VOLUME SIXTY SIX

# ADVANCES IN ATOMIC, MOLECULAR, AND OPTICAL PHYSICS

VOLUME SIXTY SIX

# ADVANCES IN ATOMIC, MOLECULAR, AND OPTICAL PHYSICS

Edited by


**ENNIO ARIMONDO**
*Physics Department*
*University of Pisa*
*Pisa, Italy*

**CHUN C. LIN**
*Department of Physics*
*University of Wisconsin Madison*
*Madison, WI, USA*

**SUSANNE F. YELIN**
*University of Connecticut*
*Storrs, CT, USA*



ELSEVIER

ACADEMIC PRESS
An imprint of Elsevier

Academic Press is an imprint of Elsevier
50 Hampshire Street, 5th Floor, Cambridge, MA 02139, United States
525 B Street, Suite 1800, San Diego, CA 92101–4495, United States
The Boulevard, Langford Lane, Kidlington, Oxford OX5 1GB, United Kingdom
125 London Wall, London, EC2Y 5AS, United Kingdom

First edition 2017



ISBN: 978-0-12-812081-1
ISSN: 1049-250X



*Publisher:* Zoe Kruze
*Acquisition Editor:* Poppy Garraway
*Editorial Project Manager:* Shellie Bryant
*Production Project Manager:* Surya Narayanan Jayachandran
*Senior Designer:* Greg Harris

Typeset by SPi Global, India

# CONTENTS

# CONTRIBUTORS

**Alessia Allevi** (49)
Università degli Studi dell'Insubria; Istituto di Fotonica e Nanotecnologie, Consiglio Nazionale delle Ricerche, Como, Italy

**Stefan Arnold** (375)
Institut für Theoretische Informatik, Universität Ulm, Ulm, Germany

**Maria Bondani** (49)
Università degli Studi dell'Insubria; Istituto di Fotonica e Nanotecnologie, Consiglio Nazionale delle Ricerche, Como, Italy

**Steven T. Cundiff** (1)
University of Michigan, Ann Arbor, MI, United States

**Michael Eckart** (375)
Institut für Quantenphysik and Center for Integrated Quantum Science and Technology ($IQ^{ST}$), Universität Ulm, Ulm, Germany

**Samuel Eden** (545)
The Open University, Milton Keynes, United Kingdom

**Maxim A. Efremov** (375)
Institut für Quantenphysik and Center for Integrated Quantum Science and Technology ($IQ^{ST}$), Universität Ulm, Ulm, Germany

**Ilya I. Fabrikant** (545)
University of Nebraska, Lincoln, Nebraska, United States

**Fredrik K. Fatemi** (439)
Army Research Laboratory, Adelphi, MD, United States

**Juraj Fedor** (545)
J. Heyrovský Institute of Physical Chemistry v.v.i., Academy of Sciences of the Czech Republic, Prague, Czech Republic

**Kevin A. Fischer** (111)
E. L. Ginzton Laboratory, Stanford University, Stanford, CA, United States

**Barry M. Garraway** (181)
University of Sussex, Brighton, United Kingdom

**Jeffrey A. Grover** (439)
Joint Quantum Institute, University of Maryland and NIST, College Park, MD, United States

**Klemens Hammerer** (263)
Institute for Theoretical Physics, Institute for Gravitational Physics (Albert Einstein Institute), Leibniz University Hannover, Hannover, Germany

**Sebastian G. Hofer** (263)
Vienna Center for Quantum Science and Technology (VCQ), Faculty of Physics, University of Vienna, Vienna, Austria

**Jonathan E. Hoffman** (439)
Joint Quantum Institute, University of Maryland and NIST, College Park, MD, United States

**Endre Kajari** (375)
Institut für Quantenphysik and Center for Integrated Quantum Science and Technology ($IQ^{ST}$), Universität Ulm, Ulm, Germany

**Hebin Li** (1)
Florida International University, Miami, FL, United States

**Nigel J. Mason** (545)
The Open University, Milton Keynes, United Kingdom

**Matthias Meister** (375)
Institut für Quantenphysik and Center for Integrated Quantum Science and Technology ($IQ^{ST}$), Universität Ulm, Ulm, Germany

**Daniela Moll** (375)
Institut für Quantenphysik and Center for Integrated Quantum Science and Technology ($IQ^{ST}$), Universität Ulm, Ulm, Germany

**Luis A. Orozco** (439)
Joint Quantum Institute, University of Maryland and NIST, College Park, MD, United States

**Hélène Perrin** (181)
CNRS, UMR 7538, Université Paris 13, Sorbonne Paris Cité, Laboratoire de physique des lasers, Villetaneuse, France

**Marina Radulaski** (111)
E. L. Ginzton Laboratory, Stanford University, Stanford, CA, United States

**Sylvain Ravets** (439)
Joint Quantum Institute, University of Maryland and NIST, College Park, MD, United States; Laboratoire Charles Fabry, Institut d'Optique, CNRS, Univ. Paris Sud, Palaiseau Cedex, France

**Steven L. Rolston** (439)
Joint Quantum Institute, University of Maryland and NIST, College Park, MD, United States

**Wolfgang P. Schleich** (375)
Institut für Quantenphysik and Center for Integrated Quantum Science and Technology ($IQ^{ST}$), Universität Ulm, Ulm, Germany; Hagler Institute for Advanced Study at Texas A&M University, Institute for Quantum Science and Engineering (IQSE), Texas A&M University, TX, United States

**Michael Schulz** (507)
Missouri University of Science and Technology, Rolla, MO, United States

**Pablo Solano** (439)
Joint Quantum Institute, University of Maryland and NIST, College Park, MD, United States

**Jelena Vučković** (111)
E. L. Ginzton Laboratory, Stanford University, Stanford, CA, United States

**Reinhold Walser** (375)
Institut für Angewandte Physik, Technische Universität Darmstadt, Darmstadt, Germany

Upon completing this volume one of us (C.C.L.) will leave the editorial team. It is a great pleasure for him to have worked with Paul Berman, Ennio Arimondo, and Susanne Yelin. He is particularly indebted to John Boffard for his invaluable assistance over the past 14 years.

Ennio Arimondo
Chun C. Lin
Susanne Yelin

CHAPTER ONE

# 2D Coherent Spectroscopy of Electronic Transitions

**Hebin Li*, Steven T. Cundiff†,[1]**
*Florida International University, Miami, FL, United States
†University of Michigan, Ann Arbor, MI, United States
[1]Corresponding author: e-mail address: cundiff@umich.edu

## Contents



## Abstract

Optical two-dimensional coherent spectroscopy has been developed as a new method over the last 20 years. It probes the structure and dynamics of materials by exciting them with a sequence of phase-coherent pulses and recording their response as two or more delays are varied. It excels at determining if resonances are coupled, overcoming the effects of inhomogeneous broadening and disentangling congested resonances by spreading them in two dimensions. In this chapter, we review the use of two-dimensional coherent spectroscopy to probe electronic transitions in atomic vapors and semiconductor heterostructures. The atomic vapors provide an ideal model system

*Advances in Atomic, Molecular, and Optical Physics*, Volume 66
ISSN 1049-250X
http://dx.doi.org/10.1016/bs.aamop.2017.03.004

to test the technique. At the same time, it provides unexpected results that arise from interactions between atoms in the vapor. In semiconductor quantum wells, the two-dimensional coherent spectra reveal the importance of many-body interactions in the nonlinear optical response. It also provides unique insight into the role of interactions in double-quantum wells.

## 1. INTRODUCTION

The concept of multidimensional spectroscopy was first developed in nuclear magnetic resonance (NMR) by Ernst et al. (1987). Implementing it in the optical domain was proposed by Tanimura and Mukamel (1993) using a Raman excitation scheme. However, the technical challenges hindered experimental implementation so that it took almost 15 years to become a practical method (Cundiff and Mukamel, 2013). It is now well established such that many ultrafast spectroscopists use it as their primary technique. Much of the work on multidimensional coherent spectroscopy is done using infrared pulses, which has the advantage of less demanding phase stability and directly coupling to molecular vibrations. This chapter will focus on multidimensional coherent spectroscopy of electronic transitions, which generally requires the use of near-infrared to visible pulses and thus is more demanding because the shorter wavelength makes the limit on acceptable path length fluctuations much smaller.

The original work in NMR used the term "multidimensional Fourier transform spectroscopy" to describe the method, a natural extension to the revolutionary development of Fourier transform methods in the 1960s when digital computers became powerful enough to generate a spectrum by taking the numerical Fourier transform of a signal captured in the time domain. This terminology was adopted at first in the optical domain; however, the phrase "Fourier transform" is not particularly informative in this context and not entirely appropriate in some situations as well. Simply dropping "Fourier transform" leads to confusion as multidimensional spectra can be generated in many ways, for example, by simply measuring incoherent emission (fluorescence or luminescence) as a function of excitation wavelength. Thus the term "multidimensional coherent spectroscopy" is often used to capture the essence of the method and also to encompass methods that measure coherent signals, but do not necessarily do so in the time domain, and do not necessarily involve taking Fourier transforms in a computer.

In this chapter, we start with an overview of spectroscopy, its goals, and the limitations of linear spectroscopy. This discussion provides the motivation for developing nonlinear methods, which we then introduce. While an improvement on linear methods, traditional nonlinear methods, for example, photon echoes, also have significant limitations, which we describe to motivate the extra effort needed for multidimensional coherent spectroscopy. We then introduce multidimensional coherent spectroscopy, explain multidimensional spectra and how to interpret them, and describe the experimental implementations, focusing, naturally, on the active stabilization approach used by our research groups. Following this introduction, we present the first 2D coherent spectrum of the $D_1$ and $D_2$ lines in an atomic vapor, namely potassium, which serves as a "model" system that produces spectra that can be easily calculated by solving the optical Bloch equations. However, it turns out that simple atomic vapors also can produce results with some surprises, at least surprising when only the single-atom response is considered. These effects show up in the two-quantum spectra, where signals due to doubly exciting the system are present. Following this, we present 3D spectra of an atomic vapor and show that they can be used to extract the Hamiltonian describing the light–matter interactions. Shifting gears, we turn to excitonic resonances in semiconductors where 2D spectroscopy clearly shows the critical role of many-body correlation terms beyond the mean field and two-quantum spectra show that the mean-field terms can actually be separated from the higher order correlations. We also discuss coupled quantum wells, which represent an "engineered dimer" eliminating the complexity of nuclear motion in molecules. Finally, we will give an overview of the use of 2D coherent spectroscopy of electronic transitions in other system, including light harvesting.

To motivate the reader to continue, before plunging into the meat of the chapter, we would like to list a few of the advantages of multidimensional coherent spectroscopy:

- removing the effects of inhomogeneous broadening;
- determining if resonances are coupled;
- spreading congested spectra in two (or more) dimensions;
- determining the phase of the nonlinear polarization with respect the excitation phase;
- measuring variation of homogeneous linewidth and level splitting with the inhomogeneous distribution.

In addition, this discussion has described a simple spectroscopy experiment, namely sweeping the frequency of an electromagnetic field incident on a sample and measuring the transmitted intensity. If a dip in the transmitted intensity is observed at a specific frequency, it indicates the presence of resonance and the width of the resonance characterizes the damping of the resonance. However, there are important ambiguities in linear spectra; namely the inability to distinguish between homogeneous and inhomogeneous broadening and the inability to determine if two resonances are coupled or uncoupled.

### 2.1.1 Inhomogeneous Broadening

Typically the sample is an ensemble of many systems, whether they are atoms undergoing electronic transitions, molecules undergoing vibrational transitions, or nuclei flipping their spins. If all the systems in the ensemble are identical, i.e., they have the same resonant frequency and same linewidth, the ensemble is designated as being "homogeneously broadened." However, this may not be the case. In particular, there may be a distribution of resonant frequencies due to effects such as the Doppler shift in a vapor or structural disorder in a nanostructure. In this case, the linewidth of the measured resonance may have nothing to do with the linewidth of the individual members of the ensemble, but rather reflects the distribution of resonance frequencies. This case is known as "inhomogeneous broadening."

The linear spectrum of an inhomogeneously broadened ensemble will have a resonance feature (the absorption "line") that has a width that is characteristic of the inhomogeneous distribution, not the damping of the individual members of the ensemble. While both are useful to know, they provide quite different information. The width in the absence of the inhomogeneous broadening, often called the "homogeneous width," provides information about processes that interrupt the oscillations, for example, collisions and radiative decay.

### 2.1.2 Coupled Resonances

If a linear spectrum shows two spectral resonances, there are two possible situations. One possibility is that the sample is heterogeneous, i.e., a mixture of two species. The other possibility is that it is pure, i.e., a single substance, but that substance has two transitions. A good example of a substance with two transitions would be the $D_1$ and $D_2$ lines in the alkali metals, which correspond to the single outer electron making a transition to ground $S_{1/2}$ state from the $P_{1/2}$ and $P_{3/2}$ states. Linear spectroscopy cannot distinguish between these two possibilities.

## 2.2 Nonlinear Spectroscopy

These ambiguities can be resolved by using some form of nonlinear spectroscopy. In nonlinear spectroscopy, as the intensity of the excitation field is increased, the polarization is no longer linearly proportional to the incident field, but rather higher order terms must be considered, i.e.,

$$P=\chi^{(1)}E+\chi^{(2)}E^2+\chi^{(3)}E^3+\cdots, \tag{1}$$

where $\chi^{(n)}$ is the coefficient for the term that is $n$th-order in the electric field of the incident light. Linear spectroscopy corresponds to the situation where all $\chi^{(n)}$ for $n>1$ are small enough that these terms can be neglected. Of course this approximation depends on the strength of $E$ because for large enough $E$, the $E^n$ factor can make $\chi^{(n)}E^n>\chi^{(1)}E$, no matter how small the ratio $\chi^{(n)}/\chi^{(1)}$. It is possible to show that for inversion symmetric systems, $\chi^{(2)}=0$. To maintain generality, i.e., not require that the system under study has broken inversion symmetry, we will discuss methods based on the third-order term in Eq. (1).

To understand how nonlinear spectroscopy can resolve the ambiguities in a linear spectrum, it is easiest to consider a simple frequency-domain method known as "spectral hole burning." In spectral hole burning, a CW "pump" laser excites the sample, saturating its absorption. A second laser is then scanned to measure the absorption of the sample. If the sample is homogeneously broadened, the absorption of the entire line simply decreases. However, if it is inhomogeneously broadened, then only the subensemble that is resonant with the pump laser is saturated. In this case, the measured absorption spectrum is unchanged, except in the spectral region close to the pump, where the absorption is decreased, known as "burning a hole." The width of the spectral hole is the homogeneous width. Thus the observation of spectral hole burning shows that the system is inhomogeneously broadened and the width gives the homogeneous width.

Similarly, if two resonances are present in the spectrum, tuning the pump laser to one resonance and probing the other can determine if they are coupled. If they are coupled, then this situation will result in a change in the absorption, whereas if they are uncoupled it will not.

This example was based on using CW lasers. Although there are some implementations of optical multidimensional coherent spectroscopy based on this approach (Chen, 2010; Chen and Gomes, 2008; Wright, 2011), most are based on using pulses and scanning their delays.

the homogeneous and inhomogeneous linewidths are associated with the cross-diagonal and diagonal directions, respectively, and can be retrieved from the lineshape analysis for any arbitrary inhomogeneities (Siemens et al., 2010). A 2D spectrum for a heterogeneous sample has spectral peaks on the diagonal line where the absorption and emission frequencies are equal, while the coupling of resonances can be identified and characterized by cross peaks. These two advantages of 2DCS will be described in more detail in Section 3.2 and confirmed in 2D spectra of atomic vapors and semiconductor quantum wells. The examples will also demonstrate other unique capabilities of 2DCS. For example, two-quantum 2D spectroscopy provides sensitive detection to many-body interactions; the phase of the nonlinear polarization can be measured to reveal additional information. Moreover, a 2D spectrum can be further unfolded into a third dimension so that excitation quantum pathways are better isolated for analysis.

2DCS can be realized with different excitation schemes in the time domain, the frequency domain (Chen, 2010; Chen and Gomes, 2008; Wright, 2011), or the mix of the two (Pakoulev et al., 2009). We will focus on time-domain TFWM approaches in the following discussions.

## 3.1 Beyond Transient Four-Wave Mixing

The three-pulse TFWM experiment can be extended to perform 2DCS. With three excitation pulses $A^*$, $B$, and $C$ ($A^*$ is considered as conjugated) as a sequence shown in Fig. 2A, the resulting TFWM signal has the form $S(\tau, T, t)$ in the time domain. Here, $\tau$ is the delay between the first and second excitation pulses, $T$ is the delay between the second and third excitation pulses, and $t$ is the signal emission time. If the TFWM signal is measured as two time delays are varied, while the third delay is fixed, a 2D spectrum can be constructed by Fourier transforming the signal into the frequency domain with respect to the two varying time delays. For example, if $\tau$ and $t$ are varied and $T$ is fixed, the resulting 2D spectrum is $S(\omega_\tau, T, \omega_t)$, where $\omega_\tau$ and $\omega_t$ are the corresponding Fourier-transformed frequencies. Similarly, a 3D spectrum can be constructed by a 3D Fourier transform if all three delays are varied.

The exact form of the TFWM signal and the resulting 2D spectrum depends on the nonlinear response of the sample and the time ordering of the excitation pulses. As an example, we consider the simplest case where a homogeneous two-level system, as shown in Fig. 2B, is excited by the pulse sequence in Fig. 2A. The TFWM signal is measured in the phase-matched direction $\mathbf{k}_s = -\mathbf{k}_A + \mathbf{k}_B + \mathbf{k}_C$ as the first time delay $\tau$ is scanned, while $T$ is fixed. We also assume that the excitation pulses are sufficiently short so that they can be treated as delta-function pulses, i.e., $\hat{E}_i = \mathcal{E}_i \delta(t - t_i)$, where $\mathcal{E}_i$ is the field amplitude, $\delta$ is the Dirac delta function, and $t_i$ is the pulse arrival time. The time-domain TFWM signal is

$$S(\tau, T, t) \propto \mu_{10}^4 \mathcal{E}_a^* \mathcal{E}_b \mathcal{E}_c e^{-(\Gamma_{10} - i\omega_{10})\tau} e^{-(\Gamma_{10} + i\omega_{10})t} \Theta(\tau)\Theta(t), \tag{5}$$

where $\mu_{10}$ is the dipole moment between excited and ground states, $\Gamma_{10}$ is the dephasing rate, $\omega_{10}$ is the resonant frequency of the transition from $|0\rangle$ to $|1\rangle$, and $\Theta$ is the Heaviside step function. The amplitude of Eq. (5) is plotted in the time domain as a function of $\tau$ and $t$, as shown in Fig. 3A. The signal amplitude decays exponentially at the rate $\Gamma_{10}$ in both $\tau$ and $t$ directions. The Heaviside step functions enforce the time causality so that the signal appears only in the first quadrant ($\tau, t \geq 0$). The frequency-domain spectrum can be obtained by Fourier transforming Eq. (5), yielding

$$S(\omega_\tau, T, \omega_t) \propto \mu_{10}^4 \mathcal{E}_a^* \mathcal{E}_b \mathcal{E}_c \frac{1}{\omega_\tau + \omega_{10} + i\Gamma_{10}} \times \frac{1}{\omega_t - \omega_{10} + i\Gamma_{10}}. \tag{6}$$

The amplitude of the spectrum is plotted as contours, as shown in Fig. 3B, with frequency axes $\omega_\tau$ and $\omega_t$. The absorption frequency $\omega_\tau$ is negative because the first pulse $A^*$ is conjugated. The spectrum features a star-shape peak on the diagonal line since an isolated two-level atom absorbs and emits energy at the same frequency. The star shape, instead of a circularly symmetric 2D Lorentzian profile, is due to the truncation of the signal at the two time axes. This is the simplest 2D spectrum.

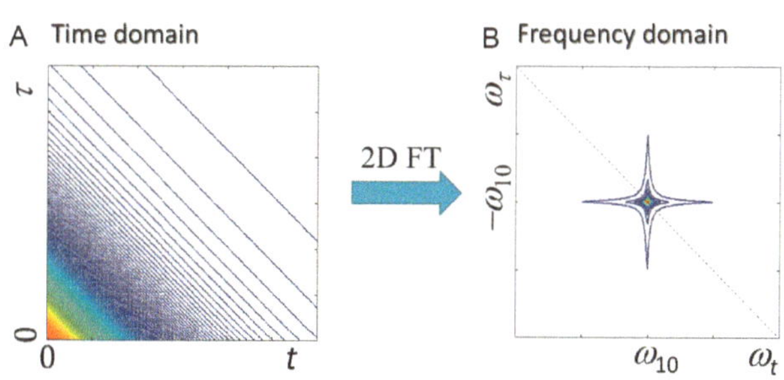


**Fig. 3** TFWM signal plotted (A) in the time domain and (B) in the frequency domain. Note that $\omega_\tau$ is negative because pulse $\mathbf{k}_A$ is conjugated in determining the signal direction.

## 3.2 Primer on 2D Spectra

The 2D spectrum in Fig. 3B reveals the fundamental parameters of the system under study. The central frequency of the peak is the resonance frequency $\omega_{10}$. The transition dipole moment $\mu_{10}$ can be calculated if the strength of the signal field and the excitation fields are measured. The dephasing rate $\Gamma_{10}$ can be determined from the linewidth analysis. The population relaxation rate $\Gamma_1$ can also be measured by scanning $T$. These parameters together fully characterize this two-level system. Although it does not surpass 1D techniques in determining these parameters for a homogeneous two-level system, 2DCS excels in more complicated cases where 1D measurements can be ambiguous.

### 3.2.1 Inhomogeneous Broadening

One can identify the inhomogeneous broadening and measure both the homogenous and inhomogeneous linewidths in a single 2D spectrum (Siemens et al., 2010). As shown in Fig. 4, the spectral peak has a symmetric star shape if the resonance is homogeneous and is elongated in the diagonal direction if the resonance is inhomogeneously broadened. In the case of inhomogeneous broadening, resonances have different frequencies satisfying a particular distribution such as a Gaussian profile. The elongation is a result of the overlapping individual resonances aligned on the diagonal line within the inhomogeneous distribution. The inhomogeneous and homogeneous linewidths are primarily associated with the slices along the diagonal and cross-diagonal directions, respectively.

Theoretically, the inhomogeneous broadening can be accounted for by integrating the result in Eq. (5) over the distribution of the frequency $\omega_{10}$. Assuming a Gaussian distribution with a linewidth $\sigma$, the time-domain TFWM signal is

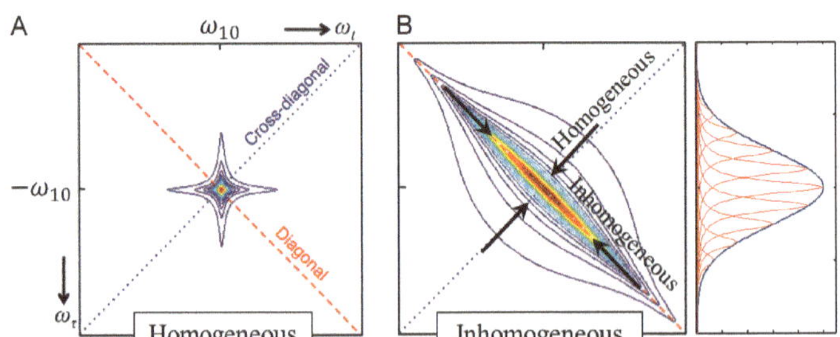


**Fig. 4** 2D spectra of (A) a homogeneous system and (B) an inhomogeneous system.

$$S(\tau, T, t) \propto \mu_{10}^4 \mathcal{E}_a^* \mathcal{E}_b \mathcal{E}_c e^{-\Gamma_{10}(t+\tau)-\sigma^2(t-\tau)^2/2} e^{-i\omega_{10}(t-\tau)} \Theta(\tau)\Theta(t). \qquad (7)$$

This result shows that the linewidths $\sigma$ and $\Gamma_{10}$ are mixed in the $t$ and $\tau$ directions but separated in the diagonal $(t-\tau)$ and cross-diagonal $(t+\tau)$ directions. Based on the projection-slice theorem of 2D Fourier transforms (Ernst et al., 1987), a projection onto a line in the time domain is equivalent to a slice in the corresponding direction in the frequency domain, so the slices in the diagonal and cross-diagonal directions in a 2D spectrum are directly associated with the linewidths $\sigma$ and $\Gamma_{10}$, respectively. Analytical expressions have been developed (Siemens et al., 2010) to fit the diagonal and cross-diagonal slices to retrieve both linewidths in a 2D spectrum with an arbitrary inhomogeneous broadening including purely homogeneous, strongly inhomogeneous, and any intermediate cases. These fundamental principles are applicable to lineshape analyses of other 2D spectra.

### 3.2.2 Coupled Resonances

A variety of couplings can be identified and characterized using optical 2DCS. From a 1D spectrum, say an absorption spectrum with two spectral peaks, one can tell that there are two resonances but cannot determine whether or how the two resonances are coupled. In 2D spectra, the coupling is revealed by the spectral pattern. Fig. 5 illustrates 2D spectra for three different scenarios: (a) if the system under study consists of two isolated two-level resonances, the spectrum has only two peaks on the diagonal; (b) if the two resonances are coupled through an incoherent relaxation channel from

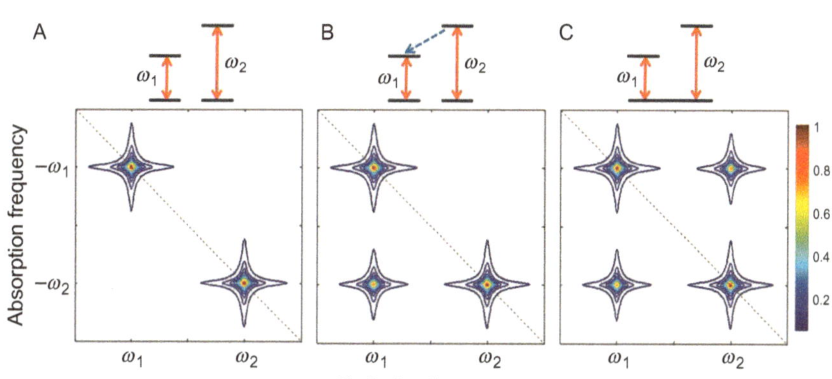


**Fig. 5** 2D spectra reveal couplings in the system. The spectra are shown for (A) two independent two-level systems, (B) two transitions coupled by an incoherent relaxation channel, and (C) two transitions coupled through a common ground state.

the higher to lower energy excited states, one cross peak is present; (c) if the two resonances are coupled through a common ground state, i.e., a three-level *V* system, there are two cross peaks in the spectrum. The cross peak in (b) is only observable in a finite waiting time ($T > 0$) since it takes time for the relaxation process to occur, while the cross peaks in (c) are present even at $T = 0$ due to the coherent nature of the couplings. The dynamics of the coupling can be studied to reveal the coupling mechanism by tracking the cross peaks as the waiting time $T$ varies. If the two excited states are coupled through a population decay, the amplitude of the cross peak decreases monotonically as the population decays. If the two excited states are coupled coherently, i.e., there is a quantum coherence between the two states, the amplitude of the cross peaks oscillates at the resonant frequency of the coherence. The dynamics during $T$ can also be investigated in the frequency domain in a 3D spectrum (Li et al., 2013a).

In additional, couplings between resonances are evident in the so-called zero-quantum and two-quantum 2D spectra (Moody et al., 2014; Nardin et al., 2014), depending on the nature of coupling. We will discuss this in the examples in Section 5 and show that a complete set of zero-, one-, and two-quantum 2D spectra provide a unique and comprehensive understanding of the coupling in the system.

## 3.3 Experimental Implementation

Experimental implementation of 2DCS is based on measuring the nonlinear response of a sample to multiple excitation laser fields. The excitation beams can be arranged in a collinear geometry where all beams are copropagating, or in the box geometry where parallel beams are aligned to the three corners of a square and converge on the sample through a lens. In the box geometry, the nonlinear signal in a particular phase-matching direction can be spatially selected so that the background including scattering of the excitation beams and signals in other directions is rejected to achieve a good signal-to-noise ratio. The collinear geometry requires a process such as phase cycling (Shim et al., 2007; Tekavec et al., 2007; Wagner et al., 2005) and frequency tagging (Nardin et al., 2013; Tekavec et al., 2007) to select the excitation pathways of interest. On the other hand, a collinear geometry benefits from a longer range of interaction, the possibility of a tight focus through an objective lens, and the flexibility of measuring higher order signals.

For time-domain approaches of 2DCS, Fourier analysis requires that the excitation pulses are phase locked and the time delays are scanned with equal steps. One way to ensure the phase stability is to have all beams propagate through common optics (Brixner et al., 2004; Cowan et al., 2004; Selig et al., 2008; Tian et al., 2003; Turner et al., 2011) so that they are subject to the same beam path fluctuations. Alternatively, the phases can be actively stabilized (Bristow et al., 2009a; Volkov et al., 2005; Zhang et al., 2005) by using interferometric error signals as a feedback.

### *3.3.1 Active Stabilization*

Our implementation of active stabilization is based on an ultrastable platform dubbed the "multidimensional optical nonlinear spectrometer" (MONSTR) (Bristow et al., 2009a). It consists of three nested Michelson interferometers, as shown in Fig. 6A, and splits a laser pulse into four copies that go through the dichroic mirrors (DCMs). The output of a CW, typically HeNe, laser copropagates with the pulses and is reflected by the DCMs to form the interferometers. The interferometric outputs of the CW laser beams at the three beam splitters are monitored by photodetectors and used as error signals for the beam path length fluctuations. The error signals are inputs to electronic servos to actively stabilize the beam path lengths by mirrors mounted on piezoelectric transducers. There are three feedback loops to stabilize the relative phases between the four pulses. This active stabilization is able to compensate small fluctuations ($<1/4$th of the CW laser wavelength) in the beam path at a frequencies up to about 10 kHz. Therefore, the whole platform has to be passively stabilized to a sufficiently low level of fluctuations. To do so, each arm (which is a Michelson interferometer itself) of the main interferometer is constructed on a separate aluminum block. One of the blocks is flipped upside down and placed on spacers above the other to form a double-deck configuration whose output beams are aligned in the box geometry. The entire setup is situated on an optical table with vibration isolators. With both the passive and active stabilization, the MONSTR has a stability of greater than 1/100th of the CW laser wavelength.

The four output pulses from the MONSTR are used to perform TFWM experiments in the box geometry as shown in Fig. 6B. Three pulses $A^*$, $B$, and $C$ are incident on the sample to generate a TFWM signal in the propagation direction of the fourth pulse. During the experiment, the fourth pulse is blocked from the sample and used as a reference pulse, which is routed around the sample and combined with the TFWM signal for spectral interferometry. Since it deviates from the common optics of other pulses, the reference pulse needs to be relocked to one of the other pulses with

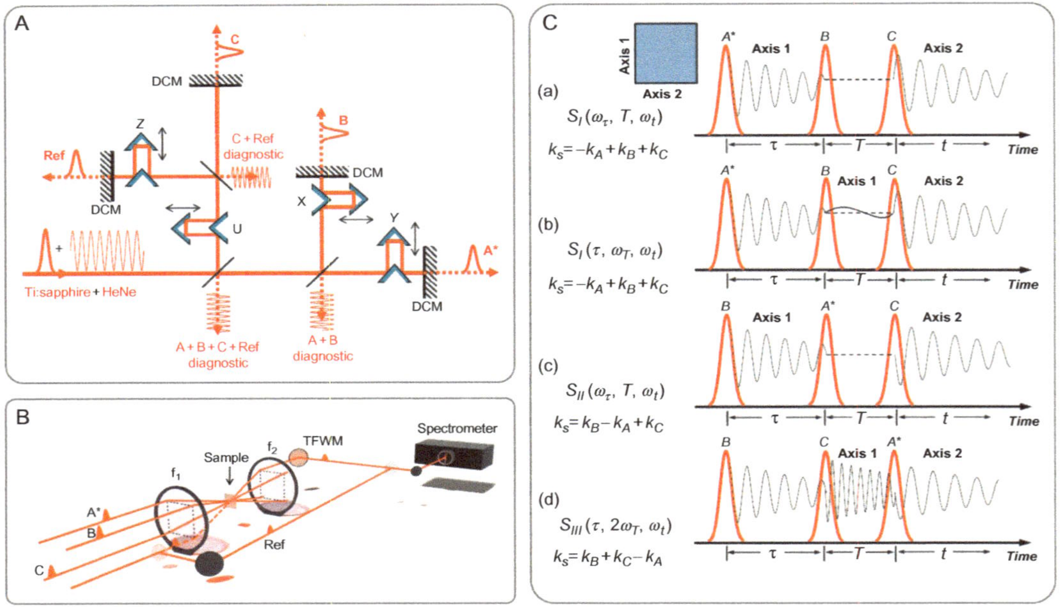


**Fig. 6** Diagrams showing the active stabilization approach of 2DCS. (A) Schematic of nested interferometers with four output femtosecond pulses for experiment and three error signals for stabilization. (B) TFWM experiment in the box geometry and the TFWM signal is measured by spectral interferometry. (C) Possible time orderings of the excitation pulses to generate different types of 2D spectra. *Panels (A) and (C): Adapted from Bristow, A.D., Karaiskaj, D., Dai, X., Zhang, T., Carlsson, C., Hagen, K.R., et al. (2009a). A versatile ultrastable platform for optical multidimensional Fourier-transform spectroscopy. Rev. Sci. Instrum. 80 (7), 73108 with permission.*

an additional feedback loop to ensure the phase stability. A spectrometer analyzes the interference of the TFWM signal and the reference pulse, and the resulting interferograms are recorded by a CCD camera. The time delays of the excitation pulses are controlled by a set of high precision translational stages. A series of interferograms are recorded as a function of time delays to generate a 2D/3D spectrum.

A variety of 2DCS experiments can be performed depending on the time ordering of the pulses and the time delay that is scanned, as shown in Fig. 6C. If pulse $A^*$ arrives first followed by $B$ and $C$, the sequence excites a rephasing process that generates a photon echo if the sample is inhomogeneous. If the time delay $\tau$ is scanned, the resulting 2D spectrum is called the "one-quantum spectrum" since the dynamics during $\tau$ involves a one-photon excitation. If the time delay $T$ is scanned, involving an upside-down Raman process, the resulting 2D spectrum is named the "zero-quantum spectrum." A pulse sequence with $A^*$ arriving second is nonrephasing and cannot generate a photon echo. This configuration can be considered as 2DCS experiments with negative $\tau$. The nonrephasing excitation does not provide extra information compared to the rephasing sequence, but can offer a better lineshape when combined with rephasing 2D spectra. If pulse $A^*$ arrives last and the time delay $T$ is scanned, the sequence can excite the two-quantum coherence between a ground state and a doubly excited state, resulting in a so-called "two-quantum 2D spectrum."

Spectral interferometry is capable of measuring the field, in addition to intensity, and the phase of the signal. The phase of the TFWM signal is determined by the phases of the sample's nonlinear response and the excitation fields. Therefore, the phase of nonlinear response can be retrieved if the excitation pulses' relative phases are measured. This can be done by analyzing spatial interference patterns and spectral interferograms of the pulses (Bristow et al., 2008).

Based on the MONSTR, 2DCS has also been implemented in the reflection geometry (Li et al., 2013b), which is necessary for studying optically thick samples or surface states. As a versatile platform, the MONSTR can be used to perform other ultrafast spectroscopic experiments than 2DCS, such as pump probe, time-integrated FWM, and coherent control.

### 3.3.2 Other Approaches

A number of approaches to implement 2DCS have been developed in the past 15 years or so. Each approach is suitable for a particular group of applications depending on their advantages and disadvantages.

The "common optics" approach has been realized in several schemes. Diffractive optics (Brixner et al., 2004; Cowan et al., 2004) can work as beam splitters to split two pulses into four and their delays can be controlled with in-beam glass slides or prims. In these schemes, the phases are only locked pairwise, while the second time delay is not stabilized. Therefore, it does not allow 2DCS that requires scanning *T* with interferometric precision. Moreover, diffraction optics and slides/prims introduce significant dispersion for ultrashort pulses, and the time delays are limited to a few ps. Several groups (Grumstrup et al., 2007; Tian et al., 2003; Turner et al., 2011) used femtosecond pulse shapers to split a pulse and control the time delays, in which case all pulses propagate through the same optical path and are phase locked. The pulse-shaper approaches enable various types of 2DCS and provide opportunities to vary temporal and spectral pulse shapes. However, some disadvantages include limited time delays, sideband pulses, and a reduced energy throughput. There is also an elegant approach (Selig et al., 2008) that does not use diffraction optics or pulse shapers but only conventional optics to achieve pairwise phase stabilization. The limitations are that the second time delay is not stabilized and it lacks the flexibility in controlling time delays. In comparison, the "active stabilization" approach phase-locks all pulses, enables various types of 2DCS, and allows long time delays up to ns.

Another remarkable approach is based on acousto-optic modulators (AOMs) (Nardin et al., 2013; Tekavec et al., 2007), in which the phases are not locked, but the phase fluctuations are monitored and removed from the final signal. In this approach, four excitation pulses are derived from conventional beam splitters and delayed with translation stages. These pulses are incident on the sample and produce an excited-state population. Each pulse is modulated by an AOM and the first-order deflection is used. The AOMs are driven at slightly different frequencies so that the pulses are frequency tagged. The phase fluctuations are monitored from the beating signal of the pulses (Tekavec et al., 2007) or copropagating CW laser beams (Nardin et al., 2013). The beat frequencies are electronically mixed to provide a reference frequency for a lock-in amplifier to demodulate the nonlinear response for a selective process. This method inherently accounts for the phase fluctuations since they are present in both the reference and the nonlinear signal.

Experiments of 2DCS can also be done in the frequency domain (Chen, 2010; Chen and Gomes, 2008; Wright, 2011). This approach uses a combination of narrowband CW laser and broadband ns pulse laser to excite a sample and measures the signal with a spectrometer. The use of a CW laser offers a better frequency resolution, which allows mapping molecular rovibronic patterns in 2D spectra. This frequency approach has been extended to perform 3D spectroscopy (Chen et al., 2013; Wells et al., 2015).

## 4. MULTIDIMENSIONAL COHERENT SPECTROSCOPY OF AN ATOMIC VAPOR

A potassium (K) atomic vapor is an ideal model system for 2DCS. This three-level "V" system formed by the $D_1$ and $D_2$ transitions is appropriately simple and well understood to interpret and verify the results and yet sufficiently complex to demonstrate the unique advantages of 2D/3D spectra. Alkali metal atomic vapors have been used in different implementations of 2DCS (Dai et al., 2010; Tekavec et al., 2007; Tian et al., 2003). One- and zero-quantum 2D spectra can be calculated from the optical Bloch equations and confirm the basic principles of implementing and understanding 2DCS. Two-quantum 2D spectra show a surprising signal that cannot be explained by considering individual atoms and requires collective resonances induced by interatomic interactions. By extension into a third dimension, a 3D spectrum unravels excitation quantum pathways that are inseparable in 1D or even 2D spectra, allowing direct characterization of each pathway to construct the system's full Hamiltonian.

In our experiments, the atomic vapor is contained in a special vapor cell that consists of a titanium body and two sapphire windows. The cell can operate at temperatures up to 800°C. The two sapphire windows form a narrow gap that can be adjusted from ~10 μm to a few mm for experiments. A short optical path of the cell is critical to ensure that the optical density is smaller than 1 to avoid the spectral distortion in 2D spectra due to pulse propagation effects (Li et al., 2013c; Spencer et al., 2015). An argon buffer gas is introduced to broaden the transitions such that the hyperfine and Zeeman sublevels are not resolvable or relevant in our measurements.

### 4.1 One- and Zero-Quantum 2D Spectra of $D_1$ and $D_2$ Lines

One- and zero-quantum 2D spectra of a K vapor have been demonstrated experimentally (Dai et al., 2010). One-quantum spectra were obtained by scanning the first time delay $\tau$ in either a rephasing or a nonrephasing excitation pulse sequence. Fig. 7C shows the amplitude of a nonrephasing (upper panel) and a rephasing (lower panel) one-quantum 2D spectra. Both spectra have two resonances (NA, NB, RA, and RB) on the diagonal line

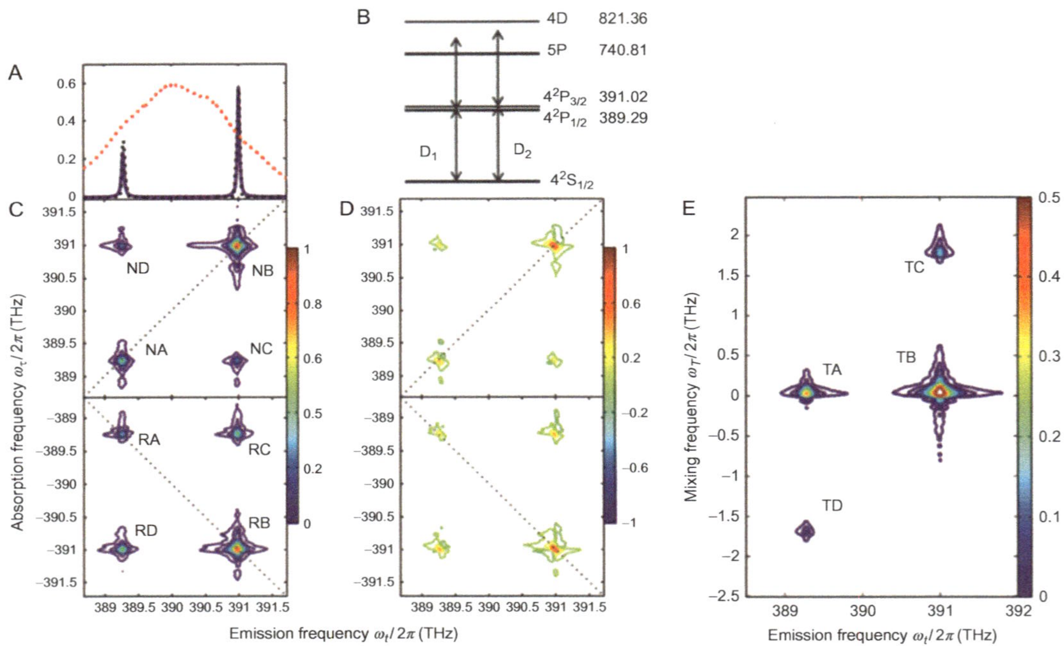


**Fig. 7** Experimental 2D spectra of a K vapor. (A) Absorption and laser spectra. (B) Relevant energy-level diagram of K atom. (C) One-quantum 2D amplitude spectra. The *upper* (*lower*) *spectrum* is a nonrephasing (rephasing) spectrum. (D) The corresponding real part of the one-quantum 2D spectra. (E) Zero-quantum 2D amplitude spectrum. *Adapted from Dai, X., Bristow, A.D., Karaiskaj, D., Cundiff, S.T., 2010. Two-dimensional Fourier-transform spectroscopy of potassium vapor. Phys. Rev. A 82 (5), 52503 with permission.*

corresponding to the $D_1$ and $D_2$ transitions, and two cross resonances (NC, ND, RC, and RD) due to the coupling of the two transitions. The experiment also measured the phase so that the real part of the spectra can be plotted as shown in Fig. 7D. The real part of the resonances has an absorptive lineshape, which is expected for this simple three-level system. Zero-quantum spectra were obtained by scanning the second time delay $T$ in either a rephasing or a nonrephasing excitation pulse sequence. A zero-quantum 2D spectrum is shown in Fig. 7E. The mixing frequency dimension is associated with the dynamics during the time period $T$. The two resonances (TC and TD) with nonzero mixing frequencies involve coherences oscillating at these frequencies, while the resonances (TA and TB) with a zero mixing frequency include a population decay during $T$.

A more rigorous interpretation of 2D spectra requires that each resonance be associated with one or more excitation quantum pathways. An effective method is to use double-sided Feynman diagrams. These diagrams have two parallel vertical lines to track the time evolution with time increasing upward. Density matrix elements are represented with symbols placed between the lines with left side indicating the ket and the right side for the bra. Diagonal arrows represent photon interactions and can act on either the left or the right side. An arrow pointing inward (outward) means the absorption (emission) of a photon. By tracking the evolution of the density matrix during each time period, the expected peak locations can be determined in a 2D spectrum.

For the 2D spectra in Fig. 7, all relevant excitation quantum pathways are represented by the double-sided Feynman diagrams in Fig. 8. The three states are $|0\rangle = |4^2S_{1/2}\rangle$, $|1\rangle = |4^2P_{1/2}\rangle$, and $|2\rangle = |4^2P_{3/2}\rangle$.

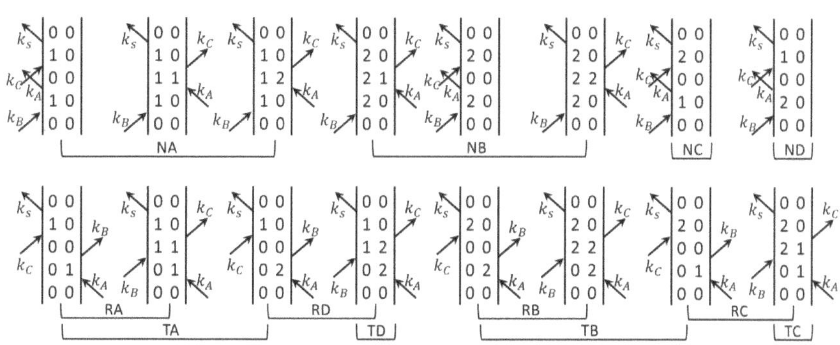


**Fig. 8** Double-sided Feynman diagrams representing all excitation quantum pathways that contribute to the 2D spectra in Fig. 7. The diagrams are grouped and labeled by the corresponding 2D peaks.

$T$ oscillates at twice the frequency of the single-quantum coherence. In this case, the spectrum features a peak at a two-quantum frequency that is twice of its emission frequency, i.e., the peak is on the diagonal line $\omega_T = 2\omega_t$. For individual atoms, the presence of a two-quantum signal relies on the existence of a doubly excited state.

However, for K atoms, within the laser bandwidth tuned to excite the D-lines, there are no atomic energy levels at twice the frequency of the laser. As shown in Fig. 9C, the $D_1$ and $D_2$ lines have transition frequencies of 389.29 and 391.02 THz, respectively. The closest higher-lying states 5P and 4D, at frequencies 740.81 and 821.36 THz, are both well outside the spectral range at the double frequency of the laser. Thus, no two-quantum signal is expected due to the absence of a doubly excited state. Surprisingly, the observed two-quantum spectrum in Fig. 10A shows clear resonances, even though there are no atomic states at these energies. The observed resonances are at exactly twice the frequencies of the $D_1$ and $D_2$ lines and their sum frequency in the two-quantum frequency axis and at the $D_1$ and $D_2$ lines in the emission frequency axis. A rephasing one-quantum spectrum is also shown in Fig. 10B for reference. The real part is plotted in these spectra.

This observation can be explained using a simple picture by combining the Hamiltonians of two individual atoms into a Hamiltonian describing both simultaneously, as shown in Fig. 9E, and working in the joint Hilbert space of both atoms. There are no doubly excited states for two isolated noninteracting two-level atoms. However, the joint space of two atoms has four levels consisting of a ground state, where both atoms are in the ground states; two singly excited states, where one of the two atoms is excited; and a doubly excited state, where both atoms are excited (two-atom Dicke states). In this four-level system, the two-quantum coherences can occur between the ground state and the doubly excited state. For K atoms (a three-level V system), the two-atom collective states are shown in Fig. 9D. However, the change in description should not alter the underlying physics. It seems contrary that the nonlinear susceptibility calculated for the four-level system has terms that are not present for the two isolated two-level systems. In fact, the two-quantum signals associated with the two pathways shown in Fig. 9B interfere destructively and exactly cancel each other if the ground state to the single excited-state transitions are identical to the singly excited-state to doubly excited-state transitions. The situation changes if there is a dipole–dipole interaction between the two atoms. As shown in Fig. 9E, the interaction between the two atoms leads to an energy splitting between the singly excited states $|10\rangle$ and

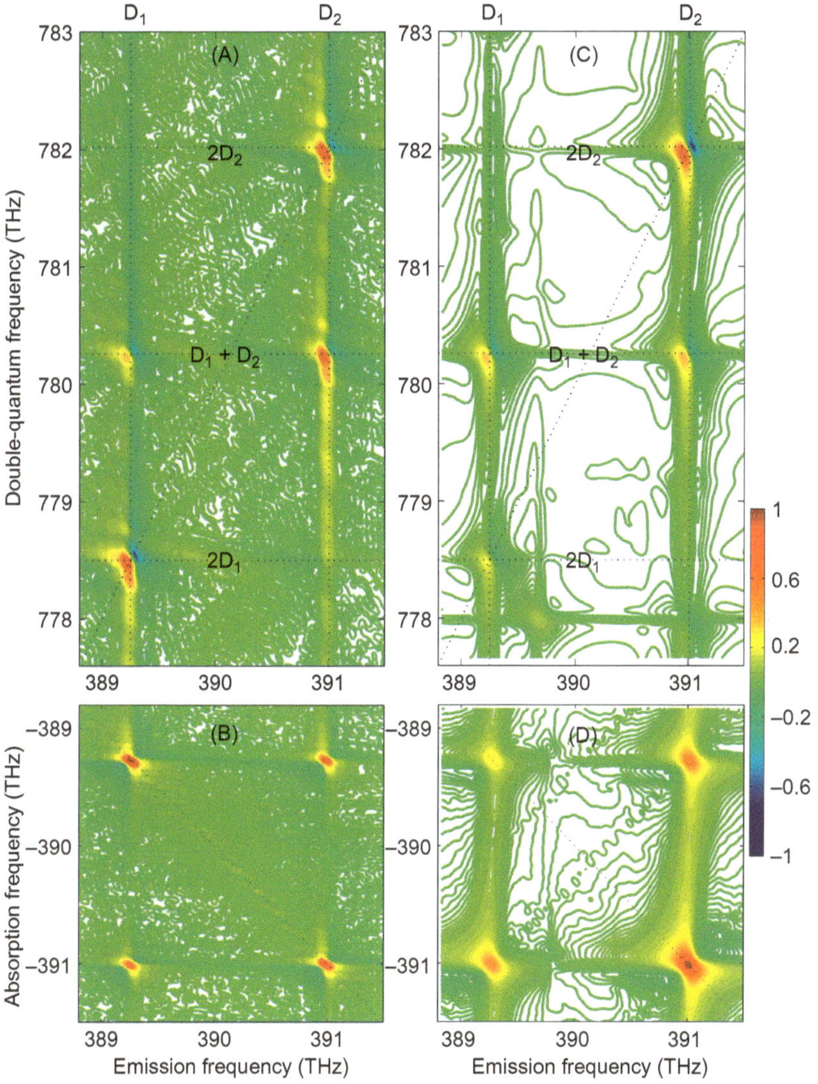


**Fig. 10** Experimental (A) two-quantum and (B) one-quantum rephasing 2D spectra and simulated (C) two-quantum and (D) one-quantum rephasing 2D spectra. The real part of the spectra is shown. *Adapted from Dai, X., Richter, M., Li, H., Bristow, A.D., Falvo, C., Mukamel, S., et al., 2012. Two-dimensional double-quantum spectra reveal collective resonances in an atomic vapor. Phys. Rev. Lett. 108 (19), 193201 with permission.*

$|01'\rangle$, an energy shift to the doubly excited state $|11'\rangle$, and a different dephasing rate for the upper transitions. In this case, the symmetry in the four-level system is broken and the interference cancellation of the available pathways will be incomplete, resulting in a nonzero two-quantum signal.

Since a slight mismatch gives rise to incomplete cancellation, the two-quantum signal provides an extremely sensitive and background-free detection to weak atom–atom interactions. The fact that the two-quantum resonances in the spectrum show no observable energy shift from the $D_1$ and $D_2$ lines indicates that the interaction is weak.

This simple model qualitatively explains how the interaction of two atoms results in a two-quantum coherence. A more complete theoretical treatment based on the exciton picture is needed to consider an ensemble of interacting atoms. The dipole–dipole interaction is considered. In the simulation, the time-dependent interatomic couplings are calculated based on the simulated trajectories of 20 potassium atoms. The 2D spectra are generated using the calculated interatomic couplings. The simulated two- and one-quantum 2D spectra are shown in Fig. 10C and D, respectively. They are in good agreement with the experimental spectra, matching the two-quantum resonances and the dispersive lineshape. The two-quantum signal disappears when the dipole–dipole interaction is turned off. This study shows that the observed two-quantum signal is attributed to a collective resonance induced by weak interatomic dipole–dipole interactions, which cannot be ignored even at a low atomic density of $10^{12}\,cm^{-3}$.

Two-quantum 2D spectroscopy has also been used to study the dipole–dipole interaction in a rubidium (Rb) vapor (Gao et al., 2016). Different from K, the energy-level structure of Rb atoms provides a doubly excited state for individual atoms. If the laser spectrum is centered at half the energy difference between states $5^2S_{1/2}$ and $5^2D$, the single-photon excitation covers the $D_2$ transition, while the two-photon excitation covers the single-atom doubly excited state $5^2D$ and the two-atom doubly excited state with two atoms in state $5^2P_{3/2}$. Therefore, the system can simultaneously produce the two-quantum signals due to the doubly excited state of individual atoms and the dipole–dipole interactions between atoms. This coincidence provides an opportunity to quantitatively study the dipole–dipole interaction by using the single-atom signal as a reference, since the two-quantum signal from individual atoms is well understood. For example, it will be interesting to study the density dependence to determine the densities for which the dipole–dipole interaction needs to be accounted for and how many atoms should be included in the interaction.

## 4.3 Three-Dimensional Coherent Spectroscopy

The technique of 2DCS can be extended into a third frequency dimension to perform three-dimensional coherent spectroscopy (3DCS) (Cundiff, 2014).

In a 2D spectrum, the excitation quantum pathways are not completely separated and thus the system's optical response is not fully revealed. A 2D spectrum has two frequency dimensions corresponding to two time delays, while the third time delay is fixed. Therefore, information about correlations associated with the fixed delay is lost. Moreover, a 3D spectrum is not equivalent to taking peak amplitudes of a series of 2D spectra taken as a function of the third time delay. It has been shown (Li et al., 2013a) that 3DCS can isolate excitation quantum pathways and reveal the full information about the system's optical response.

A 3D spectrum can be obtained by scanning both $\tau$ and $T$, and Fourier transforming the signal with respect to all three time delays. A rephasing 3D spectrum of a K vapor is shown in Fig. 11A. The spectrum is visualized with isosurfaces, a 3D analogue of 2D contour lines. The figure also shows the projections on three 2D planes. Compared to a 2D spectrum, the 3D spectrum further unfolds the nonlinear response in a third dimension, so the pathways with a zero-quantum coherence are isolated from the cross peaks.

A 3D spectrum contains complete and yet well-isolated information of the system's third-order optical response. First, various 2D spectra can be

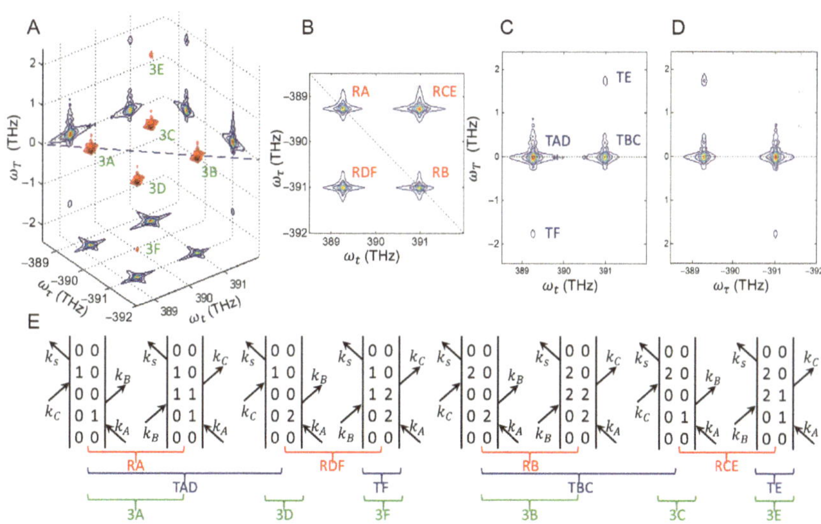


**Fig. 11** (A) Experimental rephasing 3D spectrum with 2D projections on three planes. (B) Projection on the bottom plane. (C) Projection on the left back plane. (D) Projection on the right back plane. (E) Double-sided Feynman diagrams representing all contributing quantum pathways and the corresponding peaks in the spectra. *Adapted from Li, H., Bristow, A.D., Siemens, M.E., Moody, G., Cundiff, S.T., 2013a. Unraveling quantum pathways using optical 3D Fourier-transform spectroscopy. Nat. Commun. 4, 1390 with permission.*

retrieved by projecting a 3D spectrum onto a 2D plane according to the projection-slice theorem of Fourier analysis. As shown in Fig. 11A, the projection onto the bottom plane ($\omega_\tau, \omega_t$) is the rephasing 2D spectrum at $T=0$, and the spectrum at any given $T$ can be retrieved by applying an appropriate phase shift according to the Fourier shift theorem. Similarly, the projection onto the left back plane ($\omega_t, \omega_T$) gives zero-quantum 2D spectra. These two projections, as shown in Fig. 11B and C, are the spectra that can be acquired by 2DCS. One can also retrieve 2D spectra that are not accessible by conventional 2DCS techniques. For example, the projection on the right back plane ($\omega_\tau, \omega_T$) is a unique 2D spectrum that reveals the nonlinear response at an instantaneous emission time as a function of absorption frequency $\omega_\tau$ and mixing frequency $\omega_T$, as shown in Fig. 11D.

Most importantly, the nonlinear response in a 3D spectrum is well isolated according to the excitation quantum pathways. For the rephasing 3D spectrum in Fig. 11A, the contributions from different pathways are isolated in six different peaks labeled 3A–F. There are eight possible pathways, as shown in Fig. 11E, for the given pulse sequence and phase-matching condition. In the rephasing 2D spectrum in Fig. 11B, each peak includes two different pathways and each cross peak can be divided into two in the 3D spectrum. The zero-quantum 2D spectrum in Fig. 11C isolates the zero-quantum coherence terms, but combines the diagonal and cross peaks in the 3D spectrum. In contrast, the excitation quantum pathways are most isolated in the 3D spectrum. Each of the four peaks (3C—F) is contributed by a single pathway. Two peaks 3A and 3B include two pathways each, which describe equivalent processes. Therefore, each peak in the 3D spectrum can be independently analyzed to determine the information on a particular physical process.

For example, peak 3E can be isolated and analyzed, as shown in Fig. 12, for the information about the corresponding pathway. The amplitude of peak 3E is as follows:

$$|S_{3E}(\omega_\tau, \omega_T, \omega_t)| \propto \mu_{01}^2 \mu_{02}^2 \frac{1}{\sqrt{(\omega_\tau - \omega_{01})^2 + \Gamma_{10}^2}} \cdot \frac{1}{\sqrt{(\omega_T - \omega_{21})^2 + \Gamma_{21}^2}} \cdot \frac{1}{\sqrt{(\omega_t - \omega_{20})^2 + \Gamma_{20}^2}}. \tag{8}$$

Eq. (8) describes the lineshape profile of peak 3E, which is the square root of a Lorentzian along each frequency dimension. One can fit the slices in the

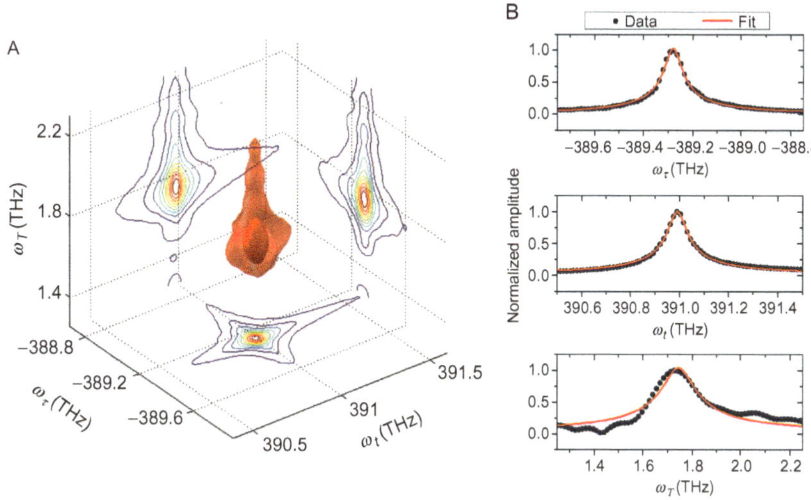


**Fig. 12** (A) A single 3D peak (3E) representing a single-quantum pathway and its projections on three planes. (B) Experimental data (*dotted lines*) and fit (*solid lines*) of the profiles of peak 3E along three frequency directions. *Adapted from Li, H., Bristow, A.D., Siemens, M.E., Moody, G., Cundiff, S.T., 2013a. Unraveling quantum pathways using optical 3D Fourier-transform spectroscopy. Nat. Commun. 4, 1390 with permission.*

three directions, as shown in Fig. 12B, to determine the parameters in this pathway including the resonance frequencies and the relaxation rates.

Overall, the spectral pattern, the strength, and the lineshape of each peak in a 3D spectrum can be used to determine the system's energy-level scheme and the parameters needed to fully characterize the Hamiltonian, including the transition energy, dipole moment, and relaxation rate (Li et al., 2013a). This ability to experimentally measure the Hamiltonian can have profound influence on some fields, since quantitative knowledge of a system's Hamiltonian can enable quantitative prediction and control of quantum phenomena. One example is coherent control, where a prior knowledge of the quantum pathways used to construct the Hamiltonian is essential for designing an effective learning algorithm, suitable initial conditions, and even a deterministic control scheme.

Other forms of 3DCS have been developed for different objectives. For example, experiments have been reported to study fifth-order susceptibility (Ding and Zanni, 2007; Fidler et al., 2010; Garrett-Roe and Hamm, 2009; Garrett-Roe et al., 2011), two-quantum coherence (Turner et al., 2009), excitons in double-quantum wells (Hall et al., 2013), molecular energy structures (Chen et al., 2013; Wells et al., 2015).

with the interaction between excitons and free carriers. A 2D spectrum separates the contributions from different many-body processes in distinct spectral features, allowing isolated studies of each process. For example, the time evolution of each process can be measured by tracking the corresponding spectral feature as the waiting time $T$ is varied (Turner et al., 2012).

### 5.2.2 Theory

Although some many-body features in the 2D spectra can be explained by using the semiconductor optical Bloch equations and including phenomenological terms of EID and EIS (Li et al., 2006), the phenomenological theory does not explain all observations in a consistent and unified manner. A more rigorous and complete description requires a full microscopic theory (Zhang et al., 2007) that starts from the Hamiltonian of interacting electrons in the appropriate band structure.

The full microscopic theory accounts for all many-body interactions/correlations and biexciton contributions. To isolate different many-body effects, the model can include all coherent contributions to the third-order optical response, include only terms of first-order Coulomb interactions (Hartree–Fock), or include Pauli blocking while neglecting the Coulomb interactions altogether.

One-quantum 2D spectra were calculated for cocircularly polarized excitation pulses. The real parts of the spectra are shown in Fig. 15. The results from three different calculations are presented. For the spectra in the first column, only Pauli-blocking terms are considered and the Coulomb interaction beyond exciton formation is neglected. No cross peaks appear in the spectra and the diagonal peak lineshape is absorptive. If the Coulomb interaction is included using the Hartree–Fock approximation, as shown in the second column, the resulting spectra display some many-body features observed in experiment, including cross peaks and the dispersive lineshape for diagonal peaks. However, the relative strengths of cross peaks to diagonal peaks do not match the experiment, and the vertical stripe does not show up at the *lh* exciton energy. The full theory includes correlation terms beyond the Hartree–Fock approximation. The results, as shown in the third column, are in good agreement with the experimental spectra. The full calculation successfully reproduces the cross peaks with proper relative strengths, the dispersive lineshape, and the vertical stripes.

With the ability to separate specific many-body contributions and phase-resolve the signal, 2D spectra provide data with stronger constraints

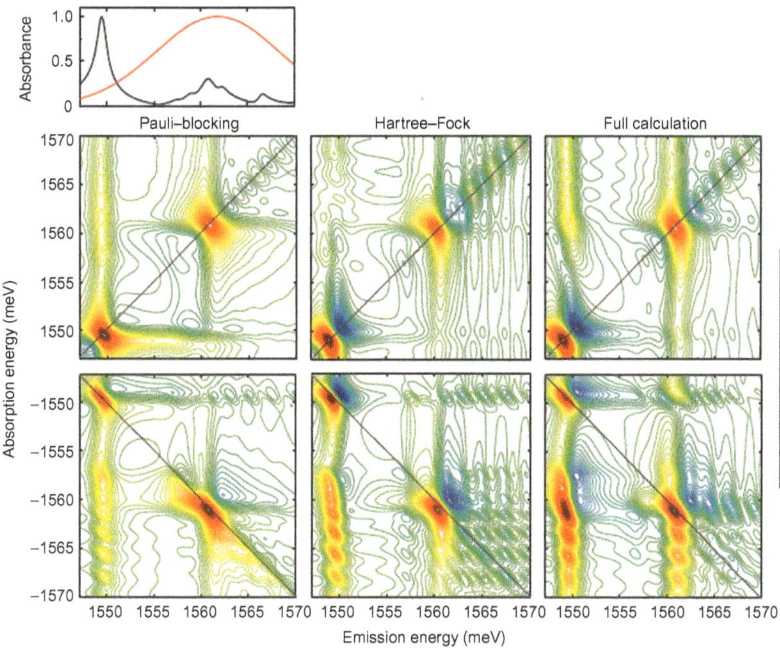


**Fig. 15** Theoretical one-quantum 2D spectra for cocircular excitation. The 2D spectra in the *upper* (*lower*) *row* are nonrephasing (rephasing) spectra. Calculations are done for three cases that include only Pauli-blocking (*left*), the Hartree–Fock approximation (*center*), and the full theory (*right*). The corresponding absorption spectrum and the laser spectrum are shown in the *top panel. Adapted from Zhang, T., Kuznetsova, I., Meier, T., Li, X., Mirin, R.P., Thomas, P., et al., 2007. Polarization-dependent optical 2D Fourier transform spectroscopy of semiconductors. Proc. Natl. Acad. Sci. U.S.A. 104 (36), 14227–14232 with permission.*

and pose a more stringent test of the theory. This helps to establish a full understanding of many-body physics in semiconductors.

## 5.3 Two-Quantum 2D Spectra of Excitons in GaAs Quantum Wells

Two-quantum 2D spectra are sensitive to two-exciton correlations in semiconductors (Yang and Mukamel, 2008a,b). In GaAs quantum wells, two excitons can form a bound state, known as the biexciton, when the two excitons have opposite electron spins, or an unbound two-exciton state when the spins are the same. The energy-level diagram in the exciton picture, as shown in Fig. 16, displays one- and two-exciton states of the *hh* and *lh* excitons. The two-exciton states have an energy that is the sum of the individual exciton energies plus a shift. The biexciton states are shifted lower

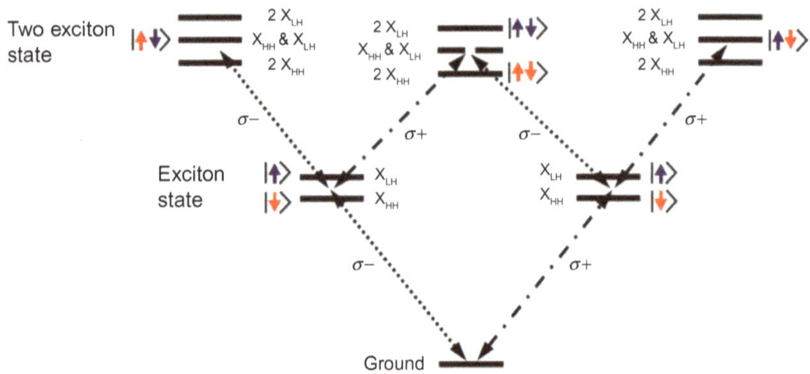


**Fig. 16** Energy-level diagram showing one-exciton and two-exciton states in GaAs quantum wells. Electron spins are labeled for one-exciton states and bound biexciton states.

by the biexciton binding energy. The unbound two-exciton states are shifted higher by the scattering energy. Two-quantum 2D spectra of these two-exciton states have been observed and shown that the dominant signal is due to mean-field many-body interactions, rather than higher order correlations terms, including biexcitons (Karaiskaj et al., 2010). This conclusion differs from the one-quantum spectra and shows that the different types of spectra are sensitive to different many-body effects.

### *5.3.1 Experiment*

The pulse sequence with the conjugated pulse $A^*$ arriving last is used for the two-quantum 2DCS experiment on GaAs quantum wells at 6K. The excitation pulses are either cross-circularly ($\sigma^-\sigma^+\sigma^+$) or cocircularly ($\sigma^+\sigma^+\sigma^+$) polarized. In the former case, the accessible two-exciton states are the bound biexcitons with two *hh* or *lh* excitons and the unbound states with mixed *hh* and *lh* excitons. In the latter case, the accessible biexcitons are a mixture of *hh* and *lh* excitons and the unbound states each include two *hh* or *lh* excitons. Moreover, the *hh* and *lh* exciton resonances can be selectively excited by tuning the laser spectrum.

Experimental two-quantum 2D spectra are shown in Fig. 17. Spectra (A) and (B) are the results when the laser spectrum is centered at the *hh* exciton resonance. For the cross-circular excitation ($\sigma^-\sigma^+\sigma^+$), the spectrum has two peaks: an absorptive peak (labeled "A") corresponding to a biexciton state, and a dispersive peak (labeled "B") due to an unbound two-exciton

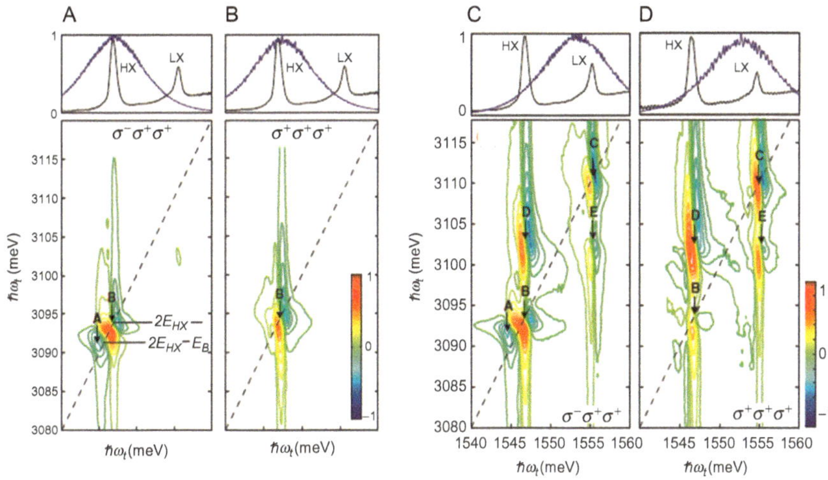


**Fig. 17** Experimental two-quantum real 2D spectra for (A) cross-circular and (B) cocircular excitations with the laser spectrum centered at the *hh* exciton, and spectra for (C) cross-circular and (D) cocircular excitations with the laser spectrum centered between the *hh* and *lh* excitons. The corresponding absorption spectra and laser spectra are also shown in the *top panel. Adapted from Karaiskaj, D., Bristow, A.D., Yang, L., Dai, X., Mirin, R.P., Mukamel, S., et al., 2010. Two-quantum many-body coherences in two-dimensional Fourier-transform spectra of exciton resonances in semiconductor quantum wells. Phys. Rev. Lett. 104 (11), 117401 with permission.*

state. The spectrum with the cocircular polarizations ($\sigma^+\sigma^+\sigma^+$) has only a dispersive peak associated with an unbound two-exciton state. These spectral peaks were predicted by previous theoretical calculations (Yang and Mukamel, 2008a,b), and the bound biexciton contribution was expected to dominate the cross-circular spectra. However, the dominant feature in experimental spectra is peak B due to the unbound two-exciton state. Peak B in both spectra has a dispersive lineshape, indicating its origin of many-body interactions. This result shows that the two-quantum signal is primarily contributed by mean-field many-body interactions, rather than bound biexcitons.

Both *hh* and *lh* exciton resonances can be excited if the laser spectrum is centered between the *hh* and *lh* excitation energies. The resulting two-quantum 2D spectra are shown in Fig. 17C and D for the cross-circular ($\sigma^-\sigma^+\sigma^+$) and cocircular ($\sigma^+\sigma^+\sigma^+$) polarizations, respectively. Besides peaks A and B from *hh* excitons, there are also contributions from *lh* excitons and mixed states of *hh* and *lh* excitons. Peak C is contributed by the two spectrum. Moreover, carrier percolation through the barrier can be ruled out as a coupling mechanism in InGaAs/GaAs quantum wells (Kim et al., 1996).

In the study of InGaAs/GaAs double-quantum wells, two samples were used and each consisted of an 8-nm narrow well (NW) and a 10-nm wide well (WW). The two wells were separated by a 30-nm GaAs barrier in the first sample and a 10-nm barrier in the second. The 2D spectrum of the first sample includes only two separate diagonal peaks corresponding to the NW and WW excitons. There was no coupling between the NW and WW with the 30-nm barrier. However, the one-quantum 2D spectrum of the second sample, as shown in Fig. 19A, has a cross peak in addition to the two diagonal peaks. One possible explanation of the cross peak is the coupling between the WW exciton and the continuum that happens to overlap with the NW exciton. This possibility can be ruled out since the cross peak does not occur for the first sample and the vertical stripes due to the continuum are not present. Therefore, the cross is clear evidence for the coupling between the WW and NW.

To understand the coupling mechanism, a set of different 2D spectra are acquired, including a phase-resolved one-quantum 2D spectrum, a zero-quantum 2D spectrum, and a two-quantum 2D spectrum. They are shown in Fig. 19B–D. The dispersive lineshape of the cross peak in Fig. 19B indicates that many-body interactions play a role in the coupling mechanism. However, the zero-quantum 2D spectrum, as shown in Fig. 19C, includes two peaks at the zero mixing frequency but no peaks at nonzero mixing frequency. A coherence between the excited states would give rise to a peak at a nonzero mixing frequency in a zero-quantum 2D spectrum. The absence of such peaks suggests that the WW and NW excitons are not coupled through a common ground state. The coupling mechanism is better understood from the two-quantum 2D spectrum, as shown in Fig. 19D. The two peaks on the diagonal line arise from the two-exciton coherence in the same quantum well, while the two cross peaks require coherent interactions between the WW and NW excitons.

To determine the role of many-body effects in the coupling, 2D spectra are reproduced in a simulation based on the optical Bloch equations for a six-level system. The energy-level structure consists of a ground state, two singly excited states with one exciton in either WW or NW, two doubly excited states with two WW excitons or two NW excitons, and one additional doubly excited state that has two mixed excitons, i.e., one from each quantum well. Many-body effects are introduced by breaking the symmetry in the

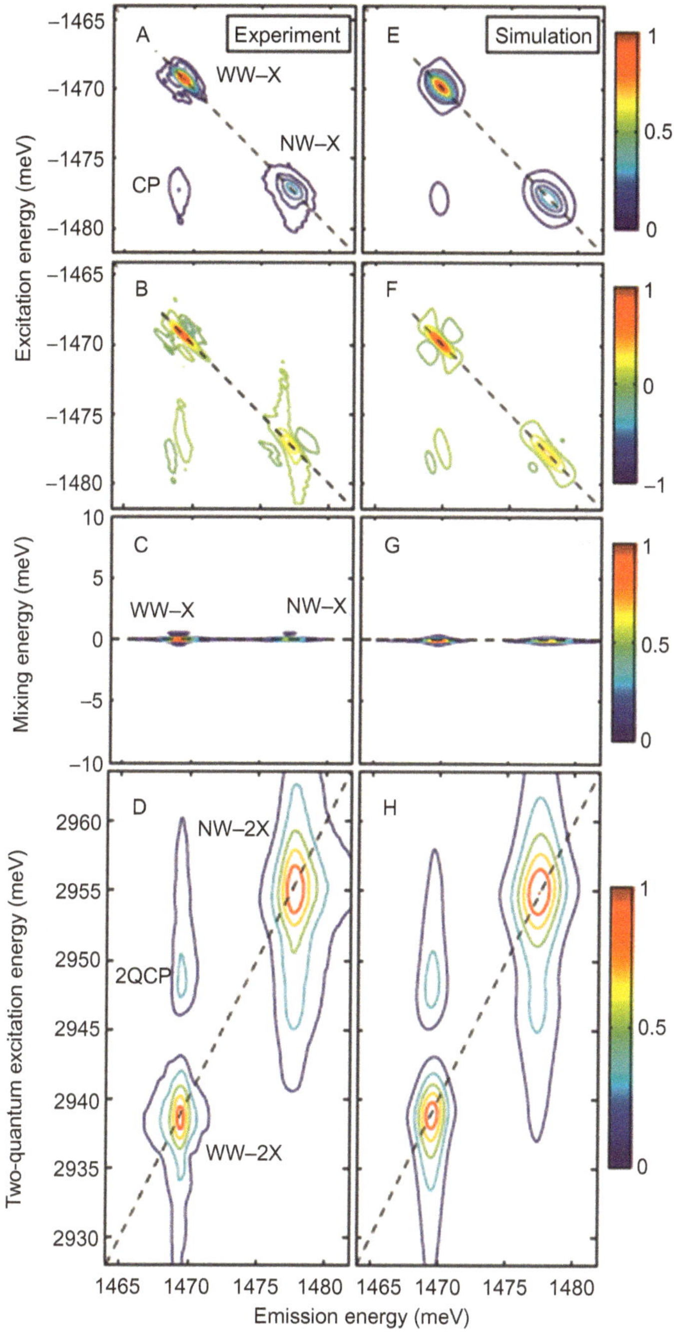


**Fig. 19** Experimental (*left*) and simulated (*right*) 2D spectra of the asymmetric double-quantum wells with a 10-nm barrier. The spectra include (A, E) the amplitude and (B, F) the real part of one-quantum rephasing 2D spectra, (C, G) the amplitude of zero-quantum 2D spectra, and (D, H) the amplitude of two-quantum 2D spectra. *Adapted from Nardin, G., Moody, G., Singh, R., Autry, T.M., Li, H., Morier-Genoud, F., et al., 2014. Coherent excitonic coupling in an asymmetric double InGaAs quantum well arises from many-body effects. Phys. Rev. Lett. 112 (4), 46402 with permission.*

energy-level structure with interaction-induced changes in transition energy and dephasing rate (Bott et al., 1993). The spectra calculated with a single set of parameters, as shown in Fig. 19E–H, reproduce all features in the experimental 2D spectra. Without the inclusion of many-body effects, the cross peaks in Fig. 19E and H would not appear. This study shows that many-body interactions between the WW and NW excitons play an important role and lead to the coherent coupling of the two quantum wells.

## 6. OTHER SYSTEMS

The examples reviewed here are only a small sampling of 2DCS studies on various systems. This technique has been used to study a variety of problems in molecular systems, such as anharmonicity in molecular vibrations (Khalil et al., 2003, 2004), formation and dissociation of solute-solvent complexes (Zheng et al., 2005), hydrogen bonding in water and proteins, and energy transfer in photosynthesis. Its applications in condensed matters also include carrier dynamics and coupling in semiconductor quantum dots (Kasprzak et al., 2010; Moody et al., 2013a,b,c,d), and valley coherence in newly emerging 2D semiconductors (Hao et al., 2016; Moody et al., 2015).

One of the most notable influences of 2DCS is in the field of light harvesting where the quantum mechanical aspect of energy transfer processes is studied. A crucial step of the light-harvesting process in photosynthesis is the conversion of absorbed photons into charge-separated states at the reaction center. This energy transfer process is found to be remarkably efficient at nearly 100%. Understanding the fundamental mechanism responsible for the near-unity efficiency is of great interest due to its potential to improve the performance of artificial light-harvesting materials. Numerous 2DCS studies of photosynthetic complexes have provided evidence for long-lasting coherence between two electronic states, suggesting that the energy transfer should be described by wavelike coherent motion instead of incoherent hopping (Engel et al., 2007). The efficiency improvement can be understood as a result of the quantum interference of multiple pathways between the electronic states involved in the energy transfer.

In these studies, 2DCS experiments were performed on light-harvesting proteins. The coupling between the chromophores in the protein gives rise to a cross peak in the 2D spectrum. Tracking the cross-peak amplitude as a function of the waiting time $T$ shows that the cross-peak amplitude is oscillating with time. This is a signature of the quantum coherence between the two involved chromophores. This quantum behavior was initially observed in protein samples that were held at a cryogenic temperature and excited by coherent light. Later experiments were also done at ambient temperature (Collini et al., 2010) and with incoherent light (Fassioli et al., 2012).

Despite intensive research, the underlying mechanisms of quantum coherence and its role in energy transfer are still not fully understood. For example, the roles played by electronic and vibrational couplings in quantum coherence are under debate (Butkus et al., 2012; Tiwari et al., 2013). Future studies are needed to answer some open questions: how is quantum coherence directly related to and improve the energy transfer efficiency? What conditions are required to support quantum coherence? How to utilize quantum coherence to improve the efficiency in artificial light-harvesting materials?

## 7. SUMMARY AND OUTLOOK

In summary, 2DCS has been developed into a robust and powerful technique for studying complex dynamics and couplings in various systems in the fields of physics, chemistry, and biology. We have reviewed the fundamental principles, experimental approaches, and data interpretations of 2DCS, and its applications in atomic vapors and semiconductor quantum wells.

The field of 2DCS is facing great challenges and opportunities to develop new capabilities for a broader scope of applications. One future direction is to extend the technique into new spectral regimes. There are efforts by a number of groups to realize 2DCS in the UV and THz. This can enable 2DCS studies of DNA molecules, which have resonances in the UV, and 2D materials and topological materials where the THz spectroscopy might be more suitable. Moreover, recent advances in coherent X-ray sources, including free-electron lasers and tabletop X-ray lasers, may enable X-ray 2DCS that provides both attosecond and nanometer resolution.

Another direction is to develop 2DCS with high spatial resolution and/or imaging capability. Most current implementations of 2DCS have a spatial resolution close to the diffraction limit or worse. Some applications such as studies of domain/domain wall on surfaces and single molecules require nanometer resolution. A promising approach is to integrate 2DCS with near-field techniques. For example, 2DCS excitation pulses can be coupled to an AFM tip to achieve a high spatial resolution.

Finally, one critical challenge in new applications of 2DCS is to improve the frequency resolution while maintaining its other advantages including multidimensional capability and temporal resolution. In molecular spectroscopy, resolving rovibronic features requires a frequency resolution better than 3 GHz. In cold atoms, a frequency resolution of MHz is necessary to resolve hyperfine levels and measurements of the energy shift due to dipole–dipole interaction demand an even higher resolution at the kHz level. A possible solution is to implement 2DCS with femtosecond frequency combs, which can simultaneously provide broad bandwidth, femtosecond temporal resolution, and high-frequency resolution. Frequency comb-based 2DCS opens new avenues in molecular spectroscopy and many-body physics in cold atoms/molecules.

## ACKNOWLEDGMENTS

We acknowledge all the members of the Cundiff research group at JILA and University of Michigan who have participated in the development and application of 2D coherent spectroscopy over the last 15 years.

## REFERENCES

Bott, K., Heller, O., Bennhardt, D., Cundiff, S.T., Thomas, P., Mayer, E.J., et al., 1993. Influence of exciton-exciton interactions on the coherent optical response in GaAs quantum wells. Phys. Rev. B 48 (23), 17418–17426.

Bristow, A.D., Karaiskaj, D., Dai, X., Cundiff, S.T., 2008. All-optical retrieval of the global phase for two-dimensional Fourier-transform spectroscopy. Opt. Express 16 (22), 18017.

Bristow, A.D., Karaiskaj, D., Dai, X., Zhang, T., Carlsson, C., Hagen, K.R., et al., 2009a. A versatile ultrastable platform for optical multidimensional Fourier-transform spectroscopy. Rev. Sci. Instrum. 80 (7), 73108.

Bristow, A.D., Karaiskaj, D., Dai, X., Mirin, R.P., Cundiff, S.T., 2009b. Polarization dependence of semiconductor exciton and biexciton contributions to phase-resolved optical two-dimensional Fourier-transform spectra. Phys. Rev. B 79 (16), 161305.

Brixner, T., Stiopkin, I.V., Fleming, G.R., 2004. Tunable two-dimensional femtosecond spectroscopy. Opt. Lett. 29 (8), 884.

Butkus, V., Zigmantas, D., Valkunas, L., Abramavicius, D., 2012. Vibrational vs. electronic coherences in 2D spectrum of molecular systems. Chem. Phys. Lett. 545, 40–43.

Chemla, D.S., Shah, J., 2001. Many-body and correlation effects in semiconductors. Nature 411 (6837), 549.

Chen, P.C., 2010. High resolution coherent 2D spectroscopy. J. Phys. Chem. A 114, 11365–11375.

Chen, P.C., Gomes, M., 2008. Two-dimensional coherent double resonance electronic spectroscopy. J. Phys. Chem. A 112 (14), 2999–3001.

Chen, P.C., Wells, T.A., Strangfeld, B.R., 2013. High-resolution coherent three-dimensional spectroscopy of Br 2. Chem. A Eur. J. 117 (29), 5981–5986.

Collini, E., Wong, C.Y., Wilk, K.E., Curmi, P.M.G., Brumer, P., Scholes, G.D., 2010. Coherently wired light-harvesting in photosynthetic marine algae at ambient temperature. Nature 463 (7281), 644–647.

Cowan, M.L., Ogilvie, J.P., Miller, R.J.D., 2004. Two-dimensional spectroscopy using diffractive optics based phased-locked photon echoes. Chem. Phys. Lett. 386, 184–189.

Cundiff, S.T., 2008. Coherent spectroscopy of semiconductors. Opt. Express 16 (7), 4639.

Cundiff, S.T., 2012. Optical two-dimensional Fourier transform spectroscopy of semiconductor nanostructures [Invited]. J. Opt. Soc. Am. B 29 (2), A69.
Cundiff, S.T., 2014. Optical three-dimensional coherent spectroscopy. Phys. Chem. Chem. Phys. 16, 8193–8200.
Cundiff, S.T., Mukamel, S., 2013. Optical multidimensional coherent spectroscopy. Phys. Today 66 (7), 44.
Cundiff, S.T., Bristow, A.D., Siemens, M., Li, H., Moody, G., Karaiskaj, D., et al., 2012. Optical 2-d Fourier transform spectroscopy of excitons in semiconductor nanostructures. IEEE J. Sel. Top. Quantum Electron. 18 (1), 318–328.
Dai, X., Bristow, A.D., Karaiskaj, D., Cundiff, S.T., 2010. Two-dimensional Fourier-transform spectroscopy of potassium vapor. Phys. Rev. A 82 (5), 52503.
Dai, X., Richter, M., Li, H., Bristow, A.D., Falvo, C., Mukamel, S., et al., 2012. Two-dimensional double-quantum spectra reveal collective resonances in an atomic vapor. Phys. Rev. Lett. 108 (19), 193201.
Davis, J.A., Hall, C.R., Dao, L.V., Nugent, K.A., Quiney, H.M., Tan, H.H., et al., 2011. Three-dimensional electronic spectroscopy of excitons in asymmetric double quantum wells. J. Chem. Phys. 135 (4), 44510.
Ding, F., Zanni, M.T., 2007. Heterodyned 3D IR spectroscopy. Chem. Phys. 341, 95–105.
Engel, G.S., Calhoun, T.R., Read, E.L., Ahn, T.-K., Mancal, T., Cheng, Y.-K., et al., 2007. Evidence for wavelike energy transfer through quantum coherence in photosynthetic systems. Nature 446, 782–786.
Ernst, R., Bodenhausen, G., Wokaun, A., 1987. Principles of Nuclear Magnetic Resonance in One and Two Dimensions. Oxford Science Publications, Oxford, UK.
Fassioli, F., Olaya-Castro, A., Scholes, G.D., 2012. Coherent energy transfer under incoherent light conditions. J. Phys. Chem. Lett. 3 (21), 3136–3142.
Fidler, A.F., Harel, E., Engel, G.S., 2010. Dissecting hidden couplings using fifth-order three-dimensional electronic spectroscopy. J. Phys. Chem. Lett. 1, 2876–2880.
Gao, F., Cundiff, S.T., Li, H., 2016. Probing dipole–dipole interaction in a rubidium gas via double-quantum 2D spectroscopy. Opt. Lett. 41 (13), 2954–2957.
Garrett-Roe, S., Hamm, P., 2009. Purely absorptive three-dimensional infrared spectroscopy. J. Chem. Phys. 130, 164510.
Garrett-Roe, S., Perakis, F., Rao, F., Hamm, P., 2011. Three-dimensional infrared spectroscopy of isotope-substituted liquid water reveals heterogeneous dynamics. J. Phys. Chem. B 115, 6976–6984.
Grumstrup, E.M., Shim, S.-H., Montgomery, M.A, Damrauer, N.H., Zanni, M.T., 2007. Facile collection of two-dimensional electronic spectra using femtosecond pulse-shaping Technology. Opt. Express 15 (25), 16681–16689.
Hall, C.R., Tollerud, J.O., Quiney, H.M., Davis, J.A., 2013. Three-dimensional electronic spectroscopy of excitons in asymmetric double quantum wells. New J. Phys. 15 (4), 45028.
Hao, K., Moody, G., Wu, F., Dass, C.K., Xu, L., Chen, C.-H., et al., 2016. Direct measurement of exciton valley coherence in monolayer WSe2. Nat. Phys. 12 (7), 677–682.
Hu, Y.Z., Binder, R., Koch, S.W., Cundiff, S.T., Wang, H., Steel, D.G., 1994. Excitation and polarization effects in semiconductor four-wave-mixing spectroscopy. Phys. Rev. B 49 (20), 14382–14386.
Karaiskaj, D., Bristow, A.D., Yang, L., Dai, X., Mirin, R.P., Mukamel, S., et al., 2010. Two-quantum many-body coherences in two-dimensional Fourier-transform spectra of exciton resonances in semiconductor quantum wells. Phys. Rev. Lett. 104 (11), 117401.
Kasprzak, J., Patton, B., Savona, V., Langbein, W., 2010. Coherent coupling between distant excitons revealed by two-dimensional nonlinear hyperspectral imaging. Nat. Photonics 5 (1), 57–63.

Khalil, M., Demirdöven, N., Tokmakoff, A., 2003. Obtaining absorptive line shapes in two-dimensional infrared vibrational correlation spectra. Phys. Rev. Lett. 90 (4), 47401.

Khalil, M., Demirdöven, N., Tokmakoff, A., 2004. Vibrational coherence transfer characterized with Fourier-transform 2D IR spectroscopy. J. Chem. Phys. 121 (1), 362.

Kim, D.-S., Shah, J., Damen, T.C., Schäfer, W., Jahnke, F., Schmitt-Rink, S., et al., 1992. Unusually slow temporal evolution of femtosecond four-wave-mixing signals in intrinsic GaAs quantum wells: direct evidence for the dominance of interaction effects. Phys. Rev. Lett. 69 (18), 2725–2728.

Kim, D.S., Ko, H.S., Kim, Y.M., Rhee, S.J., Hohng, S.C., Yee, Y.H., et al., 1996. Percolation of carriers through low potential channels in thick AlxGa1 − xAs (x < 0.35) barriers. Phys. Rev. B 54 (20), 14580–14588.

Leo, K., Wegener, M., Shah, J., Chemla, D., Göbel, E., Damen, T., et al., 1990. Effects of coherent polarization interactions on time-resolved degenerate four-wave mixing. Phys. Rev. Lett. 65 (11), 1340–1343.

Li, H., Bristow, A.D., Siemens, M.E., Moody, G., Cundiff, S.T., 2013a. Unraveling quantum pathways using optical 3D Fourier-transform spectroscopy. Nat. Commun. 4, 1390.

Li, H., Moody, G., Cundiff, S.T., 2013b. Reflection optical two-dimensional Fourier-transform spectroscopy. Opt. Express 21, 1687–1692.

Li, H., Spencer, A.P., Kortyna, A., Moody, G., Jonas, D.M., Cundiff, S.T., 2013c. Pulse propagation effects in optical 2D Fourier-transform spectroscopy: experiment. J. Phys. Chem. A 117, 6279–6287.

Li, X., Zhang, T., Borca, C., Cundiff, S., 2006. Many-body interactions in semiconductors probed by optical two-dimensional Fourier transform spectroscopy. Phys. Rev. Lett. 96 (5), 57406.

Li, X., Zhang, T., Mukamel, S., Mirin, R.P., Cundiff, S.T., 2009. Investigation of electronic coupling in semiconductor double quantum wells using coherent optical two-dimensional Fourier transform spectroscopy. Solid State Commun. 149 (9–10), 361–366.

Lindberg, M., Binder, R., Koch, S.W., 1992. Theory of the semiconductor photon echo. Phys. Rev. A 45 (3), 1865–1875.

Marzin, J.-Y., Charasse, M.N., Sermage, B., 1985. Optical investigation of a new type of valence-band configuration in InxGa1 − xAs-GaAs strained superlattices. Phys. Rev. B 31 (12), 8298–8301.

Moody, G., Singh, R., Li, H., Akimov, I.A., Bayer, M., Reuter, D., et al., 2013a. Biexcitons in semiconductor quantum dot ensembles. Phys. Status Solidi B 250 (9), 1753–1759.

Moody, G., Singh, R., Li, H., Akimov, I.A., Bayer, M., Reuter, D., et al., 2013b. Influence of confinement on biexciton binding in semiconductor quantum dot ensembles measured with two-dimensional spectroscopy. Phys. Rev. B 87 (4), 41304.

Moody, G., Singh, R., Li, H., Akimov, I.A., Bayer, M., Reuter, D., et al., 2013c. Correlation and dephasing effects on the non-radiative coherence between bright excitons in an InAs QD ensemble measured with 2D spectroscopy. Solid State Commun. 163, 65–69.

Moody, G., Singh, R., Li, H., Akimov, I.A., Bayer, M., Reuter, D., et al., 2013d. Fifth-order nonlinear optical response of excitonic states in an InAs quantum dot ensemble measured with two-dimensional spectroscopy. Phys. Rev. B 87, 45313.

Moody, G., Akimov, I.A., Li, H., Singh, R., Yakovlev, D.R., Karczewski, G., et al., 2014. Coherent coupling of excitons and trions in a photoexcited CdTe/CdMgTe quantum well. Phys. Rev. Lett. 112 (9), 97401.

Moody, G., Kavir Dass, C., Hao, K., Chen, C.-H., Li, L.-J., Singh, A., et al., 2015. Intrinsic homogeneous linewidth and broadening mechanisms of excitons in monolayer transition metal dichalcogenides. Nat. Commun. 6, 8315.

Moran, M., Dawson, P., Moore, K.J., 1998. The nature of the light hole potential profile in GaAs-InGaAs double quantum well structures. Solid State Commun. 107 (3), 119–123.
Nardin, G., Autry, T.M., Silverman, K.L., Cundiff, S.T., 2013. Multidimensional coherent photocurrent spectroscopy of a semiconductor nanostructure. Opt. Express 21 (23), 28617.
Nardin, G., Moody, G., Singh, R., Autry, T.M., Li, H., Morier-Genoud, F., et al., 2014. Coherent excitonic coupling in an asymmetric double InGaAs quantum well arises from many-body effects. Phys. Rev. Lett. 112 (4), 46402.
Pakoulev, A.V., Rickard, M.A., Kornau, K.M., Mathew, N.A., Yurs, L.A., Block, S.B., et al., 2009. Mixed frequency-/time-domain coherent multidimensional spectroscopy: research tool or potential analytical method? Acc. Chem. Res. 42 (9), 1310–1321.
Schäfer, W., Jahnke, F., Schmitt-Rink, S., 1993. Many-particle effects on transient four-wave-mixing signals in semiconductors. Phys. Rev. B 47 (3), 1217–1220.
Selig, U., Langhojer, F., Dimler, F., Löhrig, T., Schwarz, C., Gieseking, B., et al., 2008. Inherently phase-stable coherent two-dimensional spectroscopy using only conventional optics. Opt. Lett. 33 (23), 2851.
Shacklette, J., Cundiff, S., 2002. Role of excitation-induced shift in the coherent optical response of semiconductors. Phys. Rev. B 66 (4), 1–6.
Shim, S.-H., Strasfeld, D.B., Ling, Y.L., Zanni, M.T., 2007. Automated 2D IR spectroscopy using a mid-IR pulse shaper and application of this technology to the human islet amyloid polypeptide. Proc. Natl. Acad. Sci. U.S.A. 104, 14197–14202.
Siemens, M.E., Moody, G., Li, H., Bristow, A.D., Cundiff, S.T., 2010. Resonance lineshapes in two-dimensional Fourier transform spectroscopy. Opt. Express 18, 17699–17708.
Spencer, A.P., Li, H., Cundiff, S.T., Jonas, D.M., 2015. Pulse propagation effects in optical 2D Fourier-transform spectroscopy: theory. Chem. A Eur. J. 119 (17), 3936–3960.
Tanimura, Y., Mukamel, S., 1993. Two-dimensional femtosecond vibrational spectroscopy of liquids. J. Chem. Phys. 99 (12), 9496.
Tekavec, P.F., Lott, G.A., Marcus, A.H., 2007. Fluorescence-detected two-dimensional electronic coherence spectroscopy by acousto-optic phase modulation. J. Chem. Phys. 127, 2143078.
Tian, P.F., Keusters, D., Suzaki, Y., Warren, W.S., 2003. Femtosecond phase-coherent two-dimensional spectroscopy. Science 300 (5625), 1553–1555.
Tiwari, V., Peters, W.K., Jonas, D.M., 2013. Electronic resonance with anticorrelated pigment vibrations drives photosynthetic energy transfer outside the adiabatic framework. Proc. Natl. Acad. Sci. U.S.A. 110 (4), 1203–1208.
Turner, D.B., Stone, K.W., Gundogdu, K., Nelson, K.A., 2009. Three-dimensional electronic spectroscopy of excitons in GaAs quantum wells. J. Chem. Phys. 131, 144510.
Turner, D.B., Stone, K.W., Gundogdu, K., Nelson, K.A., 2011. Invited article: the coherent optical laser beam recombination technique (COLBERT) spectrometer: coherent multidimensional spectroscopy made easier. Rev. Sci. Instrum. 82.
Turner, D., Wen, P., Arias, D., Nelson, K., Li, H., Moody, G., et al., 2012. Persistent exciton-type many-body interactions in GaAs quantum wells measured using two-dimensional optical spectroscopy. Phys. Rev. B. 85, 201303.
Volkov, V., Schanz, R., Hamm, P., 2005. Active phase stabilization in Fourier-transform two-dimensional infrared spectroscopy. Opt. Lett. 30 (15), 2010.
Wagner, W., Li, C., Semmlow, J., Warren, W.S., 2005. Rapid phase-cycled two-dimensional optical spectroscopy in fluorescence and transmission mode. Opt. Express 13 (10), 3697.
Wang, H., Ferrio, K., Steel, D.G., Hu, Y.Z., Binder, R., Koch, S.W., 1993. Transient nonlinear optical response from excitation induced dephasing in GaAs. Phys. Rev. Lett. 71 (8), 1261–1264.

Wegener, M., Chemla, D.S., Schmitt-Rink, S., Schäfer, W., 1990. Line shape of time-resolved four-wave mixing. Phys. Rev. A 42 (9), 5675–5683.
Weiss, S., Mycek, M.-A., Bigot, J.-Y., Schmitt-Rink, S., Chemla, D.S., 1992. Collective effects in excitonic free induction decay: do semiconductors and atoms emit coherent light in different ways? Phys. Rev. Lett. 69 (18), 2685–2688.
Wells, T.A., Muthike, A.K., Robinson, J.E., Chen, P.C., 2015. High resolution coherent three dimensional spectroscopy of NO2. J. Chem. Phys. 142 (21), 212426.
Wright, J.C., 2011. Multiresonant coherent multidimensional spectroscopy. Annu. Rev. Phys. Chem. 62 (1), 209–230.
Yajima, T., Taira, Y., 1979. Spatial optical parametric coupling of picosecond light-pulses and transverse relaxation effect in resonant media. J. Phys. Soc. Jpn. 47 (5), 1620–1626.
Yang, L., Mukamel, S., 2008a. Revealing exciton-exciton couplings in semiconductors using multidimensional four-wave mixing signals. Phys. Rev. B 77 (7), 075335.
Yang, L., Mukamel, S., 2008b. Two-dimensional correlation spectroscopy of two-exciton resonances in semiconductor quantum wells. Phys. Rev. Lett. 100 (5), 57402.
Zhang, T., Borca, C.N., Li, X., Cundiff, S.T., 2005. Optical two-dimensional Fourier transform spectroscopy with active interferometric stabilization. Opt. Express 13 (19), 7432–7441.
Zhang, T., Kuznetsova, I., Meier, T., Li, X., Mirin, R.P., Thomas, P., et al., 2007. Polarization-dependent optical 2D Fourier transform spectroscopy of semiconductors. Proc. Natl. Acad. Sci. U.S.A. 104 (36), 14227–14232.
Zheng, J., Kwak, K., Asbury, J., Chen, X., Piletic, I.R., Fayer, M.D., 2005. Ultrafast dynamics of solute-solvent complexation observed at thermal equilibrium in real time. Science 309 (5739), 1338–1343.

CHAPTER TWO

# Nonlinear and Quantum Optical Properties and Applications of Intense Twin-Beams

**Alessia Allevi*[,†], Maria Bondani*[,†,1]**
*Università degli Studi dell'Insubria, Como, Italy
[†]Istituto di Fotonica e Nanotecnologie, Consiglio Nazionale delle Ricerche, Como, Italy
[1]Corresponding author: e-mail address: maria.bondani@uninsubria.it

## Contents



## Abstract

In this review we present the main results of some experimental investigations that we performed in the last 10 years on optical twin-beam states. We explore twin-beam statistical properties and spatio-spectral coherence in different intensity regimes, from the mesoscopic to the macroscopic domain, also including pump depletion. We also characterize the quantum nature of the mesoscopic twin-beam states in terms of different nonclassicality criteria, all of them written for detected photons and demonstrate the

*Advances in Atomic, Molecular, and Optical Physics*, Volume 66
ISSN 1049-250X
http://dx.doi.org/10.1016/bs.aamop.2017.02.001

conditional nonclassical states (Allevi et al., 2010; Lamperti et al., 2014). For this reason, it is interesting to exploit the dependence of the number of modes on the interaction parameters to optimize the tailoring of the states.

This chapter reviews some recent results on generation, characterization, and application of TWBs in the high-intensity regime up to pump depletion, experimentally achieved in the laboratory of Quantum Optics at University of Insubria. Such results have been obtained by means of a ps-pulsed laser source regeneratively amplified at 500 Hz, which delivers pulses with high-peak intensities (up to some GW/cm$^2$) at 349 nm (third harmonics). All over the years, this source allowed us to produce bright TWBs at each pump pulse. Indeed, ps-pulsed lasers represent the best sources to observe intense TWBs, as they can produce pulses with high-peak intensities in a limited pulse spectrum (up to 1 nm), thus preventing the occurrence of the dispersion effects typical of femtosecond laser sources. In our research activity, the same laser was also used to investigate the almost unexplored mesoscopic domain. Actually, this regime represents an interesting compromise between the single-photon level, where single-photon pairs are produces, and the macroscopic one, in which bright TWB states are generated. In the mesoscopic domain, where sizeable numbers of photons per pulse are produced, the TWB states are more robust than single-photon pairs with respect to losses, but still preserve, and easily exhibit, the nonclassical properties that make TWBs useful for several applications to quantum technologies. At the moment, the complete investigation of the mesoscopic regime is limited by the performance of the available detectors. Although during the last 15 years different kinds of photon-number resolving detectors have been developed and tested, the ideal solution is still far to come. In this review, we present the characterization of mesoscopic TWBs achieved by means of hybrid photodetectors, which are a particular class of commercial photon-number resolving detectors operating at room temperature. Such a characterization is not only limited to the investigation of the statistical properties, but also extended to the nonclassical features, which represent a useful resource in quantum state engineering for applications to quantum information and metrology. For instance, it is well known that TWBs can be used to produce nonclassical states by means of conditional measurements based on the existence of correlations between signal and idler. In particular, when a given number of photons is detected in the idler, the state of the signal is irreversibly changed from classical to nonclassical and from Gaussian to non-Gaussian. In this review, we show that sub-Poissonian states can be generated by performing conditional measurements on mesoscopic TWBs, where the multimode nature of TWBs represents a crucial resource. The same property is of paramount importance for the other application, we present in this review, namely the implementation of a ghost-imaging protocol. Indeed, the visibility and the signal-to-noise ratio, which are the fundamental indicators of the quality of the imaging protocols, are strongly affected by the number of both spatial and temporal modes characterizing the TWB states. Finally, we remark that the investigations of TWBs at different intensities regimes gave us a better knowledge of the nonlinear phenomenon and of the evolution of its properties, and also allowed us to understand in which way the TWB features can be optimized and tailored for the different applications to quantum technologies. The chapter is organized as follows: Section 2 introduces the multimode description of TWB states. Section 3 describes some experimental results obtained in different regimes and with different apparatuses. An initial subsection presents the well-characterized laser source, which represents a key issue for the realization of many experiments. Section 4 shows two relevant applications exploiting different properties of TWB states. Section 5 closes the chapter.

# 2. THEORY

## 2.1 Twin-Beam Generation

The process of multimode PDC is described by the interaction Hamiltonian

$$\hat{H} = \sum_{j}^{\mu} \hat{H}_j \tag{1}$$

which is the sum of $\mu$-independent contributions

$$\hat{H}_j \propto \hat{a}^{\dagger}_{si,j} \hat{a}^{\dagger}_{id,j} \hat{a}_{pu} + h.c., \tag{2}$$

each coupling three monochromatic field modes: pump (*pu*), signal (*si*), and idler (*id*), identified by their frequencies and wavevectors. $\hat{a}^{\dagger}_{l,j}$ are the field-mode operators satisfying the commutation rules $[\hat{a}_{l,j}, \hat{a}^{\dagger}_{l,i}] = \delta_{j,i}$ for $l = si, id$. In order to obtain the parametric interaction, energy and momentum conservation must be fulfilled

$$\omega_{pu} = \omega_{si,j} + \omega_{id,j} \tag{3}$$

$$\boldsymbol{k_{pu}}(\omega_{pu}) = \boldsymbol{k_{si,j}}(\omega_{si,j}) + \boldsymbol{k_{id,j}}(\omega_{id,j}). \tag{4}$$

The unitary evolution operator in the interaction picture reads

$$\hat{U} = \exp\left[-i\tau\left(\hat{a}^{\dagger}_{si,j}\hat{a}^{\dagger}_{id,j}\hat{a}_{pu} + \hat{a}_{si,j}\hat{a}_{id,j}\hat{a}^{\dagger}_{pu}\right)\right], \tag{5}$$

where $\tau$ is a rescaled interaction time.

The usual description of PDC in the low-intensity regime is performed within the so-called parametric approximation: considering that the generated TWBs have negligible intensity with respect to the pump, the pump field is assumed as nonevolving during the interaction. Mathematically speaking, the parametric approximation consists in replacing the pump-mode operator $\hat{a}_{pu}$ with the complex amplitude $A_{pu}$ of the corresponding coherent state. In phase-matching conditions, we get the evolution operator:

$$\hat{U} = \exp\left[\kappa(\boldsymbol{k_{pu}}, \boldsymbol{k_{si,j}})\hat{a}^{\dagger}_{si,j}\hat{a}^{\dagger}_{id,j} - \kappa^{*}(\boldsymbol{k_{pu}}, \boldsymbol{k_{si,j}})\hat{a}_{si,j}\hat{a}_{id,j}\right], \tag{6}$$

where the coupling function $\kappa(\boldsymbol{k_{pu}}, \boldsymbol{k_{si,j}})$ depends on the interaction parameters describing the phase-matching geometry in the specific material and on pump intensity.

The state generated from the vacuum by the evolution in Eq. (6) is the single-mode TWB state

$$|\psi\rangle_j = \sqrt{1 - |\lambda(\boldsymbol{k_{pu}}, \boldsymbol{k_{si,j}})|^2} \sum_n \lambda(\boldsymbol{k_{pu}}, \boldsymbol{k_{si,j}})^n |n\rangle|n\rangle, \tag{7}$$

$|n\rangle$ being the state containing $n$ photons, $\phi_0$ being the phase of the pump field, and $\lambda(\boldsymbol{k_{pu}}, \boldsymbol{k_{si,j}}) \equiv \tanh\left[\kappa(\boldsymbol{k_{pu}}, \boldsymbol{k_{si,j}})\right]\exp(i\phi_0)$. The mean photon-number value in the generated signal and idler is $n_{PDC,j} = \sinh^2\left[\kappa(\boldsymbol{k_{pu}}, \boldsymbol{k_{si,j}})\right]$ and the photon-number statistics is a thermal distribution

$$p_{n,j} = \frac{n^n_{PDC,j}}{(1 + n_{PDC,j})^{n+1}}, \tag{8}$$

so that the TWB state can be rewritten as

$$|\psi\rangle_j = \sum_n \sqrt{p_{n,j}} \exp(i\phi_0)|n\rangle|n\rangle. \tag{9}$$

We note that the single-mode TWB displays perfect photon-number correlations and that the wavefunction is written as a nonfactorable superposition of number states, meaning that the TWB state is entangled.

In the case of multimode PDC, we assume that all the contributions in the Hamiltonian in Eq. (1) are equal, so that the generated state is a tensor product of $\mu$ identical twin-beam states (Allevi et al., 2010), differing from each other for the wavevector. The multimode state becomes

$$|\psi_\mu\rangle = \bigotimes_{j=1}^{\mu} |\psi\rangle_j, \tag{10}$$

where $|\psi\rangle_j$ has the form in Eq. (7) and the mean photon number of all $j$ contributions is a constant, i.e., $n_{\mathrm{PDC},j} \equiv n_{\mathrm{PDC}}$. The mean total number of photons in the multimode state is $N = \mu n_{\mathrm{PDC}}$ and the statistics of photons in each arm of the TWB becomes the multimode thermal distribution

$$p_n^\mu = \frac{(n+\mu-1)!}{n!(\mu-1)!(n_{\mathrm{PDC}}+1)^\mu(1/n_{\mathrm{PDC}}+1)^n}. \tag{11}$$

We can thus write the multimode TWB in the following compact form

$$|\psi_\mu\rangle = \sum_{n=0}^{\infty} \sqrt{p_n^\mu}\,|n^\otimes\rangle|n^\otimes\rangle, \tag{12}$$

where $|n^\otimes\rangle = \delta(n - \sum_{h=1}^{\mu} n_h) \otimes_{k=1}^{\mu} |n\rangle_k$ represents the overall $n$ photons coming from the $\mu$ modes that impinge on the detector and $p_n^\mu$ is the multimode thermal distribution in Eq. (11). Note that also for the multimode TWB, perfect correlations in the number of photons are expected.

Eqs. (8) and (11) describe the statistical distributions for the number of photons in the states as they would be detected by a perfect detector. This is obviously not the case for real photodetectors, which reveal light with a nonunit detection efficiency. In Appendix we remind some basics of the description of detectors. Here we only note that the functional form of the statistics of detected photons could in principle be different from that of the photons. Actually, it can be demonstrated (Bondani et al., 2009b; Casini and Martinelli, 2013) that all the classical statistics, obtainable as superpositions of coherent states, are invariant under detection when the positive-operator valued measure (POVM) is Bernoullian (see Appendix). If instead the statistics is nonclassical, such as in the case of Fock or sub-Poissonian states, it will change upon detection.

**(3)** High-order inequality $B > 1$, introduced by Bondani in Allevi et al. (2012), where

$$B = \langle m_{si} \rangle \langle m_{id} \rangle \frac{g_{\hat{m}}^{22} - [g_{\hat{m}}^{13}]_s}{g_{\hat{m}}^{11}} + \sqrt{\langle m_{si} \rangle \langle m_{id} \rangle} \frac{[g_{\hat{m}}^{12}]_s}{g_{\hat{m}}^{11}}, \tag{18}$$

in which we have used the symmetrized quantities $[g_{\hat{m}}^{hk}]_s = (g_{\hat{m}}^{hk} + g_{\hat{m}}^{kh})/2$.

We observe that in terms of detected quantities all the above criteria depend on the intensity of the fields, but in different ways. In fact, $S \to 1$ at increasing number $N$ of photons in the TWB, regardless the values of quantum efficiencies (we remind that $\langle m_{si,\,id} \rangle = \eta_{si,id} N$). The Schwarz inequality is thus verified at any intensity, but it becomes more and more difficult to experimentally test at increasing intensities. As to $R$, this is the most widely used parameter to test TWB nonclassicality, and it is well known that the inequality is hard to violate at high intensity. The reason is that, while $R < 1$ for any value of $N$ if $\eta_{si} = \eta_{id}$, $R > 1$ for $N < 2\mu\eta_{si}\eta_{id}/(\eta_{si} - \eta_{id})^2$ if $\eta_{si} \neq \eta_{id}$, which means that there is a threshold in the detectability of entanglement in TWB that depends on the number of modes (the largest, the best) and on the perfect balancing of the efficiencies (the most similar, the best). At variance with $S$ and $R$, $B$ is not sensitive to unbalancing and increases at high intensities.

## 2.3 Multimode Twin Beam

In general, the generated TWBs are intrinsically multimode. However, in the existing literature the TWB states are usually treated as single-mode states, as the calculations are simpler and the nonclassicality criteria useful to characterize such states can be easily defined. In particular, for some physical quantities, such as the quantum discord and the non-Gaussianity, not only the theoretical extension to the multimode case is hardly achievable, but also the experimental measurements are not always easily accessible. Nevertheless, in Section 4 we will show that the multimode nature of TWBs is an important resource for the implementation of some protocols, such as for the generation of sub-Poissonian states. That is why in the following we will consider two different approaches to the description of the multimode nature of TWBs. In the first scenario, the pump beam is treated as a plane-wave undepleted field, whereas in the second one it is allowed to evolve together with signal and idler.

### 2.3.1 Effect of Phase Mismatch

In order to explore the effects of phase matching on the generation of multimode TWB, we consider the realization of the PDC nonlinear process in a uniaxial crystal, in which the pump field propagates along the normal to the crystal entrance face (Bondani et al., 2012). In Fig. 1, a generic phase-matching condition in space is depicted, along with the definitions of angles that identify the propagation directions inside the crystal. The phase-matching conditions can be written as:

$$k_{si}\sin\beta_{si} + k_{id}\sin\beta_{id} = 0 \tag{19}$$

$$k_{si}\cos\beta_{si}\sin\theta_{si} + k_{id}\cos\beta_{id}\sin\theta_{id} = 0 \tag{20}$$

$$k_{si}\cos\beta_{si}\cos\theta_{si} + k_{id}\cos\beta_{id}\cos\theta_{id} = k_{pu}. \tag{21}$$

The frequency- and phase-matching conditions allow the generation of TWBs on a continuum of frequencies and propagation directions linked to each other by nontrivial relations (Jedrkiewicz et al., 2004). Moreover, the possibility to generate light in conditions of phase-mismatch produces an output field characterized by the simultaneous presence of many modes, both spatial and temporal (Brida et al., 2009b,c). In the far field, the angular and spectral dependence of the output field can be factorized and the multimode character of the states can be considered in time and space separately. In particular, we observe that single-shot images of the output

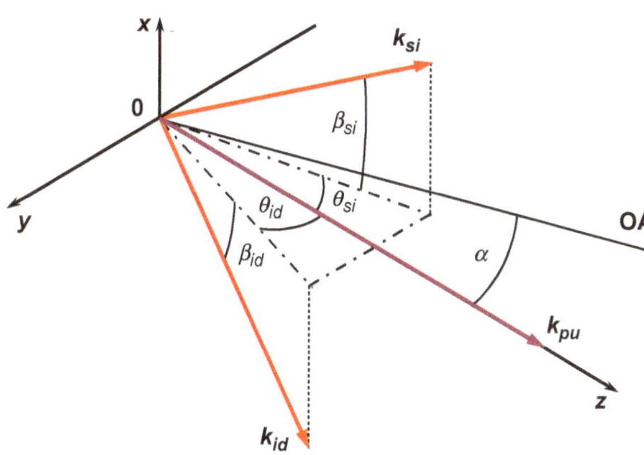


**Fig. 1** Scheme of the phase-matched interaction: $(x,y)$-plane coincides with the crystal entrance face; $\alpha$, tuning angle; $\beta_j$, angles to $(y,z)$-plane; $\theta_j$, angles on the $(y,z)$-plane; *OA*, optical axis. *Adapted from Bondani, M., Allevi, A., Andreoni, A., 2012. Ghost imaging by intense multimode twin beam. Eur. Phys. J. Spec. Top. 203 (1), 151–161. ISSN 1951-6401. http://dx.doi.org/10.1140/epjst/e2012-01540-4.*

display a "speckle" pattern that can be interpreted as a particular realization of the nonlinear process: the center of each speckle individuates the direction of one of the phase-matched wavevectors, whereas the size of the speckle depends on the angular bandwidth allowed by phase mismatch. Similarly, we can also consider the spectral bandwidth allowed by the frequency mismatch and introduce the concept of temporal modes, which, in the case of a pulsed field, can be viewed as the ratio of the spectral bandwidth of the process to the spectral bandwidth of the pump field (Paleari et al., 2004). The presence of temporal modes reflects on the photon-number statistics of the light inside a single coherence area.

In Fig. 2 we plot a typical single-shot far-field pattern generated in achievable experimental conditions (Bondani et al., 2012). To characterize the multimodal nature of the PDC output, in Mikhailov et al. (2003) we have studied the dependence of the gain of the parametric process, i.e., of the number of photons, of a single-mode TWB as a function of the interaction parameters by analogy with the classical gain of the nonlinear parametric process. We can write the gain for a given choice of mode index $j$ as:

$$\begin{aligned}\Gamma &= \frac{4\gamma^2|A_{pu}|^2}{(\boldsymbol{\Delta k}\cdot\hat{\boldsymbol{k}}_{si})^2}\left[\frac{4\gamma^2|A_{pu}|^2}{(\boldsymbol{\Delta k}\cdot\hat{\boldsymbol{k}}_{si})(\boldsymbol{\Delta k}\cdot\hat{\boldsymbol{k}}_{id})}-1\right]^{-1}\\ &\times\sinh^2\left[\sqrt{\frac{4\gamma^2|A_{pu}|^2}{(\boldsymbol{\Delta k}\cdot\hat{\boldsymbol{k}}_{si})(\boldsymbol{\Delta k}\cdot\hat{\boldsymbol{k}}_{id})}-1}\;\frac{\boldsymbol{\Delta k}\cdot\boldsymbol{r}}{2}\right],\end{aligned} \tag{22}$$

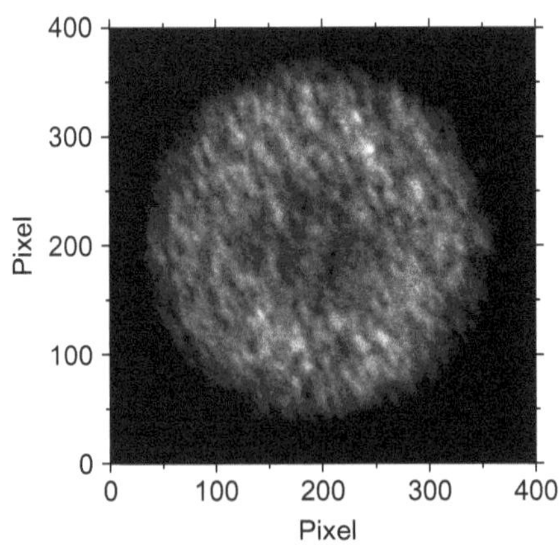


**Fig. 2** Single-shot image of the down-conversion cones at frequency degeneracy ($\lambda_{si,\,id}$ = 698 nm) in the far field. *Adapted from Bondani, M., Allevi, A., Andreoni, A., 2012. Ghost imaging by intense multimode twin beam. Eur. Phys. J. Spec. Top. 203 (1), 151–161. ISSN 1951-6401. http://dx.doi.org/10.1140/epjst/e2012-01540-4.*

where "^" here indicates the versors of the different vectors and $\gamma$ is the coupling coefficient of the nonlinear process. Mikhailov et al. (2003) have demonstrated that the direction of the phase mismatch is aligned to the bisector between signal and probe so that $\Delta\boldsymbol{k}\cdot\hat{\boldsymbol{k}}_{si}=\Delta\boldsymbol{k}\cdot\hat{\boldsymbol{k}}_{id}=|\Delta\boldsymbol{k}|\cos(\psi_{si,id}/2)$, $\psi_{si,id}$ being the angle between signal and probe. We assume $4|A_{pu}|^2/[(\Delta\boldsymbol{k}\cdot\hat{\boldsymbol{k}}_{si})(\Delta\boldsymbol{k}\cdot\hat{\boldsymbol{k}}_{id})]\gg 1$ and expand the expression in Eq. (22), obtaining:

$$\Gamma\simeq\frac{1}{4}\exp\left(2\gamma|A_{pu}|r\right)\exp\left(-\frac{|\Delta\boldsymbol{k}|^2\cos^2(\psi_{si,id}/2)}{4|A_{pu}|}r\right). \tag{23}$$

Note that the value of the gain in phase matching ($\Delta\boldsymbol{k}=0$) is $\sinh^2(\gamma|A_{pu}|r)$, that coincides with $n_{PDC}=\sinh^2(\kappa)$ upon identification of $\kappa=\gamma|A_{pu}|r$.

We can now use Eq. (23) to calculate the value of $|\Delta\boldsymbol{k}|^2$ corresponding to the full width at half maximum (FWHM) of the gain

$$|\Delta\boldsymbol{k}|^2_{\mathrm{FWHM}}=\frac{4\ \ln 2\gamma|A_{pu}|}{r\cos^2(\psi_{si,id}/2)}(\equiv D). \tag{24}$$

Since $|\Delta\boldsymbol{k}|^2$ is a function of the frequencies and of the angles of the parametric emission, we can expand it about the phase-matching condition and find the corresponding values of angular and temporal bandwidths.

First of all, we consider the angles fixed at their values in phase matching and expand $|\Delta\boldsymbol{k}|^2$ in a Taylor's series up to the second order of the expansion of the wavevectors about $\overline{\omega}_{si}$ and $\overline{\omega}_{id}$, $\overline{\omega}_j$ being the central frequency of signal and idler fields. By assuming frequency- and phase-matching for the central frequencies of the interacting fields ($\overline{\omega}_{pu}=\overline{\omega}_{si}+\overline{\omega}_{id}$ and $\boldsymbol{k}_{pu}(\overline{\omega}_{pu})-\boldsymbol{k}_{si}(\overline{\omega}_{si})-\boldsymbol{k}_{id}(\overline{\omega}_{id})=0$), we get $\delta\omega_{si}=-\delta\omega_{id}(\equiv\delta\omega)$ and

$$|\Delta\boldsymbol{k}|^2\simeq\frac{1}{v_{g,si}^2}+\frac{1}{v_{g,id}^2}-2\frac{1}{v_{g,si}}\frac{1}{v_{g,id}}\cos(\psi_{si,id})(\delta\omega)^2(\equiv A(\delta\omega)^2), \tag{25}$$

where $v_{g,j}=[(\partial k/\partial\omega)_{\overline{\omega}_j}]^{-1}$ is the group velocity. By jointly using Eqs. (24) and (25) we obtain

$$\Delta\omega_{\mathrm{FWHM}}=2\sqrt{D/A}, \tag{26}$$

and the number of temporal modes

$$\mu\equiv\frac{\tau}{\Delta t_{\mathrm{FWHM}}}=\frac{\tau\Delta\omega_{\mathrm{FWHM}}}{4\ \ln 2}. \tag{27}$$

but also the photon-number correlations between signal and idler (Allevi et al., 2012; Avenhaus et al., 2010; Bartley et al., 2013; Dovrat et al., 2013; Harder et al., 2016; Kalashnikov et al., 2012). To this aim, most investigations were produced in the single-photon regime by means of single-photon detectors, such as single-photon avalanche diodes (SPADs), photomultipliers and intensified CCD (iCCD) cameras, or electron multiplying CCD (EMCCD) cameras (Migdall et al., 2013). In the opposite intensity domain, multi-photon TWB states have been investigated by means of CCD cameras, operated either at room temperature or with a proper cooling (Brida et al., 2009a; Jedrkiewicz et al., 2004), or of pin photodiodes (Agafonov et al., 2010; Bondani et al., 2007). In the last two decades also TWB states generated in the intermediate regime, i.e., the so-called mesoscopic domain, have been measured and characterized by means of different classes of photon-number resolving detectors (Allevi et al., 2010; Lamperti et al., 2014). Among them, it is worth mentioning the hybrid photodetector (HPD), which combines a photocathode with a diode structure operated below the breakdown threshold (Andreoni and Bondani, 2009; Bondani et al., 2009a). At variance with the traditional photomultipliers, in such a detector the amplification process occurs in a single step, so that the excess noise is small enough to allow photon-number resolution (up to six detected photons). The quantum efficiency of the detector, mainly given by the photocathode, can reach good values (up to 0.5 in the visible range), whereas the dark-count rate can be neglected. As an alternative, the accomplishment of photon-counting capability can be obtained by splitting the light to be measured either in space or in time prior to detection so that at most one photon at a time hits the detector sensitive area. Among these detectors, we mention the visible light photon counter (VLPC) (Kim et al., 1999); the fiber-loop detector, which is a time-multiplexed detector based on one (Řeháček et al., 2003) or more (Fitch et al., 2003) SPADs; the silicon photomultiplier (SiPM), that is, constituted by a matrix of SPADs with a common output (Akindinov et al., 1997). Due to their composite structure, SiPMs have a good photon-counting capability, even if their large dark-count rate (100 kHz in the new generation) and their not negligible cross-talk probability (Afek et al., 2009; Ramilli et al., 2010) have till now prevented their use in the investigation of the nonclassical character of TWB states, except at very low-intensity levels. In the last decade, a tremendous progress has been achieved in the field of superconductors, so that new types of detectors have been developed, such as the transition-edge sensors (TES) (Lita et al., 2008) and the superconducting nanowires (Gol'tsman et al., 2001). Despite having a good quantum efficiency, these detectors must operate at cryogenic temperatures and thus their operation is rather cumbersome. As of today, the ideal detector has yet to appear and the optimal choice is application specific.

For what concerns the characterization of TWB states, it is worth noting that most of the experimental works were conducted in a limited intensity regime by using a specific kind of detector. Nevertheless, TWB states could in principle change their properties (mean number of photons, size of the spectral and spatial areas, photon-number correlation coefficient, number of Schmidt modes, etc.) as a function of pump power and a full investigation would be necessary. Indeed, a complete investigation of the nature of TWBs at different PDC gains has received only limited attention due to the difficulty of finding a proper detection apparatus able to cover a sufficiently wide intensity range. For instance, iCCD and EMCCD cameras can be in principle operated both at single-photon level and in the macroscopic domain. However, they cannot be used in the transient region, where the noise due to the spread of the gain strongly limits the signal-to-noise ratio. On the contrary, in the following Section 3.3 we show a different strategy that we recently adopted to investigate the statistical properties of TWBs in different intensity regimes.

## 3.2 The Laser System

All the experimental results we will discuss in the following were obtained by exploiting a few-ps-pulsed mode-locked Nd:YLF laser (High-Q Laser, Austria), regeneratively amplified at 500 Hz, as the laser source. The emitted fundamental wavelength is at 1047 nm, and second and third harmonics are built in. This source produces pulses having high-peak intensities (up to some GW/cm$^2$ at 349 nm) and a quite narrow spectral bandwidth (less than 1 nm at 349 nm). Such features make it possible the generation of bright pulsed TWBs, thus allowing the investigation of the PDC process at different intensity levels. In fact, the ps-pulses are at the same time less damaging than ns-pulses of the same intensities and narrower in spectrum than fs-pulses. As to the pump field, most of the results presented in this review have been obtained with the third harmonics of the laser at 349 nm. Frequency-degenerate TWB states at 698 nm can still be detected by cameras and photon-number resolving detectors operating in the visible range. In order to match the maximum quantum efficiency of the detectors, we also used the fourth harmonics at 262 nm, which allows the generation of TWBs of the apparatus. In fact, it has been demonstrated in (Bondani et al., 2009a) that the Fano factor $\mathfrak{F}_v = \sigma_v^2/\langle v \rangle$ for the output voltages is a linear function of the mean value of the output voltages. In particular, the limit of this function for the mean values approaching zero gives the value of $\gamma$. In the case of HPD, such a value also coincides with the distance between two consecutive valleys in the pulse-height spectrum of the detector output. The number of detected photons is thus obtained by the following procedure: each output voltage is subtracted of the mean value of the electronic noise measured in the absence of light, then the resulting values are divided by the value of $\gamma$, determined in one of the two ways described above, and rebinned in unitary bins. To perform a systematic characterization of the optical states under investigation, a proper data sample of consecutive laser pulses is usually acquired, thus allowing the study of the statistical properties of light. In the present case, for each choice of the pump power each experimental run was repeated 100,000 times. A set of neutral-density filters was used to suitably attenuate the light and keep it within the detector dynamics. For the data presented here we used filters from 0 to 6 optical densities, whose transmittance was measured at the specific wavelength. In such a way we could measure a signal beam in the range from 0.6 to $1.75 \times 10^6$ mean detected photons. The errors of the various quantities plotted in the following were calculated as the standard deviations of their mean values evaluated for blocks of 20,000 consecutive shots. The power was changed by means of a half-wave plate, followed by a polarizing cube beam splitter located in front of the BBO crystal, in order to keep the same spatial profile of the pump.

### 3.3.2 Results

In Fig. 4A we plot the experimental detected-photon distribution (dots) corresponding to different choices of pump power (black: 18.7 mW, red: 22.7 mW, green: 27 mW, blue: 58.9 mW, and magenta: 89 mW). As anticipated above, the neutral-density filters kept the light value within the dynamic range of the HPD detectors. For this reason, there is no direct connection between the pump power values and the measured mean values of the TWB. Since the light measured on either signal or idler separately is classical, the presence of the filters does not change the photon-number distribution and, in particular, does not affect the number of independent modes. Fig. 4A reports also the multimode thermal fitting curves (solid lines) for the experimental data obtained by exploiting Eq. (11), in which the mean value, $\langle m \rangle$, is fixed to the values experimentally measured and the number of modes, $\mu$, is the only fitting parameter. The $\mu$ values obtained from those

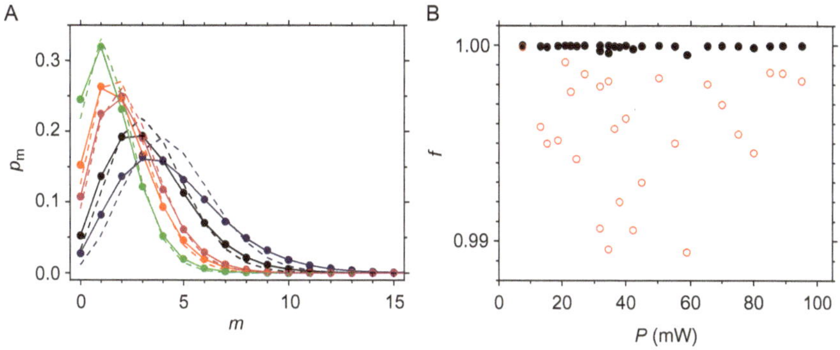


**Fig. 4** (A) Examples of experimental detected-photon number distributions (*dots*), multimode-thermal fit (*solid lines*), and Poissonian fit (*dashed lines*). Results obtained at different pump powers: 8.7 mW in *black*, 22.7 mW in *red*, 27 mW in *green*, 58.9 mW in *blue*, and 89 mW in *magenta*. (B) Fidelity for the fit to a multimode-thermal distribution (*full dots*) and fidelity for the fit to a Poissonian distribution (*empty circles*). *Adapted from Allevi, A., Bondani, M., 2014. Statistics of twin-beam states by photon-number resolving detectors up to pump depletion. J. Opt. Soc. Am. B 31 (10), B14–B19. http://dx.doi.org/10.1364/JOSAB.31.000B14.*

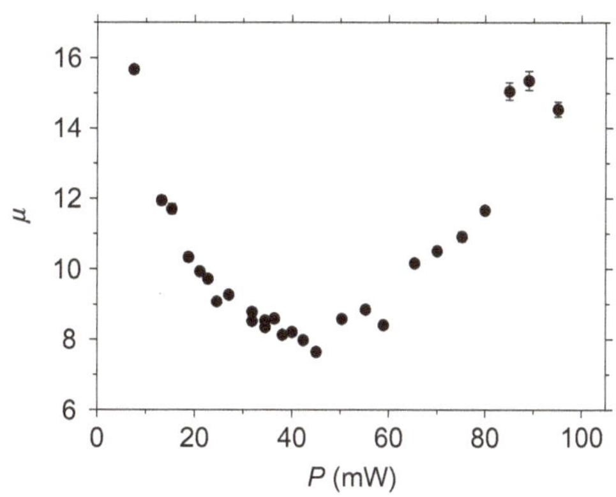


**Fig. 5** Number $\mu$ of spatio-spectral modes as a function of the pump mean power calculated from the statistics of the number of photons detected in the signal arm (*black dots*). *Adapted from Allevi, A., Bondani, M., 2014. Statistics of twin-beam states by photon-number resolving detectors up to pump depletion. J. Opt. Soc. Am. B 31 (10), B14–B19. http://dx.doi.org/10.1364/JOSAB.31.000B14.*

fits are plotted in Fig. 5. The very good agreement between data and theory is attested by the very high values of the fidelity shown in Fig. 4B (full circles). As a comparison, in Fig. 4A, we also plot the Poissonian curves (dashed lines) calculated for the experimental mean values. As shown by the fidelity values in Fig. 4B (empty circles), the superposition to the experimental data is definitely worse. We conclude that the multimode thermal nature of the photon-number statistics does not depend on pump power. On the contrary, the number of modes is strongly influenced by the complex evolution of the system, as it appears from the data of Fig. 5. In particular, it is worth noting the appearance of a minimum in the evolution of $\mu$ as a function of the input power, which appears for a specific value of the pump mean value. In order to understand what really happens inside the crystal, it is instructive to also consider the evolution of the mean number of detected photons in one of the TWB arm as a function of the pump mean power.

In Fig. 6 we plot the mean number of detected photons in the signal arm as a function of the pump mean power. The plotted values were calculated by taking into account the transmittance of the filters. First of all, we observe that for low pump power values the mean number of photons increases exponentially, as expected for high-gain PDC under the hypothesis of undepleted pump beam. This initial behavior is emphasized in the inset of the picture, where the experimental data corresponding to the lowest pump power values are presented. In the main figure, we see that the occurrence of a progressive pump depletion at very high-gain PDC processes prevents the exponential growth of the mean number of photons. On the

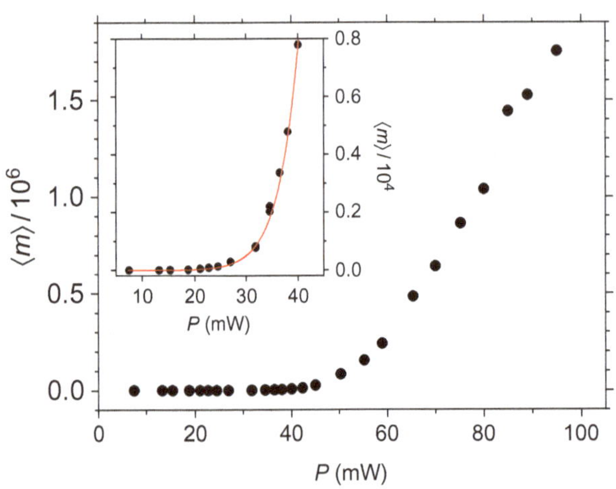


**Fig. 6** Evolution of the mean number of photons detected in the signal arm as a function of the pump mean power. *Inset*: first part of the data shown in the main figure exhibiting an exponential growth according to undepleted pump approximation. *Adapted from Allevi, A., Bondani, M., 2014. Statistics of twin-beam states by photon-number resolving detectors up to pump depletion. J. Opt. Soc. Am. B 31 (10), B14–B19. http://dx.doi.org/10.1364/JOSAB.31.000B14.*

contrary, a linear dependence on the pump power begins. We notice that the change in the slope roughly occurs at the value of pump mean power, where the minimum in the number of modes of Fig. 5 occurs.

## 3.4 Spatio-Spectral Properties

The evolution of the statistical properties of TWB in terms of photon-numbers is strictly connected to the evolution of spatial and spectral properties. Indeed, it is well known that TWB exhibits spatial and spectral correlations. As already anticipated in Section 1, the first experimental measurements of spatial correlations in TWB, aimed at determining the size of the coherence areas in a transverse plane, were performed at single-photon level on TWB states by means of scanned single-photon detectors (Blanchet et al., 2008; Haderka et al., 2005b; Hamar et al., 2010). The experimental results can be compared to a well-established theory properly working in this regime, in which, for instance, the pump beam can be considered as nonevolving during the interaction (undepleted pump approximation). On the other hand, TWB can also be generated in a much higher intensity regime (Agafonov et al., 2010; Bondani et al., 2007; Brida et al., 2009a; Jedrkiewicz et al., 2004), in which coherence areas become visible in single-shot images taken at the output of the nonlinear crystal. More recently, also the spectral features of macroscopic TWBs have been investigated in the collinear interaction geometry close to frequency degeneracy (Spasibko et al., 2012). Moreover, in the high-gain regime, the X-shaped coherence of the PDC output field (Spasibko et al., 2012) and the X-shaped spatiotemporal TWB near-field correlations (Gatti et al., 2009; Jedrkiewicz et al., 2006, 2012), originating from the space–time coupling in the phase matching, have been demonstrated.

Here, on the base of Allevi et al. (2014a), we show that the use of an EMCCD camera, combined to an imaging spectrometer that resolves emission angles and wavelengths simultaneously, represents an interesting solution to investigate the evolution of spatial and spectral properties in the high-gain regime, where pump depletion can suddenly occur (Allevi et al., 2014a).

The experimental setup used for the measurement of the PDC light structure in the angular and spectral $(\theta, \lambda)$ domain is shown in Fig. 7A. A type-I 8 mm-long BBO crystal (cut angle = 37 degree) was pumped by the third-harmonic pulses of the laser source described in Section 3.2. The FWHM of the pump beam, collimated by means of a telescope in front of the BBO, was $\sim 380\ \mu m$ at the lowest pump power. Indeed, the pump mean power was

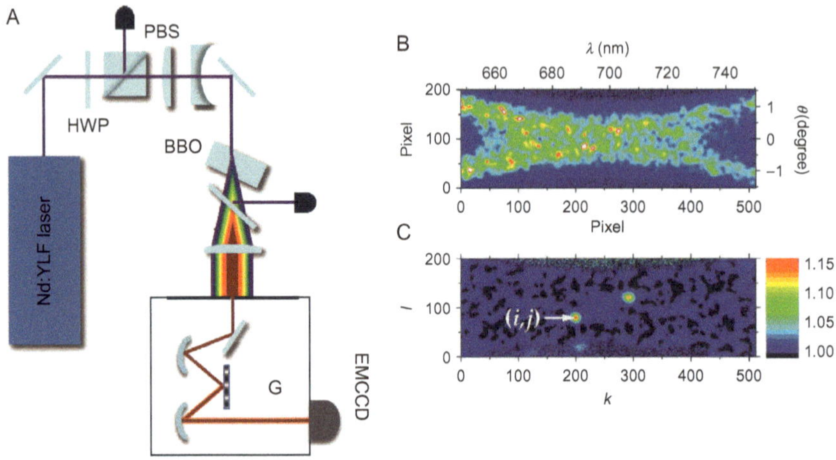


**Fig. 7** (A) Experimental setup used for the spatio-spectral measurements of the TWB. *BBO*, nonlinear crystal; *EMCCD*, electron multiplying camera; *G*, grating; *HWP*, half-wave plate; *PBS*, polarizing cube beam splitter. (B) Single-shot image recorded by the EMCCD camera, in which the typical speckle-like pattern of PDC in the spatio-spectral domain is clearly evident. (C) Typical contour plot of the correlation matrix Γ, in which the single pixel (*i*, *j*) is chosen on the left side. The intensity autocorrelation area is on the same side, whereas the cross-correlation area is on the right side. *Adapted from Allevi, A., Jedrkiewicz, O., Brambilla, E., Gatti, A., Peřina, J., Haderka, O., Bondani, M., 2014a. Coherence properties of high-gain twin beams. Phys. Rev. A 90, 063812. http://dx.doi.org/10.1103/PhysRevA.90.063812; Perina Jr., J., Haderka, O., Allevi, A., Bondani, M., 2016. Internal dynamics of intense twin beams and their coherence. Sci. Rep. 6. http://dx.doi.org/10.1038/srep22320.*

changed during the experiment by a half-wave plate followed by a polarizing cube beam splitter. The crystal was tuned to have phase-matching at frequency degeneracy in slightly noncollinear configuration. The broadband PDC light was collected by a 60-mm focal length lens and focused on the plane of the vertical slit of the already mentioned imaging spectrometer. The angularly dispersed far-field radiation was then recorded in single shot by a synchronized EMCCD camera (iXon Ultra 897, Andor), operated at full frame resolution (512 × 512 pixels, 16-μm pixel size). The resulting resolution of the system composed of the imaging spectrometer and the EMCCD camera was 0.2 nm in spectrum and 0.015 degree in angle. A typical speckle-like pattern is shown in Fig. 7B, in which the horizontal axis is connected to spectrum and the vertical axis to the angular dispersion. The existence of intensity correlations between the signal and idler portions of the TWB is well supported by the presence of symmetrical speckles around the degenerate wavelength and the collinear direction.

The evolution of the patterns at different pump mean powers *P*, and hence at different PDC gains, was investigated by calculating the intensity correlation coefficient between a single pixel at coordinates $(i, j)$ and all the pixels $(k, l)$ contained in a single image

$$\Gamma_{k,l}^{(i,j)} = \frac{\langle I_{i,j} I_{k,l} \rangle}{\langle I_{i,j} \rangle \langle I_{k,l} \rangle}, \tag{41}$$

where $I$ is the intensity value of each pixel expressed in digital numbers and upon subtraction of the mean value of the noise measured with the camera in perfect dark, whereas $\langle \ldots \rangle$ indicates the averaging over a sequence of 1000 subsequent images. The procedure was applied to a set of pixels having the abscissa $i$ close to frequency degeneracy and the ordinate $j$ in the quasi-collinear direction.

The function $\Gamma_{k,l}^{(i,j)}$ defined in Eq. (41) is a matrix having the same size as the original images, in which the horizontal axis, running on the index $k$, is connected to spectrum, while the vertical axis, running on the index $l$, is connected to the angular dispersion. As an example, in Fig. 7C the contour plot of the correlation matrix, obtained by choosing the pixel at coordinates $(i, j)$ on the left side of the images, is shown. The intensity autocorrelation area is on the same side, whereas the cross-correlation area is on the right side.

Fig. 8 shows the behaviors of the spectral (panel A) and spatial, i.e., in angular domain, (panel B) widths, FWHM, of the intensity autocorrelation and cross-correlation areas, as functions of the input pump mean power. In both panels, we can observe an initial growth that reaches the maximum at a pump power of about 30 mW and then decreases. As shown in the figure, only the first part of the data is well described by a fourth-root square function of pump power. This kind of evolution is predicted by the theory of coherence areas under the assumption of undepleted pump beam (Bondani et al., 2012; Brambilla et al., 2004; Brida et al., 2009a). The second part of our experimental results (including the peak and the decrease in the FWHM) clearly indicates that for high pump power values the assumption of undepleted pump beam does not hold anymore. Indeed, in this situation also the pump beam evolves nontrivially and the corresponding equations of motion for the three-mode interaction can be solved only numerically. Indeed, the behavior shown in Fig. 8 can be qualitatively reproduced by means of a numerical simulation of the PDC process, as already shown in Allevi et al. (2014a). In the numerical modeling the propagation equations

initially slightly lateral with respect to the center because of the pump beam walk-off inside the crystal, as shown in the bottom panel (D).

### 3.4.1 Far-Field Coherence Areas

As already described in Section 2, the evolution of the size of intensity auto- and cross-correlation areas, due to pump depletion, can be explained in terms of the modes that describe the radiation field. When the PDC process occurs at gain values leading to depletion, also the pump beam evolves in the nonlinear interaction and the dynamics of the system becomes more complex. In particular, there is a dependence of the number of effectively populated signal and idler radiation modes on the pump power. As the pump power increases, the PDC gain profile becomes narrower and narrower, and thus signal and idler fields are dominantly emitted into a smaller and smaller number of modes that gain energy to the detriment of the others (Pérez et al., 2014; Wasilewski et al., 2006). For sufficiently high values of the pump power, the process of mode selection reverts as the pump profile undergoes depletion. For this reason, the gain of the high-populated low-order modes is on the one side reduced, whereas the gain of low-populated higher-order modes is on the other side supported. Such a behavior explains the narrowing of the intensity auto- and cross-correlation areas shown in Fig. 8. The description in terms of populated radiation modes also explains the slight discrepancy between auto- and cross-correlation intensity functions plotted in Fig. 8. In fact, the cross-correlation function reflects the mutual coherence between signal and idler and originates in the pairwise PDC emission, whereas the autocorrelation function expresses the internal coherence due to the presence of three evolving fields. As such, it is more sensitive to losses in the mode selection.

With the goal of investigating how the variation of the coherence area is affected by a variation in the number of effectively populated modes, we evaluated the peak of the intensity correlation as a function of the pump power. More precisely, we evaluated the autocorrelation coefficient in Eq. (41) over a single pixel, $\Gamma_{i,j}^{(i,j)}$, so that we can exploit the best resolution of our detection system, namely 0.2 nm in spectrum and 0.015 degree in angle (Laiho et al., 2011). The procedure was applied to a set of pixels close to the frequency degeneracy and in a nearly collinear direction. In Fig. 10A we present the dependence of the maximum values of intensity autocorrelation coefficient on increasing values of the pump mean power. As already observed in Fig. 8 for the size of intensity auto- and cross-correlation areas,

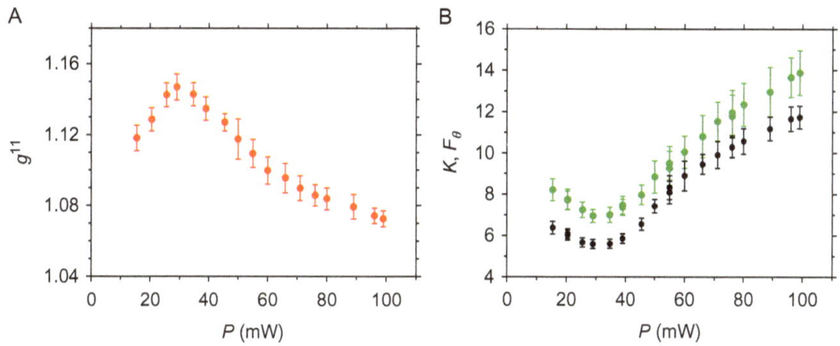


**Fig. 10** (A) Intensity autocorrelation coefficient as a function of the pump mean power. (B) Number $K$ (*green dots*) of spatio-spectral modes determined from the intensity autocorrelation coefficient and high-intensity Fedorov ratio (*black dots*), both as functions of the pump mean power. *Adapted from Allevi, A., Jedrkiewicz, O., Brambilla, E., Gatti, A., Peřina, J., Haderka, O., Bondani, M., 2014a. Coherence properties of high-gain twin beams. Phys. Rev. A 90, 063812.http://dx.doi.org/10.1103/PhysRevA.90.063812.; Allevi, A., Lamperti, M., Machulka, R., Jedrkiewicz, O., Brambilla, E., Gatti, A., Peřina, J., Haderka, O., Bondani, M., 2015. Effects of pump depletion on spatial and spectral properties of parametric down-conversion. Proc. SPIE 9505, 950508–950508-7. http://dx.doi.org/10.1117/12.2178469.*

the plot exhibits a peak for the pump power of 30 mW that lies at the beginning of the pump depletion regime. The number $K$ of modes is obtained according to Eq. (40), which formally represents the macroscopic analogous of the Schmidt number. The experimental values of $K$ are shown as green dots in Fig. 10B. The comparison between this panel and the plots in Fig. 8 confirms the complementary behavior for the size of the intensity auto- and cross-correlation areas and the numbers of modes. Note also that the behavior of $K$ is really similar to what we found for the evolution of the number of modes from statistics as presented in Section 3.3.

The evolution of the system at different pump powers can be alternatively investigated in terms of the quantity $F_\vartheta$ defined by Eq. (36). As already explained in Section 2, such a quantity, as well as its spectral counterpart, represents the analogous at high-intensity level of the Fedorov ratio, defined by Eq. (34) and commonly used to quantify the entanglement degree of the TWB state at single-photon level.

In order to calculate the values of $F_\vartheta$, first of all we experimentally determined the behavior of the spatial bandwidth of the PDC as a function of pump power (Allevi et al., 2015). The bandwidth was calculated as the FWHM of the vertical section of the average far-field spectrum in correspondence to a number of different pixels close to frequency degeneracy. The second quantity required to evaluate $F_\vartheta$ is the size of the cross-correlation area, already shown in Fig. 8. In Fig. 10B we plot the value of $F_\vartheta$ (black dots) as a function of the pump mean power. The behavior of the experimental data shows a trend with a minimum corresponding to the onset of pump depletion. From a direct comparison between $K$ and $F_\vartheta$, we notice that the two quantities are not perfectly equal as expected from the theory in the single-photon regime and for double-Gaussian biphoton function. Nevertheless, they display the same trend and a nearly proportional behavior.

### 3.4.2 Transition From Near-Field to Far-Field Coherence Areas

The results presented above were obtained in the far-field configuration. In order to get a more complete knowledge of the system, the investigation of the spatio-spectral properties of TWB in the transient region from near to far field is also interesting. In this context, most investigations have been performed in the spatial domain, either in the near-field or in the far-field configurations (D'Angelo et al., 2004; Hamar et al., 2010; Howell et al., 2004), as well as in both (Edgar et al., 2012). In addition, attention has been devoted also to the transient area between the two extremes (Chan et al., 2007; Dyakonov et al., 2015; Just et al., 2013; Tasca et al., 2009). While in the near-field position intensity cross-correlations have been confirmed in Edgar et al. (2012), momentum anticorrelations have been observed in the far field (Edgar et al., 2012; Hamar et al., 2010), as a result of the phase-matching conditions. In the transient regime, cross-correlations get blurred and, at a certain position, they cannot be observed at all (Chan et al., 2007). At this position, entanglement is entirely transferred to the phase of the two-photon amplitude (Chan et al., 2007), thus becoming hidden to intensity observations. The evolution of cross-correlations from near field to far field has been experimentally observed at the single-photon level (Just et al., 2013) and the Fedorov ratio (Fedorov et al., 2005) has been determined. In the experiment of Just et al. (2013), the Fedorov ratio equal to 1 has been measured at the position where cross-correlations were spread over the whole PDC beam. We also note that nearly all these investigations were carried out in the single-photon regime, in which a well-established theory based on the biphoton function exists. Here, we summarize the results recently published in Haderka et al. (2015), where we experimentally demonstrated the physical mechanism ruling the behavior of both spatial and spectral autocorrelations and cross-correlations of macroscopic TWB in the transition from the near field to the far field.

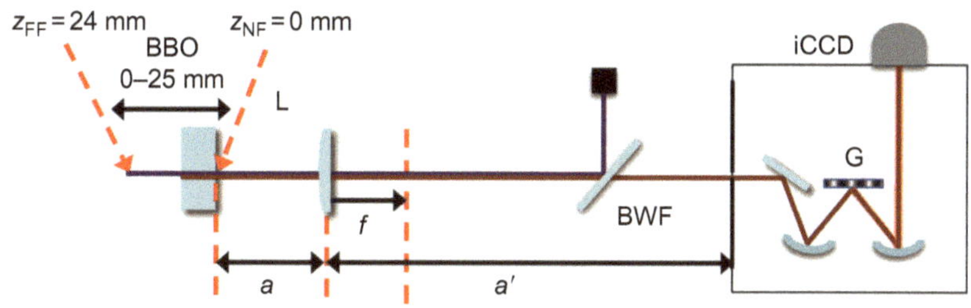


**Fig. 11** Sketch of the experimental setup. The nonlinear crystal is placed on a roto-translation stage. In the near-field position (NF, $z = 0$ mm), the lens with focal length $f$ and the imaging spectrometer are placed (distances $a, a'$) so that the output face of the crystal is imaged to the plane of the input slit of the spectrometer. Without changing $a$ and $a'$, the crystal is moved in the range 0–25 mm, so that different propagation planes of PDC are always imaged to the input slit of the spectrometer. The bandwidth filter (BWF) deflects most of the pumping beam power. *Adapted from Haderka, O., Machulka, R., Perina, J., Allevi, A., Bondani, M., 2015. Spatial and spectral coherence in propagating high-intensity twin beams. Sci. Rep. 5. http://dx.doi.org/10.1038/srep14365.*

The experimental setup is depicted in Fig. 11. The third harmonics of the already described Nd:YLF laser was used to pump a type-I BBO crystal (4 mm long, cut $\Theta = 33.8$ degree) in a nearly collinear configuration and placed on a roto-translation stage. First of all, the output face of the crystal was precisely set (distances $a, a'$ in Fig. 11) to be imaged to the input slit of an imaging spectrometer (Andor Shamrock 303i) by a lens with focal length $f = 60$ mm. The input slit was centered at the residual THG beam after most of the pumping energy was steered away with a bandwidth filter (BWF). After a precise setting of the distances $a$ and $a'$, the components were fixed to the optical table during the entire experiment. The chosen position corresponds to the near-field configuration with magnification $M = a'/a = 10$. During the experiment, the crystal was moved, in the opposite direction with respect to the spectrometer, from the near-field position up to a distance of 24 mm toward the far-field configuration. Few millimeters are enough to change the character of the spatial correlations from near-field-like to far-field-like. Since the pumping beam is well collimated, this translation along the $z$-axis does not significantly change the parameters of the interaction. The rotation stage is used to precisely set the opening angle of the PDC process so that the cone fits into the height of the input slit of the spectrometer. The light from the PDC cone entering the input slit of the monochromator gets dispersed at the grating (1200 rules per mm) in the horizontal plane. A portion of the spectrum between 683.4 and

712.5 nm has been registered by means of an iCCD camera (Andor iStar DH334T, 13 × 13 μm$^2$ pixel size). The spectral resolution of the system was 0.03 nm/pixel. The spectrum was centered approximately around the PDC degenerate wavelength (698 nm). A 50 μm-wide slit was used throughout the experiment. The camera was operated at the maximum gain and maximum A/D conversion speed (5 MHz). The gating window of the camera was set to 5 ns, synchronous with the laser pulses, to ensure the detection of single shots of the PDC in each image. At each $z$-axis setting, a sequence of 10000 camera frames (with full-resolution) was taken. We performed a scan of the spatial and spectral correlations as functions of the $z$-axis position. For each $z$-axis position, 100 points in each image were taken and processed in the same way explained above to calculate the intensity autocorrelation and cross-correlation functions.

In panels (A)–(C) of Fig. 12 we show the spatial positions of the cross-correlation peaks against the spatial positions of the corresponding autocorrelation peaks. Note that, to take into account the different divergences of the PDC beam at the transition from the near-field to the far-field configurations, in each panel the real positions on the camera were normalized to the size (FWHM) of the corresponding PDC beam. From the figure, we note a sharp diagonal around the position $z = 0$ mm, which corresponds to the near field. The location of the near field at this position is also confirmed by the direct observation of the crystal output face at the plane of the input slit of the monochromator when the monochromator is switched to the imaging mode. Moving the crystal away from the monochromator leads to a blurring of the diagonal. Around position $z = 4$ mm, the diagonal is completely lost, so there is no spatial correlation in the positions of the peaks. Moving the crystal even further to higher $z$ values, an antidiagonal character of the correlations (anticorrelation) is gradually established. Toward position $z = 24$ mm, the antidiagonal becomes sharper, thus indicating the approach to the far-field momentum correlations. We are not exactly in the far-field configuration reached at $z = \infty$, but the far-field character of the correlations is clearly established at a distance exceeding 1 cm from the crystal. Note that in the transient area (between $z = 4$ mm and $z = 9$ mm), the cross-correlations in the PDC field are nearly lost only in space. On the contrary, spectral cross-correlations remain unchanged as documented in panels (D)–(F) of Fig. 12. Here the tight anticorrelations emerging from the energy conservation are preserved at all $z$ positions.

In addition to the position of the peaks and in analogy with the investigation at different pump powers, Fig. 13A reports the spatial FWHM

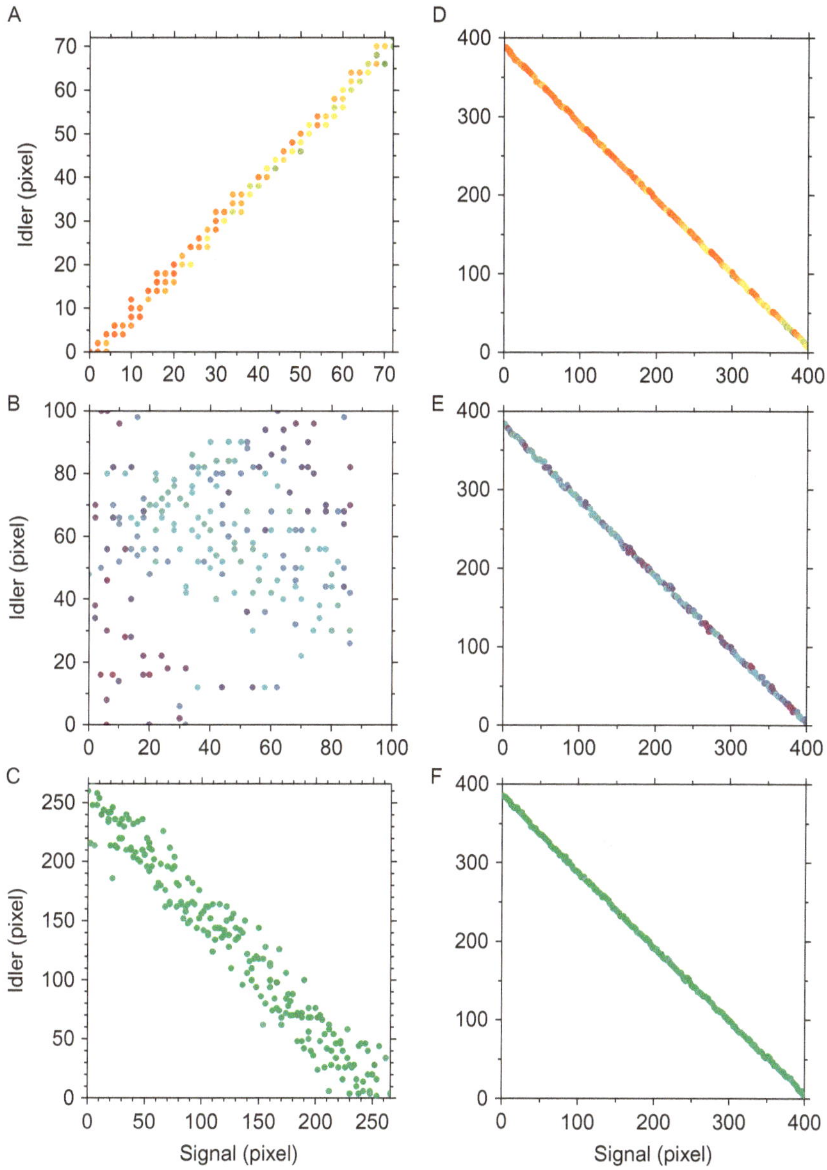


**Fig. 12** Spatial (panels A–C) and spectral (panels D–F) positions of the autocorrelation peak vs the corresponding cross-correlation peak for different *z*-axis positions: (A) and (D), $z = 0$ mm (near field); (B) and (E), $z = 5$ mm; (C) and (F), $z = 24$ mm (far field). (A)–(C) Normalized to the size of the corresponding PDC beam FWHM width to scale them to the divergence of the beam. (D)–(F) The whole registered bandwidth (683.4–712.5 nm) in both coordinates. *Adapted from Haderka, O., Machulka, R., Perina, J., Allevi, A., Bondani, M., 2015. Spatial and spectral coherence in propagating high-intensity twin beams. Sci. Rep. 5. http://dx.doi.org/10.1038/srep14365.*

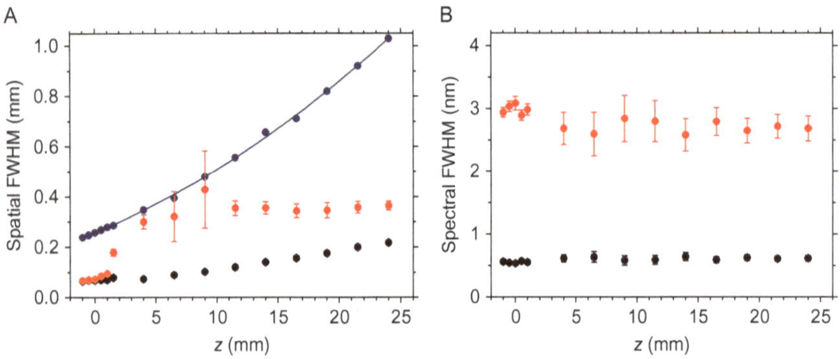


**Fig. 13** (A) Spatial intensity autocorrelation (*black, circles*) and cross-correlation (*red, circles*) peak widths (FWHM) as functions of the *z*-axis position. The *blue solid curve* shows the PDC spatial beam width (FWHM) at the input slit of the spectrometer. (B) Spectral intensity autocorrelation (*black, circles*) and cross-correlation (*red, circles*) peak widths (FWHM) as functions of the *z*-axis position. *Adapted from Haderka, O., Machulka, R., Perina, J., Allevi, A., Bondani, M., 2015. Spatial and spectral coherence in propagating high-intensity twin beams. Sci. Rep. 5. http://dx.doi.org/10.1038/srep14365.*

widths of the autocorrelation and cross-correlation areas as functions of the $z$-axis position. We note that the near-field configuration occurs at the left edge of the plots. In the near-field configuration, the autocorrelation and cross-correlation functions have nearly identical spatial widths. Moving toward the far field ($z = 24$ mm) the spatial autocorrelation width exhibits a mild growth, while the spatial cross-correlation width grows rapidly up to the PDC beam size, which was determined from the spatial extent of the beam as obtained from a large number of accumulated images at each $z$-position. As expected, the increase of the spatial cross-correlation width happens at the distances at which we find the completely blurred spatial diagonals (see for instance Fig. 12B). Going further toward the far field, momentum anticorrelations get established resulting in a gradual decrease of the spatial cross-correlation width. At variance with space, the autocorrelation and cross-correlation spectral FWHM widths shown in Fig. 13B do not exhibit any kind of evolution at different $z$-positions.

The behavior of the spatial cross-correlation widths can be also used to investigate the Fedorov ratio plotted in Fig. 14. As presented previously, at macroscopic level this ratio is again defined as the ratio of the width of the whole PDC beam to the cross-correlation width. While the spatial Fedorov ratio gives the number of paired degrees of freedom in the near field and far field, the blurring of the intensity cross-correlations prevents the determination of this number in the transition region. Nevertheless, it can be used to quantify the strength of blurring. We remark that the evolution of the spatial

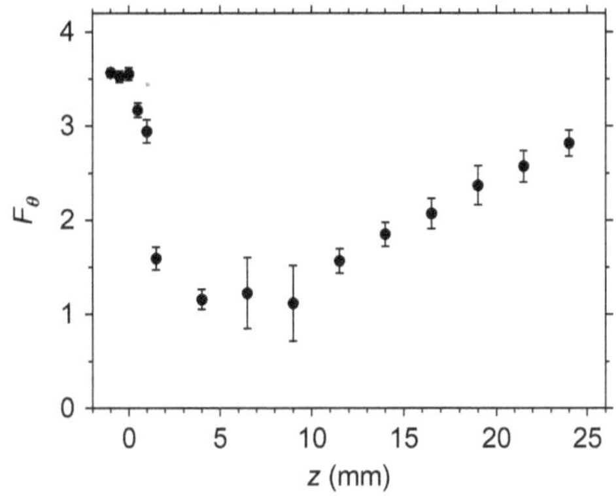


**Fig. 14** Spatial Fedorov ratio as a function of the *z*-axis position. *Adapted from Haderka, O., Machulka, R., Perina, J., Allevi, A., Bondani, M., 2015. Spatial and spectral coherence in propagating high-intensity twin beams. Sci. Rep. 5. http://dx.doi.org/10.1038/srep14365.*

Fedorov ratio is very similar to that reported by Just et al. (2013) in the single-photon regime.

## 3.5 Measurement of Nonclassical Correlations

As already mentioned in Section 2, the nonclassical character of TWB states lies in the existence of strong correlations between several degrees of freedom, such as linear position and momentum, number of photons, frequency, polarization, angular momentum, etc. Among them, photon-number correlations have been the subject of intensive investigations for state characterization. In many cases, the detection was performed by means of single or arrays of avalanche photodiodes so that the possibility to recover the correlation of the number of photons was quite straightforward (Avenhaus et al., 2010; Ivanova et al., 2006), although the intensity range actually investigated by these systems was limited to much less than one mean photon (Kalashnikov et al., 2011). Also the generation and characterization of more intense TWBs have received a lot of attention, especially in view of possible applications to quantum technology. Indeed, pulsed optical states endowed with sizeable numbers of photons represent a useful resource in the field of quantum information, as they are robust with respect to losses.

Among the nonclassicality criteria for TWB states that can be expressed in terms of photon numbers, the existence of subshot-noise photon-number correlations is routinely investigated. As already explained in Section 2, this criterion is based on the calculation of the noise reduction factor, whose values attain lower-than-1 numbers in the case of nonclassical correlations.

Indeed, subshot-noise photon-number correlations have been revealed in macroscopic TWBs by means of different kinds of detectors, such as pin photodiodes, CCD cameras, but also amplified pin-photodiodes in a lower intensity regime (more than $10^3$ detected photons). However, it is worth noting that the use of these macroscopic detectors to test nonclassicality is fragile, as a reliable a priori calibration is required and the contribution of electronic noise is not negligible and must be subtracted from the signal. On the contrary, photon-number resolving detectors, such as HPDs, can be easily calibrated by means of self-consistent methods and are endowed with a lower level of noise, which in many cases can be neglected thus making the evaluation of the nonclassicality criteria straightforward and definitely more reliable. The main drawback given by this kind of detectors is that in general their quantum efficiency is lower than 50% and thus reconstruction algorithms to describe the nonclassicality of TWBs cannot be applied. However, as already anticipated in Section 2, it is possible to express the same quantities defined for incident photons, such as correlation coefficient and noise reduction factor, in terms of detected photons. This makes the analysis of the experimental data more straightforward and less time demanding. In our research activity, we adopted such a strategy several times to characterize the nonclassicality of multimode TWB states generated in the mesoscopic domain. In particular, here we present three different nonclassicality criteria based on detected-photon-number correlations, namely the Schwarz inequality in Eq. (16), the noise reduction factor in Eq. (17) and the high-order correlation inequality in Eq. (18), and we discuss the conditions suitable for their application.

### 3.5.1 Experimental Setup

The experimental setup is sketched in Fig. 15. The fourth harmonics (at 262.2 nm) of the already described Nd:YLF laser was sent to a 6-mm-long BBO crystal (cut angle = 46.7 degree) to produce PDC in a quasi-collinear interaction geometry. Two twin portions around frequency degeneracy (i.e., at 523 nm) were spatially and spectrally filtered by a variable iris and a bandpass filter, respectively. In both arms the light was then focused into a 600-μm-core multimode fiber by means of an achromatic doublet and delivered to a PNR detector. As the PNR detectors we used two HPDs, whose operational mechanism has been already described in Appendix. We acquired sequences of 100,000 shots at different mean values of the fourth harmonics, whose energy was changed by means of a half-wave plate followed by a polarizing cube BS.

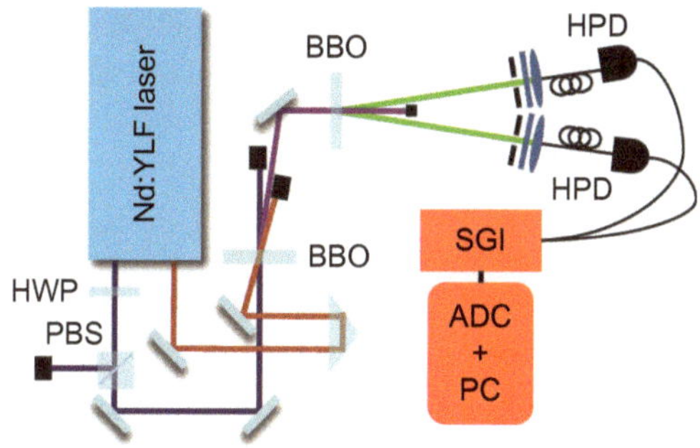


**Fig. 15** Sketch of the experimental setup for the measurement of the nonclassical correlations.

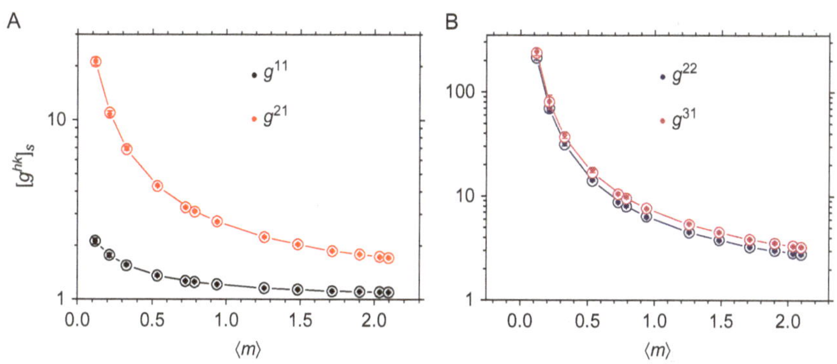


**Fig. 16** High-order correlations as functions of the mean number of photons per arm. (A) $g^{11}$ (*black symbols*) and $g^{21}$ (*red symbols*); (B) $g^{22}$ (*blue symbols*) and $g^{31}$ (*magenta symbols*). Experimental data are plotted as *dots*, whereas theoretical expectations are shown as empty circles + line.

### *3.5.2 Results*

In Section 3.3, we have shown how our detection apparatus determines the number of photons at each laser shot correctly enough to properly reconstruct the statistics of detected photons. The possibility to have access to the number of photons is fundamental to reveal shot-by-shot photon-number correlations between the parties. To this aim, in Fig. 16 we show the experimental data (colored dots) obtained by evaluating the high-order correlation functions up to the fourth order in the symmetrized form $[g^{hk}]_s = 1/2(g^{hk} + g^{kh})$. The same figure reports the theoretical expectations (colored open symbols) calculated according to Eqs. (15) by using the experimental values of the mean number of detected photons, of the number of modes $\mu$, and of the overall detection efficiency $\eta$. In particular, the mean number of photons has been directly determined by means of the self-consistent method of analysis (see Appendix). $\mu$ has been obtained as

$$\mu = \frac{\langle m \rangle}{\sigma^2(m) - \langle m \rangle} \tag{42}$$

from the first two moments of statistics. Finally $\eta$ has been calculated by the experimental value of the noise reduction factor, that is, by imposing

$$R = \frac{\sigma^2(m_1 - m_2)}{\langle m_1 \rangle + \langle m_2 \rangle} = 1 - \eta. \tag{43}$$

Notice that in both panels of Fig. 16, which correspond to the different correlation orders, the experimental data exhibit a very good agreement with theory, thus testifying the usefulness of high-order correlations for state characterization.

However, photon-number correlations are not sufficient markers of nonclassicality. As anticipated above, other quantities based on the number of photons and satisfying boundary conditions are usually considered. In addition to the already existing criteria, Allevi et al. (2012) have introduced the $B$ parameter inequality of Eq. (18), based on high-order photon-number correlations for detected photons and thus only in terms of measurable quantities.

Fig. 17 reports the noise reduction factor, the Schwarz inequality and the high-order inequality all together, in order to get a fair comparison between them. All the inequalities are calculated for detected photons and are plotted as functions of the mean number of photons per arm. The experimental data are shown as dots, whereas the theoretical predictions are plotted as empty circles + line. Notice the good agreement between data and theory for all the considered criteria. The worse superposition in the case of the $B$-parameter high-order inequality arises because such a criterion involves high-order correlations, and it is thus much more sensitive to noise in the experimental system. However, it is remarkable that the values of the high-order correlation criterion are reasonably above 1, that is, the classical boundary, thus testifying the nonclassical character of TWB. In particular, we notice that such $B$-parameter criterion well emphasizes the presence of nonclassicality for high mean photon numbers, at variance with Schwarz inequality $S$ that better discriminates at low photon numbers. It is worth also noting that in

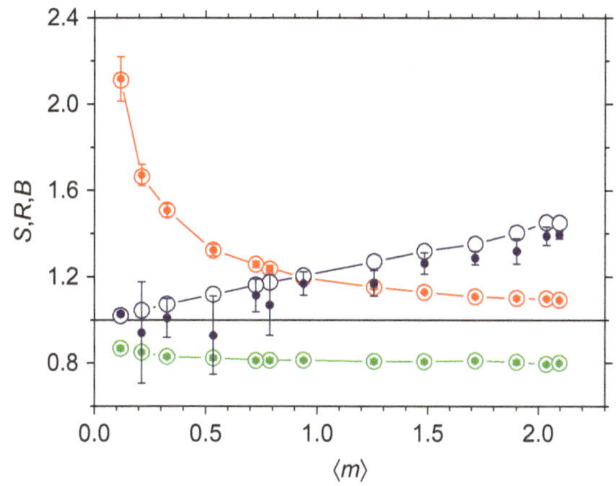


**Fig. 17** Nonclassicality criteria as functions of the mean number of photons per arm: the Schwarz inequality, *S*, (*red symbols*), the noise reduction factor, *R*, (*green symbols*), and the high-order inequality, *B* (*blue symbols*). Experimental data are plotted as *dots*, whereas theoretical expectations are shown as empty circles + line. The *horizontal black line* represents the classical boundary.

the explored intensity regime the noise reduction factor is almost constant and attains values around 0.8.

# 4. APPLICATIONS

## 4.1 Conditional States

Due to their nonclassical character, TWB states represent a useful resource in quantum state engineering for applications to quantum information and metrology, both at the single-photon level and in the mesoscopic and macroscopic domains. For instance, it is well known that such states can be used to produce nonclassical states by means of conditional measurements exploiting the existence of correlations between signal and idler. In particular, when a given number of photons is detected in the idler, the state of the signal is irreversibly modified. As the original state (the TWB) is quantum correlated, the generated conditional states can show sub-Poissonian statistics, which is nonclassical. This feature has been experimentally demonstrated in different intensity regimes, but mostly at single-photon level. However, it is worth noting that the use of photon-counting detectors instead of single-photon detectors offers the possibility not only to enhance the heralding of single-photon states by suppressing higher photon-number components (Christ and Silberhorn, 2012; Haderka et al., 2004;

Peřina et al., 2001), but also to perform multiple photon-counting operations in order to produce quasi-Fock states endowed with a number of photons larger than 1 (Peřina et al., 2013).

The extension of the experimental results presented so far to a more mesoscopic photon-number domain, where the states are more robust with respect to losses, is thus desirable and still subject to active research. Moreover, we notice that, since the conditioning operation is non-Gaussian (Allevi et al., 2010) the resulting conditional states are non-Gaussian, thus acquiring an additional interesting feature for quantum information applications. Here we show that conditional states endowed with both sub-Poissonian statistics and non-Gaussian character can be generated by exploiting the TWB states already described in Section 3.5. We performed conditional measurements by selecting 0, 1, 2, or 3 photons in the idler beam, thus obtaining the conditional states $\varrho_{m_2=0}$, $\varrho_{m_2=1}$, $\varrho_{m_2=2}$, and $\varrho_{m_2=3}$ in the signal beam, respectively (Peřina et al., 2013). This choice gives us the possibility to show that our detection apparatus can be used to perform conclusive photon subtractions by exploiting the photon-counting capability of HPDs. Among the conditioning values, $\varrho_{m_2=0}$ represents a sort of reference level. In fact, in this case the conditional states remain super-Poissonian and Gaussian. Indeed such an operation is Gaussian, at variance with all the other conclusive operations.

We also notice that the results obtained in the case $\varrho_{m_2=1}$ are definitely different from those that can be achieved by employing single-photon detectors operated in Geiger ON/OFF mode, as we do not need to assume that the output states contain 1 photon at most. Moreover, the conditioning operations presented here are useful to investigate in which way the nonclassical nature of the conditional states depends on the different experimental parameters involved in their production. In order to quantify the nonclassicality of the conditional states, here we consider their Fano factor, which is defined as $\mathfrak{F}_n = \sigma_n^2/\langle n\rangle$. Actually, in the following we will consider the analogous quantity written in terms of detected photons. Indeed, it is possible to demonstrate that $\mathfrak{F}_m$ can be easily linked to the corresponding quantity for incident photons, namely $\mathfrak{F}_m = \eta\mathfrak{F}_n + (1-\eta)$. Both quantities are equal to 1 in the case of Poissonian light. If $\mathfrak{F}_n < 1$, the light is nonclassical and is called sub-Poissonian. Note that the value 1 for the boundary between classical and nonclassical behaviors still holds for detected photons. It is interesting to notice that the minimum value of this expression coincides with the minimum value of the noise reduction factor $R = 1 - \eta$. In Fig. 18 we plot

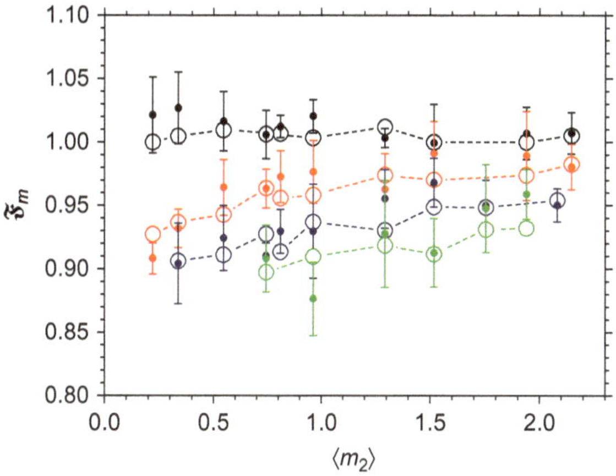


**Fig. 18** Fano factor $\mathfrak{F}_m$ of the conditional states as a function of the $\langle m_2 \rangle$ mean number of photons detected in the signal arm before conditioning. Different *colors* correspond to different conditioning values. *Black*: $m_2 = 0$, *red*: $m_2 = 1$, *blue*: $m_2 = 2$, and *green*: $m_2 = 3$. Experimental data are shown as *dots*, whereas theoretical expectations as empty circles + line.

the Fano factor of the conditional states as a function of the mean number of photons detected in the signal arm before conditioning. Different colors correspond to different conditioning values. In particular, it is worth noting that the Fano factor attains its lowest values either at increasing conditioning values or at decreasing the mean values of the unconditioned states. In general, the agreement between the experimental data and the theoretical expectations is good. For some points the discrepancies can be ascribed to possible fluctuations due to the low number of experimental data giving the conditional states.

Besides being sub-Poissonian, and hence nonclassical, the conditional states generated from TWB are also non-Gaussian, which means that they are not described by a Gaussian Wigner function. There are different ways to quantify the non-Gaussianity of a state. Here we calculate the $\delta_\rho$ relative entropy of non-Gaussianity introduced by Genoni et al. (2008) in the case of detected photons. For a generic state, this quantity is defined as the difference between the von Neumann entropy $S[\sigma] = -Tr[\sigma ln(\sigma)]$ of a reference Gaussian state $\sigma$ and that of the state under investigation, namely

$$\delta_\rho = S[\sigma] - S[\rho]. \tag{44}$$

Note that the Gaussian reference state is chosen to have the same mean value and covariance matrix as the state under investigation. According to this definition, the estimation of the non-Gaussianity implies the complete knowledge of the density matrix of the state. However, there are some cases in which the calculation is simpler, such as for diagonal states, whose von Neumann entropy (and hence non-Gaussianity) depends only on the photon-number statistics. In the case of TWB, which is in principle a diagonal state, this simplification does not hold because of its multimode nature. Thus, to calculate the non-Gaussianity amount the only possibility is to evaluate the expected density matrix of the states on the basis of the experimental values. In particular, in this calculation the mean number of photons of the conditional states, the number of modes and the quantum efficiency are involved.

Fig. 19 reports the non-Gaussianity of the conditional states as a function of the mean number of photons detected in the signal arm before conditioning. The dots represent the values of non-Gaussianity obtained by using the experimental mean values of the conditional states, whereas full circles + line have been obtained by calculating the expected mean values on the basis of the other experimental parameters, namely the mean number of photons of the unconditioned states, the number of modes, and the quantum efficiency. The agreement between the two sets of points is really good, thus remarking the reliability of the performed conditioning operations. Moreover, by comparing the behavior of non-Gaussianity shown in Fig. 19 to that

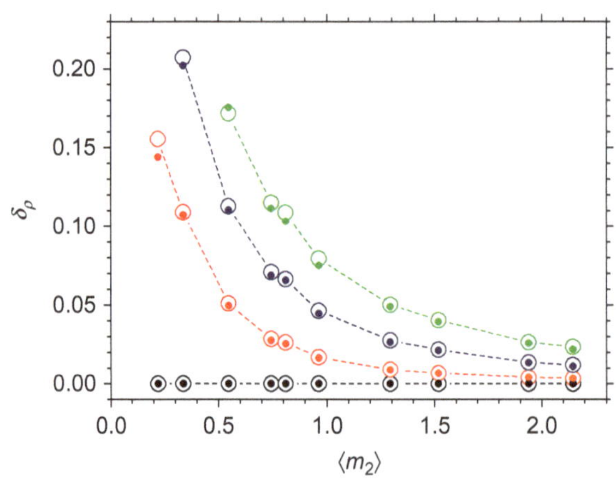


**Fig. 19** Non-Gaussianity of the conditional states as a function of the mean number of photons detected in the signal arm before conditioning. Different *colors* correspond to different conditioning values. *Black*: $m_2 = 0$, red: $m_2 = 1$, blue: $m_2 = 2$ and *green*: $m_2 = 3$. Experimental data are shown as *dots*, whereas theoretical expectations as empty circles + line.

of the Fano factor plotted in Fig. 18, it is quite clear that their evolutions are complementary. In fact, the higher values of non-Gaussianity appear for the smaller values of Fano factor.

## 4.2 Ghost Imaging

Ghost imaging is a technique based on the correlated nature of both classical and quantum fields, which allows the retrieval of the image of an object not interacting with the measured light (for a review see Gatti et al., 2008). The ghost-imaging protocols rely on the properties of correlation in the propagation direction and in the intensity between two spatially multimode fields (D'Angelo and Shih, 2005; Gatti et al., 2006, 2007; Pittman et al., 1995; Shih, 2007). The technique was originally introduced by Klyshko (1988) with the idea of exploiting quantum entanglement in signal-idler photon pairs generated by PDC. According to the original scheme, the photons of a pair are sent to two different imaging systems: the signal illuminates an object and then is detected by a "bucket" detector, whereas the idler is revealed by a position-sensitive detector. Information about the object is obtained from the coincidences of signal-idler photon pairs as functions of the transverse position of detector in the idler-arm (Abouraddy et al., 2001). By changing the optical elements in the two arms, it is possible to reconstruct either the image of the object (ghost image) or its diffraction pattern (ghost diffraction) (Brambilla et al., 2004; Ferri et al., 2005). These results were theoretically extended to the macroscopic realm, where a large number of photon pairs form entangled beams and the information about the object is retrieved by the correlation of the intensity fluctuations (Gatti et al., 2003). From the experimental point of view, the correlated light required for ghost imaging has been generated by several physical systems, starting from PDC in single-photon regime (D'Angelo and Shih, 2005; Pittman et al., 1995), passing through chaotically seeded down-conversion (Bondani et al., 2008; Puddu et al., 2007), and ending with macroscopic thermal beams (Ferri et al., 2005; Valencia et al., 2005; Zhang et al., 2005). With a suitable choice of the optical setup, all these systems lead to the production of a ghost image. Of course, according to the chosen conditions, the processes perform differently in terms of visibility, resolution, and signal-to-noise ratio (Erkmen and Shapiro, 2009, 2010; Ferri et al., 2005; O'Sullivan et al., 2010).

Bondani et al. (2012) presented the first experimental implementation of the ghost-imaging scheme obtained by means of an intense TWB, as proposed by Gatti et al. (2003), both by means of intensity correlations (as usual) and of high-order correlations. That paper demonstrated that, in analogy with classically correlated multimode light, the visibility of the ghost image increases at increasing the order of correlation, whereas the signal-to-noise ratio is expected to have a maximum at a certain value of the order of correlation (Chan et al., 2009; Chen et al., 2010; Iskhakov et al., 2011). The obtained results are here summarized.

### 4.2.1 Experimental Setup

Let's consider a ghost-imaging scheme in which a 2D-object is located in the plane $\mathbf{x}_o$ in the signal-arm at a distance $d_1$ from the crystal and an imaging lens is located in the idler-arm at a distance $d_2$ from the crystal and $d_3$ from the detector (see Fig. 20). The light transmitted by the object in the signal-arm is integrated by a "bucket" detector in the plane $\mathbf{x}_{si}$, while that propagating in the idler-arm is detected by a position-sensitive detector in the plane $\mathbf{x}_{id}$. As the sensor Bondani et al. (2012) used a 16-bit EMCCD (ImagEM, Hamamatsu, Japan) synchronized to the laser, to measure both arms of the setup. We decided to calculate the "bucket" values by integrating the light passing through the object and imaged on the sensor by lens $L_3$ in Fig. 20 with a magnification factor $M = 0.47$. This strategy allowed the authors to have a cross-check for the ghost image with a conventional image. As the object we used a mask containing four small holes of about 450 μm diameter separated by about 450 μm from each other. We chose the distances $d_1 = 30$ cm, $d_2 = 30$ cm, and $d_3 = 60$ cm in order to satisfy the thin

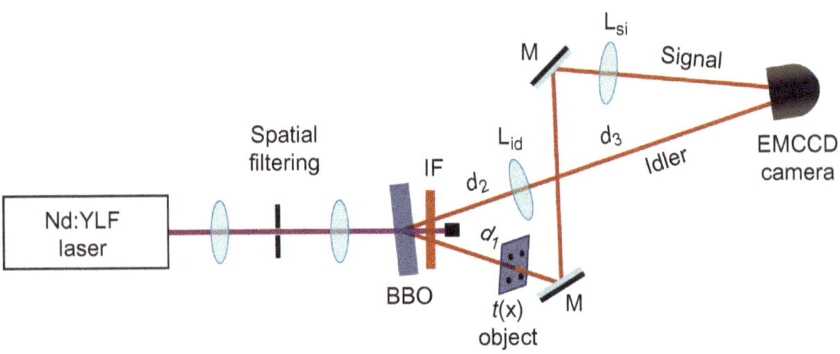


**Fig. 20** Scheme of the experimental setup for the ghost-image investigation. *BBO*, nonlinear crystal; *IF*, interference filter; *M*, mirror; $L_j$, lens; $t(x)$, transmissivity of the object. *Adapted from Bondani, M., Allevi, A., Andreoni, A., 2012. Ghost imaging by intense multimode twin beam. Eur. Phys. J. Spec. Top. 203 (1), 151–161. ISSN 1951-6401. http://dx.doi.org/10.1140/epjst/e2012-01540-4.*

lens equation $M = d_3/(d_1 + d_2)$ for the $M$ magnification factor leading to $M = 1$. Signal and idler beams were then detected by two distinct regions of the EMCCD.

### 4.2.2 Images

The reconstruction of the ghost image is achieved by calculating the normalized intensity correlation function,

$$g^{(2)}(\mathbf{x}_{id}) = \frac{\int d\mathbf{x}_{si} \langle (\hat{I}_{id}(\mathbf{x}_{id}) - \langle \hat{I}_{id}(\mathbf{x}_{id}) \rangle) (\hat{I}_{si}(\mathbf{x}_{si}) - \langle \hat{I}_{si}(\mathbf{x}_{si}) \rangle) \rangle}{\int d\mathbf{x}_{si} \langle \hat{I}_{si}(\mathbf{x}_{si}) \rangle \int d\mathbf{x}_{id} \langle \hat{I}_{id}(\mathbf{x}_{id}) \rangle}, \tag{45}$$

where $\hat{I}_i(\mathbf{x}_i) = \hat{c}_i^\dagger(\mathbf{x}_i)\hat{c}_i(\mathbf{x}_i)$ ($i = si$, $id$) is the intensity operator of the $i$-th beam at the detection plane. The field operators at the detection planes are connected to those ($\hat{b}_i(\mathbf{x})$) at the output face of the crystal by the propagators $h_{id}(\mathbf{x}_{id}, \mathbf{x}'_{id})$ and $h_{si}(\mathbf{x}_{si}, \mathbf{x}'_{si})$ describing the optical setup (Goodman, 1996)

$$\hat{c}_i(\mathbf{x}_i) = \int d\mathbf{x}'_i h_i(\mathbf{x}_i, \mathbf{x}'_i)\hat{b}_i(\mathbf{x}'_i). \tag{46}$$

The explicit expression of the propagators for the setup in Fig. 20 is calculated in Bondani et al. (2008). Introducing $1/f = 1/(d_1 + d_2) + 1/d_3$, the final expression of the correlation function becomes

$$g^{(2)}(\mathbf{x}_{id}) \propto \left| t\left(-\frac{\mathbf{x}_{id}}{M}\right)\right|^2 \int d\mathbf{x}_{si} n_{\mathrm{PDC}}\left(\frac{2\pi \mathbf{x}_{si}}{\lambda f}\right)\left[1 + n_{\mathrm{PDC}}\left(\frac{2\pi \mathbf{x}_{si}}{\lambda f}\right)\right] \tag{47}$$

In order to reconstruct the ghost image by computing the normalized second-order correlation function in Eq. (47), we recorded many EMCCD frames, each one containing both the single-shot map in the idler-arm and the corresponding bucket value in the signal-arm.

In Fig. 21A we show the image of the object obtained by the EMCCD on the signal-side and in Fig. 21B the ghost image recovered by cross-correlation. What we can see is that the object size is almost correctly reproduced, although the resolution is not very high due to the speckle size that is comparable to the object size. To explore the effects of the size and shape of the coherence areas on the image reconstruction, we can calculate the correlation of the bucket with the portion of the EMCCD sensor containing the image of the object. The result is displayed in Fig. 21C: the image obtained from autocorrelation is very similar to the ghost image, due to the

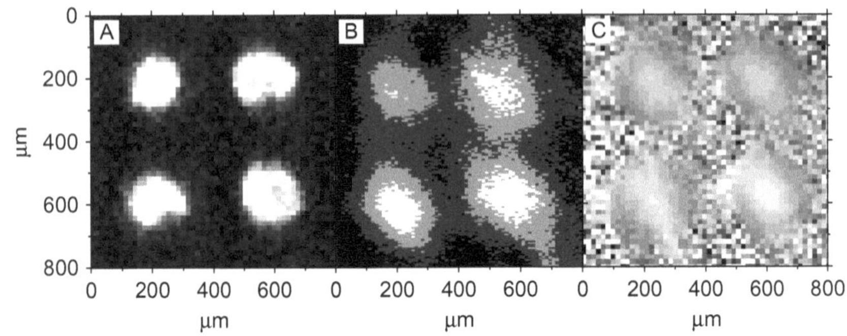


**Fig. 21** (A) Image, demagnified by a factor of 0.47, of the object used to perform the ghost-imaging experiment. (B) Ghost image recovered by cross-correlation. (C) Autocorrelation of the object obtained by correlating the bucket detector with the side of the camera containing the image of the object. *Adapted from Bondani, M., Allevi, A., Andreoni, A., 2012. Ghost imaging by intense multimode twin beam. Eur. Phys. J. Spec. Top. 203 (1), 151–161. ISSN 1951-6401. http://dx.doi.org/10.1140/epjst/e2012-01540-4.*

fact that the correlation is sensitive to the speckle structure of the field and hence the resulting image can be viewed as the convolution between the hole size and the speckle size.

In order to quantify the quality of the imaging protocol, in the following we consider the visibility (VIS) and the signal-to-noise ratio (SNR). The former quantity is defined as

$$\mathrm{VIS} = \frac{\left(g_{\mathrm{IN}}^{(2)} - g_{\mathrm{OUT}}^{(2)}\right)}{\left(g_{\mathrm{IN}}^{(2)} + g_{\mathrm{OUT}}^{(2)}\right)}, \tag{48}$$

where $g_{\mathrm{IN}}^{(2)}$ is the mean value of $g^{(2)}$ inside the image and $g_{\mathrm{OUT}}^{(2)}$ is the mean value in a portion of the matrix that does not contain the image. The latter quantity is defined as the "contrast" of the image divided by its noise, namely

$$\mathrm{SNR} = \frac{\left(g_{\mathrm{IN}}^{(2)} - g_{\mathrm{OUT}}^{(2)}\right)}{\sqrt{\mathrm{Var}\left(g_{\mathrm{IN}}^{(2)} - g_{\mathrm{OUT}}^{(2)}\right)}}. \tag{49}$$

We note that in the case presented above, the visibility of the ghost image is actually very low, due to the high total number of modes (temporal and spatial) involved in the calculation of the correlations. In fact, we expect that VIS in Eq. (48) scales as $1/\mu_{tot}$ (Gatti et al., 2004; Lopaeva and Chekhova, 2010). A way to increase the visibility is to calculate ghost images with high-order correlations, as we will see in the next section.

### 4.2.3 High-Order Ghost Imaging

Recently, there has been an increasing interest in the evaluation of high-order correlation functions applied to the calculation of ghost images. In the case of classically correlated multimode pseudo-thermal light it has been demonstrated that the visibility of the ghost image increases at increasing the order of correlations, while the signal-to-noise ratio is expected to have a maximum at a certain value of the order of correlation (Chan et al., 2009; Chen et al., 2010; Iskhakov et al., 2011).

The same kind of considerations can be applied also to our multimode TWB that represents the case of intense nonclassically correlated light. By calculating high-order correlations for TWBs, Brida et al. (2011) and Allevi et al. (2012) have observed that those correlations depend on the mean number of photons in the states. Nevertheless, the dependence disappears at high mean photon numbers and the correlation functions reduce to those of multimode thermal light (Allevi et al., 2012). In particular, in the limit of high numbers of photons we can write:

$$g^{n,m} = \frac{\langle \hat{I}_{si}^{n} \hat{I}_{id}^{m} \rangle}{\langle \hat{I}_{si} \rangle^{n} \langle \hat{I}_{id} \rangle^{m}} \simeq \prod_{k=1}^{n+m-1} \left(1 + \frac{k}{\mu}\right), \tag{50}$$

where we have assumed $\mu$ equally populated modes in the field. By following the results for classical light, from Eq. (50) we expect that also for the multimode TWBs the VIS calculated for high-order correlations increases at increasing the order of correlations and that the corresponding SNR has a maximum at a certain value of the order of correlation. Fig. 22 reports the

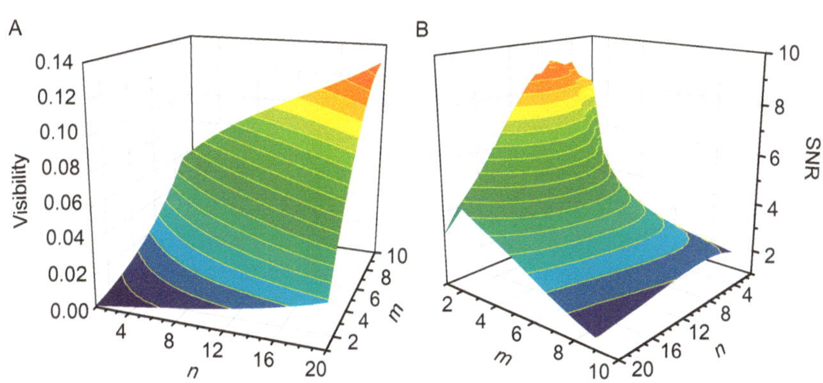


**Fig. 22** (A) VIS for ghost images of the four-hole object as a function of the orders $m$ and $n$ from Eq. (50). (B) SNR of the same results in the *left panel. Adapted from Bondani, M., Allevi, A., Andreoni, A., 2012. Ghost imaging by intense multimode twin beam. Eur. Phys. J. Spec. Top. 203 (1), 151–161. ISSN 1951-6401. http://dx.doi.org/10.1140/epjst/e2012-01540-4.*

values of VIS (panel A) and SNR (panel B) calculated from our experimental data. The experimental behavior qualitatively fits the theoretical expectations, even if the SNR is rather noisy. This problem could be overcome by increasing the number of images employed for correlations. Note that, since in the intensity regime of our experiment the calculated high-order ghost images have the same VIS and SNR as the classically correlated light, to have a real improvement in the protocols by the use of nonclassical light, one should employ a different approach, such as the subshot-noise ghost imaging introduced by Brida et al. (2010) or the ghost imaging based on the variance of the difference of detected photons introduced by Lopaeva and Chekhova (2010) and Brida et al. (2011). To this aim, an exact calibration of the camera is required to obtain a reliable estimation of the subshot-noise reduction factor, as already performed in Jedrkiewicz et al. (2004) and Brida et al. (2009a).

## 5. CONCLUSION

In this review, we presented the investigations performed on optical TWBs over the last 10 years in the laboratory of Quantum Optics at the University of Insubria. Such states were generated and characterized both in the mesoscopic and macroscopic intensity regimes, also including pump depletion. The achievement of the high-intensity levels was made possible by the availability of a pulsed laser source capable of producing few-ps UV pulses with high-peak intensity values (up to some GW/cm$^2$). Thanks to this source, we were able to produce bright TWBs, whose coherence properties have been easily characterized by means of an imaging spectrometer and two different kinds of amplified cameras. Moreover, thanks to our long-lasting experience with the use of photon-number resolving detectors, the statistical and nonclassical features of mesoscopic TWBs were extensively investigated. The results presented in the chapter are not limited to the characterization of TWB states, but also testify their possible application to the fields of quantum state engineering, quantum information, and quantum imaging. In particular, we showed that mesoscopic multimode TWBs are useful to produce sub-Poissonian and non-Gaussian states by means of conditional operations and, finally, we described the first implementation of a ghost-imaging protocol based on macroscopic TWBs.

# APPENDIX

## A.1 Detection Distributions

For the photon-number-resolving detectors described above we can assume that photodetection is performed with quantum efficiency $\eta$ and no dark counts. The probability operator-valued measure (POVM) of each detector, describing the number of detected photons $m$ as a function of the number of incident photons $n$ is thus given by a Bernoullian convolution of the ideal number operator spectral measure $\hat{P}_{n_j} = |n_j\rangle\langle n_j|$

$$\hat{\Pi}_{m_j} = \eta_j^{m_j} \sum_{n_j=m_j}^{\infty} (1-\eta_j)^{n_j-m_j} \binom{n_j}{m_j} \hat{P}_{n_j}, \tag{A.1}$$

with $j = 1, 2$. The moments of the distribution are evaluated by means of the operators

$$\widehat{m_j^p} = \sum_{m_j} m_j^p \, \hat{\Pi}_{m_j} = \sum_{n_j=0}^{\infty} (1-\eta_j)^n \, G_{\eta_j}(n_j) \, \hat{P}_{n_j}, \tag{A.2}$$

where

$$G_\eta(n) = \sum_{m=0}^{n} \binom{n}{m} \left(\frac{\eta}{1-\eta}\right)^m m^p . \tag{A.3}$$

Of course, since they are operatorial moments of a POVM, we have, in general, $\widehat{m_j^p} \neq \hat{m}_j^p$. The first two moments correspond to the operators

$$\begin{aligned} \hat{m}_j &= \eta_j \hat{n}_j \\ \widehat{m_j^2} &= \eta_j^2 \hat{n}_j^2 + \eta_j (1-\eta_j) \hat{n}_j. \end{aligned} \tag{A.4}$$

Thus for the variance of the detected statistics we have

$$\sigma^2(m_j) \equiv \langle \widehat{m_j^2} \rangle - \langle \hat{m}_j \rangle^2 = \eta_j^2 \sigma^2(n_j) + \eta_j (1-\eta_j) \langle \hat{n}_j \rangle . \tag{A.5}$$

As presented in Mandel and Wolf (1995) and Agliati et al. (2005), the resulting statistics of the number of electrons generated by the primary photon interaction (detected photons) that occurs in the detector, $p_m$, is linked to that of the number of photons in the light under measurement, $p_n$, by

$$p_m = \sum_{n=m}^{\infty} \binom{n}{m} \eta^m (1-\eta)^{n-m} p_n. \tag{A.6}$$

In principle, $p_m \neq p_n$, that is the functional form of the probability distribution of the detected photons is not necessarily the same as that of the photons (Casini and Martinelli, 2013). Indeed, it is not obvious that a Poissonian (or a thermal) distribution for the number of photons originates a Poissonian (or a thermal) distribution of detected photons. As a matter of fact, for all the fields described by a classical photon-number distribution (including signal and idler in the TWB) the statistics is invariant under the convolution in Eq. (A.6). This is definitely not true for purely quantum states, such as Fock states.

## A.2 Amplification Distributions

For almost any detector, after the primary detection process the number of generated photoelectrons undergoes amplification. To guarantee the linearity of the response, both the amplification internal to the detector (gain) and that of the electronics processing its output must be independent of the number of detected photons. If we can assume that the spread of the single photoelectron peak in the final electronic output is negligible as compared to its mean value, the relation linking the statistics of the photoelectrons to that of the final voltage outputs, $v$, derived by Andreoni and Bondani (2009) is

$$p_v = \frac{1}{\gamma} p_{\gamma m}, \tag{A.7}$$

being $\gamma$ a conversion coefficient describing the overall $m$-to-$v$ conversion process. The first two moments of the distribution in Eq. (A.7) reported by Agliati et al. (2005) are given by

$$\langle v \rangle = \gamma \langle m \rangle; \;\; \sigma_v^{(2)} = \gamma^2 \sigma_m^{(2)}, \tag{A.8}$$

where the symbols are defined as in Eq. (A.5).
To obtain the value of $\gamma$ we calculate the Fano factor for the output voltages, defined as $\mathfrak{F}_v = \sigma_v^{(2)} / \langle v \rangle$. By using Eq. (A.5) and (A.8) we find

$$\mathfrak{F}_v = \frac{\gamma^2 [\eta^2 \sigma_n^{(2)} + \eta(1-\eta)\langle n \rangle]}{\gamma \eta \langle n \rangle} = \gamma \eta \mathfrak{F}_n + \gamma(1-\eta) = \gamma \eta \mathrm{Q} + \gamma, \tag{A.9}$$

where we have rewritten the Fano factor for the photons, $\mathfrak{F}_n = \sigma^{(2)}(n)/\langle n\rangle$, as $\mathfrak{F}_n = 1 + Q$ in terms of the Mandel Q-factor. By multiplying and dividing Eq. (A.9) by $\langle n\rangle$ and reusing the above equations we get

$$\mathfrak{F}_\nu = \frac{Q}{\langle n\rangle}\langle \nu\rangle + \gamma, \tag{A.10}$$

in which $Q/\langle n\rangle = (\sigma_n^{(2)} - \langle n\rangle)/\langle n\rangle^2$ only depends on the specific state under measurement. Thus Eq. (A.10) shows a linear dependence of the Fano factor $\mathfrak{F}_\nu$ on $\langle \nu\rangle$. Note that the proportionality coefficient is zero for Poissonian light, positive for classical super-Poissonian light and negative for non-classical sub-Poissonian light. This linearity can be verified by repeatedly measuring a field upon inserting filters with different transmittance in front of the detector. In fact the insertion of filters modifies $\langle \nu\rangle$, leaving $Q/\langle n\rangle$ unchanged. We can thus obtain $\gamma$ from the fit of the experimental data. Once $\gamma$ is evaluated, it is possible to find the photoelectron distribution by dividing the $\nu$ output values by the experimental value of $\gamma$ and re-binning the data in unitary bins. The procedure has the advantage of being self-consistent as the value of $\gamma$ is obtained from the measurements on the same field under investigation (Andreoni and Bondani, 2009; Bondani et al., 2009a) and does not require an independent calibration.

## ACKNOWLEDGMENTS

The presented topics were realized in collaboration with several scientists and friends, most of them operating at the Insubria University. We thankfully acknowledge: Andrea Agliati, Alessandra Andreoni, Federica A. Beduini, Enrico Brambilla, Giovanni De Cillis, Justinas Galinis, Alessandra Gatti, Marco G. Genoni, Ondrej Haderka, Ottavia Jedrkiewicz, Marco Lamperti, Radek Machulka, Václav Michálek, Stefano Olivares, Matteo G. A. Paris, Jan Peřina Jr., Guido Zambra. We also thank Prof. Jan Perina for fruitful discussions.

The work presented in this review was supported by MIUR under Grant No. FIRB LiCHIS-RBFR10YQ3H.

## REFERENCES

Abouraddy, A.F., Saleh, B.E.A., Sergienko, A.V., Teich, M.C., 2001. Role of entanglement in two-photon imaging. Phys. Rev. Lett. 87, 123602. http://dx.doi.org/10.1103/PhysRevLett.87.123602.

Afek, I., Natan, A., Ambar, O., Silberberg, Y., 2009. Quantum state measurements using multipixel photon detectors. Phys. Rev. A 79, 043830. http://dx.doi.org/10.1103/PhysRevA.79.043830.

Agafonov, I.N., Chekhova, M.V., Leuchs, G., 2010. Two-color bright squeezed vacuum. Phys. Rev. A 82, 011801. http://dx.doi.org/10.1103/PhysRevA.82.011801.

Haderka, O., Machulka, R., Perina, J., Allevi, A., Bondani, M., 2015. Spatial and spectral coherence in propagating high-intensity twin beams. Sci. Rep. 5. http://dx.doi.org/10.1038/srep14365. https://www.scopus.com/inward/record.uri?eid=2-s2.0-84942794167&doi=10.1038%2fsrep14365&partnerID=40&md5=643b21453afde9d0f43cfa7838f4b550. Cited by 2.

Hamar, M., Peřina, J., Haderka, O., Michálek, V., 2010. Transverse coherence of photon pairs generated in spontaneous parametric down-conversion. Phys. Rev. A 81, 043827. http://dx.doi.org/10.1103/PhysRevA.81.043827.

Harder, G., Bartley, T.J., Lita, A.E., Nam, S.W., Gerrits, T., Silberhorn, C., 2016. Single-mode parametric-down-conversion states with 50 photons as a source for mesoscopic quantum optics. Phys. Rev. Lett. 116, 143601. http://dx.doi.org/10.1103/PhysRevLett.116.143601.

Heidmann, A., Horowicz, R.J., Reynaud, S., Giacobino, E., Fabre, C., Camy, G., 1987. Observation of quantum noise reduction on twin laser beams. Phys. Rev. Lett. 59, 2555–2557. http://dx.doi.org/10.1103/PhysRevLett.59.2555.

Howell, J.C., Bennink, R.S., Bentley, S.J., Boyd, R.W., 2004. Realization of the Einstein-Podolsky-Rosen paradox using momentum- and position-entangled photons from spontaneous parametric down conversion. Phys. Rev. Lett. 92, 210403. http://dx.doi.org/10.1103/PhysRevLett.92.210403.

Iskhakov, T., Allevi, A., Kalashnikov, D.A., Sala, V.G., Takeuchi, M., Bondani, M., Chekhova, M., 2011. Intensity correlations of thermal light. Noise reduction measurements and new ghost imaging protocols. Eur. Phys. J. Spec. Top. 199, 127–138. http://dx.doi.org/10.1140/epjst/e2011-01508-x.

Ivanova, O.A., Iskhakov, T.S., Penin, A.N., Chekhova, M.V., 2006. Multiphoton correlations in parametric down-conversion and their measurement in the pulsed regime. Quant. Electron. 36 (10), 951. http://stacks.iop.org/1063-7818/36/i=10/a=A08.

Jedrkiewicz, O., Jiang, Y.-K., Brambilla, E., Gatti, A., Bache, M., Lugiato, L.A., Di Trapani, P., 2004. Detection of sub-shot-noise spatial correlation in high-gain parametric down conversion. Phys. Rev. Lett. 93, 243601. http://dx.doi.org/10.1103/PhysRevLett.93.243601.

Jedrkiewicz, O., Picozzi, A., Clerici, M., Faccio, D., Di Trapani, P., 2006. Emergence of $x$-shaped spatiotemporal coherence in optical waves. Phys. Rev. Lett. 97, 243903. http://dx.doi.org/10.1103/PhysRevLett.97.243903.

Jedrkiewicz, O., Gatti, A., Brambilla, E., Di Trapani, P., 2012. Experimental observation of a skewed x-type spatiotemporal correlation of ultrabroadband twin beams. Phys. Rev. Lett. 109, 243901. http://dx.doi.org/10.1103/PhysRevLett.109.243901.

Joobeur, A., Saleh, B.E.A., Teich, M.C., 1994. Spatiotemporal coherence properties of entangled light beams generated by parametric down-conversion. Phys. Rev. A 50, 3349–3361. http://dx.doi.org/10.1103/PhysRevA.50.3349.

Joobeur, A., Saleh, B.E.A., Larchuk, T.S., Teich, M.C., 1996. Coherence properties of entangled light beams generated by parametric down-conversion: theory and experiment. Phys. Rev. A 53, 4360–4371. http://dx.doi.org/10.1103/PhysRevA.53.4360.

Jost, B.M., Sergienko, A.V., Abouraddy, A.F., Saleh, B.E.A., Teich, M.C., 1998. Spatial correlations of spontaneously down-converted photon pairs detected with a single-photon-sensitive CCD camera. Opt. Express 3 (2), 81–88. http://dx.doi.org/10.1364/OE.3.000081. http://www.opticsexpress.org/abstract.cfm?URI=oe-3-2-81.

Just, F., Cavanna, A., Chekhova, M.V., Leuchs, G., 2013. Transverse entanglement ofbiphotons. New J. Phys. 15 (8), 083015. http://stacks.iop.org/1367-2630/15/i=8/a=083015.

Kalashnikov, D.A., Tan, S.H., Chekhova, M.V., Krivitsky, L.A., 2011. Accessing photon bunching with a photon number resolving multi-pixel detector. Opt. Express 19 (10), 9352–9363. http://dx.doi.org/10.1364/OE.19.009352. http://www.opticsexpress.org/abstract.cfm?URI=oe-19-10-9352.

Kalashnikov, D.A., Tan, S.H., Iskhakov, T.S., Chekhova, M.V., Krivitsky, L.A., 2012. Measurement of two-mode squeezing with photon number resolving multipixel detectors. Opt. Lett. 37 (14), 2829–2831. http://dx.doi.org/10.1364/OL.37.002829. http://ol.osa.org/abstract.cfm?URI=ol-37-14-2829.
Kim, J., Takeuchi, S., Yamamoto, Y., Hogue, H., 1999. Multiphoton detection using visible light photon counter. Appl. Phys. Lett. 74 (7), 902–904. https://www.scopus.com/inward/record.uri?eid=2-s2.0-0005586932&partnerID=40&md5=277ac36b413f4e26b4057138d90bc2aa. Cited by 253.
Klyshko, D.N., 1988. The effect of focusing on photon correlation in parametric light scattering. Zh. Eksp. Teor. Fiz. 94, 82–90.
Klyshko, D.N., 1996. The nonclassical light. Phys. Usp. 39 (6), 573. http://stacks.iop.org/1063-7869/39/i=6/a=R02.
Klyshko, D.N., Chekhova, M., Kulik, S., 2011. Physical Foundations of Quantum Electronics. World Scientific, Singapore.
Laiho, K., Christ, A., Cassemiro, K.N., Silberhorn, C., 2011. Testing spectral filters as Gaussian quantum optical channels. Opt. Lett. 36 (8), 1476–1478. http://dx.doi.org/10.1364/OL.36.001476. http://ol.osa.org/abstract.cfm?URI=ol-36-8-1476.
Lamperti, M., Allevi, A., Bondani, M., Machulka, R., Michálek, V., Haderka, O., Peřina, J., 2014. Optimal sub-Poissonian light generation from twin beams by photon-number resolving detectors. J. Opt. Soc. Am. B 31 (1), 20–25. http://dx.doi.org/10.1364/JOSAB.31.000020. http://josab.osa.org/abstract.cfm?URI=josab-31-1-20.
Law, C.K., Eberly, J.H., 2004. Analysis and interpretation of high transverse entanglement in optical parametric down conversion. Phys. Rev. Lett. 92, 127903. http://dx.doi.org/10.1103/PhysRevLett.92.127903.
Lewis, G.N., Lipkin, D., Magel, T.T., 1941. Reversible photochemical processes in rigid media. A study of the phosphorescent state. J. Am. Chem. Soc. 63 (11), 3005–3018. http://dx.doi.org/10.1021/ja01856a043.
Lita, A.E., Miller, A.J., Nam, S.W., 2008. Counting near-infrared single-photons with 95% efficiency. Opt. Express 16 (5), 3032–3040. http://dx.doi.org/10.1364/OE.16.003032. http://www.opticsexpress.org/abstract.cfm?URI=oe-16-5-3032.
Lopaeva, E.D., Chekhova, M.V., 2010. Ghost imaging with the use of the variance of the difference photocurrent. JETP Lett. 91 (9), 447–451. ISSN 1090-6487. http://dx.doi.org/10.1134/S0021364010090031.
Lvovsky, A.I., Raymer, M.G., 2009. Continuous-variable optical quantum-state tomography. Rev. Mod. Phys. 81, 299–332. http://dx.doi.org/10.1103/RevModPhys.81.299.
Maker, P.D., Terhune, R.W., Nisenoff, M., Savage, C.M., 1962. Effects of dispersion and focusing on the production of optical harmonics. Phys. Rev. Lett. 8, 21–22. http://dx.doi.org/10.1103/PhysRevLett.8.21.
Malygin, A., Penin, A., Sergienko, A., 1985. Spatial-temporal photon bunching in spontaneous parametric light scattering. In: Akademiia Nauk SSSR Doklady. vol. 281, pp. 308–313.
Mandel, L., Wolf, E., 1995. Optical Coherence and Quantum Optics. Cambridge University Press, Cambridge.
Mauerer, W., Avenhaus, M., Helwig, W., Silberhorn, C., 2009. How colors influence numbers: photon statistics of parametric down-conversion. Phys. Rev. A 80, 053815. http://dx.doi.org/10.1103/PhysRevA.80.053815.
Migdall, A., Polyakov, S.V., Fan, J., Bienfang, J.C., 2013. Single-Photon Generation and Detection: physics and Applications. vol. 45. Academic Press, Cambridge, MA.
Mikhailov, V.N., Bondani, M., Paleari, F., Andreoni, A., 2003. Optical phase conjugation in difference-frequency generation. J. Opt. Soc. Am. B 20 (8), 1715–1723. http://dx.doi.org/10.1364/JOSAB.20.001715. http://josab.osa.org/abstract.cfm?URI=josab-20-8-1715.

Mikhailova, Y.M., Volkov, P.A., Fedorov, M.V., 2008. Biphoton wave packets in parametric down-conversion: spectral and temporal structure and degree of entanglement. Phys. Rev. A 78, 062327. http://dx.doi.org/10.1103/PhysRevA.78.062327.

Molina-Terriza, G., Minardi, S., Deyanova, Y., Osorio, C.I., Hendrych, M., Torres, J.P., 2005. Control of the shape of the spatial mode function of photons generated in noncollinear spontaneous parametric down-conversion. Phys. Rev. A 72, 065802. http://dx.doi.org/10.1103/PhysRevA.72.065802.

Monken, C.H., Ribeiro, P.H.S., Pádua, S., 1998. Transfer of angular spectrum and image formation in spontaneous parametric down-conversion. Phys. Rev. A 57, 3123–3126. http://dx.doi.org/10.1103/PhysRevA.57.3123.

Mosset, A., Devaux, F., Fanjoux, G., Lantz, E., 2004. Direct experimental characterization of the Bose-Einstein distribution of spatial fluctuations of spontaneous parametric down-conversion. Eur. Phys. J. D 28 (3), 447–451. http://dx.doi.org/10.1140/epjd/e2003-00323-2.

O'Sullivan, M.N., Chan, K.W.C., Boyd, R.W., 2010. Comparison of the signal-to-noise characteristics of quantum versus thermal ghost imaging. Phys. Rev. A 82, 053803. http://dx.doi.org/10.1103/PhysRevA.82.053803.

Paleari, F., Andreoni, A., Zambra, G., Bondani, M., 2004. Thermal photon statistics in spontaneous parametric downconversion. Opt. Express 12 (13), 2816–2824. http://dx.doi.org/10.1364/OPEX.12.002816. http://www.opticsexpress.org/abstract.cfm?URI=oe-12-13-2816.

Pérez, A.M., Iskhakov, T.S., Sharapova, P., Lemieux, S., Tikhonova, O.V., Chekhova, M.V., Leuchs, G., 2014. Bright squeezed-vacuum source with 1.1 spatial mode. Opt. Lett. 39 (8), 2403–2406. http://dx.doi.org/10.1364/OL.39.002403. http://ol.osa.org/abstract.cfm?URI=ol-39-8-2403.

Perina, J., 1991. Quantum Statistics of Linear and Nonlinear Optical Phenomena. Kluwer Academic Publishers, Dordrecht.

Perina Jr., J., Haderka, O., Allevi, A., Bondani, M., 2016. Internal dynamics of intense twin beams and their coherence. Sci. Rep. 6. http://dx.doi.org/10.1038/srep22320 https://www.scopus.com/inward/record.uri?eid=2-s2.0-84988643567&doi=10.1038%2fsrep22320&partnerID=40&md5=05b7e5c53875932d9745a1a3bb912e5b. Cited by 1.

Peřina, J., 2013. Pulsed-squeezed-light generation in a waveguide with second-subharmonic generation and periodic corrugation. Phys. Rev. A 87, 013833. http://dx.doi.org/10.1103/PhysRevA.87.013833.

Peřina, J., 2016. Spatial, spectral, and temporal coherence of ultraintense twin beams. Phys. Rev. A 93, 013852. http://dx.doi.org/10.1103/PhysRevA.93.013852.

Peřina, J., Haderka, O., Soubusta, J., 2001. Quantum cryptography using a photon source based on postselection from entangled two-photon states. Phys. Rev. A 64, 052305. http://dx.doi.org/10.1103/PhysRevA.64.052305.

Peřina, J., Hamar, M., Michálek, V., Haderka, O., 2012. Photon-number distributions of twin beams generated in spontaneous parametric down-conversion and measured by an intensified CCD camera. Phys. Rev. A 85, 023816. http://dx.doi.org/10.1103/PhysRevA.85.023816.

Peřina, J., Haderka, O., Michálek, V., 2013. Sub-Poissonian-light generation by postselection from twin beams. Opt. Express 21 (16), 19387–19394. http://dx.doi.org/10.1364/OE.21.019387. http://www.opticsexpress.org/abstract.cfm?URI=oe-21-16-19387.

Pittman, T.B., Shih, Y.H., Strekalov, D.V., Sergienko, A.V., 1995. Optical imaging by means of two-photon quantum entanglement. Phys. Rev. A 52, R3429–R3432. http://dx.doi.org/10.1103/PhysRevA.52.R3429.

Puddu, E., Andreoni, A., Degiovanni, I.P., Bondani, M., Castelletto, S., 2007. Ghost imaging with intense fields from chaotically seeded parametric downconversion. Opt. Lett. 32 (9), 1132–1134. http://dx.doi.org/10.1364/OL.32.001132. http://ol.osa.org/abstract.cfm?URI=ol-32-9-1132.

Ramilli, M., Allevi, A., Chmill, V., Bondani, M., Caccia, M., Andreoni, A., 2010. Photon-number statistics with silicon photomultipliers. J. Opt. Soc. Am. B 27 (5), 852–862. http://dx.doi.org/10.1364/JOSAB.27.000852. http://josab.osa.org/abstract.cfm?URI=josab-27-5-852.

Rarity, J.G., Tapster, P.R., Walker, J.G., Seward, S., 1990. Experimental demonstration of single photon rangefinding using parametric downconversion. Appl. Opt. 29 (19), 2939–2943. http://dx.doi.org/10.1364/AO.29.002939. http://ao.osa.org/abstract.cfm?URI=ao-29-19-2939.

Ribeiro, P.H.S., Schwob, C., Maître, A., Fabre, C., 1997. Sub-shot-noise high-sensitivity spectroscopy with optical parametric oscillator twin beams. Opt. Lett. 22 (24), 1893–1895. http://dx.doi.org/10.1364/OL.22.001893. http://ol.osa.org/abstract.cfm?URI=ol-22-24-1893.

Sciarrino, F., Vallone, G., Milani, G., Avella, A., Galinis, J., Machulka, R., Perego, A.M., Spasibko, K.Y., Allevi, A., Bondani, M., Mataloni, P., 2011. High degree of entanglement and nonlocality of a two-photon state generated at 532 nm. Eur. Phys. J. Spec. Top. 199 (1), 111–125. ISSN: 1951-6401. http://dx.doi.org/10.1140/epjst/e2011-01507-y.

Sharapova, P., Pérez, A.M., Tikhonova, O.V., Chekhova, M.V., 2015. Schmidt modes in the angular spectrum of bright squeezed vacuum. Phys. Rev. A 91, 043816. http://dx.doi.org/10.1103/PhysRevA.91.043816.

Shih, Y., 2007. Quantum imaging. IEEE J. Sel. Top. Quant. Electron. ISSN: 1077-260X13 (4), 1016–1030. http://dx.doi.org/10.1109/JSTQE.2007.902724.

Short, R., Mandel, L., 1983. Observation of sub-Poissonian photon statistics. Phys. Rev. Lett. 51, 384–387. http://dx.doi.org/10.1103/PhysRevLett.51.384.

Smithey, D.T., Beck, M., Belsley, M., Raymer, M.G., 1992. Sub-shot-noise correlation of total photon number using macroscopic twin pulses of light. Phys. Rev. Lett. 69, 2650–2653. http://dx.doi.org/10.1103/PhysRevLett.69.2650.

Spasibko, K.Y., Iskhakov, T.S., Chekhova, M.V., 2012. Spectral properties of high-gain parametric down-conversion. Opt. Express 20 (7), 7507–7515. http://dx.doi.org/10.1364/OE.20.007507. http://www.opticsexpress.org/abstract.cfm?URI=oe-20-7-7507.

Tasca, D.S., Walborn, S.P., Souto Ribeiro, P.H., Toscano, F., Pellat-Finet, P., 2009. Propagation of transverse intensity correlations of a two-photon state. Phys. Rev. A 79, 033801. http://dx.doi.org/10.1103/PhysRevA.79.033801.

Valencia, A., Scarcelli, G., D'Angelo, M., Shih, Y., 2005. Two-photon imaging with thermal light. Phys. Rev. Lett. 94, 063601. http://dx.doi.org/10.1103/PhysRevLett.94.063601.

Vasilyev, M., Choi, S.-K., Kumar, P., D'Ariano, G.M., 2000. Tomographic measurement of joint photon statistics of the twin-beam quantum state. Phys. Rev. Lett. 84, 2354–2357. http://dx.doi.org/10.1103/PhysRevLett.84.2354.

Vogel, W., Welsch, D.-G., 2006. Quantum Optics. John Wiley & Sons, New York, NY.

Řeháček, J., Hradil, Z., Haderka, O., Peřina, J., Hamar, M., 2003. Multiple-photon resolving fiber-loop detector. Phys. Rev. A 67, 061801. http://dx.doi.org/10.1103/PhysRevA.67.061801.

Waks, E., Sanders, B.C., Diamanti, E., Yamamoto, Y., 2006. Highly nonclassical photon statistics in parametric down-conversion. Phys. Rev. A 73, 033814. http://dx.doi.org/10.1103/PhysRevA.73.033814.

Wasilewski, W., Lvovsky, A.I., Banaszek, K., Radzewicz, C., 2006. Pulsed squeezed light: simultaneous squeezing of multiple modes. Phys. Rev. A 73, 063819. http://dx.doi.org/10.1103/PhysRevA.73.063819.
Wasilewski, W., Radzewicz, C., Frankowski, R., Banaszek, K., 2008. Statistics of multiphoton events in spontaneous parametric down-conversion. Phys. Rev. A 78, 033831. http://dx.doi.org/10.1103/PhysRevA.78.033831.
Zhang, D., Zhai, Y.-H., Wu, L.-A., Chen, X.-H., 2005. Correlated two-photon imaging with true thermal light. Opt. Lett. 30 (18), 2354–2356. http://dx.doi.org/10.1364/OL.30.002354. http://ol.osa.org/abstract.cfm?URI=ol-30-18-2354.

CHAPTER THREE

# Nonclassical Light Generation From III–V and Group-IV Solid-State Cavity Quantum Systems

**Marina Radulaski[1], Kevin A. Fischer[1], Jelena Vučković[2]**
E. L. Ginzton Laboratory, Stanford University, Stanford, CA, United States
[2]Corresponding author: e-mail address: jela@stanford.edu

## Contents



[1] These authors contributed equally.

*Advances in Atomic, Molecular, and Optical Physics*, Volume 66
ISSN 1049-250X
http://dx.doi.org/10.1016/bs.aamop.2017.03.001

## Abstract

In this chapter, we present the state of the art in the generation of nonclassical states of light using semiconductor cavity quantum electrodynamics (QED) platforms. Our focus is on the photon blockade effects that enable the generation of indistinguishable photon streams with high purity and efficiency. Starting with the leading platform of InGaAs quantum dots in optical nanocavities, we review the physics of a single quantum emitter strongly coupled to a cavity. Furthermore, we propose a complete model for photon blockade and tunneling in III–V quantum dot-cavity QED systems. Turning toward quantum emitters with small inhomogeneous broadening, we propose a direction for novel experiments for nonclassical light generation based on group-IV color-center systems. We present a model of a multiemitter cavity QED platform, which features richer dressed-states ladder structures, and show how it can offer opportunities for studying new regimes of high-quality photon blockade.

## 1. INTRODUCTION

Many quantum technologies, including quantum key distribution and photonic-qubit-based quantum computation, require on-demand sources of light that produce pulses containing a well-defined number of photons (O'Brien et al., 2009). Such sources are expected to have high efficiencies, rarely emit a wrong number of photons, and produce indistinguishable photons that interfere with each other. The majority of scientific efforts toward creating sources of nonclassical light have so far been focused on engineering single-photon sources (Buckley et al., 2012), and the basic idea used to generate single photons on demand in state-of-the-art approaches is very simple; a single quantum emitter is excited with a pulsed source and its emission is filtered in order to isolate a single-photon with the desired properties (Sanders et al., 2005). For example, an optical or electrical pulse can generate carriers–electrons and holes–inside a quantum dot that recombine to produce several photons at different frequencies. Subsequent spectral filtering can be used to isolate a single-photon (Santori et al., 2002; Yuan et al., 2002). Although these systems are already characterized by high multiphoton probability suppression, both the efficiency and indistinguishability of such a source can be further improved by embedding the quantum emitter into a cavity that has a high-quality factor and a small mode volume, or by resonant excitation approaches such as photon blockade (Birnbaum et al., 2005; Faraon et al., 2008).

Epitaxial III–V semiconductor quantum dots in cavities have been a leading platform for nonclassical light generation experiments, but they suffer from large inhomogeneous broadening, which impedes scaling of the systems to the substantial number of quantum emitters necessary for implementation of large entangled photon states, quantum networks or quantum simulators (Carusotto and Ciuti, 2013; Greentree et al., 2006; Hartmann et al., 2008; Kimble, 2008). Namely, a stream of indistinguishable single photons from one or more quantum emitters could be employed to generate larger photon number states or entangled photon states by quantum interference, but this requires photons to be indistinguishable and thus the emitters must be indistinguishable. Additionally, these quantum dots require operation at cryogenic temperatures. Recently, color-centers in group-IV semiconductors have emerged as alternative quantum emitters, with significantly smaller inhomogenous broadening and the ability to operate at room temperature (Beveratos et al., 2002b; Neu et al., 2011; Rogers et al., 2014; Widmann et al., 2015).

In this chapter, we present state-of-the-art demonstrations of nonclassical light generation in these semiconductor systems, and we propose a model of a system that improves the quality of the sources (efficiency, indistinguishability) through the multiemitter cavity QED. We believe that photon blockade effects will continue to play an important role in high-throughput generation of nonclassical states of light. These effects, which have been initially developed in atomic systems (Birnbaum et al., 2005), have been demonstrated only in quantum dot-based systems among solid-state optical platforms (Faraon et al., 2008; Volz et al., 2012). However, ensembles of color-centers in group-IV semiconductor cavities present a very promising platform for the implementation of new regimes of photon blockade and new quantum light sources.

## 1.1 InAs Quantum Dots in GaAs

InAs quantum dots in GaAs have served as a leading platform for solid-state quantum optics and cavity QED experiments for the past 20 years (Dietrich et al., 2016; Michler, 2009). Such quantum dots are formed by self-assembly during the growth process called molecular beam epitaxy (MBE), as a result of lattice mismatch between InAs and GaAs. Since quantum dot islands are formed by self-assembly, their locations are random and they have a

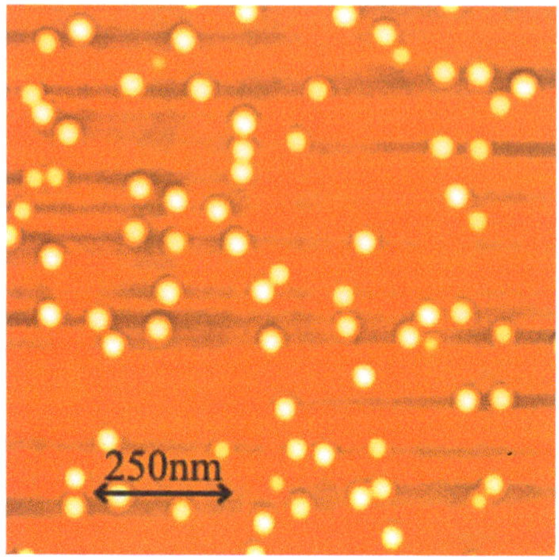


**Fig. 1** AFM image of a 1 × 1 micron square array of uncapped, self-assembled InGaAs quantum dots.

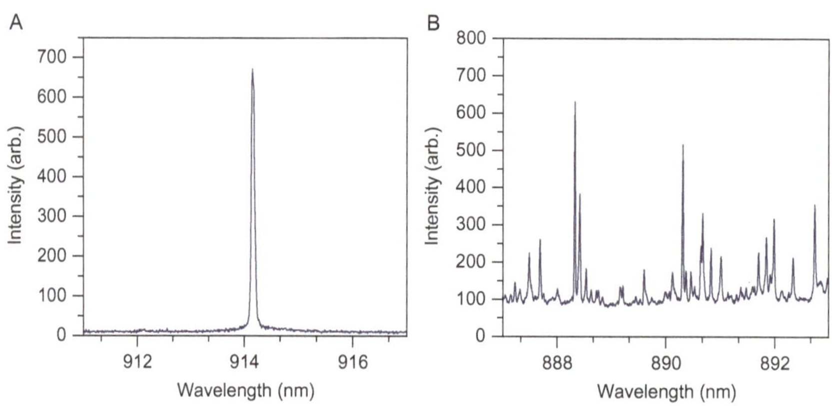


**Fig. 2** Spectra of quantum dots of (A) a neutral excitonic transition and (B) of an ensemble of excitonic transitions from many quantum dots that feature a large inhomogeneous broadening.

distribution of sizes and shapes (Fig. 1), leading to variation in the transition energies of different quantum dots on the same wafer (inhomogeneous broadening) (Fig. 2). Because of a shallow confining potential, these quantum dots require operation at cryogenic temperatures, lower than 70 K and typically around 4 K.

## 1.2 Color-Centers in Group-IV Materials

Color-centers are emerging solid-state photonics systems that hold great promise for nonclassical light generation. They can serve both as bright single-photon emitters (Aharonovich et al., 2009; Castelletto et al., 2014) with fast GHz generation rates and as spin qubits (Childress et al., 2006; Widmann et al., 2015) with demonstrated lifetimes of close to a second (Bar-Gill et al., 2013). Structurally, color-centers are point-defects in the semiconductor lattice whose localized electronic orbitals have optical transitions (Weber et al., 2010). They can occur naturally or be generated through electron, neutron, or ion irradiation (Fuchs et al., 2015; Kurtsiefer et al., 2000; Widmann et al., 2015), or doping during growth (Zhang et al., 2015), where masking, dose, and beam energy can be used to control defect density and position. In addition, the angstrom-scale size of defect centers opens opportunities to interface an ensemble of emitters with a nanocavity (Cui et al., 2015; Radulaski et al., 2016b; Sipahigil et al., 2016). Moreover, the inhomogeneous broadening of color-center optical transitions can be as narrow as four lifetime-limited linewidths (Sipahigil et al., 2014), making these emitters nearly identical. This is an advantageous property compared to other semiconductor emitters which can be utilized to access novel many-body studies in the solid-state. Additional flexibility in designing photonic systems is granted by the variety of operating wavelengths, as well as the difference in optical, mechanical, and electronic properties between the host materials.

Diamond and silicon carbide have prominently served as hosts of various vacancy-related defects (Castelletto et al., 2014; Falk et al., 2013; Kurtsiefer et al., 2000; Sipahigil et al., 2014; Widmann et al., 2015). These substrates are chemically inert which makes them biocompatible for interfacing with or within living cells as dyes and nanoscale sensors (Kucsko et al., 2013; Saddow, 2012). Color-centers also have a unique set of properties as nearly identical solid-state emitters with strong dipole moment, which we will explore for novel opportunities in cavity QED.

## 2. OVERVIEW OF CAVITY QED WITH A SINGLE QUANTUM EMITTER

Cavity quantum electrodynamical systems hold great promise for exploring fundamental light–matter interactions, generating novel quantum states of light, and playing integral roles in quantum information networks (O'Brien et al., 2009). Originally, such investigations were centered on atomic systems where both the quantum matter and light confining components of the cavity QED system possessed intrinsically narrow linewidths (Birnbaum et al., 2005). As a result, the hallmark signature of strong light–matter interaction with the emergence of new hybridized states of light and matter or *strong coupling* was more easily achievable. Such hybridized states are anharmonic, resulting in giant nonlinearities at the single-photon level. However, these atomic systems will unlikely serve as practical elements in communication networks due to their slow interaction rates and lack of scalability. In the 2000s, major advances toward the study of light–matter interaction in systems based on InGaAs quantum dots and photonic crystal nanocavities enabled the observation of strong coupling in an optical solid-state system (Englund et al., 2007; Yoshie et al., 2004). Since then, the boundaries of quantum dot-based cavity QED work have been pushed from ultra-low threshold lasers (Ellis et al., 2011) and ultra-fast single-photon sources (Müller et al., 2015b), to photon blockade and single-photon phase gates (Fushman et al., 2008). More recently, group-IV solid-state quantum emitters, in the form of the popular nitrogen (Englund et al., 2010; Wolters et al., 2010) and silicon (Radulaski et al., 2016b; Riedrich-Möller et al., 2014) vacancy-related complexes, have generated strong interest for future cavity QED experiments with their demonstrations of Purcell enhancement, single-photon emission, and exceptionally small inhomogeneous broadening. In this section, we provide an introduction to the basic concepts relevant to nonclassical light generation with a single-emitter cavity QED system.

## 2.1 Single-Photon Emission With Color-Centers

For emission purposes, color-centers can be described as two-level quasiatoms whose single-photon emission can be employed for optical networks and quantum cryptography. Similarly to quantum dots, the excited state lifetimes are in the nanosecond range. Due to the presence of vibronic sublevels of the ground and excited states, the emission spectra consist of a zero-phonon line (ZPL), a narrow feature corresponding to the energetically highest transition, and a phonon sideband, which comprises phonon-assisted transitions (Fitchen and Fowler, 1968). The ratio of the emission into the ZPL varies between color-centers, often based on the symmetry of the defect system, and is higher at low temperatures.

The negatively charged nitrogen-vacancy (NV) center in diamond has, until recently, been the dominant color-center system in photonics (Aharonovich et al., 2011). Its ZPL at 637 nm contains 3%–5% of the emission. Single-photon generation rates are in the tens of kHz and have been enhanced by the use of nanopillar structures to 168 kHz (Babinec et al., 2010). The NV center in diamond has been utilized for cryptography

(Beveratos et al., 2002a), and integrated systems with silicon nitride waveguides have been demonstrated (Mouradian et al., 2015). One of the challenges in working with the NV center is its spectral instability (Siyushev et al., 2013), which has motivated a search for novel color-center systems (Weber et al., 2010).

The negatively charged silicon vacancy (SiV) in diamond has emerged as an emitter with a large (70%) emission ratio into its ZPL at 738 nm. In addition, the ensemble linewidths are as narrow as 400 MHz at cryogenic temperatures (Sipahigil et al., 2014), which qualifies these systems as nearly identical solid-state quasiatoms. An order of magnitude increase in the ZPL extraction and two to three orders of magnitude reduction in the ZPL linewidth compared to the NV center makes SiV a more attractive center for scalable systems. Additionally, reactive ion etching can be performed while maintaining the SiV's optical properties (Fig. 3), which has been demonstrated with an array of nanopillars (Zhang et al., 2015) and in nanobeam cavities (Sipahigil et al., 2016). These recent results pave a promising path for integrating nearly identical quantum emitters in the suspended photonic structures needed for cavity QED.

Silicon carbide (SiC) has also been discovered as an alternative host of color-centers to diamond. Across its many polytypes (such as 3C, 4H, or 6H), SiC offers a variety of quantum emitters. Emitters in 3C-SiC, such as the carbon-antisite-vacancy ($Si_CV_{Si}$) (Castelletto et al., 2014) and

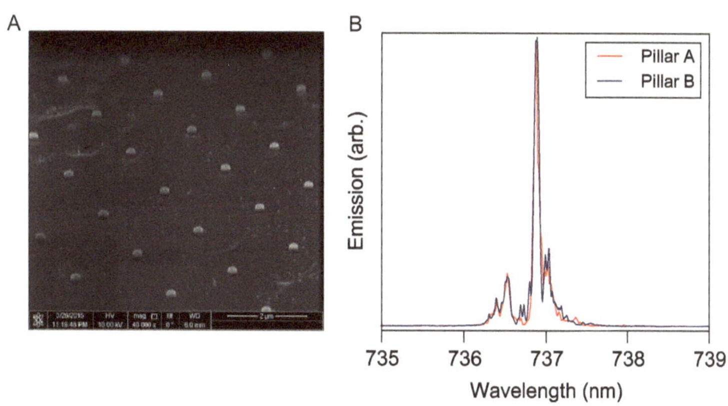


**Fig. 3** (A) Etched nanopillars in color-center rich diamond substrate. (B) Silicon-vacancy emission from two different pillars indicating low strain in the fabricated structures. *Data from Zhang, J.L., Ishiwata, H., Babinec, T.M., Radulaski, M., Müller, K., Lagoudakis, K.G., Dory, C., Dahl, J., Edgington, R., Soulière, V., et al., 2015. Hybrid group IV nanophotonic structures incorporating diamond silicon-vacancy color centers. Nano Lett. 16 (1), 212–217.*

oxidation-induced centers (Lohrmann et al., 2016), have exceptional brightness of around 1 MHz count rates. The divacancy ($V_C V_{Si}$) in 4H-SiC gives rise to six ZPL transitions in the range of 1100–1200 nm, two of which are active at room temperatures (Koehl et al., 2011). The silicon vacancy ($V_{Si}$) in 4H-SiC and 6H-SiC emits into 25–30 GHz narrow ZPLs capturing several percent of emission at cryogenic temperatures in their two and three lines (Sörman et al., 2000), respectively, in the wavelength region 860–920 nm. In particular, the 916 nm ZPL from $V_{Si}$ in 4H-SiC has been incorporated in nanopillars where nanofabrication processing left the optical properties unaltered but increased the collection efficiency of single photons by several times (Radulaski et al., 2017).

An increase in single-photon emission rate can be achieved through Purcell enhancement where a color-center is coupled to a cavity with small mode volume and high-quality factor. This has been achieved with Purcell factor as high as $F = 26$ in a hybrid approach where NV and SiV centers in diamond were coupled to GaP (Gould et al., 2016; Wolters et al., 2010) and SiC (Radulaski et al., 2016b) microresonators, as well as in monolithic approaches with $V_{Si}$ in 4H-SiC (Bracher et al., 2017).

For an overview of the established field surrounding single-photon emission from quantum dots we refer the reader to Buckley et al. (2012).

## 2.2 Strong Coupling With Quantum Dots

One of the most successful and established platforms for single-emitter cavity QED is the InGaAs quantum dot embedded within an L3 photonic crystal cavity (Akahane et al., 2003; Lodahl et al., 2015). Self-assembled InGaAs quantum dots are widely used as quantum emitters due to their excellent optical quality and are particularly useful for strong coupling experiments due to their narrowband optical transitions and ability to be integrated into planar photonic crystal cavities (Aharonovich et al., 2016). Planar photonic crystal resonators provide extremely small mode volumes ($V$) that enable a large enhancement of the light–matter interaction strength with any embedded quantum emitters, $g \propto 1/\sqrt{V}$ (Vahala, 2003). A schematic depiction of such a system is shown in Fig. 4, with the L3 photonic crystal cavity shown in the upper left (characterized by energy decay rate $\kappa$) and the quantum dot shown in the bottom right (characterized by energy decay rate $\Gamma$). The quantum dot, illustrated as the red half-dome, is either probabilistically or deliberately positioned at the center of the crystal where it can maximally couple to the fundamental mode of the cavity for optimal interaction

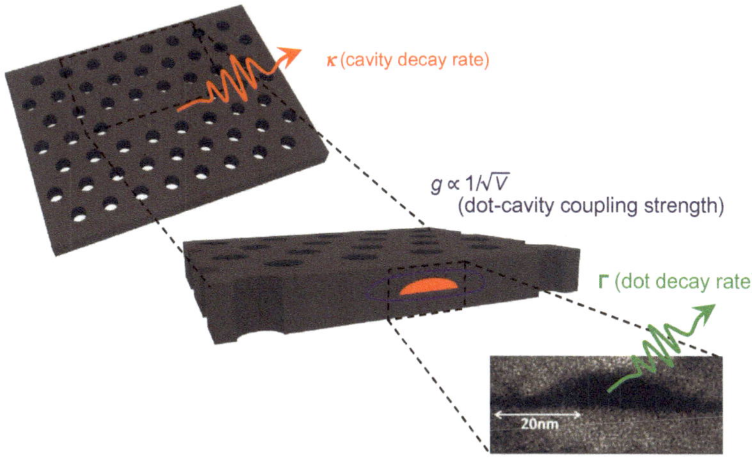


**Fig. 4** Schematic depiction of a strongly coupled system based on a single InGaAs quantum dot embedded within an L3 photonic crystal cavity. The red partial-dome in the cross-section (*middle*) represents the quantum dot and a TEM image is shown of a typical dot in the *bottom right. TEM adapted with permission from Krenner, H.J., Stufler, S., Sabathil, M., Clark, E.C., Ester, P., Bichler, M., Abstreiter, G., Finley, J.J., Zrenner, A., 2005. Recent advances in exciton-based quantum information processing in quantum dot nanostructures. New J. Phys. 7 (1), 184.*

rate $g$ (Kuruma et al., 2016). Ideally, the quantum dot behaves like a quantum two-level system, and it can be placed inside of a diode structure to control which specific excitonic transition forms the two-level system (Carter et al., 2013; Warburton et al., 2000). However, most of the work on strongly coupled systems in the solid-state has been performed without such a structure, and usually either the neutral or singly charged exciton couples to the cavity mode.

## 2.3 The Jaynes–Cummings Model

At its simplest, a strongly coupled system is modeled by the Jaynes–Cummings (JC) Hamiltonian (Shore and Knight, 1993)

$$H_{\mathrm{JC}} = \omega_a a^\dagger a + (\omega_a + \Delta_e)\sigma^\dagger\sigma + g\left(a^\dagger\sigma + a\sigma^\dagger\right), \quad (1)$$

which represents a single cavity mode coupled to a single quantum two-level system. Here, $a$ is the operator which represents the cavity's mode, $\omega_a$ is the mode's frequency, $\sigma$ is the dipole operator representing a single two-level subsystem of a quantum emitter, and $\Delta_e$ is the detuning of the quantum emitter's transition frequency from that of the cavity. As we will discuss in subsequent sections, specific solid-state considerations will have important implications in exactly modeling such a system, but for now we will consider the idealized JC model.

Because of the coherent interaction between the bosonic cavity mode and fermionic emitter mode, a new set of eigenstates arise (Shore and Knight, 1993). The optimal basis is no longer the *bare* states of the system, the direct product of individual emitter (ground $|g\rangle$ or excited $|e\rangle$) and cavity (photon number $|n\rangle$) states, but rather a so-called *dressed state* basis that comprises states known as polaritons. These polaritons are entangled states of light and matter, with eigenstates

$$|n,+\rangle = \cos(\alpha_n/2)|n-1\rangle|e\rangle + \sin(\alpha_n/2)|n\rangle|g\rangle \tag{2}$$

and

$$|n,-\rangle = -\sin(\alpha_n/2)|n-1\rangle|e\rangle + \cos(\alpha_n/2)|n\rangle|g\rangle, \tag{3}$$

where $\alpha_n = \tan^{-1}\left(2g\sqrt{n+1}/\Delta_e\right)$. These states have eigenenergies

$$E_{\pm}^{n} = n\omega_a + \Delta_e/2 \pm \sqrt{(\sqrt{n}g)^2 + (\Delta_e/2)^2}. \tag{4}$$

Consider the first two polaritons of nonzero energy ($n = 1$), called the first upper and lower polaritons: when the cavity and emitter are tuned on resonance ($\Delta_e = 0$), then the polaritons each contain equal contributions of a single electronic or a single photonic excitation. However, as the emitter is detuned from the cavity (by increasing $\Delta_e$) then the polaritons trend toward either the electronic or photonic characters.

## 2.4 Observing Strong Coupling

The evolution of the system dynamics is governed by a Liouville equation $\partial_t\rho(t) = \mathcal{L}\rho(t)$, which accounts for the nonunitary evolution induced by the cavity and emitter dissipation (Laussy et al., 2012). Specifically,

$$\mathcal{L}_{JC}\rho(t) = \mathrm{i}\left[\rho(t), H_{JC}\right] + \frac{\kappa}{2}\mathcal{D}[a]\rho(t) + \frac{\Gamma}{2}\mathcal{D}[\sigma]\rho(t), \tag{5}$$

where $\mathcal{D}[c]\rho(t) = 2c\rho(t)c^{\dagger} - c^{\dagger}c\rho(t) - \rho(t)c^{\dagger}c$ is a superoperator called the Lindblad dissipator and $c$ is an arbitrary system operator.

Now, we can investigate how strong coupling behavior can be experimentally explored. The most common way to initially observe the effects of strong coupling is to directly measure the splitting of the two lowest-energy

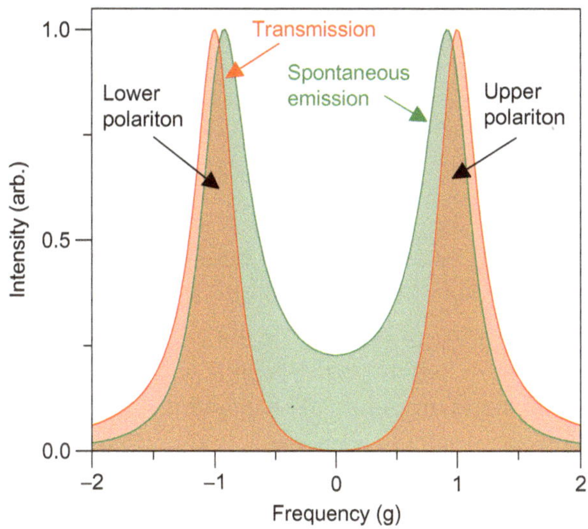


**Fig. 5** Observation of strong coupling by measuring either transmission or spontaneous emission spectra for zero emitter detuning ($\Delta_e = 0$). Typical solid-state cavity QED parameters were used, where $\kappa \approx 3/4g$ and $\Gamma$ is small enough to be insignificant ($\Gamma \ll g, \kappa$).

polaritons either in a transmission experiment (Englund et al., 2007) or spontaneous emission experiment (Reithmaier et al., 2004), as shown in Fig. 5. First, we discuss a spontaneous emission experiment. Here, the emitter is incoherently pumped by exciting charge carriers in the semiconductor to higher energy than the emitter's transition [e.g., in a quantum dot's multiexcitonic states, quasiresonant levels, or above the bandgap (Santori, 2003)], where they relax into the transition coupled to the cavity. Subsequently, the system relaxes through cavity emission. This process can easily be modeled by adding the term $\frac{P_\sigma}{2}\mathcal{D}[\sigma^\dagger]\rho(t)$ to $\mathcal{L}_{JC}\rho(t)$, where $P_\sigma$ is the incoherent pumping rate of the quantum emitter (Laussy et al., 2008).

To get an idea of whether the emission occurs through the cavity or the emitter decay channel, consider their typical emission rates. Typical rates for a well-performing cavity QED system are approximately $\Gamma/2\pi \approx 0.2$ GHz (which is often further suppressed by a photonic bandgap) and $\kappa/2\pi \approx 10$ GHz (Müller et al., 2015b). Because solid-state systems work in the bad-cavity limit where $\kappa \gg \Gamma$, the photons are almost exclusively emitted by the cavity. Therefore, to obtain the measured spectrum of spontaneous emission (e.g., through an ideal spectrometer or scanning Fabry–Perot cavity), we compute the spectrum of the cavity mode operator $a$ (Laussy et al., 2008)

$$S_a(\omega) = \lim_{t\to\infty} \int_{-\infty}^{\infty} d\tau \, \langle a^\dagger(t+\tau)a(t)\rangle e^{-i\omega\tau}. \tag{6}$$

The astute reader may also notice that it would be possible to incoherently excite the cavity mode using the term $\frac{P_a}{2}\mathcal{D}[a^\dagger]\rho(t)$, where $P_a$ is the cavity's incoherent pumping rate. While this configuration is difficult to realize experimentally, we note that the resulting incoherent spectrum of emission is formally equivalent to a transmission spectrum for arbitrarily small excitation powers. When the incoherent excitation rate is much slower than the decay rates of the first-rung polaritons, $\mathcal{D}[a^\dagger]$ randomly initializes the system with a maximum of one photon in the cavity mode. Thus, calculating a spectrum under incoherent cavity excitation is equivalent to calculating a one-photon spectra (Diniz et al., 2011), which represents the linear impulse response of the system and hence its transmission spectrum. Additionally, the computational complexity of this approach for calculating a transmission spectrum is much lower, especially for multiemitter cavity QED systems, so we use this approach for modeling transmission spectra in Section 5.

On the other hand in a transmission *experiment*, a weak continuous-wave (CW) laser incident on the cavity is scanned in frequency and the transmitted light is measured. Now, as opposed to adding the excitation source as a dissipator, the input light is modeled as a coherent state coupled to the cavity (Laussy et al., 2012). This is represented by a Hamiltonian driving term $H_{\text{drive}} = \mathcal{E}\left(a\,e^{i\omega_L t} + a^\dagger e^{-i\omega_L t}\right)$, where $\mathcal{E}$ is the real-valued driving strength (proportional to the incident field) and $\omega_L$ is the coherent state frequency. For simplicity, a rotating frame transformation $\tilde{H} = UHU^\dagger + i(\partial_t U)U^\dagger$ with $U = e^{i\omega_L\left(a^\dagger a + \sigma^\dagger\sigma\right)t}$ is used to remove the time-dependence of the excitation term (Majumdar et al., 2013) so that $\tilde{H}_{\text{drive}} = \mathcal{E}\left(a + a^\dagger\right)$, and written as a superoperator

$$\tilde{\mathcal{L}}_{\text{drive}}\,\tilde{\rho}\,(t) = i\left[\tilde{\rho}(t), \tilde{H}_{\text{drive}}\right]. \tag{7}$$

Similarly, the Jaynes–Cummings Liouvillian transforms according to

$$\tilde{\mathcal{L}}_{\text{JC}}\,\tilde{\rho}\,(t) = i\left[\tilde{\rho}(t), \tilde{H}_{\text{JC}}\right] + \frac{\kappa}{2}\mathcal{D}[a]\,\tilde{\rho}\,(t) + \frac{\Gamma}{2}\mathcal{D}[\sigma]\,\tilde{\rho}\,(t) \tag{8}$$

with

$$\tilde{H}_{\mathrm{JC}} = (\omega_a - \omega_{\mathrm{L}})a^\dagger a + (\omega_a + \Delta_{\mathrm{e}} - \omega_{\mathrm{L}})\sigma^\dagger\sigma + g\left(a^\dagger\sigma + a\sigma^\dagger\right). \tag{9}$$

Calculating the transmission spectra is then performed first by calculating the steady-state density matrix through solving

$$\left(\tilde{\mathcal{L}}_{\mathrm{JC}}(\omega_{\mathrm{L}}) + \tilde{\mathcal{L}}_{\mathrm{drive}}\right)\tilde{\rho}_{\mathrm{ss}}(\omega_{\mathrm{L}}) = 0 \tag{10}$$

and then using $\tilde{\rho}_{\mathrm{ss}}(\omega_{\mathrm{L}})$ to obtain the transmission spectrum $S_{\mathrm{T}}(\omega_{\mathrm{L}}) = \langle a^\dagger a\rangle(\omega_{\mathrm{L}}) = \mathrm{Tr}\{a^\dagger a\tilde{\rho}_{\mathrm{ss}}(\omega_{\mathrm{L}})\}$. In this way, we only simulated light transmitted through the cavity rather than including interference effects of the reflected light. Although we focus on numerical techniques in this chapter, we note that analytic solutions to the spectra under incoherent and coherent pumping may be found in Tian and Carmichael (1992) and Waks and Vuckovic (2006), respectively. From these analytic forms with dissipation, one can definitively establish a border for the strong coupling regime: when are the upper and lower branches of the polaritons nondegenerate ($g > |\kappa - \Gamma|/4$) (Andreani et al., 1999). In the incoherently excited spectra, this is slightly different than the Rayleigh criterion for when the two emission peaks are resolvable ($g > |\kappa - \Gamma|/2$).

Returning to the traces in Fig. 5, one can immediately identify that both the transmission and emission spectra clearly identify the correct splitting of the first two polaritons as $2\,g$. However, both the correct linewidths of the polaritons and their precise locations (which are slightly shifted by the effects of loss) are only captured in the spontaneous emission spectrum; we will derive this effect in Section 3.1. In transmission, because the two polaritons have cavity components that are shifted by $\pi$ phase, they interfere destructively at frequencies between the polaritons and artificially decrease the apparent polariton linewidths. In fact, this interference occurs for systems not even in strong coupling and is referred to as dipole-induced transparency (Waks and Vuckovic, 2006). Nevertheless, transmission experiments are more popular for identifying strong coupling due to the fact that incoherently exciting the quantum emitters causes experimental nonidealities. For instance, it generates excess carriers that can induce effects such as field noise (Kuhlmann et al., 2015), and specifically in quantum dot samples with randomly positioned emitters, the carriers can cause nearby dots to indirectly pump the cavity (Majumdar et al., 2012).

Now that we have discussed transmission experiments, we compare with experiments performed in a configuration called cross-polarized reflectivity

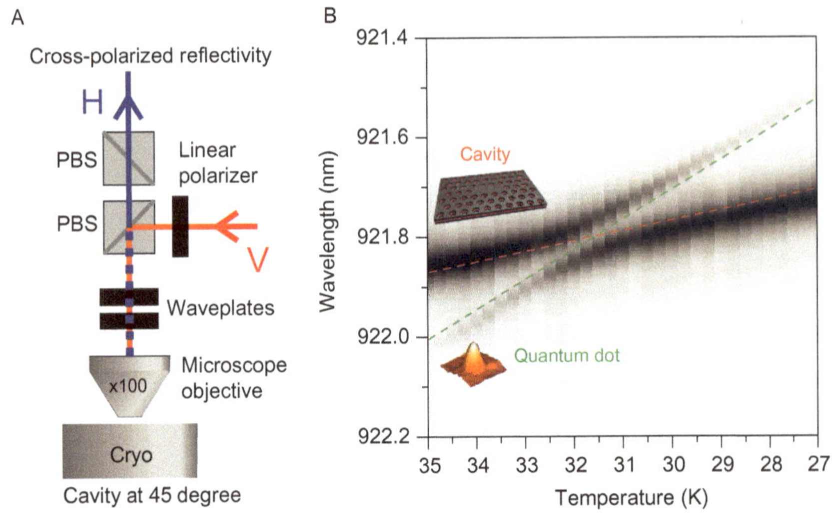


**Fig. 6** A typical reflectivity measurement used to observe strong coupling. (A) Experimental setup of a cross-polarized reflectivity measurement that is equivalent to transmission. The linear polarizer and two polarizing beamsplitter cubes perform the polarization selection, while the waveplates allow for polarization rotation and correction for bi-refringence in the optical elements (Kuhlmann et al., 2013). (B) Typical reflectivity experiment on a strongly coupled system, where the emitter detuning relative to the cavity frequency is controlled by tuning the temperature of the sample. An anticrossing between the bare cavity (*red*) and dot (*green*) states is observed. *Data from Müller, K., Rundquist, A., Fischer, K.A., Sarmiento, T., Lagoudakis, K.G., Kelaita, Y.A., Sánchez Muñoz, C., del Valle, E., Laussy, F.P., Vučković, J., 2015. Coherent generation of nonclassical light on chip via detuned photon blockade. Phys. Rev. Lett. 114 (23), 233601.*

(Englund et al., 2007) (Fig. 6A). Here, both the excitation and detection light traverse along (approximately) the same physical path into and out of the cavity. However, this setup is still mathematically equivalent to a transmission experiment due to the polarization degree of freedom and the configuration of polarizers. The linearly polarized cavity is rotated 45 degrees relative to the linear polarization of the incident light. This way, the incident light projects onto the cavity polarization with 1/2 efficiency. Meanwhile, the orthogonal polarization of the output light is filtered via the polarizing beamsplitters. Therefore, the input and output channels are strictly orthogonal and the reflectivity experiment is equivalent to transmission, to within a scaling factor of 1/2. This scaling factor does not matter in realistic experiments because the collection losses are typically at least an order of magnitude larger (Englund et al., 2007). As a result, normalized quantities are typically used to study the systems.

A typical cross-polarized reflectivity measurement revealing the strong coupling of an InGaAs-based system is shown in Fig. 6B. The emitter detuning $\Delta_e$ is controlled relative to the cavity frequency $\omega_a$ by changing the lattice temperature. The quantum dot has a stronger dependence on temperature than the cavity due to the bandgap's quadratic temperature dependence compared to the cavity's linear change in permittivity (Faraon et al., 2007). A clear avoided crossing can be seen between the two polaritons near 32.5 K, as compared to the bare dot (green) and cavity (red) states. Here, the character of each of the polaritons switches between electronic and photonic, providing clear evidence for strong coupling between the cavity and dot. This crossing matches the change in character present in the eigenstate equations for $|n, \pm\rangle(\Delta_e)$.

## 2.5 Basics of Nonclassical Light Generation

All of the above dynamics in weak excitation regimes can be fully captured by linear theories; however, the generation of nonclassical light is inherently nonlinear. In a cavity QED system, nonclassical light is generated by filtering a stream of incident coherent light through a single strongly coupled system (Faraon et al., 2008; Müller et al., 2015b; Reinhard et al., 2011). Owing to the highly nonlinear character of the interaction between the input light and a strongly coupled system, the admission of a single-photon into the cavity may enhance (photon tunneling) or diminish (photon blockade) the probability for a second photon to enter the cavity. Ideal photon blockade is depicted schematically in Fig. 7A, where the upward blue arrows represent the laser tuned for photon blockade (left side). Absorption of a photon into state $|1, +\rangle$ blocks the admission of a second photon because no state is present to absorb the second photon. Meanwhile, in photon tunneling (depicted by the upward red arrows on the right side), two photons are absorbed together in a multiphoton transition to directly excite the state $|2, +\rangle$. Following these interactions, a light beam exits the cavity with the nonlinear action imprinted on its quantum character.

The quality of the nonclassical light is typically characterized by the measured degree of second-order coherence $g^{(2)}[0]$, as discussed in Fischer et al. (2016a). If a given pulse results in a photocount distribution $P_n$, then

$$g^{(2)}[0] = \langle n(n-1)\rangle / \langle n\rangle^2, \quad (11)$$

which is a normalized second-order factorial moment. Notably, $g^{(2)}[0] = 1$ for a coherent pulse with Poissonian counting statistics, while $g^{(2)}[0] < 1$ for

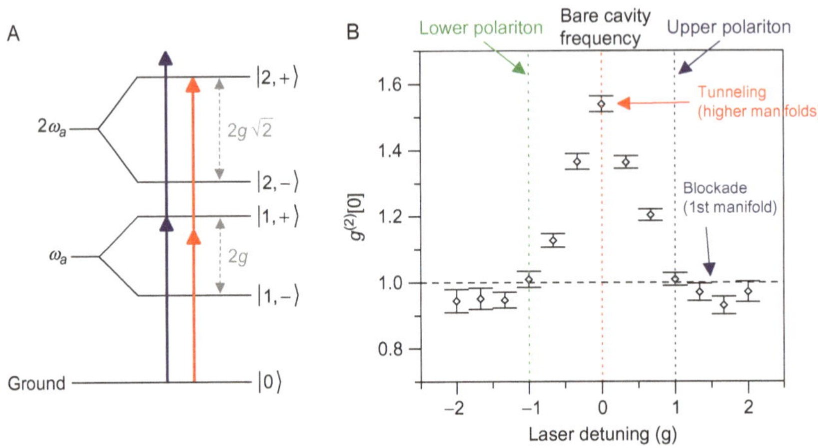


**Fig. 7** Photon blockade and photon tunneling. (A) Schematic depiction of the Jaynes–Cummings ladder in the dressed basis $|n, \pm\rangle$ when the emitter and cavity are resonant. The *upward blue arrows* (on the left side) depict photon blockade, while the *upward red arrows* (on the right side) depict photon tunneling. (B) Measured degree of second-order coherence $g^{(2)}[0]$ for an InGaAs-based resonant system, illustrating tunable nonclassical light generation from a strongly coupled system (with fitted system parameters $g/2\pi =$ 10.9 GHz and $\kappa/2\pi =$ 10 GHz). *Data from Müller, K., Fischer, K.A., Rundquist, A., Dory, C., Lagoudakis, K.G., Sarmiento, T., Kelaita, Y.A., Borish, V., Vučković, J., 2015. Ultrafast Polariton-Phonon dynamics of strongly coupled quantum Dot-Nanocavity systems. Phys. Rev. X 5 (3), 031006.*

sub-Poissonian or $g^{(2)}[0] > 1$ for super-Poissonian counting statistics; for ideal photon blockade $g^{(2)}[0] = 0$. Importantly, this statistic is completely independent of collection losses so it gives an accurate representation of the internal system dynamics in almost any experiment. Although complete characterization of the photocount distribution would require measurement of all factorial moments of the photocount distribution, $g^{(2)}[0]$ is an important measure in determining whether the source is acting as a single- or multiphoton source and is the most readily accessible in experiments (Rundquist et al., 2014). Therefore, we will focus on the measured degree of second-order coherence in this chapter.

In Fig. 7B, we present experimental evidence of photon blockade and photon tunneling in a series of typical experiments measuring $g^{(2)}[0]$ on emission from an InGaAs-based strongly coupled system. When the laser detuning is near the upper and lower polaritons (states $|1, \pm\rangle$), $g^{(2)}[0] < 1$ antibunches for photon blockade. The precise location of photon blockade occurs slightly detuned from the polaritons due to the strong dissipation in III–V systems (Müller et al., 2015b), which will be thoroughly discussed in the next section. Meanwhile, when the laser is tuned in-between the polaritonic rungs, multiphoton transitions are emphasized such that photon tunneling occurs and the light bunches causing $g^{(2)}[0] > 1$. The bunching occurs because a large component of the vacuum state is present in conjunction with super-Poissonian statistics (Rundquist et al., 2014). However, because of the strong dissipation in the system (since $g \approx \kappa$), the second-order coherence statistics do not deviate much from the laser statistics of $g^{(2)}[0] = 1$, as discussed in Müller et al. (2015b). In the next section, we will explore several nonidealities that result in this relatively poor performance and ways to leverage these nonidealities for interesting physics and better nonclassical light generation.

## 3. SINGLE-EMITTER CAVITY QED BEYOND THE JAYNES–CUMMINGS MODEL

There are several key considerations that alter the performance of a solid-state cavity QED system from that of an ideal Jaynes–Cummings system:

1. Strong dissipation caused by cavity loss rates comparable to the coherent interaction strength; dissipation decreases the fidelity of nearly all cavity QED effects (Müller et al., 2015a,b).
2. Detuning of the emitter from the cavity; this is an important tool in improving the nonclassical light generated from a dissipative JC system (Müller et al., 2015b).
3. Pulse-wise experiments, which are potentially more interesting for applications in quantum networks and are regardless required by the timing resolution of most single-photon detectors (Fischer et al., 2016a); the pulse length and pulse shape have a strong influence on the fidelity of nonclassical light generation (Müller et al., 2015a).
4. The solid-state environment results in an important interaction with phonons (Roy and Hughes, 2011); this dissipation can both be a detrimental and a positive influence toward nonclassical light generation (Müller et al., 2015a).
5. Temporal variation of the ground state of a quantum emitter, a phenomenon known as blinking (Santori et al., 2004); when the emitter blinks, it either decouples from the cavity or is so far spectrally detuned that the system is effectively no longer strongly coupled (Reinhard et al., 2011).
6. Interferometric effects owing to the photonic crystal's background density of states; the light transmitted through the cavity cannot simply be

modeled as a single cavity mode, but also requires the modeling of a continuum scattering channel (Fischer et al., 2016b, 2017; Müller et al., 2016).

In carefully studying these effects, we have learned to significantly improve both the modeling of a solid-state strongly coupled system and its ability to generate high-purity states of nonclassical light. Although this section details these nonidealities using investigations with III–V quantum emitters due to their technological maturity, the physics and the theory will be equally applicable to future experiments with group-IV systems.

## 3.1 Dissipative Structure of a Jaynes–Cummings System

The first nonideality that we explore in detail is the large dissipation present in solid-state cavity QED systems. Due to limitations in current material fabrication technologies for III–V and group-IV systems, the cavity dissipation rates are comparable to the coherent coupling rates. We explore this point through the language of non-Hermitian effective Hamiltonians (Garraway, 1997a,b). By using a non-Hermitian Hamiltonian, the effects of decay may be incorporated in a linear manner, but at the cost of generating an evolution equation that leaks probability. This non-Hermitian evolution plays an important role in the Monte-Carlo wavefunction or trajectory approach to quantum simulation (Carmichael, 2009; Steck, 2017); it governs the evolution of the wavefunction in-between photon emission events. Thus, the non-Hermitian Hamiltonian is useful because after a state has been prepared, its initial evolution and hence decay rate is set by the effective Hamiltonian. For a given system, such a Hamiltonian is

$$H_{\mathrm{EFF}} = H_0 - \mathrm{i}\sum_k \frac{\gamma_k}{2} c_k^\dagger c_k \tag{12}$$

where $c_k$ is an arbitrary system operator and $\gamma_k$ is its decay rate. For the Jaynes–Cummings system specifically, this takes the form

$$H_{\mathrm{EFF}} = H_{\mathrm{JC}} - \mathrm{i}\frac{\kappa}{2} a^\dagger a - \mathrm{i}\frac{\Gamma}{2}\sigma^\dagger \sigma. \tag{13}$$

In a similar manner to diagonalizing $H_{\mathrm{JC}}$ to obtain the dressed states in Section 2.3, $H_{\mathrm{EFF}}$ can be diagonalized to additionally obtain the loss rates of the dressed states (Laussy et al., 2012). Diagonalization yields the complex eigenenergies for the $n$-th rung of the system to be

$$E_{\pm}^{n} = n\omega_a + \frac{\Delta_e}{2} - i\frac{(2n-1)\kappa + \Gamma}{4} \pm \sqrt{\left(\sqrt{n}g\right)^2 - \left(\frac{\kappa - \Gamma}{4} + i\frac{\Delta_e}{2}\right)^2}, \quad (14)$$

where the full-width half-maxes of the linewidths are given by the imaginary parts $\pm 2\,\mathrm{Im}\{E_{\pm}^{n}\}$. These complex eigenenergies are depicted in Fig. 8A, with the bounding lines showing $\mathrm{Re}\{E_{+}^{n}\} \pm \mathrm{Im}\{E_{+}^{n}\}$ and $\mathrm{Re}\{E_{-}^{n}\} \pm \mathrm{Im}\{E_{-}^{n}\}$. One can derive the bounds for strong coupling by solving for when $\Delta_e = 0$ and $\mathrm{Re}\{E_{+}^{n}\} \neq \mathrm{Re}\{E_{-}^{n}\}$, which occurs when $g > |\kappa - \Gamma|/4$.

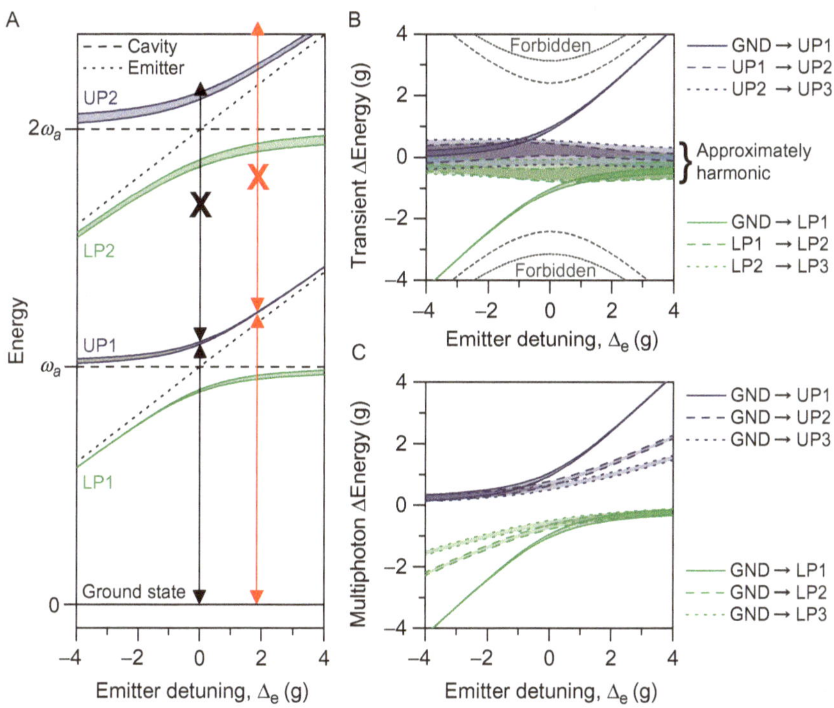


**Fig. 8** Energetic structure of a strongly coupled system. (A) Dressed states of the Jaynes–Cummings ladder and their linewidths. *Note*: the separation between the rungs of the dressed-states ladder is not to scale. (B) Transient energies to climb the Jaynes–Cummings ladder rung-by-rung. (C) Resonant laser frequencies for multiphoton transitions. (A–C) The heights of the bounding regions represent the full-width half-maxes of the linewidths for each transition. The $g/\kappa = 5$ ratio was used in order to cleanly illustrate the trends in each plot (which is now becoming achievable with state-of-the-art parameters for solid-state systems). UP*n* and LP*n* label the upper and lower polaritons, respectively.

For large emitter detunings $\Delta_e$, the statement made in Section 2.3 that the polaritons take on primarily an electronic or photonic character can be understood even more intuitively; here, one polariton has approximately the emitter lifetime $\Gamma$ and the other the cavity lifetime $\kappa$. At zero detuning, the upper (UP) polaritons have equivalent character to the lower polaritons (LP). From this plot alone, we can already suspect a way to improve photon blockade in highly dissipative systems. Considering the black arrows for photon blockade at zero detuning where the laser is tuned in resonance with UP1, then the second photon is nearly resonant with UP2 due to the finite linewidths. Notably, this problem arises because the anharmonicity of the Jaynes–Cummings ladder scales as $\sqrt{n}$, while the decay rates scale as $n$. However, by introducing a small detuning, the effective nonlinearity of the system is much higher. Now consider the red arrows for detuned photon blockade, where UP2 is much further off resonance in comparison. Thus, we would expect better quality photon blockade to occur for the detuned system.

This information can also be visualized in an alternative manner, as shown in Fig. 8B. Here, the energies to climb the dressed-states ladder one-by-one (Laussy et al., 2012) are shown, with their linewidths, by plotting

$$\Delta E^n_{++} = \mathrm{Re}\{E^n_+\} - \mathrm{Re}\{E^{n-1}_+\} \pm \left(\mathrm{Im}\{E^n_+\} + \mathrm{Im}\{E^{n-1}_+\}\right) \tag{15}$$

and

$$\Delta E^n_{--} = \mathrm{Re}\{E^n_-\} - \mathrm{Re}\{E^{n-1}_-\} \pm \left(\mathrm{Im}\{E^n_-\} + \mathrm{Im}\{E^{n-1}_-\}\right). \tag{16}$$

(Note, we define $E^0_\pm = 0$.) The linewidths add for the eigenstates involved in $E^n_{++}$ and $E^n_{--}$ because the transient energies are used to consider a resonant process that excites the system rung-by-rung. For small detunings, the jumps to the second (dashed) and third (dotted) lines are almost on top of the first jump from the ground state. Thus, we would not expect good quality nonclassical light generation to come from the system. On the other hand, for nontrivial detunings the effective anharmonicity increases linearly because the jump to the first rung differentially increases relative to the second rung with detuning. After the first jump, however, the ladder is relatively harmonic since the higher jumps are centered around $\Delta E = 0$. Importantly, these jumps occur along either the upper or lower branches, but not between branches. Jumps between branches are disallowed for the same reason the transmission dips to zero in-between the upper and lower polaritons (as seen in Fig. 5). Specifically, the transitions between branches, $\Delta E^n_{+-} = \mathrm{Re}\{E^n_+\} - \mathrm{Re}\{E^{n-1}_-\}$ and $\Delta E^n_{-+} = \mathrm{Re}\{E^n_-\} - \mathrm{Re}\{E^{n-1}_+\}$, are forbidden because the states have cavity components that are shifted by $\pi$ phase and hence zero dipolar overlap between $\langle n-1, \mp|a|n, \pm\rangle = 0$. We have plotted the forbidden transitions as dashed/dotted lines with no bounding regions. These transitions occur through emission by the quantum emitter, i.e., $\langle n-1, \mp|\sigma|n, \pm\rangle \neq 0$, but at a negligible rate due to typical solid-state cavity QED parameters.

Finally, we discuss the multiphoton structure of the dissipative Jaynes–Cummings ladder. Just like for the photon blockade argument where the detuning increased the effective nonlinearity between the first and second rungs of the dressed-states ladder, increasing the emitter detuning has the effect to separate the multiphoton transitions. This can be seen in Fig. 8C, where the multiphoton transitions fan out with detuning. Interestingly, the absorption linewidths for multiphoton emission are all $\kappa$ after the first blockaded rung, because the multiphoton transitions occur between the ground state and an upper dressed state in an idealized model that assumes the intermediate levels remain unpopulated. (Certainly, this approximation breaks down when the multiphoton transitions strongly overlap, but then it is difficult to identify linewidths for distinctive multiphoton processes anyway.) In the idealized model, the target dressed state determines the linewidth since the ground state has approximately zero dephasing. Although the dephasing rate of the $n$th dressed state scales with $n$, the $n$-photon absorption linewidth scales with $1/n$ because the laser detuning is compounded by the number of photons involved in the process. Thus, the multiphoton absorption linewidths are left at $\kappa$, so we plot

$$\mathrm{Re}\{E^n_+\}/n - \omega_a \pm \kappa/2 \tag{17}$$

and

$$\mathrm{Re}\{E^n_-\}/n - \omega_a \pm \kappa/2. \tag{18}$$

We note that it is possible to derive an effective multiphoton absorption Hamiltonian based on adiabatic elimination of the unpopulated states which rigorously supports this analysis (Linskens et al., 1996).

## 3.2 Emitter Cavity Detuning

In this section, we experimentally and numerically consider the effects of photon blockade by measuring $g^{(2)}[0]$ as a function of the excitation laser frequency, for multiple emitter cavity detunings. We excited an InGaAs-based strongly coupled system with short Gaussian pulses and experimentally measured the degree of pulse-wise second-order coherence, as shown in Fig. 9A. [For the tuned system (red), the correlation statistics are similar to the ones presented in Fig. 7B.] However, when the emitter detuning $\Delta_e$ is increased to 4 $g$, the quality of the photon blockade is increased dramatically. This can be seen in the significantly decreased value of $g^{(2)}[0]$ at the vertical red line compared to the vertical blue line. This provides

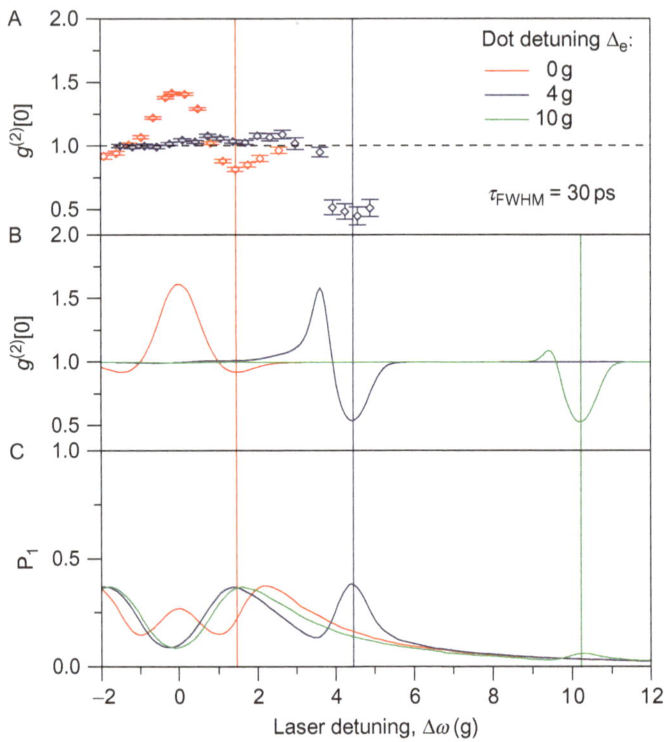


**Fig. 9** Detuned photon blockade, experiment and theory. (A) Measured second-order coherence $g^{(2)}[0]$ as a function of laser detuning for both a tuned and detuned strongly coupled system based on an InGaAs quantum dot. *Horizontal black dashed line* represents statistics of the incident laser pulses. (B) Theoretical $g^{(2)}[0]$ as a function of laser detuning for one resonant and two detuned strongly coupled systems. (C) Single-photon detection probability $P_1$ of the photocount distribution $P_n$, calculated using a quantum trajectory approach. The theoretical model for the simulations is a pure Jaynes–Cummings system with a time-dependent driving term. *Data in (A) from Müller, K., Rundquist, A., Fischer, K.A., Sarmiento, T., Lagoudakis, K.G., Kelaita, Y.A., Sánchez Muñoz, C., del Valle, E., Laussy, F.P., Vučković, J., 2015. Coherent generation of nonclassical light on chip via detuned photon blockade. Phys. Rev. Lett. 114 (23), 233601.*

experimental evidence for our discussion in Section 3.1 of how the emitter detuning modulates the effective anharmonicity of the system. [We note that in this set of experiments the cross-polarized suppression was not optimized for large detunings so in the tunneling region, where less light is emitted, the statistics were dominated by unwanted coherently scattered light (Müller et al., 2015b)].

To ensure that we fully understand the experimental behaviors, we discuss a numerical model for capturing the observed trend in photon blockade. As briefly mentioned in Section 2.5, the measurements of the degree of second-order coherence are performed in a pulsed manner, and hence we must adjust our master equation accordingly. Specifically, the driving Hamiltonian changes to $H_{\text{drive}} = \mathcal{E}(t)\left(a\,\mathrm{e}^{\mathrm{i}\omega_{\mathrm{L}}t} + a^{\dagger}\mathrm{e}^{-\mathrm{i}\omega_{\mathrm{L}}t}\right)$, where $\mathcal{E}(t)$ is the time-dependent driving strength. As before, we use the same rotating frame transformation to remove the time-dependence of the excitation term so that $\tilde{H}_{\text{drive}}(t) = \mathcal{E}(t)\left(a + a^{\dagger}\right)$ and written as a superoperator

$$\tilde{\mathcal{L}}_{\text{drive}}(t)\tilde{\rho}(t) = \mathrm{i}\left[\tilde{\rho}(t), \tilde{H}_{\text{drive}}(t)\right]. \tag{19}$$

Typically, a Gaussian pulse shape represents experiment, where $\mathcal{E}(t) = \mathcal{E}_0\mathrm{e}^{-t/2\tau_p^2}$ and $\tau_p = \tau_{\text{FWHM}}/2\sqrt{\ln 2}$ is the Gaussian pulse parameter. The overall system evolution is governed by the Liouvillian $\tilde{\mathcal{L}}(t) = \tilde{\mathcal{L}}_{\text{JC}} + \tilde{\mathcal{L}}_{\text{drive}}(t)$.

This Liouvillian can be used to calculate the measured degree of second-order coherence in two ways. The first is to unravel $\tilde{\mathcal{L}}(t)\tilde{\rho}(t)$ into a quantum trajectory equation, approximate the expected photocount distribution $P_n$ over the entire pulsed emission from an ensemble of trajectories, and estimate $g^{(2)}[0] = \sum_n P_n n(n-1)/\left(\sum_n P_n n\right)^2$ (Carmichael, 2009). The second way is to use a time-dependent form of the quantum regression theorem to calculate

$$g^{(2)}[0] = \frac{\int_0^T\int_0^T \mathrm{d}t\,\mathrm{d}t' \,\langle \mathcal{T}_-[a^{\dagger}(t)a^{\dagger}(t')]\mathcal{T}_+[a(t')a(t)]\rangle}{\left(\int_0^T \mathrm{d}t\,\langle a^{\dagger}(t)a(t)\rangle\right)^2} \equiv \frac{G^{(2)}[0]}{\langle N\rangle^2}, \tag{20}$$

where the time range $0 \to T$ encompasses the entire emission pulse and the operators $\mathcal{T}_{\pm}$ indicate the time-ordering required of a physical measurement (operators with higher time indices toward the center of the expression; Fischer et al., 2016a). The second method is typically more computationally efficient and has an intellectually satisfying connection to instantaneous two-time correlations.

Using this method and a best fit to the Jaynes–Cummings model with no dephasing (yielding $\{g/2\pi, \kappa/2\pi\} = \{10.9 \text{ GHz}, 10 \text{ GHz}\}$), we use the driving strength $\mathcal{E}_0$ as a fitting parameter for the observed photon blockade regions (Fig. 9B). Just as experimentally measured, the blockade dip grows with increasing detuning from $\Delta_e = 0\,g$ to $\Delta_e = 4\,g$, which further supports the pictorial description of detuned blockade from Section 3.1. The simulations do show a difference compared to this intuition for the very large detuning of $\Delta_e = 10\,g$: the blockade dip saturates. With increasing detuning, the oscillator strength of the emitter-like polariton decreases until its emission strength is comparable to off-resonant transmission through the cavity-like polariton. The light from the cavity-like polariton then begins to destroy the photon blockade.

Because the emitter-like polariton has a smaller oscillator strength with increasing detuning, one might be concerned with how the efficiency of single-photon generation is affected by changing the emitter detuning (Müller et al., 2015b). To answer this question, we simulated the probabilities of single-photon generation $P_1$ using quantum Monte-Carlo techniques and present these data in Fig. 9. Surprisingly, the probability for single-photon generation actually increases for small emitter detunings by approximately a factor of two over the case in resonant photon blockade (compare $P_1$ at the red and blue lines). This occurs because the highly dissipative nature of the strongly coupled system spoils the blockade so completely on resonance that with increasing detuning there are plenty of multiphoton photocounts that can be suppressed and converted into single-photon counts. Of course, this effect wears out with large enough detuning and the $P_1$ for $10\,g$ is noticeably less efficient. Thus, emitter detuning has been shown both intuitively, experimentally, and theoretically to be a valuable mechanism for enhancing nonclassical light generation.

We make two brief comments on the figure:

1. In the photon tunneling regions, the bunching values of $g^{(2)}[0]$ are largest when the minimum amount of light is transmitted. This can easily be seen by comparing the maximum simulated bunching in the blue curve to the local minimum value of $P_1$ near the blockade region. Such bunching behavior is consistent with having a photocount distribution that has multiphoton components that are emphasized over a coherent state, but additionally has a large vacuum component (Rundquist et al., 2014). For instance, although the photocount distribution $P_2 = 1$

antibunches with $g^{(2)}[0] = 0.5$, the distribution $\{P_0, P_2\} = \{0.75, 0.25\}$ bunches with $g^{(2)}[0] = 2$. This discussion further suggests that a highly dissipative Jaynes–Cummings system by itself is not necessarily ideal for the generation of multiphoton states.

2. The modeled photon blockade is actually *weaker* for the ideal Jaynes–Cummings system than experimentally measured, even for arbitrarily low powers. Thus, even though the theory and experiment do not perfectly match, we trust the strength of our general arguments in this section. This additionally suggests missing physics from the model of the solid-state strongly coupled system, which will be thoroughly discussed in the latter sections of this chapter.

## 3.3 Excitation Pulse Length

As has recently been shown in depth (Müller et al., 2015a), the length of an excitation pulse is critically important to optimizing emission from photon blockade. Specifically, optimizing photon blockade in a detuned strongly coupled system by changing the pulse length is a careful trade-off between avoiding excitation of the higher rungs for short pulses or reexcitation of the first polariton for long pulses. Here, we show experimental results from an InGaAs-based system in Müller et al. (2015a) (Fig. 10A and C) and discuss their comparison with simulated photon blockade in a pulsed regime from a Jaynes–Cummings system (Fig. 10B and D). The blockade dip has two important characteristics: its depth and its width. Here, the better pulse length of the two is seen to be 110 ps since this pulse length better optimizes between higher rung excitation and lower rung reexcitation. Regarding the blockade dip, it is widest for short pulses but narrowest for long pulses; because the detuned polariton has a relatively long lifetime, the width of the blockade dip is roughly determined by the spectral width of the laser pulse. These trends are clearly observable in both experiment and theory.

Although the trends of the blockade dips are again reproduced well by the pulsed Jaynes–Cummings model, the agreement is only qualitative. For instance, the blockade dips are again smaller in simulation than in experiment, but now shown for multiple pulse lengths. This suggests that the disparity is not simply an error in the pulse length but rather additional physics. Especially in this set of experimental data where the cross-polarized setup was better optimized for suppression than in Fig. 9A, the simulated tunneling regions quite poorly agree with the experimental data. While we supposed initially that the experimental difference was potentially an imprecision

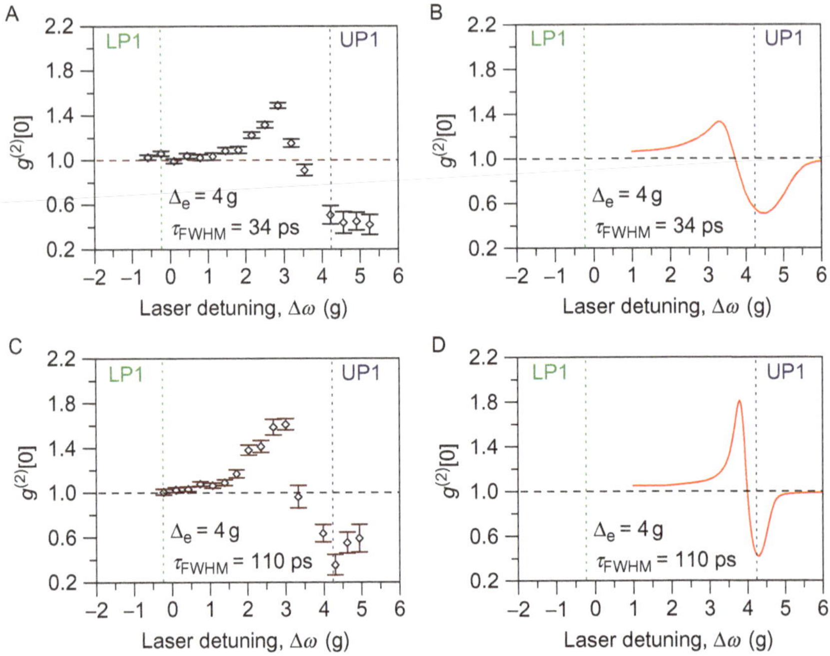


**Fig. 10** Pulse-length dependence of photon blockade. Experimentally measured degrees of second-order coherence $g^{(2)}[0]$ as a function of laser detuning for a detuned strongly coupled system based on an InGaAs quantum dot under short (A) and long (C) pulses. Theoretical $g^{(2)}[0]$ as a function of laser detuning for a detuned strongly coupled system under (B) short and (D) longer pulses. The theoretical model for the simulations is a pure Jaynes–Cummings system with a time-dependent driving term. *Horizontal black dashed lines* represent statistics of the incident laser pulses. *Data from Müller, K., Fischer, K.A., Rundquist, A., Dory, C., Lagoudakis, K.G., Sarmiento, T., Kelaita, Y.A., Borish, V., Vučković, J., 2015. Ultrafast Polariton-Phonon dynamics of strongly coupled quantum Dot-Nanocavity systems. Phys. Rev. X 5 (3), 031006.*

in our ability to accurately determine the laser detuning or an experimental drift, further investigation has shown the difference to be the result of an unexpected interference effect. We will discuss this effect later.

## 3.4 Electron–Phonon Interaction

In this section we discuss the first true nonideality to our model for the strongly coupled systems; the effect of phonons on solid-state cavity QED platforms. Any subsystem embedded within a solid-state environment, at any temperature, has the potential to feel the effects of interactions with the phonon bath. In particular, the electron–phonon interaction between solid-state emitters and their environment can lead to a variety of incoherent phenomena. Due to the differing sizes of the III–V quantum dots and group-IV color-centers, however, the types of phononic excitations they couple to are different. The physically large size of quantum dots lends well toward coupling of only acoustic phonons (Roy and Hughes, 2011), while the small size of color-centers lends well toward coupling of both acoustic and optical phonons (Davies, 1974). In this section, we discuss how phonon coupling manifests in single-emitter cavity QED systems using an InGaAs-based device as an example.

First, consider the avoided crossing spectrum of a typical InGaAs-based system (Müller et al., 2015a), reproduced in Fig. 11A. Now, the spectra are annotated according to the state lifetimes. The measured cavity lifetime is $1/\kappa = 8$ ps, while the suppressed dot lifetime is expected to be approximately 10 ns, and hence the maximally entangled polariton has a lifetime of 16 ps since it is half electronic and half photonic character. As a function of the emitter detuning, the expected lifetimes of the LP1 and UP1 polaritons from the Jaynes–Cummings model are plotted in Fig. 11B. However, from experimental measurements performed on a streak camera the lifetimes are an order of magnitude shorter for large detunings (Fig. 11C). This difference results from the electron–phonon interaction in solid-state cavity QED systems.

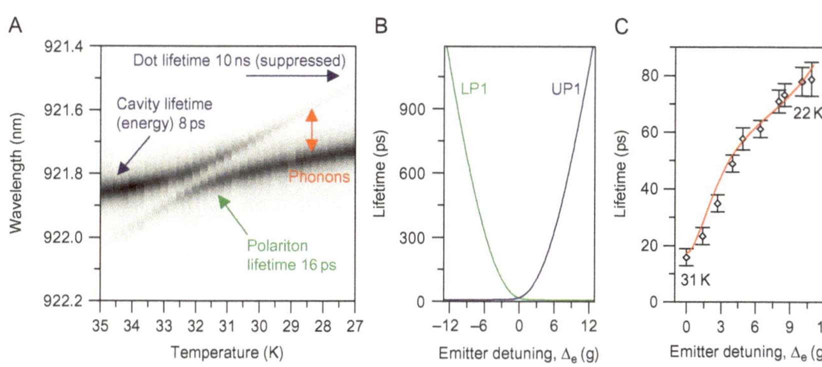


**Fig. 11** Effects of electron–phonon interaction on polariton lifetimes. (A) Annotated avoided crossing plot that matches transmission spectra to measured lifetimes. (B) Lifetimes as calculated from the cavity and dot dissipation rates alone in the Jaynes–Cummings model. (C) Measured lifetimes from a typical solid-state cavity QED system, which are an order of magnitude shorter for large detunings than those predicted by a pure Jaynes–Cummings model. *Data from Müller, K., Fischer, K.A., Rundquist, A., Dory, C., Lagoudakis, K.G., Sarmiento, T., Kelaita, Y.A., Borish, V., Vučković, J., 2015. Ultrafast Polariton-Phonon dynamics of strongly coupled quantum Dot-Nanocavity systems. Phys. Rev. X 5 (3), 031006.*

While the primary effect of electron–phonon interaction for quantum dots embedded in bulk GaAs is to generate a power-dependent damping, the effect in a cavity QED system is dramatically different. Because only acoustic phonons couple to the dots, there is an arbitrarily small density of phononic states to couple to under weak driving in bulk. However, in a cavity QED system the dressed ladder provides a constant energy difference between polaritons whereby the electron–phonon interaction samples the density of states, resulting in constant phonon emission and absorption. Three models have primarily been used to explore this effect in the solid-state:

1. Nonmarkovian models such as a path-integral form of the system dynamics (Vagov et al., 2011). This formalism includes phonon effects to all orders.
2. The polaron master equation, which is a powerful formalism for including phonon effects to all orders in a Markovian model (Roy and Hughes, 2011).
3. The effective phonon master equation, which only includes first-order phonon effects (Müller et al., 2015a; Roy et al., 2012). This model is by far the simplest, but readily captures most experimental effects observed thus-far in solid-state cavity QED systems.

For all of the nonclassical light generation phenomenon we have observed thus far, we have found the effective phonon master equation sufficient, and we will briefly review its findings here. For a cavity-driven system only (driven by $H_{\text{drive}}$), then the effects of phonons are captured by the addition of two incoherent dissipators $\frac{\Gamma_{\text{f}}}{2}\mathcal{D}[a^{\dagger}\sigma]$ and $\frac{\Gamma_{\text{r}}}{2}\mathcal{D}[\sigma^{\dagger}a]$, i.e., with the addition of the new Liouvillian term

$$\tilde{\mathcal{L}}_{\text{phonon}} = \frac{\Gamma_{\text{f}}}{2}\mathcal{D}[a^{\dagger}\sigma] + \frac{\Gamma_{\text{r}}}{2}\mathcal{D}[\sigma^{\dagger}a] \tag{21}$$

where $\Gamma_{\text{f}}$ and $\Gamma_{\text{r}}$ stand for forward and reverse phonon transfer rates, respectively. At the operating temperatures of 25 K and small emitter detunings used in our experiments, these rates are each approximately $\Gamma_{f,r} \approx (80\ \text{ps})^{-1}$. Importantly, these rates vary with detuning and fall off rapidly at large detunings; full details for the extracted rates can be found in Müller et al. (2015a).

Because the dissipators in Eq. (21) transfer excitations between the cavity and the dot, at large detunings they result in an effective transfer of population between polaritons. The transfer rates then dominate the lifetime of the emitter-like polariton for large detunings because the system emits through a phonon-induced transfer (Fig. 11A, red arrows) and cavity emission over dot emission. Specifically, the phonon-involved pathway has a lifetime of $\approx$ 80 ps + 8 ps, which is much shorter than the dot's spontaneous lifetime of $\approx$10 ns.

## 3.5 Blinking of the Quantum Emitter

While we hope the quantum emitter behaves as an ideal two-level system and is described by a single dipole operator $\sigma$, in reality it possesses a very complicated level structure. Consider InGaAs quantum dots, when placed in a high-quality electrical diode to control their precise ground states through controlling the local charge environment, they have been shown to behave as nearly ideal two-level systems (Kuhlmann et al., 2015). However, this technology was later introduced into devices with planar photonic crystal cavities (Carter et al., 2013; Laucht et al., 2009; Warburton et al., 2000), and therefore, much of the work on solid-state strongly coupled systems still shows residual effects of a slightly unstable charge environment. Because the different charge configurations of the quantum dot have different binding energies, and hence emission frequencies, when the quantum dot sequentially absorbs a charge from the environment (Santori et al., 2004) its resonant frequency is highly detuned from the cavity and the strong coupling is no longer visible (Reinhard et al., 2011; Rundquist et al., 2014). During this time, the system behaves like the bare cavity by itself and coherently scatters the incident light with a Lorentzian lineshape. When this process periodically occurs, slowly modulating the transmission spectrum between that of a resonant strongly coupled system and a bare cavity, the phenomenon is referred to as blinking. Although the experiments we discuss are with III–V quantum dots, group-IV color-centers have also been shown to blink (Bradac et al., 2010; Castelletto et al., 2013). Thus, the general physics of blinking in strongly coupled systems, which we will elaborate in this section, applies to future realizations of cavity QED systems with group-IV quantum emitters.

### *3.5.1 Effects on Transmission Spectra*

While it is certainly possible to provide a complete description of the charge states of a quantum emitter, in many situations it is sufficient to model the system as behaving as the bare cavity for some fraction of the time $f_{\text{blink}}$ and like a strongly coupled system for the rest. The statistical independence of these two situations holds because the blinking timescales are at least 10s of nanoseconds (Davanço et al., 2014), while the emission timescales are 10s of picoseconds. When the system blinks, we calculate the steady-state density matrix by solving

$$\left(\tilde{\mathcal{L}}_{\text{blink}}(\omega_{\text{L}}) + \tilde{\mathcal{L}}_{\text{drive}}\right)\tilde{\rho}'_{ss}(\omega_{\text{L}}) = 0, \tag{22}$$

where the bare cavity dynamics are represented by

$$\tilde{\mathcal{L}}_{\text{blink}}\tilde{\rho}'_{ss} = \text{i}\left[\tilde{\rho}'_{ss}, (\omega_a - \omega_{\text{L}})a^{\dagger}a\right]. \tag{23}$$

Then, the transmission spectrum while blinking is $S_{\text{blink}}(\omega_{\text{L}}) = \langle a^{\dagger}a\rangle(\omega_{\text{L}})$, using $\tilde{\rho}'_{ss}(\omega_{\text{L}})$.

To incorporate the experimental strongly coupled system that is often modeled as a Jaynes–Cummings system with phonons, we calculate the steady-state density matrix by solving

$$\left(\tilde{\mathcal{L}}_{\text{JC}}(\omega_{\text{L}}) + \tilde{\mathcal{L}}_{\text{phonon}} + \tilde{\mathcal{L}}_{\text{drive}}\right)\tilde{\rho}_{\text{ss}}(\omega_{\text{L}}) = 0. \tag{24}$$

Using $\tilde{\rho}_{\text{ss}}(\omega_{\text{L}})$, we calculate the transmission spectrum of the strongly coupled system $S_{\text{sc}}(\omega_{\text{L}}) = \langle a^{\dagger}a\rangle(\omega_{\text{L}})$.

Finally, we weight and combine the two transmission profiles to obtain

$$S_{\text{T}}(\omega_{\text{L}}) = (1 - f_{\text{blink}})S_{\text{sc}}(\omega_{\text{L}}) + f_{\text{blink}}S_{\text{blink}}(\omega_{\text{L}}). \tag{25}$$

With this quantum-optical model, we are ready to fit realistic transmission spectra with $S_{\text{T}}(\omega_{\text{L}})$. Experimental transmission spectra from an InGaAs quantum dot strongly coupled to an L3 photonic crystal cavity mode are shown in Fig. 12; in Fig. 12A the system is tuned nearly in resonance, and in Fig. 12B the system is tuned significantly off resonance. In this model, we used the phonon-induced dot-cavity transfer rates extracted from the data in Fig. 11, while $g$, $\kappa$, and $f_{\text{blink}}$ were taken as fitting parameters. From the fits, values of $\{g/2\pi, \kappa/2\pi\} = \{9.2\text{ GHz}, 12.3\text{ GHz}\}$ and $f_{\text{blink}} = 0.09$ were extracted; by fitting one spectra on-resonance and one spectra off-resonance $\kappa$ and $g$ can be extracted almost independently of one another.

We discuss several interesting features in these transmission spectra. Building the spectra component by component, we first show the blinking spectra and pure Jaynes–Cummings spectra as the dashed green and blue lines, respectively. The simulated spectra were convolved with the response function of the spectrometer (linewidth $\Gamma_{\text{FWHM}}/2\pi = 4.5$ GHz), which is why the spectra for pure Jaynes–Cummings transmission do not dip to zero

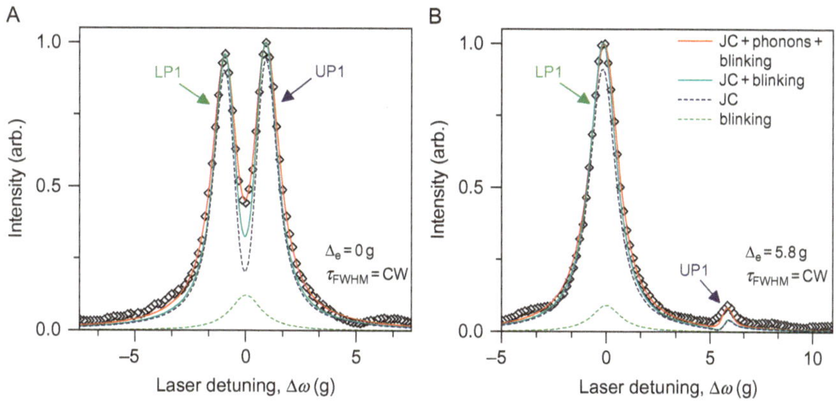


**Fig. 12** Transmission spectra from an InGaAs quantum dot strongly coupled to an L3 photonic crystal cavity mode, taken in cross-polarized reflectivity. (A) Spectrum at nearly zero dot detuning. (B) Spectrum at a large dot detuning. The spectra were taken using weak broad-band diode, which is equivalent to transmission in the linear regime of cavity QED. Simulated spectra were convolved with the response function of the spectrometer (linewidth $\Gamma_{FWHM}/2\pi = 4.5$ GHz), which is why the spectra for pure Jaynes–Cummings transmission do not dip to zero between LP1 and UP1.

between LP1 and UP1. Then, the two spectra were added together (cyan lines) to show the effects of the blinking term: blinking decreases the visibility of the strong coupling dip in the resonant case and increases the height of the cavity-like polariton in the off-resonant case. Finally, we added the effects of electron–phonon interaction to fully model our strongly coupled system. The effect of phonons is to decrease the depth of the transmission dip in the resonant case due to the additional incoherent dephasing. With this dephasing also comes a small increase in the linewidths of the polaritons. Meanwhile in the off-resonant case, the effect of the electron–phonon interaction is to reduce the lifetime of the emitter-like polariton (small peak) so that the polariton can emit at a faster rate, resulting in a higher count rate than without the interaction.

### *3.5.2 Effects on Nonclassical Light Generation*

Although the effects of blinking resulted in simply adding the strongly coupled and blinking spectra together, this approach is not sufficient for photon statistics. To understand the effect of blinking on nonclassical photon statistics, consider the standard setup for measuring pulsed $g^{(2)}[0]$, as thoroughly discussed in Fischer et al. (2016a). Using a Hanbury–Brown and Twiss interferometer operated in a pulsed manner a binned temporal coincidence histogram is built up, with peaks separated by the pulse repetition rate $\tau_r$, i.e., $h_{\text{HBT}}[n\tau_r]$ where $n \in \{\mathbb{Z} \geq 0\}$. The measured degree of second-order coherence is estimated by taking the ratio

$$\hat{g}^{(2)}[0] = \frac{h_{\text{HBT}}[0]}{h_{\text{HBT}}[\tau_r]}, \tag{26}$$

where each histogrammed time-bin $h_{\text{HBT}}[n\tau_r]$ is an independent and binomially distributed random variable. Here, we additionally consider the effects of blinking. Often the blinking time-scale is very long compared to $\tau_r$, where the counts due to transmission through the strongly coupled system or the blinking system are modeled by adding their individual histograms together, i.e., $h_{\text{HBT}}[n\tau_r] = h_{\text{SC}}[n\tau_r] + h_{\text{blink}}[n\tau_r]$. Hence,

$$\hat{g}^{(2)}[0] = \frac{h_{\text{SC}}[0] + h_{\text{blink}}[0]}{h_{\text{SC}}[\tau_r] + h_{\text{blink}}[\tau_r]}. \tag{27}$$

Because the statistics of the transmitted light are directly inherited from the laser when the system blinks, $h_{\text{blink}}[n\tau_r] \propto \langle N_{\text{blink}} \rangle^2 = \left(\int_0^T dt \, \langle a^\dagger(t) a(t) \rangle\right)^2$ and is calculated using the Liouvillian $\mathcal{L}(t) = \tilde{\mathcal{L}}_{\text{blink}} + \tilde{\mathcal{L}}_{\text{drive}}(t)$. Then, in terms of the instantaneous correlations of the system and by extension of Eq. (20)

$$g^{(2)}[0] = \frac{(1 - f_{\text{blink}}) g_{\text{sc}}^{(2)}[0] \langle N_{\text{sc}} \rangle^2 + f_{\text{blink}} \langle N_{\text{blink}} \rangle^2}{(1 - f_{\text{blink}}) \langle N_{\text{sc}} \rangle^2 + f_{\text{blink}} \langle N_{\text{blink}} \rangle^2}, \tag{28}$$

where $\langle N_{\text{sc}} \rangle$ and $g_{\text{sc}}^{(2)}[0]$ are calculated using the Liouvillian

$$\mathcal{L}_{\text{sc}}(t) = \tilde{\mathcal{L}}_{\text{JC}} + \tilde{\mathcal{L}}_{\text{phonon}} + \tilde{\mathcal{L}}_{\text{drive}}(t). \tag{29}$$

The strongly coupled system examined in this chapter had a relatively low fraction of blinking time, at only $f_{\text{blink}} = 9\%$. However, many quantum dots may have higher blinking fractions due to variation in their local charge environments. Here, we explore the effects of stronger blinking, both on the transmission spectrum and on the second-order coherence statistics, as shown in Fig. 13.

Consider the transmission spectra (Fig. 13A), the doublet that signifies strong coupling disappears with increasing blinking. Such a result is easily controlled in theory via changing $f_{\text{blink}}$, however, here we modified the local charge environment with a broad-band diode. While collecting information about the transmission spectrum, the diode power was increased in order to encourage blinking in the system. *Note*: this effect is different than saturation

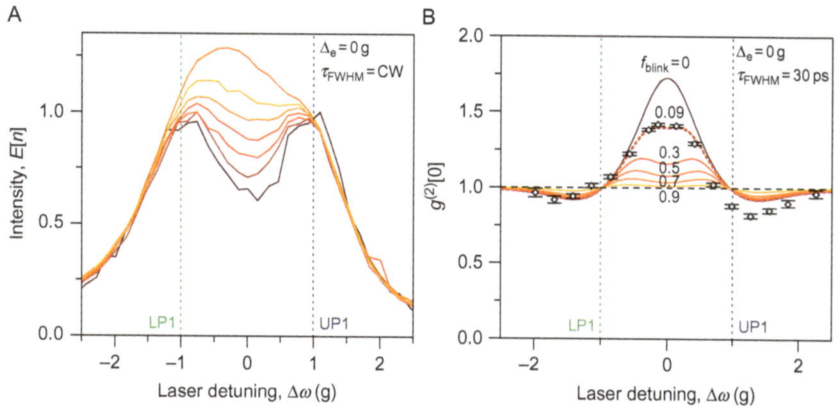


**Fig. 13** Effects of blinking on a resonant strongly coupled system. (A) Normalized CW transmission spectra with increasing blinking rate (*black* to *orange*). (B) Pulse-wise second-order coherence $g^{(2)}[0]$ as a function of laser detuning with increasing blinking rate. *Horizontal black dashed line* represents statistics of the incident laser pulses. *Experimental data for* g$^{(2)}$*[0] from Müller, K., Rundquist, A., Fischer, K.A., Sarmiento, T., Lagoudakis, K.G., Kelaita, Y.A., Sánchez Muñoz, C., del Valle, E., Laussy, F.P., Vučković, J., 2015. Coherent generation of nonclassical light on chip via detuned photon blockade. Phys. Rev. Lett. 114 (23), 233601.*

of the strongly coupled system, which also destroys the signature of strong coupling. As a Jaynes–Cummings system is driven into saturation, the separation between the doublet decreases until the spectra is comparable to that of the bare cavity, while no such linewidth narrowing is present here (Fushman et al., 2008). Of course, if the $g/\kappa$ ratio were larger then the addition of blinking would be easy to identify as a triplet in the transmission spectrum (Ota et al., 2009).

Meanwhile in the second-order coherence statistics (Fig. 13B), blinking most strongly affects the tunneling region because this is where the blinking spectrum is maximized, i.e., at the bare cavity frequency. Here, the effect to push down the $g^{(2)}[0]$ near zero laser detuning manifests itself as a broader tunneling peak and even a slight dip at zero detuning. For the case of the resonant system, the agreement between experiment and theory is excellent. Blinking has a large influence due to the spectral proximity of the strongly coupled system's emission to the bare cavity peak. By extension, it has negligible effect on the detuned spectra and cannot be used to explain the disparity between the detuned tunneling experiments and theory in Section 3.2. Finally, we note that the disparity in the blockade region of Fig. 13B will be addressed in the next section.

## 3.6 Self-Homodyne Interference

Previously, we discussed how both the experimental blockade and tunneling data seemed to reveal stronger nonclassical correlations than an ideal Jaynes–Cummings model would predict. In this chapter, we explain the missing physics that allows for better performance through enhancement of the nonclassical light emission (Fischer et al., 2016b, 2017). It is very tempting to model the physics of an L3 photonic crystal cavity (Fig. 14A) as a single mode of a harmonic oscillator, i.e., with a single Heisenberg mode operator

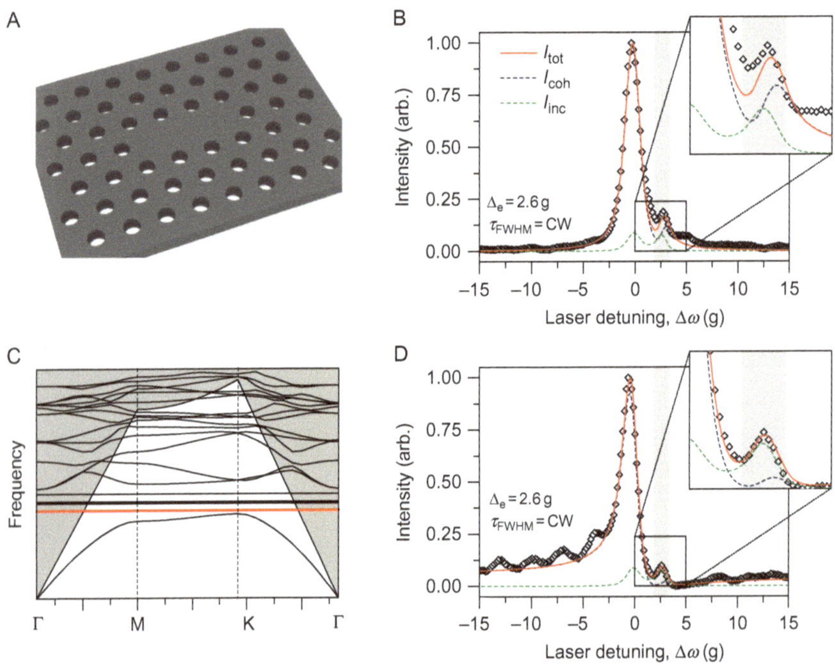


**Fig. 14** Origin of self-homodyne interference. (A) Schematic of a planar L3 photonic crystal cavity. (B) Transmission spectra under excitation conditions that produce a Lorentzian-like profile, fit to a quantum-optical model. (C) Complicated mode structure of a planar L3 Photonic crystal cavity, calculated using the MIT Photonic-Bands (MPB) package. *Red* and *black horizontal lines* depict the cavity's fundamental and higher-order modes, respectively. *Curved lines* represent photonic crystal guided modes. *Gray region* indicates leaky modes that are above the *light line*. (D) Transmission spectra under excitation conditions that produce a Fano-like profile fit to a quantum-optical model. Simulated spectra were convolved with the response function of the spectrometer (linewidth $\Gamma_{FWHM}/2\pi = 4.5$ GHz). *Transmission data in (B) and (D) from Fischer, K.A., Müller, K., Rundquist, A., Sarmiento, T., Piggott, A.Y., Kelaita, Y., Dory, C., Lagoudakis, K.G., Vučković, J., 2016b. Self-homodyne measurement of a dynamic Mollow triplet in the solid state. Nat. Photonics 10, 163–166.*

$a(t)$ which we will refer to as the discrete scattering channel. Under certain cross-polarized reflectivity conditions, such as those used in Fig. 14B, this model accurately captures the system dynamics. In this scenario, the detuned transmission profile nearly resembles the bare cavity's Lorentzian profile at high excitation powers. Here, an experimental transmission spectrum through a strongly coupled system is shown and fitted with the same quantum-optical model as in Fig. 12 (red line). The green and blue decompositions will be discussed later.

However, the L3 photonic crystal has a rich mode structure, as shown in Fig. 14C, that is not necessarily well approximated by a single mode operator $a(t)$. Instead, a better approximation is to consider an additional scattering channel that is due to a roughly constant background density of photonic states; this scattering pathway is referred to as the continuum channel. The discrete and continuum channels can interfere with one another to generate a lineshape that is closely related to a Fano resonance, and this is modeled with the operator $A(t) \rightarrow a(t) + \alpha(t)$ instead of just $a(t)$ where $\alpha(t) = \alpha\mathcal{E}(t)$, $\mathcal{E}(t)$ is the Gaussian pulse shape, and $\alpha$ is a c-number (Fischer et al., 2016b, 2017). The operator $\alpha(t)$ represents the laser light reflected into the cross-polarized output channel via the continuum modes of the photonic crystal. The lineshape can be changed between Lorentzian-like or Fano-like in cross-polarized reflectivity by altering the focal spot size and the excitation/detection polarizations, an effect which has been theoretically verified in L3 photonic crystal cavities through a rigorous scattering-matrix formalism (Vasco et al., 2013).

While the difference between these two transmission profiles may initially appear small from a comparison between these two lineshapes, the effects on nonclassical light generation are dramatic and manifest in the gray boxed regions that represent the frequency of the emitter-like polariton. In fact, the mixing action of combining the reflected laser light with light scattered by the strongly coupled system is a type of homodyning, which has the power to emphasize the incoherent or nonclassical portion of the scattered light over the coherent or classical portion. Because the mixing occurs at the level of the photonic crystal, this effect is named a self-homodyne interference (SHI) (Fischer et al., 2016b). To further explore this point, the transmission profiles in Fig. 14B and D were decomposed into their incoherent and coherent portions of emission. The coherent portions (blue) are primarily due to the classically scattered light from a subset of almost harmonically spaced dressed states or the continuum modes and hence look predominantly like the Lorentzian or Fano lineshapes. This light is due to the mean of the electric field, i.e., $I_{\mathrm{coh}} \propto \langle a^\dagger \rangle \langle a \rangle$. Meanwhile, the incoherent portions of the emissions (green) are the result of the nonlinearity in the Jaynes–Cummings system and hence from the nonclassically scattered light. This light is due to the fluctuations of the electric field, i.e., $I_{\mathrm{inc}} \propto \langle a^\dagger a \rangle - \langle a^\dagger \rangle \langle a \rangle$.

Now, we revisit Fig. 14B and D and compare the coherent and incoherent portions in the insets. Under the Lorentzian-like conditions (Fig. 14B), the coherent portion of the transmitted light (blue) dominates the incoherent portion (green). However, under the Fano-like conditions (Fig. 14D), the incoherent portion of the transmitted light dominates and over 90% of the coherently scattered portion is suppressed at the frequency of the emitter-like polariton. In this way, the effect of the quantum nonlinearity in the Jaynes–Cummings ladder is emphasized, with great potential to allow highly dissipative systems to still exhibit robust signatures of nonclassical light generation.

### *3.6.1 Effects on Emission Spectra*

As our first example of the dramatic influence of self-homodyne interference on emphasizing nonclassical light generation, we review data from Fischer et al. (2016b). In Fig. 15, we consider the effect of self-homodyne interference on the spectra of pulsed resonance fluorescence from a solid-state strongly coupled system. The ideal spectrum of pulsed resonance fluorescence for an operator $A(t)$ is calculated with

$$S(\omega) = \iint_{\mathbb{R}^2} \mathrm{d}t\, \mathrm{d}\tau\, \mathrm{e}^{-\mathrm{i}\omega\tau} \langle A^\dagger(t+\tau) A(t) \rangle. \tag{30}$$

The spectrum for resonance fluorescence from a Jaynes–Cummings system (even with phonons) is given when $A(t) \to a(t)$ and the spectrum with self-homodyne interference is given when $A(t) \to a(t) + \alpha(t)$. The two-time correlations are again computed with a time-dependent version of the quantum regression theorem, using the system Liouvillian

$$\mathcal{L}_{\mathrm{sc}}(t) = \tilde{\mathcal{L}}_{\mathrm{JC}} + \tilde{\mathcal{L}}_{\mathrm{phonon}} + \tilde{\mathcal{L}}_{\mathrm{drive}}(t). \tag{31}$$

*Note*: although we do not include blinking in this section, it has very minimal effects on the results since the dot is detuned from the bare cavity.

In the experiment, the focal conditions were first tuned to the Lorentzian-like conditions (Fig. 15A) and we then excited the emitter-like polariton (denoted by the gray box) with a high-power $\tau_{\mathrm{FWHM}} = 100$ ps pulse. The spectrum of resonance fluorescence was measured on a

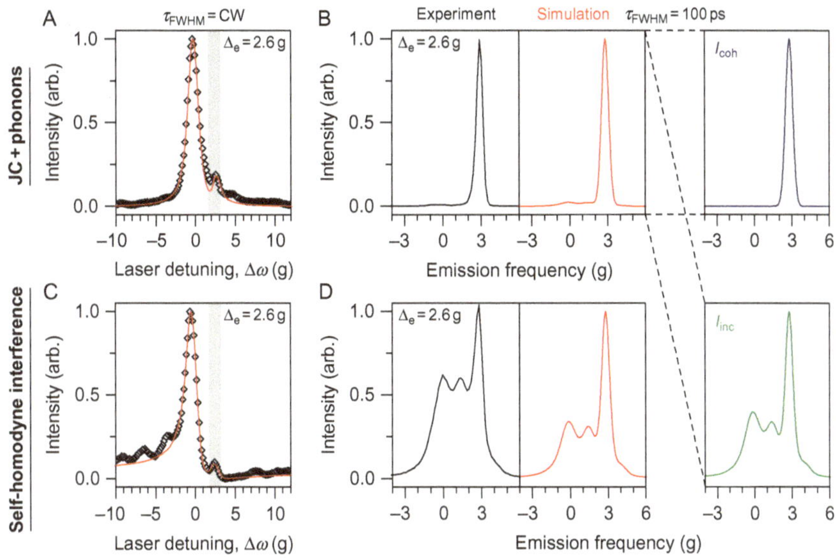


**Fig. 15** Effect of self-homodyne interference on resonance fluorescence from a dissipative strongly coupled system. (A) Transmission profile under excitation conditions that produce a Lorentzian-like profile. (B) Experimental and simulated resonance fluorescence under drive on the emitter-like polariton [*gray box* in (A)] by a high-power $\tau_{FWHM}$ = 100 ps pulse. (C) Transmission profile under excitation conditions that produces a Fano-like profile. (D) Experimental and simulated resonance fluorescence under drive on the emitter-like polariton [*gray box* in (C)] by a high-power $\tau_{FWHM}$ = 100 ps pulse, including self-homodyne interference. (A–D) Quantum-optical fits in *red*, and decomposition into coherent ($I_{coh}$) and incoherent ($I_{inc}$) components in *blue* and *green*, respectively. All simulated spectra were convolved with the response function of the spectrometer (linewidth $\Gamma_{FWHM}/2\pi$ = 4.5 GHz). *Data and simulations from Fischer, K.A., Müller, K., Rundquist, A., Sarmiento, T., Piggott, A.Y., Kelaita, Y., Dory, C., Lagoudakis, K.G., Vučković, J., 2016b. Self-homodyne measurement of a dynamic Mollow triplet in the solid state. Nat. Photonics 10, 163–166.*

spectrometer and is shown in Fig. 15B. Notably, the spectrum simply shows a single peak with a linewidth determined by the laser pulse (black). Simulating the spectrum with a Jaynes–Cummings model including phonon effects, i.e., with the Liouvillian $\mathcal{L}_{sc}(t)$, the singly peaked spectrum can be reproduced (red). Like the transmission spectra, the fluorescence spectrum can also be decomposed into the coherent and incoherent potions by making the replacement of $\langle A^\dagger(t+\tau)A(t)\rangle \rightarrow \langle a^\dagger(t+\tau)\rangle\langle a(t)\rangle$ and $\langle A^\dagger(t+\tau)A(t)\rangle \rightarrow \langle a^\dagger(t+\tau)a(t)\rangle - \langle a^\dagger(t+\tau)\rangle\langle a(t)\rangle$, respectively, in Eq. (30). These decompositions are shown in the blue and green lineshapes, and they show that the coherently scattered light completely dominates the emission spectrum at high powers. *Again, we emphasize that this is a standard feature of highly dissipative but strongly coupled Jaynes–Cummings systems, that well-known cavity QED effects can be unobservable!* On the other hand, the incoherent portion shows a very interesting quadruplet structure that is closely related to the Mollow triplet, a hallmark of quantum-mechanically scattered light. This quadruplet structure was discussed thoroughly in Fischer et al. (2016b) and its origins will not be discussed here; suffice to say that it is a structure arising from the nonclassical light emission of the cavity QED system.

In order to experimentally observe the quadruplet, self-homodyne interference was critical. We next tuned to the Fano-like conditions (Fig. 15B) and again excited the emitter-like polariton (denoted by the gray box) with the same high-power $\tau_{\mathrm{FWHM}} = 100$ ps pulse. Now, the optimally tuned self-homodyne interference experimentally reveals the interesting quadruplet structure, which is completely unobservable otherwise. The interference was also included theoretically in simulation to reveal the quadruplet (Fig. 15). We make two practical notes here regarding the modeling of these experiments. First, an entire power-series of spectra was necessary to accurately fit excitation powers [see fig. 3 in Fischer et al. (2016b)]. Second, the experimental data show a better defined quadruplet, with more energy in the sidebands than the model. We expect this difference is due to an inaccuracy of the effective phonon master equation at large driving powers.

### 3.6.2 Effects on Nonclassical Light Generation

As our second application of self-homodyne interference, we consider its effect on nonclassical light generation. As we saw in Section 3.6.1, an optimally tuned self-homodyne interference (SHI) has the ability to remove unwanted coherently scattered light from the system. On a similar principle, we would expect the interferometric technique to be capable of enhancing nonclassical light generation by isolation of the quantum-mechanical signal. In prior work, we explored the precise mechanism for how SHI is capable of enhancing both single- and multiphoton emission in dissipative Jaynes–Cummings systems, but under continuous-wave excitation (Fischer et al., 2017). In this section, we will examine the enhancement under pulsed excitation and in doing so will almost perfectly fit the experimental data from Müller et al. (2015b) and Dory et al. (2017).

First, we discuss our complete model that we believe captures nearly all experimental effects relevant for photon blockade and photon tunneling. It is again based off of the system Liouvillian

$$\mathcal{L}_{\mathrm{sc}}(t) = \tilde{\mathcal{L}}_{\mathrm{JC}} + \tilde{\mathcal{L}}_{\mathrm{phonon}} + \tilde{\mathcal{L}}_{\mathrm{drive}}(t), \tag{32}$$

but now we calculate $g^{(2)}[0]$ with a self-homodyne interference. Specifically, that means to calculate

$$g^{(2)}[0] = \frac{\int_0^T \int_0^T \mathrm{d}t\mathrm{d}t' \langle \mathcal{T}_{-}[A^{\dagger}(t)A^{\dagger}(t')]\mathcal{T}_{+}[A(t')A(t)]\rangle}{\left(\int_0^T \mathrm{d}t \langle A^{\dagger}(t)A(t)\rangle\right)^2} \equiv \frac{G_{AA}^{(2)}[0]}{\langle N_A\rangle^2}, \tag{33}$$

where $A(t) \to a(t) + \alpha(t)$ and again $\alpha(t) = \alpha\mathcal{E}(t)$. To incorporate blinking, we simply change Eq. (28) into

$$g^{(2)}[0] = \frac{(1 - f_{\mathrm{blink}})G_{AA}^{(2)}[0] + f_{\mathrm{blink}}\langle N_{\mathrm{blink}}\rangle^2}{(1 - f_{\mathrm{blink}})\langle N_{\mathrm{A}}\rangle^2 + f_{\mathrm{blink}}\langle N_{\mathrm{blink}}\rangle^2}. \tag{34}$$

Now, $\langle N_{\mathrm{blink}}\rangle = \int_0^T \mathrm{d}t \langle A^{\dagger}(t)A(t)\rangle$ and is calculated using the Liouvillian $\mathcal{L}(t) = \tilde{\mathcal{L}}_{\mathrm{blink}} + \tilde{\mathcal{L}}_{\mathrm{drive}}(t)$; again, $A(t) \to a(t) + \alpha(t)$.

With the necessary theoretical machinery established, we are now finally ready to fully model the $g^{(2)}[0]$ vs laser detuning scans. First, we revisit the data from Fig. 13, but with self-homodyne interference, in Fig. 16A. The optimal fit is given by the dashed brown line, which almost perfectly matches the data with the addition of an optimized SHI. The exact same trends are visible for blinking with or without the interference, where the tunneling regions are affected first with increasing blinking while the blockade regions are relatively unaffected. Now, we additionally show the simulated transmission plots for the incident pulses (Fig. 16B). We emphasize here that when the system blinks the SHI still occurs, and hence the system blinks with a Fano lineshape.

Next, we consider detuned blockade and tunneling with an emitter detuning of $\Delta_{\mathrm{e}} = 3.2\,g$. Comparing the resonant and detuned cases in Fig. 17, one can again observe the general trend of enhanced photon blockade with increasing detuning, as discussed in Section 3.2. (We provide a brief technical note that, just as with the resonance fluorescence experiments in Section 3.6.1, before performing any $g^{(2)}[0]$ scan the photon blockade region was optimized. At that point, a transmission experiment would reveal a Fano-like lineshape, incorporating SHI; these interference conditions were held constant over the course of the $g^{(2)}[0]$ scan.) Back to both subfigures, the quantum-optical model is broken down into several different lines, to

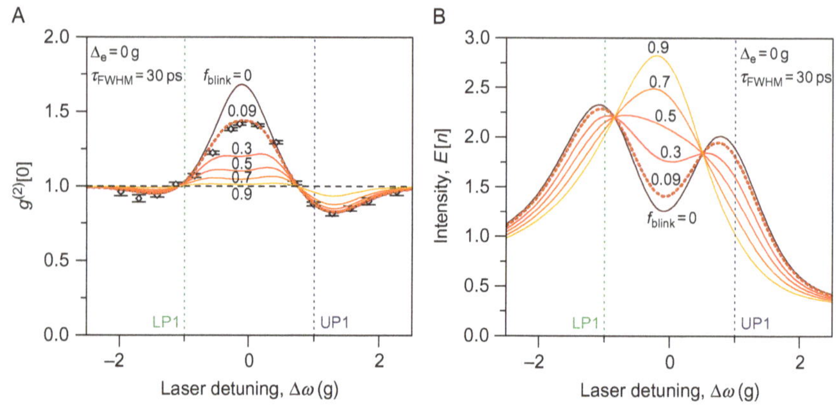


**Fig. 16** Effects of strong emitter blinking on a resonant strongly coupled system. (A) Second-order coherence $g^{(2)}[0]$ as a function of laser detuning. The ideal fit to $g^{(2)}[0]$ is given by the *brown dashed line*. *Horizontal black dashed line* represents statistics of the incident laser pulses. (B) Pulsed transmission spectrum, with quantum simulation only. In both subfigures, the statistics or intensities are given for different blinking fractions. *Experimental data in (A) from Müller, K., Rundquist, A., Fischer, K.A., Sarmiento, T., Lagoudakis, K.G., Kelaita, Y.A., Sánchez Muñoz, C., del Valle, E., Laussy, F.P., Vučković, J., 2015. Coherent generation of nonclassical light on chip via detuned photon blockade. Phys. Rev. Lett. 114 (23), 233601.*

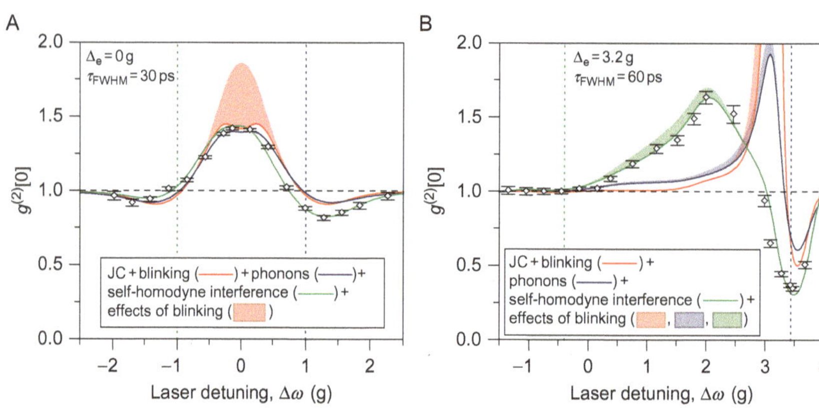


**Fig. 17** Perfectly modeled photon blockade and tunneling under pulsed excitation. Second-order coherence $g^{(2)}[0]$ as a function of laser detuning on an (A) resonant and (B) detuned system. The legends discuss the complete decomposition of the quantum-optical model into its constituent effects. *Green* and *blue dashed lines* represent the LP1 and UP1, respectively. *Horizontal black dashed lines* represent statistics of the incident laser pulses. A laser power of 5 nW was used. *Experimental data in (A) and (B) from Müller, K., Rundquist, A., Fischer, K.A., Sarmiento, T., Lagoudakis, K.G., Kelaita, Y.A., Sánchez Muñoz, C., del Valle, E., Laussy, F.P., Vučković, J., 2015. Coherent generation of nonclassical light on chip via detuned photon blockade. Phys. Rev. Lett. 114 (23), 233601; Dory, C., Fischer, K.A., Müller, K., Lagoudakis, K.G., Sarmiento, T., Rundquist, A., Zhang, J.L., Kelaita, Y., Sapra, N.V., Vučković, J., 2017. Tuning the photon statistics of a strongly coupled nanophotonic system. Phys. Rev. A 95, 023804, respectively.*

separate out the effects of blinking, electron–phonon interaction, and self-homodyne interference. The primary effects are as follows:

1. Blinking has the strongest effect in the zero dot detuning case and minimal effect in the detuned case. For zero detuning, the effect of blinking is not just to decrease the maximum achievable value of $g^{(2)}[0]$ in the tunneling region, but also to broaden the width of the tunneling region. We note that interestingly, none of the other tunable parameters are able to control this effect—without blinking, the resonant tunneling region could never be fitted properly. Additionally, the blinking fraction was taken from the fit in Section 3.5.1.
2. Electron–phonon interaction has almost negligible effects on photon blockade for either the resonant or detuned system, because the effect of the dissipation is only to change the frequency and coherence of any emitted single photons (Müller et al., 2016). In the tunneling region, phonon-induced transfers have the largest effect on the detuned system since the $g^{(2)}[0]$ values in the tunneling region are much larger due to the decrease in photon transmission, as discussed in Section 3.2.
3. Self-homodyne interference has the strongest effect on the detuned system, though it is certainly important in the resonant case as well. For both system configurations, SHI enables much lower values of $g^{(2)}[0]$ in the blockade regions, thus significantly improving the quality of single-photon emission. Because SHI occurs on one side of the cavity profile or the other, in the resonant case the LP1 blockade is worsened while the UP1 blockade is improved. This evidence of enhanced photon blockade fully suggests that in some previous experiments with strongly coupled systems where the simulations were surprisingly unable to perform as well as the experiments [e.g., those in Müller et al. (2015b), Reinhard et al. (2011), Müller et al. (2015a), and Kim et al. (2014)], researchers may have unknowingly utilized SHI.

To further discuss the effect of SHI, we note that the detuned tunneling region is dramatically different with SHI, where the peak of the $g^{(2)}[0]$ scan no longer occurs at the point of minimal photon transmission. Because the unwanted coherently scattered light is removed with SHI, the tunneling region is now a much stronger indicator of the multiphoton processes occurring in the detuned strongly coupled system. Additionally, we note that driving the quantum dot term directly (as opposed to through the cavity) was proposed as one method to increase the single-photon purity of photon blockade and could cause asymmetries like the one observed in the resonant case (Tang et al., 2015); however, this process only decreases the amplitude

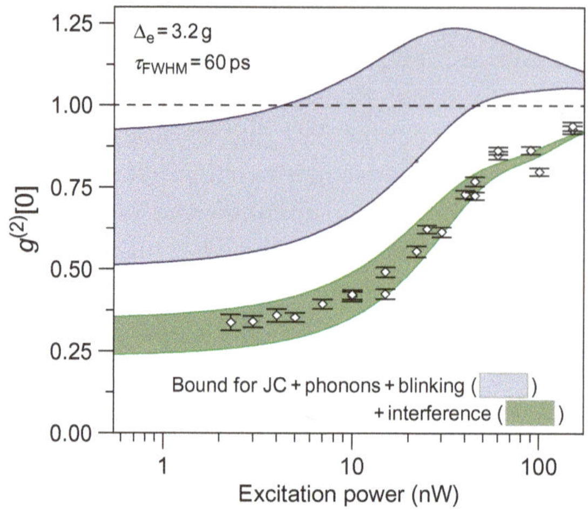


**Fig. 18** Detuned photon blockade as a function of excitation power. Data and fit parameters identical to those in Fig. 17. *Horizontal black dashed line* represents statistics of the incident laser pulses.

in the tunneling region and does not shift its maximum point of $g^{(2)}[0]$. Since no fitting parameter other than SHI was able to enhance the tunneling region, it is clear that SHI plays an important role in nonclassical light generation.

Next, we explore detuned photon blockade as a function of excitation power, both by presenting new experimental data and with quantum-optical fits (Fig. 18). This power-dependent data and fitting is important in verifying that we have not overfit our experimental data and in reaffirming the strength of the self-homodyne interference to improve photon blockade. In performing the power scan of $g^{(2)}[0]$, the experimental uncertainty in the precise laser detuning is given by the spectrometer linewidth of $\Gamma_{FWHM} \approx g/3$ (though there is little uncertainly in the laser detuning for the $g^{(2)}[0]$ vs laser detuning scans due to the high *relative* precision of the experimental pulse shaper). Therefore, we have simulated both the minimum $g^{(2)}[0]$ values in photon blockade and the values under a system that was imprecisely tuned by $\Gamma_{FWHM}/2$. This procedure pictorially shows the potential uncertainty in the correct laser detuning for the simulated values. Using the optimal fit from Fig. 17B, the green simulated blockade region matches almost perfectly with the experimental values. Meanwhile if SHI is excluded, then the blockade values are not just worse at the minimum, but much more sensitive to any possible imprecision in laser detuning. Because both the blockade and tunneling regions are sensitive to the excitation power, the strong fit with experiment helps confirm our complete model of an experimental strongly coupled system.

With such a complex model, one must be wary of overfitting the data. However, each of the elements has been independently verified and fitted through a large series of the experimental data to extract the Jaynes–Cummings dissipation, electron–phonon effects, blinking, and self-homodyne interference. Taken as a whole, *each of the nonideal effects almost independently tunes different aspects of the emission statistics and hence we believe we have identified the appropriate number of model parameters.*

## 3.7 Outlook for Single-Emitter Cavity QED

In summary, we proposed a complete model for photon blockade and tunneling in III–V quantum dot-cavity QED systems. We found that the pure Jaynes–Cummings model was incapable of accurately modeling either the spectra or photon statistics in transmission or emission from a strongly coupled system based on InGaAa quantum dots. However, by including dissipation, dot-cavity detuning, pulsed dynamics, effects of phonons, blinking, and a new effect called self-homodyne interference we were able to almost perfectly model the nonclassical light generation from our cavity QED system.

By incorporating frequency filtering of the emission, we recently showed highly indistinguishable photon generation and evidence for two-photon generation from such a strongly coupled system (Dory et al., 2017; Müller et al., 2016). Looking toward future experiments, by adapting the self-homodyne interference technique to on-chip photonic crystal waveguide devices, we expect that this work could easily become a standard feature of optical solid-state platforms in enhancing nonclassical light generation (Fischer et al., 2017). We believe this technique should also enable the first direct observation of a solid-state system's higher-order Jaynes–Cummings structure and the efficient generation of N-photon states.

Moving toward more interesting and complex level structures will allow for a much richer set of dynamics and possibilities for nonclassical light generation. For instance, charged III–V quantum dots in magnetic field (Carter et al., 2013), III–V quantum dot molecules (Vora et al., 2015), and group-IV color-centers (Riedrich-Möller et al., 2014) may allow for the exploration of cavity QED with a single, multi*level* quantum emitter coupled to a cavity (Bajcsy et al., 2013). These posses untapped level structures for improving photon blockade and studying multiphoton transitions, and they may allow experimentalists to more readily probe quantum nonlinearities in a solid-state environment. One of their most promising applications is to realize arbitrary single-photon generation in a solid-state nanocavity (Santori et al., 2009; Sweeney et al., 2014). In this scheme, the cavity mediates the generation or annihilation of an arbitrarily shaped single-photon through a Raman transition. Such devices form the backbone of spin-photon interfaces in some theoretically proposed quantum networks (Cirac et al., 1997). Unlike the already-demonstrated spin-photon interfaces that rely only on weak cavity coupling, it is possible for a flying photonic qubit to be perfectly absorbed by a cavity QED device operating in the Raman single-photon regime.

Fundamentally, however, single-emitter systems with multiple levels are also limited in their potential for nonclassical light emission. Instead, we expect the future of cavity QED, utilizing rich level structures for studying quantum light and computation, lies in multi*emitter* cavity QED. In the next section, we cover an emerging field whereby multiple color-centers couple to a single cavity mode.

## 4. OVERVIEW OF MULTIEMITTER CAVITY QED

While Section 3 illustrated the limit of what has been experimentally demonstrated so far, here we propose a direction for the future cavity QED experiments. The nonlinearity in a cavity QED system can be significantly increased through the collective coupling of multiple emitters to the resonant mode (Fig. 19). In this picture, $N$ emitters are effectively described as a single-emitter strongly coupled to the cavity with an increased interaction rate by a factor of $\sqrt{N}$. Such coupling is achievable in systems where the inhomogeneous broadening is comparable to the collective coupling rate (Diniz et al., 2011). Collective coupling is possible even when the inhomogeneous broadening exceeds the cavity linewidth, and the polariton width is dominantly defined by cavity and emitter linewidths, due to an effect called *cavity protection*. This phenomenon was previously investigated with rare-earth ions in a solid-state cavity for a very large ensemble of emitters approximating a continuum (Zhong et al., 2016). In contrast, we are interested in a regime of several emitters coupled to a nano-optical cavity and giving rise to a discretized energy ladder suitable for demonstrations of advanced photon blockade effects. Similar emitter cavity QED systems have been demonstrated in atomic (Neuzner et al., 2016; Thompson et al., 1992) and

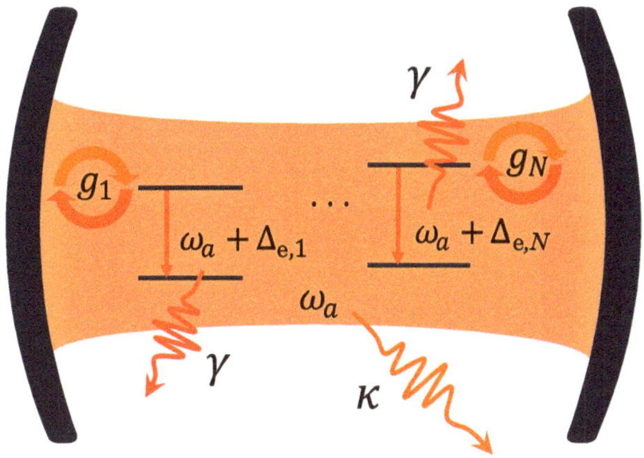


**Fig. 19** Illustration of $N$ nonidentical emitters coupled to a cavity mode.

superconducting circuit systems (Fink et al., 2009). With the developments in techniques of substrate growth and processing, the implementation of these systems is expected with color-centers in solid-state nanocavities as well, which will have an impact on the development of GHz-speed optical switches and high-quality integrated sources of single photons.

## 4.1 The Tavis–Cummings Model

We first discuss the original Tavis–Cummings model for atoms, which have negligible inhomogeneous broadening, and then extend the model to include the inhomogeneous broadening of solid-state quantum emitters. The Tavis–Cummings model was developed to describe an ensemble of atoms that strongly couple to a cavity mode (Tavis and Cummings, 1968) and gives rise to the Hamiltonian

$$H_{\mathrm{TC}} = \omega_a a^\dagger a + \sum_{n=1}^{N} \left[ \omega_a \sigma_n^\dagger \sigma_n + g_n \left( \sigma_n^\dagger a + a^\dagger \sigma_n \right) \right]. \tag{35}$$

Here, a single cavity mode individually couples to each of the $N$ emitters that otherwise do not interact with one another. As before, $a$ and $\omega_a$ represent the cavity operator and frequency, while the atoms are characterized by their dipole operator $\sigma_n$ and cavity coupling rate $g_n$. The variables $g_n$ account for nonequal positioning of emitters relative to the cavity field intensity.

Coherent interactions in the system give rise to a new set of eigenfrequencies that form a dressed ladder of states. The first rung contains $N + 1$ states, corresponding to an additional excitation of the cavity mode

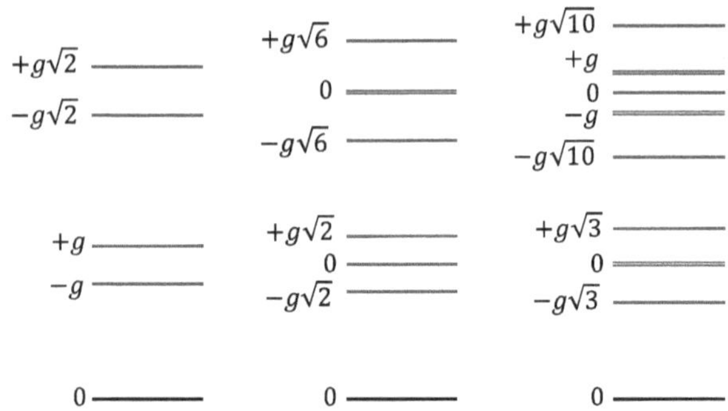


**Fig. 20** Dressed ladder of states for $N = 1, 2, 3$ identical emitters coupled to a single cavity mode.

or one of the emitters. The second rung contains $\frac{(N+1)(N+2)}{2}$ states, representing either an excitation of one of the previously nonexcited emitters or a new photon in the cavity mode.

For a system with equally coupled atoms ($g_n = g$), we illustrate these levels in Fig. 20. In particular, the levels in the first rung represent two polaritonic states and $N - 1$ degenerate states. The degenerate states are at the cavity frequency and their eigenvectors have no cavity component; therefore, the states do not couple to the environment and are referred to as the subradiant states. For the first rung, the eigenenergies are given by

$$E_1 = \omega_c - g\sqrt{N}, \tag{36}$$

$$E_{2,\ldots,N} = \omega_a, \tag{37}$$

$$E_{N+1} = \omega_a + g\sqrt{N}. \tag{38}$$

The splitting between the polaritonic states $E_1$ and $E_{N+1}$ is a result of the collective coupling with an effective coupling rate of $G_N = g\sqrt{N}$. When the emitters are unequally coupled (i.e., $g_n \neq g_m$), then the collective coupling rate can be calculated as

$$G_N = \sqrt{\sum_{n=1}^{N} g_n^2}. \tag{39}$$

*However, the degeneracy between the other $N - 1$ states occurs only for identical emitters and is lifted with the introduction of any nonidentical quantum emitters.* Compared to atoms, color-centers are nonidentical quantum emitters, and therefore we need to expand the original Hamiltonian to include a set of emitter frequencies that capture the inhomogeneous broadening in the ensemble, i.e., with

$$H'_{\mathrm{TC}} = \omega_a a^\dagger a + \sum_{n=1}^{N} \left[ (\omega_a + \Delta_{\mathrm{e},n}) \sigma_n^\dagger \sigma_n + g_n \left( \sigma_n^\dagger a + a^\dagger \sigma_n \right) \right]. \tag{40}$$

Specifically, the inhomogeneous broadening is represented by the emitter detunings $\Delta_{\mathrm{e},n}$. In the following sections, we will explore how the inhomogeneous broadening, resulting in $\Delta_{\mathrm{e},n} \neq \Delta_{\mathrm{e},m}$, affects the ability to observe collective oscillations, subradiant states, and quantum coherent phenomena.

## 4.2 Strong-Coupling Cavity QED With an Ensemble of Color-Centers

In our CQED model of an ensemble of color-centers in a cavity, the following factors are considered:

1. The loss of photons from the cavity to the environment. The loss is defined by the quality factor of the mode $Q = \omega_a/\kappa$, and is often practically imposed by fabrication limitations.
2. The loss of photons from the emitters to the environment. This mechanism is governed by the ratio of color-center emission that is directed into the cavity mode and is affected by the density of states. A high density of photonic states results in a low photon loss rate, which can be encouraged by high-quality factor and small mode volume cavities, as well as by the good positioning of the emitters within the cavity.
3. The distribution of emission frequencies in the ensemble. This factor is usually influenced by the local strain in the lattice and can be especially pronounced in nanoparticles or heteroepitaxial layers, compared to bulk substrates.
4. The variable positioning of individual emitters in the cavity. Some control over the depth of the color-centers in the substrates can be gained through selective doping during substrate growth or by defining the ion energy during irradiation of the pregrown sample. Laterally, the use of a focused ion beam or masked apertures can provide a degree of localization.
5. The ratio of emission into the zero-phonon line (ZPL). This is an intrinsic property of each color-center, and it is the property that motivates the search for new systems with high Debye-Waller factor.

We now discuss how these effects are incorporated into our quantum-optical model. The cavity and emitter losses can be modeled through the Liouville's equation as presented in Section 2.4. Now, the superoperator that describes the system dynamics has the form

$$\mathcal{L}_{\mathrm{TC}}\rho(t) = \mathrm{i}\left[\rho(t), H'_{\mathrm{TC}}\right] + \frac{\kappa}{2}\mathcal{D}[a]\rho(t) + \sum_{n=1}^{N}\frac{\Gamma}{2}\mathcal{D}[\sigma_n]\rho(t), \tag{41}$$

where $\kappa$ represents the cavity energy decay rate, and $\Gamma$ represents the individual emitter linewidth which is assumed to be constant within the ensemble.

The emission frequencies are sampled from a Gaussian distribution centered at $\omega_a + \Delta_e$ with standard deviation of $\frac{\delta}{2}$, where $\delta$ is the inhomogeneous linewidth of the ensemble. It is worth noting that the shape of individual color-center emission is a Lorentzian (homogeneous broadening), however, the distribution of central frequencies in the ensemble is a Gaussian (inhomogeneous broadening).

The coupling strength is limited by the maximal value $g_{\max}$ imposed by the system parameters: ZPL extraction ratio (Debye-Waller factor) $\rho_{\mathrm{ZPL}}$, emitter lifetime $\tau$, index of refraction $n$, cavity mode volume $V$ and operating frequency $\omega$ (McCutcheon and Lončar, 2008)

$$g_{\max} = \sqrt{\frac{3\pi c^3 \rho_{\mathrm{ZPL}}}{2\tau\omega^2 n^3 V}}. \tag{42}$$

The coupling rate is reduced by imperfect positioning of emitters $r_n$ relative to the resonant field maximum $E_{\max}$ and the angle between the dipole and field orientation $\phi_n$

$$g_n = g_{\max}\left|\frac{E(r_n)}{E_{\max}}\cos(\phi_n)\right|. \tag{43}$$

The distribution of the intensity in range $[0, g_{\max}]$ is also specific to the system, mainly to the spatial positioning of emitters and the resonant field intensity distribution. Here, we will sample the coupling $g_n$ from a uniform distribution.

We choose a realistic set of parameters

$$\{\kappa/2\pi, \Gamma/2\pi, g_{\max}/2\pi\} = \{25\ \mathrm{GHz}, 0.1\ \mathrm{GHz}, 10\ \mathrm{GHz}\}$$

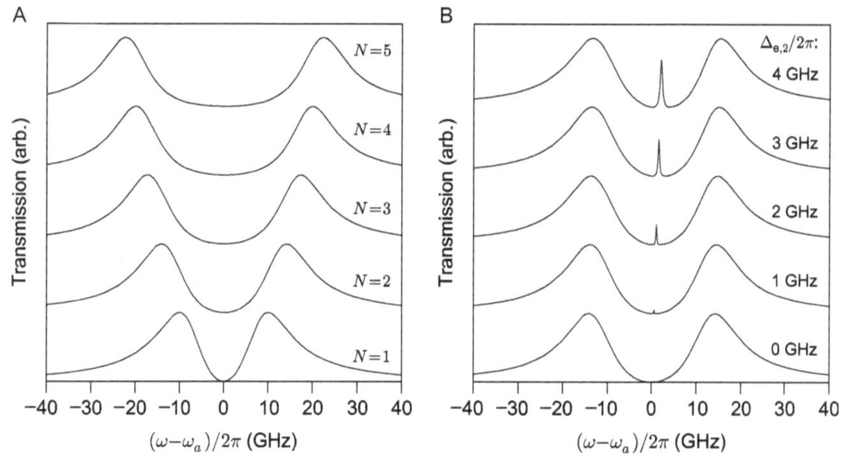


**Fig. 21** (A) Transmission spectra of $N$ identical and equally coupled emitters. (B) Transmission spectra showing the emerging subradiant peak as the second of $N = 2$ emitters becomes off-resonant.

to model the dynamics of the multiemitter cavity system. In order to numerically investigate the transmission spectra of the multiemitter cavity system, we use the incoherent *cavity* pumping technique discussed in Section 2.4. From these simulations, in Fig. 21A we plot the increasing polaritonic spacing with an increased number of identical and equally coupled emitters. The trend is consistent with the lossless case analysis where the collective coupling is given as $G_N = g\sqrt{N}$. As discussed previously, nonidenticality of emitters lifts the degeneracy of the new states in the ladder and makes them visible in the spectrum. With increasing emitter–emitter detuning, the subradiant states centered between the emitter frequencies become more visible. This is illustrated with the emergence of an intermediate peak in the transmission spectrum for $N = 2$ nonidentical emitters in Fig. 21B.

Next, we analyze the influence of the inhomogeneous broadening to the transmission spectrum and present one of our most important findings, that the inhomogenous broadening will not obscure the observation of collective many-body physics in systems comprising color-centers with small inhomogeneous broadening. First, we investigate this phenomenon in Fig. 22, which shows randomly generated spectra with $N = 4$ emitters for variable inhomogeneous broadening with

$$\delta/2\pi \in \{1\text{ GHz}, 10\text{ GHz}, 100\text{ GHz}\}.$$

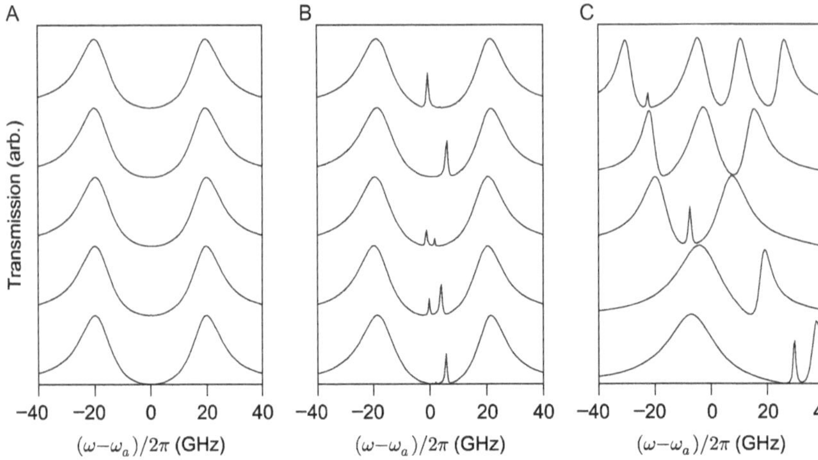


**Fig. 22** Randomly generated transmission spectra for $N = 4$, $\{\kappa/2\pi, \Gamma/2\pi, g_{max}/2\pi\} =$ {25 GHz, 0.1 GHz, 10 GHz} and $\delta/2\pi =$ (A) 1 GHz, (B) 10 GHz, and (C) 100 GHz.

The first two sets of spectra show a small or a moderate perturbation to the system with identical emitters, featuring two polariton peaks and several subradiant peaks, suggesting that silicon-vacancy-related complexes could be used in multiemitter cavity QED systems. In the case of large (100 GHz) inhomogeneous broadening, however, the spectrum is significantly perturbed and the identification of the polariton peaks is difficult or impossible (Fig. 22C). Therefore, a multiemitter systems with such broadening may not show the signs of collective strong coupling.

To further analyze the character of the collective strong coupling, we generate 100 systems ($G_N > \kappa/2$) for variable inhomogeneous broadening with

$$\delta/2\pi \in \{1\ \text{GHz}, 10\ \text{GHz}, 20\ \text{GHz}\}$$

and plot the spacing between the emitters against the expected collective coupling $G_N$. We use the same formula as in the case of identical emitters: $G_N = \sqrt{\sum_{n=1}^{N} g_n^2}$, so that we benchmark the results with inhomogeneous broadening against the ideal multiemitter cavity systems. As seen in Fig. 23 the effective coupling rate is fit well by $2G_N$, implying that the collective coupling rate of nonidentical emitters is comparable to identical emitters for small inhomogeneous broadening. Crucially, because the numerically investigated polariton splittings closely match the $G_N$ for the cases of diamond SiV ($\delta \sim 1$ GHz) and SiC silicon vacancy ($\delta \sim 20$ GHz) systems, we expect these

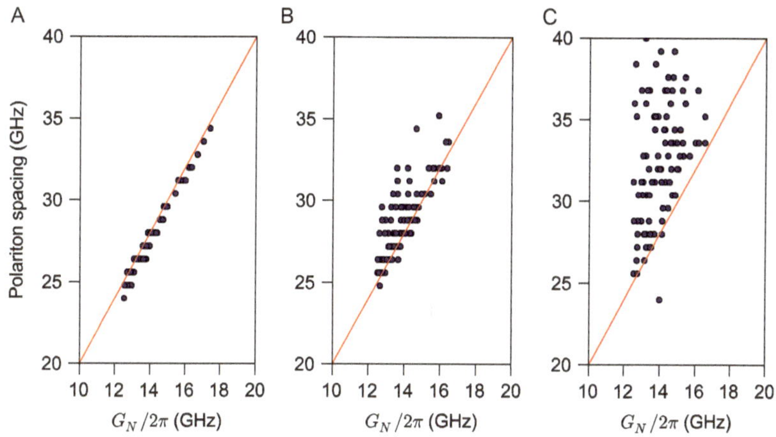


**Fig. 23** Polariton spacing of randomly generated strongly coupled systems vs expected collective coupling rate for $N=4$, $\{\kappa/2\pi, \Gamma/2\pi, g_{\text{max}}/2\pi\} = \{25\ \text{GHz}, 0.1\ \text{GHz}, 10\ \text{GHz}\}$ and $\delta/2\pi =$ (A) 1 GHz, (B) 10 GHz, and (C) 20 GHz. The *red line* corresponds to the system with identical emitters.

emitters to exhibit multiemitter cavity QED phenomenon in spite of their inhomogeneous broadening.

## 4.3 Effective Hamiltonian Approach to Multiemitter Cavity QED

The quantum master equation provides a full numerical treatment of the system dynamics in transmission, however, the diagonalization of the density matrix makes this approach computationally challenging. This practically limits the number of emitters in the system to $N \lesssim 5$. The effective Hamiltonian approach introduced in Section 3.1 can provide a significant speedup. For example, in the modeling of the transmission spectra, computation is effectively reduced to the diagonalization of a $(N+1) \times (N+1)$ matrix, from the initial $2^{(N+1)} \times 2^{(N+1)}$ size.

Extended from the single-emitter case presented in Section 3.1, the effective Hamiltonian for multiemitter system is

$$H_{\text{EFF}} = H'_{\text{TC}} - \mathrm{i}\frac{\kappa}{2}a^\dagger a - \mathrm{i}\sum_{n=1}^{N}\frac{\Gamma}{2}\sigma_n^\dagger \sigma_n. \tag{44}$$

Its eigensystem can be represented as a set of eigenenergies and eigenvectors $\{E_n^{\text{EFF}}, \psi_n^{\text{EFF}}\}$. Bearing in mind that $\mathrm{Re}\{E_n^{\text{EFF}}\}$ gives the frequency of an energy level and $2\,\mathrm{Im}\{E_n^{\text{EFF}}\}$ represent its linewidth, we can reconstruct some of the spectral information obtained through the quantum master equation approach. We again generate 100 random systems for

$$\delta/2\pi \in \{1\ \text{GHz}, 10\ \text{GHz}, 20\ \text{GHz}\}$$

and plot the frequency difference between the two eigenstates with the highest linewidths against the expected strong coupling rate $G_N$ (Fig. 24). Here, the eigenstates with the highest linewidths represent the two states of strongest cavity emission since we operate in the bad-cavity limit. The correlation between the simulated polariton splitting for nonidentical emitters to the one expected for identical emitters is excellent. The latter can be analytically calculated by approximating the multiemitter system as a single-emitter system with effective coupling $G_N$. Using the derivation presented in Section 3.1, the polariton splitting is expressed as

$$E_{N+1} - E_1 = 2\sqrt{G_N^2 - \left(\frac{\kappa - \Gamma}{4}\right)^2}. \tag{45}$$

While here we confirm that the nonidentical multiemitter systems behave similarly to identical ones in the effective Hamiltonian approximation, additional steps are needed to derive a close approximation to the transmission spectrum. To do this, we first introduce a driving term to the effective Hamiltonian

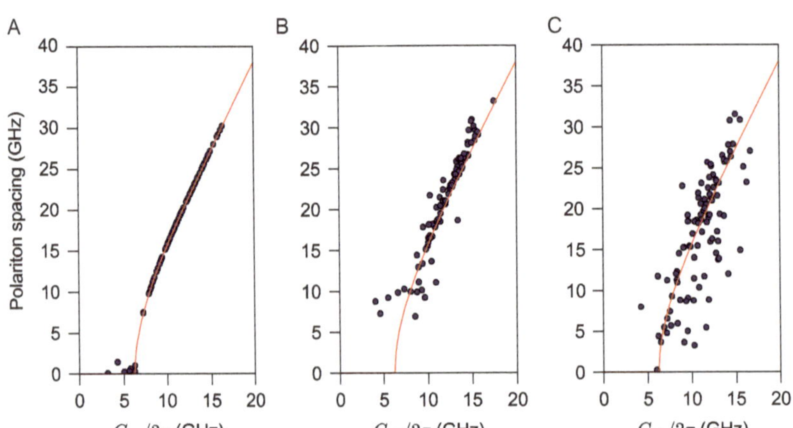


**Fig. 24** Polariton spacing calculated using the effective Hamiltonian diagonalization vs expected collective coupling rate for $N = 4$, $\{\kappa/2\pi, \Gamma/2\pi, g_{max}/2\pi\} = \{25\ \text{GHz}, 0.1\ \text{GHz}, 10\ \text{GHz}\}$ and $\delta/2\pi =$ (A) 1 GHz, (B) 10 GHz, and (C) 20 GHz. The *red line* corresponds to the system with identical emitters.

$$H_{\text{drive}}(\omega_{\text{L}}) = \mathcal{E}\left(a e^{i\omega_{\text{L}}t} + a^{\dagger} e^{-i\omega_{\text{L}}t}\right), \tag{46}$$

where $\omega_{\text{L}}$ is the laser frequency and $\mathcal{E}$ is proportional to the laser field intensity. Transforming the Hamiltonian into the rotating-wave frame (as discussed in Section 2.4) and diagonalizing the effective Hamiltonian, we obtain the eigensystem $\{E_n^{\text{EFF}}(\omega_{\text{L}}), \psi_n^{\text{EFF}}(\omega_{\text{L}})\}$. Now, we assume that each energy level emits light with Lorentzian intensity distribution centered at $\text{Re}\,\{E_n^{\text{EFF}}(\omega)\}$ and with linewidth of $2\text{Im}\,\{E_n^{\text{EFF}}(\omega)\}$. To combine the individual levels into a transmission spectrum, we weight each of the eigenstates by their cavity occupation terms $\langle \psi_n^{\text{EFF}}(\omega_{\text{L}}) | a^{\dagger} a | \psi_n^{\text{EFF}}(\omega_{\text{L}}) \rangle$. From there, we derive our first effective Hamiltonian approximation to the spectrum $S^{\text{EFF}}(\omega)$ as

$$S_I^{\text{EFF}}(\omega) = \sum_{n=1}^{N} \langle \psi_n^{\text{EFF}}(\omega) | a^{\dagger} a | \psi_n^{\text{EFF}}(\omega) \rangle L\left(\omega; \text{Re}\,\{E_n^{\text{EFF}}(\omega)\}, \text{Im}\,\{E_n^{\text{EFF}}(\omega)\}\right), \tag{47}$$

where $L(\omega;\, \omega_0,\, \eta)$ represents Lorentzian distribution defined as

$$L(\omega; \omega_0, \eta) = \frac{1}{\pi\eta\left[1 + \left(\dfrac{\omega - \omega_0}{\eta}\right)^2\right]}. \tag{48}$$

In Fig. 25 we compare the spectra obtained through the effective Hamiltonian approach to the ones calculated by the quantum master equation. The qualitative agreement is very good, while the quantitative match between peak locations and intensities is close, but not complete, because it does not capture the interference effects between the polaritons. To improve on this, we derive another approximation which models the light field interference more reliably.

Because we consider the transmission spectra when the emitters are all weakly excited (as discussed in Section 2.4), they may be considered to primarily emit coherent radiation (Steck, 2017). Thus, we can ignore their incoherent portions of emission and use the classical limit

$$\langle a^{\dagger} a \rangle \approx \langle a^{\dagger} \rangle \langle a \rangle. \tag{49}$$

Now, we can take the fields ($\propto \langle a \rangle$) rather than intensities ($\propto \langle a^{\dagger} a \rangle$) to combine the transmission profiles in the spectrum, thereby incorporating the interference effects between the different polaritons and subradiant states.

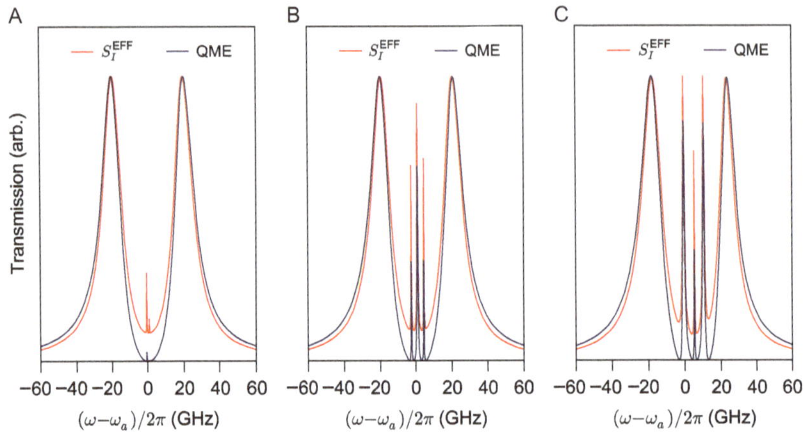


**Fig. 25** A comparison between the transmission spectra calculated using quantum master equation (QME) and the first effective Hamiltonian approximation ($S_I^{\text{EFF}}$) for a $N=4$ multiemitter system $\{\kappa/2\pi, \Gamma/2\pi, g_{\text{max}}/2\pi\} = \{25\text{ GHz}, 0.1\text{ GHz}, 10\text{ GHz}\}$ and emitter frequencies sampled for $\delta/2\pi =$ (A) 1 GHz, (B) 10 GHz, and (C) 20 GHz.

Then, in our second effective Hamiltonian approximation, the transmission spectrum is calculated as

$$S_{II}^{\text{EFF}}(\omega) = \left| \sum_{n=1}^{N} \langle \psi_n^{\text{EFF}}(\omega) | a | \psi_n^{\text{EFF}}(\omega) \rangle \sqrt{L\left(\omega; \text{Re}\left\{E_n^{\text{EFF}}(\omega)\right\}, \text{Im}\left\{E_n^{\text{EFF}}(\omega)\right\}\right)} \right|^2 \tag{50}$$

where now the terms inside the summation represent field contributions that are interfered before the final modulus squared is taken to calculate the intensity. This result is formally very similar to that obtained by diagonalizing a set of coupled Heisenberg–Langevin equations and using the approximation that the emitters are all in their ground states (Waks and Vuckovic, 2006). While these two methods arrive at similar answers, we believe the intuitive connection to the complex eigenstates of the non-Hermitian effective Hamiltonian provides additional insight. Fig. 26 shows that this approximation provides a close fit to the quantum master equation results.

With these approximations at hand, we can conclude that the effective Hamiltonian approach can provide an excellent insight into transmission properties of a cavity QED system. This is especially valuable for systems with a large number of emitters where the solving of the quantum master equation would require unrealistic computational resources. We analyze one such system based on $N = 100$ emitters for $\delta/2\pi \in \{10\text{ GHz}, 40\text{ GHz}, 80\text{ GHz}\}$. Fig. 27 shows the extension of the cavity protection effects to systems with larger inhomogeneous broadening, granted by the

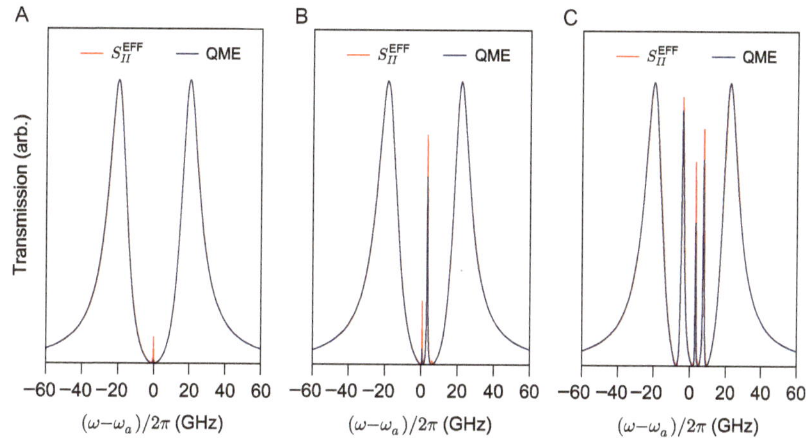


**Fig. 26** A comparison between the transmission spectra calculated using quantum master equation (QME) and the second effective Hamiltonian approximation ($S_{II}^{EFF}$) for a $N = 4$ multiemitter system $\{\kappa/2\pi, \Gamma/2\pi, g_{max}/2\pi\} = \{25\text{ GHz}, 0.1\text{ GHz}, 10\text{ GHz}\}$ and emitter frequencies sampled for $\delta/2\pi =$ (A) 1 GHz, (B) 10 GHz, and (C) 20 GHz.

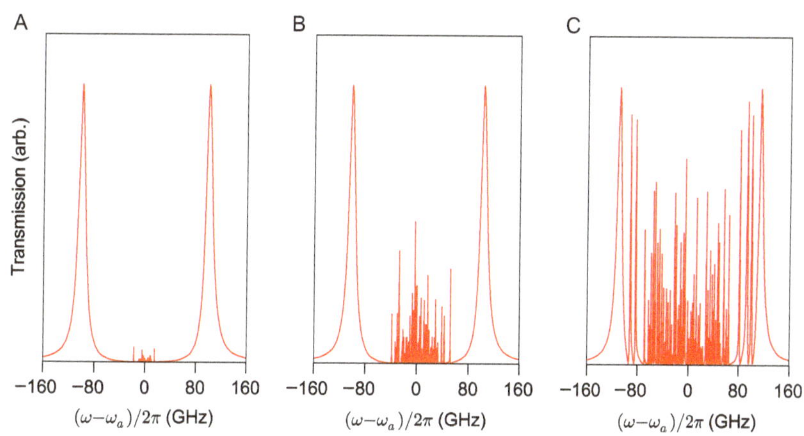


**Fig. 27** Transmission spectra calculated by the second effective Hamiltonian approximation for a $N = 100$ multiemitter system $\{\kappa/2\pi, \Gamma/2\pi, g_{max}/2\pi\} = \{25\text{ GHz}, 0.1\text{ GHz}, 10\text{ GHz}\}$ and emitter frequencies sampled for $\delta/2\pi =$ (A) 10 GHz, (B) 40 GHz, and (C) 80 GHz.

increase in the collective coupling amounting from an increased number of emitters in the system. Despite the large inhomogeneous broadening, collective oscillations as the highest-energy and lowest-energy states are still readily apparent. These collective excitations manifest as two polaritons, each with approximately half the cavity linewidth, that are separated by $2G_N$.

In the next section, we will analyze new opportunities for nonclassical light generation in multiemitter cavity QED systems.

# 5. NONCLASSICAL LIGHT GENERATION WITH MULTIEMITTER CAVITY QED SYSTEMS

As discussed in Section 3, nonclassical light generation in cavity QED systems finds its origin in the discretized and anharmonic character of the energy states. The differences in the dressed ladder of states for an increasing number of emitters are presented in Fig. 20 and compared to the Jaynes–Cummings ladder. These more complicated level structures for multiemitter systems bring in new opportunities for $n$-photon emission.

## 5.1 Resonant Photon Blockade

In previous sections, we only considered photon blockade operated in a pulsed regime. Here, we consider it under continuous-wave (CW) excitation. Emerging technologies for photodetection such as superconducting single-photon detectors are beginning to provide the timing resolution required for the study of CW photon correlations. Because these correlations do not involve averaging over an entire pulse, they can act as a more sensitive probe of the underlying system dynamics.

For the CW case, the second-order coherence $g^{(2)}(0)$ is defined as

$$g^{(2)}(0) = \frac{\langle a^\dagger a^\dagger a a \rangle}{\langle a^\dagger a \rangle^2}, \tag{51}$$

and it is related to the statistics of simultaneously emitting two (or more) photons (Glauber, 1963). Just like for the pulsed statistic $g^{(2)}[0]$, when the CW statistic $g^{(2)}(0)$ is less than one (ideally zero), the light has a sub-Poissonian nonclassical character that indicates its potential for single-photon emission. However, to confirm the single-photon character of the emitted light in the CW case, higher-order coherences need also to be taken into account (Carreño et al., 2016).

Now, we use this tool to study transmission through the collective polaritonic excitations of multiemitter cavity QED systems. Considering the transmission through polaritonic states, we naively expect that the photon blockade effect should become even more pronounced with a $\sqrt{N}$ increase in coupling strength. However, this seems to be the case only while $G_N < \kappa$—we attribute this somewhat unexpected finding to the addition of energy states in the second rung (see Fig. 20) with an increasing number

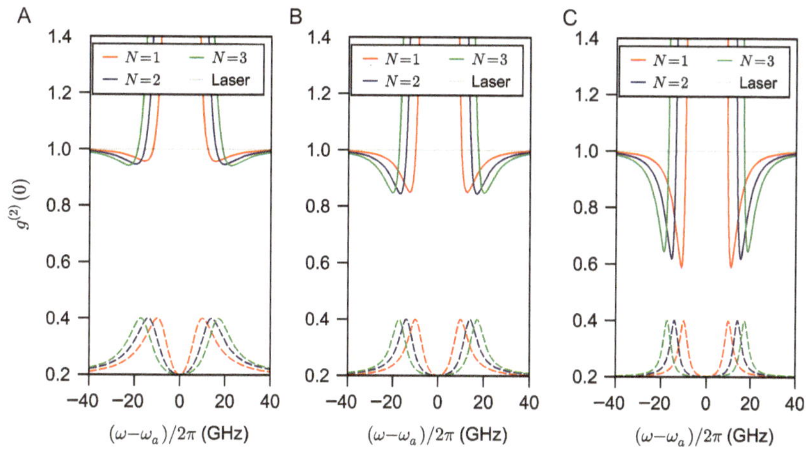


**Fig. 28** Photon blockade trends with an increasing $N$ for parameters $\{\Delta_{e,n}/2\pi, \Gamma/2\pi, g_n/2\pi\}$ = {0 GHz, 0.1 GHz, 10 GHz} and $\kappa/2\pi$ = (A) 25 GHz ($G_N < \kappa$), (B) 15 GHz ($G_N \sim \kappa$), and (C) 8 GHz ($G_N > \kappa$). *Dashed lines* illustrate the transmission spectra.

of emitters. While a detailed investigation is needed for full understanding of this phenomenon, we suggest that the photon blockade effects are greatly influenced by the dipolar coupling mechanisms between the two energy rungs. We present these effects on a system of identical emitters, resonantly and equally coupled to the cavity. Fig. 28 compares second-order coherence $g^{(2)}(0)$ values with an increasing $N$ for three systems whose parameters capture the change in the trend around $G_N \approx \kappa$. As with the single-emitter photon blockade, the points of best single-photon emission are located around the frequencies of the polariton transmission peaks. When $G_N < \kappa$, the photon blockade improves with the increasing $N$, while for $G_N > \kappa$ the behavior reverses. Hence, we have identified a potential fundamental limit in multiemitter photon blockade under resonant conditions, in that high-quality factor cavities may not necessarily yield the best photon blockade in a multiemitter cavity QED system. Notably, this limit is relaxed for detuned photon blockade.

Next, we analyze the effect of unequal coupling to the photon blockade, and compare the second-order coherence values for systems with variable coupling of three emitters (Fig. 29). Starting with a system of identically coupled emitters, we gradually decrease coupling rates of the second and the third emitter, which corresponds to the decrease of the collective coupling rate $G_N$. We observe that the quality of the photon blockade decreases ($g^{(2)}_{\min}(0)$ increases) with the lowering of the collective coupling rate.

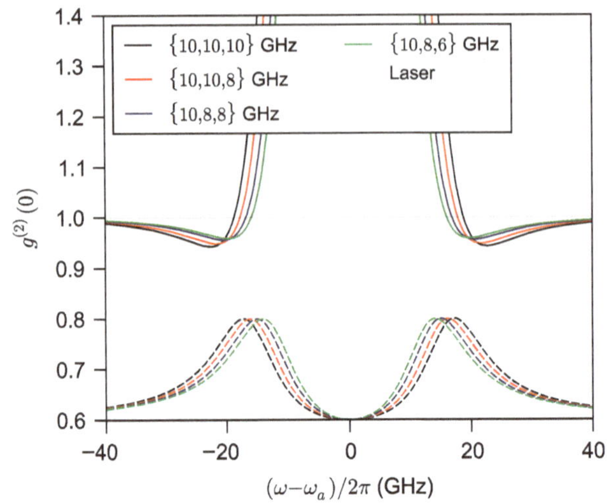


**Fig. 29** Photon blockade trends for $N = 3$, $\{\kappa/2\pi, \Gamma/2\pi, \Delta_{e,n}/2\pi\} = \{25\text{ GHz}, 0.1\text{ GHz}, 0\text{ GHz}\}$ and the variable coupling parameters $\{g_1/2\pi, g_2/2\pi, g_3/2\pi\}$ noted in the legend.

## 5.2 Detuned Photon Blockade in the Multiemitter System

As in the single-emitter case, we expect an enhanced photon blockade for a system where multiple emitters are detuned from the cavity. Here, we focus on a two-emitter system and analyze the minima of $g^{(2)}(0)$ as a function of the emitters' detuning from the cavity. In Fig. 30 we see that the photon blockade improves as the emitters detune from the cavity. For unequally coupled emitters, the trend becomes less pronounced for the lowered $G_N = \sqrt{g_1^2 + g_2^2}$. Peculiarly, we also see an effect that additionally lowers $g_{\min}^{(2)}(0)$ and occurs when both emitters are highly detuned from the cavity and a little detuned from one another. This behavior becomes even more pronounced for unequally coupled emitters providing a lower second-order coherence value, and shifting the $g_{\min}^{(2)}(0)$ further away with the emitters' detuning from the cavity. The effect is asymmetric and favors systems where the most strongly coupled emitter is also the highest detuned one.

Let us look more closely into what causes this additional second-order coherence reduction. Fig. 31A shows the $g^{(2)}(0)$ dependence on the transmission wavelength. We see that the minima of the function occur around the frequencies of the transmission peaks. When the emitters are identical there are only two local minima corresponding to transmission near the polariton peaks. For the nonidentical emitters the second-order coherence has three local minima, with the middle one corresponding to the frequency

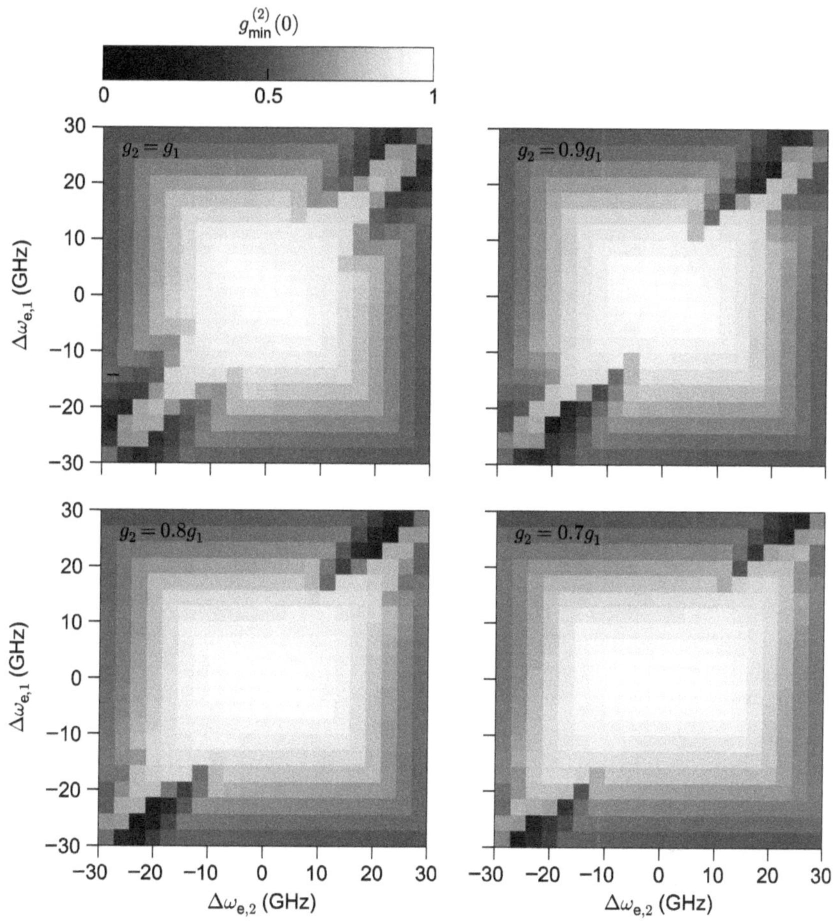


**Fig. 30** Zero-time second-order coherence minima for a detuned system with $N = 2$ emitters and $\{\kappa/2\pi, \Gamma/2\pi, g_1/2\pi\} = \{25\text{ GHz}, 0.1\text{ GHz}, 10\text{ GHz}\}$; $g_2$ is quoted in each panel.

of the emerging peak. For large detuning, transmission through this intermediate peak gains an advantageous second-order coherence value, which corresponds exactly to the peculiar minima in $g^{(2)}(0)$ seen in Fig. 30. The origin of this effect has been traced back to the specifics of the dressed ladder of states (Radulaski et al., 2016a). Time evolutions of the second-order coherence function (for details see Radulaski et al., 2016a) show that the dynamics of the single-photon emission from either of the photon blockade frequencies evolves at the 100 ps scale, which represents a significant speedup in the single-photon emission relative to the expected 10 ns lifetime of individual color-centers.

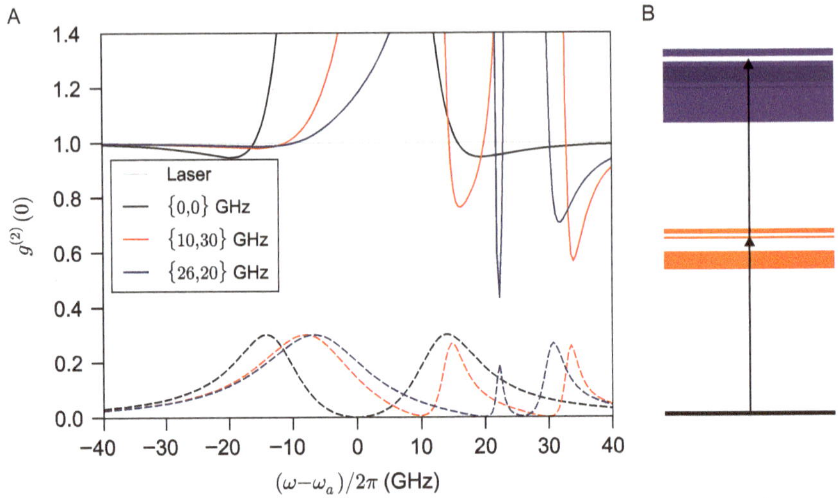


**Fig. 31** (A) Zero-time second-order coherence for $N = 2$, $\{\kappa/2\pi, \Gamma/2\pi, g_{1,2}/2\pi\} = \{25$ GHz, 0.1 GHz, 10 GHz$\}$ and the different emitter detunings $\{\Delta_{e,1}/2\pi, \Delta_{e,2}/2\pi\}$ noted in the legend; plotted with transmission spectra (*dashed lines*). (B) Dressed ladder of states calculated with our effective Hamiltonian approach for $\{\Delta_{e,1}/2\pi, \Delta_{e,2}/2\pi\} = \{26$ GHz, 20 GHz$\}$, plotted with their linewidths; *black vertical arrows* illustrate first and second excitation at the point of the best photon blockade. Energy spacing between the rungs is significantly reduced for clarity.

## 5.3 Effective Hamiltonian Approach to Photon Blockade

The effective Hamiltonian approach can provide an insight into the photon blockade properties of the system. The frequency spacing between the eigenstates in the first and the second rung is calculated by the $H_{\mathrm{EFF}}$ diagonalization previously discussed in Section 4.3. Fig. 31B illustrates the energy levels of the system with 26 and 20 GHz emitter detunings, plotted in red in Fig. 31A. The levels of the first and the second rung are depicted in red and blue, respectively. Each state's width corresponds to its full width at half maximum. The spacing between the rungs is not true to size. The black vertical arrows indicate the frequencies of the best photon blockade (22.6 GHz), owing their performance to the highest frequency clearance from a second photon absorption. Due to the variable linewidth of the states, it is important to consider the widths of the individual levels in the analysis of their frequency overlap. When there is less frequency overlap between subsequent jumps up our extended Tavis–Cummings ladder, the quality of the photon blockade increases.

This approach can be used in more complex systems, where the calculation of quantum master equation is too lengthy, to indicate potentially advantageous parameters for high-quality single-photon emission.

## 5.4 Effective Hamiltonian Approach to N-Photon Generation

Multiemitter cavity QED systems can generate light in the so-called *unconventional photon blockade* regime. Here, the second-order coherence has sub-Poissonian character, while the third-order coherence has super-Poissonian statistics, therefore promoting the generation of three-photon states.

Presenting data from Radulaski et al. (2016a), we show how the effective Hamiltonian approach can be used to identify such regimes. We focus on the system with $N = 2$ emitters,

$$\{\kappa/2\pi, \Gamma/2\pi, g_{1,2}/2\pi, \Delta_{e,1}/2\pi\} = \{25\ \text{GHz}, 0.1\ \text{GHz}, 10\ \text{GHz}, 30\ \text{GHz}\}$$

and variable detuning of the second emitter $\Delta_{e,2}/2\pi \in [30\ \text{GHz}, 40\ \text{GHz}]$. Fig. 32A shows the frequency overlap of the first three rungs of the dressed ladder calculated using $H_{\text{EFF}}$, while Fig. 32B and C shows the second-order and the third-order coherences calculated using the quantum-optical master equation. For the region corresponding to $\omega/2\pi = 35$ GHz, we see that the eigenstate of the third rung has no frequency overlap with states of the first two rungs and should promote three-photon emission. This is confirmed by the coherence functions which show a suppression of the two-photon

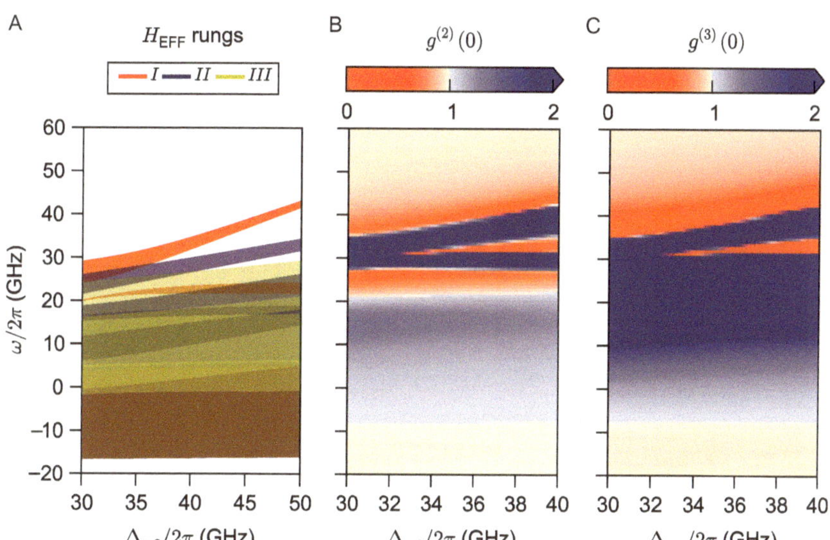


**Fig. 32** (A) Frequency overlap $(E_n - n\omega_a)/n$ where $E_n$ are eigenenergies of $n$-th rung, between three rungs of the dressed ladder for $N = 2$ emitters and $\{\kappa/2\pi, \Gamma/2\pi, g_{1,2}/2\pi, \Delta_{e,1}/2\pi\} = \{25\ \text{GHz}, 0.1\ \text{GHz}, 10\ \text{GHz}, 30\ \text{GHz}\}$. (B) Second-order coherence function and (C) third-order coherence function for the same system. *Data from Radulaski, M., Fischer, K.A., Lagoudakis, K.G., Zhang, J.L., Vuckovic, J., 2016. Nonclassical light generation in two-emitter-cavity systems. arXiv: 1612.03261.*

emission, but an enhancement in the three photon emission as discussed in Radulaski et al. (2016a). These results show that effective Hamiltonian approach can be used to identify parameters advantageous for higher-order $n$-photon generation.

### 5.5 Outlook for Multiemitter Cavity QED

Multiemitter cavity QED at the level of several emitters is close to its first demonstrations in a solid-state platform. Recent developments in low-strain material growth, nanofabrication in bulk substrates, and controlled implantation of color-centers are setting the path for achieving high-quality factor, small mode volume cavities with preferentially positioned nearly identical quasiemitters.

We have set forth an extended Tavis–Cummings model that captures the core dynamics of a multiemitter cavity QED system. The initial theoretical findings on improved photon blockade and photon bundle generation are expected to be expanded and unveil rich physics, even with the inhomogeneous broadening of emerging color-centers. Robust regimes of operation have been predicted, with promising applications in reliable quantum-optical networks. Experimental data will set this theory to a test, and based on the experience with the single-emitter systems, we envision additional inclusion of factors in the quantum model explicitly, such as dephasing and phonon interaction.

The systems with lossy cavities will require a higher number of emitters to reach the strong coupling regime. Our new effective Hamiltonian approach to estimating transmission spectra could provide insights into the dynamics of such systems by finding the best fit to the transmission spectra, identifying individual emitter frequencies, and establishing regimes of operation. From there, the same framework can be used to evaluate frequency overlap between the rungs of excited states and target operating frequencies for nonclassical light generation.

## 6. CONCLUSIONS

In this chapter, we presented the state of the art in the generation of nonclassical states of light using semiconductor cavity QED platforms, focusing in particular on the photon blockade effects that enable the generation of indistinguishable photon streams with high purity and efficiency. InGaAs quantum dots in optical nanocavities have been the leading platform for such experiments for many years and have enabled exciting fundamental science demonstrations (including the first demonstration of photon blockade in solid-state), with direct applications to nonclassical light generation. However, the limitations of these quantum dots, particularly in terms of their random positioning and large inhomogeneous broadening, are impeding their employment in systems requiring more than one emitter. This in turn has generated much interest in systems based on color-centers in group-IV semiconductors—diamond and SiC, which feature very small inhomogenous broadening and even room temperature operation. This facilitates interference of photons emitted from different quantum emitters, and it enables the implementation of multiemitter cavity QED systems that feature richer dressed-states ladder structures and offers opportunities for studying new regimes of photon blockade (Radulaski et al., 2016a). However, the remaining piece of the puzzle is the demonstration of a strongly coupled cavity QED platform based on a few color-centers coupled to a cavity, which should be within reach with today's technologies (Burek et al., 2012; Radulaski et al., 2016a). Once this milestone is achieved, not only will it open the door to implementation of new quantum light sources, but also will create opportunities beyond nonclassical light, including quantum many-body physics simulation (Carusotto and Ciuti, 2013; Greentree et al., 2006; Hartmann et al., 2008).

## ACKNOWLEDGMENTS

The authors thank Kai Müller for productive and helpful discussions in framing the context of this work, and for contributing experimental data to Figs. 13 and 18. The authors are grateful to Tomas Sarmiento and Jingyuan Linda Zhang for providing data to Figs. 1 and 3.

This material is based upon work supported by the Air Force Office of Scientific Research under award number FA9550-17-1-0002 and the National Science Foundation Division of Materials Research, Grant Numbers 1406028 and 1503759. K.A.F. acknowledges support from the Lu Stanford Graduate Fellowship and the National Defense Science and Engineering Graduate Fellowship.

## REFERENCES

Aharonovich, I., Castelletto, S., Simpson, D.A., Stacey, A., McCallum, J., Greentree, A.D., Prawer, S., 2009. Two-level ultrabright single photon emission from diamond nanocrystals. Nano Lett. 9 (9), 3191–3195.

Aharonovich, I., Greentree, A.D., Prawer, S., 2011. Diamond photonics. Nat. Photonics 5 (7), 397–405.

Aharonovich, I., Englund, D., Toth, M., 2016. Solid-state single-photon emitters. Nat. Photonics 10 (10), 631–641.

Akahane, Y., Asano, T., Song, B.S., Noda, S., 2003. High-Q photonic nanocavity in a two-dimensional photonic crystal. Nature 425 (6961), 944–947.

Andreani, L.C., Panzarini, G., Gérard, J.M., 1999. Strong-coupling regime for quantum boxes in pillar microcavities: theory. Phys. Rev. B Condens. Matter 60 (19), 13276–13279.

Babinec, T.M., Hausmann, B.J., Khan, M., Zhang, Y., Maze, J.R., Hemmer, P.R., Lončar, M., 2010. A diamond nanowire single-photon source. Nat. Nanotechnol. 5 (3), 195–199.

Bajcsy, M., Majumdar, A., Rundquist, A., Vučković, J., 2013. Photon blockade with a four-level quantum emitter coupled to a photonic-crystal nanocavity. New J. Phys. 15 (2), 025014.

Bar-Gill, N., Pham, L.M., Jarmola, A., Budker, D., Walsworth, R.L., 2013. Solid-state electronic spin coherence time approaching one second. Nat. Commun. 4, 1743.

Beveratos, A., Brouri, R., Gacoin, T., Villing, A., Poizat, J.P., Grangier, P., 2002. Single photon quantum cryptography. Phys. Rev. Lett. 89 (18), 187901.

Beveratos, A., Kühn, S., Brouri, R., Gacoin, T., Poizat, J.P., Grangier, P., 2002. Room temperature stable single-photon source. Eur. Phys. J. D 18 (2), 191–196.

Birnbaum, K.M., Boca, A., Miller, R., Boozer, A.D., Northup, T.E., Kimble, H.J., 2005. Photon blockade in an optical cavity with one trapped atom. Nature 436 (7047), 87–90.

Bracher, D.O., Zhang, X., Hu, E.L., 2017. Selective Purcell enhancement of two closely linked zero-phonon transitions of a silicon carbide color center. Proc. Natl. Acad. Sci. 114 (16), 4060.

Bradac, C., Gaebel, T., Naidoo, N., Sellars, M.J., Twamley, J., Brown, L.J., Barnard, A.S., Plakhotnik, T., Zvyagin, A.V., Rabeau, J.R., 2010. Observation and control of blinking nitrogen-vacancy centres in discrete nanodiamonds. Nat. Nanotechnol. 5 (5), 345–349.

Buckley, S., Rivoire, K., Vučković, J., 2012. Engineered quantum dot single-photon sources. Rep. Prog. Phys. 75 (12), 126503.

Burek, M.J., de Leon, N.P., Shields, B.J., Hausmann, B.J., Chu, Y., Quan, Q., Zibrov, A.S., Park, H., Lukin, M.D., Loncar, M., 2012. Free-standing mechanical and photonic nanostructures in single-crystal diamond. Nano Lett. 12 (12), 6084–6089.

Carmichael, H., 2009. An Open Systems Approach to Quantum Optics: Lectures Presented at the Université Libre de Bruxelles, October 28 to November 4, 1991. Lecture Notes in Physics Monographs, Springer, Berlin, Heidelberg.

Carreño, J.C.L., Casalengua, E.Z., del Valle, E., Laussy, F.P., 2016. Criterion for single photon sources. arXiv: 1610.06126.

Carter, S.G., Sweeney, T.M., Kim, M., Kim, C.S., Solenov, D., Economou, S.E., Reinecke, T.L., Yang, L., Bracker, A.S., Gammon, D., 2013. Quantum control of a spin qubit coupled to a photonic crystal cavity. Nat. Photonics 7 (4), 329–334.

Carusotto, I., Ciuti, C., 2013. Quantum fluids of light. Rev. Mod. Phys. 85 (1), 299.

Castelletto, S., Johnson, B.C., Ivády, V., Stavrias, N., Umeda, T., Gali, A., Ohshima, T., 2013. A silicon carbide room-temperature single-photon source. Nat. Mater. 13 (2), 151–156.

Castelletto, S., Johnson, B.C., Zachreson, C., Beke, D., Balogh, I., Ohshima, T., Aharonovich, I., Gali, A., 2014. Room temperature quantum emission from cubic silicon carbide nanoparticles. ACS Nano 8 (8), 7938–7947.

Childress, L., Dutt, M.G., Taylor, J., Zibrov, A., Jelezko, F., Wrachtrup, J., Hemmer, P., Lukin, M., 2006. Coherent dynamics of coupled electron and nuclear spin qubits in diamond. Science 314 (5797), 281–285.

Cirac, J., Zoller, P., Kimble, H., Mabuchi, H., 1997. Quantum state transfer and entanglement distribution among distant nodes in a quantum network. Phys. Rev. Lett. 78 (16), 3221–3224.

Cui, S., Zhang, X., Liu, T., Lee, J., Bracher, D., Ohno, K., Awschalom, D., Hu, E.L., 2015. Hybrid plasmonic photonic crystal cavity for enhancing emission from near-surface nitrogen vacancy centers in diamond. ACS Photonics 2 (4), 465–469.

Davanço, M., Hellberg, C.S., Ates, S., Badolato, A., Srinivasan, K., 2014. Multiple time scale blinking in InAs quantum dot single-photon sources. Phys. Rev. B Condens. Matter 89 (16), 161303.

Davies, G., 1974. Vibronic spectra in diamond. J. Phys. C Solid State Phys. 7 (20), 3797.

Dietrich, C.P., Fiore, A., Thompson, M.G., Kamp, M., Höfling, S., 2016. GaAs integrated quantum photonics: towards compact and multi-functional quantum photonic integrated circuits. Laser Photonics Rev. 10 (6), 857–1061.

Diniz, I., Portolan, S., Ferreira, R., Gérard, J., Bertet, P., Auffeves, A., 2011. Strongly coupling a cavity to inhomogeneous ensembles of emitters: potential for long-lived solid-state quantum memories. Phys. Rev. A 84 (6), 063810.

Dory, C., Fischer, K.A., Müller, K., Lagoudakis, K.G., Sarmiento, T., Rundquist, A., Zhang, J.L., Kelaita, Y., Sapra, N.V., Vučković, J., 2017. Tuning the photon statistics of a strongly coupled nanophotonic system. Phys. Rev., A 95, 023804.

Ellis, B., Mayer, M.A., Shambat, G., Sarmiento, T., Harris, J., Haller, E.E., Vučković, J., 2011. Ultralow-threshold electrically pumped quantum-dot photonic-crystal nanocavity laser. Nat. Photonics 5 (5), 297–300.

Englund, D., Faraon, A., Fushman, I., Stoltz, N., Petroff, P., Vucković, J., 2007. Controlling cavity reflectivity with a single quantum dot. Nature 450 (7171), 857–861.

Englund, D., Shields, B., Rivoire, K., Hatami, F., Vučković, J., Park, H., Lukin, M.D., 2010. Deterministic coupling of a single nitrogen vacancy center to a photonic crystal cavity. Nano Lett. 10 (10), 3922–3926.

Falk, A.L., Buckley, B.B., Calusine, G., Koehl, W.F., Dobrovitski, V.V., Politi, A., Zorman, C.A., Feng, P.X.L., Awschalom, D.D., 2013. Polytype control of spin qubits in silicon carbide. Nat. Commun. 4, 1819.

Faraon, A., Englund, D., Fushman, I., Vučković, J., Stoltz, N., Petroff, P., 2007. Local quantum dot tuning on photonic crystal chips. Appl. Phys. Lett. 90 (21), 213110.

Faraon, A., Fushman, I., Englund, D., Stoltz, N., Petroff, P., Vučković, J., 2008. Coherent generation of non-classical light on a chip via photon-induced tunnelling and blockade. Nat. Phys. 4 (11), 859–863.

Fink, J., Bianchetti, R., Baur, M., Göppl, M., Steffen, L., Filipp, S., Leek, P., Blais, A., Wallraff, A., 2009. Dressed collective qubit states and the Tavis-Cummings model in circuit QED. Phys. Rev. Lett. 103 (8), 083601.

Fischer, K.A., Müller, K., Lagoudakis, K.G., Vučković, J., 2016a. Dynamical modeling of pulsed two-photon interference. New J. Phys. 18 (11), 113053.

Fischer, K.A., Müller, K., Rundquist, A., Sarmiento, T., Piggott, A.Y., Kelaita, Y., Dory, C., Lagoudakis, K.G., Vučković, J., 2016b. Self-homodyne measurement of a dynamic Mollow triplet in the solid state. Nat. Photonics 10, 163–166.

Fischer, K.A., Kelaita, Y.A., Sapra, N.V., Dory, C., Lagoudakis, K.G., Müller, K., Vučković, J., 2017. On-chip architecture for self-homodyned nonclassical light. Phys. Rev. A 7, 044002.

Fitchen, D., Fowler, W.B., 1968. Physics of Color Centers. Academic Press, New York, London, p. 293.

Fuchs, F., Stender, B., Trupke, M., Simin, D., Pflaum, J., Dyakonov, V., Astakhov, G., 2015. Engineering near-infrared single-photon emitters with optically active spins in ultrapure silicon carbide. Nat. Commun. 6, 7578.

Fushman, I., Englund, D., Faraon, A., Stoltz, N., Petroff, P., Vuckovic, J., 2008. Controlled phase shifts with a single quantum dot. Science 320 (5877), 769–772.

Garraway, B.M., 1997a. Decay of an atom coupled strongly to a reservoir. Phys. Rev. A 55 (6), 4636–4639.

Garraway, B.M., 1997b. Nonperturbative decay of an atomic system in a cavity. Phys. Rev. A 55 (3), 2290–2303.

Glauber, R.J., 1963. The quantum theory of optical coherence. Phys. Rev. 130 (6), 2529–2539.

Gould, M., Schmidgall, E.R., Dadgostar, S., Hatami, F., Fu, K.M.C., 2016. Efficient extraction of zero-phonon-line photons from single nitrogen-vacancy centers in an integrated GaP-on-Diamond platform. Phys. Rev. A 6, 011001.

Greentree, A.D., Tahan, C., Cole, J.H., Hollenberg, L.C., 2006. Quantum phase transitions of light. Nat. Phys. 2 (12), 856–861.

Hartmann, M.J., Brandao, F.G., Plenio, M.B., 2008. Quantum many-body phenomena in coupled cavity arrays. Laser Photonics Rev. 2 (6), 527–556.

Kim, H., Shen, T.C., Roy-Choudhury, K., Solomon, G.S., Waks, E., 2014. Resonant interactions between a Mollow triplet sideband and a strongly coupled cavity. Phys. Rev. Lett. 113 (2), 027403.

Kimble, H.J., 2008. The quantum internet. Nature 453 (7198), 1023–1030.

Koehl, W.F., Buckley, B.B., Heremans, F.J., Calusine, G., Awschalom, D.D., 2011. Room temperature coherent control of defect spin qubits in silicon carbide. Nature 479 (7371), 84–87.

Kucsko, G., Maurer, P., Yao, N.Y., Kubo, M., Noh, H., Lo, P., Park, H., Lukin, M.D., 2013. Nanometre-scale thermometry in a living cell. Nature 500 (7460), 54–58.

Kuhlmann, A.V., Houel, J., Brunner, D., Ludwig, A., Reuter, D., Wieck, A.D., Warburton, R.J., 2013. A dark-field microscope for background-free detection of resonance fluorescence from single semiconductor quantum dots operating in a set-and-forget mode. Rev. Sci. Instrum. 84 (7), 073905.

Kuhlmann, A.V., Prechtel, J.H., Houel, J., Ludwig, A., Reuter, D., Wieck, A.D., Warburton, R.J., 2015. Transform-limited single photons from a single quantum dot. Nat. Commun. 6, 8204.

Kurtsiefer, C., Mayer, S., Zarda, P., Weinfurter, H., 2000. Stable solid-state source of single photons. Phys. Rev. Lett. 85 (2), 290.

Kuruma, K., Ota, Y., Kakuda, M., Takamiya, D., Iwamoto, S., Arakawa, Y., 2016. Position dependent optical coupling between single quantum dots and photonic crystal nanocavities. Appl. Phys. Lett. 109 (7), 071110.

Laucht, A., Hofbauer, F., Hauke, N., Angele, J., Stobbe, S., Kaniber, M., Böhm, G., Lodahl, P., Amann, M., Finley, J., 2009. Electrical control of spontaneous emission and strong coupling for a single quantum dot. New J. Phys. 11 (2), 023034.

Laussy, F.P., del Valle, E., Tejedor, C., 2008. Strong coupling of quantum dots in microcavities. Phys. Rev. Lett. 101 (8), 083601.

Laussy, F.P., del Valle, E., Schrapp, M., Laucht, A., Finley, J.J., 2012. Climbing the Jaynes–Cummings ladder by photon counting. J. Nanophotonics 6 (1), 061803–061803.

Linskens, A.F., Holleman, I., Dam, N., Reuss, J., 1996. Two-photon Rabi oscillations. Phys. Rev. A 54 (6), 4854–4862.

Lodahl, P., Mahmoodian, S., Stobbe, S., 2015. Interfacing single photons and single quantum dots with photonic nanostructures. Rev. Mod. Phys. 87 (2), 347.

Lohrmann, A., Castelletto, S., Klein, J., Ohshima, T., Bosi, M., Negri, M., Lau, D., Gibson, B., Prawer, S., McCallum, J., et al., 2016. Activation and control of visible single defects in 4H-, 6H-, and 3C-SiC by oxidation. Appl. Phys. Lett. 108 (2), 021107.

Majumdar, A., Bajcsy, M., Rundquist, A., Kim, E., Vučković, J., 2012. Phonon-mediated coupling between quantum dots through an off-resonant microcavity. Phys. Rev. B Condens. Matter 85 (19), 195301.

Majumdar, A., Kaer, P., Bajcsy, M., Kim, E.D., Lagoudakis, K.G., Rundquist, A., Vučković, J., 2013. Proposed coupling of an electron spin in a semiconductor quantum dot to a nanosize optical cavity. Phys. Rev. Lett. 111 (2), 027402.

McCutcheon, M.W., Lončar, M., 2008. Design of a silicon nitride photonic crystal nanocavity with a Quality factor of one million for coupling to a diamond nanocrystal. Opt. Express 16 (23), 19136–19145.

Michler, P., 2009. Single Semiconductor Quantum Dots. Springer, Berlin.

Mouradian, S.L., Schröder, T., Poitras, C.B., Li, L., Goldstein, J., Chen, E.H., Walsh, M., Cardenas, J., Markham, M.L., Twitchen, D.J., et al., 2015. Scalable integration of long-lived quantum memories into a photonic circuit. Phys. Rev. X 5 (3), 031009.

Müller, K., Fischer, K.A., Rundquist, A., Dory, C., Lagoudakis, K.G., Sarmiento, T., Kelaita, Y.A., Borish, V., Vučković, J., 2015a. Ultrafast Polariton-Phonon dynamics of strongly coupled quantum Dot-Nanocavity systems. Phys. Rev. X 5 (3), 031006.

Müller, K., Rundquist, A., Fischer, K.A., Sarmiento, T., Lagoudakis, K.G., Kelaita, Y.A., Sánchez Muñoz, C., del Valle, E., Laussy, F.P., Vučković, J., 2015b. Coherent generation of nonclassical light on chip via detuned photon blockade. Phys. Rev. Lett. 114 (23), 233601.

Müller, K., Fischer, K.A., Dory, C., Sarmiento, T., Lagoudakis, K.G., Rundquist, A., Kelaita, Y.A., Vučković, J., 2016. Self-homodyne-enabled generation of indistinguishable photons. Optica 3 (9), 931–936.

Neu, E., Steinmetz, D., Riedrich-Möller, J., Gsell, S., Fischer, M., Schreck, M., Becher, C., 2011. Single photon emission from silicon-vacancy colour centres in chemical vapour deposition nano-diamonds on iridium. New J. Phys. 13 (2), 025012.

Neuzner, A., Körber, M., Morin, O., Ritter, S., Rempe, G., 2016. Interference and dynamics of light from a distance-controlled atom pair in an optical cavity. Nat. Photonics 10 (5), 303–306.

O'Brien, J.L., Furusawa, A., Vučković, J., 2009. Photonic quantum technologies. Nat. Photonics 3 (12), 687–695.

Ota, Y., Kumagai, N., Ohkouchi, S., Shirane, M., Nomura, M., Ishida, S., Iwamoto, S., Yorozu, S., Arakawa, Y., 2009. Investigation of the spectral triplet in strongly coupled quantum dot–nanocavity system. Appl. Phys. Express 2 (12), 122301.

Radulaski, M., Fischer, K.A., Lagoudakis, K.G., Zhang, J.L., Vuckovic, J., 2016a. Nonclassical light generation in two-emitter-cavity systems. arXiv: 1612.03261.

Radulaski, M., Tzeng, Y.K., Zhang, J.L., Lagoudakis, K.G., Ishiwata, H., Dory, C., Alassaad, K., Ferro, G., Shen, Z.X., Melosh, N., Chu, S., Vuckovic, J., 2016b. Hybrid microresonator enhanced emission from silicon-vacancy and chromium-related color centers in diamond. arXiv: 1610.03183.

Radulaski, M., Widmann, M., Niethammer, M., Zhang, J.L., Lee, S.Y., Rendler, T., Lagoudakis, K.G., Son, N.T., Janzen, E., Ohshima, T., et al., 2017. Scalable quantum photonics with single color centers in silicon carbide. Nano Lett. 17 (3), 1782–1786.

Reinhard, A., Volz, T., Winger, M., Badolato, A., Hennessy, K.J., Hu, E.L., Imamoğlu, A., 2011. Strongly correlated photons on a chip. Nat. Photonics 6 (2), 93–96.

Reithmaier, J.P., Sek, G., Löffler, A., Hofmann, C., Kuhn, S., Reitzenstein, S., Keldysh, L.V., Kulakovskii, V.D., Reinecke, T.L., Forchel, A., 2004. Strong coupling in a single quantum dot-semiconductor microcavity system. Nature 432 (7014), 197–200.

Riedrich-Möller, J., Arend, C., Pauly, C., Mücklich, F., Fischer, M., Gsell, S., Schreck, M., Becher, C., 2014. Deterministic coupling of a single silicon-vacancy color center to a photonic crystal cavity in diamond. Nano Lett. 14 (9), 5281–5287.

Rogers, L.J., Jahnke, K.D., Teraji, T., Marseglia, L., Müller, C., Naydenov, B., Schauffert, H., Kranz, C., Isoya, J., McGuinness, L.P., et al., 2014. Multiple intrinsically identical single-photon emitters in the solid state. Nat. Commun. 4739.

Roy, C., Hughes, S., 2011. Influence of Electron–Acoustic-Phonon scattering on intensity power broadening in a coherently driven Quantum-Dot–Cavity system. Phys. Rev. X 1 (2), 021009.

Roy, C., Kim, H., Waks, E., Hughes, S., 2012. Anomalous phonon-mediated damping of a driven quantum dot embedded in a high-Q semiconductor microcavity. Photonics Nanostruct. Fundam. Appl. 10 (4), 359–368.
Rundquist, A., Bajcsy, M., Majumdar, A., Sarmiento, T., Fischer, K., Lagoudakis, K.G., Buckley, S., Piggott, A.Y., Vučković, J., 2014. Nonclassical higher-order photon correlations with a quantum dot strongly coupled to a photonic-crystal nanocavity. Phys. Rev. A 90 (2), 023846.
Saddow, S.E., 2012. Silicon Carbide Biotechnology: A Biocompatible Semiconductor for Advanced Biomedical Devices and Applications. Elsevier, Boston.
Sanders, B., Vuckovic, J., Grangier, P., 2005. Single photons on demand. Europhys. News 36 (2), 56–58.
Santori, C., 2003. Generation of Nonclassical Light Using Semiconductor Quantum Dots (Ph.D. thesis). Stanford University.
Santori, C., Fattal, D., Vučković, J., Solomon, G.S., Yamamoto, Y., 2002. Indistinguishable photons from a single-photon device. Nature 419 (6907), 594–597.
Santori, C., Fattal, D., Vučković, J., Solomon, G.S., Waks, E., Yamamoto, Y., 2004. Submicrosecond correlations in photoluminescence from InAs quantum dots. Phys. Rev. B 69 (20), 205324.
Santori, C., Fattal, D., Fu, K.M.C., Barclay, P.E., Beausoleil, R.G., 2009. On the indistinguishability of Raman photons. New J. Phys. 11 (12), 123009.
Shore, B.W., Knight, P.L., 1993. The Jaynes-Cummings model. J. Mod. Opt. 40 (7), 1195–1238.
Sipahigil, A., Jahnke, K.D., Rogers, L.J., Teraji, T., Isoya, J., Zibrov, A.S., Jelezko, F., Lukin, M.D., 2014. Indistinguishable photons from separated silicon-vacancy centers in diamond. Phys. Rev. Lett. 113 (11), 113602.
Sipahigil, A., Evans, R.E., Sukachev, D.D., Burek, M.J., Borregaard, J., Bhaskar, M.K., Nguyen, C.T., Pacheco, J.L., Atikian, H.A., Meuwly, C., Camacho, R.M., Jelezko, F., Bielejec, E., Park, H., Lončar, M., Lukin, M.D., 2016. An integrated diamond nanophotonics platform for quantum-optical networks. Science 354 (6314), 847–850. ISSN 0036–8075.
Siyushev, P., Pinto, H., Vörös, M., Gali, A., Jelezko, F., Wrachtrup, J., 2013. Optically controlled switching of the charge state of a single nitrogen-vacancy center in diamond at cryogenic temperatures. Phys. Rev. Lett. 110 (16), 167402.
Sörman, E., Son, N., Chen, W., Kordina, O., Hallin, C., Janzén, E., 2000. Silicon vacancy related defect in 4H and 6H SiC. Phys. Rev. B 61 (4), 2613.
Steck, D.A., 2017. Quantum and Atom Optics. available online at http://steck.us/teaching (revision 0.11.6, 24 February 2017).
Sweeney, T.M., Carter, S.G., Bracker, A.S., Kim, M., Kim, C.S., Yang, L., Vora, P.M., Brereton, P.G., Cleveland, E.R., Gammon, D., 2014. Cavity-stimulated Raman emission from a single quantum dot spin. Nat. Photonics 8 (6), 442–447.
Tang, J., Geng, W., Xu, X., 2015. Quantum interference induced photon blockade in a coupled single quantum dot-cavity system. Sci. Rep. 5, 9252.
Tavis, M., Cummings, F.W., 1968. Exact solution for an N-molecule radiation-field Hamiltonian. Phys. Rev. 170 (2), 379.
Thompson, R., Rempe, G., Kimble, H., 1992. Observation of normal-mode splitting for an atom in an optical cavity. Phys. Rev. Lett. 68 (8), 1132.
Tian, L., Carmichael, H.J., 1992. Incoherent excitation of the Jaynes-Cummings system. Quantum Opt. 4 (2), 131.
Vagov, A., Croitoru, M.D., Glässl, M., Axt, V.M., Kuhn, T., 2011. Real-time path integrals for quantum dots: quantum dissipative dynamics with superohmic environment coupling. Phys. Rev. B Condens. Matter 83 (9), 094303.
Vahala, K.J., 2003. Optical microcavities. Nature 424 (6950), 839–846.

Vasco, J.P., Vinck-Posada, H., Valentim, P.T., Guimãraes, P.S.S., 2013. Modeling of Fano resonances in the reflectivity of photonic crystal cavities with finite spot size excitation. Opt. Express 21 (25), 31336–31346.

Volz, T., Reinhard, A., Winger, M., Badolato, A., Hennessy, K.J., Hu, E.L., Imamoğlu, A., 2012. Ultrafast all-optical switching by single photons. Nat. Photonics 6 (9), 605–609.

Vora, P.M., Bracker, A.S., Carter, S.G., Sweeney, T.M., Kim, M., Kim, C.S., Yang, L., Brereton, P.G., Economou, S.E., Gammon, D., 2015. Spin-cavity interactions between a quantum dot molecule and a photonic crystal cavity. Nat. Commun. 6, 7665.

Waks, E., Vuckovic, J., 2006. Dipole induced transparency in drop-filter cavity-waveguide systems. Phys. Rev. Lett. 96 (15), 153601.

Warburton, R.J., Schaflein, C., Haft, D., Bickel, F., Lorke, A., Karrai, K., Garcia, J.M., Schoenfeld, W., Petroff, P.M., 2000. Optical emission from a charge-tunable quantum ring. Nature 405 (6789), 926–929.

Weber, J., Koehl, W., Varley, J., Janotti, A., Buckley, B., Van de Walle, C., Awschalom, D.D., 2010. Quantum computing with defects. Proc. Natl. Acad. Sci. 107 (19), 8513–8518.

Widmann, M., Lee, S.Y., Rendler, T., Son, N.T., Fedder, H., Paik, S., Yang, L.P., Zhao, N., Yang, S., Booker, I., et al., 2015. Coherent control of single spins in silicon carbide at room temperature. Nat. Mater. 14 (2), 164–168.

Wolters, J., Schell, A.W., Kewes, G., Nüsse, N., Schoengen, M., Döscher, H., Hannappel, T., Löchel, B., Barth, M., Benson, O., 2010. Enhancement of the zero phonon line emission from a single nitrogen vacancy center in a nanodiamond via coupling to a photonic crystal cavity. Appl. Phys. Lett. 97 (14), 141108.

Yoshie, T., Scherer, A., Hendrickson, J., Khitrova, G., Gibbs, H.M., Rupper, G., Ell, C., Shchekin, O.B., Deppe, D.G., 2004. Vacuum Rabi splitting with a single quantum dot in a photonic crystal nanocavity. Nature 432 (7014), 200–203.

Yuan, Z., Kardynal, B.E., Stevenson, R.M., Shields, A.J., Lobo, C.J., Cooper, K., Beattie, N.S., Ritchie, D.A., Pepper, M., 2002. Electrically driven single-photon source. Science 295 (5552), 102–105.

Zhang, J.L., Ishiwata, H., Babinec, T.M., Radulaski, M., Müller, K., Lagoudakis, K.G., Dory, C., Dahl, J., Edgington, R., Soulière, V., et al., 2015. Hybrid group IV nanophotonic structures incorporating diamond silicon-vacancy color centers. Nano Lett. 16 (1), 212–217.

Zhong, T., Kindem, J.M., Rochman, J., Faraon, A., 2017. On-chip storage of broadband photonic qubits in a cavity-protected rare-earth ensemble. Nat. Commun. 8, 14107.

CHAPTER FOUR

# Trapping Atoms With Radio Frequency Adiabatic Potentials

**Hélène Perrin*,[1], Barry M. Garraway†**
*CNRS, UMR 7538, Université Paris 13, Sorbonne Paris Cité, Laboratoire de physique des lasers, Villetaneuse, France
†University of Sussex, Brighton, United Kingdom
[1]Corresponding author: e-mail address: helene.perrin@univ-paris13.fr

## Contents



*Advances in Atomic, Molecular, and Optical Physics*, Volume 66
ISSN 1049-250X
http://dx.doi.org/10.1016/bs.aamop.2017.03.002

## Abstract

In this chapter we review the field of radio frequency dressed atom trapping. We emphasize the role of adiabatic potentials and give simple but generic models of electromagnetic fields that currently produce traps for atoms at microkelvin temperatures and below. This chapter aims to be didactic and starts with general descriptions of the essential ingredients of adiabaticity and magnetic resonance. As examples of adiabatic potentials we pay attention to radio frequency dressing in both the quadrupole trap and the Ioffe–Pritchard trap. We include a description of the effect of different choices of radio frequency polarization and orientations or alignment. We describe how the adiabatic potentials, formed from radio frequency fields, can themselves be probed and manipulated with additional radio frequency fields including multiphoton-effects. We include a description of time-averaged adiabatic potentials. Practical issues for the construction of radio frequency adiabatic potentials are addressed including noise, harmonics, and beyond rotating wave approximation effects.

## 1. INTRODUCTION

The manipulation of the external motion of atoms and molecules with light or magnetic fields represents one of the great successes of atomic physics in the past decades (Cohen-Tannoudji and Guéry-Odelin, 2011) and has led to major discoveries acknowledged by several Nobel prizes. Laser cooling and trapping (Chu, 1998; Cohen-Tannoudji, 1998; Phillips, 1998) gives access to a sample of billions of neutral particles, isolated from the environment, at temperatures in the microkelvin range. This results in a suppression of the Doppler effect, and cold atom samples have found their first applications in high precision measurements, in particular atomic clocks (Bize et al., 2005) and atom interferometry (Cronin et al., 2009), including acceleration and rotation sensors (Tin, 2014). Thus the subject also now plays its role in the emergence of quantum technologies.

At these low temperatures the laser cooled atoms may then be confined in trapping potentials resulting from optical or magnetic fields. While they would be immediately lost should they touch a normal physical container, we can avoid this problem by using, for example, magnetic fields, which allows us to keep the atoms confined for up to several minutes. Starting from this point, and applying the technique of evaporative cooling, which consists in the elimination of the most energetic particles confined in such conservative traps, has led to the first observation of Bose–Einstein condensation in dilute atomic gases by Cornell and Wieman (2002) and Ketterle (2002). This achievement has opened the now mature research field of ultracold quantum gases.

The study of ultracold atomic gases has given rich insights into many physical phenomena including superfluidity (Leggett, 2006), low-dimensional physics such as the Berezinskii–Kosterlitz–Thouless transition, many-body physics (Bloch et al., 2008), and much more. This is made possible by the control over most of the relevant physical parameters. These include the temperature, the interaction strength, the dimension of the system, and the confinement geometry. A famous example is the confinement of ultracold atoms in optical lattices (Bloch, 2005), which opens up the way to the quantum simulation of the electronic properties in solid state systems with ultracold atoms in periodic lattices as a model system (Bloch et al., 2012).

Both in the contexts of quantum sensors and quantum simulators, it becomes crucial to control precisely the confinement geometry of the sample to something far from the simple case of the harmonic trap. Optical lattices provide a well-established way to realize periodic potentials, but there is also a need for other geometries such as double-wells, flat traps for two-dimensional gases, or ring traps. Moreover, the dynamical control of the trapping potential is essential, for example, in the study of out-of-equilibrium dynamics in these systems. Adiabatic potentials arising from the dressing of atoms with a radio frequency field provide widely tunable potentials for these purposes.

Radio frequency (rf) fields are already routinely used in quantum gas experiments for the evaporative cooling stage in magnetic traps (Ketterle and van Druten, 1996). An rf "knife" is applied over a time of order a few seconds to induce a change in the atomic internal state and remove the most energetic atoms so that, on average, the remaining atoms have a lower average energy, leading to lower temperatures. The transfer of an energetic atom to an untrapped state is efficient if the rf amplitude is large enough, and this can be seen as a direct modification of the trapping potential due to the presence of the rf field. This rf field can then have its frequency continuously changed during the process, to adapt to the situation of fewer and colder trapped atoms. In effect then, we use our control of rf radiation to in turn control a microscopic system at ultracold temperatures. These basic ideas are also used for a next-level control of atomic systems through adiabatic potentials.

In this chapter we give an overview of an approach that broadens the scenario of rf evaporation to a much wider range of situations with rf adiabatic potentials that allow more complex trapping potentials needed for quantum simulation, atomtronics (Amico et al., 2016), and sensing systems. We will recall that we can control frequency, amplitude, polarization, and the timing of rf radiation. We can have multiple rf fields. All this control will allow us to fine-tune atomic potentials, such as barriers, without negative effects. It will allow us to split Bose–Einstein condensates for interferometry, even when they are close (microns) to noisy surfaces such as those found on atom chips. It will allow us to create ring traps for rotation physics and for sensing, dumb-bell traps, bubble traps, 2D traps, and other exotic atom traps for the study of cold atomic gas in new topologies.

In order to get to grips with the adiabatic potentials in this chapter, we first introduce the basic ideas of adiabaticity in Section 2. This will present the ideas that lead to a robust potential that can be modified by the experimentalist, and which the atoms follow as they move through space. To illustrate the approach to adiabaticity, we will include a classic nonadiabatic scenario: the Landau–Zener crossing. Then, in Section 3, we will discuss magnetic resonance in order to give the fundamentals of the interaction of our atoms with the rf field. In Section 4 we combine the concepts of Sections 2 and 3 to introduce the adiabatic potentials that are the core of the chapter. This includes a discussion of polarization and "loading" (i.e., the method used to transfer cold atoms into the trap). At this point the reader will have the needed basic theory and Section 5 discusses significant examples of adiabatic potentials arising from some standard field configurations: i.e., from the Ioffe–Pritchard trap (Section 5.1) and from the 3D quadrupole trap (Section 5.2). We include here (Section 5.3) a discussion on ring traps as this is an area of intriguing variety and potential applications.

The basic ideas and some examples are covered before the next Sections 6 and 7 which look at some more complex ways of creating adiabatic potentials with even more variety. Section 6 introduces the idea of modifying adiabatic potentials by moving and oscillating them in space at a frequency faster than the mechanical response of the atom. The resulting effective potential is governed by a wider set of parameters and has a new form which can allow us to make potentials that would not be allowed by Maxwell's equations applied to a static magnetic field. We give a didactic example in the section. Our second expansion of the basic ideas examines possibilities with multiple rf fields in Section 7. This can involve one rf field used to make an adiabatic potential, and another one probing it. This can be used for spectroscopy of the trap. Equally, a second rf field can be used to cool down an rf-dressed trap (rf evaporative cooling again). Or, three different frequencies can be used to make a double-well potential (or a double shell), and this idea can be extended to more potential wells with additional fields at different frequencies. Finally, we look at some practical matters in Section 8, that is, issues of trap lifetime, strong rf and alignment of polarization directions, and then we conclude in Section 9.

# 2. BASIC CONCEPTS

## 2.1 Concept of Adiabaticity

If you have ever heard a violin or a guitar make a sliding note just from the slide of the fingers on the neck of the instrument you will have an idea of a classical version of adiabatic following. In the instrument case, an eigenmode (or modes) of the string are excited, and subsequently the properties of that mode are changed by moving the finger at the boundary. In particular the frequency of the mode changes, and provided the change in eigenmode pitch is not too rapid, the excitation, e.g., of a fundamental, *follows* the mode. That is, if we started with a fundamental excitation at low frequency, we obtain a fundamental excitation at a higher frequency after shortening the string. This would not be true if the slide is too quick: in that case, starting with only a fundamental excitation, we would obtain excited harmonics after the slide. The situation is very similar for a quantum system as we see in Section 2.3. This is best exemplified with a two-state quantum system such as found for many simplified atomic systems, or for a spin-half particle undergoing an interaction. (We will examine systems with more than two levels in the next sections where atoms interacting with magnetic fields naturally provide multilevel systems with an angular momentum quantum number $F = 1/2, 1, 3/2, 2, 5/2, 3, \ldots$.) The two-level system then can be parameterized by a Hamiltonian of the form

$$\hat{H}(t) = \begin{bmatrix} -\alpha(t) & \beta(t) \\ \beta(t) & \alpha(t) \end{bmatrix}. \tag{1}$$

Here the parameter $\beta(t)$ represents the coupling between two energy levels which have energies $\mp\,\alpha(t)$. Everything is allowed to be time dependent in the Hamiltonian (which immediately makes the problem hard to solve exactly, apart from some special cases). We do not include here a constant energy term; if we do, it affects both levels in the same way and it just adds a time-dependent phase factor to all the states. In the example of the magnetic spin problem, the parameter $\alpha(t)$ might arise from the interaction of the magnetic dipole with a (slowly) time-varying magnetic field, and the parameter $\beta(t)$ might arise from the additional interaction with an rf field, which causes a coupling of the Zeeman states with energies $\pm\,\alpha(t)$. The parameter $\beta(t)$ might also have a time-varying amplitude.

As in the case of the sliding guitar string, we want to start with a particular eigenenergy (frequency) and then see if we can "follow" it as the Hamiltonian parameters change. At an instant of time $t$ the eigenvalues of the Hamiltonian of Eq. (1) are found by diagonalization with a unitary operator $\hat{U}$ such that

$$\hat{H}_A(t) = \begin{bmatrix} -\mathcal{E}(t) & 0 \\ 0 & \mathcal{E}(t) \end{bmatrix} = \hat{U}^\dagger(t)\hat{H}(t)\hat{U}(t). \tag{2}$$

Here, the instantaneous eigenenergies are given by $\pm\mathcal{E}$ where we easily find that

$$\mathcal{E}(t) = \sqrt{\alpha^2(t) + \beta^2(t)}. \tag{3}$$

Then the adiabatic principle states that the system state "follows" the eigenstate of Eq. (1), with instantaneous energies $\pm\mathcal{E}$, provided the changes in time of $\alpha(t)$ and $\beta(t)$ are slow enough (Messiah, 1966). But how slow is "slow enough"? We can get a more precise idea of "slowness" by looking at the actual Schrödinger dynamics in the adiabatic basis. That is, if the original Hamiltonian of Eq. (1) satisfies the Schrödinger equation

$$i\hbar\frac{\partial}{\partial t}\Psi(t) = \hat{H}(t)\Psi(t) \tag{4}$$

where $\Psi(t)$ is a two-component spinor, then to obtain the equivalent, exact, equation in the adiabatic basis we substitute for $\hat{H}(t)$ from Eq. (2), such that $\hat{H}(t) = \hat{U}(t)\hat{H}_A(t)\hat{U}^\dagger(t)$, and transform the states to the adiabatic basis through

$$\Psi(t) = \hat{U}(t)\Psi_A(t). \tag{5}$$

Then we obtain

$$i\hbar\frac{\partial}{\partial t}\Psi_A(t) = \hat{H}_A(t)\Psi_A(t) - i\hbar\hat{U}^\dagger\frac{\partial}{\partial t}\hat{U}\Psi_A(t) \tag{6}$$

so that we see that the actual, effective Hamiltonian in the adiabatic basis, is

$$\hat{H}_A(t) - i\hbar \hat{U}^\dagger \frac{\partial}{\partial t} \hat{U},$$

which consists of the adiabatic Hamiltonian $\hat{H}_A(t)$, as intended, as well as an additional term, to be regarded (here) as a *correction*. In the case of the two-component Hamiltonian of Eq. (1) we can find that the correction term has the form

$$\begin{bmatrix} 0 & \gamma(t) \\ \gamma^*(t) & 0 \end{bmatrix} \tag{7}$$

which plays the role of the *nonadiabatic coupling* between the eigenstates. The nonadiabaticity parameter $\gamma(t)$ is given by

$$\gamma(t) = -i\frac{\hbar}{2} \frac{\dot{\alpha}(t)\beta(t) - \alpha(t)\dot{\beta}(t)}{\mathcal{E}^2(t)}. \tag{8}$$

Thus, given that the exact Hamiltonian in the adiabatic basis has the eigenenergy $\mathcal{E}(t)$ on the diagonal and the nonadiabatic coupling $\gamma(t)$ on the off-diagonals, it is common to use as a simple measure of adiabaticity the requirement that (Messiah, 1966)

$$|\gamma(t)| \ll \mathcal{E}(t). \tag{9}$$

This is then a simple measure of how slow is "slow enough."

## 2.2 The Landau–Zener Paradigm

At present we imagine that the time dependence of the Hamiltonian of Eq. (1) is due to an external controlling parameter such as a magnetic field strength. However, we shall shortly see that the time dependence can also arise from the dynamics of an atom in a potential (Section 2.3), i.e., the motion of an atom can itself generate the time dependence in the parameters. In essence, if the position of an atom is given by some dynamics as $x_0(t)$, and a potential for the atom is given by $U_0(x)$ we can make a semiclassical approximation by letting (see Eq. (1)) $\alpha(t) \to U_0(x_0(t))$, which implies $\dot{\alpha} = \dot{x}_0 \partial_x U_0$. As a result, we note that Eq. (9) now appears as an adiabaticity condition depending on the atomic velocity $\dot{x}_0$. This is a half-way house to a full quantum treatment to be considered in Section 8.1.

A simple example of a specific Hamiltonian in the form of Eq. (1) in the basis $\{|1\rangle, |2\rangle\}$ of the bare states is the Landau–Zener model, where the Hamiltonian of Eq. (1) takes the form

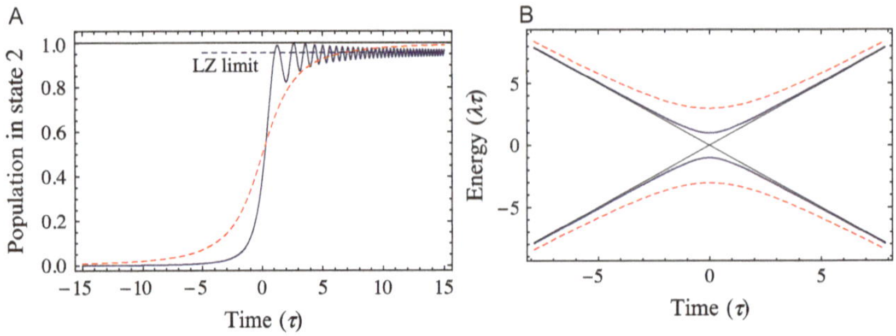


**Fig. 1** Hamiltonian evolution during a linear sweep, starting from state $|1\rangle$ at $t=-\infty$, for two different values of the coupling: $V=\lambda\tau$ corresponding to $\Lambda=1$ (*blue full line*) and $V=3\lambda\tau$ (*red dashed line*, $\Lambda=9$). The time unit is $\tau=\sqrt{\hbar/\lambda}$. (A) Evolution of the population $P_2$ in state $|2\rangle$. The expected $t\to\infty$ Landau–Zener limit, $P_2=1-\exp(-\pi\Lambda)$, is indicated as a *dashed blue line* in the case $V=\lambda\tau$. (B) Energies of the bare states (*thin black line*) and of the adiabatic states $\pm\mathcal{E}(t)$.

$$\hat{H}(t)=\begin{bmatrix}-\lambda t & V\\ V & \lambda t\end{bmatrix} \tag{10}$$

where the constant $\lambda$ describes the rate of change of potential with time and the coupling $V$ between states $|1\rangle$ and $|2\rangle$ is assumed to be a constant. This is an example of Eq. (1) with $\alpha(t)\to\lambda t$ and $\beta\to V$. This model can be an approximation to the dynamics of a particle passing through a region of rf resonance at approximately constant speed. The adiabatic energies of Eq. (10) are $\mathcal{E}(t)=\sqrt{(\lambda t)^2+V^2}$ as illustrated in Fig. 1. The related non-adiabaticity parameter $\gamma(t)$, Eq. (8), is given by the expression

$$\gamma(t)=-i\frac{\hbar}{2}\frac{\lambda V}{\left(\lambda^2t^2+V^2\right)}. \tag{11}$$

We saw in Eq. (9) that the ratio $|\gamma(t)|/\mathcal{E}(t)$ should be small for an adiabatic process. In the case of the Landau–Zener model we find that $|\gamma(t)|/\mathcal{E}(t)=\hbar\lambda V/[2(\lambda^2t^2+V^2)^{3/2}]$. We see that this ratio has its largest value at $t=0$, where $|\gamma(0)|/\mathcal{E}(0)=\hbar\lambda/(2V^2)$. This, in turn, implies that to be adiabatic we would like a strong coupling $V$ and/or a "low" rate of in the potential, $\lambda$ so that $\hbar\lambda/(2V^2)\ll 1$. Conventionally, we define a parameter $\Lambda$ such that (Suominen et al., 1991)

$$\Lambda=\frac{V^2}{\hbar\lambda} \tag{12}$$

and then we have $\Lambda\gg 1$ for highly adiabatic behavior.

The Landau–Zener model (Landau, 1932; Zener, 1932) is a nice example of an exactly solvable and nontrivial time-dependent quantum mechanical problem. The time-dependent solution to Eq. (10) involves the parabolic cylinder, or Weber functions and is quite complex (Vitanov and Garraway, 1996; Zener, 1932). A typical time-evolution is shown in Fig. 1A for a medium value of adiabaticity parameter, $\Lambda = 1$, where most, but not all, of the probability of the initial state is transferred to the other state of the system around $t = 0$. However, despite the complex time-evolution, the probability amplitudes in the long-time limit (starting from $t \to -\infty$) have a very simple form which can be used as a model for practical applications. The probability for the system to remain in its initial eigenstate is given by

$$P = \exp(-\pi\Lambda). \tag{13}$$

When the adiabaticity parameter $\Lambda$ is high, this probability becomes exponentially small, which happens because of adiabatic transfer out of the bare state at the point of the crossing of bare levels.

The adiabatic transfer to the other state is understood by looking at the bare energies $\alpha(t)$ and adiabatic energies $\mathcal{E}(t)$ which are illustrated in Fig. 1B. Although the bare state energies $\pm\,\alpha(t)$ cross over at $t = 0$, the adiabatic energies do not, and a state following the adiabatic path ends up changing its bare state by avoiding the bare state "crossing" at $t = 0$.

## 2.3 Application to Adiabatic Traps

This situation can be applied to magnetic spin problems in several ways. For example, an atomic or nuclear spin in a magnetic field can be subjected to an rf field far below resonance. If the frequency is "slowly" swept through the resonance to far above resonance the spin population is transferred from one spin state to the opposite state in a process known as adiabatic rapid passage (ARP). The situation is analogous to Fig. 1 with the change in bare state energy mapped to the frequency of the exciting field. This means that the measure of "slowly" is given by Eq. (9), which for a linear chirp requires a big value of $\Lambda$ as given by Eq. (12).

Another example, which is highly relevant to this article, occurs if the rf field has constant frequency (and amplitude) in time, but the "static" magnetic field slowly varies in time so that a passage through resonance again takes place. Again, this can lead to the so-called adiabatic rapid passage, but a point of special interest is the case where the magnetic field is *spatially* varying. This realizes a potential, $U_0(x)$, see Section 3.2, down which the atom can accelerate, until it reaches a resonance region such as that found in Fig. 1B. Indeed the time coordinate of the original Landau–Zener model can now be mapped to space through $t \to x/v$. This means that an atom traveling from the left, following the upper adiabatic (red) curve of Fig. 1B will find that provided its velocity $v$ is not too high in the critical Landau–Zener crossing region, it remains on the adiabatic potential after the crossing and starts decelerating on the upward potential. Eventually, the kinetic energy of the atom is lost and it turns around and starts accelerating back to the left side. This cycle will repeat and thus the atom is trapped, adiabatically, by the magnetic resonance location, see Section 4. The condition for remaining trapped, over one cycle of oscillation, is that the probability given by Eq. (13) should be very small. To evaluate this, the parameter $\lambda$ in Eq. (12) should be replaced by $v\dfrac{\partial U_0}{\partial x}$ where the gradient of potential $\dfrac{\partial U_0}{\partial x}$ is to be evaluated at the resonance location. This will be discussed more precisely in Section 8.1.

# 3. INTRODUCTION TO MAGNETIC RESONANCE IN CLASSICAL AND QUANTIZED DESCRIPTIONS

## 3.1 Angular Momentum Operators and Rotation

In this section, for the sake of fixing the notations, we recall simple results on rotation of angular momentum operators, which we will use in this review, and which play a key role in finding and utilizing the adiabatic states. The example we have in mind is an atom having a nuclear spin $\hat{\mathbf{I}}$, an orbital angular momentum $\hat{\mathbf{L}}$ and an electronic spin $\hat{\mathbf{S}}$. The total angular momentum operator relevant to the interaction with a weak magnetic field is $\hat{\mathbf{F}} = \hat{\mathbf{J}} + \hat{\mathbf{I}} = \hat{\mathbf{L}} + \hat{\mathbf{S}} + \hat{\mathbf{I}}$. For example, for a rubidium 87 atom in its $5S_{1/2}$ ground state, we have $I = 3/2$, $L = 0$ and $S = 1/2$, such that $J = 1/2$ and $F \in \{|I-J|, \ldots |I+J|\}$: $F = 1$ or $F = 2$, which are the two hyperfine states of the atomic ground state. For the purpose of this review, where spins will interact with static magnetic fields or rf fields, we consider a fixed value $F$ of the angular momentum quantum number. In the following, in order to lighten up the writing, we will extend the term "spin" to the total angular momentum $\hat{\mathbf{F}}$.

### 3.1.1 Spin Operators

A spin operator, which by convention here we will take as the total atomic angular momentum $\hat{\mathbf{F}}$, is a vector operator (dimension $\hbar$) associated to the quantum number $F$. $F \geq 0$ is an integer for bosonic particles, or a half integer for fermions. The projection of $\hat{\mathbf{F}}$ along any axis, represented by a unit vector $\mathbf{u}$, is denoted as $\hat{F}_{\mathbf{u}} = \hat{\mathbf{F}} \cdot \mathbf{u}$, and is an operator in the space of spin vectors.

Given a quantization axis $\mathbf{e}_z$, we can find a basis where both $\hat{\mathbf{F}}^2$ and $\hat{F}_z$ are diagonal. The spin eigenstates are labeled $|F, m\rangle$ where $m \in \{-F, -F+1, \ldots, F-1, F\}$, with eigenvalues given by

$$\hat{\mathbf{F}}^2|F,m\rangle = F(F+1)\hbar^2|F,m\rangle, \tag{14}$$

$$\hat{F}_z|F,m\rangle = m\hbar|F,m\rangle. \tag{15}$$

We note that the quantum numbers have the labels $F$ and $m$ here. The label $m_F$ is often chosen instead of $m$, but here we choose the latter label in order to simplify some of our later expressions.

We also introduce the rising and lowering operators $\hat{F}_+$ and $\hat{F}_-$, defined as

$$\hat{F}_\pm = \hat{F}_x \pm i\hat{F}_y. \tag{16}$$

It is clear from their definition that $[\hat{F}_\pm]^\dagger = \hat{F}_\mp$. Their commutation relations with $\hat{F}_z$ are

$$[\hat{F}_z, \hat{F}_\pm] = \pm\hbar\hat{F}_\pm. \tag{17}$$

From these relations, we can deduce their effect on $|F, m\rangle$, which is, in the case of $\hat{F}_+$, to increase $m$ by one unit, whereas in the case of $\hat{F}_-$ the quantum number $m$ is decreased by one unit. That is

$$\hat{F}_\pm|F,m\rangle = \hbar\sqrt{F(F+1) - m(m\pm 1)}|F, m\pm 1\rangle. \tag{18}$$

### 3.1.2 Rotation Operators

From now on, as we will concentrate on operators which do not change the value of $F$, we will further simplify the state notation and use $|m\rangle$, where the number $F$ is implicit, and we may also use $|m\rangle_z$ to emphasize that the quantization axis is chosen along $z$. Conversely, an eigenstate of $\hat{F}_{\mathbf{u}}$ will be labeled $|m\rangle_{\mathbf{u}}$.

The operator which allows us to transform $|m\rangle_z$ into $|m\rangle_{z'}$ where $z'$ is a new quantization axis is a rotation operator. The rotation around any unit vector $\mathbf{u}$ by an angle $\alpha$ is described by the unitary operator

$$\hat{R}_{\mathbf{u}}(\alpha) = \exp\left[-\frac{i}{\hbar}\alpha\hat{\mathbf{F}}\cdot\mathbf{u}\right]. \tag{19}$$

This derives in fact from the fact that $\hat{\mathbf{F}}$ is the generator of infinitesimal rotations in the Hilbert space, see, for example, Cohen-Tannoudji et al. (1991) or Walraven (2016). The inverse rotation, by an angle $-\alpha$, is described by its Hermitian conjugate: $\hat{R}_{\mathbf{u}}(-\alpha) = [\hat{R}_{\mathbf{u}}(\alpha)]^{-1} = [\hat{R}_{\mathbf{u}}(\alpha)]^{\dagger}$. Starting from an eigenstate $|m\rangle_z$ of $\hat{F}_z$, the effect of $\hat{R} = \hat{R}_{\mathbf{u}}(\alpha)$ is to give the corresponding eigenstate $|m\rangle_{z'} = \hat{R}^{\dagger}|m\rangle_z$ of the rotated operator $\hat{F}_{z'} = \hat{R}^{\dagger}\hat{F}_z\hat{R}$:

$$|m\rangle_z = \hat{R}|m\rangle_{z'} \Rightarrow \hat{F}_{z'}|m\rangle_{z'} = \hat{R}^{\dagger}\hat{F}_z|m\rangle_z = m\hbar\hat{R}^{\dagger}|m\rangle_z = m\hbar|m\rangle_{z'}.$$

The rotation by the sum of two angles is simply the product of the two rotations around the same axis:

$$\hat{R}_{\mathbf{u}}(\alpha+\beta) = \hat{R}_{\mathbf{u}}(\alpha)\hat{R}_{\mathbf{u}}(\beta).$$

However, as the spin projections onto different axes do not commute, the composition of rotations around different axes does not commute. A useful formula is the decomposition of a rotation around any vector $\mathbf{u}$ in terms of rotations around the basis axes $(x, y, z)$. If the spherical angles describing the direction of the unit vector $\mathbf{u}$ are $(\theta, \phi)$ such that

$$\mathbf{u} = \sin\theta\left(\cos\phi\,\mathbf{e}_x + \sin\phi\,\mathbf{e}_y\right) + \cos\theta\,\mathbf{e}_z, \tag{20}$$

we can write

$$\hat{R}_{\mathbf{u}}(\alpha) = \hat{R}_z(\phi)\hat{R}_y(\theta)\hat{R}_z(\alpha)\hat{R}_y(-\theta)\hat{R}_z(-\phi), \tag{21}$$

where $\hat{R}_i$ stands for $\hat{R}_{\mathbf{e}_i}$. Starting from the right-hand side, the two first rotations put $\mathbf{u}$ on top of $z$, the central operator makes the rotation by $\alpha$ around $z$, and the two last operators bring back $\mathbf{u}$ to its original position.

### 3.1.3 Rotation of Usual Spin Operators

In the following, we will need to transform hamiltonians $\hat{H}$ through rotations, where the Hamiltonian is a sum of spin operators, and calculate operators such as $\hat{R}^{\dagger}\hat{F}_{\mathbf{u}'}\hat{R}$, where $\hat{R} = \hat{R}_{\mathbf{u}}(\alpha)$ and the direction $\mathbf{u}'$ is in general different from $\mathbf{u}$. Here we give its effect in simple cases. Let us first consider rotations by $\alpha$ around the quantization axis $z$, such that $\hat{R} = \hat{R}_z(\alpha)$.

$$\left[\hat{R}_z(\alpha)\right]^{\dagger}\hat{F}_z\hat{R}_z(\alpha) = \hat{F}_z, \tag{22}$$

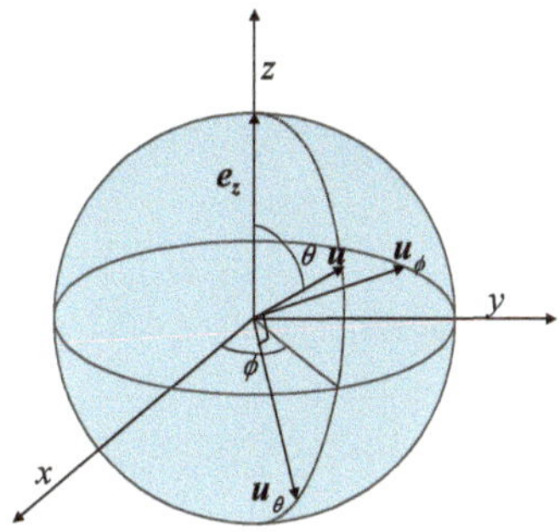

**Fig. 2** Orientation of the basis ($\mathbf{u}$, $\mathbf{u}_\theta$, $\mathbf{u}_\phi$) relative to the basis ($\mathbf{e}_z$, $\mathbf{e}_x$, $\mathbf{e}_y$).

$$\left[\hat{R}_z(\alpha)\right]^\dagger \hat{F}_\pm \hat{R}_z(\alpha) = e^{\pm i\alpha}\hat{F}_\pm, \tag{23}$$

$$\left[\hat{R}_z(\alpha)\right]^\dagger \hat{F}_x \hat{R}_z(\alpha) = \cos\alpha\,\hat{F}_x - \sin\alpha\,\hat{F}_y. \tag{24}$$

The general expression for the rotated operator $\hat{R}^\dagger_{\mathbf{u}}\hat{F}_{\mathbf{u}'}\hat{R}_{\mathbf{u}}$ can be deduced from these equations.

### 3.1.4 Time-Dependent Rotations

For the description of an atom moving in an inhomogeneous magnetic field, we will need to deal with time-dependent rotation angles, and with the derivatives of the $\hat{R}$ operators. After using Eq. (21) and some straightforward arithmetic, we come to the following expression involving $\partial_t\hat{R}_{\mathbf{u}}(\alpha)$:

$$\begin{aligned} i\hbar\hat{R}^\dagger_{\mathbf{u}}(\alpha)\partial_t\hat{R}_{\mathbf{u}}(\alpha) = \dot{\alpha}\hat{F}_{\mathbf{u}} + (1-\cos\alpha)\left[-\dot{\theta}\hat{F}_{\mathbf{u}_\phi} + \dot{\phi}\sin\theta\hat{F}_{\mathbf{u}_\theta}\right] \\ + \sin\alpha\left[\dot{\phi}\sin\theta\hat{F}_{\mathbf{u}_\phi} + \dot{\theta}\hat{F}_{\mathbf{u}_\theta}\right], \end{aligned} \tag{25}$$

where ($\mathbf{u}$, $\mathbf{u}_\theta$, $\mathbf{u}_\phi$) form an orthonormal basis, see Fig. 2, with

$$\mathbf{u}_\theta = \cos\theta\left(\cos\phi\,\mathbf{e}_x + \sin\phi\,\mathbf{e}_y\right) - \sin\theta\,\mathbf{e}_z \tag{26}$$

and

$$\mathbf{u}_\phi = -\sin\phi\,\mathbf{e}_x + \cos\phi\,\mathbf{e}_y. \tag{27}$$

The time variation of $\hat{R}$ can be recast using the rising and lowering operators $\hat{F}_{\mathbf{u},\pm} = \hat{F}_{\mathbf{u}_\theta} \pm i\hat{F}_{\mathbf{u}_\phi}$ with respect to the eigenstates of $\hat{F}_{\mathbf{u}}$, under the form

$$i\hbar\hat{R}^\dagger_{\mathbf{u}}(\alpha)\partial_t\hat{R}_{\mathbf{u}}(\alpha) = \dot{\alpha}\hat{F}_{\mathbf{u}} + \sin(\alpha/2)\left[(\dot{\theta} - i\dot{\phi}\sin\theta)\,e^{i\frac{\alpha}{2}}\hat{F}_{\mathbf{u},+} + h.c.\right]. \tag{28}$$

This last expression makes clear that, if the direction $\mathbf{u}$ around which the rotation is performed varies with time, the time derivative of the rotation operator now involves also spin projections along directions orthogonal to $\mathbf{u}$

$$(\dot{\theta} - i\dot{\phi}\sin\theta)\sin\frac{\alpha}{2}\,e^{i\frac{\alpha}{2}}\hat{F}_{\mathbf{u},+}, \tag{29}$$

which couple different eigenstates of $\hat{F}_{\mathbf{u}}$ with an amplitude

$$\sqrt{\dot{\theta}^2 + \dot{\phi}^2\sin^2\theta}\,\sin\frac{\alpha}{2}. \tag{30}$$

## 3.2 Magnetic Interaction

Consider an atom with a total spin $F$ in its ground state. The Landé factor is labeled as $g_F$. In the presence of a static and homogeneous magnetic field $\mathbf{B}_0 = B_0\,\mathbf{e}_z$ which we use to set the quantization axis $z$, i.e., $B_0$ is positive, the magnetic dipolar interaction between the atomic spin and the field reads

$$\hat{H}_0 = \frac{g_F\mu_B}{\hbar}\mathbf{B}_0\cdot\hat{\mathbf{F}} = \frac{g_F\mu_B}{\hbar}B_0\hat{F}_z. \tag{31}$$

Here, $\mu_B$ is the Bohr magneton. From Eq. (15), we get the first-order Zeeman energy of each magnetic sublevel $|F, m\rangle_z$:

$$E_m = mg_F\mu_B B_0. \tag{32}$$

We see that because $m$ can have different signs the Zeeman energies $E_m$ can also have different signs. This means that when we later introduce a field gradient, we find that some $m$ state, with a positive energy, lead to forces which tend to return the atom to regions of weaker field: these states are called low field seeking states as a result. The low field seeking states, used for magnetic trapping in an inhomogeneous magnetic field, are the states such that $m > 0$ for a positive $g_F$, or $m < 0$ if $g_F < 0$. We introduce

$$s = g_F/|g_F| = \pm 1 \tag{33}$$

as being the sign of $g_F$. The low field seeking states are those with $sm > 0$, with a positive energy $E_m = |m|\hbar\omega_0$, where the frequency spacing between Zeeman levels, the Larmor frequency, is defined as

$$\omega_0 = |g_F|\mu_B B_0/\hbar. \tag{34}$$

## 3.3 Theory of Magnetic Resonance and Coupling to a Classical rf Field

Let us now consider the coupling of these Zeeman states with a near resonant rf field, i.e., one of frequency $\omega$ close to $\omega_0$. In a first approach, we will concentrate on a linearly polarized rf field. Other polarization configurations will be discussed in Section 3.3.2. While the electric field coupling is negligible at rf frequencies, the magnetic field couples to the atomic spin through the magnetic dipolar interaction.

The rf field is usually produced by a coil or a wire in the vicinity of the atomic sample. Its amplitude is usually larger than a mG, such that the atomic spin interacts with many rf photons. For this reason, the rf field can be described by a classical coherent field. We will first present in Section 3.3.1 a semiclassical analysis where a classical field couples to a quantum spin. A full quantum treatment for the field and the spin will be derived in Section 3.4. We will see in Section 7.3 that a quantum description for the rf field provides a simple interpretation for the possible transitions between dressed states. In this section, we will assume that both the static magnetic field and the rf field are spatially homogeneous. In this case, the atom–field coupling term does not depend on the external variables of the atom, whose evolution is thus decoupled from the spin variables. The case of spatially dependent fields is essential for the realization of adiabatic potentials, though. It will be treated in Section 4.

### 3.3.1 *A Spin Coupled to a Classical Field*

In this section, we consider a linearly polarized, homogeneous, classical rf field, along some direction $\boldsymbol{\epsilon}$, where $\boldsymbol{\epsilon}$ is a real unit vector. The magnetic component of the classical rf field is $\mathbf{B}_1(t) = B_1 \cos(\omega t + \phi)\, \boldsymbol{\epsilon}$, where $\phi$ is the field phase at initial time. It couples to the atomic spin through the Hamiltonian

$$\hat{V}_1 = \frac{g_F \mu_B}{\hbar} \mathbf{B}_1 \cdot \hat{\mathbf{F}} \cos(\omega t + \phi) \tag{35}$$

as in Eq. (31). The total Hamiltonian describing the atomic state is then

$$\hat{H} = \hat{H}_0 + \hat{V}_1 = \frac{g_F \mu_B}{\hbar} \mathbf{B}_0 \cdot \hat{\mathbf{F}} + \frac{g_F \mu_B}{\hbar} \mathbf{B}_1 \cdot \hat{\mathbf{F}} \cos(\omega t + \phi). \tag{36}$$

We now use the direction of the static field $\mathbf{B}_0 = B_0\, \mathbf{e}_z$ as quantization axis $z$ ($B_0 > 0$). The presence of a component of the rf field along $z$ only slightly modifies the result and will be discussed in Section 8.4. In this section, we will consider instead the special, and effective case where

$\mathbf{B}_1$ is polarized linearly and in a direction orthogonal to $z$ (known as $\sigma$ polarization), such that we can set the axis $x$ parallel to the rf field: $\mathbf{B}_1(t) = B_1 \cos(\omega t + \phi)\,\mathbf{e}_x$ $(B_1 > 0)$. In the end this turns out to be an optimum choice for a linearly polarized rf field, see Section 8.4. The time-dependent Hamiltonian now reads

$$\hat{H} = s\left[\omega_0 \hat{F}_z + 2\Omega_1 \cos(\omega t + \phi)\hat{F}_x\right]. \tag{37}$$

We have introduced the Rabi frequency

$$\Omega_1 = \frac{|g_F|\mu_B B_1}{2\hbar}. \tag{38}$$

Using the $\hat{F}_\pm$ operators, the Hamiltonian can be written as

$$\hat{H} = s\omega_0 \hat{F}_z + s\frac{\Omega_1}{2}\left[e^{-i(\omega t+\phi)}\,\hat{F}_+ + e^{i(\omega t+\phi)}\,\hat{F}_- + e^{-i(\omega t+\phi)}\,\hat{F}_- + e^{i(\omega t+\phi)}\,\hat{F}_+\right]$$

where we recall that $s = \pm 1$, see Eq. (33).

The evolution of the atomic spin state $|\psi\rangle$ through this Hamiltonian is given by the Schrödinger equation $i\hbar\partial_t|\psi\rangle = \hat{H}|\psi\rangle$. The first term of the Hamiltonian is responsible for a spin precession around the $z$-axis at frequency $\omega_0$, a in direction determined by $s$. The other terms couple different $|m\rangle_z$ states and induce transitions. These transitions will be resonant for $\omega = \omega_0$. To emphasize this point, it is useful to write the Hamiltonian in the basis rotating at the frequency $s\omega$ around $z$. We introduce the rotated state

$$|\psi_{\rm rot}\rangle = \hat{R}^\dagger|\psi\rangle \quad \text{or} \quad |\psi\rangle = \hat{R}|\psi_{\rm rot}\rangle, \quad \text{where} \quad \hat{R} = \hat{R}_z[s(\omega t + \phi)]. \tag{39}$$

This rotated state now obeys a Schrödinger equation $i\hbar\partial_t|\psi_{\rm rot}\rangle = \hat{H}_{\rm rot}|\psi_{\rm rot}\rangle$, where

$$\hat{H}_{\rm rot} = -i\hbar\hat{R}^\dagger\left[\partial_t\hat{R}\right] + \hat{R}^\dagger\hat{H}\hat{R}. \tag{40}$$

Using Eqs. (22), (23), and (25), we obtain

$$\begin{aligned}\hat{H}_{\rm rot} = -s\delta\hat{F}_z &+ s\frac{\Omega_1}{2}\left[e^{i(s-1)\omega t}\,\hat{F}_+ + e^{-i(s-1)\omega t}\,\hat{F}_-\right]\\ &+ s\frac{\Omega_1}{2}\left[e^{i(s+1)\omega t}\,\hat{F}_+ + e^{-i(s+1)\omega t}\,\hat{F}_-\right].\end{aligned} \tag{41}$$

where we have introduced the detuning $\delta = \omega - \omega_0$.

Depending on the sign of $s$, either the two first terms (for $s=1$) or the two last terms (for $s=-1$) in the brackets become static in this rotating frame. On the other hand, the two other terms evolve at high frequency $\pm\, 2\omega$.

We are interested in the limit of near resonant rf coupling, so that the detuning $\delta$ is very small as compared to the rf frequency $\omega$. If we also assume that the rf coupling $\Omega_1$ is much smaller than $\omega$, we can apply the rotating wave approximation (RWA). In this approximation, the time-dependent terms of Eq. (41), which evolve at a frequency much larger than the other frequencies in the problem, are effectively time-averaged to zero. The choice of which terms are dropped from Eq. (41) depends on the sign $s$. We are then left with a time-independent Hamiltonian

$$\hat{H}_{\text{eff}} = -s\delta\hat{F}_z + s\frac{\Omega_1}{2}\left(\hat{F}_+ + \hat{F}_-\right) = s\left(-\delta\hat{F}_z + \Omega_1\hat{F}_x\right), \tag{42}$$

which can be written in terms of a very simple spin Hamiltonian, in the spirit of Eq. (31):

$$\hat{H}_{\text{eff}} = \Omega\hat{F}_\theta. \tag{43}$$

Here, we have defined the frequency splitting through the *generalized Rabi frequency*

$$\Omega = \sqrt{\delta^2 + \Omega_1^2}, \tag{44}$$

and the spin projection $\hat{F}_\theta = \hat{\mathbf{F}}\cdot\mathbf{e}_\theta$, where $\mathbf{e}_\theta = \cos\theta\,\mathbf{e}_z + \sin\theta\,\mathbf{e}_x$, is

$$\hat{F}_\theta = \cos\theta\,\hat{F}_z + \sin\theta\,\hat{F}_x, \quad \text{with} \quad \theta = \arccos\left(-\frac{\delta}{\Omega}\right) + \frac{s-1}{2}\pi. \tag{45}$$

$\hat{H}_{\text{eff}}$ is the Hamiltonian of a spin interacting with a static, effective, magnetic field, pointing in the direction $s\mathbf{e}_\theta$

$$\mathbf{B}_{\text{eff}} = \frac{\hbar\Omega}{g_F\mu_B}\mathbf{e}_\theta. \tag{46}$$

$\hat{F}_\theta$ is linked to $\hat{F}_z$ by the rotation operator of angle $\theta$ around $y$:

$$\left[\hat{R}_y(\theta)\right]^\dagger\hat{F}_\theta\hat{R}_y(\theta) = \hat{F}_z, \quad \text{or} \quad \hat{F}_\theta = \hat{R}_y(\theta)F_z\hat{R}_y(-\theta). \tag{47}$$

The eigenstates of $\hat{H}_{\text{eff}}$ are spin states $|m\rangle_\theta$ in the new basis of quantization axis set by the angle $\theta$. They can be deduced from the initial states in the basis with quantization axis $z$ through the same rotation:

$$|m\rangle_\theta = \hat{R}_y(\theta)|m\rangle_z = e^{-i\theta\hat{F}_y/\hbar}|m\rangle_z. \tag{48}$$

The two states $|m\rangle_\theta$ and $|m\rangle_z$ are mapped and have the same spin projection $m\hbar$ in their respective basis:

$$\hat{F}_\theta|m\rangle_\theta = m\hbar|m\rangle_\theta \quad \text{and} \quad \hat{F}_z|m\rangle_z = m\hbar|m\rangle_z. \tag{49}$$

Their eigenenergies thus read

$$E'_m = m\hbar\Omega. \tag{50}$$

With our convention of the orientation of the new quantization axis $\mathbf{e}_\theta$ with respect to the effective magnetic field $\mathbf{B}_{\text{eff}}$, the low field seeking states, with the highest energies, are always those with $m > 0$, whatever the sign of $g_F$. As a result, for detunings $\delta$ very large as compared to $\Omega_1$ and negative, the extreme low field seeking state $|m{=}sF\rangle_z$ is connected to (i.e., has the largest projection onto) the low field seeking state $|m{=}F\rangle_\theta$ in the new basis, while for $\delta \gg \Omega_1$ and $\delta > 0$, $|m{=}F\rangle_\theta$ is connected to $|m{=}-sF\rangle_z$.

The system can be flipped from $|sF\rangle_z$ to $|-sF\rangle_z$ with a linear sweep of the rf frequency by following adiabatically the state $|F\rangle_\theta$, in a way similar to the two-level system presented in the introduction, see Fig. 1. The adiabaticity criterion given at Eq. (9) can be recast under a very simple formula:

$$|\dot{\theta}| \ll \Omega. \tag{51}$$

This condition expresses that changes in the orientation $\theta$ of the adiabatic state must be slower than the effective Rabi frequency $\Omega$. This is analog to the case of a spin in a static magnetic field, whose direction must change slower than the Larmor frequency for the spin to follow adiabatically.

### *3.3.2 Generalization of Magnetic Resonance Theory to Any rf Polarization*

We now discuss the case of an rf polarization which is no longer linear. The rf field can be written very generally as

$$\mathbf{B}_1 = B_x\cos(\omega t + \phi_x)\,\mathbf{e}_x + B_y\cos(\omega t + \phi_y)\,\mathbf{e}_y + B_z\cos(\omega t + \phi_z)\,\mathbf{e}_z \tag{52}$$

$z$ being the direction of the static magnetic field. In principle, the amplitudes $B_i$ (with $i = x, y, z$) and phases $\phi_i$ could depend on position. To start with, we consider a homogeneous rf field.

We now use a complex notation $\boldsymbol{\mathcal{B}}_1$ for the field amplitude, with $\mathbf{B}_1 = \boldsymbol{\mathcal{B}}_1\, e^{-i\omega t} + c.c.$ and

$$\mathcal{B}_1 = \frac{B_x}{2} e^{-i\phi_x} \mathbf{e}_x + \frac{B_y}{2} e^{-i\phi_y} \mathbf{e}_y + \frac{B_z}{2} e^{-i\phi_z} \mathbf{e}_z. \tag{53}$$

The coupling between the $|m\rangle_z$ states due to the $z$ component $B_z$ of the rf field, aligned along the quantization axis, is extremely small if the corresponding Rabi frequency $\Omega_z$ verifies $|\Omega_z| \ll \omega$, which is the case if the RWA holds. Its effect on the resulting adiabatic energy will be discussed specifically in Section 8.4 and be ignored elsewhere.

We introduce the spherical basis ($\mathbf{e}_+$, $\mathbf{e}_-$, $\mathbf{e}_z$) where

$$\mathbf{e}_+ = -\frac{1}{\sqrt{2}}\left(\mathbf{e}_x + i\mathbf{e}_y\right), \quad \mathbf{e}_- = \frac{1}{\sqrt{2}}\left(\mathbf{e}_x - i\mathbf{e}_y\right). \tag{54}$$

Using this basis, we can write the rf field restricted to the $xy$-plane as

$$\mathcal{B}_1 = B_+ \mathbf{e}_+ + B_- \mathbf{e}_-. \tag{55}$$

The complex projections $B_+$ and $B_-$ are given by the scalar products $\mathbf{e}_\pm^* \cdot \mathcal{B}_1$:

$$B_+ = \frac{1}{2\sqrt{2}}\left(-B_x e^{-i\phi_x} + iB_y e^{-i\phi_y}\right), \quad B_- = \frac{1}{2\sqrt{2}}\left(B_x e^{-i\phi_x} + iB_y e^{-i\phi_y}\right)$$

and we note that $B_+ \neq B_-^*$. We see that $\hat{\mathbf{F}} \cdot \mathbf{e}_+ = -\frac{1}{\sqrt{2}}\hat{F}_+$ and $\hat{\mathbf{F}} \cdot \mathbf{e}_- = \frac{1}{\sqrt{2}}\hat{F}_-$. If we define the complex coupling amplitudes as

$$\Omega_\pm = \mp\sqrt{2}\frac{|g_F|\mu_B}{\hbar} B_\pm, \tag{56}$$

the total spin Hamiltonian reads

$$\hat{H} = s\omega_0 \hat{F}_z + s\left[\frac{\Omega_+}{2} e^{-i\omega t} \hat{F}_+ + \frac{\Omega_+^*}{2} e^{i\omega t} \hat{F}_- + \frac{\Omega_-}{2} e^{-i\omega t} \hat{F}_- + \frac{\Omega_-^*}{2} e^{i\omega t} \hat{F}_+\right]. \tag{57}$$

In order to emphasize the resonant terms and apply RWA, we will write the rotated Hamiltonian, in the frame rotating at frequency $s\omega$. Depending on the sign of $s$, either the two first terms (for $s > 0$) or the two last terms (for $s < 0$) of Eq. (57) will be static, as in the previous case of linear polarization, see Eq. (41). We thus define the effective Rabi coupling $\Omega_1 = |\Omega_1| e^{-i\phi}$ as

$$\Omega_1 = \Omega_s = \begin{cases} \Omega_+ & \text{for } s > 0, \\ \Omega_- & \text{for } s < 0. \end{cases} \tag{58}$$

Note that in any case $\Omega_1 = -\sqrt{2} g_F \mu_B B_s / \hbar$. After a rotation around $z$ of angle $s(\omega t + \phi)$, and application of the rotating wave approximation, only the term in $\Omega_s$ remains and the effective Hamiltonian is

$$\hat{H}_{\text{eff}} = -s\delta \hat{F}_z + s\frac{|\Omega_1|}{2}\left(\hat{F}_+ + \hat{F}_-\right) = s\left(-\delta \hat{F}_z + |\Omega_1|\hat{F}_x\right). \tag{59}$$

We recover Eq. (42), and thus the eigenenergies $m\hbar\sqrt{\delta^2 + |\Omega_1|^2}$ and the eigenstates, deduced from $|m\rangle_z$ by a rotation of $\theta$ around $y$, where $\theta$ is given by Eq. (45).

It is clear from Eq. (58) that the relevant coupling is only the $\sigma^+$ polarized part of the rf field for $s = 1$ (or the $\sigma^-$ component for $s = -1$). It is related to the $x$ and $y$ projections of the rf field through

$$|\Omega_1| = \frac{|g_F|\mu_B}{2\hbar}\sqrt{B_x^2 + B_y^2 + 2sB_xB_y\sin\left(\phi_x - \phi_y\right)}. \tag{60}$$

For a linearly polarized field in a plane perpendicular to $z$, for example, with $B_y = 0$, we recover the amplitude of Eq. (38). The coupling is maximum for purely circular polarization $\sigma^s$ with respect to $z$, which we obtain when $\phi_x - \phi_y = s\pi/2$ and $B_x = B_y$. The amplitude is then twice as large as in the case of the linear field along $x$. For a linear transverse polarization with $\phi_x = \phi_y$ and $B_x = B_y = B_1$, the coupling is smaller by $\sqrt{2}$ than for the circular case. Finally, the coupling totally vanishes in the case of a $\sigma^{-s}$ polarization ($\sigma^-$ for $s = 1$, and vice versa).

We must emphasize that all this reasoning has been done with the direction of the static magnetic field chosen as the quantization axis. Should the direction of this field change in space, the relevant amplitude would be the $\sigma^s$ component along the new, local direction of the magnetic field. This will be illustrated in Section 5.

## 3.4 Dressed Atom Approach Applied to Magnetic Resonance With an rf Field

Although the rf field is classical in the sense that the mean photon number $\langle N \rangle$ interacting with the atoms is very large, and its relative fluctuations $\Delta N / \langle N \rangle$ negligible, it gives a deeper insight in the coupling to use a quantized description for the rf field (Cohen-Tannoudji and Reynaud, 1977). This will make much clearer the interpretation of rf spectroscopy, see

Section 7.3, or the effect of strong rf coupling, beyond RWA, see Section 8.3. Instead of writing the rf field as a classical field, it can be written as a quantum field with creation and annihilation operators $\hat{a}$ and $\hat{a}^\dagger$ of a photon in the rf mode of frequency $\omega$. The effect of $\hat{a}$ and $\hat{a}^\dagger$ on a field state $|N+n\rangle$ where $n < \Delta N$ is then approximately

$$\hat{a}|N+n\rangle \simeq \sqrt{\langle N\rangle}|N+n-1\rangle \quad \hat{a}^\dagger|N+n\rangle \simeq \sqrt{\langle N\rangle}|N+n+1\rangle. \tag{61}$$

We start from the expression Eq. (55) of the classical field in the spherical basis. The quantum rf magnetic field operator can be described as follows:

$$\hat{\mathbf{B}}_1 = (b_+\,\mathbf{e}_+ + b_-\,\mathbf{e}_-)\,\hat{a} + h.c. \tag{62}$$

where $b_\pm = B_\pm/\sqrt{\langle N\rangle}$. The field Hamiltonian is then

$$\hat{H}_{\mathrm{rf}} = \hbar\omega\left(\hat{a}^\dagger\hat{a} + \frac{1}{2}\right). \tag{63}$$

Keeping the definition of Eq. (56) for the Rabi coupling, the one-photon Rabi coupling is $\Omega_\pm^{(0)} = \mp\sqrt{2}|g_F|\mu_B b_\pm/\hbar = \Omega_\pm/\sqrt{\langle N\rangle}$. The operator which couples the atom and the field now reads

$$\hat{V}_1 = s\left[\frac{\Omega_+^{(0)}}{2}\hat{a}\hat{F}_+ + \frac{\Omega_+^{(0)*}}{2}\hat{a}^\dagger\hat{F}_- + \frac{\Omega_-^{(0)}}{2}\hat{a}\hat{F}_- + \frac{\Omega_-^{(0)*}}{2}\hat{a}^\dagger\hat{F}_+\right]. \tag{64}$$

The total Hamiltonian $\hat{H}_0 + \hat{H}_{\mathrm{rf}} + \hat{V}_1$ for the spin and the field in interaction thus reads

$$\hat{H} = s\omega_0\hat{F}_z + \hbar\omega\hat{a}^\dagger\hat{a} + s\frac{1}{2\sqrt{\langle N\rangle}}\left[\Omega_+\hat{a}\hat{F}_+ + \Omega_+^*\hat{a}^\dagger\hat{F}_- + \Omega_-\hat{a}\hat{F}_- + \Omega_-^*\hat{a}^\dagger\hat{F}_+\right], \tag{65}$$

where we have set the origin of energy so as to include the zero photon energy $\hbar\omega/2$.

### 3.4.1 Uncoupled States

In the following, we will chose $s = +1$ for simplicity, the other choice simply changing the role of the two polarizations. We also assume that $\Omega_+$ is real, without lack of generality. In the absence of coupling (for $\Omega_\pm = 0$), the eigenstates of the {atom + photons} system $\hat{H}_0 = \omega_0\hat{F}_z + \hbar\omega\hat{a}^\dagger\hat{a}$ are

$|m, N\rangle_z = |m\rangle_z |N\rangle$, where $|m\rangle_z$ is an eigenstate of $\hat{F}_z$ and $|N\rangle$ an eigenstate of $\hat{a}^\dagger \hat{a}$, with respective eigenvalues $m\hbar$ and $N$:

$$\hat{H}_0 |m, N\rangle_z = E^0_{m,N} |m, N\rangle_z, \qquad E^0_{m,N} = m\hbar\omega_0 + N\hbar\omega.$$

Let us write this energy in terms of the detuning $\delta = \omega - \omega_0$:

$$E^0_{m,N} = -m\hbar\delta + (N + m)\hbar\omega.$$

From this expression, we see that for each fixed $N$ the states in the manifold $\mathcal{E}_N = \{|m, N - m\rangle, m = -F \ldots F\}$ have an energy

$$E^0_{m,N-m} = -m\hbar\delta + N\hbar\omega.$$

For a frequency $\omega$ close to $\omega_0$, that is if $|\delta| \ll \omega$, the energy splitting inside a manifold, of order $\hbar\delta$, is very small as compared to the energy splitting between neighboring manifolds, which is $\hbar\omega$.

### *3.4.2 Effect of the rf Coupling*

The Hamiltonian contains four-coupling terms. The two first terms, proportional to $\Omega_+$, act inside a given $\mathcal{E}_N$ manifold, between two states split by a small frequency $\pm\,\delta$:

$$\left\langle m \pm 1, N - m \mp 1 \left| \frac{\Omega_+}{2\sqrt{\langle N \rangle}} \left( \hat{a}\,\hat{F}_+ + \hat{a}^\dagger\,\hat{F}_- \right) \right| m, N - m \right\rangle$$
$$\simeq \frac{\Omega_+}{2} \sqrt{F(F+1) - m(m \pm 1)} = \Omega_+ \;\; {}_z\langle m \pm 1 | F_x | m \rangle_z.$$

Here we have assumed, following Eq. (61), $N - m \simeq \langle N \rangle$ when applying the operators $\hat{a}$ and $\hat{a}^\dagger$, which is valid because $\langle N \rangle \gg 1$.

The two last coupling terms, proportional to $\Omega_-$, couple states of the $\mathcal{E}_N$ manifold to states of the $\mathcal{E}_{N\pm 2}$ manifold, split by a much larger frequency $\omega_0 + \omega$:

$$\left\langle m \pm 1, N - m \pm 1 \left| \frac{\Omega_-}{2\sqrt{\langle N \rangle}} \left( \hat{a}^\dagger\,\hat{F}_+ + \hat{a}\,\hat{F}_- \right) \right| m, N - m \right\rangle \simeq \Omega_- \;\; {}_z\langle m \pm 1 | F_x | m \rangle_z.$$

An estimation of the effect of these two terms on the energy with perturbation theory, valid in the limit $|\delta| \gg |\Omega_+|$, will lead to a shift of order $\hbar|\Omega_+|^2/\delta$ for the $\Omega_+$ terms, and $\hbar|\Omega_-|^2/(\omega_0 + \omega)$ for the $\Omega_-$ term. The rotating wave approximation, which applies if $|\delta|$, $|\Omega_\pm| \ll \omega$, consists in neglecting the effect of the $\Omega_-$ terms, and to concentrate on the states belonging to a given multiplicity. An estimation of this effect in the case of large rf coupling is given in Section 8.3.

### 3.4.3 Dressed States in the Rotating Wave Approximation

Within the rotating wave approximation, we just need to find the eigenstates in a given manifold. The result is given by using a generalized spin rotation, with the angle given at Eq. (45), where the photon number is changed accordingly to stay in the $\mathcal{E}_N$ manifold. The eigenstates are the *dressed states* $|m, N\rangle_\theta$ given by (see also Garraway, 2011)

$$\begin{aligned} |m,N\rangle_\theta &= \sum_{m'=-F}^{F} {}_z\langle m'|m\rangle_\theta |m',N-m'\rangle_z \\ &= \sum_{m'=-F}^{F} {}_z\langle m'|\hat{R}_y(\theta)|m\rangle_z |m',N-m'\rangle_z \end{aligned} \tag{66}$$

with $\theta$ given by Eq. (45) and $|m\rangle_\theta = \hat{R}(\theta)|m\rangle_z$ as in Eq. (48). Their energies are

$$E_{m,N} = E'_m + N\hbar\omega, \quad \text{where} \quad E'_m = m\hbar\sqrt{\delta^2 + |\Omega_+|^2}. \tag{67}$$

The spin states are dressed by the rf field, in such a way that the eigenstates are now combining different spin and field states, and cannot be written as a product state spin$\otimes$field. The dressed states are connected to the uncoupled states for $|\delta| \gg |\Omega_+|$. The effect of the coupling is to repel the states inside the multiplicity: the frequency splitting increases from $|\delta|$ to $\sqrt{\delta^2 + |\Omega_+|^2}$.

## 4. ADIABATIC POTENTIALS FOR RF-DRESSED ATOMS

In the previous sections, we have assumed uniform static and rf fields, so that the spin Hamiltonian fully decouples from the external degrees of freedom. Here, we will consider instead that either the static field $\mathbf{B}_0(\mathbf{r}) = B_0(\mathbf{r})\mathbf{u}(\mathbf{r})$ or the rf field

$$\mathbf{B}_1(\mathbf{r},t) = \boldsymbol{\mathcal{B}}_1(\mathbf{r})\, e^{-i\omega t} + c.c. = \mathcal{B}_1(\mathbf{r})\, \boldsymbol{\epsilon}(\mathbf{r})\, e^{-i\omega t} + c.c., \tag{68}$$

or both, are space dependent. $\mathcal{B}_1(\mathbf{r})$ is the complex amplitude[a] and $\boldsymbol{\epsilon}(\mathbf{r})$ is the complex polarization of the rf field. The local Larmor frequency is $\omega_0(\mathbf{r}) = |g_F|\mu_B B_0(\mathbf{r})/\hbar$, the local detuning is $\delta(\mathbf{r}) = \omega - \omega_0(\mathbf{r})$.

[a] In the particular case of a linearly polarized field $\mathbf{B}_1(\mathbf{r},t) = B_1\cos(\omega t + \phi)\,\boldsymbol{\epsilon}$ (with $\boldsymbol{\epsilon}$ real) introduced in Section 3.3.1, the complex amplitude would read $\mathcal{B}_1 = B_1/2$.

Within the semiclassical treatment of Section 3.3.1, the total Hamiltonian for the external variables and spin reads

$$\hat{H} = \frac{\hat{\mathbf{P}}^2}{2M} + \frac{g_F\mu_B}{\hbar} B_0(\hat{\mathbf{R}})\,\hat{\mathbf{F}}\cdot\mathbf{u}(\hat{\mathbf{R}}) + \left\{\frac{g_F\mu_B}{\hbar}\mathcal{B}_1(\hat{\mathbf{R}})\,e^{-i\omega t}\hat{\mathbf{F}}\cdot\boldsymbol{\epsilon}(\hat{\mathbf{R}}) + h.c.\right\}. \tag{69}$$

Here, $\hat{\mathbf{R}}$ and $\hat{\mathbf{P}}$ are the atomic position and momentum operators, respectively, and $M$ is the atomic mass. We can recast this expression under the form

$$\hat{H} = \hat{T} + \hat{H}_{\text{spin}}(\hat{\mathbf{R}}, t) \tag{70}$$

where the kinetic part is $\hat{T} = \hat{\mathbf{P}}^2/(2M)$ and $\hat{H}_{\text{spin}}(\hat{\mathbf{R}}, t)$ contains all the angular momentum operators.

## 4.1 Adiabatic Potentials

The approach of Section 3.3.1 still holds to diagonalize the spin part of the Hamiltonian $\hat{H}_{\text{spin}}(\hat{\mathbf{R}}, t)$, although the unitary transformation now depends on the position operator $\hat{\mathbf{R}}$, as $\hat{H}_{\text{spin}}(\hat{\mathbf{R}}, t)$ commutes with $\hat{\mathbf{R}}$. For each value $\mathbf{r} = \langle\hat{\mathbf{R}}\rangle$ of the average position, the eigenstates are $|m\rangle_{\theta(\mathbf{r})}$, where $\theta(\mathbf{r})$ is given with respect to the local quantization axis $\mathbf{u}(\mathbf{r})$. The generalized Rabi frequency $\Omega(\mathbf{r}) = \sqrt{\delta(\mathbf{r})^2 + |\Omega_1(\mathbf{r})^2|}$ gives the eigenenergies at point $\mathbf{r}$, see Eq. (44). Here, $\Omega_1(\mathbf{r}) = |\Omega_1(\mathbf{r})|e^{-i\phi(\mathbf{r})}$ is the local component on the efficient local circular polarization $\sigma^s$ (defined with respect to the quantization axis $\mathbf{u}$). Its derivation will be given in Section 4.2.

In principle, the unitary transformation applied to diagonalize the spin Hamiltonian does not commute with the kinetic energy. However, this effect can be ignored if the spatial dependence of the static and rf fields are weak enough, at small atomic velocities and large rf splitting $\Omega$. This is called the *adiabatic theorem* (Messiah, 1966). In the following, we will assume that the adiabatic condition holds. Nonadiabatic transitions between the adiabatic states will be discussed in Section 8.1.

In this limit, an atom prepared in a state $|m\rangle_{\theta(\mathbf{r})}$ will follow adiabatically this local eigenstate, called an *adiabatic state*, as $\mathbf{r}$ varies due to the atomic motion. The local eigenenergy thus acts as a potential for the atomic motion, indexed by $m$:

$$V_m(\mathbf{r}) = m\hbar\Omega(\mathbf{r}) = m\hbar\sqrt{\delta(\mathbf{r})^2 + |\Omega_1(\mathbf{r})|^2}. \tag{71}$$

Fig. 3 illustrates the adiabatic potentials (for an angular momentum $F = 1$) in the simple case where the magnetic field increases linearly with the spatial

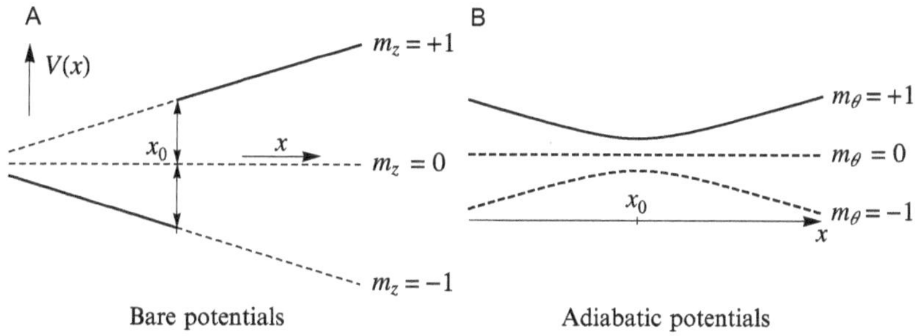


**Fig. 3** Simple example of an adiabatic potential in the case $F = 1$, where the static magnetic field varies linearly with position $x$, the Rabi frequency $\Omega_1$ is position independent and the rf resonance occurs at $x = x_0$. (A) Magnetic potentials or bare potentials, corresponding to the energies of the uncoupled (or bare) states $|m\rangle_z$. (B) Adiabatic potentials, corresponding to the energies of the three dressed states $|m\rangle_\theta$. The frequency splitting at the avoided crossing is $\Omega_1$.

coordinate $x$, such that $\omega_0(x) = \alpha x$ where $\alpha$ is a constant gradient. The rf field is resonant at a position $x_0 = \omega/\alpha$, where the adiabatic potentials present an avoided crossing, with a frequency splitting equal to the Rabi coupling. Far from this region, the adiabatic state $|m\rangle_\theta$ connects to one of the bare states $|\pm m\rangle_z$, with opposite sign on both ends, as illustrated by the bold lines of Fig. 3A, which correspond to the two far ends of the bold line in Fig. 3B.

In general, the maximally polarized state $|m{=}F\rangle_{\theta(\mathbf{r})}$ is used to trap atoms. For a spin 1 or 1/2, this is the only trappable state around the resonance point. For $F > 1$, this choice allows one to suppress inelastic collisions: the spin of the atoms initially in a nonfully polarized state can flip in a collision, and cause losses, whereas such a spin-flip is prevented by conservation of total angular momentum when both atoms are fully polarized (Ketterle et al., 1999). The same is also true in a magnetic trap. One may wonder if this stays true for dressed spin, as the direction of the quantization field rotates at frequency $\omega$. However, all the spins at the same location rotate in phase, such that they can be considered as polarized in the same extremal state. Polarizing the atoms in the extremal state $|m{=}F\rangle_\theta$ is thus an efficient way to prevent inelastic collisions in an adiabatic potential (Moerdijk et al., 1996). In this adiabatic state $|F\rangle_\theta$, the potential energy is always positive

$$V_F(\mathbf{r}) = F\hbar\sqrt{\delta(\mathbf{r})^2 + |\Omega_1(\mathbf{r})|^2}. \tag{72}$$

## 4.2 Local Coupling

The rf field frequency $\omega$ is chosen in the typical range of the Larmor frequency $\omega_0(\mathbf{r})$. At some particular positions, the Larmor frequency is exactly $\omega$. The locus of the points such that $\omega_0(\mathbf{r}) = \omega$ is a surface in space, characterized by a given value of the magnetic field $B_0 = \hbar\omega/(|g_F|\mu_B)$, which we refer to as the *resonance surface*. This surface is determined by the choice of $\omega$. It is often topologically equivalent to a sphere.

The system is described by the Hamiltonian of Eq. (69). Assuming that the adiabatic approximation is valid, we will take a semiclassical approach to describe the atom and replace the positions and momentum operators by their average value $\mathbf{r}$ and $\mathbf{p}$. At each fixed position $\mathbf{r}$, we can apply the procedure described in Section 3.3.1 to get the spin eigenstates and the eigenenergies: apply a rotation of angle $s\omega t$ around the axis $\mathbf{u}(\mathbf{r})$, apply the rotating wave approximation to remove counter-rotating terms, and diagonalize the effective, time-independent Hamiltonian.

An important point is that the $\sigma^s$ polarization, which is the only efficient component to couple the spin states in the RWA, must be defined with respect to the local quantization axis $\mathbf{u}(\mathbf{r})$. Using the *local* spherical basis $(\mathbf{e}_+(\mathbf{r}), \mathbf{e}_-(\mathbf{r}), \mathbf{u}(\mathbf{r}))$, the relevant rf coupling is

$$\Omega_1(\mathbf{r}) = \Omega_s(\mathbf{r}) = -\sqrt{2}\frac{g_F\mu_B}{\hbar}\mathcal{B}_1(\mathbf{r})\,\mathbf{e}_s^*(\mathbf{r})\cdot\boldsymbol{\epsilon}(\mathbf{r}). \tag{73}$$

It is clear from this expression that, even if the rf complex amplitude $\mathcal{B}_1$ and the polarization $\boldsymbol{\epsilon}$ are homogeneous, the Rabi frequency is position dependent, because of the position-dependent direction $\mathbf{u}(\mathbf{r})$ of the static magnetic field.

Using the properties of the spherical basis, in particular $\mathbf{e}_\pm \times \mathbf{u} = \pm i\mathbf{e}_\pm$, or equivalently $\mathbf{e}_s \times \mathbf{u} = is\mathbf{e}_s$ where $s = \pm 1$ is the sign of $g_F$, we find $\mathbf{e}_s^* \cdot \boldsymbol{\epsilon}$ as follows:

$$(\boldsymbol{\epsilon} - is\boldsymbol{\epsilon}\times\mathbf{u})\times\mathbf{u} = \boldsymbol{\epsilon}\times\mathbf{u} + is\,\mathbf{u}\times(\boldsymbol{\epsilon}\times\mathbf{u}) = 2is(\mathbf{e}_s^*\cdot\boldsymbol{\epsilon})\mathbf{e}_s, \tag{74}$$

$$\left|\mathbf{e}_s^*\cdot\boldsymbol{\epsilon}\right| = \frac{1}{2}\left|\boldsymbol{\epsilon}\times\mathbf{u} + is\,\mathbf{u}\times(\boldsymbol{\epsilon}\times\mathbf{u})\right|. \tag{75}$$

This expression makes clear that the rf coupling comes from the projections $\boldsymbol{\epsilon} \times \mathbf{u}$ and $\mathbf{u} \times (\boldsymbol{\epsilon} \times \mathbf{u})$ of the rf field in a plane orthogonal to the direction $\mathbf{u}$ of the static magnetic field. From this expression comes finally, for any complex rf field $\mathcal{B}_1(\mathbf{r})\boldsymbol{\epsilon}(\mathbf{r})$ and any nonvanishing static field $\mathbf{B}_0(\mathbf{r})$ such that $\mathbf{u}(\mathbf{r}) = \mathbf{B}_0(\mathbf{r})/B_0(\mathbf{r})$,

$$|\Omega_1(\mathbf{r})| = \frac{\sqrt{2}\left|g_F\mu_B\mathcal{B}_1(\mathbf{r})\,\mathbf{e}_s^*(\mathbf{r})\cdot\boldsymbol{\epsilon}(\mathbf{r})\right|}{\hbar} = \frac{|g_F\mu_B\mathcal{B}_1|}{\hbar\sqrt{2}}\left|\boldsymbol{\epsilon}\times\mathbf{u} + is\,\mathbf{u}\times(\boldsymbol{\epsilon}\times\mathbf{u})\right|. \tag{76}$$

Another convenient expression for the calculation of the rf coupling reads

$$|\Omega_1(\mathbf{r})| = \frac{|g_F \mu_B \mathcal{B}_1|}{\hbar\sqrt{2}} \sqrt{1 - |\boldsymbol{\epsilon}\cdot\mathbf{u}|^2 + |\boldsymbol{\epsilon}\times\mathbf{u}|^2 + 2is\,\mathbf{u}\cdot(\boldsymbol{\epsilon}\times\boldsymbol{\epsilon}^*)}. \tag{77}$$

Let us discuss two important particular choices of the rf polarization, namely circular or linear. For a polarization $\boldsymbol{\epsilon}$ which is circular $\sigma^s$ with respect to some given axis $z$, we remark that $(\boldsymbol{\epsilon}, \boldsymbol{\epsilon}^*, \mathbf{e}_z)$ form a basis. The last cross product reads $\boldsymbol{\epsilon}\times\boldsymbol{\epsilon}^* = -is\,\mathbf{e}_z$, such that the argument of the square root in Eq. (77) simplifies to $1 - |\boldsymbol{\epsilon}\cdot\mathbf{u}|^2 + |\boldsymbol{\epsilon}\times\mathbf{u}|^2 + 2u_z$ with $u_z = \mathbf{u}\cdot\mathbf{e}_z$. Moreover, using the decomposition of $\mathbf{u}$ in the $(\boldsymbol{\epsilon}, \boldsymbol{\epsilon}^*, \mathbf{e}_z)$ basis

$$\mathbf{u} = (\mathbf{u}\cdot\boldsymbol{\epsilon}^*)\boldsymbol{\epsilon} + (\mathbf{u}\cdot\boldsymbol{\epsilon})\boldsymbol{\epsilon}^* + u_z\mathbf{e}_z,$$

we have

$$\boldsymbol{\epsilon}\times\mathbf{u} = (\mathbf{u}\cdot\boldsymbol{\epsilon})\boldsymbol{\epsilon}\times\boldsymbol{\epsilon}^* + u_z\boldsymbol{\epsilon}\times\mathbf{e}_z = is[u_z\boldsymbol{\epsilon} - (\mathbf{u}\cdot\boldsymbol{\epsilon})\mathbf{e}_z].$$

As $\boldsymbol{\epsilon}$ and $\mathbf{e}_z$ are orthogonal, the modulus square of the two components add and we get finally

$$\begin{aligned} 1 - |\boldsymbol{\epsilon}\cdot\mathbf{u}|^2 + |\boldsymbol{\epsilon}\times\mathbf{u}|^2 + 2is\,\mathbf{u}\cdot(\boldsymbol{\epsilon}\times\boldsymbol{\epsilon}^*) &= 1 - |\boldsymbol{\epsilon}\cdot\mathbf{u}|^2 + 2u_z + u_z^2 + |\mathbf{u}\cdot\boldsymbol{\epsilon}|^2 \\ &= (1+u_z)^2. \end{aligned}$$

The local coupling for an rf polarization $\sigma^s$ circular of sign $s$ along the axis $z$ is

$$|\Omega_1(\mathbf{r})| = \frac{|g_F \mu_B \mathcal{B}_1(\mathbf{r})|}{\hbar\sqrt{2}}[1 + u_z(\mathbf{r})]. \tag{78}$$

It is maximum and equal to $\sqrt{2}|g_F\mu_B\mathcal{B}_1(\mathbf{r})|/\hbar$ where the field is aligned along $+z$ and vanishes at the places where the static field points toward $-z$.

In the case of a linear polarization $\boldsymbol{\epsilon}$, the cross product $\boldsymbol{\epsilon}\times\boldsymbol{\epsilon}^*$ vanishes. Moreover, $1 - |\boldsymbol{\epsilon}\cdot\mathbf{u}|^2$ and $|\boldsymbol{\epsilon}\times\mathbf{u}|^2$ are equal and correspond to the square of the projection of $\mathbf{u}$ in the direction orthogonal to $\boldsymbol{\epsilon}$. The factor under the square root is then $2(1 - |\boldsymbol{\epsilon}\cdot\mathbf{u}|^2)$. The local coupling for a linear rf polarization along some axis $z$ is thus

$$|\Omega_1(\mathbf{r})| = \frac{|g_F \mu_B \mathcal{B}_1(\mathbf{r})|}{\hbar} \sqrt{1 - u_z(\mathbf{r})^2}. \tag{79}$$

It is maximum and equal to $|g_F\mu_B\mathcal{B}_1(\mathbf{r})|/\hbar$ at the positions in space where the field is orthogonal to the $z$-axis, and vanishes where the static field and the rf field are parallel.

## 4.3 Trap Geometry: Role of the Isomagnetic Surface

In order to understand better the spatial shape of the adiabatic potential, let us first assume that the rf effective coupling $|\Omega_1(\mathbf{r})|$ is homogeneous and equal to $\Omega_1$ while $\delta$ varies with position in Eq. (72). This is relevant if the direction of the static field varies only slightly close to the potential minimum of the adiabatic potential. It is then clear that, in the absence of gravity, the potential is minimum where the detuning vanishes, $\delta(\mathbf{r}) = 0$, that is on the isomagnetic resonant surface defined by $\omega_0(\mathbf{r}) = \omega$. This provides us with very anisotropic traps, where one direction transverse to the isomagnetic surface is confined, and may be strongly confined by the avoided crossing of the adiabatic potential, while the directions parallel to the surface are free to move.

How much are the atoms indeed confined to an isomagnetic surface $\delta(\mathbf{r}) = 0$? By definition, the direction normal to the surface is given by the gradient of the Larmor frequency $\boldsymbol{\nabla}\omega_0$. Along this direction, locally, the variations of $\omega_0$ are linear, and so are the variations of $\delta$: $\delta(\mathbf{r} + \mathbf{dr}) \simeq \delta(\mathbf{r}) + \mathbf{dr} \cdot \boldsymbol{\nabla}\omega_0$. We can expand $V_F$ around $\mathbf{r}$ to determine the oscillation frequency in the harmonic approximation in the direction $\boldsymbol{\nabla}\omega_0$ normal to the surface. We find

$$\omega_{\text{transverse}} = \alpha\sqrt{\frac{F\hbar}{M\Omega_1}}, \tag{80}$$

where $\alpha = |\boldsymbol{\nabla}\omega_0|$ is the local magnetic gradient in units of frequency. This formula is similar to the one giving the largest of the two frequencies of a cigar-shape Ioffe–Pritchard (IP) trap, provided the Larmor frequency at the trap bottom is replaced by the Rabi frequency. As the Rabi frequency is in general much smaller than the Larmor frequency, as required for the RWA, for a given magnetic gradient, the confinement to the isomagnetic surface in an adiabatic potential is thus significantly larger than what is obtained in a IP trap. The transverse confinement frequency is typically in the range of a few hundred Hz to a few kHz. We can thus have a good idea of the trap geometry by assuming that the atoms will be confined to the isomagnetic surface. The effect of the position dependence of $|\Omega_1(\mathbf{r})|$ or of gravity is essentially to shape a refined landscape inside this surface. This will

**Fig. 4** Spatial density distribution of cold rf-dressed atoms confined in an adiabatic potential. The static magnetic field comes from a cigar-shape Ioffe–Pritchard trap, see Section 5.1. Atoms are spread around an ellipsoid of radii 160 μm×1.3 mm, isomagnetic surface of the magnetic static field. The atomic density is higher at the bottom of the ellipsoid, where gravity causes the atoms to collect. *From Morizot, O., 2007. Pièges radiofréquence très anisotropes pour un condensat de Bose–Einstein. Université Paris 13. http://tel.archives-ouvertes.fr/tel-00142650/fr/, reprinted by courtesy of the author.*

be discussed in more detail in Section 5, where we will consider various examples of interesting configurations.

If the static magnetic field $B(\mathbf{r})$ has a local minimum $B_{\min}$, as is the case in a magnetic trap, the isomagnetic surfaces close to this minimum are typically ellipsoids (a trap is generally harmonic close to its minimum). This is the basic idea for a bubble trap, as proposed by Zobay and Garraway (2001, 2004) and first realized by Colombe et al. (2004), see Fig. 4.

When gravity is included, the isomagnetic surface is no longer an isopotential of the total potential $V_{\text{tot}}(\mathbf{r}) = F\hbar\sqrt{\delta(\mathbf{r})^2 + \Omega_1^2} + Mgz$. There is a single minimum, at the bottom of the ellipsoid. Depending on the energy of the cloud, which for thermal atoms is set by their temperature, the atoms will fill the bubble up to a certain height given by the barometric energy: $h_{\max} \sim k_B T/(Mg)$. For a Bose–Einstein condensate, the relevant energy scale is the chemical potential $\mu$ and $h_{\max} \sim \mu/(Mg)$.

The radii of the resonant ellipsoid can be adjusted with the choice of the rf frequency $\omega$. Increasing $\omega$ from the minimum Larmor frequency in the trap center $\omega_{0,\min}$ makes the bubble inflate, so that the $z$ radius becomes larger than $h_{\max}$ at some point. Above this frequency, the atoms are confined to a curved plane. Fig. 5 shows ultracold atoms confined in such an anisotropic trap, for various values of the dressing frequency $\omega$.

## 4.4 Loading From a Magnetic Trap

Adiabatic potentials could in principle be loaded from a magneto-optical trap, like magnetic traps. However, their trapping volume is in general much smaller, as the interesting feature of the large anisotropy also comes with a small volume. Moreover, spin flips can occur at high temperature (a few tens of μK) because the relevant effective splitting frequency is $\Omega_1$, of order

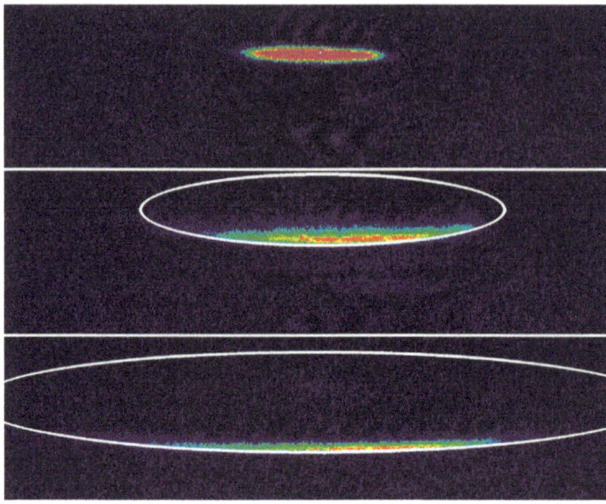

**Fig. 5** Cold rf-dressed rubidium atoms in hyperfine state $F = 2$ confined in an adiabatic potential, for different values of the dressing frequency (Colombe et al., 2004). *Top*: no rf dressing, atoms are trapped in a Ioffe–Pritchard trap with a cigar shape. The Larmor frequency at the trap bottom is 1.3 MHz. *Middle*: rf dressing field at 3 MHz. The atoms occupy the lower part of a bubble. The field of view is 4.5 mm (horizontally) $\times$ 1.2 mm (vertically). The center of mass is shifted vertically by 160 μm. *Lower picture*: dressing frequency 8 MHz. The isomagnetic bubble is larger, the atoms are more shifted vertically (by 490 μm) and the cloud is even more anisotropic. The resonant isomagnetic surfaces are marked with a *white line*.

100 kHz, in general smaller than the minimum Larmor frequency $\omega_{0,\min}$ in a magnetic trap, of order 1 MHz. Adiabatic potentials are very well adapted to trap ultracold atoms or condensates, precooled by evaporative cooling in a magnetic trap.

The loading procedure from a magnetic trap presenting a nonzero Larmor frequency $\omega_{0,\min}$ at its bottom is straightforward and is sketched in Fig. 6. The idea is to operate a frequency sweep from an initial frequency below $\omega_{0,\min}$, up to the desired final value above $\omega_{0,\min}$. The rf field is switched on at a fixed negative detuning $\delta = \omega - \omega_{0,\min}$, in a time sufficiently long to ensure the adiabatic condition Eq. (51), which for a variation in $\Omega_1$ only writes $|\dot{\Omega}_1| \ll \delta^2$. The initial, trapped, upper magnetic state $|sF\rangle_z$ is connected to the upper dressed state $|F\rangle_{\mathbf{u}}$. When the rf frequency is subsequently increased, the atoms stay in this upper adiabatic state if the ramp is slow, so that $|\dot{\delta}| \ll \Omega_1^2$ and we used again Eq. (51). Once $\omega$ reaches $\omega_{0,\min}$, the atoms have reached the resonant surface, and they remain at this surface, which is now a potential minimum, as the rf frequency is further increased.

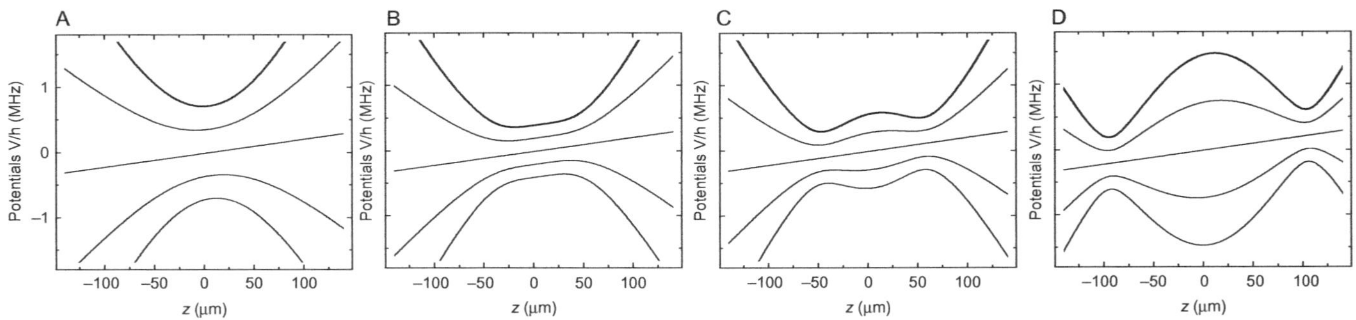


**Fig. 6** Loading procedure from a magnetic trap with nonzero minimum, corresponding to the experimental situation of Fig. 5. The adiabatic potentials are represented for the five states of a $F = 2$ spin, along the vertical axis. Gravity is included. From (A) to (D), the rf frequency is ramped from below resonance to the final value. The atoms start in the upper potential, which is the same as the magnetic trapping potential in (A). (B) corresponds to the resonance at the bottom of the trap, $\omega = \omega_{0,\mathrm{min}}$. Above $\omega_{0,\mathrm{min}}$, see (C) and (D), there is a single trap minimum at the bottom of the ellipsoid, due to gravity, which position $z$ changes with the dressing frequency. *From Colombe, Y., 2004. Condensat de Bose–Einstein, champs évanescents et piégeage bidimensionnel (Ph.D. thesis), Université Paris 13, reprinted by courtesy of the author.*

We note that apart from a condition on the adiabatic following of the adiabatic *spin* state, the loading procedure involves a trap deformation, which could lead to heating if this deformation is not adiabatic with respect to the *mechanical* degrees of freedom. In case of rapid loading resulting in unwanted heating, evaporative cooling can be readily applied in the final adiabatic trap, see Section 7.2.

# 5. EXAMPLES OF ADIABATIC POTENTIALS

We will now present two important examples of adiabatic potentials. The reasoning followed in this section can then be generalized to any kind of trapping configuration with rf-induced adiabatic potentials. Other trap configurations are reviewed in Garraway and Perrin (2016), including lattices (Courteille et al., 2006; Lin et al., 2010; Morgan et al., 2011), adiabatic traps combined with optical lattices allowing to reach subwavelength periods (Lundblad et al.,2008, 2014; Shotter et al., 2008; Yi et al., 2008), or using inductively coupled currents (Griffin et al., 2008; Pritchard et al., 2012; Sinuco-León et al., 2014; Vangeleyn et al., 2014).

Here, we discuss two configurations based on Ioffe–Pritchard and quadrupole static magnetic traps, respectively, for various choices of the rf polarization, which can substantially impact on the resulting adiabatic potential. We start with traps based on a static Ioffe–Pritchard configuration, which were the first to be implemented (Colombe et al., 2004; Schumm et al., 2005). We then discuss how a quadrupole field can be used to tailor a very anisotropic adiabatic trap for confining quasi two-dimensional gases (De Rossi et al., 2016; b), and eventually lead to an annular trap in combination with an optical potential (Heathcote et al., 2008; Morizot et al., 2006).

In this section, we will consider for simplicity a uniform rf magnetic field, in amplitude and polarization, which reads

$$\mathbf{B}_1 = \mathcal{B}_1 \boldsymbol{\epsilon}\, e^{-i\omega t} + c.c.$$

where $\boldsymbol{\epsilon}$ is a complex polarization of unit modulus and $\mathcal{B}_1$ a complex amplitude.

## 5.1 The Dressed Ioffe–Pritchard Trap

The first rf-dressed trap was obtained from dressing atoms initially confined in a Ioffe–Pritchard magnetic trap (Colombe et al., 2004). We will first recall the static field configuration in this case, then discuss the effect of the rf field depending on the rf polarization and amplitude.

### *5.1.1 Isomagnetic Surfaces*

The static magnetic field we start from is the one of a Ioffe–Pritchard magnetic trap (Ketterle et al., 1999). Let us write explicitly the magnetic field in this trap near the trap minimum, up to second order in the coordinates:

$$\begin{aligned}\mathbf{B}_0(\mathbf{r}) &= \left(B_{\min} + \frac{b''}{2}z^2\right)\mathbf{e}_z + b'(x\,\mathbf{e}_x - y\,\mathbf{e}_y) - \frac{b''}{2}z(x\,\mathbf{e}_x + y\,\mathbf{e}_y)\\ &= B_z(z)\,\mathbf{e}_z + b'(x\,\mathbf{e}_x - y\,\mathbf{e}_y) - \frac{b''}{2}z(x\,\mathbf{e}_x + y\,\mathbf{e}_y).\end{aligned} \tag{81}$$

Its modulus up to second order in $x$, $y$, $z$ is

$$B_0(\mathbf{r}) = B_{\min}\left[1 + \frac{b''}{2B_{\min}}z^2 + \frac{b'^2}{2B_{\min}^2}(x^2+y^2)\right] = B_0(\rho, z)$$

giving rise to a cylindrically symmetric cigar-shape harmonic trapping near the center, with frequencies $\omega_x = \omega_y \gg \omega_z$. Away from the $z$-axis, specifically when $\rho = \sqrt{x^2+y^2} \gg B_{\min}/b'$, the trap is closer to a linear potential, with $B_0(\mathbf{r}) \simeq b'\rho$. The quantization axis is defined as

$$\mathbf{u} = \frac{B_z(\mathbf{r})}{B_0(\mathbf{r})}\mathbf{e}_z + \frac{b'x - b''xz/2}{B_0(\mathbf{r})}\mathbf{e}_x - \frac{b'y + b''yz/2}{B_0(\mathbf{r})}\mathbf{e}_y. \tag{82}$$

For discussing adiabatic potentials, frequency units are often more convenient. The Larmor frequency is denoted $\omega_0(\rho, z)$, with a minimum value $\omega_{0,\min}$ at the center, and the gradient in frequency units is $\alpha = |g_F|\mu_B b'/\hbar$. We also define $\eta = |g_F|\mu_B b''/\hbar$, the curvature in units of frequency. We can then also write, up to second order in the coordinates,

$$\omega_0(\rho, z) = \omega_0(0, z) + \frac{\alpha^2}{2\omega_{0,\min}}\rho^2, \quad \text{with} \quad \omega_0(0, z) = \omega_{0,\min} + \frac{1}{2}\eta z^2.$$

The resonant surface is defined by $\omega_0(\rho_{\mathrm{res}}(z), z) = \omega$. At each longitudinal position $z$, the resonant radius $\rho_{\mathrm{res}}$ is given by

$$\rho_{\mathrm{res}}(z) = \frac{1}{\alpha}\sqrt{\omega^2 - \left(\omega_{0,\min} + \frac{\eta z^2}{2}\right)^2}. \tag{83}$$

It depends only slowly on $z$. For $\omega \sim \omega_{0,\min}$, the maximum radius $\rho_0 = \rho_{\rm res}(0)$ scales like $\sqrt{\omega - \omega_{0,\min}}$. For rf frequencies much larger than $\omega_{0,\min}$, the dependence becomes linear:

$$\rho_0 \simeq \frac{\omega}{\alpha} \quad \text{for} \quad \omega \gg \omega_{0,\min}.$$

Tuning the frequency is thus a natural and efficient way to increase the radius of the isomagnetic surface.

### 5.1.2 Circular rf Polarization in the Ioffe–Pritchard Case

We consider here a circular rf polarization along $z$, in a direction chosen according to the sign of $g_F$, $\boldsymbol{\epsilon} = -\frac{1}{\sqrt{2}}\left(s\mathbf{e}_x + i\mathbf{e}_y\right)$. This choice corresponds to a maximum coupling on the $z$-axis of the trap. Defining $\Omega_0$ as the maximum coupling, corresponding to $\hbar\Omega_0 = \sqrt{2}|g_F|\mu_B|\mathcal{B}_1|/\hbar$, we deduce the local coupling from Eq. (78)

$$|\Omega_1(\mathbf{r})| = \frac{\Omega_0}{2}(1 + u_z) = \frac{\Omega_0}{2}\left[1 + \frac{B_z(z)}{B_0(\rho,z)}\right] = \frac{\Omega_0}{2}\left[1 + \frac{\omega_0(0,z)}{\omega_0(\rho,z)}\right]. \tag{84}$$

The rf coupling reaches its maximum $\Omega_{\max}$ on the $z$-axis and is reduced as $\rho$ increases. The coupling does not depend on the polar angle around $z$, and the rotational invariance of the Ioffe–Pritchard potential is preserved. On the resonance surface $\omega_0 = \omega$, the rf coupling is

$$|\Omega_1(\rho_{\rm res}(z), z)| = \frac{\Omega_0}{2}\left[1 + \frac{\omega_0(0,z)}{\omega}\right].$$

The bubble geometry described at Section 4.3 is not much changed, see Fig. 7, left. Because of the reduction of the coupling with increasing $\rho$, the potential minimum is simply shifted very slightly to a larger radius with respect to the resonant radius $\rho_{\rm res}(z)$. Along $z$, on the resonance surface, the minimum coupling is obtained for $z = 0$.

In the presence of gravity, the trap minimum is at the bottom of the resonant surface, as was the case in the first demonstration (Colombe et al., 2004). If, however, one can manage to compensate for gravity, the region of minimum total potential is ring shaped, or more precisely tubular, because it is very elongated along $z$ (Lesanovsky et al., 2006). Such a gravity compensation can be approximately obtained by using a position varying rf amplitude, as is the case with atom chips producing large magnetic gradients, including for the rf field, as recently demonstrated (Kim et al., 2016).

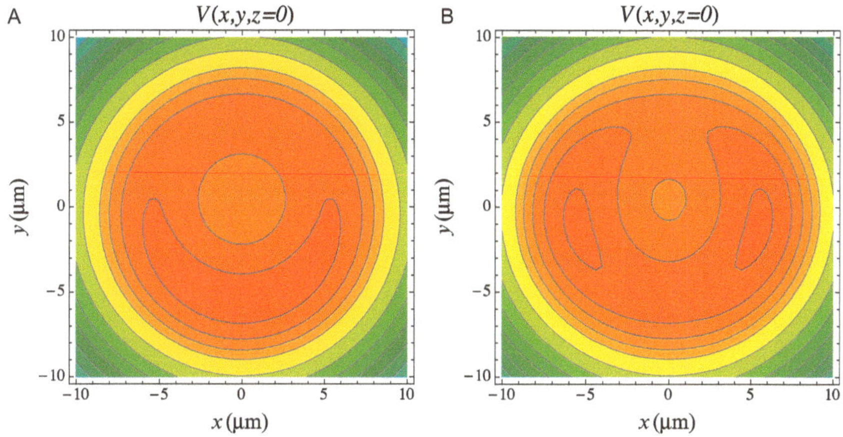


**Fig. 7** Isopotential lines in the "ring" trap (*left*) and the double-well trap (*right*), seen in the *xy*-plane. Gravity is along *y* and is taken into account. These traps are elongated in the *z*-direction leading to a tubular trap on the left and two long traps on the right. Parameters for both plots: $\omega_{0,\mathrm{min}}/(2\pi) = 1000$ kHz, $\alpha/(2\pi) = 100$ kHz μm$^{-1}$, $\omega/(2\pi) = 1120$ kHz, $\Omega_0/(2\pi) = 200$ kHz, $F = 1$. *Left*: circular polarization. *Right*: linear polarization.

### *5.1.3 Linear rf Polarization*

The situation is quite different for a linear polarization. If gravity is along the $y$-axis, let us consider an rf field polarized along $x$: $\boldsymbol{\epsilon} = \mathbf{e}_x$. From Eq. (79), we get for the local coupling

$$|\Omega_1(\mathbf{r})| = \Omega_0\sqrt{1-u_x^2} = \Omega_0\left[1-\frac{b'^2x^2}{B_0^2}\right]^{1/2} = \Omega_0\left[1-\frac{\alpha^2x^2}{\omega_0(\rho,z)^2}\right]^{1/2} \tag{85}$$

up to second order in the coordinates. $\Omega_0 = |g_F|\mu_B|\mathcal{B}_1|/\hbar$ is the maximum coupling, reached in the $x = 0$ plane. Note that, for identical rf power, it is smaller than the maximum in the case of a circular polarization by a factor $\sqrt{2}$. The coupling is lower on the $x$-axis than on the $y$-axis, by a factor $B_z/B_0$. On the resonant surface $\rho = \rho_{\mathrm{res}}(z)$, the effective Rabi coupling is thus

$$|\Omega_1| = \Omega_0\left[1-\frac{\alpha^2x^2}{\omega^2}\right]^{1/2}, \quad |x| \leq \rho_{\mathrm{res}}(z).$$

$|\Omega_1|$ is minimum for $y = 0$ and $x = \rho_{\mathrm{res}}(z)$, for which $|\Omega_1| = \Omega_0\omega_0(0, z)/\omega$. The absolute coupling minimum within the resonant surface is thus $|\Omega_1| = \Omega_0\omega_{0,\mathrm{min}}/\omega$, obtained at the two equatorial positions $(\pm\rho_0, 0, 0)$ where $\rho_0 = \rho_{\mathrm{res}}(z=0) = \alpha^{-1}\sqrt{\omega^2-\omega_{0,\mathrm{min}}^2}$.

Let us look for the potential minimum inside the resonant surface. Because of the reduced coupling at these points, in the absence of gravity, the potential has two minima, located at $x = \pm\rho_0$, $y = 0$, $z = 0$, see Fig. 7, right. The energy at these points is $F\hbar\Omega_0\sqrt{1-(\alpha\rho_0/\omega)^2} = F\hbar\Omega_0\,\omega_{0,\mathrm{min}}/\omega$, to be compared to the highest energy points at $x = 0$, $y = \pm\rho_{\mathrm{res}}(z)$ where the energy is $F\hbar\Omega_0$. The energy difference due to the inhomogeneity in the rf coupling is equal to $F\hbar\Omega_0\left(1-\omega_{0,\mathrm{min}}/\omega\right)$.

As discussed in Section 5.1.1, the resonant radius $\rho_0$ is tuned by changing the rf frequency. For $\omega \sim \omega_{0,\mathrm{min}}$, the maximum radius $\rho_0$ scales like $\sqrt{\omega-\omega_{0,\mathrm{min}}}$. For rf frequencies much larger than $\omega_{0,\mathrm{min}}$, $\rho_0 \simeq \omega/\alpha$. It is then straightforward to adjust at will the distance $2\rho_0$ between the two wells by tuning the rf frequency, which has been done to perform atom interferometry on a chip (Jo et al., 2007; Schumm et al., 2005), see Fig. 8, or to realize species selective trapping (Extavour et al., 2006). These double waveguides have also been used to split a cloud in two parts during its propagation (van Es et al., 2008).

We must now discuss the effect of gravity, which in the case of the tubular potential significantly changes the shape of the minimum if it is not compensated. Here, gravity will not affect qualitatively the double-well trap if the gravitational energy difference between the equator and the bottom of the resonant surface is less pronounced than the rf coupling difference, that is if

$$Mg\rho_0 < F\hbar\Omega_0\left(1-\omega_{0,\mathrm{min}}/\omega\right). \tag{86}$$

To discuss if this relation is fulfilled, we introduce the parameter $\beta = F\dfrac{\hbar\alpha}{Mg}$, which is the ratio between the magnetic force and the gravity force. It has to

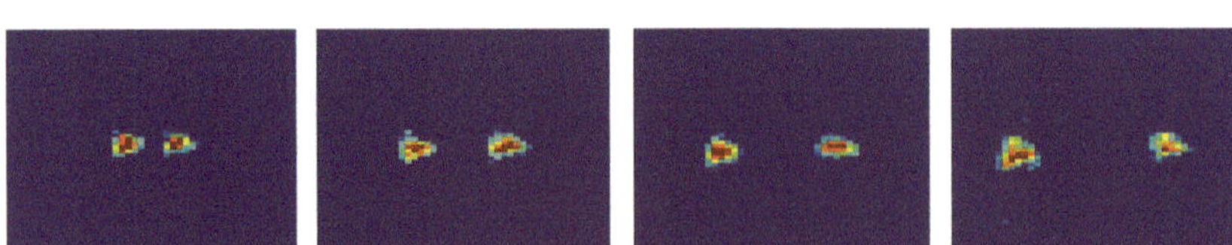

**Fig. 8** First experimental realization of the rf double-well trap (Schumm et al., 2005). The distance between two Bose–Einstein condensates in elongated traps is adjusted with the rf frequency. Distance between wells, from *left* to *right*: 20 μm, 36 μm, 45 μm, and 60 μm. *Unpublished figure, courtesy of Thorsten Schumm.*

be larger than one for the Ioffe–Pritchard trap to confine atoms against gravity. The condition given by Eq. (86) can be written using $\beta$ as:

$$1<\beta\frac{\Omega_0}{\omega}\sqrt{\frac{\omega-\omega_{0,\min}}{\omega+\omega_{0,\min}}}.$$

The ratio under the square root is always less than one and reaches one in the limit where $\omega \gg \omega_{0,\min}$. A necessary condition for Eq. (86) to be fulfilled for all double-well distances is thus

$$1<\beta\frac{\Omega_0}{\omega}, \quad \text{or} \quad \omega<\beta\Omega_0. \tag{87}$$

If we want to stay in the RWA where $\Omega_0 \ll \omega$, the magnetic gradient must be much stronger than gravity, $\beta \gg 1$. This explains why atom chips are ideal devices for these double-well traps, as they provide us with large magnetic gradients.

### 5.1.4 Trapping Away From Resonance

The existence of a resonant surface is not necessary for an adiabatic trapping to occur. If, as is the case in a Ioffe–Pritchard configuration, the magnetic field presents a local minimum, the rf frequency can be less than the Larmor frequency everywhere, provided that $\omega < \omega_{0,\min}$. The magnetic potential in the vicinity of the magnetic minimum is nonetheless modified if $|\omega-\omega_{0,\min}| \lesssim |\Omega_1|$. A double-well potential with a small separation between the wells, enabling tunnel coupling between the wells, has been produced with the Ioffe–Pritchard magnetic field and a linear rf polarization discussed in the previous section but with a frequency below the minimum Larmor frequency (Schumm et al., 2005).

Let us write the condition to obtain a double-well. The adiabatic potential for the adiabatic state $|m=F\rangle_\theta$ reads

$$V_F(\mathbf{r}) = F\hbar\sqrt{(\omega_0(\rho,z)-\omega)^2+\Omega_0^2\left(1-\frac{\alpha^2x^2}{\omega_0(\rho,z)^2}\right)}.$$

We ignore gravity here for simplicity. It only has a small effect on the atom chip experiment where the magnetic gradients are large ($\beta \gg 1$). The minima will lie on the $x$-axis by symmetry. Along this axis, the expression of the potential simplifies as

$$V_F(x) = F\hbar\omega_{0,\min}\sqrt{\left(\frac{X^2}{2} - \tilde{\delta}_{\min}\right)^2 + \tilde{\Omega}_0^2\frac{1 - X^2/2}{1 + X^2/2}} \tag{88}$$

where $X = \alpha x/\omega_{0,\min}$, $\tilde{\Omega}_0 = \Omega_0/\omega_{0,\min} < 1$, and $\tilde{\delta}_{\min} = (\omega - \omega_{0,\min})/\omega_{0,\min} < 0$.

For small values of $X$, the Taylor expansion of the potential is of the form

$$V_F(x) \simeq V_0 + a_2X^2 + a_4X^4 + O(X^6).$$

We can show that $a_4$ is always positive. A double-well occurs as soon as $a_2 < 0$. As $a_2 \propto -(\tilde{\Omega}_0^2 + \tilde{\delta}_{\min})$, see Eq. (88), the condition on the rf coupling for a double-well to appear is

$$\tilde{\Omega}_0^2 > -\tilde{\delta}_{\min} \quad \text{or} \quad \Omega_0^2 > \omega_{0,\min}|\delta_{\min}|.$$

The separation between the two wells scales as $\Delta x \propto \sqrt{\Omega_0^2 - \omega_{0,\min}|\delta_{\min}|}$ and can thus be adjusted by a control on the rf amplitude at fixed detuning.

The trap loading is simplified in this case, as the rf detuning can remain fixed at $\delta_{\min} < 0$, while the rf amplitude is ramped up to its final value. The single initial minimum of the Ioffe–Pritchard trap splits into two minima as the rf amplitude increases and approaches $|\omega - \omega_{0,\min}|$. This method allows in particular to prepare two copies of an initial Bose–Einstein condensate, which can then interfere after a given waiting time, enabling the study of the phase diffusion in elongated BECs with interferometric methods (Betz et al., 2011; Schumm et al., 2005).

## 5.2 The Dressed Quadrupole Trap

After the dressed Ioffe–Pritchard trap, another interesting case, which allows very flat traps for 2D quantum gases, is the adiabatic potential obtained from dressing atoms in a quadrupole field (Merloti et al., 2013b; Morizot et al., 2007). We will discuss this trap here.

### *5.2.1 Magnetic Field Geometry in the Quadrupole Trap*

In a quadrupole field, which can be used as a magnetic trap (Migdall et al., 1985), the magnetic field is linear in the position, for example:

$$\mathbf{B}_0(\mathbf{r}) = b'(x\,\mathbf{e}_x + y\,\mathbf{e}_y - 2z\,\mathbf{e}_z). \tag{89}$$

The factor 2 in the $z$ gradient ensures a vanishing divergence of the magnetic field. The corresponding Larmor frequency is

$$\omega_0(\mathbf{r}) = \alpha\sqrt{x^2 + y^2 + 4z^2}, \tag{90}$$

where $\alpha = |g_F|\mu_B b'/\hbar$.

The isomagnetic surfaces is this case are ellipsoids, with a radius smaller in the vertical direction by a factor of two. For a given rf frequency $\omega$, the equation of the resonant ellipsoid is

$$x^2 + y^2 + 4z^2 = r_0^2,$$

where the ellipsoid horizontal radius is related to the frequency through

$$r_0 = \frac{\omega}{\alpha}. \tag{91}$$

In the following, we will use the generalized distance to the center of the quadrupole,

$$\ell(\rho, z) = \sqrt{\rho^2 + 4z^2}, \tag{92}$$

where $\rho$ is the polar coordinate in the $xy$-plane, $\rho\ \mathbf{e}_\rho = x\ \mathbf{e}_x + y\ \mathbf{e}_y$.

The direction of the magnetic field is given by

$$\mathbf{u} = \frac{\rho\ \mathbf{e}_\rho - 2z\ \mathbf{e}_z}{\ell(\rho, z)}. \tag{93}$$

The naive potential, forgetting polarization issues, is thus a bubble defined by $\ell(\rho, z) = r_0$, the atoms being attracted to its bottom by gravity. This explains why this configuration is well adapted to the trapping of two-dimensional gases.

### *5.2.2 Circular Polarization in the Quadrupole Trap*

The magnetic field at the bottom of the ellipsoid $(0, 0, -r_0/2)$ is aligned along the $+\ z$-axis. It is then natural to consider first a circularly polarized field, To maximize the coupling at the bottom, we will hence chose a circular polarization $\sigma^s$ of sign $s$ aligned with $z$:

$$\boldsymbol{\epsilon} = -\frac{1}{\sqrt{2}}\left(s\ \mathbf{e}_x + i\mathbf{e}_y\right).$$

Using Eq. (78), the efficient Rabi component is then

$$|\Omega_1(\mathbf{r})| = \frac{\Omega_0}{2}(1 + u_z) = \frac{\Omega_0}{2}\left(1 - \frac{2z}{\ell}\right). \tag{94}$$

$\Omega_0$ is the maximum Rabi frequency, obtained as expected on the negative side of the $z$-axis, where $\ell = -2z$. On the other hand, the effective coupling vanishes on the positive side of the $z$-axis, where $\ell = 2z$. The polarization here is $\sigma^{-s}$ with respect to the local orientation of the magnetic field, which points downward. This results in a half axis of zero coupling. The trap minimum must lie away from this $+z$-axis, in order to prevent spin flips.

The total potential, including gravity, in the extreme adiabatic state $|m{=}F\rangle_\theta$ is

$$V_F(\mathbf{r}) = F\hbar\sqrt{[\alpha\ell(\rho,z) - \omega]^2 + \frac{\Omega_0^2}{4}\left[1 - \frac{2z}{\ell(\rho,z)}\right]^2} + Mgz. \tag{95}$$

We can immediately notice that the potential is rotationally invariant around $z$, and depends only on $\rho$ and $z$. The first term will be minimum when both the detuning term and the coupling term vanish, at the point $(0, 0, r_0/2)$. However, this is the point of the resonant surface where gravity is maximum, and gravity will help bringing the atoms to the bottom of the ellipsoid, away from the region of vanishing Rabi frequency.

Again, it is reasonable to assume that the minimum will lie on the resonance surface, which allows one to set $\delta$ to zero. Let us write the value of the potential for atoms living on the two-dimensional resonance surface $\ell(\rho, z) = r_0$. It depends only on $z$ now, as $\rho$ is given by $z$ and $r_0$:

$$V_{F,\,\mathrm{surf}}(z) = F\hbar\frac{\Omega_0}{2}\left(1 - \frac{2z}{r_0}\right) + Mgz, \quad |z| \leq \frac{r_0}{2}.$$

The potential on the surface is linear in $z$.

From this expression, it is clear that there will be a competition between the gradient of rf coupling and gravity. Let us use the same parameter $\beta = F\dfrac{\hbar\alpha}{Mg}$ as in Section 5.1.3. It has to be larger than 1/2 if the quadrupole field itself is supposed to confine atoms against gravity. If $Mgr_0 > F\hbar\Omega_0$, that is for $\omega > \beta\Omega_0$, gravity is dominant and the minimum is at $z = -r_0/2$, at the bottom of the ellipsoid where the coupling is maximum. On the other hand, if $\omega < \beta\Omega_0$, the atoms are pushed upward to the point of the ellipsoid where the coupling vanishes, and the trap is unstable with respect to Landau–Zener losses. For a given coupling, this sets a minimum frequency which should be used

$$\omega > \beta\Omega_0. \tag{96}$$

In contrast to the requirement for a double well, this is now easily compatible with RWA, which requires $\omega \gg \Omega_0$. The trap is thus normally at the bottom of the ellipsoid, even for moderate gradients.

### 5.2.3 Isotropic Trap for a 2D Gas Using Quadrupole Fields

Let us assume that we have indeed $\omega > \beta\Omega_0$. In the vicinity of the bottom of the resonant surface, we can develop the full potential to find the oscillation frequencies. In the vertical direction, the trap is similar to the radial direction of a Ioffe–Pritchard magnetic trap:

$$\begin{aligned} V_F(0,0,z) &= F\hbar\sqrt{\alpha^2(r_0+2z)^2+\Omega_0^2}+Mgz \\ &\simeq F\hbar\Omega_0+Mgz+F\hbar\frac{4\alpha^2}{2\Omega_0}\left(z+\frac{r_0}{2}\right)^2. \end{aligned}$$

The vertical oscillation frequency is thus (Merloti et al., 2013b)

$$\omega_z = 2\alpha\sqrt{\frac{F\hbar}{M\Omega_0}}\left(1-\frac{1}{4\beta^2}\right)^{3/4}, \tag{97}$$

$$\omega_z \simeq 2\alpha\sqrt{\frac{F\hbar}{M\Omega_0}}. \tag{98}$$

The $1/\beta^2$ correction arises because the minimum does not strictly belong to the resonant surface, due to gravity (Merloti et al., 2013b). The analogy with the Ioffe–Pritchard trap is immediate: the Larmor frequency is just replaced by the Rabi frequency.

In the horizontal direction, the trap is isotropic. The oscillation frequency is imposed by the geometry of the ellipsoid and by gravity: the motion is pendulum-like, with an oscillation frequency in the harmonic approximation of order $\sqrt{g/(2r_0)}$. More precisely, defining the absolute value of the vertical equilibrium position as

$$R = \frac{r_0}{2}\left(1+\frac{1}{\sqrt{4\beta^2-1}}\frac{\Omega_0}{\omega}\right), \tag{99}$$

the expression of the horizontal frequency is (Merloti et al., 2013b)

$$\omega_\rho = \sqrt{\frac{g}{4R}}\left[1 - \frac{F\hbar\Omega_0}{2MgR}\sqrt{1 - \frac{1}{4\beta^2}}\right]^{1/2}, \tag{100}$$

$$\omega_\rho \simeq \sqrt{\frac{g}{2r_0}}\left[1 - \frac{\beta\Omega_0}{\omega}\right]^{1/2}. \tag{101}$$

The correction in $\beta\Omega_0/\omega$ comes from the vertical dependence of the Rabi frequency. The exact expression Eq. (100) is slightly affected by the gravitational sag. The two values given by Eqs. (98) and (101), however, are a very good estimate of the oscillation frequencies. The trap is very anisotropic, and the aspect ratio is approximately

$$\frac{\omega_z}{\omega_\rho} \simeq 2\alpha\sqrt{\frac{2F\hbar r_0}{Mg\Omega_0}} = 2\sqrt{\frac{2F\hbar\alpha\omega}{Mg\Omega_0}} = 2\sqrt{\frac{2\beta\omega}{\Omega_0}}. \tag{102}$$

It is bounded from above and from below:

$$2\sqrt{2}\beta < \frac{\omega_z}{\omega_\rho} < 2\sqrt{2}\frac{\omega}{\Omega_0}.$$

The last inequality is the condition in Eq. (96) for the trap minimum to be at the bottom of the ellipsoid. We see here that as soon as the underlying quadrupole trap compensates gravity, the trap is naturally anisotropic. Moreover, if the rf field fulfills the RWA $\omega \gg \Omega_0$, the upper bound is very high. In a regime where the aspect ratio reaches about 100, this trap has been used to prepare quasi two-dimensional quantum gases (Merloti et al., 2013b).

### 5.2.4 Linear Polarization

If the polarization is chosen to be linear, the best choice is a horizontal polarization, to ensure a maximum coupling at the bottom. Let us chose a polarization $\mathbf{e}_x$ along the $x$-axis. From Eq. (79), the rf coupling is

$$|\Omega_1(\mathbf{r})| = \Omega_0\sqrt{1 - u_x^2} = \Omega_0\sqrt{1 - \frac{x^2}{\ell^2}}.$$

$\Omega_0$ is defined as the maximum coupling. If we again assume that the potential minimum lies on the resonance surface, we end up with

$$V_{F,\mathrm{surf}}(\mathbf{r}) = F\hbar\Omega_0\sqrt{1 - \frac{x^2}{r_0^2}} + Mgz, \quad |z| \le \frac{r_0}{2}, \quad |x| \le r_0, \quad \ell(\rho, z) = r_0.$$

Now, the rotational symmetry is broken by the choice of the rf polarization. Two "holes" with a vanishing coupling appear on the resonant surface, at positions $(\pm r_0, 0, 0)$ opposite on the equator of the resonant ellipsoid. Spin-flips can occur if the atoms reach these points (Morizot et al., 2007), which can, however, be prevented thanks to gravity.

Indeed, there is again a competition between gravity and the coupling gradient, in the $y = 0$ plane where $V_{\mathrm{surf}}(z) = 2F\hbar\Omega z/r_0 + Mgz$. Gravity wins if $\omega > 2\beta\Omega_0$. This condition is almost the same as in the previous case and is fulfilled within the RWA for moderate magnetic gradients. For large magnetic gradients, the rf frequency must be increased to fulfill the condition on gravity.

The vertical oscillation frequency at the trap bottom is unchanged with respect to the previous case of Eq. (97). As the coupling strength is now uniform in the $yz$-plane, the oscillation frequency along $y$ is the bare pendulum frequency:

$$\omega_y = \sqrt{\frac{g}{4R}} \simeq \sqrt{\frac{g}{2r_0}}. \tag{103}$$

The $x$ oscillation frequency is lowered by the attraction to the holes (Merloti et al., 2013b):

$$\omega_x = \sqrt{\frac{g}{4R}}\left[1 - \frac{F\hbar\Omega_0}{MgR}\sqrt{1 - \frac{1}{4\beta^2}}\right]^{1/2} \simeq \sqrt{\frac{g}{2r_0}}\left[1 - \frac{2\beta\Omega_0}{\omega}\right]^{1/2}. \tag{104}$$

This effect is twice as large as in the circularly polarized case, because the holes are at half the height. The condition on gravity is clear from the expression of the oscillation frequency, which would vanish at the limit $\omega = 2\beta\Omega_0$ where the single minimum at the bottom disappears, and the two minima at the equator appear—with zero coupling and strong spin flips.

Here, the important point is that, while the trap is still extremely anisotropic in the vertical versus horizontal directions, it also becomes anisotropic in-plane, with an approximate aspect ratio $\sqrt{1 - 2\beta\Omega_0/\omega}$ controlled by the rf amplitude, opening the way to trapping two-dimensional anisotropic gases. The direction of this anisotropy is controlled by the polarization axis.

In short, modulating the direction of the polarization axis or the rf amplitude allows one to set the gas into rotation or to excite quadrupole oscillations, respectively. This has been used recently to study collective modes of a two-dimensional superfluid and probe its superfluidity, with an anisotropy of about 4/3 (De Rossi et al., 2016; Dubessy et al., 2014; Merloti, 2013). The smooth character of the adiabatic potential (Merloti et al., 2013b; van Es et al., 2008) is very well suited for such studies.

How much can we increase the in-plane anisotropy while keeping a strong vertical-horizontal anisotropy necessary for a 2D gas? We remark that the two aspect ratios $\omega_z/\omega_y$ and $\omega_x/\omega_y$ are related:

$$\frac{\omega_y}{\omega_z} \simeq \sqrt{\frac{\Omega_0}{8\beta\omega}} \quad \Rightarrow \quad \frac{\omega_x}{\omega_y} \simeq \left[1 - \frac{2\beta\Omega_0}{\omega}\right]^{1/2} \simeq \left[1 - 16\beta^2\left(\frac{\omega_y}{\omega_z}\right)^2\right]^{1/2}.$$

For an important in-plane anisotropy, $\omega_y/\omega_z$ should approach $1/(4\beta)$. In order to stay in the 2D regime, which requires in particular a strong anisotropy $\omega_z \gg \omega_y$, very large gradients $4\beta \gg 1$ are necessary. Experimentally, the maximum in-plane anisotropy which can be obtained in a 2D gas thus depends on the maximal feasible gradient.

## 5.3 Ring Traps

A proper choice of inhomogeneous static and rf fields can lead to an annular confining potential, as proposed by Lesanovsky et al. (2006) and demonstrated by Kim et al. (2016) with an atom chip in the regime of negative detuning ($\omega < \omega_{0,\min}$) as in Section 5.1.4: see Fig. 9. This geometry has been discussed briefly at Section 5.1.2 and is presented in more detail in Garraway and Perrin (2016). The gravitational sag can be compensated by a gradient in the rf amplitude. This trap can provide a strong radial annular confinement (with frequencies in the kHz range), as it relies on the confinement to a resonant surface. The axial confinement is typically weaker, and the resulting geometry is tubular.

In order to increase the axial confinement and get a ring with comparable trapping frequencies in all directions, adiabatic potentials can be combined with optical dipole potentials (Grimm et al., 2000) which widens the range of possible trapping geometries. Starting from a dressed quadrupole trap with a circular polarization, see Section 5.2.3, and taking advantage of the strong confinement to the ellipsoidal resonant surface, a ring trap is obtained by cutting the ellipsoid by a horizontal plane. The confinement

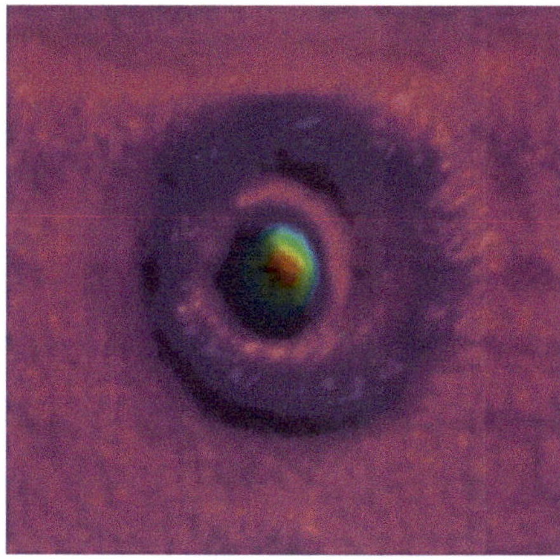

**Fig. 9** Rubidium atoms confined in an annular adiabatic potential analog to Fig. 7, left, but with a negative detuning. The central spot corresponds to atoms trapped at the minimum of the static magnetic field in the adiabatic states with $m_\theta < 0$. The experimental parameters are $\omega_{0,\text{min}}/2\pi = 700$ kHz, $\alpha = 75$ kHz μm$^{-1}$, $\omega/(2\pi) = 650$ kHz, $\Omega_0/(2\pi) \simeq 550$ kHz, $F = 2$. Note that the rf amplitude is beyond the RWA regime. Picture size: $4 \times 4$ μm$^2$. *Adapted with permission from Kim, S.J., Yu, H., Gang, S.T., Anderson, D.Z., Kim, J.B., 2016. Controllable asymmetric double well and ring potential on an atom chip. Phys. Rev. A 93, 033612. Copyrighted by the American Physical Society.*

to a given horizontal plane can be done with a standing laser wave (Morizot et al., 2006) or of a pair of blue-detuned light sheets (Heathcote et al., 2008), as illustrated in Fig. 10A–C where a possible loading mechanism is sketched.

The ring radius is easily adjustable dynamically, for example, with the rf frequency $\omega$ or the magnetic gradient of the static quadrupole field (see Fig. 10D–E). If the confinement plane is at the equator of the ellipsoid, the radial confinement is purely due to the adiabatic potential and is given by Eq. (98) (provided that the vertical gradient $2\alpha$ is replaced by the horizontal gradient $\alpha$), while the vertical confinement is purely set by the optical trap, allowing an independent tuning of these two parameters.

A circular polarization should be used to ensure the rotational symmetry of the adiabatic potential. On the other hand, using a noncircular polarization, with an additional component on $x$, for example, induces a deformation of the ring along $x$. Rotating the direction of this additional rf component allows one to induce a rotation of the atoms in the ring and create a superflow (Heathcote et al., 2008).

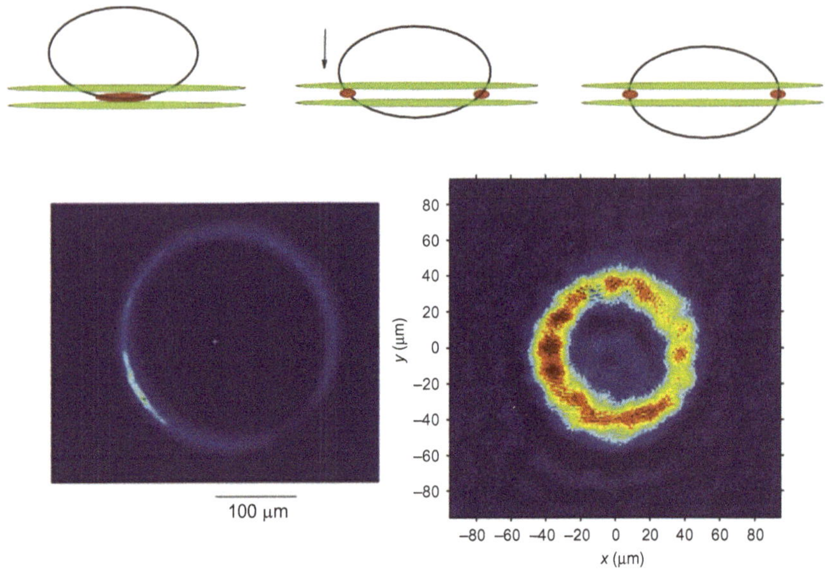


**Fig. 10** *Top row*: loading of a ring trap starting from atoms confined to a dressed quadrupole trap. (A) The two blue-detuned light sheets (in *green*) are switched on in order to confine the initial oblate atomic cloud (in *red*) between the two light maxima. (B) The resonant surface in then translated downward thanks to an additional static magnetic field aligned along the vertical axis. The atoms stay at fixed height due to the light sheets and a ring forms. (C) The maximum ring radius—and maximum radial trapping frequency—is obtained when the atoms are in the equatorial plane. *Bottom row*: atoms confined in the ring trap. (D) Radius 130 μm obtained for a low magnetic gradient of $b' = 55.4\ \mathrm{G\ cm^{-1}}$ and an rf frequency of 1 MHz. (E) Radius 40 μm corresponding to a larger magnetic gradient $b' = 218\ \mathrm{G\ cm^{-1}}$ and a dressing frequency of 600 kHz (image LPL). *Adapted from De Rossi, C., 2016. Gaz de Bose en dimension deux: modes collectifs, superfluidité et piège annulaire (Ph.D. thesis), Université Paris 13. https://tel.archives-ouvertes.fr/tel-01420085, reprinted by courtesy of the author.*

## 6. TIME-AVERAGED ADIABATIC POTENTIALS

Time-averaged potentials have been used for some time in atomic physics. An example, in a field which is close to this review, is the Time Orbiting Potential, otherwise known as a TOP trap (Petrich et al., 1995). This is a pure magnetic trap (with no rf radiation) which is often based on a quadrupole configuration. The quadrupole field (e.g., $[b'x,\ b'y,\ -2b'z]$, Section 5.2.1) has a field zero at its center where the spin direction is undefined. As a result atoms are easily lost from the trap. This problem can be solved for the magnetic trap by adding an additional potential (e.g., an optical "plug" using different, light induced forces, as done by Davis et al., 1995; Naik and Raman, 2005; and Dubessy et al., 2012). Alternatively, the time-orbiting potential approach can be used to make the TOP trap. In this case the center of the quadrupole field is rotated about another point by using a rotating bias field. Provided the rotation of the field is fast compared to the atomic dynamics, the atoms "see" an average field: the resulting time-averaged potential is a harmonic trapping potential. This requires that the harmonic trap frequency is less than the TOP rotation frequency which is less than the Larmor frequency $\omega_0$, Eq. (34).

The same approach can be used with adiabatic traps, with the only major complication being that although the atoms should experience a time-averaged potential, requiring fast changes in the potential, the changes should not be so fast as to cause nonadiabatic transitions. The first example was proposed in a theoretical paper by Lesanovsky and von Klitzing (2007) who proposed a double-well and a ring trap by this method.

Although several different types of time-averaged adiabatic potentials (TAAP) are possible and have now been realized (Gildemeister et al., 2010, 2012; Navez et al., 2016; Sherlock et al., 2011), we will here give one of the original proposals of Lesanovsky and von Klitzing as a particular example. We have already seen in Section 4.3 how resonant dressing with a quadrupolar field tends to make a "bubble" potential with the atoms trapped on the resonance surface. Lesanovsky and von Klitzing add an oscillating bias field which shifts the center of the bubble up and down (see Fig. 11). In this example they also oscillate the frequency and the strength of the rf to achieve the ring trap potential after time-averaging over all these additional oscillations.

We start by imposing an oscillating bias field on the quadrupole trap of Section 5.2.1, that is, in Eq. (89) so that

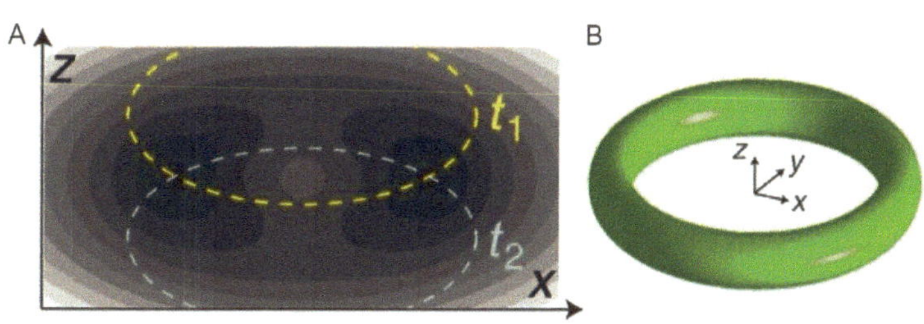


**Fig. 11** Use of time-averaged adiabatic potential, or TAAP, to change a bubble potential to a ring potential. (A) Vertical slice through the time-averaged ring potential with the averaged adiabatic potential shown with contours. Two extreme positions of the instantaneous adiabatic potential are shown with the *dashed lines* which indicate the instantaneous minimum in that potential. (B) A 3D visualization of an equipotential shown in section in (A). *From Lesanovsky, I., von Klitzing, W., 2007. Time-averaged adiabatic potentials: versatile matter-wave guides and atom traps. Phys. Rev. Lett. 99 (8), 083001. Copyrighted by the American Physical Society.*

$$\mathbf{B}_0(\mathbf{r},t) = b'(x\,\mathbf{e}_x + y\,\mathbf{e}_y - 2z\,\mathbf{e}_z) + B_m \sin(\omega_m t)\mathbf{e}_z, \tag{105}$$

where $B_m$ is the amplitude of the oscillating (modulated) field which is taken to have a vector direction along $z$ and an oscillation frequency $\omega_m$. The immediate effect of this oscillation is that the center of the quadrupole field can be regarded as oscillating up and down with a new center given by

$$z_0(t) = \frac{B_m}{2b'} \sin(\omega_m t). \tag{106}$$

At a fixed location (in cylindrical coordinates where $\rho \equiv \sqrt{x^2 + y^2}$) the Larmor frequency, see Eq. (90), is now given by

$$\omega_0(\rho, z, t) = \alpha \sqrt{\rho^2 + 4[z - z_0(t)]^2}, \tag{107}$$

where, as before, we have the frequency gradient $\alpha = |g_F|\mu_B b'/\hbar$ in a radial direction of the magnetic field. The system interacts with an rf field which is taken to be uniform and with a linear polarization in $z$-direction, i.e.,

$$\mathbf{B}_1(t) = B_1 \cos(\omega_{\mathrm{rf}} t)\,\mathbf{e}_z \tag{108}$$

which is also $\mathbf{B}_1(t) = \mathcal{B}_1\, e^{-i\omega_{\mathrm{rf}} t}\,\mathbf{e}_z + \mathcal{B}_1^*\, e^{i\omega_{\mathrm{rf}} t}\,\mathbf{e}_z$, Eq. (68) with a real value of the complex vector amplitude $\mathcal{B}_1 = B_1/2$. In Eq. (108) we use the notation $\omega_{\mathrm{rf}}$ for the frequency of the rf, rather than the usual $\omega$, as in the development below this frequency will itself become time dependent.

The linear polarization could be replaced with a circular polarization around $z$; the main point is that at the location of the proposed ring the interaction of the rf radiation with the atoms should be strong, given the approximately radial field at that point, and that the rf coupling should not break the rotational symmetry around the $z$-axis. Two aspects of the rf field have to be considered carefully: the amplitude and the frequency. Both of these quantities can also be time dependent as we see below.

The instantaneous dressed potential is given by a modified (i.e., time dependent) version of Eq. (72)

$$V_F(\rho, z, t) = F\hbar \sqrt{[\omega_0(\rho, z, t) - \omega_{\mathrm{rf}}(t)]^2 + \Omega_1(\rho, z, t)^2}, \tag{109}$$

where the local Rabi frequency $\Omega_1$ is given by Eq. (79)

$$\Omega_1(\rho, z, t) = \left|\frac{g_F \mu_B}{2\hbar}\right| B_1 \sqrt{1 - u_z(\rho, z, t)^2} = \Omega_0 \sqrt{1 - u_z(\rho, z, t)^2}. \tag{110}$$

The maximum Rabi frequency for a linear polarization $\Omega_0 = |g_F \mu_B B_1|/(2\hbar)$ is also given in Eq. (38), while, for rf polarized in the $z$-direction,

$$u_z(\rho, z, t) = \frac{\mathbf{B}_0(\rho, z, t) \cdot \mathbf{e}_z}{|\mathbf{B}_0(\rho, z, t)|} = \frac{-2[z - z_0(t)]}{\sqrt{\rho^2 + 4[z - z_0(t)]^2}} \tag{111}$$

reduces the Rabi frequency as the fields become less orthogonal. Note that we neglect the effect of gravity in this calculation for simplicity. The gravitational term will simply cause some sag in the final ring potential.

Now we suppose that we will form a ring trap centered on the origin and with a ring radius $\rho_0$, and we examine the time-dependent fields acting on an atom located at $\rho = \rho_0$, $z = 0$. The aim is to make this location a stable and static circle so that the surrounding fields, when time-averaged, form a trapping potential. The stable ring would naturally be a circle on the magnetic iso-surface given by $\omega_0 = \omega_{\mathrm{rf}}$, as this would already have a trapping potential in the radial direction. However, we note that the oscillation in $\omega_0$ would make the iso-surface expand and contract at twice the modulation frequency $\omega_m$. This would not imply a static potential point at $\rho = \rho_0$, $z = 0$, but we can compensate for this by simultaneously modulating the rf frequency so that $\omega_0(\rho_0,\ z{=}0,\ t) - \omega_{\mathrm{rf}}(t) = 0$. This implies that

$$\omega_{\mathrm{rf}}(t) = \alpha \sqrt{\rho_0^2 + [2 z_0(t)]^2}, \tag{112}$$

and then defining the lowest rf frequency as $\omega_{\mathrm{rf}}^0 = \alpha \rho_0$ we can also write this time-dependent frequency as

$$\omega_{\mathrm{rf}}(t) = \omega_{\mathrm{rf}}^0 \sqrt{1 + \beta_m^2 \sin^2(\omega_m t)}, \tag{113}$$

where $\beta_m = B_m/(b'\rho_0)$ gives the scale of vertical shaking relative to the size of the resonance surface.

This nearly leads to a stable ring in Eq. (109), but we still have oscillations in energy at $\rho = \rho_0$, $z = 0$ because the oscillating magnetic field $\mathbf{B}_0(\rho, z, t)$ causes the Rabi frequency, Eq. (110), to oscillate. This is because as the quadrupole field center moves out of the $z = 0$ plane, the field changes its angle in the $z = 0$ plane so that it points slightly out of the plane with its magnitude increasing. If the rf field is in the fixed direction $z$, as above, the Rabi frequency around the ring circle will oscillate as

$$\Omega_1(\rho_0, 0, t) = \frac{\Omega_0}{\sqrt{1 + \beta_m^2 \sin^2(\omega_m t)}} \tag{114}$$

from Eq. (110) and using $2z_0(t) = \rho_0 \beta_m \sin(\omega_m t)$. This Rabi frequency would not oscillate at the proposed ring location $\rho = \rho_0$, $z = 0$ if the rf magnetic field amplitude $B_1$ is itself modulated by replacing

$$B_1 \rightarrow B_1 \sqrt{1 + \beta_m^2 \sin^2(\omega_m t)}, \tag{115}$$

so that the Rabi frequency around the trapping circle is fixed at $\Omega_0 = |g_F \mu_B B_1|/(2\hbar)$ although it will oscillate at other locations (away from $\rho = \rho_0$, $z = 0$). In this case, i.e., away from the trapping circle, using Eq. (111) we find

$$\Omega_1(\rho, z, t) = \Omega_0 \frac{(\rho/\rho_0)\sqrt{1 + \beta_m^2 \sin^2(\omega_m t)}}{\sqrt{(\rho/\rho_0)^2 + [2z/\rho_0 - \beta_m \sin(\omega_m t)]^2}}. \tag{116}$$

So far we have engineered the three oscillating field parameters to produce a nontime-oscillating point around $\rho = \rho_0$, $z = 0$. To demonstrate that it is stable we must expand the dressed potential, Eq. (109), about this point and then time-average it to obtain the TAAP.

Thus we expand Eq. (109), firstly in $\rho$, and we let $\rho = \rho_0 + \Delta\rho$. Then for the first part in the square root of Eq. (109) we find that

$$\omega_0(\rho, z = 0, t) - \omega_{\mathrm{rf}}(t) = \omega_{\mathrm{rf}}^0 \frac{\Delta\rho}{\rho_0} \frac{1}{\sqrt{1 + \beta_m^2 \sin^2(\omega_m t)}}. \tag{117}$$

This expression is only needed in first order in $\Delta\rho$ to expand $V_F(\rho, z, t)$ to second order in $\Delta\rho$, which then becomes

$$V_F(\rho_0 + \Delta\rho, z, t) \simeq F\hbar\Omega_0 \left\{ 1 + \frac{1}{2}\left(\frac{\omega_{\mathrm{rf}}^0}{\Omega_0}\right)^2 \left(\frac{\Delta\rho}{\rho_0}\right)^2 \frac{1}{1 + \beta_m^2 \sin^2(\omega_m t)} + \cdots \right\}. \tag{118}$$

Here we have neglected terms from the expansion of Eq. (116) about $\rho = \rho_0$, $z = 0$. This is because they are smaller by a factor of $(\Omega_0/\omega_{\mathrm{rf}}^0)^2$ which makes only a slight difference to the second-order term in Eq. (118), which will be the main result. There is also a first-order term from the spatial dependence of the Rabi frequency, Eq. (116), but this simply slightly shifts the radial location of the ring trap.

Finally, for the radial direction, we time-average the approximate potential given by Eq. (118). By using

$$\int_0^{2\pi/\omega_m} \frac{1}{1+\beta_m^2 \sin^2(\omega_m t)} dt = \frac{1}{\sqrt{1+\beta_m^2}}, \tag{119}$$

we obtain the approximate second-order potential, expanded in the $\rho$-direction, as

$$V_F(\rho_0 + \Delta\rho, z) \simeq F\hbar\Omega_0 \left\{ 1 + \frac{1}{2}\left(\frac{\omega_{\mathrm{rf}}^0}{\Omega_0}\right)^2 \left(\frac{\Delta\rho}{\rho_0}\right)^2 \frac{1}{\sqrt{1+\beta_m^2}} \right\}, \tag{120}$$

where, again, we have neglected the slight effects of the spatial variation of the Rabi frequency in Eq. (116). The potential of Eq. (120) implies a vibrational frequency through

$$V_F = V_0 + \frac{1}{2} M\omega_\rho^2 (\rho - \rho_0)^2, \tag{121}$$

where $V_0$ is the potential at the ring trap bottom (where $\rho = \rho_0$, $z = 0$), $M$ is the mass of the atom and $\omega_\rho$ is the vibrational frequency. By comparing Eq. (120) to Eq. (121) we find the radial trapping frequency

$$\omega_\rho = \omega_{\mathrm{trans}} (1 + \beta_m^2)^{-1/4}. \tag{122}$$

This result has been referred to the standard transverse trapping frequency $\omega_{\mathrm{trans}}$, Eq. (80), which would be the dressed trapping frequency in the radial direction if there was no modulation $B_m$ and with the atoms confined by other means to the plane $z = 0$. In this situation we can obtain $\omega_{\mathrm{trans}}$ from the limit of Eq. (122) as $\beta_m \to 0$, or from Eq. (80) as $\omega_{\mathrm{trans}} = |g_F \mu_B b'| \sqrt{F/(M\hbar\Omega_0)}$. We note that, as expected, this result, Eq. (122), does not depend on the modulation frequency $\omega_m$.

A similar calculation can be made in the $z$-direction. This time, the position is set to $(\rho, z) \to (\rho_0, \Delta z)$, which has a small displacement $\Delta z$ in the $z$-direction. From Eqs. (107) and (112) we find the approximate detuning

$$\omega_0(\rho_0, \Delta z, t) - \omega_{\mathrm{rf}}(t) \simeq -2\omega_{\mathrm{rf}}^0 \frac{\Delta z}{\rho_0} \frac{\beta_m \sin(\omega_m t)}{\sqrt{1+\beta_m^2 \sin^2(\omega_m t)}}. \tag{123}$$

This becomes squared inside $V_F$, Eq. (109), so that if we make the same approximation of a constant Rabi frequency, because of the modulation given by Eq. (115) and small displacements, we can obtain the approximated adiabatic potential

$$V_F(\rho_0, \Delta z, t) \simeq F\hbar\Omega_0 \left\{ 1 + 2\left(\frac{\omega_{\rm rf}^0}{\Omega_0}\right)^2 \left(\frac{\Delta z}{\rho_0}\right)^2 \frac{\beta_m^2 \sin^2(\omega_m t)}{1 + \beta_m^2 \sin^2(\omega_m t)} + \cdots \right\}. \tag{124}$$

Then after time-averaging over a modulation period $2\pi/\omega_m$ we will find the $z$-direction vibrational frequency:

$$\omega_z = 2\omega_{\rm trans}\sqrt{1 - (1 + \beta_m^2)^{-1/2}}. \tag{125}$$

Eq. (122) shows that the effect of increased modulation depth is to slowly weaken the confinement of the atom in the radial direction. However, in the $z$-direction, Eq. (125) shows that the opposite happens. With no modulation, there is no confinement in this direction, but as the modulation depth is increased the trapping frequency monotonically increases (in the harmonic approximation). The choice $\beta_m = 3/4$ gives equal oscillation frequencies along $z$ and $\rho$.

An argument based on the symmetry of the vertical oscillation, see Eq. (105), and the expansion oscillation, see Eq. (113), which has half the period, indicates that when neglecting the spatial variation of the Rabi frequency in Eq. (110), the local $\rho$- and $z$-axes form the principal axes in the potential.

The full 3D TAAP calculation (Lesanovsky and von Klitzing, 2007) produces the ring trap seen in Fig. 11. Here the potentials are calculated exactly, i.e., beyond the harmonic regime. The resonance surface is shown at two extreme positions as dashed lines in Fig. 11A, where it has an expanded size due to Eq. (113).

As mentioned above, the TAAP ring was realized by Sherlock et al. (2011), and Navez et al. (2016). The TAAP concept has been applied to other trapping situations involving adiabatic traps. Lesanovsky and von Klitzing (2007) proposed "dumbbell" traps and other exotic structures. Vangeleyn et al. (2014) proposed an rf ring trap based on fields induced in a metal ring. The bias field required for rf dressing necessarily introduced an asymmetry which was removed by rotating the bias potential at a low frequency and time-averaging it (see discussion in Garraway and Perrin, 2016): i.e., by essentially forming a TAAP. In another example, Garraway and Perrin (2010) obtained a TAAP by rotating a polarization direction which affected the couplings in the magnetic resonance problem.

# 7. MULTIPLE RF FIELDS

## 7.1 Well-Separated rf Frequencies

Up to this point in this review we have focused on single frequency rf fields, although it could be argued that the time-dependent fields of the previous section on TAAPs (Section 6) present a spread of frequencies to the atom. In this section we examine in detail the physics and features of multiple discrete rf frequencies with a particular emphasis on two fields with frequencies $\omega_1$ and $\omega_2$.

It would seem clear that if an atom interacts with two rf fields which are "well" separated the fields will not interfere with each other and at a particular point in space the atom will interact with the dominant field at that location. (Here we consider that, because of experimental preparation, only a single hyperfine multiplet $F$ is involved.) That is, given a spatially inhomogeneous Larmor frequency $\omega_0(\mathbf{r})$, the conditions $\omega_0(\mathbf{r}) = \omega_1$ and $\omega_0(\mathbf{r}) = \omega_2$ define different regions of space when $\omega_1$ and $\omega_2$ are different. This means that, for resonant rf trapping, we could have atoms trapped in adiabatic traps in two locations with the two rf frequencies defining the two distinct traps. But then we have the question of what is meant by "well" separated, and what happens if the regions of space overlap? A plausible criterion is that the frequency separation $\Delta = \omega_2 - \omega_1$ should be much larger than a characteristic Rabi frequency, $\Omega_1$ or $\Omega_2$, at all places visited by the atom for this simple, single field, picture to be valid. An interesting aspect of even this simple picture is that two such adiabatic traps are not simply connected if the magnetic field strength changes monotonically from one place to another. Following the adiabatic potential from the minimum at $\omega_0(\mathbf{r}) = \omega_1$ we will find we arrive at a potential maximum at $\omega_0(\mathbf{r}) = \omega_2$. Essentially, this is because adiabatic potentials cannot cross.

This is not to preclude the formation of double-well potentials with a single frequency, or more, because of spatial variations in magnetic field strength, or rf polarization, which are not monotonic. However, by using three rf fields it is possible to create two connected resonant rf atom traps, and with five rf fields three traps can be created, and so on. This approach has been suggested to make an rf-dressed atom lattice by Courteille et al. (2006). Such a lattice would be highly controllable as individual lattice sites could have their position adjusted by changing their associated rf frequency and the height of the barriers between lattice sites could be adjusted, to some extent, by controlling the strength of the Rabi coupling at those locations

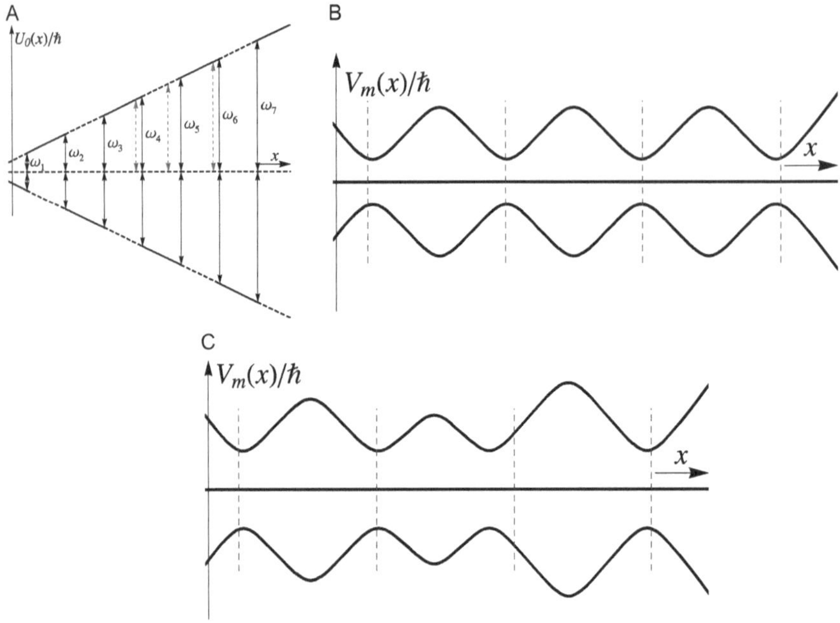


**Fig. 12** A one-dimensional lattice obtained with a rf frequency comb (here an example with seven frequencies such that $\omega_n = n\omega_1$, $n = 1\ldots7$ for a $F = 1$ state). (A) The bare magnetic potential $U_0(x)$ is linear, and the resonance points (*full line arrows*) occur at evenly spaced places. (B) Resulting adiabatic potentials $V_m(x)$. The correction given by Eq. (126) has been included. (C) Adiabatic potential where $\omega_5$ is shifted to a lower value, while $\omega_4$ and $\omega_6$ are shifted twice as less, see the *red dashed arrows* in (A). As a consequence, the third well is shifted to the left.

(while maintaining adiabaticity). Fig. 12 presents an example with seven frequencies forming evenly separated wells. By modifying three of these frequencies, it is possible to shift spatially one of the wells among the others in a controlled way, for example, to modify the tunneling between sites.

However, even this picture of well-separated traps is not so simple as the atom trap at one location will be perturbed by the field of another even if we have the condition $|\Delta| \gg \Omega_1$. For example, Eq. (72) can be considered in the limit where $|\omega_1 - \omega_0(\mathbf{r})| \gg \Omega_1$ in order to examine perturbatively the effect of the field $\omega_1$ at the resonant location of the field $\omega_2$ (where $\omega_2 = \omega_0(\mathbf{r})$) under the condition $|\Delta| \gg \Omega_{1,2}$. Thus, letting $\omega_1 - \omega_0(\mathbf{r}) \to \Delta$ we find

$$V_{F,1}(\mathbf{r}) = F\hbar\sqrt{[\omega_1 - \omega_0(\mathbf{r})]^2 + |\Omega_1(\mathbf{r})|^2} \quad \to \quad F\hbar|\Delta| + \frac{1}{2}F\hbar\frac{|\Omega_1(\mathbf{r})|^2}{|\Delta|}. \tag{126}$$

In perturbation theory this acts as a shift in the resonant frequency required for $\omega_2$, or alternatively, if a perfect spatial lattice is required then offsets such as Eq. (126) must be compensated for to obtain resonance in the correct location. The effects are pervasive because the correction, Eq. (126), depends on the inverse of the frequency difference $\omega_1 - \omega_2$ which vanishes slowly with frequency difference. Furthermore, if many rf frequencies are involved, such as in the rf lattice proposed by Courteille et al. (2006), then the shifts of all other rf frequencies have to be included, as a sum, at any particular lattice site. An example with seven frequencies is shown in Fig. 12. In addition, from a practical point-of-view, care has to be taken over the presence of harmonics from any of the chosen frequencies as harmonic contamination could produce couplings in unexpected places (see Section 8.2).

The above picture is, however, essentially one-dimensional and may be much more complex in three-dimensions. For example, using an atom chip lattice of wires one of us has been able to design a 2D rf-dressed lattice with a single field where the spatial variation is produced by near-fields to the chip rather than different frequencies (Sinuco-León and Garraway, 2015).

Despite these complexities, trapping with well-separated multifrequency fields has been analyzed using the quadrupole trap (Bentine et al., 2017). In this case, two rf fields which are well separated in frequency space create two well-separated resonance surfaces (see Section 5.2). The same work uses four rf frequencies to create a two-species trap with a double well for one species and a single well for the other.

Returning to the case of two rf frequencies, we realize that if one of the fields is strong, and the other is weak, we can consider the case of a probe field (the second rf field ) performing spectroscopy on a dressed rf atom trap. Such a probe, which typically induces atom loss at particular locations from the atom trap can also be used as a tool for evaporative cooling of the adiabatic trap. This system is also amenable to analytic treatment, even when the two fields are not too well separated, and we turn in the next sections to this problem in more detail.

## 7.2 Double-Dressing and rf Evaporative Cooling

The situation when a second, weaker rf field is added to the main dressing rf field is of important practical interest, as the second rf field can be used to evaporatively cool the trapped atoms or to probe the Rabi splitting between the dressed states.

In this section, to simplify the discussion, we take $g_F > 0$ ($s = 1$) and consider a homogeneous static magnetic field aligned along $z$ with a Larmor frequency $\omega_0$. The generalization to the inhomogeneous case proceeds as in Section 4. The spin states are coupled by a first, strong rf field of frequency

$\omega_1$ and Rabi frequency $\Omega_{1,+}$, real and optimally ($\sigma^+$) polarized, with $\Omega_{1,+} \ll \omega_1$ to ensure the application of RWA. The purpose of this section is to describe the effect of a second field of frequency $\omega_2$ with Rabi frequencies $\Omega_{2,z}$ and $\Omega_{2,+}$ in the $\pi$ (linear along $z$) and $\sigma^+$ (circularly polarized around $z$) directions, respectively. We do not consider a $\sigma^-$ polarization, as its effect is neglected within the rotating wave approximation.

We start with the time-dependent Hamiltonian

$$\begin{aligned}\hat{H}(t) = \omega_0 \hat{F}_z + \Omega_{1,+}\left(\hat{F}_x \cos\omega_1 t + \hat{F}_y \sin\omega_1 t\right) \\ + \Omega_{2,+}\left(\hat{F}_x \cos\omega_2 t + \hat{F}_y \sin\omega_2 t\right) + 2\Omega_{2,z}\hat{F}_z \cos\omega_2 t.\end{aligned} \tag{127}$$

The states dressed by the first field are obtained by first applying a rotation of frequency $\omega_1$ around $z$, then a rotation of angle $\theta$ around $y$, as in Section 3.3.1. We introduce sequentially the state $|\psi_{\text{rot}}\rangle$ defined by Eq. (39) following the rotated Hamiltonian of Eq. (40), and the state $|\tilde{\psi}\rangle$, related to the initial state $|\psi\rangle$ through $|\psi\rangle = \hat{R}_z(\omega_1 t)\hat{R}_y(\theta)|\tilde{\psi}\rangle$. These unitary transformations diagonalize the first line of Eq. (127), but also modify the terms describing the second field. The transformed Hamiltonian for $|\tilde{\psi}\rangle$ reads

$$\begin{aligned}\hat{\tilde{H}} &= \left[\hat{R}_y(\theta)\right]^\dagger\left(-i\hbar\hat{R}_z^\dagger\left[\partial_t\hat{R}_z\right] + \left[\hat{R}_z(\omega_1 t)\right]^\dagger\hat{H}\hat{R}_z(\omega_1 t)\right)\hat{R}_y(\theta) \\ &= \Omega\hat{F}_z + \Omega_{2,+}\left[\cos\theta\cos\Delta t\hat{F}_x + \sin\theta\cos\Delta t\hat{F}_z + \sin\Delta t\hat{F}_y\right] \\ &\quad + 2\Omega_{2,z}\cos\omega_2 t\left(\cos\theta\hat{F}_z - \sin\theta\hat{F}_x\right)\end{aligned} \tag{128}$$

$$\hat{\tilde{H}} = \hat{H}_{\|} + \hat{H}_+ + \hat{H}_- + \hat{H}_z, \quad \text{with}$$

$$\hat{H}_{\|} = \Omega\hat{F}_z + \Omega_{2,+}\sin\theta\cos\Delta t\hat{F}_z + 2\Omega_{2,z}\cos\omega_2 t\cos\theta\hat{F}_z \tag{129}$$

$$\hat{H}_+ = \Omega_{2,+}\cos^2(\theta/2)\left(\cos\Delta t\hat{F}_x + \sin\Delta t\hat{F}_y\right) \tag{130}$$

$$\hat{H}_- = -\Omega_{2,+}\sin^2(\theta/2)\left(\cos\Delta t\hat{F}_x - \sin\Delta t\hat{F}_y\right) \tag{131}$$

$$\hat{H}_z = -2\Omega_{2,z}\sin\theta\cos\omega_2 t\hat{F}_x \tag{132}$$

where $\Omega = \sqrt{\delta_1^2 + |\Omega_{1,+}|^2}$, $\delta_1 = \omega_1 - \omega_0$, and $\Delta = \omega_2 - \omega_1$.

This new Hamiltonian is analogous to the one of a spin in a static magnetic field and an rf field. The terms in $\hat{H}_{\|}$, Eq. (129), proportional to $\hat{F}_z$ do not couple the states dressed by the first field, but the terms along $\hat{F}_x$ or $\hat{F}_y$ do. They oscillate either at $\pm\Delta$ or at $\omega_2$, and will be resonant when one of these three frequencies approaches the splitting $\Omega$ between states.

Specifically, for a frequency $\omega_2$ close to $\omega_1 + \varepsilon\Omega$ with $\varepsilon = \pm 1$, either $\hat{H}_+$, Eq. (130), or $\hat{H}_-$, Eq. (131), is nearly resonant, and we can define a detuning $\delta_2 = \omega_2 - \omega_1 - \varepsilon\Omega = \Delta - \varepsilon\Omega$. We repeat the procedure described in Section 3.3.1: another rotation around $\hat{F}_z$ at a frequency $\varepsilon\Delta$ and the application of the rotating wave approximation yields the rotated Hamiltonian

$$\hat{H}_{2,+} = -\delta_2\hat{F}_z + \Omega_{2,+}\cos^2(\theta/2)\hat{F}_x \quad \text{in the case} \quad \varepsilon = 1, \tag{133}$$

$$\hat{H}_{2,-} = \delta_2\hat{F}_z - \Omega_{2,+}\sin^2(\theta/2)\hat{F}_x \quad \text{in the case} \quad \varepsilon = -1. \tag{134}$$

This expression is valid in the situation where $|\delta_2|, \Omega_{2,+} \ll |\Delta|, \Omega \ll \omega_1$. This ensures the application of RWA first to the main $\omega_1$ dressing field, then to the second, weaker $\omega_2$ field.

We note that the effective coupling $\Omega_{2,\text{eff}} = \Omega_{2,+}\cos^2(\theta/2)$, or $\Omega_{2,+}\sin^2(\theta/2)$, of the second rf field is reduced by a factor $\cos^2(\theta/2)$ or $\sin^2(\theta/2)$. This reduced coupling, asymmetric with respect to the sign of $\delta_1$, is easily understood in the limit $|\delta_1| \gg \Omega_{1,+}$, where $\Omega \simeq |\delta_1|$. For $\delta_1 < 0$, $\theta \simeq 0$ and only the coupling at $\omega_2 \simeq \omega_1 + \Omega \simeq \omega_1 - \delta_1 = \omega_0$ is nonvanishing. This simply corresponds to the direct coupling of $\omega_2$ to the initial undressed spin, at the Larmor frequency $\omega_0$. In the same way, for large and positive $\delta_1$, $\theta \simeq \pi$ and only the coupling at $\omega_2 \simeq \omega_1 - \Omega \simeq \omega_1 - \delta_1 = \omega_0$ remains. The apparition of the other coupling is always due to the dressing by the first rf field, such that $\delta_1$ should not be too large for the second resonance to occur. We come back to the interpretation of the different couplings at the end of Section 7.3.

In the case of a low frequency $\omega_2$, close to $\Omega$, the last term $\hat{H}_z$, Eq. (132), is resonant. A rotation at $\omega_2$ around $z$ yields the transformed Hamiltonian within RWA

$$\hat{H}_{2,z} = -\delta_2'\hat{F}_z - \Omega_{2,z}\sin\theta\hat{F}_x, \quad \text{with} \quad \delta_2' = \omega_2 - \Omega. \tag{135}$$

Resonance occurs at $\omega_2 = \Omega$, with again a reduced coupling amplitude $\Omega_{2,\text{eff}} = \sin\theta\,\Omega_{2,z}$, which is now symmetric with respect to the sign of $\delta_1$, and maximum for $\delta_1 = 0$. This shows that coupling is only relevant when the first rf field is nearly resonant, and $\theta$ is different from both 0 and $\pi$.

In any of these three cases, if now the Larmor frequency depends on position, the application of the second rf field can lead to doubly dressed states, with a modified adiabatic potential

$$V_{\text{dd},m}(\mathbf{r}) = m\hbar\sqrt{\delta_2(\mathbf{r})^2 + \Omega_{2,\text{eff}}(\mathbf{r})^2}. \tag{136}$$

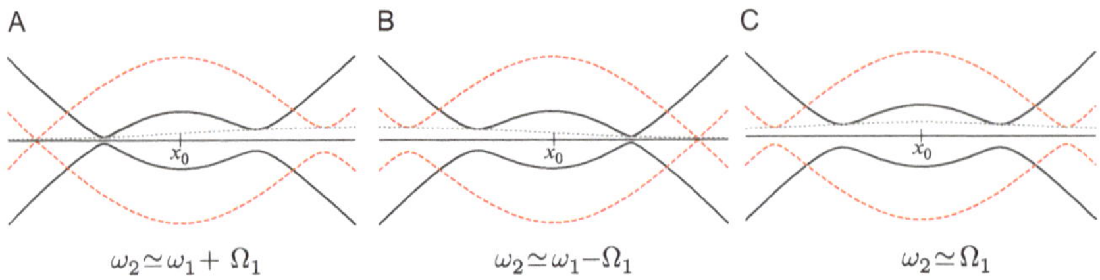


**Fig. 13** Effect of a second rf field on the adiabatic potentials sketched in Fig. 3 ($F = 1$), for the three cases $\omega_2 \simeq \omega_1 \pm \Omega_1$ and $\omega_2 \simeq \Omega_1$, each time for two values of $\omega_2$. In each plot, the *dotted line* represents the Rabi splitting due to the second field, which is asymmetric in (A) and (B) and symmetric in (C). (A) $\omega_2 = \omega_1 + 1.2\,\Omega_1$ (*full black line*) and $\omega_2 = \omega_1 + 1.6\,\Omega_1$ (*dashed red line*). (B) $\omega_2 = \omega_1 - 1.2\,\Omega_1$ (*full black line*) and $\omega_2 = \omega_1 - 1.6\,\Omega_1$ (*dashed red line*). (C) $\omega_2 = 1.2\,\Omega_1$ (*full black line*) and $\omega_2 = 1.6\,\Omega_1$ (*dashed red line*).

Due to the asymmetry in the effective coupling when $\omega_2 \simeq \omega_1 \pm \Omega_1$, the doubly dressed potentials also are asymmetric, as seen in Fig. 13 for the simple case of a linear static magnetic field (compare with Fig. 3 sketching the adiabatic potential with a single frequency in the same situation). Atoms initially trapped in the adiabatic potential resulting from the first field $\omega_1$ now see a reduced trap depth as a result of the second rf field $\omega_2$, which allows one to perform rf evaporation in the adiabatic potential (Garrido Alzar et al., 2006). In this context, the asymmetry in the cases depicted in Fig. 13A and B means that, far from the trap bottom (for large $|\delta_2|$), evaporation will occur essentially on a resonance surface around (or inside) the trapping surface of the first adiabatic potential. On the other hand, the low-frequency resonance at $\omega_2 \simeq \Omega_1$ provides an evaporation on both sides, but is less efficient at larger detuning $|\delta_2'|$, where the effective Rabi frequency decreases.

## 7.3 Trap Spectroscopy

When the second rf field $\omega_2$ is very weak, the adiabatic condition is not fulfilled any more and the effect of the second field is to induce nonadiabatic transitions between the states dressed by the first rf field (Cohen-Tannoudji and Reynaud, 1977). The second field can thus be used as a probe, by measuring the nonadiabatic losses as a function of the frequency $\omega_2$, which allows us to characterize the effect of the first field. To describe the effect of the probe, the quantum approach gives more insight in the elementary processes at work. In the quantum field approach the fields $\omega_1$ and $\omega_2$ are described by annihilation and creation operators. The full Hamiltonian in the presence of the second field at $\omega_2$, with the same notations as in the previous section, reads

$$\hat{H} = \hat{H}_0 + \hat{V}_+ + \hat{V}_z, \tag{137}$$

where

$$\hat{H}_0 = \hat{H}_1 + \hbar\omega_2 \hat{a}_2^\dagger \hat{a}_2, \tag{138}$$

$$\hat{H}_1 = \hbar\omega_1 \hat{a}_1^\dagger \hat{a}_1 + \omega_0 \hat{F}_z + \frac{\Omega_{1,+}^{(0)}}{2}(\hat{a}_1 \hat{F}_+ + \hat{a}_1^\dagger \hat{F}_-), \tag{139}$$

$$\hat{V}_+ = \frac{\Omega_{2,+}^{(0)}}{2}(\hat{a}_2 \hat{F}_+ + \hat{a}_2^\dagger \hat{F}_-), \tag{140}$$

$$\hat{V}_z = \Omega_{2,z}^{(0)}(\hat{a}_2 + \hat{a}_2^\dagger)\hat{F}_z. \tag{141}$$

$\Omega_i^{(0)}$ with $i = 1, 2+, 2z$ are the single-photon Rabi frequencies, corresponding to average Rabi frequencies $\Omega_i = \sqrt{\langle N_i \rangle}$, taken as real for simplicity. The effect of the probe field $\omega_2$ is to couple the eigenstates of the "unperturbed" Hamiltonian $\hat{H}_0$, describing the effect of the dressing field $\omega_1$.

The states dressed by the first field, eigenstates of $\hat{H}_1$, are $|m, N_1\rangle_\theta$, given by Eq. (66). The eigenstates of $\hat{H}_0$ are then simply $|m, N_1\rangle_\theta |N_2\rangle_2$, where $|N_2\rangle_2$ is a Fock state for the probe field $\omega_2$, with the same decomposition on the bare product states than Eq. (66):

$$|m, N_1\rangle_\theta |N_2\rangle_2 = \sum_{m'=-F}^{F} \langle m'|_z \hat{R}_y(\theta)|m\rangle_z |m'\rangle_z |N_1 - m'\rangle_1 |N_2\rangle_2 \tag{142}$$

where $|N\rangle_i$ represents a Fock states with $N$ photons for the field $\omega_i$.

These eigenstates of $\hat{H}_0$ are coupled by $\hat{V}_+$ and $\hat{V}_z$. When the frequency $\omega_2$ is close to $\omega_1 \pm \Omega_{1,+}$, the eigenstates of $\hat{H}_0$ are resonantly coupled by $\hat{V}_+$, while they are resonantly coupled by $\hat{V}_z$ when $\omega_2 \simeq \Omega_{1,+}$. After simple algebra, writing the matrix element between two states $|m, N_1\rangle_\theta |N_2\rangle$ and $|m', N_1'\rangle_\theta |N_2'\rangle$ induced by $\hat{V}_+$ or $\hat{V}_z$ allows us to recover the effective couplings derived in the last section. The matrix elements are given below:

$$\langle m-1, N_1 - 1|_\theta \langle N_2 + 1| \hat{V}_+ |m, N_1\rangle_\theta |N_2\rangle = \frac{\Omega_{2,+}}{2} \cos^2\frac{\theta}{2} \langle m-1|_z \hat{F}_- |m\rangle_z$$

$$\langle m-1, N_1 + 1|_\theta \langle N_2 - 1| \hat{V}_+ |m, N_1\rangle_\theta |N_2\rangle = -\frac{\Omega_{2,+}}{2} \sin^2\frac{\theta}{2} \langle m-1|_z \hat{F}_- |m\rangle_z$$

$$\langle m-1, N_1|_\theta \langle N_2 + 1| \hat{V}_z |m, N_1\rangle_\theta |N_2\rangle = \frac{\Omega_{2,z}}{2} \sin\theta \langle m-1|_z \hat{F}_- |m\rangle_z$$

**Table 1** Effective Coupling of the Probe for the Various One-Photon Resonances

| Resonant $\omega_2$ | Probe Polarization | Coupling $\Omega_{\text{eff}}$ | $\theta \to 0$ | $\theta \to \pi$ |
|---|---|---|---|---|
| $\omega_1 + \Omega_{1,+}$ | $\sigma_+$ | $\Omega_{2,+}\cos^2(\theta/2)$ | $\Omega_{2,+}$ | $\dfrac{\Omega_{1,+}\Omega_{2,+}^2}{2\Omega^2}$ |
| $\omega_1 - \Omega_{1,+}$ | $\sigma_+$ | $\Omega_{2,+}\sin^2(\theta/2)$ | $\dfrac{\Omega_{1,+}\Omega_{2,+}^2}{2\Omega^2}$ | $\Omega_{2,+}$ |
| $\Omega_{1,+}$ | $\pi$ | $\Omega_{2,z}\sin\theta$ | $\dfrac{\Omega_{1,+}\Omega_{2,+}}{\Omega}$ | $\dfrac{\Omega_{1,+}\Omega_{2,+}}{\Omega}$ |

and the corresponding effective coupling $\Omega_{\text{eff}}$, where $\Omega_{\text{eff}}/2$ appears in front of $\hat{F}_\pm$, is given in Table 1. We recover the asymmetric couplings also valid for rf-dressed atoms coupled to probe fields in the optical or microwave range (Cohen-Tannoudji and Haroche, 1969). The corresponding absorption or emission of probe photons is sketched in Fig. 14.

In the limit where $\theta$ is close to 0 or $\pi$, which corresponds to probing regions away from the resonance of the strong dressing field ($|\delta_1| \gtrsim \Omega_{1,+}$), the coupling amplitudes are proportional to $\Omega_{1,+}^n\Omega_{2,+}$ (see Table 1). The transitions can then be interpreted in terms of single, two- or three-photon processes, with one photon of the probe and $n$ photons of the dressing field, depending on the power $n$ of $\Omega_{1,+}$ in the expression of the coupling. Fig. 15 shows the three resonances in the basis of the bare states, which is relevant for large $\delta_1$, as a decomposition into multiphoton processes implying one or several photons of the dressing field. This picture makes apparent the reason of the asymmetry in the coupling of the resonances at $\omega_1 \pm \Omega_{1,+}$, and the symmetry of the resonance at $\Omega_{1,+}$. The weak resonance is a three-photon process, whereas the strong resonance results from the direct coupling of the probe $\Omega_{2,+}$ between bare states.

## 7.4 Higher-Order Probe in the Dressed Atom Trap

Even higher-order processes can contribute if a probe $\omega_2$ is used with a strength that is too great. For example, an $\omega_2$ excitation is possible between dressed states $|m_\theta = +1, N_1{+}1\rangle_\theta$ and $|m_\theta = 0, N_1{-}1\rangle_\theta$ (see Fig. 14) via a point half-way between $|m_\theta = +1, N_1\rangle_\theta$ and $|m_\theta = 0, N_1\rangle_\theta$, under the condition $\Delta = \omega_2 - \omega_1 = \Omega/2$. Therefore, unlike the probe process of the previous section, this process takes place through an intermediate point which is nonresonant (as in a nonresonant Raman process in three-level systems).

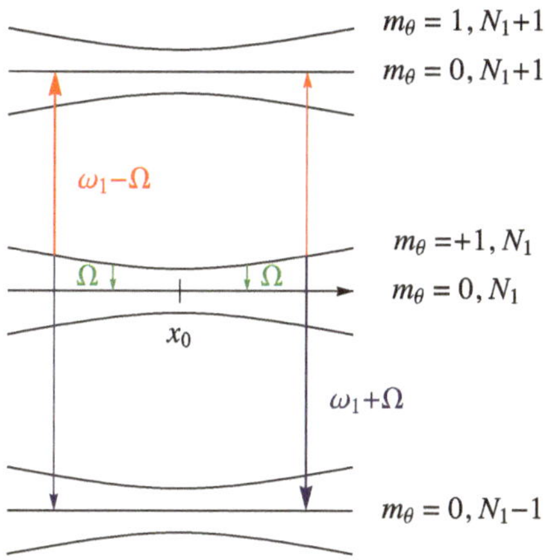


**Fig. 14** Effect of a second, weak rf field on the adiabatic potentials sketched in Fig. 3 ($F = 1$), represented in the dressed picture for three cases: $\omega_2 = \omega_1 + \Omega$ (*blue arrows*) and $\omega_2 = \omega_1 - \Omega$ (*red arrows*) with $\Omega = 1.6\Omega_{1,+}$, which couple different manifolds; and $\omega_2 = \Omega$ (*green arrows*) with $\Omega = 1.2\Omega_{1,+}$, within a given manifold. The stronger resonances in the cases $\omega_2 \simeq \omega_1 \pm \Omega_{1,+}$ are indicated by a *bolder arrow*.

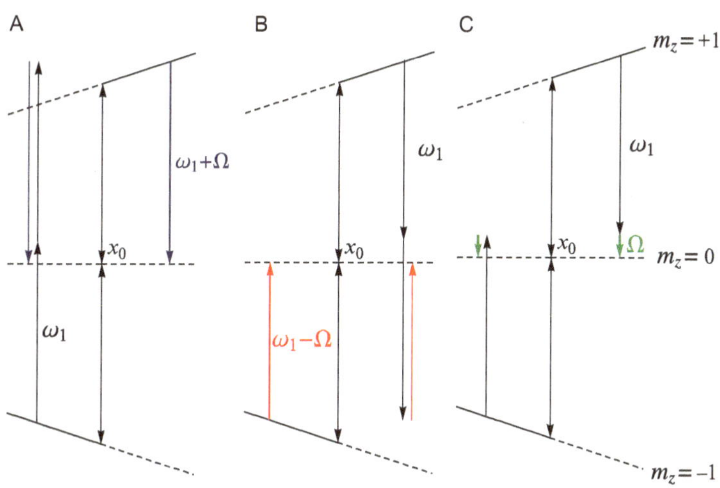


**Fig. 15** Probe resonances as seen in the basis of the bare states, for a linear static magnetic field (same field as in Fig. 14) and away from the resonance point at $x_0$. The processes at frequency $\omega_2 \simeq \omega_1 \pm \Omega_{1,+}$, resonant at the two positions $x$ such that $|\omega_2 - \omega_1| = \Omega(x)$, involve either one probe photon alone or three photons, with one probe photon and two photons from the dressing field. The resonances at frequency $\Omega_{1,+}$ always involve a probe photon and a dressing photon. (A) $\omega_2 \simeq \omega_1 + \Omega_1$. (B) $\omega_2 \simeq \omega_1 - \Omega_1$. (C) $\omega_2 \simeq \Omega_1$

We will call this a two-photon multiphoton probe process and examine the conditions under which it can take place, and explain when it should be avoided, in this section. The more general case of this kind of probe resonance requires $\Delta = \Omega/n$, where $n$ is an integer ($n \neq 0$) characterizing the order of the process.

Our starting point for this analysis is the Hamiltonian in the dressed frame, Eq. (128). However, for simplicity, and because it does not contribute significantly to the resonance $\Delta = \Omega/n$, we will drop the last term which is proportional to $\Omega_{2,z}$. (We can treat it separately, however, to obtain an equivalent "low" $\omega_2$ resonance under the different condition $\omega_2 = \Omega/n$.) Thus, from Eq. (128) we have

$$\hat{\tilde{H}} = \Omega \hat{F}_z + \Omega_{2,+}\left[\cos\theta \cos\Delta t \hat{F}_x + \sin\theta \cos\Delta t \hat{F}_z + \sin\Delta t \hat{F}_y\right]. \tag{143}$$

This Hamiltonian is dominated by the diagonal $\Omega\hat{F}_z$ term under the condition that $\Omega_{2,+}$ is rather weak. To proceed, we follow the argument of Pegg and Series (1970) (see also Allegrini and Arimondo, 1971, Pegg, 1972, Garraway and Vitanov, 1997) and we further transform the Hamiltonian to an interaction representation by performing a rotation about $z$ that completely removes the diagonal term. That is

$$\begin{aligned}\hat{\tilde{H}} \to \hat{\tilde{H}}' &= e^{i\frac{\phi(t)}{\hbar}\hat{F}_z}\hat{\tilde{H}}\, e^{-i\frac{\phi(t)}{\hbar}\hat{F}_z} - \hbar\phi'(t)\hat{F}_z \\ &= \Omega_{2,+}\cos(\Delta t)\cos\theta\left[\hat{F}_x\cos\phi(t) - \hat{F}_y\sin\phi(t)\right] \\ &\quad + \Omega_{2,+}\sin(\Delta t)\left[\hat{F}_y\cos\phi(t) + \hat{F}_x\sin\phi(t)\right]\end{aligned} \tag{144}$$

where $\phi(t)$ comes from integrating the coefficient of $\hat{F}_z$ in Eq. (143), i.e.,

$$\phi(t) = \int_0^t dt'(\Omega + \Omega_{2,+}\sin\theta\cos\Delta t') = \Omega t + \frac{\Omega_{2,+}\sin\theta}{\Delta}\sin(\Delta t). \tag{145}$$

We now inspect the trigonometric functions in $\hat{\tilde{H}}'$, Eq. (144). To deal with the time-dependence, involving a trigonometric function of a trigonometric function, we follow Pegg and Series (1970) and use a Bessel function expansion to rewrite the exponential of $\phi(t)$, Eq. (145): i.e., we observe that we have the Fourier expansion

$$e^{i\phi(t)} = e^{i\Omega t}\cdot e^{ib\sin(\Delta t)} = e^{i\Omega t}\sum_{n=-\infty}^{\infty} e^{in\Delta t}J_n(b), \tag{146}$$

where we let $b = \sin\theta\,\Omega_{2,+}/\Delta$, which we expect to have a small value. The real and imaginary parts of Eq. (146) provide the needed trigonometric functions in Eq. (144) as infinite sums.

Under the $n$-photon resonance condition $n\Delta = \Omega$ we find that the Hamiltonian of Eq. (144) contains purely harmonic terms and we can now make a second rotating wave approximation (Garraway and Vitanov, 1997; Pegg, 1973) where we keep only the static terms and drop all the rotating terms. After doing this we are left with a new Hamiltonian in the 2nd RWA where the infinite sums are now lost, i.e.,

$$\hat{\tilde{H}}'_{n,\mathrm{RWA}} = (-1)^{n+1}\frac{\Omega_{2,+}}{2}[(\cos\theta - 1)J_{n+1}(b) + (\cos\theta + 1)J_{n-1}(b)]\hat{F}_x. \tag{147}$$

This coupling is proportional to the weak probe strength $\Omega_{2,+}$, and couples the dressed states when $n\Delta = \Omega$. The strength of this coupling can also be shown to be a measure of the width of the resonance.

From Eq. (147) we can read off an effective Rabi frequency $\Omega_{n,\mathrm{eff}}$ for the coupling (i.e., the coefficient of $\hat{F}_x$). Given that the parameter $b = \sin\theta\,\Omega_{2,+}/\Delta$ is so small when $n\Delta = \Omega$, we simplify this for practical purposes by using the Bessel function expansion

$$J_n(x) \simeq \frac{1}{n!}\left(\frac{x}{2}\right)^n, \quad n \in \mathbb{N} \tag{148}$$

and keeping only the most significant terms (which means neglecting $J_{n+1}(b)$ compared to $J_{n-1}(b)$), so that from Eq. (147), and for $n \geq 1$,

$$\begin{aligned}\Omega_{n,\mathrm{eff}} &= (-1)^{n-1}\frac{\Omega_{2,+}}{2}[(\cos\theta + 1)J_{n-1}(b) + (\cos\theta - 1)J_{n+1}(b)] \\ &\simeq (-1)^{n-1}\frac{\Omega_{2,+}}{2}(\cos\theta + 1)\frac{1}{(n-1)!}\left(\frac{b}{2}\right)^{n-1} \\ &= (-1)^{n-1}\frac{\Omega_{2,+}}{2}(\cos\theta + 1)\frac{1}{(n-1)!}\left(\frac{n\Omega_{2,+}\sin\theta}{2\Omega}\right)^{n-1}\end{aligned} \tag{149}$$

where in the last line we have used the expression for $b$ and the value of $\Delta$ on resonance.

We note that when $n = 1$ we obtain from Eq. (149) the previous result of Eq. (133), i.e.,

$$\Omega_{n=1,\mathrm{eff}} = \frac{\Omega_{2,+}}{2}(\cos\theta + 1). \tag{150}$$

For $n = -1$, we can use the relation

$$J_{-n}(x) = (-1)^n J_n(x) \tag{151}$$

in Eq. (147) and then we find that the other Bessel function dominates so that $(\cos\theta + 1)$ is replaced by $(\cos\theta - 1)$ in Eq. (150), in agreement with Eq. (134).

The case discussed at the start of this section occurs when $n = 2$. Then

$$\Omega_{n=2,\mathrm{eff}} = -\frac{\Omega_{2,+}^2}{2\Omega}\sin\theta(\cos\theta + 1). \tag{152}$$

For $n = -2$ a similar result emerges, but again with $(\cos\theta + 1)$ replaced by $(\cos\theta - 1)$.

Eq. (147) shows a dependence on $\Omega_{2,+}^n$ which means that the resonances where $\Delta = \Omega/n$ will become extremely weak for high order $n$. In addition the resonances are correspondingly narrow, and the $\sin\theta$ dependence in Eq. (149) may mean that the gravitational sag, which reduces $\sin\theta$, will reduce the coupling strength yet further. As discussed above, the case $n = \pm 1$ is used in evaporative cooling of rf-dressed atom traps and for rf trap spectroscopy. The case $n = \pm 2$ requires a higher probe strength to become visible and can be regarded as a consequence of a probe that is too strong. This means that stray rf fields must be avoided, not only close to the usual probe resonance at $|\Delta| = \Omega_{1,+}$, but also close to other frequencies and especially $\Omega_{1,+}/2$. This has recently been observed at LPL (see Fig. 16).

Finally, we note that if we keep the $\Omega_{2,z}$ terms in Eq. (128), but as part of an approximation drop the $\Omega_{2,+}$ terms; then the analysis above produces resonances under the condition $\omega_2 = \Omega/n$ at low frequency. In this case, Eq. (149) becomes (for $n \geq 1$)

$$\Omega_{n,\mathrm{eff}} = (-1)^n \Omega_{2,z}\sin\theta \frac{1}{(n-1)!}\left(\frac{\cos\theta\,\Omega_{2,z}}{\omega_2}\right)^{n-1}. \tag{153}$$

This result also agrees with Section 7.3 when $n = 1$.

# 8. PRACTICAL ISSUES WITH RF-DRESSED ATOM TRAPS

## 8.1 Estimates for the Decay of rf Traps

In this section, we discuss in more detail the validity of the adiabatic following of the adiabatic states. With the notations of Section 4.1, the unitary

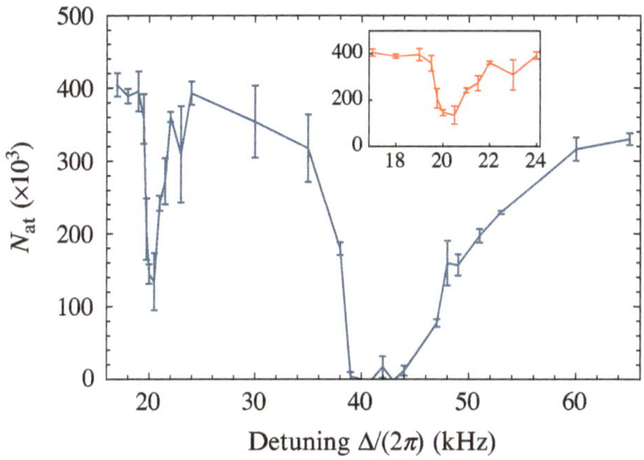


**Fig. 16** Recent data taken at LPL for the $n = \pm 2$ multiphoton process in a rf-dressed quadruple trap, with an rf dressing frequency $\omega_1/(2\pi) = 1200$ kHz and a Rabi frequency $\Omega_{1,+}/(2\pi) = 41$ kHz. A probe at frequency $\omega_2$ is applied for 1 s to the trapped atoms, and the number of remaining atoms is measured. The probe Rabi frequency is of order 2 kHz, much larger than the amplitude normally used for the observation of the single-photon resonance. The probe is scanned between 1216 and 1266 kHz, corresponding to a detuning $\Delta/(2\pi)$ scanned between 16 and 66 kHz. Two resonances occur: a broad resonance (because of the large probe amplitude) at 1241 kHz, i.e., $\Delta/(2\pi) = 41$ kHz, corresponding to the single-photon resonance at $\omega_2 = \omega_1 + \Omega_{1,+}$; and a narrow resonance at 1220.5 kHz, i.e., $\Delta/(2\pi) = 20.5$ kHz, corresponding to the two-photon resonance at $\omega_2 = \omega_1 + \Omega_{1,+}/2$. *Inset*: zoom around the two-photon resonance.

operator $\hat{U}(\mathbf{r},t)$ applied to $\hat{H}_{\text{spin}}(\mathbf{r},t)$ is a position-dependent rotation operator, of the form

$$\hat{U}(\mathbf{r},t) = \hat{R}_{\mathbf{u}}[s\omega t + s\phi(\mathbf{r})]\hat{R}_{\mathbf{u}_\perp}[\theta(\mathbf{r})]. \tag{154}$$

The $\mathbf{r}$ dependence has been omitted in $\mathbf{u}(\mathbf{r})$ and in $\mathbf{u}_\perp(\mathbf{r})$, the vector orthogonal to $\mathbf{u}(\mathbf{r})$ around which the last rotation is performed.

More precisely, the application of $\hat{U}(\hat{\mathbf{R}},t)$ on $\hat{H}_{\text{spin}}(\hat{\mathbf{R}},t)$ writes

$$\overline{\hat{U}(\hat{\mathbf{R}},t)^\dagger \hat{H}_{\text{spin}}(\hat{\mathbf{R}},t)\hat{U}(\hat{\mathbf{R}},t)} - i\hbar\overline{\hat{U}^\dagger\left[\partial_t \hat{U}\right]} = \Omega(\hat{\mathbf{R}})\hat{F}_z \tag{155}$$

where the bar represents the time-averaging applied within the rotating wave approximation.

As $\hat{U}(\hat{\mathbf{R}},t)$ now depends on the position operator, the diagonalization procedure, when applied to the total Hamiltonian of Eq. (70), gives rise to extra terms due to the noncommutation of $\hat{\mathbf{R}}$ and $\hat{\mathbf{P}}$, and the full transformed Hamiltonian reads

$$\hat{H}' = \hat{T} + \Omega(\hat{\mathbf{R}})\hat{F}_z + \Delta\hat{T} = \hat{H}_{\text{adia}} + \Delta\hat{T}. \tag{156}$$

In this expression $\hat{\mathbf{T}} = \hat{\mathbf{P}}^2/2M$ and $\hat{H}_{\text{adia}} = \hat{T} + \Omega(\hat{\mathrm{R}})\hat{F}_z$ is called the adiabatic Hamiltonian. The internal and external degrees of freedom are decoupled, such that for each spin state, it describes the motion of a particle in a different *adiabatic potential*, see Section 4.1, without any coupling between spin states.

The correction to this Hamiltonian is given by

$$\Delta\hat{T} = \overline{\hat{U}(\hat{\mathbf{R}},t)^{\dagger}\hat{T}\hat{U}(\hat{\mathbf{R}},t)} - \hat{T}. \tag{157}$$

Its effect is to couple the spin eigenstates of $\hat{H}_{\text{adia}}$, in a way which depends on the atomic velocity. The adiabatic approximation consists in neglecting this correction $\Delta\hat{T}$ in front of the energy splitting $\hbar\Omega$ induced by the rf coupling. The purpose of this section is to discuss the effect of $\Delta\hat{T}$ and give a condition for the application of the adiabatic approximation.

Let us first write the transform of the momentum operator $\hat{\mathbf{P}}$ under $\hat{U}(\hat{\mathbf{R}},t)$:

$$\hat{U}(\hat{\mathbf{R}},t)^{\dagger}\hat{\mathbf{P}}\hat{U}(\hat{\mathbf{R}},t) = \hat{\mathbf{P}} - i\hbar\hat{U}(\hat{\mathbf{R}},t)^{\dagger}\nabla\hat{U}(\hat{\mathbf{R}},t) = \hat{\mathbf{P}} + \hat{\mathbf{A}}(\hat{\mathbf{R}},t). \tag{158}$$

The transformed kinetic operator thus reads

$$\begin{aligned}\hat{T}' &= \frac{1}{2M}\left[\hat{\mathbf{P}} + \hat{\mathbf{A}}(\hat{\mathbf{R}},t)\right]^2 = \hat{T} + \frac{1}{2M}\left[\hat{\mathbf{P}}\cdot\hat{\mathbf{A}} + \hat{\mathbf{A}}\cdot\hat{\mathbf{P}}\right] + \frac{\hat{\mathbf{A}}^2}{2M} \\ &= \hat{T} + \hat{\mathbf{A}}\cdot\frac{\hat{\mathbf{P}}}{M} - i\frac{\hbar}{2M}\nabla\cdot\hat{\mathbf{A}} + \frac{\hat{\mathbf{A}}^2}{2M}.\end{aligned}$$

After time-averaging we get

$$\Delta\hat{T} = \overline{\hat{\mathbf{A}}(\hat{\mathbf{R}},t)}\cdot\frac{\hat{\mathbf{P}}}{M} - i\frac{\hbar}{2M}\nabla\cdot\overline{\hat{\mathbf{A}}(\hat{\mathbf{R}},t)} + \frac{\overline{\hat{\mathbf{A}}(\hat{\mathbf{R}},t)^2}}{2M}. \tag{159}$$

As $\hat{U}$ is a rotation operator, $\hat{\mathbf{A}}$ is a combination of the angular momentum operators $\hat{F}_z$, $\hat{F}_{\pm}$, with position-dependent vectorial weights oscillating at $\omega$:

$$\hat{\mathbf{A}}(\hat{\mathbf{R}},t) = \mathbf{A}_z(\hat{\mathbf{R}},t)\hat{F}_z + \mathbf{A}_+(\hat{\mathbf{R}},t)\hat{F}_+ + \mathbf{A}_-(\hat{\mathbf{R}},t)\hat{F}_-. \tag{160}$$

The effect of $\Delta\hat{T}$ is thus twofold: (i) it produces an energy shift due to the terms in $\hat{F}_z$ or $\hat{F}_z^2$, and (ii) it couples the adiabatic state $|m\rangle_{\theta(\mathbf{r})}$ to $|m\pm1\rangle_{\theta}$

(terms in $\hat{F}_{\pm}$) or to $|m{\pm}2\rangle_\theta$ (terms in $\hat{F}_{\pm}^2$). This second effect (ii) is responsible for nonadiabatic Landau–Zener losses.

Computing the coefficients in front of $\hat{F}_{\pm}$ and $\hat{F}_{\pm}^2$ in the general case is beyond the scope of this review. A recent derivation of loss rates in some particular cases can be found in Burrows et al. (2017). Instead, we will consider a simple toy model where the static magnetic field has a fixed direction $z$ and a spatial gradient along the $x$-direction:

$$\mathbf{B}_0(\mathbf{r}) = b'x\,\mathbf{e}_z = \frac{\hbar\alpha}{\mu}x\,\mathbf{e}_z. \tag{161}$$

It is a "dummy field" in the sense that it does verify Maxwell's equation $\boldsymbol{\nabla}\times\mathbf{B}_0 = \mathbf{0}$, but it is enough to give an insight in the loss processes. We also assume that the rf phase $\phi$ and Rabi frequency $\boldsymbol{\Omega}_1$ are homogeneous, and we chose $\phi = 0$. $\mathbf{u}_\perp$ is then equal to $\mathbf{e}_y$. In this case, $\hat{U}$ is the product of a position-independent rotation and a rotation which depends on position through $\theta(\mathbf{r})$:

$$\hat{U}(\hat{\mathbf{R}},t) = \hat{R}_z(s\omega t)\hat{R}_y[\theta(\hat{\mathbf{R}})]. \tag{162}$$

Using the analog of Eq. (28) with gradients, we find that

$$\begin{aligned}\hat{\mathbf{A}} &= -i\hbar\hat{U}(\hat{\mathbf{R}},t)^\dagger\nabla\hat{U}(\hat{\mathbf{R}},t) = -i\hbar\hat{R}_y^\dagger(\theta)\nabla\hat{R}_y(\theta)\\ \hat{\mathbf{A}} &= -\nabla\theta\hat{F}_y = i\frac{\nabla\theta}{2}\left(\hat{F}_+ - \hat{F}_-\right).\end{aligned} \tag{163}$$

The coupling term outside the adiabatic states is thus proportional to $\boldsymbol{\nabla}\theta$. The situation is analog to the case where the rf or the Rabi frequency varies in time, which produces a coupling outside the dressed states of the form $\dot{\theta}\hat{F}_+$, see Eq. (29) and the adiabatic condition Eq. (51).

Let us consider atoms initially in the extreme $|m{=}F\rangle_{\theta(\mathbf{r})}$ adiabatic state. The matrix element from $|F\rangle_{\theta(\mathbf{r})}$ to $|F{-}1\rangle_{\theta(\mathbf{r})}$ is thus of order

$$V_- = i\hbar\sqrt{2F}\,\hat{\mathbf{v}}\cdot\nabla\theta + \sqrt{2F}\frac{\hbar^2}{2M}\triangle\theta \tag{164}$$

where $\hat{\mathbf{v}} = \hat{\mathbf{P}}/M$, while the matrix element from $|F\rangle_{\theta(\mathbf{r})}$ to $|F{-}2\rangle_{\theta(\mathbf{r})}$ is of order

$$V_{2-} \simeq F\frac{\hbar^2}{2M}|\nabla\theta|^2. \tag{165}$$

In the example of a linear magnetic field, we have

$$\tan\theta = \frac{-\delta}{\Omega_1} = \frac{\alpha x - \omega}{\Omega_1} \Rightarrow \nabla\theta = \frac{\alpha\Omega_1}{\Omega_1^2 + (\alpha x - \omega)^2}\mathbf{e}_x. \tag{166}$$

$\boldsymbol{\nabla}\theta$ is largest at the resonance surface $x = \omega/\alpha$, where it reaches $\alpha/\Omega_1$. On the other hand, $\triangle\theta$ cancels on the resonance surface $x = \omega/\alpha$ and is largest at positions $x = (\omega \pm\Omega_1)/\alpha$ where it reaches $\alpha^2/(2\Omega_1)^2$.

The adiabatic potential confines particles along the $x$-direction, with an oscillation frequency at the bottom of the extreme $m_\theta = F$ adiabatic potential given as in Eq. (80) by

$$\omega_x = \alpha\sqrt{\frac{\hbar F}{M\Omega_1}}. \tag{167}$$

The matrix elements can then be recast under the form

$$V_- \lesssim \hbar\sqrt{2F}\frac{\alpha|v_x|}{\Omega_1} + \frac{1}{\sqrt{2F}}\left(\frac{\omega_x}{\Omega_1}\right)^2\frac{\hbar\Omega_1}{2}, \qquad V_{2-} \lesssim \left(\frac{\omega_x}{\Omega_1}\right)^2\frac{\hbar\Omega_1}{2}. \tag{168}$$

Let us discuss first the terms in $(\omega_x/\Omega_1)^2\Omega_1$. They have to remain very small as compared to the oscillation at the Rabi frequency $\Omega_1$ responsible for the rf dressing and the existence of the adiabatic states. This requires $\omega_x \ll \Omega_1$, which states exactly that the motion in the adiabatic potential should be adiabatic with respect to the level splitting $\Omega_1$.

On the other hand, the first term of $V_-$ is reminiscent of the Landau–Zener paradigm, see Section 2.2, and its extension to a motion in space described in Section 2.3, with a rate in energy change $\hbar\alpha v_x$. The Landau–Zener adiabatic parameter for the linear field situation reads

$$\Lambda = \frac{V^2}{\hbar^2\alpha|v_x|} = \frac{\Omega_1^2}{\alpha|v_x|} \tag{169}$$

which is equivalent to the ratio of the rf coupling $\Omega_1$ to the first term of $V_-$.

The order of magnitude of this first term $V_-$ for a particle with total energy $(n + 1/2)\hbar\omega_x$ in the trap is

$$\hbar\sqrt{2F}\frac{\alpha|v_x|}{\Omega_1} \simeq \hbar\alpha\sqrt{\frac{(2n+1)F\hbar\omega_x}{M\Omega_1^2}} = \hbar\sqrt{2n+1}\left(\frac{\omega_x}{\Omega_1}\right)^{3/2}\Omega_1. \tag{170}$$

This shows that this coupling term is the largest, even in the ground state $n = 0$. The condition for low Landau–Zener losses is again $\omega_x \ll \Omega_1$, or equivalently

$$\frac{\hbar\alpha^2}{M} \ll \Omega_1^3, \tag{171}$$

and the adiabaticity parameter for the $n$th level of the adiabatic trap is of order

$$\Lambda \simeq \frac{1}{\sqrt{2n+1}}\left(\frac{\Omega_1}{\omega_x}\right)^{3/2} = \frac{1}{\sqrt{2n+1}}\left(\frac{M\Omega_1^3}{\hbar\alpha^2 F}\right)^{3/4}. \tag{172}$$

For a given trap geometry, adiabaticity is better ensured for the lowest harmonic states. This can lead to selective losses of excited atoms, which provides a cooling mechanism, as observed experimentally at LPL (Merloti, 2013).

## 8.2 rf Stability and Other Experimental Considerations

The first experimental issue that may arise when using rf fields to dress atoms is the harmonic distortion in the rf signal, which could come from the amplification process of the rf source. The presence of rf fields at frequencies $2\omega$, $3\omega$... could lead to a deformation of the adiabatic potential or to rf evaporation, see Section 7. In particular, a field at a frequency $2\omega$, even with a small amplitude, could lead to atom losses if some atoms reach an energy $\hbar(\omega - \Omega_1)$ above the bottom of the adiabatic potential. This may be an issue during the loading phase where the dressing field is switched on from below the minimum Larmor frequency $\omega_{0,\mathrm{min}}$, see Section 4.4 and Fig. 17A. As nonlinearity is hard to avoid totally in an rf amplifier, the initial rf must be chosen in such a way that the energy distribution of the atoms in the initial magnetic field lie between $\hbar\omega_{0,\mathrm{min}}$ and $2\hbar\omega$, as in Fig. 17B.

Another possible experimental issue for adiabatic trapping with rf fields concerns noise sources in rf frequency, amplitude, polarization, or in the amplitude of the static magnetic field. In particular, phase or frequency jitter should be avoided to prevent unwanted spin flips, during the loading ramp (see Section 4.4) or while operating the trap. For this reason, rf sources based on Direct Digital Synthesis (DDS) are generally used to realize adiabatic traps. This problem has been addressed in detail by Morizot et al. (2008), together with an analysis on the possible sources of heating, related to frequency, amplitude, or phase fluctuations. In this section, we will briefly review the effect of noise in the different experimental parameters, in the simple case of a linear magnetic field as in Eq. (161).

The fluctuations in the experimental parameters can have two main effects: displacing the position of the resonance surface and thus the trap

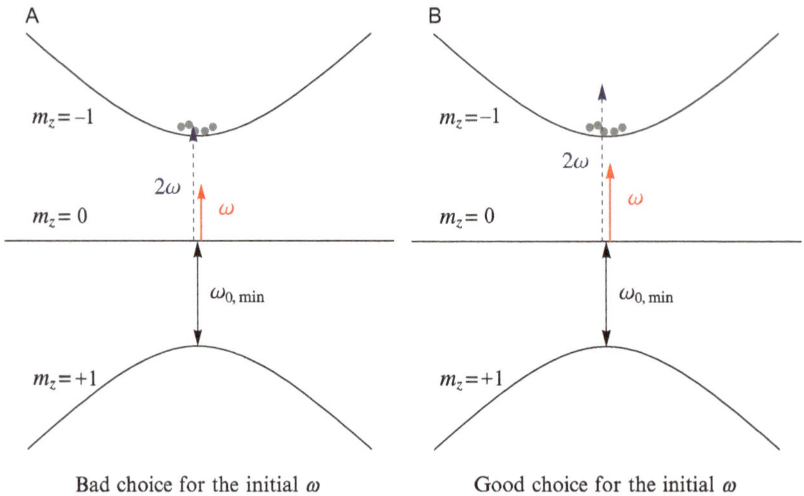


**Fig. 17** Effect of a second harmonic at $2\omega$ (*blue dashed arrow*) during the loading phase, where the rf source (*red arrow*) is switched on from below the minimum Larmor frequency $\omega_{0,\mathrm{min}}$ (*black arrow*) in the initial magnetic trap. The $F=1$ levels are those of the unperturbed bare states $|m\rangle_z$. Disks symbolize atoms. (A) The frequency $2\omega$ is resonant with the Zeeman splitting at a position where atoms are present. This leads to losses. (B) While the rf frequency $\omega$ is below $\omega_{0,\mathrm{min}}$ as it should be, the second harmonic $2\omega$ is far above $\omega_{0,\mathrm{min}}$, at an energy that no atom can reach. This avoids unwanted losses.

center, see Section 4.3, or modifying the trapping frequency Eq. (80). The first effect gives rise to linear dipolar heating, the second to exponential parametric heating.

We first start with dipolar heating. We recall the heating rate $\dot{E}$ (Gehm et al., 1998)

$$\dot{E}=\frac{1}{4}M\omega_x^4\,S_x(\nu_x) \tag{173}$$

where $\nu_x=\omega_x/(2\pi)$, $\omega_x$ is given by Eq. (167) and $S_x$ is the one-sided Power Spectral Density (PSD) of the position fluctuations $\delta x$, defined as the Fourier transform of the time correlation function (Gehm et al., 1998)

$$S_x(\nu)=4\int_0^{\infty}d\tau\,\cos(2\pi\nu\tau)\langle\delta x(t)\,\delta x(t+\tau)\rangle. \tag{174}$$

The average trap position $x_0$ is fixed by the resonance condition $\omega=\alpha x_0$. Position fluctuations can thus be due either[b] to rf frequency or static field

[b] In the case of an external force like gravity in the direction $x$, the trap position is shifted and rf amplitude or phase fluctuations can also lead to dipolar heating.

amplitude fluctuations. If these noise sources are not correlated to each other, $S_x(\nu)$ is related to the sum of the relative frequency ($\delta\omega/\omega$) PSD $S_y(\nu)$ and the relative static field ($\delta\alpha/\alpha$) PSD $S_b(\nu)$ taken at the trap frequency $\nu_x$ through

$$S_x(\nu_x) = x_0^2\,[S_y(\nu_x) + S_b(\nu_x)] = \frac{\omega^2}{\alpha^2}\,[S_y(\nu_x) + S_b(\nu_x)].$$

Replacing $\omega_x$ by its expression Eq. (167), we find that the linear heating rate is

$$\dot{E} = \frac{F}{4}\hbar\omega_x\frac{\omega_x}{\Omega_+}\omega^2\,[S_y(\nu_x) + S_b(\nu_x)] = \frac{\hbar^2F^2\alpha^2\omega^2}{4M\Omega_+^2}\,[S_y(\nu_x) + S_b(\nu_x)].$$

The contribution of the rf frequency noise can also be written in terms of the phase noise PSD $S_\varphi$ of the source:

$$\dot{E}_{\text{phase noise}} = \frac{F\,\hbar\omega_x^4}{4\,\Omega_+}S_\varphi(\nu_x).$$

Relative frequency noise of DDS sources can be very low. A typical standard figure for the phase noise is $S_\varphi = 10^{-10}\ \text{Hz}^{-1}$ in the kHz range. If we take $\Omega_+/\omega_x = 10$ to ensure adiabaticity (see Section 8.1) and $\nu_x = 1$ kHz, we find a very small heating rate of order 5 pK s$^{-1}$, which is not a problem for trapping degenerate gases for seconds in an adiabatic trap.

On the other hand, the contribution of the magnetic field fluctuations reads

$$\dot{E}_{\text{static field}} = \frac{F}{4}\frac{\hbar\omega_x^2\omega^2}{\Omega_+}S_b(\nu_x).$$

This contribution to the linear heating is expected to be larger than the previous one, because of the large $\omega^2/\omega_x^2$ factor between $\dot{E}_{\text{static field}}$ and $\dot{E}_{\text{phase noise}}$. For the low noise power supply used in our experiment[c] we have measured a PSD of the relative amplitude of $S_b = 10^{-14}\ \text{Hz}^{-1}$ around 1 kHz, which is quite low. This corresponds to a heating rate due to the magnetic fluctuation of order 5 nK s$^{-1}$ for a dressing frequency of $\omega/(2\pi) = 1$ MHz and a trapping frequency of $\nu_x = 1$ kHz. In the experiment Merloti et al. (2013b) measure indeed a heating rate of few nK s$^{-1}$. Batteries can also provide low noise current sources.

[c] SM15-100 from Delta Elektronika.

We now turn to parametric heating, which is due to fluctuations in the trapping frequency. The temperature is expected to increase exponentially, with a rate $\Gamma_{\text{param}}$ given by (Gehm et al., 1998):

$$\Gamma_{\text{param}} = \pi^2 \nu_x^2 S_k(2\nu_x) \tag{175}$$

where $S_k$ is the one-sided power spectrum of the fractional fluctuation in the spring constant $k = M\omega_x^2 = \hbar F\alpha^2/\Omega_+$, taken at twice the trap frequency. $S_k$ can be due to static magnetic field amplitude fluctuations via the $\alpha^2$ term, or to fluctuations in $\Omega_+$, in turn due to rf amplitude noise or noise in the relative orientation between the rf and static field.

Let us consider the contribution of rf amplitude noise to parametric heating. We have

$$\Gamma_{\text{rf amplitude}} = \pi^2 \nu_x^2 S_a(2\nu_x)$$

where $S_a$ is the PSD of relative rf amplitude noise ($\delta\Omega_+/\Omega_+$). For $\Gamma_{\text{rf amplitude}}$ to be lower than 0.001 $\text{s}^{-1}$ in the $\nu_x = 1$ kHz trap considered here, the PSD of the relative rf amplitude should be less than $S_a = 10^{-10}\ \text{Hz}^{-1}$ around 2 kHz. Such low noise rf sources are easily accessible, and $S_a = 10^{-11}\ \text{Hz}^{-1}$ is typical. Care must, however, be taken with the amplification chain that may be installed at the output of the rf source. In particular, Merloti et al. (2013b) found that rf attenuators can bring additional noise, and a full digital control of the rf amplitude with the DDS provides better performance.

The noise in the relative static magnetic field contributes twice more, because of the square dependence on $\alpha$ mentioned above:

$$\Gamma_{\text{static field}} = 2\pi^2 \nu_x^2 S_b(2\nu_x).$$

For the same performance, the requirement is now $S_b(2\nu_x) < 5 \times 10^{-11}$ $\text{Hz}^{-1}$. We see that this condition is less stringent than the one on linear heating.

Finally, the strongest experimental requirements for quiet adiabatic potentials are on the noise in the static magnetic field, mostly because of linear dipolar heating. While a good DDS source, together with low noise amplifiers, will be good enough to prevent heating due to the rf source, the static field power supply is required to have an excellent noise performance.

## 8.3 Beyond the Rotating Wave Approximation

In the calculation of the adiabatic potentials, we have ignored the contribution of terms rotating at a frequency $2\omega$ after the transformation to the frame rotating with the rf frequency, see, e.g., Eqs. (41) and (42). This holds either when the local polarization is purely circular, which in general is not true especially because the direction of the static field is not homogeneous, or when both the local detuning to the rf frequency and the local Rabi coupling are small as compared to the Rabi frequency: $|\delta|,\ \Omega_1 \ll \omega$.

In real experiments, this is not always true, in particular when moderate rf frequencies (below 1 MHz) are used. Counter-rotating terms have then to be included in the calculation of the adiabatic potential. The effect of beyond RWA terms has been studied in the late 1960s in the context of the development of the dressed state approach by Cohen-Tannoudji and Haroche (1969), see also Cohen-Tannoudji et al. (1998). It is also relevant to estimate the trapping potential due to far-off resonant laser beams (Grimm et al., 2000). More recently, it has been studied in the context of rf-dressed adiabatic potentials from atom chips by Hofferberth et al. (2007), in a situation where the rf frequency is chosen below the minimum Larmor frequency ($\omega < \omega_{0,\min}$).

In particular, the Bloch–Siegert shift (Bloch and Siegert, 1940; Ramsey, 1955) has been observed by Hofferberth et al. (2007) using rf spectroscopy with a weak probe field, see Section 7.3. If $\Omega_-$ becomes of order $\omega$, an exact treatment with a numerical approach is required, as done by Hofferberth et al. (2007). When, however, $\Omega_-/\omega$ is not too large, this shift can be estimated by second-order perturbation theory of the bare states due to the presence of the counter-rotating term. A simple calculation of the shift for the bare state $|m,\ N\rangle_z$ similar to Eq. (126) leads to the new energy

$$\tilde{E}_{m,N} = N\hbar\omega - m\hbar\delta + m\frac{\hbar|\Omega_-|^2}{2(\omega+\omega_0)}. \tag{176}$$

The detuning $\delta$ should thus be replaced in the equations by a shifted detuning

$$\tilde{\delta} = \delta - \frac{|\Omega_-|^2}{2(\omega+\omega_0)} = \delta - \frac{|\Omega_-|^2}{2(2\omega-\delta)}.$$

As a result, the rf resonance for the corotating field is shifted and occurs now for a detuning $\delta_{res} = \omega - \omega_0$ such that

$$\delta_{res} = \frac{|\Omega_-|^2}{2(2\omega-\delta_{res})},$$

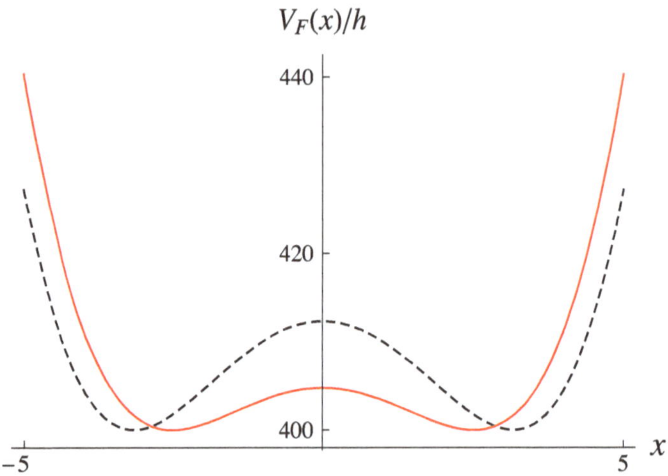


**Fig. 18** Effect of beyond RWA terms in the shape of a double-well potential arising from dressing an initially harmonic magnetic potential $V_0(x) = \hbar\omega_{0,\min} + \hbar\kappa x^2/2$ with an rf frequency $\omega > \omega_{0,\min}$ and a particularly large value of $\Omega_\pm/\omega$. *Dashed black line*: potential calculated within RWA. *Red full line*: potential $\tilde{V}_F$ calculated with Eq. (177), including non-RWA terms to second order. The shape of the double-well is significantly different. The resonance occurs at a smaller value of the magnetic field, resulting in closer wells. Parameters: $\omega_{0,\min}/(2\pi) = 1$ MHz, $\omega/(2\pi) = 1.1$ MHz, $\Omega_+/(2\pi) = \Omega_-/(2\pi) = 400$ kHz, bare trap curvature $\kappa/(2\pi) = 20$ kHz μm$^{-2}$. Units: $x$ in μm, $\tilde{V}_F/h$ in kHz.

which corresponds for an inhomogeneous magnetic field to a resonance surface defined by

$$\omega_0(\mathbf{r}) = \sqrt{\omega^2 - |\Omega_-|^2}.$$

The adiabatic potential should then be corrected and now reads

$$\tilde{V}_m(\mathbf{r}) = m\hbar\sqrt{\left[\omega - \omega_0(\mathbf{r}) - \frac{|\Omega_-(\mathbf{r})|^2}{2(\omega + \omega_0(\mathbf{r}))}\right]^2 + |\Omega_+(\mathbf{r})|^2}. \qquad (177)$$

The effect of this correction can be important when adiabatic potentials are used to produce double-wells, as the tunnel coupling between the two wells depends exponentially on the barrier height. Fig. 18 shows the effect of beyond RWA terms on the potential, in the case of dressed magnetic trap with a nonzero minimum, for a very strong Rabi frequency.

## 8.4 Misalignment Effects of the rf Polarization

In the beginning of Section 3.3.2 we stated that the component of the rf field aligned with the static magnetic field does not couple different states of angular momentum, such that the rf polarization should always be chosen with a nonzero projection in the plane orthogonal to the static field. In fact, this is not strictly true. When the rf field has some component $\Omega_z$ along the axis of the magnetic field, taken as the $z$-axis, its effect is to modify the Landé factor, by a factor $J_0(\Omega_z/\omega)$ (Haroche et al., 1970). It can also lead to transitions at submultiples of the Larmor frequency.

The Hamiltonian for a static field along $z$ and an rf field with a nonzero projection along $z$ writes

$$\hat{H} = (\omega_0 + \Omega_z \sin\omega t)\,\hat{F}_z + \left\{\frac{\Omega_+}{2}\,e^{-i\omega t}\hat{F}_+ + h.c.\right\}. \tag{178}$$

The rf field is circularly polarized, but also has a linear component of its polarization along $z$.

To understand the effect of this $\Omega_z$ term, we will look for the solution of a spin rotated through $\hat{R} = \hat{R}_z[(n+1)\omega t - \frac{\Omega_z}{\omega}\cos\omega t]$ where $n \in \mathbb{N}$ is a fixed integer. This rotation is chosen to cancel the $\Omega_z$ term of the initial Hamiltonian. We get for the rotated Hamiltonian Eq. (40):

$$\begin{aligned}\hat{H}_{\rm rot} &= \hat{R}^\dagger \hat{H}\hat{R} - [(n+1)\omega + \Omega_z \sin\omega t]\,\hat{F}_z \\ &= -\delta_n \hat{F}_z + \left\{\frac{\Omega_+}{2}\,e^{in\omega t}e^{-i\frac{\Omega_z}{\omega}\cos\omega t}\hat{F}_+ + h.c.\right\}\end{aligned} \tag{179}$$

where $\delta_n = (n+1)\omega - \omega_0$. The exponential of the cosine may be expanded in terms of Bessel functions of the first kind. This gives

$$\hat{H}' = -\delta_n \hat{F}_z + \left\{\frac{\Omega_+}{2}\sum_{n'=-\infty}^{+\infty}(-i)^{n'}J_{n'}\left(\frac{\Omega_z}{\omega}\right)e^{i(n-n')\omega t}\hat{F}_+ + h.c.\right\}. \tag{180}$$

It is now clear that the term with $n' = n$ is the sum is stationary. This means that resonances appear at frequencies $\omega$ such that $\delta_n = 0$, with a coupling amplitude given by the Bessel function:

$$\text{coupling} \quad \Omega_+ J_n\left(\frac{\Omega_z}{\omega}\right) \quad \text{at frequency} \quad \omega = \frac{\omega_0}{n+1}. \tag{181}$$

The $n = 0$ case is the usual, expected transition. However, the rf coupling is modified from $\Omega_+$ to $\Omega_+ J_0\left(\frac{\Omega_z}{\omega}\right)$. We recover a coupling $\Omega_+$ when $\Omega_z$ vanishes. Everything happens as if the Landé factor had been modified by a factor $J_0\left(\frac{\Omega_z}{\omega}\right)$, smaller than one, which can even change sign if $\Omega_z$ is comparable with $\omega$ (Cohen-Tannoudji and Haroche, 1969; Haroche et al., 1970). As a consequence, if an accurate description of the adiabatic potential is required, the corrected value of the Landé factor must be used instead of its bare value.

The cases $n > 0$ correspond to resonances at submultiples of the Larmor frequencies (Cohen-Tannoudji and Haroche, 1969; Pegg, 1973), with smaller amplitudes $\Omega_+ J_n\left(\frac{\Omega_z}{\omega}\right)$. For $\Omega_z \ll \omega$, the coupling amplitude scales as $\left(\frac{\Omega_z}{\omega}\right)^n$ and is very small. For practical purposes in rf-dressed adiabatic potentials, the rf source often has harmonics of the frequency $\omega$ due to nonlinear amplification. This misalignment effect is another reason, apart from possible anharmonicity of the rf source, for avoiding to have atoms at a position where the Larmor frequency is close to $2\omega$ (see Section 8.2).

## 9. CONCLUSION

Magnetic resonance has a long and rich history (see, for example, Abragam, 1961). However, in combination with spatially varying fields a new set of phenomena have emerged such as the resonant and off-resonant trapping of ultracold atomic gases. By using the adiabatic potentials from magnetic resonance to manipulate atoms, rf dressing has now also become a standard tool for preparation of cold atomic systems. The approach is also applicable to dressed microwave potentials (Agosta et al., 1989; Ammar et al., 2015; Böhi et al., 2009, 2010; Spreeuw et al., 1994; Treutlein et al., 2006) where there may also be applications to atomic clocks (Kazakov and Schumm, 2015; Sárkány et al., 2014).

In this chapter we focused on generic experimental setup configurations, not discussing much the way the fields are produced, by macroscopic or microscopic arrangement of wires. However, it must be stressed that atom chips provide strong fields and gradients that make new field configurations which are strongly influenced by the geometric design of the wires on the chips. These cases include rf chips as well as microwave potentials from chips such as those mentioned above. The compactness of atom chip systems makes this an area of interest for research into Quantum Technologies. Emerging possibilities for cold atom experiments in space (Lundblad,

2015) further broaden the possibilities of unusual topologies such as the bubble states of matter (Zobay and Garraway, 2004), when atoms are released from the gravitational potential.

## ACKNOWLEDGMENTS

H.P. thanks the École de physique des Houches, where part of the text has been written for a lecture series, and the whole BEC group at LPL for their contribution to the experiment over the last 15 years. The work presented here is connected to the PhD thesis of Colombe (2004), Morizot (2007), Kollengode Easwaran (2009), Liennard (2011), Merloti (2013), De Rossi (2016), Burrows (2016), and Mathieu de Goër de Herve. B.M.G. would like to thank the Leverhulme Trust, the CNRS, the UK EPSRC (grants EP/I010394/1 and EP/M005453/1), and the University of Sussex for supporting the research that contributed to this review.

Laboratoire de physique des lasers is UMR 7538 of CNRS and Paris 13 University. LPL is member of the Institut Francilien de Recherche sur les Atomes Froids (IFRAF).

## REFERENCES

Abragam, A., 1961. The Principles of Nuclear Magnetism. Oxford University Press, Oxford.

Agosta, C.C., Silvera, I.F., Stoof, H.T.C., Verhaar, B.J., 1989. Trapping of neutral atoms with resonant microwave radiation. Phys. Rev. Lett. 62 (20), 2361–2364.

Allegrini, M., Arimondo, E., 1971. Radiofrequency transitions in a dressed atom. J. Phys. B: At. Mol. Phys. 4 (8), 1008. http://stacks.iop.org/0022-3700/4/i=8/a=004.

Amico, L., Birkl, G., Boshier, M.G., Kwek, L.C., 2016. Focus on atomtronics-enabled quantum technologies. New J. Phys. 19, 020201. (special issue). http://iopscience.iop.org/1367-2630/focus/Focus%20on%20Atomtronics-enabled%20Quantum%20Technologies.

Ammar, M., Dupont-Nivet, M., Huet, L., Pocholle, J.P., Rosenbusch, P., Bouchoule, I., Westbrook, C.I., Estève, J., Reichel, J., Guerlin, C., Schwartz, S., 2015. Symmetric microwave potentials for interferometry with thermal atoms on a chip. Phys. Rev. A 91, 053623.

Bentine, E., Harte, T.L., Luksch, K., Barker, A., Mur-Petit, J., Yuen, B., Foot, C.J., 2017. Species-selective confinement of atoms dressed with multiple radiofrequencies. Preprint arXiv:1701.05819. https://arxiv.org/abs/1701.05819.

Betz, T., Manz, S., Bücker, R., Berrada, T., Koller, C., Kazakov, G., Mazets, I.E., Stimming, H.P., Perrin, A., Schumm, T., Schmiedmayer, J., 2011. Two-point phase correlations of a one-dimensional bosonic Josephson junction. Phys. Rev. Lett. 106, 020407. http://link.aps.org/doi/10.1103/PhysRevLett.106.020407.

Bize, S., Laurent, P., Abgrall, M., Marion, H., Maksimovic, I., Cacciapuoti, L., Grünert, J., Vian, C., Pereira dos Santos, F., Rosenbusch, P., Lemonde, P., Santarelli, G., Wolf, P., Clairon, A., Luiten, A., Tobar, M., Salomon, C., 2005. Cold atom clocks and applications. J. Phys. B: At. Mol. Opt. Phys. 38, S449–S468.

Bloch, I., 2005. Ultracold quantum gases in optical lattices. Nat. Phys. 1, 23.

Bloch, F., Siegert, A., 1940. Magnetic resonance for nonrotating fields. Phys. Rev. 57, 522–527. http://link.aps.org/doi/10.1103/PhysRev.57.522.

Bloch, I., Dalibard, J., Zwerger, W., 2008. Many-body physics with ultracold gases. Rev. Mod. Phys. 80, 885–964.

Bloch, I., Dalibard, J., Nascimbène, S., 2012. Quantum simulations with ultracold quantum gases. Nat. Phys. 8 (4), 267–276. http://dx.doi.org/10.1038/nphys2259.

Böhi, P., Riedel, M.F., Hoffrogge, J., Reichel, J., Hänsch, T.W., Treutlein, P., 2009. Coherent manipulation of Bose-Einstein condensates with state-dependent microwave potentials on an atom chip. Nat. Phys. 5 (8), 592–597.
Böhi, P., Riedel, M.F., Hänsch, T.W., Treutlein, P., 2010. Imaging of microwave fields using ultracold atoms. Appl. Phys. Lett. 97, 051101.
Burrows, K.A., 2016. Non-adiabatic losses from radio frequency dressed cold atom traps (Ph.D. thesis). University of Sussex, Brighton. http://sro.sussex.ac.uk/61380/.
Burrows, K.A., Garraway, B.M., Perrin, H., 2017. Non-adiabatic losses from RF-dressed cold atom traps: beyond the Landau-Zener model. ArXiv e-prints.
Chu, S., 1998. Nobel lecture: the manipulation of neutral particles. Rev. Mod. Phys. 70, 685.
Cohen-Tannoudji, C., 1998. Nobel lecture: manipulating atoms with photons. Rev. Mod. Phys. 70, 707.
Cohen-Tannoudji, C., Guéry-Odelin, D., 2011. Advances in Atomic Physics. An Overview. World Scientific, Singapore.
Cohen-Tannoudji, C., Haroche, S., 1969. Absorption et diffusion de photons optiques par un atome en interaction avec des photons de radiofréquence. J. Phys. Fr. 30 (2–3), 153–168. https://doi.org/10.1051/jphys:01969003002–3015300.
Cohen-Tannoudji, C., Reynaud, S., 1977. Dressed-atom description of resonance fluorescence and absorption spectra of a multi-level atom in an intense laser beam. J. Phys. B: At. Mol. Opt. Phys. 10 (3), 345–363.
Cohen-Tannoudji, C., Diu, B., Laloë, F., 1991. Quantum Mechanics. Wiley, New York.
Cohen-Tannoudji, C., Dupont-Roc, J., Grynberg, G., 1998. Atom-Photon Interactions : Basic Processes and Applications. Wiley, New York.
Colombe, Y., 2004. Condensat de Bose–Einstein, champs évanescents et piégeage bidimensionnel (Ph.D. thesis), Université Paris 13.
Colombe, Y., Knyazchyan, E., Morizot, O., Mercier, B., Lorent, V., Perrin, H., 2004. Ultracold atoms confined in rf-induced two-dimensional trapping potentials. Europhys. Lett. 67 (4), 593–599.
Cornell, E.A., Wieman, C.E., 2002. Nobel lecture: Bose-Einstein condensation in a dilute gas, the first 70 years and some recent experiments. Rev. Mod. Phys. 74, 875.
Courteille, Ph.W., Deh, B., Fortágh, J., Günther, A., Kraft, S., Marzok, C., Slama, S., Zimmermann, C., 2006. Highly versatile atomic micro traps generated by multifrequency magnetic field modulation. J. Phys. B: At. Mol. Opt. Phys. 39, 1055.
Cronin, A.D., Schmiedmayer, J., Pritchard, D.E., 2009. Optics and interferometry with atoms and molecules. Rev. Mod. Phys. 81 (3), 1051–1129.
Davis, K.B., Mewes, M.O., Andrews, M.R., van Druten, N.J., Durfee, D.S., Kurn, D.M., Ketterle, W., 1995. Bose-Einstein condensation in a gas of sodium atoms. Phys. Rev. Lett. 75 (22), 3969–3973.
De Rossi, C., 2016. Gaz de Bose en dimension deux: modes collectifs, superfluidité et piège annulaire (Ph.D. thesis), Université Paris 13. https://tel.archives-ouvertes.fr/tel-01420085.
De Rossi, C., Dubessy, R., Merloti, K., de Goër de Herve, M., Badr, T., Perrin, A., Longchambon, L., Perrin, H., 2016. Probing superfluidity in a quasi two-dimensional Bose gas through its local dynamics. New J. Phys. 18 (6), 062001. http://stacks.iop.org/1367-2630/18/i=6/a=062001.
Dubessy, R., Merloti, K., Longchambon, L., Pottie, P.E., Liennard, T., Perrin, A., Lorent, V., Perrin, H., 2012. Rubidium-87 Bose-Einstein condensate in an optically plugged quadrupole trap. Phys. Rev. A 85, 013643. http://link.aps.org/doi/10.1103/PhysRevA.85.013643.
Dubessy, R., De Rossi, C., Badr, T., Longchambon, L., Perrin, H., 2014. Imaging the collective excitations of an ultracold gas using statistical correlations. New J. Phys. 16 (12), 122001. http://stacks.iop.org/1367-2630/16/i=12/a=122001.

Extavour, M.H.T., Le Blanc, L.J., Schumm, T., Cieslak, B., Myrskog, S., Stummer, A., Aubin, S., Thywissen, J.H., 2006. Dual-species quantum degeneracy of $^{40}$K and $^{87}$Rb on an atom chip. At. Phys. 20, 241.
Garraway, B.M., 2011. The Dicke model in quantum optics: Dicke model revisited. Phil. Trans. R. Soc. A 369, 1137.
Garraway, B.M., Perrin, H., 2010. An effective scalar magnetic interaction for resonantly trapped atoms. Phys. Scr. T140, 014006.
Garraway, B.M., Perrin, H., 2016. Recent developments in trapping and manipulation of atoms with adiabatic potentials. J. Phys. B: At. Mol. Opt. Phys. 49 (17), 172001. http://stacks.iop.org/0953-4075/49/i=17/a=172001.
Garraway, B.M., Vitanov, N.V., 1997. Population dynamics and phase effects in periodic level crossings. Phys. Rev. A 55 (6), 4418–4432.
Garrido Alzar, C.L., Perrin, H., Garraway, B.M., Lorent, V., 2006. Evaporative cooling in a radio-frequency trap. Phys. Rev. A 74, 053413.
Gehm, M.E., O'Hara, K.M., Savard, T.A., Thomas, J.E., 1998. Dynamics of noise-induced heating in atom traps. Phys. Rev. A 58 (5), 3914–3921. (Erratum: Phys. Rev. A 61, 029902, 2000).
Gildemeister, M., Nugent, E., Sherlock, B.E., Kubasik, M., Sheard, B.T., Foot, C.J., 2010. Trapping ultracold atoms in a time-averaged adiabatic potential. Phys. Rev. A 81, 031402. http://link.aps.org/doi/10.1103/PhysRevA.81.031402.
Gildemeister, M., Sherlock, B.E., Foot, C.J., 2012. Techniques to cool and rotate Bose–Einstein condensates in time-averaged adiabatic potentials. Phys. Rev. A 85 (5), 053401.
Griffin, P.F., Riis, E., Arnold, A.S., 2008. Smooth inductively coupled ring trap for atoms. Phys. Rev. A 77, 051402.
Grimm, R., Weidemüller, M., Ovchinnikov, Yu.B., 2000. Optical dipole traps for neutral atoms. Adv. At. Mol. Opt. Phys. 42, 95–170.
Haroche, S., Cohen-Tannoudji, C., Audoin, C., Schermann, J.P., 1970. Modified Zeeman hyperfine spectra observed in $^{1}$H and $^{87}$Rb ground states interacting with a nonresonant rf field. Phys. Rev. Lett. 24, 861–864. http://link.aps.org/doi/10.1103/PhysRevLett.24.861.
Heathcote, W.H., Nugent, E., Sheard, B.T., Foot, C.J., 2008. A ring trap for ultracold atoms in an rf-dressed state. New J. Phys. 10 (4), 043012.
Hofferberth, S., Fischer, B., Schumm, T., Schmiedmayer, J., Lesanovsky, I., 2007. Ultracold atoms in radio-frequency dressed potentials beyond the rotating-wave approximation. Phys. Rev. A 76, 013401. http://link.aps.org/doi/10.1103/PhysRevA.76.013401.
Jo, G.B., Shin, Y., Will, S., Pasquini, T.A., Saba, M., Ketterle, W., Pritchard, D.E., Vengalattore, M., Prentiss, M., 2007. Long phase coherence time and number squeezing of two Bose-Einstein condensates on an atom chip. Phys. Rev. Lett. 98 (3), 030407.
Kazakov, G.A., Schumm, T., 2015. Magic radio-frequency dressing for trapped atomic microwave clocks. Phys. Rev. A 91, 023404.
Ketterle, W., 2002. When atoms behave as waves: Bose-Einstein condensation and the atom laser. Rev. Mod. Phys. 74, 1131.
Ketterle, W., van Druten, N.J., 1996. Evaporative cooling of trapped atoms. In: Bederson, B., Walther, H. (Eds.), Advances in Atomic, Molecular and Optical Physics. vol. 37. Academic Press, San Diego, pp. 181–236.
Ketterle, W., Durfee, D.S., Stamper-Kurn, D.M., 1999. Making, probing and understanding Bose-Einstein condensates. In: Inguscio, M., Stringari, S., Wieman, C.E. (Eds.), Proceedings of the International School of Physics "Enrico Fermi", Course CXL. IOS Press, Amsterdam, pp. 67–176.
Kim, S.J., Yu, H., Gang, S.T., Anderson, D.Z., Kim, J.B., 2016. Controllable asymmetric double well and ring potential on an atom chip. Phys. Rev. A 93, 033612.

Kollengode Easwaran, R., 2009. Spectroscopy and Evaporative Cooling of Rubidium Atoms in Radiofrequency Dressed Traps (Ph.D. thesis), Université Paris 13. http://tel.archives-ouvertes.fr/tel-00376470/fr/.

Landau, L., 1932. Zur Theorie der Energieübertragung—II. Phys. Z. Sowjetunion 2, 46–51.

Leggett, A.J., 2006. Quantum Liquids: Bose Condensation and Cooper Pairing in Condensed-Matter Systems. Oxford University Press, New York.

Lesanovsky, I., von Klitzing, W., 2007. Time-averaged adiabatic potentials: versatile matter-wave guides and atom traps. Phys. Rev. Lett. 99 (8), 083001.

Lesanovsky, I., Schumm, T., Hofferberth, S., Andersson, L.M., Krüger, P., Schmiedmayer, J., 2006. Adiabatic radio-frequency potentials for the coherent manipulation of matter waves. Phys. Rev. A 73 (3), 033619.

Liennard, T., 2011. Construction d'un montage de condensation de Bose–Einstein de rubidium et étude théorique d'un superfluide en rotation dans un anneau (Ph.D. thesis), Université Paris 13. http://tel.archives-ouvertes.fr/tel-00667804.

Lin, D., Hui, Y., Jin, W., Ming-Sheng, Z., 2010. Chip-based square wave dynamic micro atom trap. Chin. Phys. Lett. 27 (5), 053201.

Lundblad, N., 2015. Nasa project. Private communication.

Lundblad, N., Lee, P.J., Spielman, I.B., Brown, B.L., Phillips, W.D., Porto, J.V., 2008. Atoms in a radio-frequency-dressed optical lattice. Phys. Rev. Lett. 100 (15), 150401.

Lundblad, N., Ansari, S., Guo, Y., Moan, E., 2014. Observations of $\lambda/4$ structure in a low-loss radio-frequency-dressed optical lattice. Phys. Rev. A 90, 053612.

Merloti, K., 2013. Condensat de Bose–Einstein dans un piège habillé: modes collectifs d'un superfluide en dimension deux (Ph.D. thesis), Université Paris 13. http://tel.archives-ouvertes.fr/tel-00949914.

Merloti, K., Dubessy, R., Longchambon, L., Olshanii, M., Perrin, H., 2013a. Breakdown of scale invariance in a quasi-two-dimensional Bose gas due to the presence of the third dimension. Phys. Rev. A 88, 061603. http://link.aps.org/doi/10.1103/PhysRevA.88.061603.

Merloti, K., Dubessy, R., Longchambon, L., Perrin, A., Pottie, P.E., Lorent, V., Perrin, H., 2013b. A two-dimensional quantum gas in a magnetic trap. New J. Phys. 15 (3), 033007. http://stacks.iop.org/1367-2630/15/i=3/a=033007.

Messiah, A., 1966. Quantum Mechanics. North Holland Publishing Company, Amsterdam; John Wiley & Sons Inc., New York

Migdall, A.L., Prodan, J.V., Phillips, W.D., Bergeman, T.H., Metcalf, H.J., 1985. First observation of magnetically trapped neutral atoms. Phys. Rev. Lett. 54, 2596–2599. http://link.aps.org/doi/10.1103/PhysRevLett.54.2596.

Moerdijk, A.J., Verhaar, B.J., Nagtegaal, T.M., 1996. Collisions of dressed ground-state atoms. Phys. Rev. A 53 (6), 4343–4351.

Morgan, T., O'Sullivan, B., Busch, T., 2011. Coherent adiabatic transport of atoms in radio-frequency traps. Phys. Rev. A 83 (5), 053620.

Morizot, O., 2007. Pièges radiofréquence très anisotropes pour un condensat de Bose–Einstein. Université Paris 13. http://tel.archives-ouvertes.fr/tel-00142650/fr/.

Morizot, O., Colombe, Y., Lorent, V., Perrin, H., Garraway, B.M., 2006. Ring trap for ultracold atoms. Phys. Rev. A 74, 023617.

Morizot, O., Garrido Alzar, C.L., Pottie, P.E., Lorent, V., Perrin, H., 2007. Trapping and cooling of rf-dressed atoms in a quadrupole magnetic field. J. Phys. B: At. Mol. Opt. Phys. 40, 4013–4022.

Morizot, O., Longchambon, L., Kollengode Easwaran, R., Dubessy, R., Knyazchyan, E., Pottie, P.E., Lorent, V., Perrin, H., 2008. Influence of the radio-frequency source properties on RF-based atom traps. Eur. Phys. J. D 47, 209.

Naik, D.S., Raman, C., 2005. Optically plugged quadrupole trap for Bose-Einstein condensates. Phys. Rev. A 71, 033617.

Navez, P., Pandey, S., Mas, H., Poulios, K., Fernholz, T., von Klitzing, W., 2016. Matter-wave interferometers using TAAP rings. New J. Phys. 18 (7), 075014. http://stacks.iop.org/1367-2630/18/i=7/a=075014.

Pegg, D.T., 1972. Semi-classical theory of radiofrequency transitions in a dressed atom. J. Phys. B: At. Mol. Phys. 5 (1), L4. http://stacks.iop.org/0022-3700/5/i=1/a=002.

Pegg, D.T., 1973. Misalignment effects in magnetic resonance. J. Phys. B: Atom. Mol. Phys. 6, 241.

Pegg, D.T., Series, G.W., 1970. Semi-classical theory of the Hanle effect with transverse static and oscillating magnetic fields. J. Phys. B: Atom. Mol. Phys. 3, L33.

Petrich, W., Anderson, M.H., Ensher, J.R., Cornell, E.A., 1995. Stable, tightly confining magnetic trap for evaporative cooling of neutral atoms. Phys. Rev. Lett. 74, 3352–3355. http://link.aps.org/doi/10.1103/PhysRevLett.74.3352.

Phillips, W.D., 1998. Nobel lecture: laser cooling and trapping of neutral atoms. Rev. Mod. Phys. 70, 721.

Pritchard, J.D., Dinkelaker, A.N., Arnold, A.S., Griffin, P.F., Riis, E., 2012. Demonstration of an inductively coupled ring trap for cold atoms. New J. Phys. 14 (10), 103047.

Ramsey, N.F., 1955. Resonance transitions induced by perturbations at two or more different frequencies. Phys. Rev. 100, 1191–1194. http://link.aps.org/doi/10.1103/PhysRev.100.1191.

Sárkány, L., Weiss, P., Hattermann, H., Fortágh, J., 2014. Controlling the magnetic-field sensitivity of atomic-clock states by microwave dressing. Phys. Rev. A 90, 053416.

Schumm, T., Hofferberth, S., Andersson, L.M., Wildermuth, S., Groth, S., Bar-Joseph, I., Schmiedmayer, J., Krüger, P., 2005. Matter wave interferometry in a double well on an atom chip. Nat. Phys. 1, 57.

Sherlock, B.E., Gildemeister, M., Owen, E., Nugent, E., Foot, C.J., 2011. Time-averaged adiabatic ring potential for ultracold atoms. Phys. Rev. A 83, 043408. http://link.aps.org/doi/10.1103/PhysRevA.83.043408.

Shotter, M., Trypogeorgos, D., Foot, C., 2008. Enhancement of on-site interactions of tunneling ultracold atoms in optical potentials using radio-frequency dressing. Phys. Rev. A 78 (5), 051602.

Sinuco-León, G.A., Garraway, B.M., 2015. Radio-frequency dressed lattices for ultracold alkali atoms. New J. Phys. 17 (5), 053037.

Sinuco-León, G.A., Burrows, K.A., Arnold, A.S., Garraway, B.M., 2014. Inductively guided circuits for ultracold dressed atoms. Nat. Commun. 5, 5289.

Spreeuw, R.J.C., Gerz, C., Goldner, L.S., Phillips, W.D., Rolston, S.L., Westbrook, C.I., Reynolds, M.W., Silvera, I.F., 1994. Demonstration of neutral atom trapping with microwaves. Phys. Rev. Lett. 72 (20), 3162–3165.

Suominen, K.A., Garraway, B.M., Stenholm, S., 1991. The adiabatic limit of level crossing models. Opt. Commun. 82 (3–4), 260–266.

Tino, G.M., Kasevich, M.A. (Eds.), 2014. Atom Interferometry. Proceedings of the International School of Physics "Enrico Fermi", vol. 188. IOS Press, Amsterdam.

Treutlein, P., Hänsch, T.W., Reichel, J., Negretti, A., Cirone, M.A., Calarco, T., 2006. Microwave potentials and optimal control for robust quantum gates on an atom chip. Phys. Rev. A 74, 022312.

van Es, J.J.P., Whitlock, S., Fernholz, T., van Amerongen, A.H., van Druten, N.J., 2008. Longitudinal character of atom-chip-based rf-dressed potentials. Phys. Rev. A 77, 063623. http://link.aps.org/doi/10.1103/PhysRevA.77.063623.

Vangeleyn, M., Garraway, B.M., Perrin, H., Arnold, A.S., 2014. Inductive dressed ring traps for ultracold atoms. J. Phys. B: At. Mol. Opt. Phys. 47 (7), 071001. http://stacks.iop.org/0953-4075/47/i=7/a=071001.

Vitanov, N.V., Garraway, B.M., 1996. Landau-Zener model: effects of finite coupling duration. Phys. Rev. A 53, 4288–4304. http://link.aps.org/doi/10.1103/PhysRevA.53.4288.

Walraven, J.T.M., 2016. Atomic physics. Lecture Notes, Course Given at the University of Amsterdam, https://staff.fnwi.uva.nl/j.t.m.walraven/walraven/Publications files/2017-Atomic-Physics.pdfYi2008.

Yi, W., Daley, A.J., Pupillo, G., Zoller, P., 2008. State-dependent, addressable subwavelength lattices with cold atoms. New J. Phys. 10 (7), 073015.

Zener, C., 1932. Non-adiabatic crossing of energy levels. Proc. R. Soc. London Ser. A 137, 696–702.

Zobay, O., Garraway, B.M., 2001. Two-dimensional atom trapping in field-induced adiabatic potentials. Phys. Rev. Lett. 86 (7), 1195–1198.

Zobay, O., Garraway, B.M., 2004. Atom trapping and two-dimensional Bose-Einstein condensates in field-induced adiabatic potentials. Phys. Rev. A 69, 023605.

CHAPTER FIVE

# Quantum Control of Optomechanical Systems

**Sebastian G. Hofer*, Klemens Hammerer†,[1]**
*Vienna Center for Quantum Science and Technology (VCQ), Faculty of Physics, University of Vienna, Vienna, Austria
†Institute for Theoretical Physics, Institute for Gravitational Physics (Albert Einstein Institute), Leibniz University Hannover, Hannover, Germany
[1]Corresponding author: e-mail address: klemens.hammerer@itp.uni-hannover.de

## Contents



*Advances in Atomic, Molecular, and Optical Physics*, Volume 66
ISSN 1049-250X
http://dx.doi.org/10.1016/bs.aamop.2017.03.003

## Abstract

In this chapter we aim at bringing together the fields of quantum control theory and quantum optomechanics, exploring the prospects of entanglement-enhanced quantum control of these systems. We first analyze in detail how the radiation pressure interaction can be used to generate entanglement between a mechanical mode and the electromagnetic field, both in continuous-wave and pulsed regimes, and introduce an optomechanical teleportation scheme to transfer an arbitrary quantum state from a traveling-wave light pulse onto the mechanical system. Making use of continuous measurement and optimal control theory, we then show how similar schemes can be implemented in a time-continuous regime; analyzed protocols include optimal optomechanical feedback cooling, time-continuous teleportation, and time-continuous entanglement swapping. Finally we discuss the implementation of a Kalman filter for an optomechanical system, representing an important first step toward the experimental realization of the discussed protocols. Additionally, elementary aspects of quantum stochastic calculus and quantum control theory are given in comprehensive appendices.

## NOTATION

In this chapter vectors are printed in a bold, slanted font, e.g., $\boldsymbol{M}$, while matrices, such as $\mathbf{M}$, are denoted by bold, upright letters. The corresponding transposed quantities are written as $\boldsymbol{M}^{\mathrm{T}}$ and $\mathbf{M}^{\mathrm{T}}$, respectively. The complex conjugate of a complex number $z$ is denoted by $z^*$, the Hermitian conjugate of an operator $c$ is $c^\dagger$. For tuples of operators we use the convention $(c_1, c_2, \ldots)^\dagger = (c_1^\dagger, c_2^\dagger, \ldots)$ which is different from $(c_1^\dagger, c_2^\dagger, \ldots)^{\mathrm{T}}$. The same symbol is used to denote the Hermitian conjugate of complex matrices (or vectors), i.e., $\mathbf{M}^\dagger = (\mathbf{M}^*)^{\mathrm{T}}$ for $\mathbf{M} \in \mathbb{C}^{n\times m}$. Quantum-mechanical mean values with respect to the initial state of a system are written as $\langle c \rangle$. A time dependence is indicated in a Heisenberg-type notation, such as $\langle c(t) \rangle$. The symbol $\mathbb{1}$ is used to denote both the identity operator in a Hilbert space or the identity matrix. In the former case a subscript identifies the (sub)system the operator acts on, in the latter case a numerical subscript denotes the matrix dimension.

Also note that we use the convention $\hbar = 1$, except in Section 2, where the explicit usage of $\hbar$ is convenient for discussing the physics underlying our optomechanical models.

# 1. INTRODUCTION

Conducting any kind of quantum experiment requires, in a general sense, excellent control over the physical system and the experimental setup. On the one hand this means that the system under study must be, to a large degree, decoupled from the environment and all classical noise sources—or from all sources of decoherence in general—in order for genuine quantum effects to emerge. On the other hand one must be able to manipulate the system's dynamics and quantum state very precisely. This was impressively realized in a broad range of atomic, molecular, and optical (AMO) systems, such as pulsed and continuous wave (CW) light fields, optical and microwave (MW) fields of cavity modes, single trapped ions or neutral atoms in optical or MW cavities, Coulomb crystals of trapped ions, neutral cold atoms in optical potentials, and even in room temperature atomic ensembles. An equally high level of quantum control has been achieved in solid-state realizations of quantum optical systems in the form of artificial atoms, such as defect centers in crystals, quantum dots, or superconducting qubits. The exquisite control over these systems is based (i) on the possibility to prepare almost pure quantum states, e.g., by appropriate cooling or optical pumping techniques, (ii) on the ability to have strong interactions between these systems for quantum state manipulation, where strong means that coherent evolution dominates over decoherence, and finally (iii) on methods to perform precise, quantum-limited measurements. This has led to remarkably precise tests of fundamental aspects of quantum theory, in particular to the preparation of nonclassical and entangled quantum states. Finally this progress has also paved the way toward exciting technological applications of quantum physics in quantum information processing.

In a remarkable parallel development quantum control over optomechanical systems has been achieved during the last years. Optomechanical systems encompass micro- or nanofabricated structures realizing high-quality mechanical oscillators coupled to optical or MW cavities where the coupling is typically achieved via radiation pressure or capacitive forces, respectively. Radiation pressure effects on mechanical oscillators have already been analyzed theoretically in the late 1960s (Braginskii and Manukin, 1967), and experimentally demonstrated using microwaves (Braginskii et al., 1970; Gozzini et al., 1985) and optical fields (Dorsel et al., 1983) several years later. In a vigorous growth during the last two decades the field of cavity optomechanics established itself as a separate field within the quantum sciences. Eventually, within the last 6 years, a number of breakthrough experiments demonstrated optomechanical dynamics in the quantum regime: first, cooling of micromechanical oscillators to the ground state has been shown in 2011 by Chan et al. (2011) and Teufel et al. (2011). Since then quantum coherent coupling among mechanical systems and light was achieved in Verhagen et al. (2012) and used to transfer quantum states from mechanics to light (Palomaki et al., 2013a) as well as to generate entangled states of light and a mechanical oscillator (Palomaki et al., 2013b). State transfer between a superconducting qubit and a mechanical oscillator was shown before by O'Connell et al. (2010). The quantum coherent nature of the optomechanical coupling may give rise to ponderomotive squeezing of light, as seen in Brooks et al., (2012); Safavi-Naeini et al. (2013), and Nielsen et al. (2017). When an optomechanical system is used as a position or force sensor entering the quantum regime implies that the measurement sensitivity will be limited by the quantum noise of measurement back action, as demonstrated in Purdy et al. (2013). The strong effect light may exert on the quantum state of a mechanical oscillator is further witnessed by the generation of mechanical squeezed states by means of reservoir engineering (Pirkkalainen et al., 2015; Wollman et al., 2015) or quantum non-demolition measurements (Lecocq et al., 2015). Very recently quantum back action evading measurements of collective modes of two mechanical oscillators (Ockeloen-Korppi et al., 2016) and of a mechanical oscillator and a collective atomic spin has been demonstrated (Møller et al., 2016). The quantum-limited measurement of mechanical oscillators by means of light present a key tool for the quantum control of these mesoscopic objects, especially when combined with fast feedback control within the decoherence time as shown by Wilson et al. (2015). The effects demonstrated in these experiments witness an optomechanical (quantum) cooperativity parameter larger than unity which is a necessary condition to observe quantum effects in optomechanics. In this regime the quantum control of optomechanical systems by means of time-continuous measurement and feedback operations becomes an exciting perspective within reach of state-of-the-art experimental platforms.

Quantum control theory combines concepts from classical control theory with concepts from quantum physics, exploiting inherently non-classical features such as coherent superpositions and entanglement. It is nowadays an active research field that plays a crucial role in modern quantum experiments across different fields (Dong and Petersen, 2010a; Gough,

2012). Prominent examples of successful application of quantum control protocols are spin squeezing (Kuzmich et al., 2000) and steady-state entanglement generation in atomic ensembles (Krauter et al., 2011), or come from the field of quantum metrology, such as nuclear magnetic resonance (Geremia et al., 2003) and atomic clocks (Bergquist et al., 2001). Strongly connected (and partly overlapping) fields are those of *quantum filtering* (Belavkin and Guta, 2008; Bouten et al., 2007) and *quantum estimation theory* (Helstrom, 1969; Paris, 2009), which concern themselves with the estimation of the quantum state itself (Belavkin, 1980; Gammelmark et al., 2013), or the estimation of classical parameters (or classical signals) from the measurements of quantum systems (Tsang, 2009; Tsang et al., 2011). Quantum estimation techniques have already been successfully applied in optomechanics (Ang et al., 2013; Iwasawa et al., 2013; Szorkovszky et al., 2012), and optimal estimation of the optomechanical quantum state has been shown in Wieczorek et al. (2015).

Despite these advancements, control of optomechanical systems has not yet been developed to the high degree attained in more mature branches of quantum physics; applications of quantum control theory in optomechanics range from feedback cooling of the mechanical motion (Cohadon et al., 1999), mechanical squeezing (Clerk et al., 2008; Woolley et al., 2008), and two-mode squeezing (Woolley and Clerk, 2013) to back-action elimination (Courty et al., 2003; Wiseman, 1995) with possible applications in gravitational-wave detection. Importantly for quantum information processing and communication, it can also be used to robustly generate entanglement between remote quantum systems, as has been demonstrated recently for spin qubits (Dolde et al., 2014). At the same time entanglement itself can be an essential component to facilitate control of quantum systems, e.g., as a resource for teleportation (Bennett et al., 1993), when employed as a means for remote state preparation. Quantum control protocols are often operated in a time-continuous fashion and rely on continuous measurements that are capable of tracking the quantum state of the controlled system. The resulting measurement record—and the so-called conditional quantum state inferred from it—is then used as a basis for the applied feedback (Wiseman and Milburn, 2009). Thus, the control protocol's success critically depends on the precision of the employed measurements. In the experiment by Wilson et al. (2015) the mechanical oscillator was monitored with a measurement strength matching its thermal decoherence rate (equivalent to a cooperativity above 1) and measurement-based feedback cooling to an occupation number of several phonons (limited by residual absorption) has been demonstrated.

In view of the achievements attained in the field of optomechanics, we believe it is the right time to make better known the framework of quantum control theory in the field. Moreover, we believe that researchers in the field of quantum control theory will benefit from a brief introduction to optomechanics summarizing its theoretical description and state of the art. These two goals we are following with the present review. We emphasize that we do not try to give a comprehensive account or review of either field, quantum control theory or optomechanics. In both fields a number of highly recommendable reviews and textbooks have been published: for optomechanics we refer to the reviews (Aspelmeyer et al., 2010, 2012, 2014b; Favero and Karrai, 2009; Favero et al., 2014; Genes et al., 2009; Kippenberg and Vahala, 2007, 2008; Marquardt and Girvin, 2009; Meystre, 2013; Schliesser and Kippenberg, 2010) and textbooks (Aspelmeyer et al., 2014a; Bowen and Milburn, 2016). Regarding the theory of quantum measurement and quantum control we refer to Dong and Petersen (2010b) and the excellent textbooks (Jacobs, 2014; Wiseman and Milburn, 2009) as well as references therein.

Our review is based on the doctoral thesis of Hofer (2015) and in parts on the articles (Hofer and Hammerer, 2015; Hofer et al., 2011; Wieczorek et al., 2015). It is organized as follows: Section 2 provides an introduction to the physics of cavity-optomechanical systems, as well as to their description we use throughout this work. We introduce the total system's description as an open quantum system in terms of (stochastic) master equations and quantum Langevin equations and discuss characteristic features of the optomechanical steady state. Section 3 is devoted to pulsed optomechanical quantum control and quantum information protocols. We discuss a protocol to create and verify continuous-variable entanglement between a mechanical oscillator and a light pulse, including an extension to a teleportation scheme. Section 4 discusses time-continuous control protocols, analyzing in detail several measurement-based feedback schemes, including time-continuous teleportation and entanglement swapping. In Section 5 we present first experimental steps toward the implementation of such protocols in optomechanical systems. Finally, Appendices A and B review important mathematical results from quantum filtering and control theory used in the main text.

# 2. CAVITY-OPTOMECHANICAL SYSTEMS

## 2.1 The Physical System

Although optomechanical systems exist in many different physical implementations, many of them can be described by the same simple physical model (Aspelmeyer et al., 2014b). To exemplify the most important features of a cavity-optomechanical system, we will use as our toy-model a Fabry–Pérot cavity with one oscillating mirror (see Fig. 1). On the most basic level this simple system consists of two coupled harmonic oscillators. In this section we will introduce its main constituents—the optical and mechanical resonator—and the physical background of the optomechanical interaction.

Although we will mainly be concerned with optomechanical setups working with optical photons, the presented description as well applies to *electromechanical* systems, where an LC resonator driven by a microwave field takes the role of the optical cavity (see Fig. 1).

### *2.1.1 Mechanical Oscillators*

Due to their size and high number of degrees of freedom, experimentally employed mechanical oscillators in general posses a multitude of mechanical eigenmodes, whose spectral features are determined by the oscillator's geometry, material properties, and the coupling to its support. For general geometries, the spatial mechanical mode shape can be arbitrarily complex and can be described by a displacement field $\boldsymbol{u}(\boldsymbol{r}, t)$ (Pinard et al., 1999). We can expand $\boldsymbol{u}(\boldsymbol{r}, t)$ in terms of the oscillator's eigenmodes $\boldsymbol{u}_n(\boldsymbol{r})$,

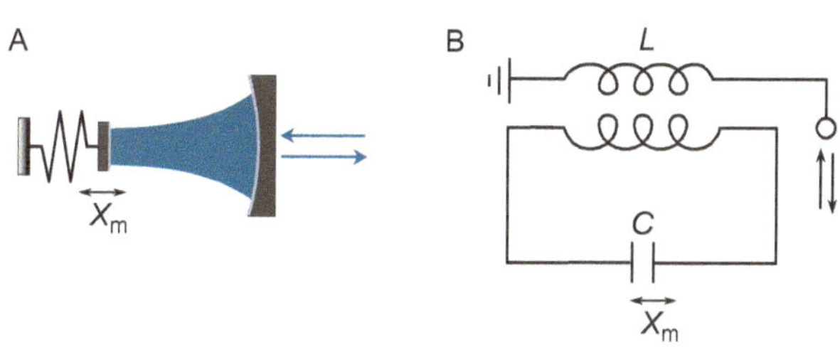


**Fig. 1** (A) Schematics of an optomechanical setup. The cavity length $L(X_m)$ is modulated by the mechanical motion. (B) Schematics of an electromechanical setup. The capacitance $C(X_m)$ of the LC resonator is modulated by the mechanical motion.

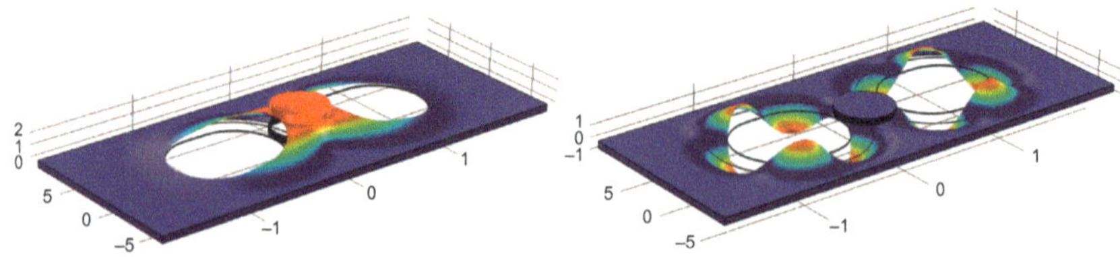


**Fig. 2** Finite-element simulation of the fundamental (*left*) and a higher-order mode (*right*) of a doubly clamped oscillating mirror as has been used in Gröblacher et al. (2009). *Courtesy of Witlef Wieczorek, University of Vienna.*

$$u(r,t)=\sum_n X_n(t)u_n(r), \tag{1}$$

with the corresponding time-dependent amplitudes $X_n(t)$. Two examples of different eigenmodes of a doubly clamped oscillating mirror pad are shown in Fig. 2. While mechanical oscillators are in general not linear (in terms of their response to applied forces), linearity poses a good approximation for the small displacements typically attained in normal operation. We therefore model the amplitudes $X_n(t)$ to follow the damped harmonic evolution given by

$$\ddot{X}_n(t)+\gamma_n\dot{X}_n(t)+\omega_n^2X_n(t)=\frac{F_{\text{ext}}}{m_{\text{eff}}^{(n)}}. \tag{2}$$

Here $\omega_n$ and $\gamma_n$ are the angular resonance frequency and the damping constant of the $n$th eigenmode and $m_{\text{eff}}^{(n)}$ denotes the corresponding effective mass determined by the mode volume of $u_n$. $F_{\text{ext}}$ denotes the sum of all external forces acting on the specific mechanical mode, e.g., radiation pressure or fluctuating Langevin forces.

In this chapter we will consider the single-mode case only, where we assume that we can experimentally address a single one of the mechanical eigenmodes. This mode is assumed to have no significant spectral overlap with its neighboring modes. In the following we will denote this mode's resonance frequency by $\omega_{\text{m}}$, and the energy damping rate as $\gamma_{\text{m}}$ [the full width at half maximum (FWHM) of the corresponding spectral peak]. The corresponding amplitude of the oscillation we will call $X_{\text{m}}(t)$. In our toy-model, the Fabry–Pérot cavity, the center-of-mass oscillation is the only available eigenmode, and the amplitude $X_{\text{m}}(t)$ is simply the mirror's displacement from its equilibrium position. The effective mass $m_{\text{eff}}$ is then approximately given by the total mass of the mirror (not accounting for the finite mass of the supporting spring).

The viscous damping term in Eq. (2) is due to the coupling of the mechanical oscillator to its support, which at the same time also acts as a thermal environment (commonly also referred to as heat bath) and represents a noise source for the mechanical oscillator. Customary this heat bath is modeled as a collection of an infinite number of harmonic oscillators in a thermal state at a temperature $T$, as first introduced by Caldeira and Leggett (1983). The mean bath occupation number follows a Bose–Einstein distribution, and we thus have $n_{\mathrm{B}}(\omega) = \left[\exp(\hbar\omega/k_{\mathrm{B}}T) - 1\right]^{-1}$ (with the reduced Planck constant $\hbar$ and the Boltzmann constant $k_{\mathrm{B}}$). Later an important figure for us will be the mean bath occupation at the mechanical frequency, i.e., $\bar{n} = n_{\mathrm{B}}(\omega_{\mathrm{m}})$. In the high-temperature limit we can approximate $n_{\mathrm{B}}(\omega_{\mathrm{m}}) \approx k_{\mathrm{B}}T/\hbar\omega_{\mathrm{m}}$. We will see in Section 2.3.1 that the effective decoherence rate (often called thermal decoherence rate) is given by $\bar{n}\gamma_{\mathrm{m}} \approx k_{\mathrm{B}}T/\hbar Q_{\mathrm{m}}$, where we defined the mechanical quality factor (Q-factor) $Q_{\mathrm{m}} = \omega_{\mathrm{m}}/\gamma_{\mathrm{m}}$. One thus sees that in order to have low thermal decoherence we need a high-Q mechanical oscillator and a low-temperature bath, which can be obtained by cryogenic cooling of the experimental setup. Apart from the Caldeira–Leggett approach above, another (phenomenological) model takes into account internal friction effects (Saulson, 1990); additionally, more realistic microscopic models and experimental studies of system–bath coupling exist (Cole et al., 2011; Gröblacher et al., 2015).

For the quantum treatment of the mechanical oscillator we introduce position and momentum operators $X_{\mathrm{m}}$, $P_{\mathrm{m}}$, which fulfill canonical commutation relations $[X_{\mathrm{m}}, P_{\mathrm{m}}] = i\hbar$. As is customary in quantum optics, we also introduce the dimensionless quadratures $x_{\mathrm{m}}$, $p_{\mathrm{m}}$ with $[x_{\mathrm{m}}, p_{\mathrm{m}}] = \mathrm{i}$. We express them in terms of creation and annihilation operators $c_{\mathrm{m}}^{\dagger}$, $c_{\mathrm{m}}$ ($[c_{\mathrm{m}}, c_{\mathrm{m}}^{\dagger}] = 1$)

$$x_{\mathrm{m}} = \frac{c_{\mathrm{m}} + c_{\mathrm{m}}^{\dagger}}{\sqrt{2}}, \qquad p_{\mathrm{m}} = \frac{c_{\mathrm{m}} - c_{\mathrm{m}}^{\dagger}}{\sqrt{2}\mathrm{i}}. \tag{3}$$

We can convert them into $X_m = \sqrt{2}x_0 x_{\mathrm{m}}$ and $P_{\mathrm{m}} = \sqrt{2}m_{\mathrm{eff}}\omega_{\mathrm{m}}x_0 p_{\mathrm{m}}$ by rescaling with the oscillator's ground state extension

$$x_0 = \sqrt{\hbar/2m_{\mathrm{eff}}\omega_{\mathrm{m}}}. \tag{4}$$

### 2.1.2 Optical Resonators

The second constituent of a cavity-optomechanical system is the optical cavity which forms a resonator for photons. A simple Fabry–Pérot cavity consists of two highly reflecting mirrors separated by a distance $L$. Such a cavity contains a sequence of equally spaced modes with resonance frequencies $\nu_n = n\Delta\nu$ $(n \in \mathbb{N})$ (Vogel and Welsch, 2006), which are separated by a frequency $\Delta\nu$, the so-called free spectral range (Fig. 3). The free spectral range is determined by the cavity length, and—in the absence of an optical medium—is given by

$$\Delta\nu = \frac{c}{2L}. \tag{5}$$

Again, in this work we will focus on the case of the mechanical oscillator interacting with a single cavity mode with a central angular frequency $\omega_c = 2\pi\nu_c$ only. In nano- and microoptomechanical setups this situation can easily be created experimentally, as the free spectral range is many orders of magnitude larger than any frequency scale of the mechanical system. Thus different cavity modes do not interact via coupling to the mirror. There are systems, however, where coupling between different optical modes can occur (deliberately or as a perturbative effect) (Favero et al., 2014).

Due to finite mirror reflectivity, absorption in the mirror substrate, or photon scattering out of the cavity mode the lifetime of photons in the cavity is limited to a finite number of round trips. This number is called the optical finesse $\mathcal{F}_{\mathrm{cav}}$ and is connected to the cavity decay rate $\kappa$ by

$$\mathcal{F}_{\mathrm{cav}} = \frac{2\pi\Delta\nu}{\kappa} = \frac{\pi c}{\kappa L}. \tag{6}$$

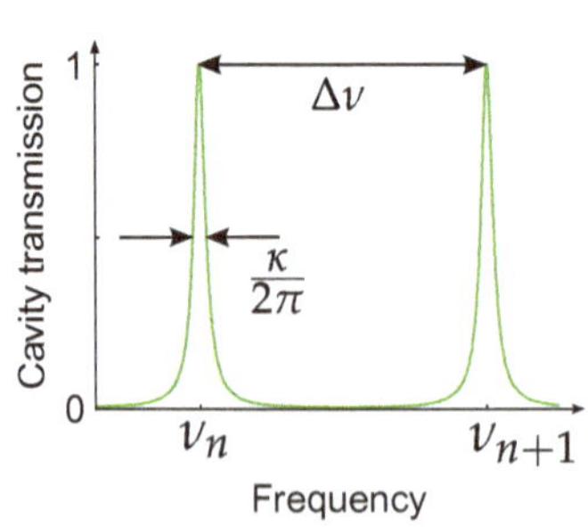


**Fig. 3** Cavity transmission (power) for $\mathcal{F}_{\mathrm{cav}} = 20$.

Throughout this review we denote by $\kappa$ the energy (FWHM) decay rate, i.e., the number of photons leaving the cavity per unit time. As we need to monitor the output light of the cavity in virtually every experiment we need to discriminate between pure photon losses (absorption, scattering) and transmission into the external field which we can eventually measure. The total decay rate is thus $\kappa = \kappa_{\text{in}} + \tilde{\kappa}$, where $\kappa_{\text{in}}$ is associated with the input-coupler of the cavity (i.e., the decay channel we can measure), whereas $\tilde{\kappa}$ collects all other loss mechanisms we cannot measure.

Let us assume we drive an optical cavity by a constant coherent laser field with a central frequency $\omega_0$ and a (appropriately rescaled) complex amplitude $\varepsilon$. This amplitude is connected to the input power $P_{\text{in}}$ by $|\varepsilon| = \sqrt{P_{\text{in}}/\hbar\omega_0}$, and therefore describes the square root of the driving laser's photon flux. The mean intracavity field $\alpha_c$ then follows the equation of motion

$$\dot{\alpha}_c = -\left(i\omega_c + \frac{\kappa}{2}\right)\alpha_c + \sqrt{\kappa_{\text{in}}}\varepsilon e^{-i\omega_0 t}. \tag{7}$$

We can get rid of the (trivial) evolution at optical frequencies by introducing $\tilde{\alpha}_c(t) = \alpha_c(t)e^{i\omega_0 t}$. After a transient period $\tilde{\alpha}_c$ will assume a constant steady-state amplitude $\tilde{\alpha}_c^{ss} = \lim_{t\to\infty}\tilde{\alpha}_c(t)$ given by

$$\tilde{\alpha}_c^{ss} = \frac{\sqrt{\kappa_{\text{in}}}\varepsilon}{i\Delta_0 - \dfrac{\kappa}{2}}, \tag{8}$$

with the detuning $\Delta_0 = \omega_0 - \omega_c$ of the laser with respect to the cavity resonance frequency. Note the distinction between the total decay rate $\kappa$ and the input-coupling rate $\kappa_{\text{in}}$, which leads to a decreased intracavity photon number $|\alpha_c|^2$ for increasing losses $\tilde{\kappa}$.

Taking the Fourier transform of Eq. (7) we obtain the cavity's susceptibility $\chi(\omega_0)$, i.e., its linear response function. The value $\chi(\omega_0)$ determines the response of the system for a constant input at a frequency $\omega_0$. It is given by the Lorentzian

$$\chi(\omega) = \frac{\sqrt{\kappa_{\text{in}}}}{\dfrac{\kappa}{2} - i(\omega - \omega_c)}. \tag{9}$$

Its modulus and its argument give the amplitude and phase response of the intracavity field, respectively.

Quantum-mechanically the intracavity field can be described as a damped harmonic oscillator (Walls and Milburn, 2008). In analogy to Section 2.1.1 we introduce creation and annihilation operators $c_c^\dagger$, $c_c$, and the corresponding (dimensionless) quadrature operators

$$x_c = \frac{c_c + c_c^\dagger}{\sqrt{2}}, \quad p_c = \frac{c_c - c_c^\dagger}{\sqrt{2}i}. \tag{10}$$

We call $x_c$ the amplitude quadrature and $p_c$ the phase quadrature of the electromagnetic field. Their equations of motion we will discuss in Section 2.3.1.

### 2.1.3 Cavity-Optomechanical Interaction

In cavity-optomechanical systems the interaction between the optical and the mechanical mode typically manifests itself in the form of a dispersive coupling, which means that the cavity resonance frequency experiences a shift depending on the mechanical oscillator's position. Physically, the interaction is mediated by radiation pressure (Einstein, 1909) in the form of momentum transfer due to reflection (Fabry–Pérot-type setups, microtoroids) (Braginskii and Manukin, 1967; Braginskii et al., 1970) or gradient forces [membrane in the middle setups (Thompson et al., 2008), levitated microobjects (Barker and Shneider, 2010; Chang et al., 2010; Romero-Isart et al., 2010)]. In addition to the dispersive regime there exists a dissipative optomechanical coupling where the cavity decay rate is dependent on the mechanical position (Elste et al., 2009; Sawadsky et al., 2015; Xuereb et al., 2011). We will not consider this kind of coupling in this work, however.

For photons impinging on a single mirror the radiation pressure is just proportional to the incoming photon flux. As this force exerted by photons on a massive object is typically very weak, cavity-optomechanical experiments employ a resonator for the photons—an optical cavity—in order to enhance the effective photon flux and therefore the radiation pressure coupling. Conversely, the intracavity photon flux now very sensitively depends on the mirror's position, as the mirror parametrically changes the cavity resonance frequency. This can be nicely illustrated by plotting the modified susceptibility of the cavity as a function of the mirror position $X_m$, which can be obtained from Eq. (9) by letting the resonance frequency

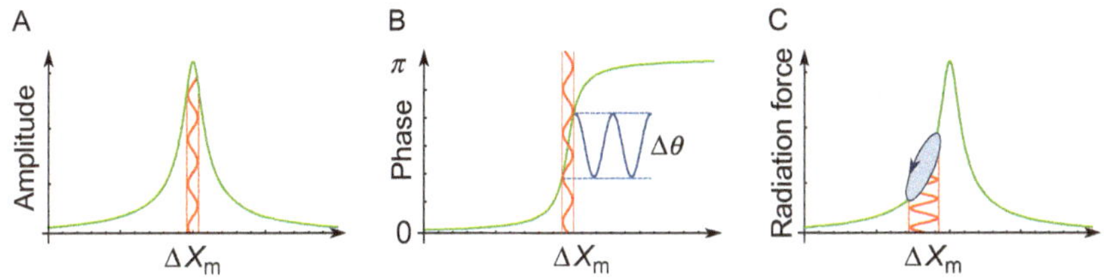


**Fig. 4** (A and B) Optical response to mechanical modulation. (C) Radiation pressure (proportional to the intracavity amplitude) as a function of the mechanical motion for a red-detuned drive.

depend linearly on the mirror position, i.e., $\omega_c = \omega_c(X_m)$ (see also the discussion of the radiation pressure Hamiltonian in the next section). Fig. 4 shows that moving the mirror modulates the intracavity-field amplitude [shown in (A)], as well as its phase [shown in (B)]. For resonant driving (as depicted in the figure) the amplitude stays constant to second order in $X_m$, while the phase is changed linearly. Hence, mechanical displacement sensing is typically operated on resonance as there the sensitivity on the mechanical displacement $\Delta\theta/\Delta X_m$ is maximal (Aspelmeyer et al., 2014b).

The finite decay time of the cavity (i.e., the average time it takes photons to leave the cavity) leads to a time-lag in the radiation pressure force on the mirror. This leads to so-called *dynamical back-action effects*, most notably the optical-spring effect and back-action cooling (or heating) of the mechanical motion, both of which we will again encounter again below. A classical picture of optomechanical cooling can be given by considering the cavity's susceptibility (see Fig. 4) (Marquardt and Girvin, 2009): as the mirror oscillates and therefore sweeps the cavity's resonance profile, it modulates the number of photons inside the cavity and thus also the radiation pressure. However, due to the time-delay associated with the cavity bandwidth, the radiation pressure is increased in one half-cycle, while it is decreased in the other half-cycle (schematically depicted by the blue ellipse in Fig. 4). Depending on the relative phase between the radiation pressure cycle and the mechanical oscillation this leads to additional damping for a red-detuned laser drive or amplification for a blue-detuned drive.

## 2.2 Hamiltonian

Here we restrict ourselves to the case where a single mechanical mode interacts with a single optical mode. The generalization to a multimode scenario is straightforward. We give a simple derivation of the cavity-optomechanical

Hamiltonian; a rigorous version was given by Law (1995). We start from the Hamiltonian of two uncoupled harmonic oscillators

$$H_0 = \hbar\omega_m c_m^\dagger c_m + \hbar\omega_c c_c^\dagger c_c, \tag{11}$$

where $\omega_m$ and $\omega_c$ denote the mechanical and (nominal) cavity resonance frequency, respectively, and we have $[c_i, c_j^\dagger] = \delta_{ij}$ for $i, j \in \{m, c\}$. The optical resonance frequency $\omega_c$ is determined by the round-trip time of photons in the cavity and thus by the effective cavity length $L$, i.e., $\omega_c/2\pi = nc/2L$, for some $n \in \mathbb{N}$. In the case of a cavity with a moving mirror, the effective length $L(X_m)$ depends on the (dimensionful) position of the mechanical oscillator $X_m$. Moving the mirror thus shifts the resonance frequency and therefore changes the energy stored inside the cavity mode. For small displacements $X_m/L \ll 1$ [assuming that $X_m = 0$ is the rest position and hence $\omega_c := \omega_c(0)$ the nominal cavity frequency] we can expand $\omega_c(X_m)$ in a Taylor series around $X_m = 0$,

$$\omega_c(X_m) = \omega_c + \frac{\partial\omega_c}{\partial X_m} X_m + \mathcal{O}(X_m)^2. \tag{12}$$

Using that $X_m = x_0(c_m + c_m^\dagger) = \sqrt{2}x_0 x_m$, the Hamiltonian for the optomechanical system is (to first order in $X_m$)

$$H'_{nl} = \hbar\omega_m c_m^\dagger c_m + \hbar\omega_c c_c^\dagger c_c + \hbar g_0 x_m c_c^\dagger c_c. \tag{13}$$

The last term is the sought-after radiation pressure interaction with the so-called single-photon optomechanical coupling strength

$$g_0 = \sqrt{2}x_0 \frac{\partial\omega_c}{\partial X_m}, \tag{14}$$

which quantifies the interaction between the mechanical oscillator and a single photon in the cavity. For the Fabry–Pérot case we choose $L(X_m) = L_0 - X_m$ and thus find $g_0 = \sqrt{2}x_0\omega_c/L_0$. Note that under these conventions a positive value for the displacement $X_m > 0$ means that the cavity length decreases, and thus the energy in the cavity increases.

Expression (13) shows that the radiation pressure coupling is nonlinear in the amplitudes $c_i$, $c_i^\dagger$ and depends on the number of photons in the cavity. However, as $g_0$ is very small in many current optomechanical systems, the generated nonlinear dynamics—and optomechanical effects on the single-photon level in general—are hard to observe. To enhance the radiation pressure one can drive the optomechanical cavity by a strong laser beam with a large coherent (time dependent) amplitude $\varepsilon(t)$. Such a drive with a center frequency $\omega_0$ can be described by adding an additional driving term

$$H_{\text{drive}} = -i\hbar\left[E^*(t)e^{i\omega_0 t}c_c - E(t)e^{-i\omega_0 t}c_c^\dagger\right] \tag{15}$$

to Eq. (13), where $E(t) = \sqrt{\kappa_{\text{in}}}\varepsilon(t)$ describes the driving strength and $\varepsilon(t) \in \mathbb{C}$ is the complex amplitude of the input field. This Hamiltonian basically arises from a beam-splitter-like interaction, where photons are scattered into the cavity mode from the external coherent field. In order to get rid of the explicit time dependence in (15) we go to a frame rotating at the optical frequency $\omega_0$, which prompts us to introduce the detuning $\Delta_0 = \omega_0 - \omega_c$. The complete optomechanical Hamiltonian including the nonlinear radiation pressure interaction then takes the form

$$H_{\text{nl}} = \hbar\omega_m c_m^\dagger c_m - \hbar\Delta_0 c_c^\dagger c_c + \hbar g_0 x_m c_c^\dagger c_c - i\hbar\left[E^*(t)c_c - E(t)c_c^\dagger\right]. \tag{16}$$

Driving the optomechanical cavity in such a way has two effects: (i) the mirror is shifted to a new equilibrium position, and (ii) the radiation pressure interaction is enhanced by the classical intracavity amplitude $\alpha_c$ created by the laser drive, effectively linearizing it given that we have $|\alpha_c| \gg 1$. (This situation can be created by employing a strong laser drive or a high-finesse cavity.) For a full description of these effects one needs to account for the open-system dynamics, as discussed in Mancini and Tombesi (1994). In this regime the optomechanical Hamiltonian (16) can be approximated by

$$H_{\text{lin}} = \hbar\omega_m c_m^\dagger c_m - \hbar\Delta_c c_c^\dagger c_c + \hbar\frac{g_0\alpha_c}{\sqrt{2}}(c_m + c_m^\dagger)(c_c + c_c^\dagger), \tag{17}$$

where we introduced an effective detuning $\Delta_c$ which is shifted with respect to $\Delta_0$ due to the shifted equilibrium position of the mirror. We can see that the coupling strength $g = g_0\alpha_c/\sqrt{2}$ of the linear interaction is enhanced by $\alpha_c$, which in the present case is the square-root of the intracavity photon number. (Here we assumed $\alpha_c \in \mathbb{R}$ without loss of generality.) For a high-finesse cavity or a strong laser drive the mean number of photons in the cavity can be large, and the interaction strength can be enhanced by several orders of magnitude.

Depending on the chosen detuning $\Delta_c$ we can identify different types of interactions which are realized by the linearized radiation pressure Hamiltonian. Expanding the interaction part of (17) we can identify two contributions: the so-called *beam-splitter* (BS) Hamiltonian

$$H_{\text{bs}} = \hbar g(c_{\text{m}}c_{\text{c}}^{\dagger} + c_{\text{m}}^{\dagger}c_{\text{c}}) \tag{18}$$

is resonant for $\Delta_{\text{c}} = -\omega_{\text{m}}$ and describes coherent exchange of energy between the mechanical oscillator and the cavity mode. This term is relevant for cooling the mechanical motion via sideband cooling (Marquardt et al., 2007; Wilson-Rae et al., 2007) (see also Section 2.4) and can be employed to generate a state swap between the two modes. The so-called two-mode squeezing (TMS) interaction

$$H_{\text{tms}} = \hbar g(c_{\text{m}}c_{\text{c}} + c_{\text{m}}^{\dagger}c_{\text{c}}^{\dagger}), \tag{19}$$

on the other hand, is resonant for $\Delta_{\text{c}} = \omega_{\text{m}}$, and describes simultaneous creation and annihilation of excitations in both modes, and is the optomechanical analog to the optical down-conversion process. It is known to create optomechanical correlations and entanglement. Customarily we call the $\Delta_{\text{c}} > 0$ "blue detuned" and $\Delta_{\text{c}} < 0$ "red detuned."

Both terms can be interpreted in terms of three-mode scattering processes involving the incoming laser beam (see Fig. 5A and B). In the beam-splitter process a phonon is annihilated in order to up-scatter a laser photon into the cavity mode (resonant for $\omega_0 = \omega_{\text{c}} - \omega_{\text{m}}$). In the two-mode squeezing process a fraction $\hbar\omega_{\text{m}}$ of a laser photon's energy is transferred to the mechanical motion creating a phonon, while the photon is down-scattered into the cavity mode (resonant for $\omega_0 = \omega_{\text{c}} + \omega_{\text{m}}$).

In the case of zero detuning $\Delta_{\text{c}} = 0$ both processes contribute equally and the full interaction Hamiltonian proportional to $x_{\text{m}}x_{\text{c}}$ is retained. This interaction is often referred to as quantum nondemolition (QND) interaction (Braginsky et al., 1980; Thorne et al., 1978). Additionally, a resonant drive is commonly employed to measure position changes of the mechanical oscillator via phase shifts of the light field; these measurements can in turn be used

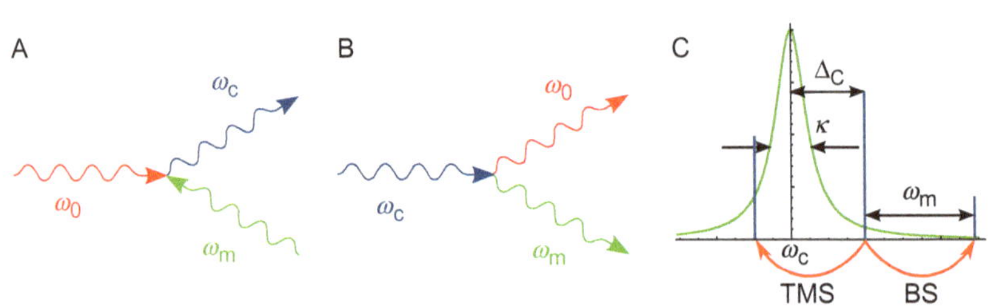


**Fig. 5** Scattering processes corresponding to (A) the BS and (B) the TMS Hamiltonian. (C) Spectral illustration of scattering processes into mechanical sidebands (here the *blue-detuned* case).

to cool the mechanical motion by active feedback (Courty et al., 2001; Genes et al., 2008b; Vitali et al., 2002).

The resonance conditions for the listed processes can be illustrated by going into an interaction picture with $\hbar\omega_{\mathrm{m}} c_{\mathrm{m}}^{\dagger} c_{\mathrm{m}} - \hbar\Delta_{\mathrm{c}} c_{\mathrm{c}}^{\dagger} c_{\mathrm{c}}$ which leads to

$$\tilde{H}_{\mathrm{int}} = \hbar g \left[ c_{\mathrm{m}} c_{\mathrm{c}}^{\dagger} e^{-\mathrm{i}(\omega_{\mathrm{m}} + \Delta_{\mathrm{c}})t} + \mathrm{H.\,c.} \right] + \hbar g \left[ c_{\mathrm{m}} c_{\mathrm{c}} e^{-\mathrm{i}(\omega_{\mathrm{m}} - \Delta_{\mathrm{c}})t} + \mathrm{H.\,c.} \right].$$

Tuning the laser to one of the mechanical sidebands, i.e., choosing $\Delta_{\mathrm{c}} = \pm\omega_{\mathrm{m}}$, makes one of the terms in $\tilde{H}_{\mathrm{int}}$ resonant while the second term is oscillating rapidly at a frequency $\pm\ 2\omega_{\mathrm{m}}$. Invoking a rotating-wave approximation (RWA), which is equivalent to taking a temporal average, then allows us to neglect the corresponding off-resonant term. The RWA is a good approximation in the sideband-resolved weak-coupling regime $g < \kappa \ll \omega_{\mathrm{m}}$, where the off-resonant scattering terms are strongly suppressed by the cavity's density of states which is strongly peaked in this regime (see Fig. 5C).

One of the most important optomechanical parameters for us in this review will be the *optomechanical cooperativity*

$$C = \frac{4g^2}{\kappa\gamma_{\mathrm{m}}(\bar{n} + 1)}, \tag{20}$$

which quantifies the strength of the unitary optomechanical interaction compared to electromagnetic and mechanical decoherence rates. Intuitively, the coherent interaction should be stronger then the decohering dynamics in order to observe quantum effects. Indeed we will see later that all discussed protocols need to be operated in the strong cooperativity regime of $C > 1$.

## 2.3 Open System Dynamics

A cavity-optomechanical system couples to two different kinds of environments: on the one hand the mechanical oscillator couples to a mechanical heat bath through its support; on the other hand, photons can leak out of the cavity into the electromagnetic environment in a finite time. Due to this interaction with the environment, cavity-optomechanical systems inherently are open quantum systems and a Hamiltonian description does not suffice to describe their full dynamics. In quantum optics, powerful methods have been developed to describe open quantum systems (Gardiner and Zoller, 2004). In the following sections we will show how to describe the dynamics of optomechanical systems in the master equation (MEQ) and the quantum Langevin equations (QLEs) approach. As is customary we will call the optomechanical system simply "system," while we will refer to the external electromagnetic field and the mechanical environment as "bath" or "environment."

### 2.3.1 Quantum Langevin Equations

The linear open-system dynamics of optomechanical systems [corresponding to the Hamiltonian Eq. (17)] can be described by the (Stratonovich) quantum Langevin equations (Gardiner and Collett, 1985)

$$\dot{x}_{\mathrm{m}} = \omega_{\mathrm{m}} p_{\mathrm{m}}, \tag{21a}$$

$$\dot{p}_{\mathrm{m}} = -\omega_{\mathrm{m}} x_{\mathrm{m}} - \gamma_{\mathrm{m}} p_{\mathrm{m}} - 2g x_{\mathrm{c}} + \sqrt{2\gamma_{\mathrm{m}}} f, \tag{21b}$$

$$\dot{x}_{\mathrm{c}} = -\Delta_{\mathrm{c}} p_{\mathrm{c}} - \frac{\kappa}{2} x_{\mathrm{c}} + \sqrt{\kappa} x_{\mathrm{in}}, \tag{21c}$$

$$\dot{p}_{\mathrm{c}} = \Delta_{\mathrm{c}} x_{\mathrm{c}} - \frac{\kappa}{2} p_{\mathrm{c}} - 2g x_{\mathrm{m}} + \sqrt{\kappa} p_{\mathrm{in}}, \tag{21d}$$

which are written in the frame rotating at the driving laser frequency $\omega_0$. The input from the external electromagnetic field to the cavity mode is described by the field operators $a_{\mathrm{in}}$ and $a_{\mathrm{in}}^{\dagger}$ (see Appendix A). In a Markov approximation they are modeled as a white noise fields and are therefore $\delta$-correlated. For optical systems the electromagnetic bath is normally assumed to be in the vacuum state, which is due to the fact that at room temperature the mean occupation number of the bath at relevant frequencies is negligible, i.e., $n_{\mathrm{B}}(\omega_{\mathrm{c}}) \approx n_{\mathrm{B}}(\omega_0) \approx 0$. We thus have

$$\langle a_{\mathrm{in}}(t)\rangle = 0, \quad \langle a_{\mathrm{in}}^{\dagger}(t)\rangle = 0, \tag{22a}$$

$$\langle a_{\mathrm{in}}^{\dagger}(t) a_{\mathrm{in}}(t')\rangle = 0, \quad \langle a_{\mathrm{in}}(t) a_{\mathrm{in}}^{\dagger}(t')\rangle = \delta(t-t'). \tag{22b}$$

Note, however, that in electromechanical systems where $\omega_{\mathrm{c}}$ is in the microwave regime one might have a finite thermal occupation, i.e., $n_{\mathrm{B}}(\omega_{\mathrm{c}}) > 0$. The same description can also be valid if the laser drive of an optomechanical system exhibits broadband amplitude or phase noise that can be approximated by classical white noise. When the external field is in vacuum, we will often refer to the noise contribution $a_{\mathrm{in}}$ as *shot noise*, which is a purely quantum-mechanical contribution. In the experiment one typically has full access to the cavity input and the corresponding output field. This means that we can (at least in principle) engineer the input state [which modifies

Eq. (22)] and monitor the state of the outgoing field. The output of the cavity is given by the input–output relation (Gardiner and Collett, 1985) (also see Appendix A.3 for a more detailed discussion)

$$a_{\text{out}}(t) = \sqrt{\kappa} c_c(t) - a_{\text{in}}(t). \tag{23}$$

This shows that by measuring quadrature operators $x_{\text{out}}$, $p_{\text{out}}$ (e.g., by homodyne detection) of the output light we can directly monitor the intracavity-field quadratures $x_c$ and $p_c$ (plus a noise contribution).

In contrast to the treatment of the electromagnetic field, no single accepted model exists for the coupling of the mechanical oscillator to its thermal environment (Vacchini, 2000). A common approach is to model it in the form of Brownian motion damping (Caldeira and Leggett, 1983; Giovannetti and Vitali, 2001), where $f$ denotes the corresponding Gaussian, Hermitian noise operator. Here we adopt a quantum-optical stance and treat $f$ in a Markov approximation, which is known to be valid in the high-temperature limit where $k_B T/\hbar\omega_m \gg 1$ (Giovannetti and Vitali, 2001) and for high-Q oscillators (Benguria and Kac, 1981). We thus approximate $f$ as a zero mean white noise field which is $\delta$-correlated,

$$\langle f(t)\rangle = 0, \tag{24a}$$

$$\langle f(t)f(t') + f(t')f(t)\rangle \approx (2\bar{n}+1)\delta(t-t'). \tag{24b}$$

### 2.3.2 Master Equation and Stochastic Master Equation

As an alternative to the quantum Langevin equations, which constitute the Heisenberg description of the dynamics, the evolution of the optomechanical system can also be described in the Schrödinger picture. We denote by $\rho$ the density matrix of the bipartite system consisting of the mechanical and the optical mode. The state $\rho$ evolves according to a Lindblad master equation (Lindblad, 1976) of the form (Wilson-Rae et al., 2007)

$$\dot{\rho} = \mathcal{L}\rho = -\frac{i}{\hbar}[H_{\text{lin}}, \rho] + \kappa\mathcal{D}[c_c]\rho + \gamma_m(\bar{n}+1)\mathcal{D}[c_m]\rho + \gamma_m\bar{n}\mathcal{D}[c_m^\dagger]\rho. \tag{25}$$

Its initial condition we typically choose to be $\rho_0 = \rho_{\text{th}}^{(m)} \otimes |0\rangle\langle 0|_c$, where $\rho_m^{(\text{th})}$ denotes a thermal state of the mechanical oscillator given by

$$\rho_{\text{th}}^{(m)} = \frac{1}{\bar{n}+1}\sum_{n=0}^{\infty}\left(\frac{\bar{n}}{\bar{n}+1}\right)^n |n\rangle\langle n|_m \tag{26}$$

and $|0\rangle\langle 0|_c$ is the ground state of the cavity. The master equation (25) describes the unitary evolution generated by the linearized Hamiltonian $H_{\rm lin}$ [given by Eq. (17)] together with the dissipative dynamics created by the coupling to the environment. The decoherence effects are captured by the Lindblad terms (Gardiner and Zoller, 2004)

$$\mathcal{D}[s]\rho = s\rho s^{\dagger} - \frac{1}{2}\rho s^{\dagger}s - \frac{1}{2}s^{\dagger}s\rho. \tag{27}$$

The operator $\mathcal{D}$ is trace-preserving [i.e., $\mathrm{tr}(\mathcal{D}[c]\rho) = 0$], and hence preserves the normalization of the density matrix $\rho$.

In the weak-coupling regime $g \ll \kappa$ we can derive an effective master equation for the mechanical mode from which the cavity is adiabatically eliminated (see Wilson-Rae et al., 2007; Marquardt et al., 2007). We obtain the equation (written in a frame rotating at $\omega_{\rm m}$)

$$\dot{\rho}^{(\rm m)} = -\mathrm{i}[\delta\omega_{\rm m}c_{\rm m}^{\dagger}c_{\rm m}, \rho^{(\rm m)}] + \gamma_{-}\mathcal{D}[c_{\rm m}]\rho^{(\rm m)} + \gamma_{+}\mathcal{D}[c_{\rm m}^{\dagger}]\rho^{(\rm m)}. \tag{28}$$

This equations nicely illustrates the characteristic dynamical features of optomechanical systems we touched on in Section 2.1.3: firstly we see that the mechanical frequency is shifted with respect to the nominal value $\omega_{\rm m}$ by

$$\delta\omega_{\rm m} = g^2\,\mathrm{Im}(\eta_{-} + \eta_{+}), \tag{29}$$

where we introduced the parameters

$$\eta_{\pm} = \frac{1}{\frac{\kappa}{2} + \mathrm{i}(-\Delta_c \pm \omega_{\rm m})}. \tag{30}$$

This is the *optical-spring effect* (Braginsky and Vyatchanin, 2002; Braginsky et al., 1997; Corbitt et al., 2007), which leads to a softening of the spring for a red-detuned drive ($\Delta_c < 0$) and a hardening for a blue-detuned drive ($\Delta_c > 0$). Secondly we observe a modification of the mechanical damping rate by an optomechanical contribution, which leads to the effective rates $\gamma_{-} = \Gamma_{-} + \gamma_{\rm m}(\bar{n}+1)$, $\gamma_{+} = \Gamma_{+} + \gamma_{\rm m}\bar{n}$, where

$$\Gamma_{\pm} = 2g^2\,\mathrm{Re}\,\eta_{\pm}. \tag{31}$$

This modification leads to sideband cooling for $\Delta_c < 0$, for which we have $\Gamma_{-} > \Gamma_{+}$ (Marquardt et al., 2007; Wilson-Rae et al., 2007), while for the case of $\Delta_c > 0$ we find $\Gamma_{+} > \Gamma_{-}$ and thus amplification of the mechanical motion.

Throughout the chapter we will often encounter the situation where the optomechanical system is continuously monitored via a homodyne measurement of the output light, realizing a so-called *diffusive* measurement of a given cavity quadrature [set by the local oscillator (LO) angle $\phi$]. The system can then be described by the *conditional quantum state* $\rho_c$ that encodes our knowledge of the system given a specific measurement record $I(t)$. The conditional state's evolution is conveniently described in the Schrödinger picture by the Itō stochastic master equation (SME) (Gardiner and Zoller, 2004; Wiseman and Milburn, 2009)

$$d\rho_c = \mathcal{L}\rho_c\, dt + \sqrt{\eta\kappa}\,\mathcal{H}[e^{i\phi}c_c]\rho_c\, dW, \tag{32}$$

where $0 \leq \eta \leq 1$ is the efficiency of the detection. The effect of conditioning is described by the measurement operator

$$\mathcal{H}[s]\rho_c = [s - \mathrm{tr}(s\rho_c)]\rho_c + \rho_c[s - \mathrm{tr}(s\rho_c)]^{\dagger}. \tag{33}$$

$\mathcal{H}$ is thus nonlinear in $\rho_c$, as is expected for a measurement term. The photocurrent $I(t)$ can be expressed as

$$I(t)dt = \sqrt{\eta\kappa}\langle c_c(t)e^{i\phi} + c_c^{\dagger}(t)e^{-i\phi}\rangle_c dt + dW(t), \tag{34}$$

where $W$ is a Wiener process with an increment $dW(t) := W(t + dt) - W(t)$, which has zero mean and a variance $dW(t)^2 = dt$. Here and in the following we denote by $\langle A(t)\rangle_c = \mathrm{tr}(A\rho_c(t))$ the expectation value with respect to the conditional state at time $t$. Eq. (34) shows that $I(t)$ encodes information about a generalized intracavity quadrature defined by $\phi$.

In contrast to the conditional state $\rho_c$ which solves a SME, we will call the solution of a standard MEQ [such as (25)] the *unconditional state*, which we denote by $\rho$. A general introduction to SMEs is given in Appendix B.

### *2.3.3 Phase Space Description*

In the Gaussian regime (i.e., for linearized optomechanical interaction) the system's unconditional state is fully characterized by the first and second moments of the vector $\boldsymbol{X} = (x_m, p_m, x_c, p_c)^T$, i.e., the mean values $\langle \boldsymbol{X}(t)\rangle = \mathrm{tr}(\boldsymbol{X}\rho(t))$ and the symmetric covariance matrix $\boldsymbol{\Sigma}(t)$ with elements

$$\begin{aligned}\Sigma_{ij}(t) &= \frac{1}{2}\langle X_i(t)X_j(t) + X_j(t)X_i(t)\rangle - \langle X_i(t)\rangle\langle X_j(t)\rangle \\ &= \mathrm{Re}\big(\langle X_i(t)X_j(t)\rangle\big) - \langle X_i(t)\rangle\langle X_j(t)\rangle.\end{aligned} \tag{35}$$

The linear equations of motion of $\langle \boldsymbol{X}(t)\rangle$ and $\boldsymbol{\Sigma}(t)$ are given by (see Appendix B.3.1)

$$\frac{\mathrm{d}}{\mathrm{d}t}\langle \mathbf{X}(t)\rangle = \mathbf{F}\langle \mathbf{X}(t)\rangle, \tag{36a}$$

$$\frac{\mathrm{d}}{\mathrm{d}t}\mathbf{\Sigma}(t) = \mathbf{F}\mathbf{\Sigma}(t) + \mathbf{\Sigma}(t)\mathbf{F}^{\mathrm{T}} + \mathbf{N}. \tag{36b}$$

The 4 × 4 matrices **F** and **N** describe the system's dynamics and noise properties, respectively, and can be deduced directly from the Langevin equations (21) or the corresponding master equation. For the model used in Eq. (21) they read

$$\mathbf{F} = \begin{pmatrix} 0 & \omega_{\mathrm{m}} & 0 & 0 \\ -\omega_{\mathrm{m}} & -\gamma_{\mathrm{m}} & -2g & 0 \\ 0 & 0 & -\frac{\kappa}{2} & -\Delta_{\mathrm{c}} \\ -2g & 0 & \Delta_{\mathrm{c}} & -\frac{\kappa}{2} \end{pmatrix} \tag{37}$$

and

$$\mathbf{N} = \frac{1}{2}\,\mathrm{diag}(0, 2\gamma_{\mathrm{m}}(2\bar{n}+1), \kappa, \kappa). \tag{38}$$

Gaussian conditional quantum states on the other hand are fully described by the mean vector $\hat{\boldsymbol{X}}(t) = \langle \mathbf{X}(t)\rangle_c$ and covariance matrix

$$\begin{aligned} \hat{\mathbf{\Sigma}}_{ij}(t) &= \frac{1}{2}\langle X_i(t)X_j(t) + X_j(t)X_i(t)\rangle_c - \langle X_i(t)\rangle_c\langle X_j(t)\rangle_c \\ &= \mathrm{Re}\big(\langle X_i(t)X_j(t)\rangle_c\big) - \hat{X}_i(t)\hat{X}_j(t), \end{aligned} \tag{39}$$

defined with respect to $\rho_c$ [the solution of a stochastic master equation such as (32)]. Their equations of motion are given by a linear stochastic differential equation and a (deterministic) matrix Riccati equation, respectively (Belavkin, 1980) (see Appendix B.3.1 for a derivation),

$$\mathrm{d}\hat{\boldsymbol{X}}(t) = \mathbf{F}\hat{\boldsymbol{X}}(t)\mathrm{d}t + \mathbf{K}(t)\big[I(t) - \mathbf{H}\hat{\boldsymbol{X}}(t)\big]\mathrm{d}t, \tag{40a}$$

$$\begin{aligned} \frac{\mathrm{d}}{\mathrm{d}t}\hat{\mathbf{\Sigma}}(t) &= \mathbf{F}\hat{\mathbf{\Sigma}}(t) + \hat{\mathbf{\Sigma}}(t)\mathbf{F}^{\mathrm{T}} + \mathbf{N} \\ &\quad - \big[\hat{\mathbf{\Sigma}}(t)\mathbf{H}^{\mathrm{T}} + \mathbf{M}\big]\big[\hat{\mathbf{\Sigma}}(t)\mathbf{H}^{\mathrm{T}} + \mathbf{M}\big]^{\mathrm{T}}, \end{aligned} \tag{40b}$$

where **H** describes the homodyne measurement and **M** is related to the system's noise properties [see Eqs. (B.39b) and (B.41c)]. **K**($t$) is a time-dependent gain factor

$$\mathbf{K}(t) = \hat{\mathbf{\Sigma}}(t)\mathbf{H}^{\mathrm{T}} + \mathbf{M}. \tag{41}$$

which itself depends on the state through $\hat{\boldsymbol{\Sigma}}(t)$. In the case of a single homodyne detection with a LO phase $\phi$ and detection efficiency $\eta$ [as described by Eq. (32)], $\mathbf{H}$ and $\mathbf{M}$ are

$$\mathbf{H} = \sqrt{2\eta\kappa}(0 \quad 0 \quad \cos(\phi) \quad -\sin(\phi)), \tag{42a}$$

$$\mathbf{M} = \sqrt{\frac{\eta\kappa}{2}}(0 \quad 0 \quad -\cos(\phi) \quad \sin(\phi))^{\mathrm{T}}. \tag{42b}$$

For a one-dimensional system [with a two-dimensional phase space $(x, p)$] these equations allow us to give a simple graphic interpretation of the SME (32) in terms of a phase space description (see Fig. 6): the conditional trajectory $\hat{X}$ (blue line) is determined by the measurements $I(t)$ and therefore follows a random walk in phase space. The covariance matrix $\hat{\boldsymbol{\Sigma}}$ (red ellipse) on the other hand evolves deterministically, independent of the measurement results. Averaging over all possible phase space trajectories recovers the broad Gaussian distribution described by the standard MEQ (25) [or equivalently, Eq. (36)]. For an unstable system (e.g., in the blue-detuned regime), the blue line will spiral outward, leading to a growing unconditional covariance. The conditional covariance matrix $\hat{\boldsymbol{\Sigma}}$, however, may still possess a (finite) steady state. This is due to the fact that the exponential growth is tracked by the conditional mean, with respect to which the covariance matrix is defined. The steady-state conditional covariance matrix $\hat{\boldsymbol{\Sigma}}^{\mathrm{ss}}$ can

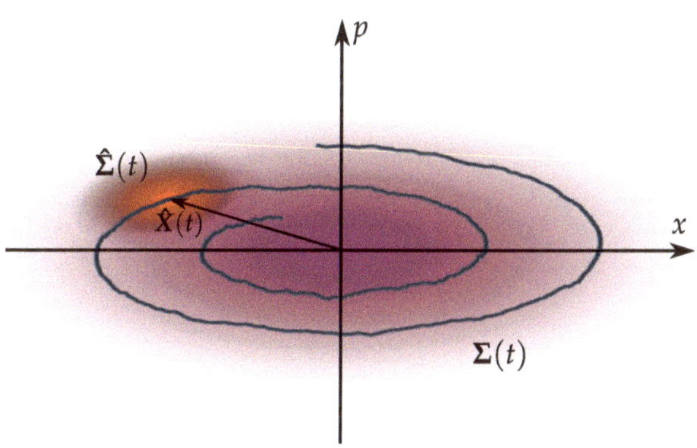


**Fig. 6** Schematic comparison of the master equation (25) and the stochastic master equation (32) for a single-mode Gaussian system in phase space. The conditional state with a covariance $\hat{\boldsymbol{\Sigma}}(t)$, depicted by the *red ellipse*, moves through phase space on a trajectory given by a realization of $\hat{X}(t)$ (*blue line*). Averaging over many sample paths recovers the broad, unconditional distribution, determined by $\boldsymbol{\Sigma}(t)$ (*violet ellipse*).

be found in analogy to $\boldsymbol{\Sigma}^{ss}$ by setting the left-hand side of Eq. (40b) to zero and by solving the resulting algebraic Riccati equation (Wiseman and Milburn, 2009).

## 2.4 The Optomechanical Phase Diagram

Provided a linear optomechanical system is stable [as determined by the Routh–Hurwitz criterion (Gradshteyn and Ryshik, 2007), see also Hofer (2015)] it will in the long-term assume a steady state, $\lim_{t\to\infty}\rho(t)=\rho^{ss}$, where $\rho^{ss}$ is determined by the condition $\mathcal{L}\rho^{ss}=0$. If a stable steady state exists the means $\langle \boldsymbol{X}\rangle^{ss}:=\mathrm{tr}(\boldsymbol{X}\rho^{ss})=0$ vanish [as is clear from Eq. (36a)], while the steady-state covariance matrix $\boldsymbol{\Sigma}^{ss}$ is given by the solution to the so-called Lyapunov equation, which is obtained from (36b) by setting its left-hand side to zero, i.e., $\mathbf{F}\boldsymbol{\Sigma}^{ss}+\boldsymbol{\Sigma}^{ss}\mathbf{F}^{T}+\mathbf{N}=0$.

The characteristic features of an optomechanical system's steady state can nicely be illustrated by plotting a phase diagram showing the mechanical steady-state occupation number $n_{m}^{ss}=\langle c_{m}^{\dagger}c_{m}\rangle^{ss}$ and optomechanical entanglement between the mechanical resonator and the intracavity field. We quantify the entanglement by the steady-state logarithmic negativity $E_{\mathcal{N}}^{ss}=E_{\mathcal{N}}(\rho^{ss})$(Plenio, 2005; Vidal and Werner, 2002), where

$$E_{\mathcal{N}}(\rho)=\log_2\left(\|\rho^{T_m}\|_1\right), \tag{43}$$

$\|\cdot\|_1$ is the trace norm, and $\rho^{T_m}$ is the partial transpose with respect to the mechanical subsystem. Fig. 7 shows such a phase diagram of an optomechanical system in the resolved-sideband regime ($\kappa<\omega_m$) for a high-Q mechanical oscillator with respect to the laser detuning $\Delta_c$ and the coupling $g$. The gray background depicts the regions of instability, where no steady state exists. The first thing to note is that the system is unstable in nearly all the right half-plane, i.e., for blue-detuned laser drive, while for red detuning the system becomes unstable only for appreciably high optomechanical coupling. Exactly on resonance $\Delta_c=0$ the system is unconditionally stable for arbitrary high coupling (as long as $Q_m>0$). Centered around the first mechanical sideband at $\Delta_c=-\omega_m$, where the beam-splitter part of the optomechanical interaction is resonant, lies the region where $n_m^{ss}<1$ (dashed red line) and thus ground state cooling of the mechanical system through optomechanical sideband cooling is possible. Optomechanical sideband cooling can be understood in analogy to sideband cooling of trapped ions (Cirac et al., 1992) by looking at the level scheme depicted in the inset: tuning the laser to the lower mechanical sideband resonantly drives the

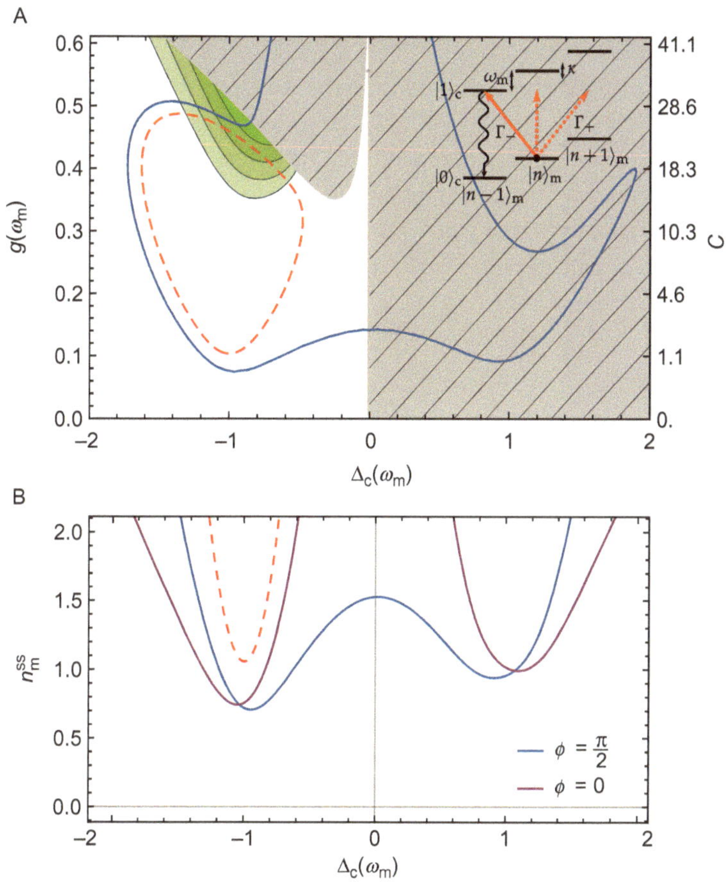


**Fig. 7** *Upper plot*: steady-state phase diagram of an optomechanical system for $\kappa = \omega_m/4$, $Q_m = 10^7$, $\bar{n} = 3.5 \cdot 10^5$. The *gray-hatched area* depicts unstable regions where no steady state exists. The *dashed red line* shows regions of ground state cooling of the mechanical oscillator, here $n_m^{ss} < 1$. The *green area* shows optomechanical entanglement (logarithmic negativity $E_{\mathcal{N}}^{ss}$), with largest values close to the instability region. The *blue line* encloses the regions where the conditional mean mechanical occupation number $\langle c_m^\dagger c_m \rangle_c^{ss} < 1$ for a measurement of the optical phase quadrature, i.e., $\phi = \pi/2$. The *right axis* shows the corresponding optomechanical cooperativity, given by $C = 4g^2/(\bar{n}+1)\gamma_m\kappa$. *Inset*: illustration of sideband cooling based on the optomechanical level scheme. *Lower plot*: cut through the phase diagram at $g = \omega_m/10$, depicting the conditional phonon number $\langle c_m^\dagger c_m \rangle_c^{ss}$ for LO phases $\phi = \pi/2$ (*blue*) and $\phi = 0$ (*purple*). The *dashed red line* again shows the mean occupation number for the unconditional state for sideband cooling.

transition $|n\rangle_{\rm m}|k\rangle_{\rm c} \rightarrow |n-1\rangle_{\rm m}|k+1\rangle_{\rm c}$ with a rate $\Gamma_-$ (solid red arrow), while in the sideband-resolved regime $\kappa \ll \omega_{\rm m}$ all other transitions (dashed red arrows) are strongly suppressed. Due to the finite cavity lifetime the systems consecutively decays into $|n-1\rangle_{\rm m}|k\rangle_{\rm c}$ and the cycle starts over. Devoid of additional mechanical decoherence (and for $\Gamma_- \gg \Gamma_+$) the mechanical system would thus eventually be driven into the ground state $|0\rangle_{\rm m}$. Assuming stability we can solve the Lyapunov equation corresponding to Eq. (28) to obtain the steady-state covariance matrix of the mechanical mode

$$\boldsymbol{\Sigma}_{\rm m}^{\rm ss} = \left(\frac{\bar{n}\gamma + \Gamma_+}{\gamma + \Gamma_- - \Gamma_+} + \frac{1}{2}\right)\mathbb{1}_2. \tag{44}$$

The first term corresponds to the effective mechanical occupation number $n_{\rm eff}$ in steady state. We can thus see that in order to achieve ground state cooling (i.e., $n_{\rm eff} < 1$) we need $\Gamma_- \gg \gamma, \Gamma_+$ and $\Gamma_- > \bar{n}\gamma$. In leading order we then have $n_{\rm eff} \approx \bar{n}\gamma/\Gamma_- \approx 1/C$. The region of ground-state cooling thus coincides with the regime of strong cooperativity $C \gtrsim 1$ (clearly visible in the plot).

Right at the border of stability, for a similar detuning, we find regions of large steady-state entanglement between the intracavity field and the mechanical resonator (colored in green) (Genes et al., 2008a). On the opposite side of the phase diagram, around $\Delta_{\rm c} = \omega_{\rm m}$, we also expect to observe optomechanical entanglement due to the effect of the optomechanical two-mode squeezing dynamics. However, there the formation of a steady state is inhibited by the optomechanical instability, which is due to parametric amplification of the amplitude of both the mechanical and the optical mode (Aspelmeyer et al., 2014b). The connection of laser cooling, entanglement generation, and the instability region has been analyzed in detail in Genes et al. (2008a).

Although no steady state exists for a blue-detuned laser drive, various alternative approaches permit to work with the resonantly enhanced two-mode squeezing dynamics of the optomechanical interaction. In Section 3 we will analyze in detail a pulsed protocol for entanglement creation, which does not require to be operated in a stable regime. Working with a continuous-wave blue-detuned laser drive on the other hand is still possible if one employs stabilizing feedback that inhibits the exponential growth of the optomechanical system's quadratures. One possible type of feedback is measurement-based feedback using homodyne detection, which we will consider in Section 4.

The measurement-based feedback protocols considered in this work rely on the evaluation of the system's conditional state, whose dynamics are described by a stochastic master equation such as (32). For Gaussian systems (such as a linear optomechanical system) the SME is easily integrated, by solving the corresponding Riccati equation. The conditional steady state's mean mechanical occupation number $\langle c_{\mathrm{m}}^{\dagger} c_{\mathrm{m}} \rangle_c^{\mathrm{ss}}$ is depicted in Fig. 7A for a measurement of the phase quadrature of the light field, i.e., $\phi = \pi/2$. We find a large region where $\langle c_{\mathrm{m}}^{\dagger} c_{\mathrm{m}} \rangle_c^{\mathrm{ss}} < 0.4$ for all detunings $-\omega_{\mathrm{m}} \lesssim \Delta_{\mathrm{c}} \lesssim \omega_{\mathrm{m}}$ (blue line). In the region around the red sideband $\Delta_{\mathrm{c}} \approx -\omega_{\mathrm{m}}$ this effect can mainly be attributed to passive sideband cooling of the mirror, which we discussed above. However, we now also find a region of low occupation on the opposite (blue) sideband at $\Delta_{\mathrm{c}} \approx \omega_{\mathrm{m}}$. In this region the reduction of the conditional phonon number, which at the same time means an increase of the mechanical state's purity, is due to correlations between the mechanical oscillator and the light field. These correlations allow us to extract information about the mechanics from the homodyne measurement. We will see in the next section that in the sideband-resolved regime $\kappa < \omega_{\mathrm{m}}$ this effect is strongest for $\Delta_{\mathrm{c}} \approx \omega_{\mathrm{m}}$ where the two-mode squeezing, entangling term of the optomechanical Hamiltonian is resonant.

To illustrate how the choice of the LO phase influences the conditional mechanical occupation we plot a cut through Fig. 7A at a fixed optomechanical coupling $g = \omega_{\mathrm{m}}/10$ in Fig. 7B. If we choose to measure the optical amplitude quadrature we find that on resonance we do not have a reduction of the conditional phonon number. For a detuned laser drive ($|\Delta_{\mathrm{c}}| \gtrsim \omega_{\mathrm{m}}$), however, we again find regions of $\langle c_{\mathrm{m}}^{\dagger} c_{\mathrm{m}} \rangle_c^{\mathrm{ss}} < 1$. This is easily explained by noting that on resonance ($\Delta_{\mathrm{c}} = 0$) only the optical phase quadrature couples to the mechanical oscillator, while the amplitude quadrature contains noise only. Measuring the amplitude quadrature therefore does not allow us to make inferences about the mechanical motion. In general there will be an optimal LO angle, depending on all system parameters (especially $g$, $\Delta_{\mathrm{c}}$, $\kappa$) at which we obtain maximal information about the mechanical motion. Thus, homodyne detection at this particular angle yields the minimal conditional occupation. Typically—especially in the weak-coupling regime where $g < \kappa$—the optimal angle corresponds to the optical quadrature which is *anti* squeezed by the optomechanical interaction and thus features the best signal-to-noise ratio. We will see later how these features of the optomechanical phase diagram connect to feedback cooling of the mechanical oscillator.

Although no steady state exists in the blue-detuned regime, it is well suited as a working point for optomechanical quantum control protocols. The entanglement created by the two-mode squeezing interaction can be exploited to obtain information about the mechanical system's state, as well as to prepare it in a certain state. We will explore several such protocols in the following sections.

## 3. PULSED ENTANGLEMENT CREATION AND VERIFICATION

Entanglement creation between a mechanical oscillator and light (Abdi et al., 2011; Galve et al., 2010; Genes et al., 2008a; Ghobadi et al., 2011b; Mari and Eisert, 2009; Miao et al., 2010; Paternostro et al., 2007; Vitali et al., 2007a,c), as well as between two mechanical oscillators (Hartmann and Plenio, 2008; Huang and Agarwal, 2009; Ludwig et al., 2010; Mancini et al., 2002; Pinard et al., 2005; Pirandola et al., 2006; Vacanti et al., 2008; Vitali et al., 2007b; Zhang et al., 2003) has been analyzed in a multitude of theoretical studies. These studies are mainly concerned with entanglement in the steady-state regime, where an optomechanical system driven by one or more continuous-wave light modes is prepared in a stationary nonseparable multipartite state.

Our discussion of the optomechanical phase diagram showed that creation of steady-state entanglement is limited to a rather small parameter regime: for a blue-detuned drive dynamical instability prevents formation of a steady state altogether, while in the red-detuned regime entanglement reaches a maximum right at the stability border. Conversely, the amount of achievable entanglement is determined by the conditions guaranteeing the existence of a stationary state. This is of course a consequence of the amplifying two-mode squeezing dynamics [Eq. (19)] which generate entanglement in the linear regime.

In this section we follow a different approach to generate optomechanical entanglement which circumvents these limitations; entanglement is created and verified by two subsequent pulses of light, as originally analyzed in Hofer et al. (2011). Working in such a pulsed regime we do not rely on the existence of a steady state and thus provides us with the benefit of not being limited by any stability requirements. Additionally, the temporal ordering of pulses provides us with an unambiguous way to test light–matter entanglement, ruling out effects such as distribution of entanglement without entanglement (Cubitt et al., 2003; Mišta and Korolkova, 2009). As the generated state will exhibit strong Einstein–Podolsky–Rosen (EPR) correlations, the protocol can directly be extended into a continuous-variable teleportation protocol. The employed strategy has first been developed in the context of atomic ensembles (Hammerer et al., 2005) and was later considered as a protocol for levitated nanospheres (Romero-Isart et al., 2011). Entanglement between a mechanical oscillator and the electromagnetic field has for the first time been created in electromechanical systems (Palomaki et al., 2013b) by employing the scheme analyzed in this section.

## 3.1 Protocol Outline and Central Results

The protocol is performed in two steps (see Fig. 8): First, one drives the cavity with a blue-detuned laser pulse at $\Delta_c = \omega_m$ for a duration $\tau$ to resonantly enhance the optomechanical two-mode squeezing interaction (19) (Fig. 8A). For a sufficiently strong coupling this generates EPR entanglement between the traveling-wave light pulse and the mechanical oscillator. In a second step a red-detuned light pulse with a detuning $\Delta_c = -\omega_m$ is sent onto the cavity (Fig. 8C). This resonantly drives the beam-splitter interaction (18) and hence swaps the mechanical state onto the light pulse. Measuring both pulses via homodyne detection and calculating their correlation function then allows us to verify entanglement between the first pulse and the mirror.

To get an intuition for the protocol, we first treat it in an idealized scenario, neglecting any perturbative dynamics. This treatment is justified under the following assumptions about the system's parameters: first, we assume to work in the weak-coupling, resolved-sideband regime, i.e., $g \ll \kappa \ll \omega_m$. In this limit off-resonant terms in the optomechanical interaction are strongly suppressed and only first-order interactions of photons with the mechanics contribute to the dynamics. This allows us to adiabatically eliminate the cavity mode and invoke a rotating-wave-approximation in the interaction term. Second, we for now neglect mechanical decoherence effects, which is justified as long the pulses are much shorter than

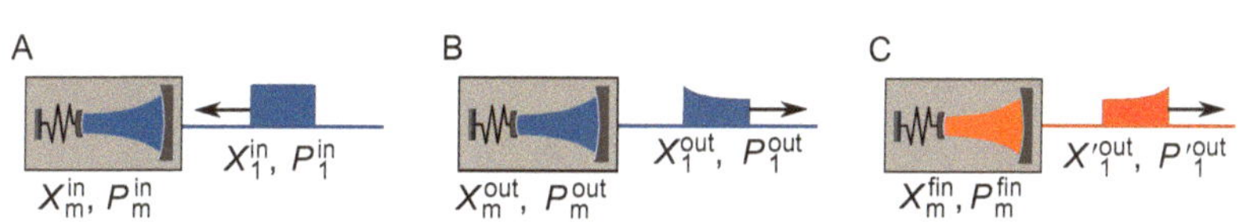


**Fig. 8** Pulsed entanglement creation and verification (Hofer et al., 2011): (A) A *blue-detuned* flat-top pulse resonantly drives the two-mode squeezing dynamics. (B) Photons down-scattered into the cavity resonance become entangled to the mechanical oscillator. (C) A *red-detuned* light pulse is used to read out the mechanical state.

the *effective* mechanical coherence time, i.e., $\tau \ll 1/\gamma_m \bar{n}$. The pulses are, on the other hand, chosen to be long on the timescale of one mechanical period, i.e., $\omega_m \tau \gg 1$. The pulses we assume to be flat-top pulses that have a constant amplitude $\varepsilon(t) \approx \sqrt{N_{ph}/\tau}$ for the largest part, where $N_{ph}$ is the total number of photons in the pulse. We can then model the optomechanical system in a quasi steady state (see Hofer et al., 2011) with a coupling constant given by (assuming a one-sided cavity with $\kappa_{in} = \kappa$)

$$g = g_0 \left( \frac{\kappa}{\Delta_c^2 + \frac{\kappa^2}{4}} \frac{N_{ph}}{2\tau} \right)^{\frac{1}{2}}. \tag{45}$$

### 3.1.1 Entanglement Creation

In the limit considered here we can adiabatically eliminate the cavity mode from the dynamics to obtain an effective equation of motion for the mechanical system alone, as is in detail discussed in Hofer (2015). Additionally we for the moment neglect mechanical damping and thermal decoherence, and thus set $\gamma_m = 0$. Under these approximations we find, for $\Delta_c = \omega_m$, and in a frame rotating at the mechanical frequency $\omega_m$,

$$\dot{\bar{c}}_m^\dagger = \frac{1}{2} G \bar{c}_m^\dagger + i\sqrt{G} \bar{a}_{in}, \tag{46}$$

where $\bar{c}_m$ and $\bar{a}_{in}^\dagger$ denote time coarse grained operators introduced in the adiabatic approximation. This equation can be formally integrated to yield

$$\bar{c}_m^\dagger(t) = e^{\frac{1}{2}Gt} \bar{c}_m^\dagger(0) + i\sqrt{G} e^{\frac{1}{2}Gt} \int_0^t ds e^{-\frac{1}{2}Gs} \bar{a}_{in}(s). \tag{47}$$

This shows that the mirror gets correlated to a light mode with a central frequency $\omega_c$ (as $\bar{a}_{in}$ here has a center frequency $\omega_0 - \omega_m = \omega_c$), with an exponentially shaped envelope $\alpha_{in}(t) \propto e^{-\frac{1}{2}Gt}$. Under the same approximations the adiabatic solution for the cavity is given by

$$\bar{c}_c(t) = -i \frac{2g}{\kappa} \bar{c}_m^\dagger(t) + \frac{2}{\sqrt{\kappa}} \bar{a}_{in}(t). \tag{48}$$

Using the input–output relations $\bar{a}_{out} = \sqrt{\kappa} \bar{c}_c - \bar{a}_{in}$ now allows us to define a set of normalized temporal light modes which carry the entanglement to the mirror. We find by using solutions (47) and (48) that

$$
\begin{aligned}
\bar{a}_{\text{out}}(t) &= -\mathrm{i}\sqrt{G}\bar{c}_{\text{m}}^{\dagger}(t) + \bar{a}_{\text{in}}(t) \\
&= -\mathrm{i}\sqrt{G}\mathrm{e}^{\frac{1}{2}Gt}c_{\text{m}}^{\dagger}(0) + G\mathrm{e}^{\frac{1}{2}Gt}\int_0^t \mathrm{d}s\mathrm{e}^{-\frac{1}{2}Gs}\bar{a}_{\text{in}}(s) + \bar{a}_{\text{in}}(t).
\end{aligned}
\tag{49}
$$

Eqs. (47) and (49) connect the state of the mirror to the state of the outgoing light field. This prompts us to introduce normalized temporal modes of the field,

$$
A_{\text{in}} = \int_0^{\tau} \mathrm{d}t\alpha_{\text{in}}(t)\bar{a}_{\text{in}}(t), \quad \alpha_{\text{in}}(t) = \sqrt{\frac{G}{1-\mathrm{e}^{-G\tau}}}\mathrm{e}^{-\frac{1}{2}Gt} \tag{50a}
$$

$$
A_{\text{out}} = \int_0^{\tau} \mathrm{d}t\alpha_{\text{out}}(t)\bar{a}_{\text{out}}(t), \quad \alpha_{\text{out}}(t) = \sqrt{\frac{G}{\mathrm{e}^{G\tau}-1}}\mathrm{e}^{\frac{1}{2}Gt} \tag{50b}
$$

which in our current approximation obey canonical commutation relations $[A_{\text{in}}, A_{\text{in}}^{\dagger}] = [A_{\text{out}}, A_{\text{out}}^{\dagger}] = 1$. Together with the definitions $B_{\text{in}} = \bar{c}_{\text{m}}(0)$ and $B_{\text{out}} = \bar{c}_{\text{m}}(\tau)$ we can derive scattering relations which relate the mechanical and optical mode at the end of the pulse $t=\tau$ to their initial state at $t=0$.

We plug (49) into the definition of $A_{\text{out}}$, and obtain

$$
\begin{aligned}
A_{\text{out}} &= -\mathrm{i}\frac{G}{\sqrt{\mathrm{e}^{G\tau}-1}}\int_0^{\tau}\mathrm{d}t\mathrm{e}^{Gt}\bar{c}_{\text{m}}^{\dagger}(0) \\
&\quad -\sqrt{\frac{G}{\mathrm{e}^{G\tau}-1}}\int_0^{\tau}\mathrm{d}t\left[G\mathrm{e}^{Gt}\int_0^t \mathrm{d}s\mathrm{e}^{-\frac{1}{2}Gs}\bar{a}_{\text{in}}(s) + \mathrm{e}^{\frac{1}{2}Gt}\bar{a}_{\text{in}}(t)\right] \\
&= -\mathrm{i}\sqrt{\mathrm{e}^{G\tau}-1}\bar{c}_{\text{m}}^{\dagger}(0) \\
&\quad -\sqrt{\frac{G}{\mathrm{e}^{G\tau}-1}}\int_0^{\tau}\mathrm{d}s\left[G\mathrm{e}^{-\frac{1}{2}Gs}\bar{a}_{\text{in}}(s)\int_s^{\tau}\mathrm{d}t\mathrm{e}^{Gt} + \mathrm{e}^{\frac{1}{2}Gs}\bar{a}_{\text{in}}(t)\right] \\
&= -\mathrm{i}\sqrt{\mathrm{e}^{G\tau}-1}B_{\text{in}}^{\dagger} + \sqrt{\frac{G}{1-\mathrm{e}^{-G\tau}}}\int_0^{\tau}\mathrm{d}s\mathrm{e}^{-\frac{1}{2}Gs}\bar{a}_{\text{in}}(s),
\end{aligned}
$$

where in the second line we used the identity

$$
\int_0^{\tau}\mathrm{d}s\int_0^{s}\mathrm{d}s'h(s,s') = \int_0^{\tau}\mathrm{d}s'\int_{s'}^{\tau}\mathrm{d}s\, h(s,s'),
$$

for an integrable function $h$. Using this and Eq. (47) we finally obtain

$$
A_{\text{out}} = \mathrm{e}^{\frac{1}{2}G\tau}A_{\text{in}} - \mathrm{i}\sqrt{\mathrm{e}^{G\tau}-1}B_{\text{in}}^{\dagger}, \tag{51a}
$$

$$
B_{\text{out}} = \mathrm{e}^{\frac{1}{2}G\tau}B_{\text{in}} - \mathrm{i}\sqrt{\mathrm{e}^{G\tau}-1}A_{\text{in}}^{\dagger}. \tag{51b}
$$

Physically, the rising pulse shape of the output mode is due to the two-mode optomechanical squeezing interaction, which amplifies the mirror motion. The photons scattered into the cavity resonance in this way contribute to the entanglement of the pulse to the mirror.

We can express Eq. (51) in terms of quadratures $X_{\mathrm{m}}^{i}=(B_i+B_i^{\dagger})/\sqrt{2}$ and $X_{\mathrm{l}}^{i}=(A_i+A_i^{\dagger})/\sqrt{2}$, $i\in\{\mathrm{in},\ \mathrm{out}\}$, and their corresponding conjugate variables, and calculate the so-called EPR variance $\Delta_{\mathrm{EPR}}$ of the system at the end of the protocol. We assume the electromagnetic field to initially be in vacuum $\left[(\Delta X_{\mathrm{l}}^{\mathrm{in}})^2=(\Delta P_{\mathrm{l}}^{\mathrm{in}})^2=\frac{1}{2}\right]$ and the mirror to be in a thermal state $[(\Delta X_{\mathrm{m}}^{\mathrm{in}})^2=(\Delta P_{\mathrm{m}}^{\mathrm{in}})^2=n_0+\frac{1}{2}$, where $n_0$ is the initial occupation number of the mechanical oscillator]. The state is then entangled if (Duan et al., 2000; Simon, 2000)

$$\begin{aligned}\Delta_{\mathrm{EPR}}&=\left[\Delta(X_{\mathrm{m}}^{\mathrm{out}}+P_{\mathrm{l}}^{\mathrm{out}})\right]^2+\left[\Delta(P_{\mathrm{m}}^{\mathrm{out}}+X_{\mathrm{l}}^{\mathrm{out}})\right]^2\\&=2(n_0+1)\left(\mathrm{e}^{r}-\sqrt{\mathrm{e}^{2r}-1}\right)^2<2,\end{aligned}\tag{52}$$

where $r=\frac{1}{2}G\tau$ is the (two-mode) squeezing parameter. In the limit of large squeezing ($r\gg 1$) the EPR variance $\Delta_{\mathrm{EPR}}\approx(n_0+1)\mathrm{e}^{-2r}/2$ is exponentially suppressed. The generated state thus approximates a two-mode-squeezed state, and, asymptotically, an EPR state; it can therefore readily be used to conduct optomechanical teleportation as described in a later section.

Solving (52) for $r$, we see that the state is entangled if

$$r>r_0=\frac{1}{2}\ln\left(\frac{(n_0+2)^2}{4(n_0+1)}\right)\overset{n_0\to\infty}{\sim}\frac{1}{2}\ln n_0.\tag{53}$$

In this scheme the required effective optomechanical interaction strength (as quantified by $r=2g^2\tau/\kappa$) thus scales logarithmically with the initial mechanical occupation number $n_0$, which eases the protocol's experimental realization, as neither $g$ nor $\tau$ can be arbitrarily increased. Note that $n_0$ can be decreased below the mean bath occupation $\bar{n}$ by laser cooling to improve the protocol's performance.

### *3.1.2 Entanglement Verification*

To verify the success of the protocol we send a red-detuned laser pulse ($\Delta_{\mathrm{c}}=-\omega_{\mathrm{m}}$) into the cavity to resonantly drive the beam-splitter interaction and to generate a state swap between the mechanical and the optical mode. The corresponding solutions are (Hofer, 2015)

$$\bar{c}_{\mathrm{m}}(t) = \mathrm{e}^{-\frac{1}{2}Gt}\bar{c}_{\mathrm{m}}(0) - \mathrm{i}\sqrt{G}\mathrm{e}^{-\frac{1}{2}Gt}\int_0^t \mathrm{d}s \mathrm{e}^{\frac{1}{2}Gs}\bar{a}_{\mathrm{in}}(s), \tag{54a}$$

$$\bar{c}_{\mathrm{c}}(t) = -\mathrm{i}\frac{1}{2}G\bar{c}_{\mathrm{m}}(t) + \frac{2}{\sqrt{\kappa}}\bar{a}_{\mathrm{in}}(t). \tag{54b}$$

Introducing slightly modified mode functions $\alpha'_{\mathrm{in(out)}} = \alpha_{\mathrm{out(in)}}$ together with the corresponding light modes $A'_{\mathrm{in(out)}}$ we can derive scattering relations in analogy to Eq. (51),

$$A'_{\mathrm{out}} = \mathrm{e}^{-\frac{1}{2}G\tau}A'_{\mathrm{in}} + \mathrm{i}\sqrt{1-\mathrm{e}^{-G\tau}}B'_{\mathrm{in}}, \tag{55a}$$

$$B'_{\mathrm{out}} = \mathrm{e}^{-\frac{1}{2}G\tau}B'_{\mathrm{in}} + \mathrm{i}\sqrt{1-\mathrm{e}^{-G\tau}}A'_{\mathrm{in}}. \tag{55b}$$

Again the pulse shape is determined by the optomechanical interaction, which here dampens the mirror motion (due to the resonant beam-splitter dynamics) and thus leads to an exponentially decreasing scattering into the cavity resonance. As the read out happens directly after the entanglement creation in the protocol we can set $B'_{\mathrm{in}} = B_{\mathrm{out}} = \bar{c}_{\mathrm{m}}(\tau)$, while $B'_{\mathrm{out}} = \bar{c}_{\mathrm{m}}(2\tau)$ is the mirror state after the second pulse. We see that this state-swap operation also features an exponential scaling with $G\tau$. In the limit $G\tau \to \infty$ Eq. (55) reduce to $A'_{\mathrm{out}} = \mathrm{i}B'_{\mathrm{in}}$ and $B'_{\mathrm{out}} = \mathrm{i}A'_{\mathrm{in}}$, showing that—apart from a phase shift—the mechanical state is perfectly transferred to the output light mode.

The state-swap operation allows us to access mechanical quadratures by measuring quadratures of light and thus to verify the created optomechanical entanglement. This can be achieved by correlating the entangling and the read-out pulse, which gives access to the EPR variance $\Delta'_{\mathrm{EPR}}$ between both pulses. Note that due to the phase shift introduced by the read-out we have to correlate a different combination of quadratures. We define

$$\Delta'_{\mathrm{EPR}} = \left[\Delta\left(X_1^{\mathrm{out}'} - X_1^{\mathrm{out}}\right)\right]^2 + \left[\Delta\left(P_1^{\mathrm{out}'} + P_1^{\mathrm{out}}\right)\right]^2, \tag{56}$$

which is constructed such that in the limit of $G\tau \to \infty$ in the read-out we have $\Delta'_{\mathrm{EPR}} \to \Delta_{\mathrm{EPR}}$.

### *3.1.3 Optomechanical Teleportation Protocol*

We showed in the previous sections that using a pulsed protocol can generate EPR-type entanglement, an entanglement resource central to many quantum information processing protocols (Braunstein and van Loock, 2005).

The scheme presented above can immediately be extended to an optomechanical continuous-variables (CV) quantum teleportation protocol. The goal of quantum state teleportation in this context is to transfer an arbitrary, unknown quantum state $|\psi_{in}\rangle$ encoded in a traveling-wavellight pulse onto the mechanical resonator, not making use of a direct interaction, but of entanglement between the two systems instead. This can be achieved in full analogy to the continuous-variable teleportation protocol for photons (Braunstein and Kimble, 1998; Vaidman, 1994). The presented scheme resembles the protocol used in atomic ensembles (Hammerer et al., 2005; Sherson et al., 2006) and it has also been proposed in the context of optomechanical experiments with levitated microspheres (Romero-Isart et al., 2011).

The protocol works as follows (see Fig. 9A for a schematic of the setup and Fig. 10 for a phase space description): the blue-detuned light pulse A is

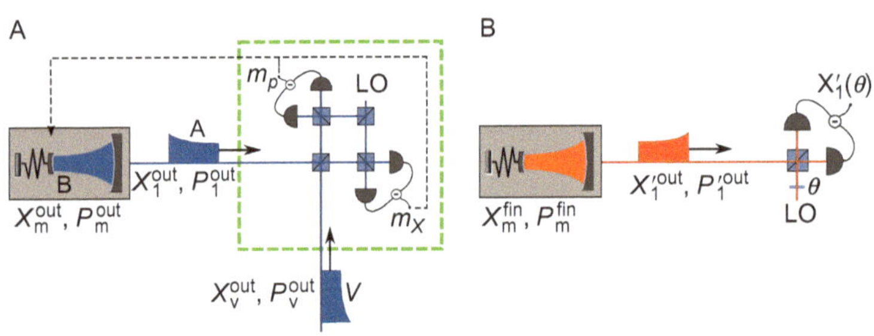


**Fig. 9** Setup for the optomechanical teleportation protocol (Hofer et al., 2011): (A) A CV Bell measurement (*green box*) is used to make a joint measurement on the output light of the optomechanical cavity and a second pulse prepared in the input state. The result $(m_X, m_P)$ is fed back to the mechanical oscillator, displacing it in phase space (see Fig. 10). (B) Homodyne tomography operated on the red mechanical sideband is used to evaluate the success of the teleportation.

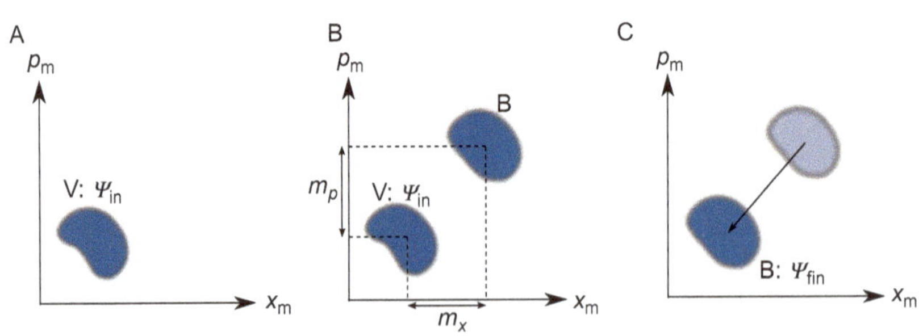


**Fig. 10** Phase space description of CV teleportation: state of the B–V system (A) before the Bell measurement, (B) after the Bell measurement, and (C) after feedback.

sent to the optomechanical cavity, becoming entangled to its mechanical mode B by the two-mode squeezing dynamics. A second pulse V is prepared in an arbitrary state $|\psi_{\rm in}\rangle$ that is to be transferred to mode B (Fig. 10A). The two pulses interfere on a beam-splitter. Behind the beam-splitter two homodyne detectors measure the joint quadratures $P_{\rm l}^{\rm out} + X_{\rm v}$ and $X_{\rm l}^{\rm out} + P_{\rm v}$, with measurement outcomes $m_X$ and $m_P$, respectively. This measurement realizes the continuous-variable analog to the Bell measurement in the case of qubit teleportation and projects previously uncorrelated systems A and V onto an EPR state (Bouwmeester et al., 1997; Furusawa et al., 1998). After the measurement the state of mode B is thus displaced in phase space with respect to $|\psi_{\rm in}\rangle$ by the measurement results $(m_x, m_p)$ (Fig. 10B). To recover the original input state, the protocol is concluded by a feedback step, displacing the mirror in phase space by $m_X$ and $m_P$ (Fig. 10C). A possible way to achieve this displacement is by means of short light pulses, applying methods described in Vanner et al. (2011) and Cerrillo et al. (2011). At the end of the protocol the state of the mechanical mode is then described by Braunstein and van Loock (2005)

$$
\begin{aligned}
X_{\rm m}^{\rm fin} &= X_{\rm m}^{\rm out} + P_{\rm l}^{\rm out} + X_{\rm v},\\
&= X_{\rm v} + \left(e^{r} - \sqrt{e^{2r}-1}\right)\left(X_{\rm m}^{\rm in} - P_{\rm l}^{\rm in}\right),
\end{aligned}
\tag{57a}
$$

$$
\begin{aligned}
P_{\rm m}^{\rm fin} &= P_{\rm m}^{\rm out} + X_{\rm l}^{\rm out} + P_{\rm v},\\
&= P_{\rm v} + \left(e^{r} - \sqrt{e^{2r}-1}\right)\left(P_{\rm m}^{\rm in} - X_{\rm l}^{\rm in}\right).
\end{aligned}
\tag{57b}
$$

This shows that the final state corresponds to the input state plus quantum noise contributions; the total noise added to both quadratures is given by $\Delta_{\rm EPR}$, which vanishes for $r \to \infty$. In this limit we therefore approach perfect teleportation fidelity, i.e., $X_{\rm m}^{\rm fin} = X_{\rm v}$ and $P_{\rm m}^{\rm fin} = P_{\rm v}$.

The success of the protocol can be verified by applying tomography of the mechanical state using red-detuned light pulses in the same manner as discussed in the previous section.

In Section 4.3 we will discuss a time-continuous version of the teleportation protocol, which allows us to prepare the mechanical oscillator in a squeezed steady state.

## 3.2 Full Dynamics

In the ideal scenario analyzed above we found that the amount of generated entanglement scales exponentially with the effective optomechanical coupling strength and the duration of the laser pulse. In the following we will consider a more realistic scenario, taking into account thermal noise effects and the full optomechanical interaction. Although these effects impede entanglement creation, we will show that one can find an optimal parameter regime where they can be suppressed sufficiently strongly.

### 3.2.1 Perturbative Dynamics

We include the following perturbative dynamics: contributions from the off-resonant beam-splitter Hamiltonian, higher-order optomechanical interactions beyond the adiabatic approximation, and mechanical decoherence due to coupling to a thermal environment. To minimize the impact of these effects the working point of the protocol should be chosen in the following parameter regime:

1. Working in the weak coupling, sideband-resolved regime $g \lesssim \kappa \ll \omega_{\mathrm{m}}$ results in the strong enhancement of the two-mode-squeezing dynamics with respect to the suppressed beam-splitter interaction, and essentially constitutes the regime where the RWA is a good approximation.
2. At the same time, however, we need to choose $g\tau \gg 1$, as the squeezing parameter $r = (g/\kappa)g\tau$ must be large to generate entanglement, while $g/\kappa$ needs to be small.
3. Most importantly, the whole protocol needs to be operated within the mechanical coherence time, i.e., $\bar{n}\gamma_{\mathrm{m}}\tau \ll 1$. As for low-frequency mechanical oscillators $\bar{n}$ can be very high even at cryogenic temperatures, this condition can pose a very strict upper limit for the pulse duration $\tau$.

This analysis leads us to consider the sequence of parameter inequalities

$$\bar{n}\gamma_{\mathrm{m}} \ll \frac{1}{\tau} \ll g \ll \kappa \ll \omega_{\mathrm{m}}, \tag{58}$$

which defines a suitable parameter regime. In Section 3.1 we assumed all these conditions to be well satisfied and we completely neglected the existence of mechanical decoherence. It is intuitively clear that low mechanical decoherence (i.e., low bath occupation number and a high mechanical quality factor) are essential for the success of the protocol. For all other parameters we expect an optimum to exist, which arises from balancing all contributions.

The sequence of inequalities (58) can also be connected to the cooperativity $C$. To do this we first write the squeezing parameter as $r = G\tau/2 \approx \bar{n}\gamma_{\mathrm{m}}\tau C/2$. Following the arguments above we require $\bar{n}\gamma_{\mathrm{m}}\tau \ll 1$, which means that we must have $r \ll C$. On the other hand we need a certain amount of squeezing in order to create entanglement, i.e., $r \gg r_0$, where $r_0$ is a monotonically increasing function of the initial mechanical occupation $n_0$. Combining these considerations to $r_0 \ll r \ll C$ we now see that we can either have a large cooperativity such that we can tolerate a large $n_0$ (and thus a large $r_0$), or we prepare an initial state with a low $n_0$ (and thus a small $r_0$). The connection between the two situations is given by the pulse length $\tau$. In the former case we work with a high-Q oscillator which allows us to increase the pulse length in order to generate entanglement without suffering from mechanical decoherence; in the latter a short pulse is sufficient to create the required amount of two-mode squeezing. For realistic parameters of $r_0$ this means that we have to operate the protocol in the strong cooperativity regime $C > 1$, as will be illustrated in the discussion of Fig. 11.

### 3.2.2 Results

To take into account the full optomechanical dynamics, including the perturbative effects discussed above, we can integrate the Langevin equations (21) in an exact manner, and use the solution together with the cavity input–output relations to obtain an expression for $\Delta_{EPR}$. The complete procedure is detailed in Hofer et al. (2011); here we only present the most important numerical results.

Following the analysis above, we expect—when keeping $\bar{n}$ and $Q_m$ fixed—to obtain optimal values for the other parameters $g$, $\kappa$, $\tau$, and thus minimize $\Delta_{EPR}$ with respect to these parameters. The maximal possible entanglement will ultimately be set by $\bar{n}$ and $Q_m$, which will also constitute hard boundaries in typical experiments.

The results of this minimization are shown in Fig. 11. In Fig. 11A we plot the resulting entanglement against the thermal occupation number of the mechanical bath for different Q-factors and initial mechanical occupation numbers. As we showed in the previous section, $\Delta_{EPR}$ depends on the initial occupation number of the mechanical oscillator. Precooling the mechanics by means of laser cooling can therefore drastically increase the amount of entanglement created during this protocol. This is illustrated in Fig. 11A, which shows that it is possible to create an entangled state even for a fairly large bath occupation. This is facilitated by a short pulse duration, during which the effective mechanical decoherence is small.

As they stand, the optimized parameters from Fig. 11A correspond to a very high optomechanical cooperativity. This is, however, not a requirement, as is illustrated in Fig. 11B. For this plot the EPR-variance is optimized with respect to the pulse length $\tau$ for fixed parameters

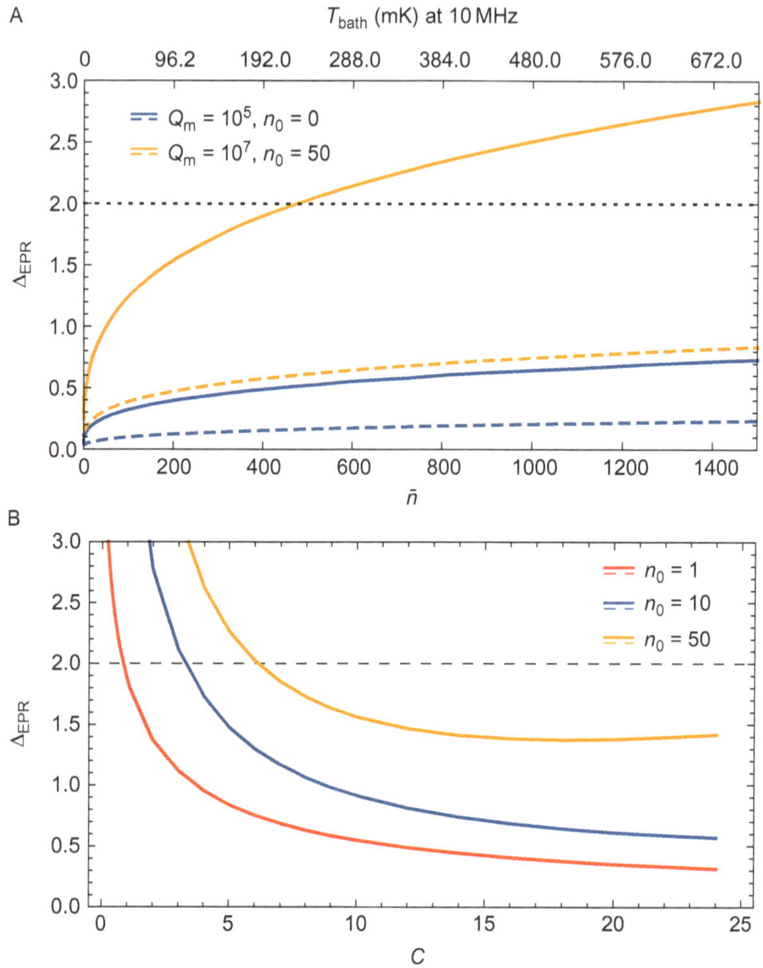


**Fig. 11** (A) Optimized entanglement (minimal $\Delta_{EPR}$) against $\bar{n}$ for $Q_m = 10^5$ and $n_0 = 0$ (*blue line*), and $Q_m = 10^7$ and $n_0 = 50$ (*yellow line*), where $n_0$ is the initial mechanical occupation and $\bar{n}$ is the mean bath occupation. This corresponds to the two cases of large $Q_m$ with moderate prelaser-cooling, and lower $Q_m$ with precooling into the ground state. The *upper axis* gives the corresponding bath temperature for a oscillator with a resonance frequency of 3.8 MHz. (B) $\Delta_{EPR}$ against cooperativity for different initial mechanical occupation numbers, optimized with respect to pulse duration $\tau$ for fixed values $\kappa/\omega_m = 2/100$, $Q_m = 3 \cdot 10^6$, $\bar{n} = 40$.

$\kappa/\omega_m = 2/100$, $Q_m = 3 \cdot 10^6$ and $\bar{n} = 40$ [which roughly corresponds to the parameters from Palomaki et al. (2013b)]. It shows that entanglement can be produced for $C \gtrsim 1$, depending on the initial mechanical occupation number $n_0$. As we discussed above a high $n_0$ requires a large squeezing parameter $r$ to create entanglement, which, for a fixed value of $C$ leads to a long pulse duration $\tau$ and thus a large contribution from mechanical decoherence effects. For low initial mechanical occupation short pulses and little squeezing suffice.

Adding to these fundamental imperfections, technical perturbations, such as mode-mismatch and detector inefficiencies, also decrease the amount of entanglement. Such effects can be collectively taken into account as beam-splitter losses adding vacuum noise to the optical signal; they will never completely break entanglement, as long as the total loss is smaller than unity.

# 4. TIME-CONTINUOUS QUANTUM CONTROL

We discussed in Section 3 how pulsed protocols can be used to achieve quantum control of optomechanical systems. These schemes do not rely on the existence of a steady state, i.e., they can also be operated in an unstable regime, and thus also in the case of a resonantly driven two-mode squeezing interaction.

In this chapter we explore *continuous-wave* protocols that exploit entanglement as a resource for measurement-based *time-continuous* control of cavity-optomechanical systems, and which allow us to engineer a steady state. These protocols were originally developed and published in Hofer et al. (2013) and Hofer and Hammerer (2015). Specifically we will discuss feedback cooling of the mechanical oscillator to its ground state for a blue-detuned laser drive (Section 4.1), preparation of a mechanical oscillator in a general Gaussian (squeezed) state via time-continuous teleportation (Section 4.3), and preparation of two remote mechanical systems in an (Einstein–Podolsky–Rosen) entangled state by time-continuous entanglement swapping (Section 4.4).

## 4.1 Optomechanical Feedback Cooling

Feedback cooling of mechanical oscillators has been analyzed in detail in the literature (Courty et al., 2001; Doherty and Jacobs, 1999; Genes et al., 2008b; Hamerly and Mabuchi, 2013; Mancini et al., 1998; Vitali et al., 2002). The setup typically considered, which is also the basis for the scheme discussed here, is depicted in Fig. 12: the output light of an optomechanical cavity is measured by a homodyne detector and the measurement results are fed back to the cavity in a suitable manner, such that the mechanical system is driven to a low-entropy steady state. The regime for feedback cooling

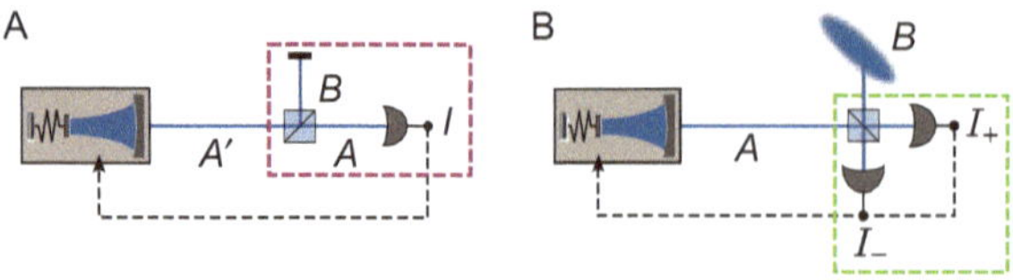


**Fig. 12** (A) Single homodyne detection for feedback cooling (nonunit efficiency is modeled by a beam-splitter before a perfect detector, marked by the *purple box*; see Appendix B.1.1). (B) Time-continuous teleportation setup. The *green box* marks the time-continuous Bell measurement.

discussed in the literature is typically restricted to a resonant drive and the bad-cavity regime $\kappa > \omega_{\mathrm{m}}$. In this section we will discuss that feedback cooling can also be effectively operated in the sideband-resolved regime $\kappa < \omega_{\mathrm{m}}$, and even on the blue sideband $\Delta_{\mathrm{c}} = \omega_{\mathrm{m}}$, which is normally affiliated with heating. We will show that we can harness the entanglement created by the optomechanical two-mode squeezing interaction for a measurement-based feedback scheme, which enables us to cool the mechanical motion to its ground state.

Feedback control typically consists of two stages: an estimation part that aims to obtain the best possible understanding of the system's state, and a control part that generates a feedback signal from the knowledge obtained in the estimation. We already discussed in Section 2.4 how measurement results from homodyne detection can be used to calculate the conditional quantum state of the mechanical system. This state can now be used as a basis for applying a linear–quadratic–Gaussian (LQG) control scheme—a tool from classical control theory that attains the optimal cooling performance for the chosen configuration—in order to prepare a cold mechanical quantum state.

Feedback onto the mechanical system can either be effected by direct driving through a piezoelectric device (Poggio et al., 2007), or by modulation of the laser input, as we will assume in the following. This type of optical feedback can be described by adding an additional time-dependent term

$$H_{\mathrm{fb}} = -\mathrm{i}\sqrt{\kappa}[\varepsilon(t)^{*}c_{\mathrm{c}} - \varepsilon(t)c_{\mathrm{c}}^{\dagger}] = \sqrt{2\kappa}[u_p(t)x_{\mathrm{c}} + u_x(t)p_{\mathrm{c}}] \tag{59}$$

to the Hamiltonian, where $\varepsilon(t) = u_x(t) + \mathrm{i}u_p(t) \in \mathbb{C}$ is the complex amplitude of the feedback signal, and $|\varepsilon(t)|^2$ accordingly is the incoming photon flux. To choose an appropriate feedback strategy we employ quantum LQG control (Belavkin, 1999), which is designed to minimize a quadratic cost function as described in Appendix B.3.2. Applied to feedback cooling the basic working principle is the following: from the measurement results of the homodyne detection we calculate the system's conditional state $\rho_c(t)$, whose evolution is described by Eq. (32). Based on this state we can then determine the optimal feedback signal $\varepsilon(t)$ that minimizes the steady-state mechanical occupation number $\langle c_{\mathrm{m}}^{\dagger} c_{\mathrm{m}} \rangle^{\mathrm{ss}} = \frac{1}{2}[\langle x_{\mathrm{m}}^2 + p_{\mathrm{m}}^2 \rangle^{\mathrm{ss}} - 1]$. This of course means that the final occupation number depends on the conditional state (more specifically on the covariance matrix $\hat{\boldsymbol{\Sigma}}$), and thus on the chosen LO angle for the homodyne detection as we discussed above. A suitable cost function for this problem is given by

$$h(x_{\mathrm{m}}(t), p_{\mathrm{m}}(t), \varepsilon(t)) = h_{\mathrm{m}} \left[ x_{\mathrm{m}}(t)^2 + p_{\mathrm{m}}(t)^2 \right] + |\varepsilon(t)|^2, \tag{60}$$

with $h_{\mathrm{m}} > 0$. Note that $h$ also includes a contribution by the feedback signal $\varepsilon$, which precludes feedback strategies with unrealistically high feedback strength. The parameter $h_{\mathrm{m}}$ therefore effects a trade-off between feedback strength and final occupation number $n_{\mathrm{m}}^{\mathrm{ss}}$: high values of $h_{\mathrm{m}}$ result in low occupation number possibly requiring large $|\varepsilon|$ and vice versa. The mean photon flux in the feedback signal can be calculated as described in Appendix B.3.2. For the parameters used in this section we find that on average $|\varepsilon|^2$ is small compared to the overall driving strength in typical experiments. Only in the region of $\kappa \rightarrow 0$—where almost no photons enter the cavity—the required $|\varepsilon|^2$ may increase dramatically. We note that in order for LQG control to work correctly, certain observability and controllability conditions need to be satisfied (Wiseman and Milburn, 2009), which is indeed the case for our system. Additionally, we assume here that the feedback onto the laser field is instantaneous. In practice this means that any feedback delay $\tau$ should be small on the typical timescales of the system, i.e., $\tau \ll 1/\omega_{\mathrm{m}}, 1/\kappa$.

The final mechanical occupation is found by first calculating the steady-state variances $(\Delta x_{\mathrm{m}})^2$ and $(\Delta p_{\mathrm{m}})^2$ for a closed feedback loop as outlined in Appendix B.3.2. $n_{\mathrm{m}}^{\mathrm{ss}}$ is then given by $n_{\mathrm{m}}^{\mathrm{ss}} = \frac{1}{2}[(\Delta x_{\mathrm{m}})^2 + (\Delta p_{\mathrm{m}})^2 - 1]$ (for $\langle x_{\mathrm{m}} \rangle^{\mathrm{ss}} = \langle p_{\mathrm{m}} \rangle^{\mathrm{ss}} = 0$).

In Fig. 13 we plot the steady-state occupation numbers of the feedback-cooled mechanical mode against the laser detuning $\Delta_{\mathrm{c}}$, for the bad-cavity regime (upper plot) and the sideband-resolved regime (lower plot), for two different coupling strengths $g$. For each detuning the homodyne phase $\phi$ is chosen such that the occupation number is minimized. Note that we keep $g$ constant while varying $\Delta_{\mathrm{c}}$ (or $\kappa$). This means that the driving laser

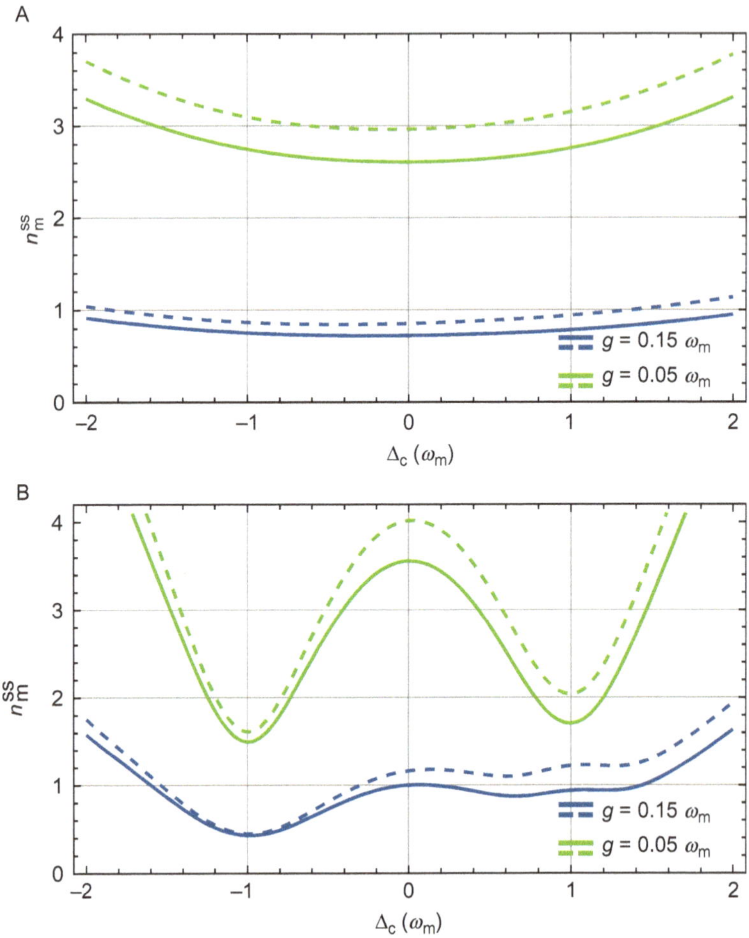


**Fig. 13** Steady-state mechanical occupation number $n_m^{ss}$ minimized with respect to the LO angle $\phi$ against detuning of the driving laser $\Delta_c$ (A) in the bad-cavity regime ($\kappa = 4\omega_m$) and (B) in the sideband-resolved regime ($\kappa = \omega_m/2$), for detection efficiencies $\eta = 1$ (*solid lines*) and $\eta = 8/10$ (*dashed lines*). *Different colors* denote different coupling strengths $g$. Other parameters: $Q = 5 \cdot 10^6$, $\bar{n} = 3.5 \cdot 10^5$, $h_m = 100$.

power has to be adjusted accordingly. In the bad-cavity regime $\kappa > \omega_m$ we find that driving on resonance is favorable for both values of $g$. In this case the optimal LO phase is $\phi = \pi/2$, as discussed in the previous section. This is the usual regime for feedback cooling (Cohadon et al., 1999; Wilson et al., 2015).

For micromechanical systems, however, the sideband-resolved regime $\kappa < \omega_m$ is typically more relevant. In this regime the picture changes completely. For weak coupling ($g < \kappa$) we find two pronounced dips at *both* mechanical sidebands ($\Delta_c = \pm\,\omega_m$), where $n_m^{ss}$ is locally minimal and clearly lies below the value on resonance. It is obvious from the figure that cooling works best on the red sideband ($\Delta_c = -\omega_m$), where we have a cumulative effect from passive sideband and feedback cooling (see also Fig. 14).

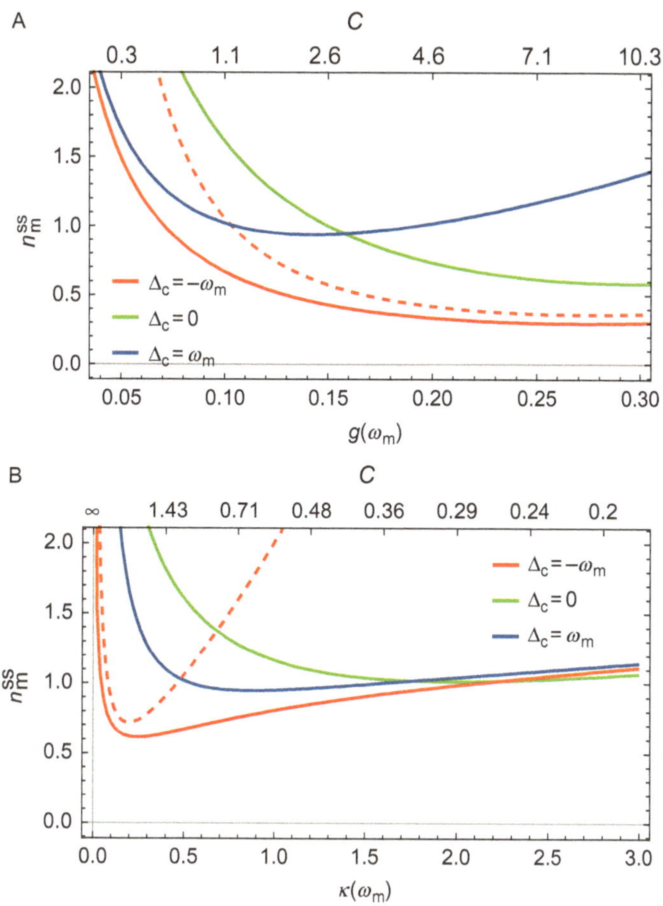


**Fig. 14** Steady-state mechanical occupation number $n_m^{ss}$ minimized with respect to $\phi$ for different driving frequencies $\Delta_c = 0, \pm\,\omega_m$, corresponding to a laser drive on the mechanical sidebands and on resonance (represented by *different colors*). *Solid lines* represent feedback cooling, while the *dashed line* (for $\Delta_c = -\omega_m$ only) corresponds to sideband cooling without feedback. (A) $n_m^{ss}$ against coupling $g$ for fixed cavity decay rate $\kappa = \omega_m/2$ (sideband-resolved regime). (B) $n_m^{ss}$ against $\kappa$ for $g = \omega_m/10$ (weak-coupling regime). Other parameters for both plots: $Q = 5 \cdot 10^6$, $\bar{n} = 3.5 \cdot 10^5$, $h_m = 100$, $\eta = 1$.

However, even on the blue sideband ($\Delta_c = \omega_m$)—which is commonly associated with heating—we find an appreciable reduction of the mechanical occupation by feedback cooling. As we discussed in Section 2.4, we can attribute this effect to large optomechanical correlations, which allow for a good read out of the mechanical motion and thus a good feedback performance. If we increase the coupling strength to $g = 0.3\kappa$ we see a peak appearing around the blue sideband (which we attribute to ponderomotive squeezing of the output fields), pushing the occupation number above the value at $\Delta_c = 0$. For both regimes we plot graphs for two different detection efficiencies $\eta = 1$ (lossless detection) and $\eta = 8/10$. Clearly, nonunit detection efficiency leads to a noticeable degradation of feedback-cooling performance. Only at the red sideband and in the sideband-resolved regime, where the effect of sideband cooling dominates, the final occupation number is virtually unaffected.

Fig. 14A shows the mechanical occupation for three detunings $\Delta_c = 0, \pm\omega_m$ plotted against $g$. For $\Delta_c = -\omega_m$ we show, additionally to the closed-loop feedback solution (red solid line), the solution for sideband cooling (red dashed line). While for $\Delta_c = 0$ and $\Delta_c = -\omega_m$ the occupation number steadily decreases—in the depicted range—for growing $g$, for $\Delta_c = \omega_m$ a clear minimum is visible in the weak-coupling regime at $g \approx \kappa/10$. This minimum lies well below the value for $\Delta_c = 0$ (but still above the value for the red sideband). This means that there exists a considerably large parameter regime where a detuned operation significantly improves the performance of feedback cooling. Note that all curves rise drastically in the strong-coupling regime, where $g/\kappa \gtrsim 1$ (not shown in the plot). In Fig. 14B we plot $n_m^{ss}$ against cavity linewidth $\kappa$ for constant coupling $g$. Again we find that feedback on the sidebands works best in the sideband-resolved regime, while in the bad-cavity regime working on resonance yields (slightly) better performance. Again, the occupation number is minimized with respect to the homodyne phase $\phi$ at each point in the plot.

In summary we illustrated that feedback cooling in the resolved-sideband regime is a viable option for cooling the mechanical oscillator into its ground state. It turns out that in this regime driving the system on the blue mechanical sideband yields a lower mechanical occupation number than operating on resonance. As an extension of this protocol we will show in Section 4.3 that a similar setup operating at the same working point can be used to remotely prepare a squeezed mechanical state via time-continuous teleportation.

## 4.2 Time-continuous Bell Measurements

We discussed in Section 3 how a continuous-variable Bell measurement can be used to teleport the quantum state of a light pulse onto a mechanical oscillator. In systems which are amenable to strong projective measurements [e.g., photons (Bouwmeester et al., 1997; Sherson et al., 2006) and atoms (Barrett et al., 2004; Riebe et al., 2004)], Bell measurements constitute a well-established, versatile tool for quantum control and state engineering. However, in many physical systems only weak, indirect, but time-continuous measurements are available. Continuous measurements, described by stochastic Schrödinger and master equations, themselves constitute a corner stone of quantum control. In this section we will combine these important concepts and establish the notion of time-continuous Bell measurements. They are realized via continuous homodyne detection of electromagnetic fields and can be applied to a great number of systems, including those which cannot be measured projectively. We derive the constitutive equations of motion—the conditional stochastic Schrödinger/master equation and the unconditional feedback master equation—of the monitored systems. In particular we analyze two generic scenarios—*time-continuous quantum teleportation* and *continuous entanglement swapping*. We will first discuss both protocols for generic systems (Sections 4.2.1 and 4.2.2); the optomechanical implementation is the subject of Sections 4.3 and 4.4.

### 4.2.1 Time-Continuous Teleportation

We consider the setup depicted in Fig. 12: a one-dimensional electromagnetic field mode $A$ [described by the field operator $a(t)$] couples to a system $S$ via a beam-splitter-like interaction

$$H_{\text{int}} = \mathrm{i}[sa^\dagger(t) + s^\dagger a(t)]. \tag{61}$$

$A$ is then mixed with a second field $B$ [with a field operator $b(t)$] on a beam-splitter and the resulting fields are sent to two homodyne detection setups which measure the EPR quadratures $x_+ = x_a + x_b$ and $p_- = p_a - p_b$ where $x_a$, $x_b$ and $p_a$, $p_b$ are the amplitude and phase quadratures of the respective fields. The field $B$ is prepared in a pure state of Gaussian squeezed white noise, which we denote by $|M\rangle$, where $M \in \mathbb{C}$ characterizes the squeezing; $|M|$ describes the absolute increase (reduction) of the (anti-)squeezed quadrature, while $\arg(M)$ determines the squeezing angle. Provided the system–field interaction creates entanglement between the system and the outgoing light field, the state of $B$ can be teleported to $S$ by applying (instantaneous) feedback proportional to the measurement results of the Bell measurement ($I_\pm$). This effectively generates dissipative dynamics which drive the system into a steady state coinciding with the input light state.

#### 4.2.1.1 Conditional Master Equation

To find the corresponding SME, describing the state of $S$ conditioned on measurements of $x_+$ and $p_-$, we apply the same reasoning as in Appendix B.1.1 to describe homodyne detection. [The presented results have also been derived in a more formal way in Dąbrowska and Gough (2016).] We start from the Schrödinger equation [Eq. (A.27)] for the full system ($S$+$A$+$B$)

$$\mathrm{d}|\Psi(t)\rangle = [-\mathrm{i}H_{\mathrm{eff}}\mathrm{d}t + s\mathrm{d}A^\dagger(t) - s^\dagger\mathrm{d}A(t)]|\Psi(t)\rangle, \tag{62}$$

with the initial condition $|\Psi_0\rangle = |\psi_0\rangle_S \otimes |\mathrm{vac}\rangle_A \otimes |M\rangle_B$, $H_{\mathrm{eff}} = H_{\mathrm{sys}} - \mathrm{i}\frac{1}{2}s^\dagger s$, and an unspecified system Hamiltonian $H_{\mathrm{sys}}$. As mode $A$ is assumed to be in vacuum, the quantum Wiener increment $\mathrm{d}A$ obeys the Itō multiplication table (A.14) in Appendix A.2. As $B$ is prepared in a squeezed state the multiplication table for $\mathrm{d}B$ on the other hand is given by

| $\times$ | $\mathrm{d}B$ | $\mathrm{d}B^\dagger$ | $\mathrm{d}t$ |
|---|---|---|---|
| $\mathrm{d}B$ | $M\,\mathrm{d}t$ | $(N+1)\,\mathrm{d}t$ | 0 |
| $\mathrm{d}B^\dagger$ | $N\,\mathrm{d}t$ | $M^*\,\mathrm{d}t$ | 0 |
| $\mathrm{d}t$ | 0 | 0 | 0 |

Due to the Heisenberg uncertainty relation $N$ and $M$ must obey $|M|^2 \leq N(N+1)$, where equality holds for a pure state. The eigenvalue equation for squeezed states can be written as

$$[\mathrm{d}B(t) - \alpha\mathrm{d}B^\dagger(t)]|M\rangle = 0, \tag{63}$$

with $\alpha = (N+M)/(N+M^*+1)$. Using this together with the fact that $\mathrm{d}A(t)|\mathrm{vac}\rangle = 0$, we can write

$$\begin{aligned}\mathrm{d}|\Psi\rangle &= \left\{-\mathrm{i}H_{\mathrm{eff}}\,\mathrm{d}t + s\left[\mathrm{d}A^\dagger - \alpha\mathrm{d}A + \mathrm{d}B - \alpha\mathrm{d}B^\dagger\right]\right\}|\Psi\rangle \\ &= \left\{-\mathrm{i}H_{\mathrm{eff}}\,\mathrm{d}t + \sqrt{1/2}\,s[\mu\,\mathrm{d}X_+ + \mathrm{i}\nu\mathrm{d}P_-]\right\}|\Psi\rangle,\end{aligned} \tag{64}$$

where $\mu = 1-\alpha$, $\nu = 1+\alpha$. We introduced $\mathrm{d}X_+(t) = x_+(t)\mathrm{d}t$, $\mathrm{d}P_-(t) = p_-(t)\,\mathrm{d}t$ and going from the first to the second line we used the fact that $a^\dagger + b = (x_+ + \mathrm{i}p_-)/\sqrt{2}$. We now note that $x_+$ and $p_-$ commute and can be measured simultaneously. We can thus directly project equation (64) onto the EPR state $|I_+I_-\rangle_{AB}$ defined by $x_+|I_+I_-\rangle_{AB} = I_+|I_+I_-\rangle_{AB}$ and $p_-|I_+I_-\rangle_{AB} = I_-|I_+I_-\rangle_{AB}$. This yields the linear stochastic Schrödinger equation

$$\mathrm{d}|\tilde{\psi}_c\rangle = \left\{-\mathrm{i}H_{\mathrm{eff}}\mathrm{d}t + \sqrt{1/2}\, s[\mu \mathrm{d}\tilde{X}_+ + \mathrm{i}\nu \mathrm{d}\tilde{P}_-]\right\}|\psi_c\rangle, \tag{65}$$

where $\mathrm{d}\tilde{X}_+(t) = I_+(t)\mathrm{d}t$ and $\mathrm{d}\tilde{P}_-(t) = I_-(t)\mathrm{d}t$ are classical processes, which possess the same statistical properties as their quantum counterparts. The photo-currents $I_\pm$ (analogous to homodyne detection discussed in Section 2.3.2) can be written as

$$I_+\mathrm{d}t = \sqrt{1/2}\langle s + s^\dagger\rangle_c\, \mathrm{d}t + \mathrm{d}W_+, \tag{66a}$$

$$I_-\mathrm{d}t = \mathrm{i}\sqrt{1/2}\langle s - s^\dagger\rangle_c\, \mathrm{d}t + \mathrm{d}W_-, \tag{66b}$$

with correlated Wiener increments $\mathrm{d}W_\pm$. The (co-)variances of $\mathrm{d}W_\pm$ are given by

$$w_1\mathrm{d}t := (\mathrm{d}W_+)^2 = [N + 1 + (M + M^*)/2]\mathrm{d}t, \tag{67a}$$

$$w_2\mathrm{d}t := (\mathrm{d}W_-)^2 = [N + 1 - (M + M^*)/2]\mathrm{d}t, \tag{67b}$$

$$w_3\mathrm{d}t := \mathrm{d}W_+\mathrm{d}W_- = -[\mathrm{i}(M - M^*)/2]\mathrm{d}t, \tag{67c}$$

and directly follow from the definition of $\mathrm{d}X_+$, $\mathrm{d}P_-$ and the multiplication tables for $\mathrm{d}A$ and $\mathrm{d}B$. To find the SME corresponding to Eq. (65) we first use Itō calculus to find an expression for $\mathrm{d}|\tilde{\psi}_c\rangle\langle\tilde{\psi}_c|$, which we can then normalize. (See Appendix B.1.1 for a more detailed explanation of this process.) It is convenient to introduce the complex increment

$$\mathrm{d}Y(t) = \mu \mathrm{d}\tilde{X}_+(t) + \mathrm{i}\nu \mathrm{d}\tilde{P}_-(t), \tag{68}$$

which obeys

$$\mathrm{d}Y^2 = 2\zeta \mathrm{d}t := -2N/M^*\mathrm{d}t, \tag{69a}$$

$$|\,\mathrm{d}Y|^2 = \left[|\mu|^2 w_1 + |\nu|^2 w_2 + 2\,\mathrm{Im}(\mu\nu^*)\right]\mathrm{d}t = 2\mathrm{d}t. \tag{69b}$$

To show this it is useful to note the relations

$$\mu = \frac{1 - 2\mathrm{i}w_3}{w_1 - \mathrm{i}w_3}, \qquad \nu = \frac{2w_1 - 1}{w_1 - \mathrm{i}w_3}, \qquad w_2 = \frac{2w_3 + w_1}{2w_1 - 1}, \tag{70}$$

which can be derived by using $|M|^2 = N(N + 1)$. With this knowledge we can immediately write the Zakai equation—the equation of motion for the unnormalized state (Wiseman and Milburn, 2009, p. 197)—corresponding to (65) as

$$\mathrm{d}\tilde{\rho}_c = \mathcal{L}\rho_c \mathrm{d}t + \sqrt{1/2}\left(s\rho_c \mathrm{d}Y + \rho_c s^\dagger \mathrm{d}Y^*\right), \tag{71}$$

with the standard Liouvillian $\mathcal{L}\rho = -\mathrm{i}[H_{\mathrm{sys}},\rho] + \mathcal{D}[s]\rho$. To normalize this equation we first calculate

$$\mathrm{tr}(\tilde{\rho}_c(t+\mathrm{d}t)) = 1 + \sqrt{1/2}\left[\langle s(t)\rangle_c \mathrm{d}Y(t) + \mathrm{H.\,c.}\right], \tag{72}$$

which we use to obtain (by expanding to second order in d$Y$)

$$\begin{aligned}\mathrm{tr}(\tilde{\rho}_c(t+\mathrm{d}t))^{-1} = 1 &- \sqrt{1/2}\left[\langle s(t)\rangle_c \mathrm{d}Y(t) + \mathrm{H.c.}\right]\\ &+ (1/2)\left\{[\langle s(t)\rangle_c]^2\zeta + [\langle s^\dagger(t)\rangle_c]^2\zeta^* + 4|\langle s(t)\rangle_c|^2\right\}\mathrm{d}t.\end{aligned} \tag{73}$$

Combining this with (71) we find after some algebra

$$\begin{aligned}\mathrm{d}\rho_c &= \mathcal{L}\rho_c \mathrm{d}t + \sqrt{1/2}\left[\mu(s - \langle s\rangle_c)\rho_c + \mathrm{H.\,c.}\right]\rho_c \mathrm{d}W_+\\ &\quad+ \sqrt{1/2}\left[\mathrm{i}\nu(s - \langle s\rangle_c)\rho_c + \mathrm{H.\,c.}\right]\rho_c \mathrm{d}W_-\\ &= \mathcal{L}\rho_c \mathrm{d}t + \sqrt{1/2}\{\mathcal{H}[\mu s]\rho_c \mathrm{d}W_+ + \mathcal{H}[\mathrm{i}\nu s]\rho_c \mathrm{d}W_-\}.\end{aligned} \tag{74}$$

It can easily be checked that due to the fact $|\mathrm{d}Y|^2 = 2\mathrm{d}t$ this is a equation of the form of Eq. (B.17) of Appendix B.1.1 and thus a valid Belavkin equation (Belavkin, 1992).

To conclude this section let us briefly discuss, as a slight variation of the above setup, the situation where instead of the squeezed state $|M\rangle_B$ we use a displaced squeezed state $|M,\beta\rangle_B = \exp(\beta b^\dagger - \beta^* b)|M\rangle_B = D_B(\beta)|M\rangle_B$ with $\beta \in \mathbb{C}$ as an initial state of mode $B$, and thus as an input state for teleportation. Transforming the Schrödinger equation (62) into a displaced frame with $D_B(\beta)$ shows that the structure of the stochastic Schrödinger equation (SSE) (65) and the SME (74) remains unchanged, if the measurement processes are replaced by appropriately displaced versions

$$\mathrm{d}\tilde{X}'_+ = \mathrm{d}\tilde{X}_+ - \sqrt{1/2}(\beta + \beta^*)\mathrm{d}t, \tag{75a}$$

$$\mathrm{d}\tilde{P}'_- = \mathrm{d}\tilde{P}_- + \mathrm{i}\sqrt{1/2}(\beta - \beta^*)\mathrm{d}t \tag{75b}$$

[which consequently is also the case for the currents $I_\pm$ in Eqs. (66)]. Note that due to Itō rules we have $\mathrm{d}\tilde{X}'^2_- = \mathrm{d}\tilde{X}^2_-$, $\mathrm{d}\tilde{P}'^2_- = \mathrm{d}\tilde{P}^2_-$.

#### 4.2.1.2 Feedback Master Equation

We follow (Wiseman, 1994) to apply Markovian (i.e., instantaneous) feedback proportional to the homodyne currents $I_\pm$ to the system $S$. (The simpler case of homodyne feedback is reviewed in detail in Appendix B.2.) We want to add a term

$$(\mathbf{S})\quad [\mathrm{d}\rho_c]_{\mathrm{fb}} = \sqrt{1/2\eta}(\mathcal{K}_+ I_+ + \mathcal{K}_- I_-)\rho_c \tag{76}$$

describing the feedback to the conditional evolution (where $\eta$ describes a nonunit detector efficiency, as discussed in Appendix B.1.1). The feedback we model as generalized forces $F_\pm = F_\pm^\dagger$ in the form of additional Hamiltonian terms, which we write as

$$\mathcal{K}_\pm \rho I_\pm = -\mathrm{i}[F_\pm I_\pm, \rho]. \tag{77}$$

As $\mathcal{K}$ is linear, Eq. (76) can easily be converted to Itō form,

$$(\mathbf{I})\quad [\mathrm{d}\rho_c]_{\mathrm{fb}} = \sqrt{1/2\eta}(\mathcal{K}_+ \mathrm{d}\tilde{X}_+ + \mathcal{K}_- \mathrm{d}\tilde{P}_-)\rho_c + (1/4\eta)(\mathcal{K}_+ \mathrm{d}\tilde{X}_+ + \mathcal{K}_- \mathrm{d}\tilde{P}_-)^2 \rho_c. \tag{78}$$

This follows from the fact that for a Stratonovich equation

$$(\mathbf{S})\ \mathrm{d}\rho = \sum \mathcal{K}_i \rho \mathrm{d}W_i \tag{79}$$

with correlated noise processes $W_i$ and linear operators $\mathcal{K}_i$, we find the equivalent Itō equation

$$(\mathbf{I})\ \mathrm{d}\rho = \sum \mathcal{K}_i \rho \mathrm{d}W_i + \frac{1}{2}\sum M_{ij}\mathcal{K}_i\mathcal{K}_j \rho\, \mathrm{d}t \tag{80}$$

with $\mathrm{d}W_i \mathrm{d}W_j = M_{ij}\,\mathrm{d}t$. We then rewrite the SME (74) in terms of the complex Wiener increment

$$\mathrm{d}W_\gamma = \mu \mathrm{d}W_+ + \mathrm{i}\nu \mathrm{d}W_-. \tag{81}$$

Additionally introducing the (nonlinear) operators $\mathcal{G}\rho = [s - \mathrm{tr}(s\rho)]\rho$ and $\mathcal{G}'\rho = (\mathcal{G}\rho)^\dagger$ this leads to

$$\mathrm{d}\rho_c = \mathcal{L}\rho_c \mathrm{d}t + \sqrt{\frac{\eta}{2}}\left(\mathcal{G}\rho_c \mathrm{d}W_\gamma + \mathcal{G}'\rho_c \mathrm{d}W_\gamma^*\right). \tag{82}$$

The evolution under continuous feedback and measurement is then described by Wiseman (1994)

$$\rho_c(t+\mathrm{d}t) = [1 + (1/2\eta)(\mathcal{K}_+\mathrm{d}\tilde{X}_+ + \mathcal{K}_-\mathrm{d}\tilde{P}_-) + (1/4\eta)(\mathcal{K}_+\mathrm{d}\tilde{X}_+ + \mathcal{K}_-\mathrm{d}\tilde{P}_-)^2] \\ \times[1 + \mathcal{L}\mathrm{d}t + \sqrt{\eta/2}\left(\mathcal{G}\mathrm{d}W_y + \mathcal{G}'\mathrm{d}W_y^*\right)]\rho_c(t), \tag{83}$$

which can be cast into the form

$$\begin{aligned} \mathrm{d}\rho_c &= \mathcal{L}\rho_c + (1/4\eta)\left(\mathcal{K}_+\mathrm{d}\tilde{X}_+ + \mathcal{K}_-\mathrm{d}\tilde{P}_-\right)^2\rho_c \\ &+ (1/2)\left(\mathcal{K}_+\mathrm{d}\tilde{X}_+ + \mathcal{K}_-\mathrm{d}\tilde{P}_-\right)\left(\mathcal{G}\mathrm{d}W_y + \mathcal{G}'\mathrm{d}W_y^*\right)\rho_c \\ &+ \sqrt{1/2\eta}\left(\mathcal{K}_+\mathrm{d}\tilde{X}_+ + \mathcal{K}_-\mathrm{d}\tilde{P}_-\right)\rho_c \\ &+ \sqrt{\eta/2}\left(\mathcal{G}\mathrm{d}W_y + \mathcal{G}'\mathrm{d}W_y^*\right)\rho_c. \end{aligned} \tag{84}$$

Note that the ordering $\mathcal{KG}$ ensures that we obtain a trace-preserving master equation (Wiseman and Milburn, 1993). We can now average over all possible measurement trajectories (i.e., over all classical processes $\tilde{X}_+$, $\tilde{P}_-$) to obtain an unconditional (deterministic) master equation. Expanding the quadratic terms yields,

$$\left(\mathcal{K}_+\mathrm{d}\tilde{X}_+ + \mathcal{K}_-\mathrm{d}\tilde{P}_-\right)^2 = [w_1\mathcal{K}_+^2 + w_2\mathcal{K}_-^2 + w_3(\mathcal{K}_+\mathcal{K}_- + \mathcal{K}_-\mathcal{K}_+)]\mathrm{d}t,$$

and

$$\begin{aligned} \left(\mathcal{K}_+\mathrm{d}\tilde{X}_+ + \mathcal{K}_-\mathrm{d}\tilde{P}_-\right)\left(\mathcal{G}\mathrm{d}W_y + \mathcal{G}'\mathrm{d}W_y^*\right)\rho_c &= \mathcal{K}_+(\mathcal{G}+\mathcal{G}')\rho_c\mathrm{d}t + \mathcal{K}_-(\mathrm{i}\mathcal{G}-\mathrm{i}\mathcal{G}')\rho_c\mathrm{d}t, \\ &= \mathcal{K}_+\mathcal{H}[s]\rho_c\mathrm{d}t + \mathcal{K}_-\mathcal{H}[\mathrm{i}s]\rho_c\mathrm{d}t, \end{aligned}$$

where we used

$$\begin{aligned} \mathrm{d}X_+\mathrm{d}W_y &= \mu w_1 + \mathrm{i}\nu w_3 = 1, \\ \mathrm{d}P_-\mathrm{d}W_y &= \mu w_3 + \mathrm{i}\nu w_2 = \mathrm{i}. \end{aligned}$$

Taking this all together leads to

$$\begin{aligned} \dot{\rho} = \mathcal{L}\rho &+ \frac{1}{2}\left\{\frac{w_1 - w_3}{\eta}\mathcal{D}[F_+] - \mathrm{i}[F_+, s\rho + \rho s^\dagger]\right\} \\ &+ \frac{1}{2}\left\{\frac{w_2 - w_3}{\eta}\mathcal{D}[F_-] - \mathrm{i}[F_-, (\mathrm{i}s)\rho + \rho(\mathrm{i}s)^\dagger]\right\} \\ &+ \frac{w_3}{2\eta}\mathcal{D}[F_+ + F_-], \end{aligned} \tag{85}$$

where we used the fact that $\frac{1}{2}(\mathcal{K}_\pm)^2 = \mathcal{D}[F_\pm]$ and $\frac{1}{2}(\mathcal{K}_+\mathcal{K}_- + \mathcal{K}_-\mathcal{K}_+) = \mathcal{D}[F_+ + F_-] - \mathcal{D}[F_-] - \mathcal{D}[F_+]$. Using the identity $\mathcal{D}[s + \mathrm{i}F_\pm]\rho = \mathcal{D}[s]\rho + \mathcal{D}[F_\pm]\rho + \mathrm{i}[F_\pm, s\rho + \rho s^\dagger] + \frac{\mathrm{i}}{2}[\rho, F_\pm s + s^\dagger F_\pm]$ this can be written in the more familiar Lindblad form

$$\begin{aligned}\dot{\rho} = -\mathrm{i}\left[H + \frac{1}{4}\{(F_+ + \mathrm{i}F_-)s + s^\dagger(F_+ - \mathrm{i}F_-)\}, \rho\right] \\ + \frac{1}{2}\Bigg\{\mathcal{D}[s - \mathrm{i}F_+]\rho + \mathcal{D}[s - F_-]\rho + \frac{w_3}{\eta}\mathcal{D}[F_+ + F_-]\rho \\ + \left(\frac{w_1 - w_3}{\eta} - 1\right)\mathcal{D}[F_+]\rho + \left(\frac{w_2 - w_3}{\eta} - 1\right)\mathcal{D}[F_-]\rho\Bigg\}.\end{aligned} \tag{86}$$

This is the main result of this section. The evolution of the system $S$ thus effectively depends on the state of the field $B$ (via $w_i$) which has never interacted with $S$, and which can, in principle, even change (adiabatically) in time. Eq. (86) can thus be viewed as a continuous "remote preparation" of quantum states.

To illustrate this point we consider the case where the target system $S$ is a bosonic mode. For a system to be amenable to continuous teleportation the system–field interaction must enable entanglement creation. We thus set $s = d^\dagger$, with $d$ a bosonic annihilation operator, and therefore obtain $H_{\text{int}} \propto d\, a(t) + d^\dagger a^\dagger(t)$ (which is just a two-mode-squeezing interaction). Additionally we choose $F_+ = \mathrm{i}(d - d^\dagger)$ and $F_- = (d + d^\dagger)$, which means that the photocurrents $I_+$, $I_-$ will be fed back to the $x \propto (d + d^\dagger)$ and $p \propto \mathrm{i}(d - d^\dagger)$ quadrature, respectively. The resulting equation can be brought into the form

$$\dot{\rho} = -\mathrm{i}[H_{\text{sys}}, \rho] + (2N + 1)\mathcal{D}[J]\rho, \tag{87}$$

where the jump operator $J$ is determined by

$$J \propto -\mathrm{i}(2N + 1 - M - M^*)x + (1 + M - M^*)p \tag{88}$$

(with an appropriate normalization). If we define as $U_{\pi/2}$ the local unitary which effects the canonical transformation $x \to p$, $p \to -x$, we find by comparison with the eigenvalue equation for $|M\rangle$,

$$[(1 - M + M^*)x + \mathrm{i}(2N + 1 + M + M^*)p]|M\rangle = 0, \tag{89}$$

that $U^\dagger_{\pi/2} J U_{\pi/2}|-M\rangle = 0$. Taking into account the relations

$$\langle x^2 \rangle_M = \frac{1}{2}(2N + 1 + M + M^*), \tag{90a}$$

$$\langle p^2 \rangle_M = \frac{1}{2}(2N + 1 - M - M^*), \tag{90b}$$

$$\mathrm{Re}(\langle xp \rangle_M) = -\frac{\mathrm{i}}{2}(M - M^*), \tag{90c}$$

one can easily see that $U_{\pi/2}|-M\rangle = |M\rangle$. For $H_{\mathrm{sys}} = 0$ equation (87) thus has the steady-state solution $\rho^{\mathrm{ss}} = |M\rangle\langle M|$, where $J|M\rangle = 0$ and hence $\mathcal{D}[J]\rho^{\mathrm{ss}} = 0$. Note that for the vacuum case $N = M = 0$ we find $J = c$, which means that, devoid of other decoherence terms, the system will be driven to its ground state.

Below, we will come back to this scenario and discuss its implementation on the basis of an optomechanical system in more detail.

If we consider again the situation where we use a displace squeezed state $|M, \beta\rangle_B$ as input, we make the replacements $\mathrm{d}\tilde{X}_+ \to \mathrm{d}\tilde{X}'_+$ and $\mathrm{d}\tilde{P}_- \to \mathrm{d}\tilde{P}'_-$ [see (75)] in Eq. (84). This only changes the third line as all products $\mathrm{d}\tilde{X}^2$, $\mathrm{d}\tilde{X}\,\mathrm{d}W_y$, etc. are unaffected. After taking the classical average this yields an additional Hamiltonian term

$$H_{\mathrm{coh}} = \sqrt{2}[\mathrm{Re}(\beta)F_+ - \mathrm{Im}(\beta)F_-] \tag{91}$$

which has to be incorporated into $\mathcal{L}$ in the feedback master equation (85), and effects a displacement of the steady state.

### *4.2.2 Time-Continuous Entanglement Swapping*

We now replace the Gaussian input state in mode $B$ with a field state emitted by a second system (see Fig. 15), which is then, together with field $A$, sent to

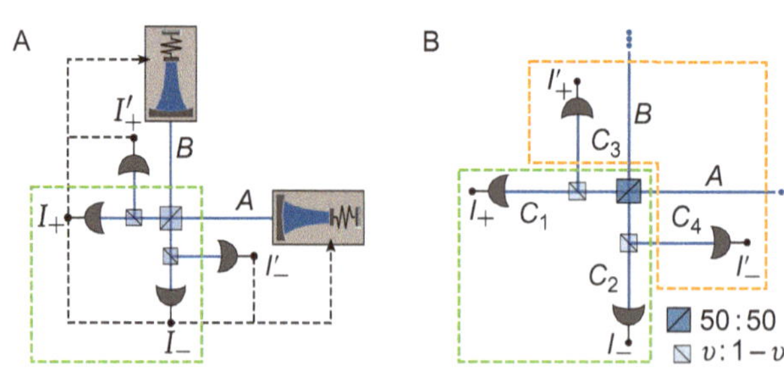


**Fig. 15** (A) Schematic of time-continuous entanglement swapping setup (*green dashed box* marks the time-continuous Bell measurement). (B) Detail of the additional stabilizing measurements (*yellow box*).

a Bell measurement as before. If before the measurement the light fields are entangled to the emitting systems, the bipartite system state will be projected into an EPR-entangled state. As this might lead to unstable system dynamics, we add two homodyne detectors to the setup, which help us to stabilize the systems (Vasilyev et al., 2013).

#### 4.2.2.1 Conditional Master Equation

We assume that the second system $S_2$ couples to the field $B$ via $H_{\text{int}} = \mathrm{i}[s_2 b^\dagger(t) - s_2^\dagger b(t)]$. $A$ and $B$ are combined on a 50:50 beam-splitter to form the combinations $a \pm b$ in the outputs. These outputs are sent to a pair of beam-splitters (with an uneven splitting ratio $\upsilon : 1 - \upsilon$) and subsequently to a total of four homodyne setups. If we label the modes incident on the homodyne detectors as $C_i$ (described by field operators $c_i$) for $i = 1, \ldots, 4$ (see Fig. 15), we find the following relations to modes $A$ and $B$,

$$a = \sqrt{\upsilon/2}(c_1 + c_2) - \sqrt{(1-\upsilon)/2}(c_3 + c_4), \tag{92a}$$

$$b = \sqrt{\upsilon/2}(c_1 - c_2) - \sqrt{(1-\upsilon)/2}(c_3 - c_4). \tag{92b}$$

We now choose the LO phases of the four homodyne setups such that they measure $x_+ = c_1 + c_1^\dagger$, $p_- = -\mathrm{i}(c_2 - c_2^\dagger)$, $x_- = c_4 + c_4^\dagger$, and $p_+ = -\mathrm{i}(c_3 - c_3^\dagger)$. These four measurements allow us to simultaneously monitor both quadratures of both systems (although with reduced efficiency). The measurement of $x_+$ and $p_-$, which we choose to have a relative strength $\upsilon$ set by the beam-splitting ratio, realize a Bell measurement as before, while the measurement of the conjugate quadratures $x_-$ and $p_+$, with a strength $1 - \upsilon$, we will need for stabilization of $S_1$ and $S_2$.

To derive the corresponding SME we apply the same logic as before. We start from the Schrödinger equation for the full system ($S_1$ + $S_2$ + field modes),

$$\mathrm{d}|\Psi\rangle = \left[-\mathrm{i}H_{\text{eff}}\mathrm{d}t + s_1(\mathrm{d}A^\dagger + \mathrm{d}A) + s_2(\mathrm{d}B^\dagger + \mathrm{d}B)\right]|\Psi\rangle, \tag{93}$$

with an initial state $|\Psi_0\rangle = |\psi_0\rangle_{S_1 S_2} \otimes |\text{vac}\rangle_{\text{field}}$ and an effective Hamiltonian $H_{\text{eff}} = H_{\text{sys}}^{(1)} + H_{\text{sys}}^{(2)} - \frac{\mathrm{i}}{2}\sum_{i=1,2} s_i^\dagger s_i$. We then rewrite this in terms of $\mathrm{d}X_\pm$ and $\mathrm{d}P_\pm$ [by plugging (92) into Eq. (93)] and project onto eigenstates corresponding to measurement outcomes $I_\pm$, $I'_\pm$. With the definition $s_\pm = s_1 \pm s_2$ we find

$$
\begin{aligned}
\mathrm{d}|\tilde{\psi}_c\rangle = &-\mathrm{i}H_{\mathrm{eff}}|\psi_c\rangle \mathrm{d}t + \sqrt{v/2}[s_+\mathrm{d}\tilde{X}_+ + \mathrm{i}s_-\mathrm{d}\tilde{P}_-]|\psi_c\rangle \\
&+ \sqrt{(1-v)/2}[\mathrm{i}s_+\mathrm{d}\tilde{X}_- + s_-\mathrm{d}\tilde{P}_+]|\psi_c\rangle
\end{aligned} \tag{94}
$$

where $|\tilde{\psi}_c\rangle$ is unnormalized. As all electromagnetic field modes are assumed to be in vacuum we find that the measurement processes have unit variance, $\mathrm{d}\tilde{X}_\pm(t)^2 = \mathrm{d}\tilde{P}_\pm(t)^2 = \mathrm{d}t$, and that they are mutually uncorrelated, i.e., $\mathrm{d}\tilde{X}_+\mathrm{d}\tilde{X}_- = \mathrm{d}\tilde{X}_+\mathrm{d}\tilde{P}_+ = 0$, etc. This can be shown by expressing $c_i$ in terms of $a$, $b$ and using Itō rules, where we have to take into account vacuum noise entering through the open ports of the second pair of beam-splitters (not explicitly introduced here). The homodyne currents are given by

$$I_+\mathrm{d}t = \sqrt{v/2}\,\langle s_+ + s_+^\dagger\rangle_c + \mathrm{d}W_+, \tag{95a}$$

$$I_-\mathrm{d}t = \mathrm{i}\sqrt{v/2}\,\langle s_- - s_-^\dagger\rangle_c + \mathrm{d}W_-, \tag{95b}$$

$$I'_+\mathrm{d}t = \mathrm{i}\sqrt{(1-v)/2}\,\langle s_+ - s_+^\dagger\rangle_c + \mathrm{d}V_+, \tag{95c}$$

$$I'_-\mathrm{d}t = \sqrt{(1-v)/2}\,\langle s_- + s_-^\dagger\rangle_c + \mathrm{d}V_-. \tag{95d}$$

where the Wiener increments $\mathrm{d}W_\pm$ and $\mathrm{d}V_\pm$ obey a multiplication table corresponding to the one of $\mathrm{d}\tilde{X}_\pm$ and $\mathrm{d}\tilde{P}_\pm$. As we now have four uncorrelated homodyne measurements we can directly apply the procedure laid out in Appendix B.1.1 and arrive at the SME

$$
\begin{aligned}
\mathrm{d}\rho_c = &\mathcal{L}\rho_c\mathrm{d}t + \sqrt{v/2}\{\mathcal{H}[s_+]\rho_c\mathrm{d}W_+ + \mathcal{H}[\mathrm{i}s_-]\rho_c\mathrm{d}W_-\} \\
&+ \sqrt{(1-v)/2}\{\mathcal{H}[\mathrm{i}s_+]\rho_c\mathrm{d}V_+ + \mathcal{H}[s_-]\rho_c\mathrm{d}V_-\},
\end{aligned} \tag{96}
$$

with $\mathcal{L}\rho = -\mathrm{i}[H_{\mathrm{sys}}^{(1)} + H_{\mathrm{sys}}^{(2)}, \rho] + \mathcal{D}[s_1]\rho + \mathcal{D}[s_2]\rho$. These results can alternatively be derived in a similar spirit but in a more formal way within the framework of quantum networks, as for example presented in Gough and James (2009).

#### 4.2.2.2 Feedback Master Equation

In this entanglement swapping scheme all four homodyne currents, $I_\pm$ (Bell measurement) and $I'_\pm$ (stabilizing measurements), are fed back to *both* systems. (For convenience we will in the following label the photocurrents by $I_i$, $i = 1, \ldots, 4$, according to the light modes $C_i$ they correspond to.) We again describe this by the operations $\mathcal{K}[F_i]\rho I_i = -\mathrm{i}[F_i I_i, \rho]$ ($i = 1, \ldots, 4$), where $F_i = F_i^\dagger$ now act on the combined Hilbert space of $S_1 + S_2$. Using the procedure from before it is straightforward to show that the corresponding FME can be written as

$$\dot{\rho} = -\mathrm{i}\left[H_{\mathrm{sys}},\rho\right] - \frac{\mathrm{i}}{2}\sum_{i=1}^{4}\left[s_i^\dagger F_i + F_i s_i,\rho\right] + \sum_{i=1}^{4}\left\{\mathcal{D}[s_i - \mathrm{i}F_i] + \frac{1-\eta}{\eta}\mathcal{D}[F_i]\right\}, \tag{97}$$

with $(s_i)_{i=1}^4 = (\sqrt{v}s_+, \mathrm{i}\sqrt{v}s_-, \mathrm{i}\sqrt{1-v}s_+, \sqrt{1-v}s_-)$. [Compare this to Eq. (B.34)]. Here we assumed that all detectors have the same efficiency $\eta$.

Let us again consider the case of bosonic modes, with jump operators (assuming a unity coupling rate) $s_1 = d_1^\dagger$, $s_2 = d_2^\dagger$ ($[d_i, d_j^\dagger] = \delta_{ij}$), not including the stabilizing measurements (i.e., $v = 1$). We thus have $(s_-, s_+) = (d_+^\dagger, d_-^\dagger)$ and choose $(F_1, F_2) = (\mathrm{i}d_+ - \mathrm{i}d_+^\dagger, d_- + d_-^\dagger)$. For $H_{\mathrm{sys}} = 0$ (and $\eta = 1$) this gives the master equation

$$\dot{\rho} = -\frac{\mathrm{i}}{2}\left[d_+^2 - d_-^2 - \mathrm{H.\,c.},\rho\right] + \mathcal{D}[d_+]\rho + \mathcal{D}[d_-]\rho.$$

The Lindblad terms damp the system, while the Hamiltonian term generates squeezing dynamics in the EPR-modes, which amplifies the system. This becomes clearer when looking at the corresponding Langevin equations for the quadratures corresponding quadratures $x_\pm \propto d_\pm + d_\pm^\dagger$, $p_\pm \propto \mathrm{i}(d_\pm - d_\pm^\dagger)$,

$$\frac{\mathrm{d}}{\mathrm{d}t}x_+ = -\left(\frac{1}{2} + 2\right)x_+ + x_+^{\mathrm{in}}, \quad \frac{\mathrm{d}}{\mathrm{d}t}p_+ = -\left(\frac{1}{2} - 2\right)p_+ + p_+^{\mathrm{in}},$$
$$\frac{\mathrm{d}}{\mathrm{d}t}x_- = -\left(\frac{1}{2} - 2\right)x_- + x_-^{\mathrm{in}}, \quad \frac{\mathrm{d}}{\mathrm{d}t}p_- = -\left(\frac{1}{2} + 2\right)p_- + p_-^{\mathrm{in}},$$

where $x_\pm^{\mathrm{in}}$, $p_\pm^{\mathrm{in}}$ denote appropriate noise contributions. These equations indicate that $x_+$ and $p_-$ will become squeezed while $x_-$ and $p_+$ will be anti-squeezed. This instability can be prevented by introducing the additional measurement (and feedback) channels; they will, however, also limit the generated squeezing and thus the amount of created entanglement.

## 4.3 Optomechanical Time-Continuous Teleportation

In Section 4.2.1 we derived the constitutive equations of motion describing time-continuous teleportation (the conditional master equation and feedback master equation) for a generic system. In this section we will focus on the optomechanical implementation (see Fig. 12). Technical details are discussed in Section 4.3.1.

In order to successfully implement continuous teleportation in optomechanical systems we need to appropriately design our measurement setup. We saw that in the regime $g \ll \kappa \ll \omega_\mathrm{m}$ and for a blue drive with $\Delta_\mathrm{c} = \omega_\mathrm{m}$ the optomechanical interaction essentially gives rise to a two-mode squeezing dynamics between the mechanical mode and the extracavity light field, which is the required interaction for teleportation. Consequently the mechanical oscillator resonantly scatters photons into the lower sideband at $\omega_\mathrm{c} = \omega_0 - \omega_\mathrm{m}$; spectrally, the photons which are correlated with the mechanical motion are therefore located at this sideband frequency. We thus set up our Bell measurement in the following way: firstly, we choose the center frequency of the squeezed input light located at the sideband frequency $\omega_\mathrm{c}$. Secondly, we now use heterodyne detection to measure quadratures at the same frequency.

In Section 4.3.1 we show that after adiabatic elimination of the cavity mode and a rotating-wave approximation, the evolution of the conditional mechanical state $\rho_c^{(\mathrm{m})}$ in a rotating frame with $\omega_\mathrm{m}$ (neglecting the mechanical frequency shift by the optical-spring effect, see Section 4.3.1) is described by the stochastic master equation

$$\begin{aligned} \mathrm{d}\rho_c^{(\mathrm{m})} &= \gamma_- \mathcal{D}[c_\mathrm{m}]\rho_c^{(\mathrm{m})}\mathrm{d}t + \gamma_+ \mathcal{D}[c_\mathrm{m}^\dagger]\rho_c^{(\mathrm{m})}\mathrm{d}t \\ &\quad - \sqrt{\frac{\eta g^2 \kappa}{2}} \Big\{ \mathcal{H}[\mathrm{i}\mu\eta_+ c_\mathrm{m}^\dagger]\rho_c^{(\mathrm{m})}\mathrm{d}W_+ - \mathcal{H}[\nu\eta_+ c_\mathrm{m}^\dagger]\rho_c^{(\mathrm{m})}\mathrm{d}W_- \Big\}. \end{aligned} \tag{98}$$

The first row corresponds to the standard adiabatic master equation (28), while the second row describes the time-continuous Bell measurement, where the squeezing parameter $M$ is encoded in $\mu = 1 - \alpha$ and $\nu = 1 + \alpha$, with $\alpha = (N + M)/(N + M^* + 1)$. $\eta$ is the detection efficiency as before. The photo-currents resulting from the Bell measurement are

$$I_+ \mathrm{d}t = -\mathrm{i}\sqrt{\eta g^2 \kappa / 2}\, \langle \eta_+ c_\mathrm{m}^\dagger - \mathrm{H.c.} \rangle \mathrm{d}t + \mathrm{d}W_+, \tag{99a}$$

$$I_- \mathrm{d}t = \sqrt{\eta g^2 \kappa / 2}\, \langle \eta_+ c_\mathrm{m}^\dagger + \mathrm{H.c.} \rangle \mathrm{d}t + \mathrm{d}W_-, \tag{99b}$$

where $\mathrm{d}W_\pm$ are correlated Wiener increments whose (co-)variances are given in Eq. (67). For the choice $\Delta_\mathrm{c} = \omega_\mathrm{m}$ we have $\eta_+ = 2/\kappa$ and $\eta_- = 1/(\frac{\kappa}{2} + 2\mathrm{i}\omega_\mathrm{m})$. Thus $I_\pm$ approximately correspond to measurements of the mechanical quadratures $p_\mathrm{m}$ and $x_\mathrm{m}$, respectively. We model the feedback as instantaneous displacements of the mechanical oscillator in phase space, where the feedback strength is proportional to the heterodyne currents

$I_\pm(t)$. This is described by Hamiltonian terms $I_\pm(t)F_\pm$, where $F_\pm = F_\pm^\dagger$ are generalized forces. The feedback operators we choose to be

$$F_+ = -\sqrt{2g^2\kappa}\,\eta_+ x_m, \tag{100a}$$
$$F_- = -\sqrt{2g^2\kappa}\,\eta_+ p_m, \tag{100b}$$

which generate a displacement in $p_m$ and $x_m$, respectively. The prefactors of $F_\pm$ (i.e., the feedback gain) we chose such that they match the measurement strength of the Bell detection. The corresponding feedback master equation (in the same rotating frame) can be written as

$$\begin{aligned}\dot{\rho}^{(m)} &= \gamma(\bar{n}+1)\mathcal{D}[c_m]\rho^{(m)} + \gamma\bar{n}\mathcal{D}[c_m^\dagger]\rho^{(m)} \\ &\quad + (4g^2/\kappa)\{\lambda_1(\epsilon)\mathcal{D}[J_1(\epsilon)] + \lambda_2(\epsilon)\mathcal{D}[J_2(\epsilon)]\}\rho^{(m)},\end{aligned} \tag{101}$$

where $\epsilon = [1+(4\omega_m/\kappa)^2]^{-1}$ quantifies the suppression of the counter-rotating interaction terms (i.e., the optomechanical beam-splitter) as before. The effective Lindblad terms are determined by $\lambda_i$ and $J_i$, which are obtained (see Section 4.3.1) from the eigenvalue decomposition of the positive matrix

$$\mathbf{\Lambda} = \begin{pmatrix} \frac{w_2}{\eta} - \frac{1}{2}(1+\epsilon) & -\frac{w_3}{\eta} + \frac{i}{2}(1+\epsilon) \\ -\frac{w_3}{\eta} - \frac{i}{2}(1+\epsilon) & \frac{w_1}{\eta} - \frac{1}{2}(1+\epsilon) \end{pmatrix}.$$

For efficient detection ($\eta = 1$) we obtain $\lambda_1 = (2N+1) + O(\epsilon)$ and $\lambda_2 = O(\epsilon)$, which means that in the sideband-resolved regime the jump operator $J_2(\epsilon)$ contributes only weakly. In zeroth order in $\epsilon$ the dominating dissipative dynamics are generated by

$$J_1(0) \propto -i(2N+1-M-M^*)x_m + (1+M-M^*)p_m, \tag{102}$$

which is just the situation we already encountered in Section 4.2.1. This means that $|M\rangle\langle M|$ is a dark state of $D[J_1(0)]$ and thus in the ideal limit of $\gamma = 0$, $\epsilon = 0$, $\eta = 1$ the steady state of (101) is

$$\lim_{t\to\infty} \rho^{(m)}(t) = |M\rangle\langle M|. \tag{103}$$

Hence, the optical input state is perfectly transferred to the mechanical mode.

Moving away from the ideal case, the protocol's performance is degraded by mechanical decoherence effects ($\gamma\bar{n} > 0$), counter-rotating terms of the optomechanical interaction which are suppressed by $\epsilon < 1$, and inefficient

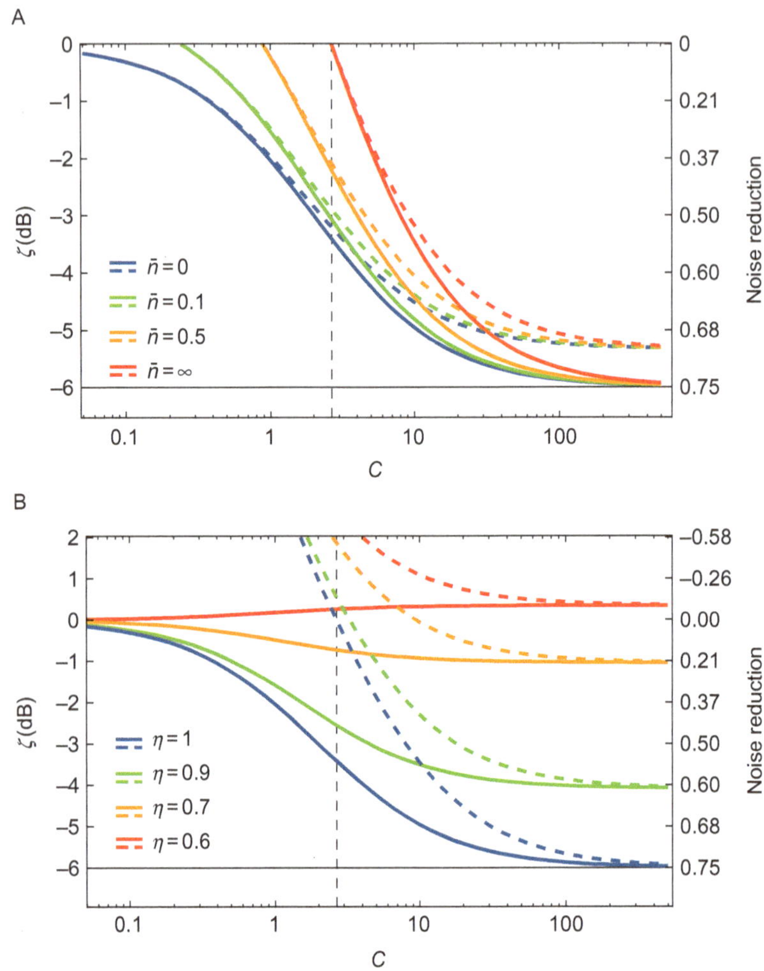


**Fig. 16** Mechanical squeezing $\zeta$ against cooperativity $C$: (A) Varying mechanical bath occupation $\bar{n} = 0,\ 1/10,\ 1/2,\ \infty$ (represented by *different colors*) and unit detection efficiency $\eta = 1$; The *solid* (*dashed*) *lines* represent a sideband resolution of $\kappa/\omega_m = 1/10$ (1). (B) Different detection efficiencies $\eta = 1,\ 9/10,\ 7/10,\ 6/10$ (represented by *different colors*) and $\kappa = \omega_m/10$; here the *solid* (*dashed*) *lines* represent $\bar{n} = 0\ \ (\infty)$. In both plots the *horizontal solid line* at $\zeta = -6$dB (corresponding to $N \approx 0.56$) shows the squeezing level of the input light and the *vertical dashed line* the critical cooperativity $C_{crit} \approx 2.7$. Note that in (B) $C_{crit}$ marks the value of $C$ above which the *blue dashed line* ($n \to \infty, \eta = 1$) falls below 0 (i.e., squeezing is transferred).

detection ($\eta < 1$) which leads to imperfect feedback. [All these effects are captured by the master equation (101).] Fig. 16 shows the steady-state squeezing $\zeta$ transmitted to the mechanical mode for different parameters plotted against the optomechanical cooperativity $C = 4g^2/(\bar{n}+1)\gamma\kappa$. In the upper plot we assume perfect detection efficiency $\eta = 1$ and find that in this case there exists a critical value $C_{\text{crit}}(N) = 1/[\sqrt{N(N+1)}]$ determined by the input squeezing $N$ above which the resulting mechanical state is squeezed for any thermal occupation number $\bar{n}$. The lower plot clearly shows that this is no longer true if we assume nonunit detection efficiency $\eta$. We find that below a certain critical value $\eta_{\text{crit}}(N, \bar{n})$ we can no longer transfer squeezing to the mechanical oscillator, but we rather heat it instead. (This is even true for a zero-temperature mechanical environment, as illustrated in the plot.) In this general case it can be beneficial to chose a modified feedback gain, i.e., use feedback operators $\widetilde{F}_{\pm} = \sigma F_{\pm}$ with $\sigma \neq 1$. In the parameter regime we consider, however, the resulting improvements are negligible.

### *4.3.1 Detailed Derivation*

Here we derive the stochastic and feedback master equations for the optomechanical teleportation setup presented in Section 4.3, following the lines of Section 4.2.1 with modifications accommodating the optomechanical implementation. The one-dimensional electromagnetic field $A$ couples to the cavity via the linear interaction $H_{\text{int}} = \mathrm{i}\sqrt{\kappa}[c_c a^\dagger(t) - c_c^\dagger a(t)]$. As before, $A$ is assumed to be in the vacuum state, while the second field $B$ is in a pure squeezed state. In this section we refer to several different rotating frames: the frame of the driving laser rotating at $\omega_0$ (which is our standard frame of reference), the squeezing frame which defines the central frequency for the squeezed input light at $\omega_s$, and the local oscillator frame at $\omega_{\text{lo}}$ in reference to which all measurements will be made. We therefore have the relations

$$a(t) = a_{\text{lo}}(t)\mathrm{e}^{-\mathrm{i}\Delta_{\text{lo}}t}, \tag{104a}$$

$$b(t) = b_s(t)\mathrm{e}^{-\mathrm{i}\Delta_s t} = b_{\text{lo}}(t)\mathrm{e}^{-\mathrm{i}\Delta_{\text{lo}}t}, \tag{104b}$$

with the definitions $\Delta_{\text{lo}} = \omega_{\text{lo}} - \omega_0$, $\Delta_s = \omega_s - \omega_0$. The squeezed input state is then, in the squeezing frame, defined by the eigenvalue equation $[b_s(t) - \alpha b_s^\dagger(t)]|M\rangle_B = 0$ with $\alpha = (N + M)/(N + M^* + 1)$ and $N(N + 1) = |M|^2$. The Schrödinger equation of the full system in the LO frame can be written as (neglecting for the moment the coupling to the mechanical bath as this can easily be added in the end)

$$\begin{aligned} \mathrm{d}|\Psi\rangle = &-\mathrm{i}H_{\text{eff}}|\Psi\rangle\mathrm{d}t + \sqrt{\kappa}\left(\mathrm{d}A_{\text{lo}}^\dagger + \alpha\mathrm{d}A_{\text{lo}}\right)\mathrm{e}^{\mathrm{i}\Delta_{\text{lo}}t}c_c|\Psi\rangle \\ &+ \sqrt{\kappa}\left[\mathrm{d}B_{\text{lo}}\mathrm{e}^{-\mathrm{i}(\Delta_{\text{lo}}-\Delta_s)t} - \alpha\mathrm{d}B_{\text{lo}}^\dagger\mathrm{e}^{\mathrm{i}(\Delta_{\text{lo}}-\Delta_s)t}\right]\mathrm{e}^{\mathrm{i}\Delta_{\text{lo}}t}c_c|\Psi\rangle, \end{aligned} \tag{105}$$

where $|\Psi\rangle$ is the state describing the complete system with an initial condition $|\Psi_0\rangle = |\psi_0\rangle_S \otimes |\text{vac}\rangle_A \otimes |M\rangle_B$. If we now choose $\Delta_s = \Delta_{\text{lo}}$, we can rewrite this as

$$\mathrm{d}|\Psi\rangle = \left[-\mathrm{i}H_{\text{eff}}\mathrm{d}t + \sqrt{\kappa/2}(\mu \mathrm{d}X_+ + \mathrm{i}\nu \mathrm{d}P_-)c_{\text{c}}\,\mathrm{e}^{\mathrm{i}\Delta_{\text{lo}}t}\right]|\Psi\rangle,$$

where $\mathrm{d}X_+ = \sqrt{1/2}(a_{\text{lo}} + a_{\text{lo}}^\dagger + b_{\text{lo}} + b_{\text{lo}}^\dagger)\mathrm{d}t$ and $\mathrm{d}P_- = \mathrm{i}\sqrt{1/2}(a_{\text{lo}} - a_{\text{lo}}^\dagger - b_{\text{lo}} + b_{\text{lo}}^\dagger)\mathrm{d}t$, and $\mu = 1 - \alpha$, $\nu = 1 + \alpha$ as before. By comparing this to Schrödinger equation (64) we can deduce that the heterodyne Bell measurement at $\omega_{\text{lo}}$ is described by SME (74) together with the expression for the measurement currents (66) if we set $s = c_{\text{c}}\,\mathrm{e}^{\mathrm{i}\Delta_{\text{lo}}t}$. Thus the master equation

$$\mathrm{d}\rho_c = \mathcal{L}\rho_c\mathrm{d}t + \sqrt{\kappa}\mathcal{H}[(\mu \mathrm{d}W_+ + \mathrm{i}\nu \mathrm{d}W_-)c_{\text{c}}\mathrm{e}^{\mathrm{i}\Delta_{\text{lo}}t}]\rho_c \tag{106}$$

together with the output equations

$$I_+\mathrm{d}t = \sqrt{1/2}\langle c_{\text{c}}\mathrm{e}^{\mathrm{i}\Delta_{\text{lo}}t} + \text{H.c.}\rangle_c\mathrm{d}t + \mathrm{d}W_+, \tag{107a}$$

$$I_-\mathrm{d}t = \mathrm{i}\sqrt{1/2}\langle c_{\text{c}}\mathrm{e}^{\mathrm{i}\Delta_{\text{lo}}t} - \text{H.c.}\rangle_c\mathrm{d}t + \mathrm{d}W_- \tag{107b}$$

provides us with a description of the conditional state of the full optomechanical system (including the cavity mode) conditioned on the heterodyne currents $I_\pm$. What we eventually seek to obtain is an effective description of the mechanical system only. Under the weak-coupling assumption $g/\kappa \ll 1$ we can eliminate the cavity mode. To make the desired expansion we must first transform (106) into the interaction picture defined by the free Hamiltonian $\omega_{\text{m}}c_{\text{m}}^\dagger c_{\text{m}} - \Delta_c c_{\text{c}}^\dagger c_{\text{c}}$,

$$\begin{aligned}\mathrm{d}\tilde{\rho}_c = &-\mathrm{i}g\left[(c_{\text{c}}\mathrm{e}^{\mathrm{i}\Delta_c t} + \text{H.c.})(c_{\text{m}}\mathrm{e}^{-\mathrm{i}\omega_{\text{m}}t} + \text{H.c.}), \tilde{\rho}_c\right]\mathrm{d}t\\ &+ \sqrt{\kappa}\mathcal{H}[(\mu \mathrm{d}W_+ + \mathrm{i}\nu \mathrm{d}W_-)c_{\text{c}}\mathrm{e}^{\mathrm{i}(\Delta_{\text{lo}} + \Delta_c)t}]\tilde{\rho}_c + \kappa\mathcal{D}[c_{\text{c}}]\tilde{\rho}_c\mathrm{d}t.\end{aligned} \tag{108}$$

(All operators marked with a tilde, e.g., $\tilde{\rho}$, are defined with respect to this rotating frame.) One can then show that the SME for the mechanical system (in the rotating frame at $\omega_{\text{m}}$) can be written as (Doherty and Jacobs, 1999)

$$\begin{aligned}\mathrm{d}\tilde{\rho}_c^{(\text{m})} = &-\sqrt{2}g^2\left[\tilde{x}_{\text{m}}, \tilde{y}\tilde{\rho}_c^{(\text{m})} - \tilde{\rho}_c^{(\text{m})}\tilde{y}^\dagger\right]\mathrm{d}t\\ &+ \sqrt{g^2\kappa}\mathcal{H}[(-\mathrm{i}\mu \mathrm{d}W_+ + \nu \mathrm{d}W_-)\,\tilde{y}\,\mathrm{e}^{\mathrm{i}\Delta_{\text{lo}}t}]\tilde{\rho}_c^{(\text{m})},\end{aligned} \tag{109}$$

where $\tilde{\rho}_c^{(\text{m})}$ denotes the conditional state of the mechanical subsystem. Eq. (109) does not give rise to a valid Lindblad equation when averaged over all possible measurement trajectories as $\tilde{y}$ is not a Hermitian operator.

In order to get a consistent equation we apply a RWA to both the dynamics generated by the first commutator term and the measurement term. Plugging the definitions of $\tilde{x}_{\mathrm{m}}$ and $\tilde{y}$ into the first term we find resonant terms of the form $c_{\mathrm{m}}\tilde{\rho}^{(\mathrm{m})}c_{\mathrm{m}}^{\dagger}$, $c_{\mathrm{m}}^{\dagger}c_{\mathrm{m}}\tilde{\rho}^{(\mathrm{m})}$, etc., and off-resonant terms oscillating at $\mathrm{e}^{\pm 2\mathrm{i}\omega_{\mathrm{m}}t}$. The resonant terms have two effects (see Section 2.3.2): first they give rise to cooling and heating, and second to a frequency shift of the mechanical resonance frequency to $\omega_{\mathrm{m}}^{\mathrm{eff}}$. We have to account for this frequency shift by changing to a different frame rotating at $\omega_{\mathrm{m}}^{\mathrm{eff}}$, which we still denote by $\tilde{\rho}^{(\mathrm{m})}$ for simplicity.

We can now introduce a time coarse graining in the form of $\mathrm{d}\overline{\rho}_c^{(\mathrm{m})} = \int_t^{t+\delta t}\mathrm{d}\tilde{\rho}_c^{(\mathrm{m})}$ (see Hofer, 2015). We assume that it can be arranged such that $\tilde{\rho}_c^{(\mathrm{m})}$ varies slowly on the timescale $\delta t$ (and can thus be pulled out from all time integrals), while we still average over many mechanical periods, i.e., $\delta t\,\omega_{\mathrm{m}}^{\mathrm{eff}} \gg 1$. In the adiabatic regime the relevant system timescales are given by $g^2/\kappa$ and $\bar{n}\gamma$, the effective interaction strength and mechanical decoherence rate, respectively. Although Eq. (109) is valid for any $\Delta_{\mathrm{lo}}$ and $\Delta_{\mathrm{c}}$, the form of the resulting equation in RWA depends on the choice of $\Delta_{\mathrm{lo}}$. As we illustrated before we drive the optomechanical cavity on the blue sideband ($\omega_0 = \omega_c + \omega_{\mathrm{m}}$), but want the LO to be resonant with the scattered photons ($\omega_{\mathrm{lo}} = \omega_0 - \omega_{\mathrm{m}}^{\mathrm{eff}}$), and thus set $\Delta_{\mathrm{lo}} = -\omega_{\mathrm{m}}^{\mathrm{eff}}$ (note the shifted resonance frequency). For the first term in (109) the RWA then simply amounts to dropping all terms oscillating with $\exp(\pm 2\mathrm{i}\omega_{\mathrm{m}}^{\mathrm{eff}}t)$, which introduces an error of order $\mathcal{O}(1/\delta t\,\omega_{\mathrm{m}}^{\mathrm{eff}})$. To treat the heterodyne measurement we introduce the coarse-grained noise increment $\mathrm{d}W_{\pm}^{(0)} = \int_t^{t+\delta t}\mathrm{d}W_{\pm}$ (see Appendix B.1.1), which obey (67). We end up with an effective SME for the mechanical system,

$$
\begin{aligned}
\mathrm{d}\tilde{\rho}_c^{(\mathrm{m})} &= \gamma_{-}\mathcal{D}[c_{\mathrm{m}}]\tilde{\rho}_c^{(\mathrm{m})}\mathrm{d}t + \gamma_{+}\mathcal{D}[c_{\mathrm{m}}^{\dagger}]\tilde{\rho}_c^{(\mathrm{m})}\mathrm{d}t \\
&\quad + \sqrt{g^2\kappa/2}\,\mathcal{H}[(-\mathrm{i}\mu\mathrm{d}W_{+}^{(0)} + \nu\mathrm{d}W_{-}^{(0)})\eta_{+}c_{\mathrm{m}}^{\dagger}]\tilde{\rho}_c^{(\mathrm{m})},
\end{aligned}
\tag{110}
$$

where we added mechanical decoherence terms; this is essentially equation Eq. (98).

To apply feedback we have to extract the modes corresponding to the filtered noise processes $W_{\pm}^{(0)}$ from the heterodyne currents $I_{\pm}$. This can be achieved by applying the coarse-graining procedure from above to (107a) and (107b), i.e., $I_{\pm}^{(0)} = \int_t^{t+\delta t} I_{\pm}\mathrm{d}t$. Together with $\langle c_{\mathrm{c}}\rangle_c = -\mathrm{i}g\langle y\rangle_c$, which results from the adiabatic elimination, we find

$$I_+^{(0)}\mathrm{d}t \approx -\mathrm{i}g\sqrt{\kappa/2}\langle\eta_+c_\mathrm{m}^\dagger - \mathrm{H.c.}\rangle\mathrm{d}t + \mathrm{d}W_+^{(0)}, \tag{111a}$$

$$I_-^{(0)}\mathrm{d}t \approx g\sqrt{\kappa/2}\langle\eta_+c_\mathrm{m}^\dagger + \mathrm{H.c.}\rangle\mathrm{d}t + \mathrm{d}W_-^{(0)}, \tag{111b}$$

where we neglected contributions from higher sidebands, introducing corrections on the order $\mathcal{O}(1/\delta t\,\omega_\mathrm{m}^\mathrm{eff})$. With the identification $s = -\mathrm{i}g\sqrt{\kappa}\,\eta_+c_\mathrm{m}^\dagger$ the set of equations (110), (111) is equivalent to the generic case discussed before. However, Eq. (110) additionally contains decoherence terms due to the coupling to the mechanical environment $(\gamma\bar{n}\mathcal{D}[c_\mathrm{m}^\dagger] + \gamma(\bar{n}+1)\mathcal{D}[c_\mathrm{m}])$ and due to optomechanical back-action $(2g^2\,\mathrm{Re}(\eta_-)\mathcal{D}[c_\mathrm{m}])$. For the choice

$$F_+ = -g\sqrt{\kappa}\eta_+(c_\mathrm{m} + c_\mathrm{m}^\dagger), \tag{112a}$$

$$F_- = \mathrm{i}g\sqrt{\kappa}\eta_+(c_\mathrm{m} - c_\mathrm{m}^\dagger), \tag{112b}$$

(where the prefactors are chosen to match the operator $s$), and after adding the appropriate decoherence terms, the FME for optomechanical teleportation can be written as

$$\begin{aligned}\dot{\tilde{\rho}}^{(\mathrm{m})} &= \{\gamma(\bar{n}+1)\mathcal{D}[c_\mathrm{m}] + \gamma\bar{n}\mathcal{D}[c_\mathrm{m}^\dagger]\}\tilde{\rho}^{(\mathrm{m})}\\ &+ \frac{4g^2}{\kappa}\bigg\{(1+\epsilon)\mathcal{D}[c_\mathrm{m}] + \frac{w_3}{\eta}\mathcal{D}[x_\mathrm{m}+p_\mathrm{m}]\\ &+ \left(\frac{w_1-w_3}{\eta}-1\right)\mathcal{D}[p_\mathrm{m}] + \left(\frac{w_2-w_3}{\eta}-1\right)\mathcal{D}[x_\mathrm{m}]\bigg\}\tilde{\rho}^{(\mathrm{m})},\end{aligned} \tag{113}$$

where $\epsilon = [1+(4\omega_\mathrm{m}/\kappa)^2]^{-1}$ and we finally set $\Delta_\mathrm{c} = \omega_\mathrm{m}$. Note that this equation is not necessarily in Lindblad form, as the decoherence terms' prefactors may in general be negative. To fix this we can apply the following diagonalization procedure. We parametrize the jump operators in terms of $\boldsymbol{X}_\mathrm{m} = (x_\mathrm{m}, p_\mathrm{m})^\mathrm{T}$ as $\boldsymbol{s} = \tilde{\Lambda}\boldsymbol{X}$ and rewrite the master equation as

$$\begin{aligned}\dot{\rho} &= \sum_k l_k\mathcal{D}[s_k] = \sum_i l_k\left[s_k\rho s_k^\dagger - \frac{1}{2}\rho s_k^\dagger s_k - \frac{1}{2}s_k^\dagger s_k\rho\right]\\ &= \sum_{ijk} l_k\tilde{\Lambda}_{ki}\tilde{\Lambda}_{kj}^*\left[X_i\rho X_j - \frac{1}{2}\rho X_jX_i - \frac{1}{2}X_jX_i\rho\right]\\ &= \sum_{ijm}\lambda_m O_{im}O_{mj}^\mathrm{T}\left[X_i\rho X_j - \frac{1}{2}\rho X_jX_i - \frac{1}{2}X_jX_i\rho\right]\\ &= \sum_m\lambda_m\left[J_m\rho J_m - \frac{1}{2}\rho J_mJ_m - \frac{1}{2}J_mJ_m\rho\right] = \sum_m\lambda_m\mathcal{D}[J_m]\rho.\end{aligned} \tag{114}$$

This shows that by virtue of the eigenvalue decomposition of the $(2 \times 2)$ Hermitian matrix $\mathbf{\Lambda}$, $\Lambda_{ij} = \sum_k l_k \widetilde{\Lambda}_{ki} \widetilde{\Lambda}^*_{kj}$, we can write $\dot{\rho} = \sum_i \lambda_i \mathcal{D}[J_i]\rho$ with $J_i = \boldsymbol{v}_i \cdot \mathbf{X}$, where $\lambda_i$ and $\boldsymbol{v}_i$ $(i = 1, 2)$ are the eigenvalues and eigenvectors of $\mathbf{\Lambda}$, respectively. In all cases considered in this review, $\mathbf{\Lambda}$ is positive definite and thus $\lambda_i > 0$; we can therefore bring the corresponding master equations into Lindblad form.

## 4.4 Optomechanical Time-Continuous Entanglement Swapping

We now replace the squeezed field mode with a second optomechanical cavity, as is depicted in Fig. 15. The goal of this scheme is to generate stationary entanglement between the two mechanical subsystems. The implementation is akin to the teleportation protocol presented above: both cavities are driven on the blue sideband to resonantly enhance the two-mode squeezing interaction, and their output light is sent to the Bell detection setup which operates at the cavity resonance frequency $\omega_c$. Feeding back the Bell detection results $I_\pm$ corresponding to the $x_+$ and $p_-$ quadratures of the optical fields to both mechanical systems dissipatively drives them toward an entangled state. There is a slight complication, however. A single Bell measurement only allows us to separately monitor two of the four variables $(x_{\mathrm{m},\,1}, p_{\mathrm{m},\,1}, x_{\mathrm{m},\,2}, p_{\mathrm{m},\,2})$ needed to describe the quantum state of the mechanical systems. Combined with the fact that we drive the system on the blue side of the cavity resonance (and thus in an unstable regime) this means that we cannot actively stabilize the system and—depending on the driving strength and sideband resolution—no steady state may exist. To compensate for this we extend the setup by two additional heterodyne detectors, measuring $x_-$ and $p_+$ with outcomes $I'_\mp$ . The effective measurement strength of this stabilizing measurements with respect to the Bell measurement is set by the transmissivity $\upsilon$ of the beam-splitter in front of the heterodyne setup (see Fig. 15). Appropriate feedback of all measurement currents $I_\pm$, $I_\pm{}'$ (for simplicity labeled $I_i$, $i = 1, \ldots, 4$, below) to both mechanical systems finally allows us to stabilize them in an entangled state. Note that this setup effectively realizes two simultaneous Bell measurements of the pairs $(x_+, p_-)$ and $(x_-, p_+)$ with detection efficiencies $\upsilon$ and $1 - \upsilon$, respectively. In the rest of this section the two optomechanical systems are assumed to be identical and all detectors to have the same quantum detection efficiency $\eta$.

In Section 4.4.1 we show that in an adiabatic approximation the conditional state of the two mechanical oscillators $\rho^{(\mathrm{m})}$ in a rotating frame can be described by the SME (setting $\Delta_c = \omega_m$)

$$\begin{aligned} \mathrm{d}\rho_c^{(\mathrm{m})} &= \epsilon\frac{4g^2}{\kappa}\left\{\mathcal{D}[c_{\mathrm{m},1}]\rho_c^{(\mathrm{m})} + \mathcal{D}[c_{\mathrm{m},2}]\rho_c^{(\mathrm{m})}\right\}\mathrm{d}t \\ &+ \sum_{i=1}^{2}\left\{\gamma(\bar{n}+1)\mathcal{D}[c_{\mathrm{m},i}]\rho_c^{(\mathrm{m})} + \gamma\bar{n}\mathcal{D}[c_{\mathrm{m},i}^\dagger]\rho_c^{(\mathrm{m})}\right\}\mathrm{d}t \\ &+ \frac{2g^2}{\kappa}\sum_{i=1}^{4}\left\{\mathcal{D}[J_i]\rho_c^{(\mathrm{m})}\mathrm{d}t + \sqrt{\eta}\mathcal{H}[J_i]\rho_c^{(\mathrm{m})}\mathrm{d}W_i\right\}, \end{aligned} \tag{115}$$

where we set $(J_1, J_2) = \sqrt{\upsilon}\left(c_{\mathrm{m},+}^\dagger, \mathrm{i}c_{\mathrm{m},-}^\dagger\right)$, $(J_3, J_4) = \sqrt{1-\upsilon}(\mathrm{i}c_{\mathrm{m},+}^\dagger, c_{\mathrm{m},-}^\dagger)$, and $c_{\mathrm{m},\pm} = c_{\mathrm{m},1} \pm c_{\mathrm{m},2}$. The Wiener processes $W_i$ are uncorrelated and thus obey $\mathrm{d}W_i\mathrm{d}W_j = \delta_{ij}\mathrm{d}t$, and correspond to the photo-currents

$$I_i\mathrm{d}t = \sqrt{4g^2/\kappa}\langle J_i + J_i^\dagger\rangle_c\mathrm{d}t + \mathrm{d}W_i. \tag{116}$$

The final steady state of this protocol depends on the feedback operators $F_i = F_i^\dagger$ we apply. In analogy to the previous section we choose $(F_1, F_2) = \sqrt{\upsilon}\sigma(\mathrm{i}c_{\mathrm{m},+} - \mathrm{i}c_{\mathrm{m},+}^\dagger, c_{\mathrm{m},-} + c_{\mathrm{m},-}^\dagger)$ and $(F_3, F_4) = \sqrt{1-\upsilon}\sigma(c_{\mathrm{m},+} + c_{\mathrm{m},+}^\dagger, \mathrm{i}c_{\mathrm{m},-} - \mathrm{i}c_{\mathrm{m},-}^\dagger)$, which can realize independent displacements of all mechanical quadratures. This time we introduced an additional gain parameter $\sigma$ which we can vary in order to optimize the amount of entanglement in the resulting steady state. With these choices the FME for optomechanical entanglement swapping takes the form

$$\begin{aligned} \dot{\rho}_c^{(\mathrm{m})} &= -\mathrm{i}[H_{\mathrm{fb}}, \rho_c^{(\mathrm{m})}] + \epsilon\frac{4g^2}{\kappa}(\mathcal{D}[c_{\mathrm{m},1}] + \mathcal{D}[c_{\mathrm{m},2}])\rho_c^{(\mathrm{m})} \\ &+ \sum_{i=1}^{2}\left\{\gamma(\bar{n}+1)\mathcal{D}[c_{\mathrm{m},i}]\rho_c^{(\mathrm{m})} + \gamma\bar{n}\mathcal{D}[c_{\mathrm{m},i}^\dagger]\rho_c^{(\mathrm{m})}\right\} \\ &+ \frac{2g^2}{\kappa}\sum_{i=1}^{4}\left\{\mathcal{D}[J_i - \mathrm{i}F_i]\rho_c^{(\mathrm{m})} + \frac{1-\eta}{\eta}\mathcal{D}[F_i]\right\}, \end{aligned} \tag{117}$$

where the dynamics generated by the feedback is described by

$$H_{\mathrm{fb}} = \mathrm{i}\frac{2g^2}{\kappa}[(1+\sigma)\upsilon - 1](c_{\mathrm{m},+}^2 + c_{\mathrm{m},-}^2 - \mathrm{H.\,c.}). \tag{118}$$

We can now analyze the stability properties of the linear feedback system by evaluating the corresponding Routh–Hurwitz criterion. In the case of no stabilizing feedback ($\upsilon = 1$) we find that the admissible optomechanical coupling is limited from above by $4g^2/\kappa < 1/(1-\epsilon)$. This condition thus only allows appreciably values of $4g^2/\kappa$ for $\epsilon \approx 1$, and thus in the bad-cavity regime. The stabilization is caused by the counter-rotating beam-splitter terms $c_{\mathrm{m},\,i}c_{\mathrm{m},i}^{\dagger}$ + H. c. of the optomechanical Hamiltonian, which cool the mechanical systems. This cooling effect, however, diminishes the amount of generated steady-state entanglement. If we switch on the stabilizing feedback and thus choose $\upsilon < 1$, we can rewrite the Routh–Hurwitz criterion in the form

$$[3 + (4g^2/\kappa)^{-1} + \epsilon] > 4\upsilon > [(1-\epsilon) - (4g^2/\kappa)^{-1}]/\sigma$$

(where we assumed $\epsilon < 1$). These inequalities are tightest in the limit $\epsilon \to 0$, $g^2/\kappa \to \infty$ where we have $3 > 4\upsilon > 1/\sigma$. For the rest of this section we choose $\upsilon = 3/4$ which ensures stability of the feedback system for any values of $g^2/\kappa$ and $\epsilon$—and consequently the sideband resolution $\kappa/\omega_{\mathrm{m}}$—as long as the feedback gain fulfills $\sigma > 1/3$. In the stable regime and for $\epsilon = 0, \eta = 1$ we find a simple analytic expression for the steady-state logarithmic negativity,

$$E_{\mathcal{N}}^{\mathrm{ss}} = \ln\left(\frac{\frac{1}{2}C(\bar{n}+1)(3\sigma-1)(4\upsilon-1)+1}{C(\bar{n}+1)[3\sigma(\sigma-1)+1]+2\bar{n}+1}\right), \tag{119}$$

where we again introduced the cooperativity $C = 4g^2/(\bar{n}+1)\gamma\kappa$. Generally we can—for a set of parameters $(C, \bar{n}, \epsilon, \upsilon, \eta)$—maximize the entanglement $E_{\mathcal{N}}$ with respect to the feedback gain $\sigma$. In Fig. 17 we plot the resulting steady-state values in terms of logarithmic negativity $E_{\mathcal{N}}$ and EPR variance

$$\Delta_{\mathrm{EPR}} = \min_{\phi_1,\phi_2}\left([\Delta(x_{\mathrm{m},1}^{\phi_1} - x_{\mathrm{m},2}^{\phi_2})]^2 + [\Delta(p_{\mathrm{m},1}^{\phi_1} + p_{\mathrm{m},2}^{\phi_2})]^2\right), \tag{120}$$

where $x_{\mathrm{m},i}^{\phi} = (c_{\mathrm{m},i}\mathrm{e}^{-\mathrm{i}\phi} + c_{\mathrm{m},i}^{\dagger}\mathrm{e}^{+\mathrm{i}\phi})/\sqrt{2}$ and $p_{\mathrm{m},\,i}^{\phi} = x_{\mathrm{m},i}^{\phi+\pi/2}$ are rotated mechanical quadratures. [Again, the criterion for entanglement is $\Delta_{\mathrm{EPR}} < 2$ (Duan et al., 2000; Simon, 2000)]. In the first plot we assume a perfect detection efficiency $\eta = 1$ and consider different bath occupation numbers $\bar{n}$. We again see that there exists a critical cooperativity $C_{\mathrm{crit}}$ above which we are able to generate entanglement regardless of $\bar{n}$. From (119) we can deduce the expression

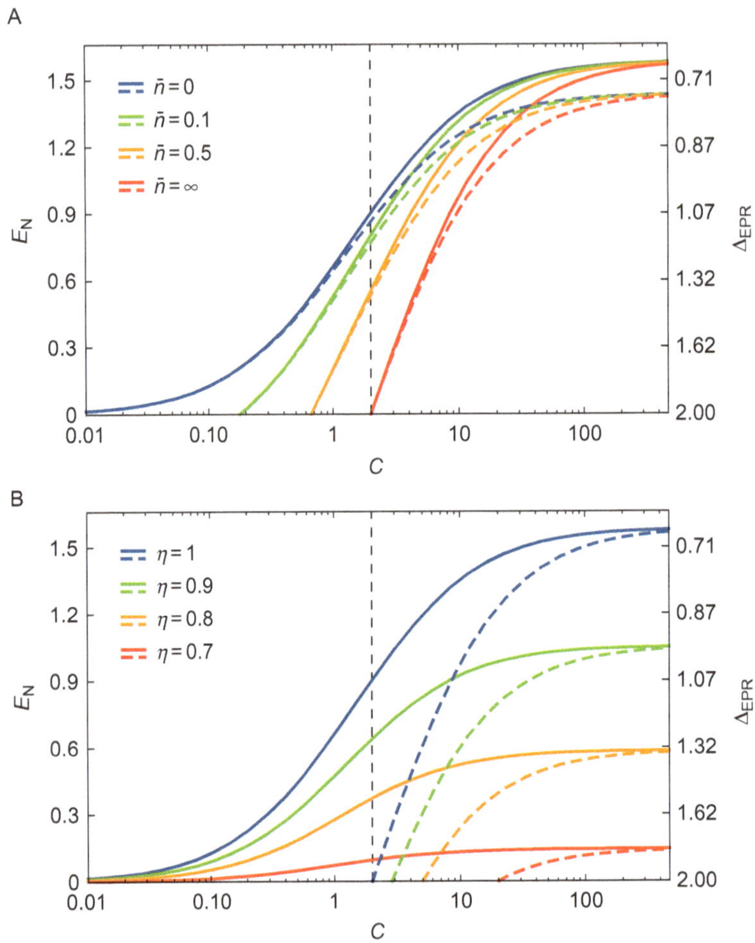


**Fig. 17** Two-mode mechanical steady-state entanglement in terms of $E_{\mathcal{N}}^{ss}$ and $\Delta_{EPR}$ against cooperativity $C$, maximized with respect to feedback gain $\sigma$. (A) Varying mechanical bath occupation $\bar{n}=0, 1/10, 1/2, \infty$ (represented by *different colors*) for unit detection efficiency $\eta=1$; the *solid* (*dashed*) *lines* represent a sideband resolution of $\kappa/\omega_m = 1$ (1/10). (B) Different detection efficiencies $\eta = 1, 9/10, 7/10, 6/10$ (represented by *different colors*) and $\kappa=\omega_m/10$; Here the *solid* (*dashed*) *lines* represent $\bar{n}=0$ $(\infty)$. The *black vertical line* shows the critical cooperativity $C_{crit}=2$.

$$C_{crit}(\upsilon,\sigma) = 4/[3\sigma(1+4\upsilon-2\sigma)-(1+4\upsilon)]. \tag{121}$$

(As is evident from the plot, the $C_{crit}$ is independent of $\epsilon$.) For the parameters used in the plot (taking into account the optimization with respect to $\sigma$) we find $C_{crit}=2$. Again, counter-rotating terms decrease entanglement but are strongly suppressed by the sideband resolution. In Fig. 17B we take into account losses and nonunit detection efficiency, $\eta < 1$, which drastically reduces the amount of achieved entanglement. As before we find a critical loss value $\eta_{\text{crit}}(\bar{n}, v)$ (for the parameters chosen in the plot slightly above 65%) below which entanglement creation is prohibited.

### 4.4.1 Detailed Derivation

In this section we derive the SME (115) and FME (117) which specify the generic case in Section 4.2.2 for the optomechanical implementation. Again, the goal is to derive equations for the mechanical systems, which we obtain by adiabatic elimination of the cavity and subsequent application of a RWA. As before, the Bell detection operates at the cavity frequency $\omega_c$ detuned by $\Delta_{\text{lo}} = \omega_{\text{lo}} - \omega_0$ with respect to the driving laser, and relations (104) still apply. Following the logic from Section 4.3.1 we thus define $s_i = \sqrt{\kappa}\, e^{i\Delta_{\text{lo}} t} c_{\text{m},i}$, which we use together with the generic entanglement SME (96) and FME (97) as the starting point for our approximations. Going to the rotating frame with $\sum_i (\omega_{\text{m}} c^\dagger_{\text{m},i} c_{\text{m},i} - \Delta_c c^\dagger_{\text{m},i} c_{\text{m},i})$ and applying the adiabatic approximation procedure from Doherty and Jacobs (1999) to (96) leaves us with

$$
\begin{aligned}
d\tilde{\rho}_c^{(\text{m})} = & -\sqrt{2} g^2 \sum_{i=1,2} \left[ \tilde{x}_{\text{m},i}, \tilde{y}_i \tilde{\rho}_c^{(\text{m})} - \tilde{\rho}_c^{(\text{m})} \tilde{y}_i^\dagger \right] dt \\
& + \sqrt{g^2 \kappa v / 2}\, \mathcal{H}[(\tilde{y}_+ dW_+ + i\tilde{y}_- dW_-) e^{i\Delta_{\text{lo}} t}] \tilde{\rho}_c^{(\text{m})} \\
& + \sqrt{g^2 \kappa (1 - v)/2}\, \mathcal{H}[(i\tilde{y}_+ dV_+ + \tilde{y}_- dV_-) e^{i\Delta_{\text{lo}} t}] \tilde{\rho}_c^{(\text{m})},
\end{aligned}
\tag{122}
$$

where $\tilde{y}_\pm = \tilde{y}_1 \pm \tilde{y}_2$ with $\tilde{y}_i = \eta_- c_{\text{m},i} e^{-i\omega_{\text{m}} t} + \eta_+ c^\dagger_{\text{m},i} e^{i\omega_{\text{m}} t}$. To apply a time coarse graining as in Section 4.3.1 we first change to the frame rotating with $\omega_{\text{m}}^{\text{eff}}$. If we choose $\Delta_{\text{lo}} = -\omega_{\text{m}}^{\text{eff}}$ we can drop the fast rotating terms in the first line of Eq. (122). For the measurement terms (second and third line) we again introduce $dW_\pm^{(0)}$ [as in Eq. (110)] and average over any sideband modes. We end up with

$$
\begin{aligned}
d\tilde{\rho}_c^{(\text{m})} = & \gamma_- \{\mathcal{D}[c_{\text{m},1}] + \mathcal{D}[c_{\text{m},2}]\} \tilde{\rho}_c^{(\text{m})} dt \\
& + \gamma_+ \{\mathcal{D}[c^\dagger_{\text{m},1}] + \mathcal{D}[c^\dagger_{\text{m},2}]\} \tilde{\rho}_c^{(\text{m})} dt \\
& + \sqrt{g^2 \kappa v / 2}\, \eta_+ \mathcal{H}[c^\dagger_{\text{m},+} dW_+^{(0)} + i c^\dagger_{\text{m},-} dW_-^{(0)}] \tilde{\rho}_c^{(\text{m})} \\
& + \sqrt{g^2 \kappa (1 - v)/2}\, \eta_+ \mathcal{H}[i c^\dagger_{\text{m},+} dV_+^{(0)} + c^\dagger_{\text{m},-} dV_-^{(0)}] \tilde{\rho}_c^{(\text{m})}
\end{aligned}
\tag{123}
$$

where we introduced $c_{\mathrm{m},\pm} = c_{\mathrm{m},1} \pm c_{\mathrm{m},2}$ and we added mechanical decoherence terms. We apply the same coarse graining to the measurement currents (95) and find, by using $s_i = \sqrt{\kappa}\mathrm{e}^{\mathrm{i}\Delta_{\mathrm{lo}}t}c_{\mathrm{m},i}$ and $\langle c_{\mathrm{m},i}\rangle_c = -\mathrm{i}g\langle y_i\rangle_c$,

$$I_+^{(0)}\mathrm{d}t = -\mathrm{i}\sqrt{g^2\kappa\upsilon/2}\,\langle\eta_+c_{\mathrm{m},+}^\dagger - \mathrm{H.c.}\rangle_c + \mathrm{d}W_+^{(0)}, \tag{124a}$$

$$I_-^{(0)}\mathrm{d}t = \sqrt{g^2\kappa\upsilon/2}\,\langle\eta_+c_{\mathrm{m},-}^\dagger + \mathrm{H.c.}\rangle_c + \mathrm{d}W_-^{(0)}, \tag{124b}$$

$$I_+'^{(0)}\mathrm{d}t = \sqrt{g^2\kappa(1-\upsilon)/2}\,\langle\eta_+c_{\mathrm{m},+}^\dagger + \mathrm{H.c.}\rangle_c + \mathrm{d}V_+^{(0)}, \tag{124c}$$

$$I_-'^{(0)}\mathrm{d}t = -\mathrm{i}\sqrt{g^2\kappa(1-\upsilon)/2}\,\langle\eta_+c_{\mathrm{m},-}^\dagger - \mathrm{H.c.}\rangle_c + \mathrm{d}V_-^{(0)}. \tag{124d}$$

One can clearly see that Eqs. (123) and (124) are equivalent to SME (96) and measurement currents (95) if we set $s_\pm = \sqrt{g^2\kappa}\eta_+c_{\mathrm{m},\pm}^\dagger$ and add appropriate decoherence terms. We can therefore use FME (97) directly, and together with the choice $(F_1, F_2) = \sqrt{\upsilon}\sigma(\mathrm{i}c_{\mathrm{m},+} - \mathrm{i}c_{\mathrm{m},+}^\dagger, c_{\mathrm{m},-} + c_{\mathrm{m},-}^\dagger)$, $(F_3, F_4) = \sqrt{1-\upsilon}(c_{\mathrm{m},+} + c_{\mathrm{m},+}^\dagger, \mathrm{i}c_{\mathrm{m},-} - \mathrm{i}c_{\mathrm{m},-}^\dagger)$ we find Eq. (117).

# 5. EXPERIMENTAL IMPLEMENTATION OF A KALMAN FILTER

In the previous chapters we have extensively discussed different stochastic master equations that describe the evolution of conditional quantum states, conditioned on the outcomes of different measurement setups. This section will be devoted to illustrating how conditional quantum states can be evaluated from experimental data.

## 5.1 General Considerations

The generic SME for homodyne detection (32) of a cavity's output field is typically written in a form such that the evolution is driven by a Wiener process $W(t)$. The connection to measured data in an experiment is given by the identity (34), which shows that the Wiener increment $\mathrm{d}W(t)$ is the difference between the expected measurement outcome [according to the conditional state $\rho_c(t)$] and the actual measurement result. In filtering theory $W(t)$ is called *innovation process* (or innovation sequence) (Heijden et al., 2005; Stengel, 1994). In order to evaluate the state $\rho_c(t)$ from a measurement trajectory $\{I(s) : s \leq t\}$ we thus have to integrate a nonlinear equation of the form (choosing to measure the amplitude quadrature and neglecting photon losses)

$$d\rho_c(t) = \mathcal{L}\rho_c(t)dt + \sqrt{\kappa}[I(t) - \sqrt{\kappa}\langle c_c(t) + c_c^\dagger(t)\rangle_c]\mathcal{H}[c_c]\rho_c(t)dt. \quad (125)$$

We show in Appendix B.3 that in the case of Gaussian systems and a linear measurement (such as homodyne detection), this equation can be mapped to the (stochastic) differential equations (40). In classical filtering theory equations (40) are known as the *Kalman filter* (Kalman, 1960; Kalman and Bucy, 1961), which presents the optimal estimator under a minimum mean square error criterion (Stengel, 1994; Xiong, 2008). Let us briefly discuss the interpretation of the classical Kalman filter in its continuous form. If we describe by $\boldsymbol{X}$ a state of a Gaussian system in a classical state space (e.g., the phase space of a harmonic oscillator), the Kalman filter ascribes to it an optimal estimate $\hat{\boldsymbol{X}}$ and an estimation-error covariance matrix $\hat{\boldsymbol{\Sigma}} = \mathbb{E}[(\boldsymbol{X} - \hat{\boldsymbol{X}})(\boldsymbol{X} - \hat{\boldsymbol{X}})^{\mathrm{T}}]$, where $\mathbb{E}[\cdot]$ denotes the expectation value with respect to the initial probability distribution describing the system. The estimate is chosen to minimize the mean square error $\mathbb{E}[\|\boldsymbol{X} - \hat{\boldsymbol{X}}\|^2] = \mathrm{tr}(\hat{\boldsymbol{\Sigma}})$. The true state $\boldsymbol{X}$ (a point in the state space) thus follows a Gaussian distribution with a covariance given by $\hat{\boldsymbol{\Sigma}}$ centered at $\hat{\boldsymbol{X}}$. The Kalman filter equations describe how the conditional state is iteratively updated (see Fig. 18): first, the estimate $\hat{\boldsymbol{X}}$ and the covariance $\hat{\boldsymbol{\Sigma}}$ are propagated for an infinitesimal time interval $dt$ according to the system dynamics. Second, the measurement outcome is incorporated as a Bayesian update that corrects the value of the estimate $\hat{\boldsymbol{X}}$ and contracts the covariance ellipse. The updated values are again propagated by $dt$ and the procedure is repeated. Note that of the Kalman equations only the one for $\hat{\boldsymbol{X}}(t)$ actually depends on the measurement trajectory $I(t)$, while the evolution of $\hat{\boldsymbol{\Sigma}}(t)$ solely depends on the linear dynamics of

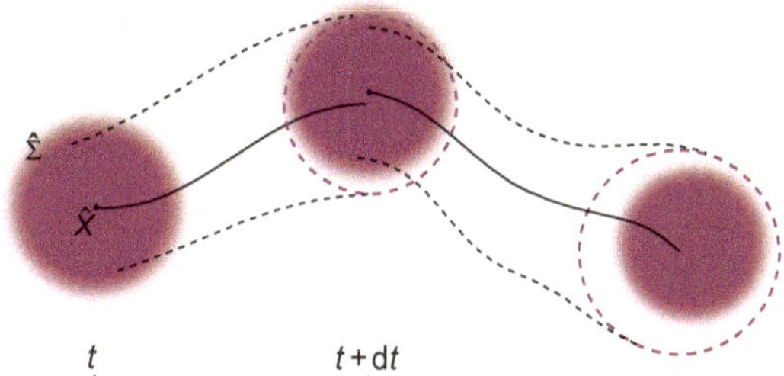


**Fig. 18** Schematic state-space representation of the Kalman evolution. The conditional state is depicted by a circle and evolves in time according to the system dynamics. After a time $dt$ a Bayesian update is applied based on the measurement outcome to find the new conditional state.

the underlying system. This has an important consequence: given an initial state and a system's state-space model (i.e., the matrices **F**, **H**, **G**, **N**, and **M**; see Appendix B.3.1) all quantities that depend on $\hat{\boldsymbol{\Sigma}}(t)$ only [e.g., the mean occupation number or the logarithmic negativity] are predetermined for the complete time evolution and are independent of the observations made in the experiment. It is therefore crucial to build an accurate model of the experimental setup and verify its accuracy against the measured data. This can be achieved by testing that the (experimentally obtained) innovation process $W(t)$ is indeed a Wiener process, i.e., that for $t > s$ it obeys $W(t) - W(s) \sim \mathcal{N}(0, t-s)$, where $\mathcal{N}(\mu, \sigma^2)$ is the normal distribution with a mean $\mu$ and a variance $\sigma^2$ (Kloeden, 1994). How this can be implemented in practice will be discussed in the next section.

## 5.2 Experimental Implementation

We now discuss the implementation of the Kalman filter for an optomechanical experiment, demonstrating optimal state estimation of the optomechanical quantum state. Although the system is currently operated in a classical regime, this experiment constitutes an important next step toward real-time quantum control of optomechanical systems. The experimental data presented in this section were taken by Jason Hölscher-Obermaier and Witlef Wieczorek in Markus Aspelmeyer's laboratories at the University of Vienna (Wieczorek et al., 2015).

### *5.2.1 The Experimental Setup*

In this setup (as schematically depicted in Fig. 19) we drive two independent optical modes of an optomechanical cavity with two laser beams, one of which (denoted by a subscript r) is resonant with the cavity (i.e., $\Delta_{\mathrm{r}} = 0$), while the other one (subscript d) is tuned to the lower mechanical sideband at $\Delta_{\mathrm{d}} = -\omega_{\mathrm{m}}$. The resonant beam is used to monitor the cavity length to lock the laser frequency with respect to the cavity resonance; the detuned beam can be used as a control beam, for example, to cool the mechanical motion. The cavity is double-sided, with a total decay rate $\kappa = \kappa_{\mathrm{in}} + \overline{\kappa}$ (as discussed in Section 2.1.2), which is the same for both optical modes. The fields reflected from the cavity are independently measured by two homodyne detection setups, yielding measurements of the generalized quadratures $y_{\mathrm{r}}$ and $y_{\mathrm{d}}$. Both beams carry classical amplitude and phase noise, and suffer from passive beam-splitter losses. To incorporate the classical noise into our description, we extend the standard Langevin equations to (Abdi et al., 2011; Ghobadi et al., 2011a; Giovannetti and Vitali, 2001; Rabl et al., 2009)

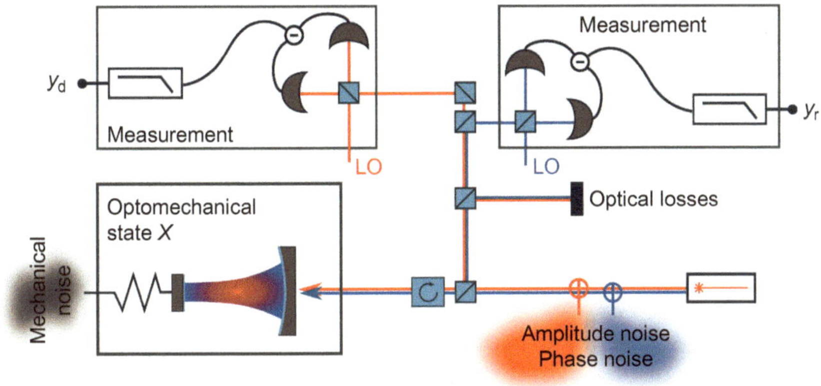


**Fig. 19** Schematic of experimental setup: the optomechanical cavity is driven by two laser fields, which are both subject to classical amplitude and phase noise. The output light is detected by two independent homodyne detectors, yielding measurement processes $y_r$ and $y_d$.

$$\dot{x}_m = \omega_m p_m, \tag{126a}$$

$$\dot{p}_m = -\omega_m x_m - \gamma_m p_m + \sum_{i=r,d} 2g_i[\cos\theta_i x_i - \sin\theta_i p_i] + f, \tag{126b}$$

$$\begin{aligned}\dot{x}_i &= -\frac{\kappa}{2}x_i + \Delta_i p_i + g_i \sin(\theta_i)x_m + \sqrt{\kappa_{in}}x_{i,1}^{in} \\ &\quad + \sqrt{\bar{\kappa}}x_{i,2}^{in} + \sqrt{2\kappa_{in}}\,\delta\beta_i + |\alpha_{0,i}|\sin\theta_i\,\dot{\phi}_i,\end{aligned} \tag{126c}$$

$$\begin{aligned}\dot{p}_i &= -\frac{\kappa}{2}p_i - \Delta_i x_i + 2g_i\cos(\theta_i)x_m + \sqrt{\kappa_{in}}p_{i,1}^{in} \\ &\quad + \sqrt{\bar{\kappa}}p_{i,2}^{in} + |\alpha_{0,i}|\cos\theta_i\,\dot{\phi}_i,\end{aligned} \tag{126d}$$

where $x_i$, $p_i$ with $[x_i, p_j] = \mathrm{i}\delta_{ij}$ (for $i, j \in \{\mathrm{r,d}\}$) describe the amplitude and phase quadrature of the two cavity modes. Vacuum noise entering through both cavity mirrors is denoted by $x_i^{in}$, $y_i^{in}$, with variances $\langle x_i^{in}(t)x_j^{in}(s)\rangle = \langle p_i^{in}(t)p_j^{in}(s)\rangle = \frac{1}{2}\delta_{ij}\delta(t-s)$ $(i, j \in \{1, 2\})$. The angles $\theta_i$ are determined by $\theta_i = \arctan(\Delta_i/\kappa)$. Terms proportional to $\delta\beta_i$ and $\dot{\phi}_i$ describe classical amplitude and phase noise of the driving lasers. The quadratures detected by the homodyne detectors are given by cavity input–output relations, yielding

$$\begin{aligned}y_i &= \sqrt{1-\eta}\left(\sqrt{\kappa_{in}}\,x_i + x_i^{in} + \delta\beta_i\right)\cos(\varphi_i) \\ &\quad + \sqrt{1-\eta}\left(\sqrt{\kappa_{in}}\,p_i + p_i^{in}\right)\sin(\varphi_i) + \sqrt{\eta}y_i^{in},\end{aligned} \tag{127}$$

where $\varphi_i$ denotes the local-oscillator phases. Additionally we introduced the overall efficiency $\eta$ (including losses and detector efficiency) and an independent vacuum noise process $y_i^{\rm in}$. The classical noise processes $\delta\beta_i$ and $\dot{\phi}_i$ are modeled as Gaussian colored noise, whose spectra has to be determined from experimental data.

Using the Langevin equations (126) as a basis, we can build a sophisticated state-space model of the complete experimental setup, see Wieczorek et al. (2015) for details. The resulting Kalman filter can then be used to estimate the state of the mechanical and the optical subsystems, as well as their correlations.

### 5.2.2 Results

We apply the Kalman filter in postprocessing to the recorded homodyne data to obtain estimates of the full optomechanical state vector $\boldsymbol{X} = (x_{\rm m}, p_{\rm m}, x_{\rm r}, p_{\rm r}, x_{\rm d}, p_{\rm d})^{\rm T}$, together with the corresponding estimation error. Short snippets of the resulting trajectories are shown in Fig. 20, for two different values of the coupling strength $g_{\rm d}$, while $g_{\rm r}$ is always fixed to $g_{\rm r} = 0.2\kappa$. The left column shows results from the weak-coupling regime ($g_{\rm d} = 0.2\kappa$), nicely displaying the harmonic mechanical oscillation in the mechanical as well as in the optical quadratures—the latter being modulated via the

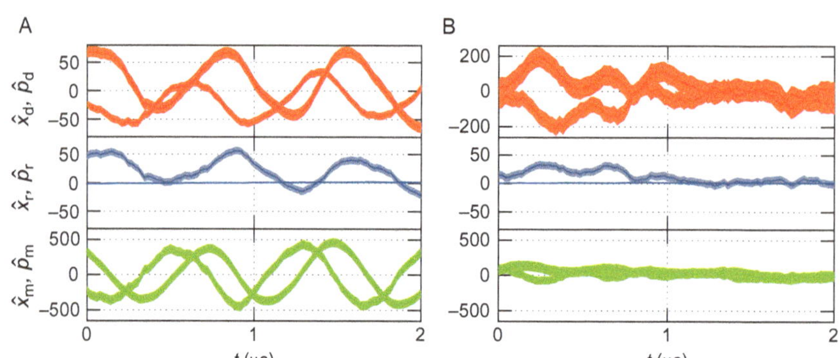


**Fig. 20** Estimated optomechanical quadratures in the (A) weak ($g_{\rm d} = 0.2\kappa$) and (B) strong ($g_{\rm d} = 1.68\kappa$) coupling regime. The *solid lines* show the amplitude (position) quadratures $\hat{x}_{\rm d}$, $\hat{x}_{\rm r}$ ($\hat{x}_{\rm m}$), the *dashed lines* the phase (momentum) quadratures $\hat{p}_{\rm d}$, $\hat{p}_{\rm r}$ ($\hat{p}_{\rm m}$). The *shaded areas* correspond to two times the estimation error (i.e., the 95% confidence interval). The experiment was performed at room temperature with a micromechanical oscillator of $\omega_{\rm m} = 2\pi \times 1.278$MHz, $\gamma_{\rm m} = 2\pi \times 265$Hz, and optomechanical parameters $\kappa = 0.34\,\omega_{\rm m}$, $g_0 = 2\pi \times 7.7$Hz. For both coupling strengths $g_{\rm d}$ of the detuned beam we use $g_{\rm r} = 0.2\,\kappa$, $\varphi_{\rm d} \approx 0$, and $\varphi_{\rm r} = \pi/2$. *Adapted from Wieczorek, W., Hofer, S.G., Hoelscher-Obermaier, J., Riedinger, R., Hammerer, K., Aspelmeyer, M., 2015. Optimal state estimation for cavity optomechanical systems. Phys. Rev. Lett. 114 (22), 223601. doi:10.1103/PhysRevLett.114. 223601.*

optomechanical interaction. In the right column the system is operated in the strong-coupling regime ($g_d = 1.68\kappa$); this leads to sideband cooling of the mechanical mode and a strong suppression of the mechanical oscillation. In both cases it is clear to see that only the phase quadrature of the resonant cavity couples to the mechanical motion, while the amplitude quadrature contains noise. Both quadratures of the detuned beam, on the other hand, are modulated by the mechanics.

Fig. 21 shows a phase space representation of the estimated optomechanical quadratures, similar to the schematic depiction in Fig. 6. In the weak-coupling regime the mechanical system—rotating in phase space—traces out a symmetric, thermal distribution (Fig. 21A). The black dashed

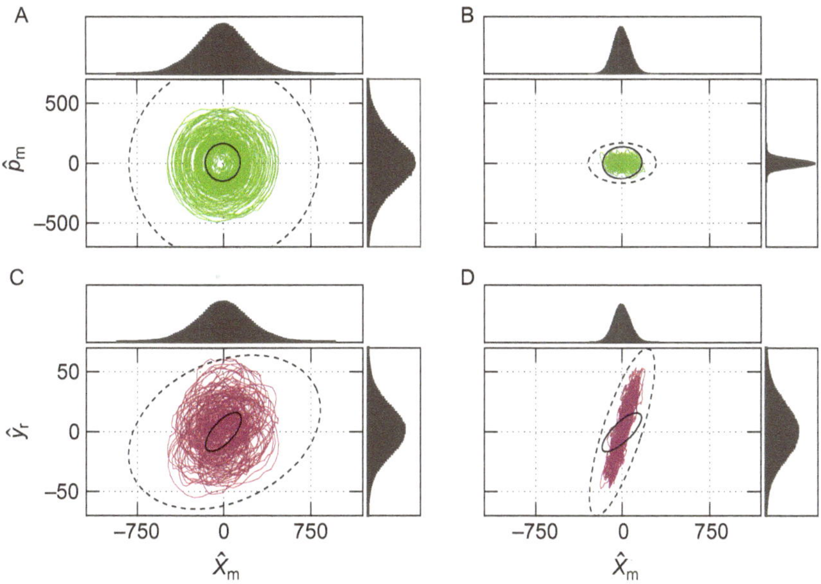


**Fig. 21** Phase space picture of the estimated quadratures. The *left column* [(A), (C)] shows the weak-coupling regime ($g_d = 0.2\kappa$), the right column [(B), (D)] the strong-coupling regime ($g_d = 1.68\kappa$). (A) and (B) depict the phase space of the mechanical oscillator, (C) and (D) depict correlations between the phase quadrature of the resonantly driven cavity mode and the mechanical position. Each trajectory covers 100 μs. The *dashed ellipses* indicate the 95% confidence region (calculated from the unconditional covariance matrix $\boldsymbol{\Sigma}$), the *solid black ellipses* show the corresponding extent of the conditional state (as determined by $\hat{\boldsymbol{\Sigma}}$). Histograms over the estimated quadratures are shown as sidepanels (averaged over 100 ms). *Adapted from Wieczorek, W., Hofer, S.G., Hoelscher-Obermaier, J., Riedinger, R., Hammerer, K., Aspelmeyer, M., 2015. Optimal state estimation for cavity optomechanical systems. Phys. Rev. Lett. 114 (22), 223601. doi:10.1103/PhysRevLett.114. 223601.*

line shows the uncertainty ellipse of the unconditional state, which is here chosen to be the 95% confidence region. The area enclosed by it is proportional to the uncertainty product $\det(\Sigma_{\mathrm{m}})$ of the mechanical state (denoting by $\Sigma_{\mathrm{m}}$ the mechanical covariance matrix), and thus inversely proportional to the state's purity

$$\mathrm{tr}(\rho_{\mathrm{m}}^2) = \frac{1}{\sqrt{\det(\Sigma_{\mathrm{m}})}}. \tag{128}$$

Comparing it to the uncertainty ellipse of the conditional state (black solid line), we note a clear reduction in uncertainty due to the conditioning procedure. This effect—the conditional reduction of the occupation number—is sometimes referred to as *cooling by measurement* (Vanner et al., 2013). In our case it leads to a reduction of the occupation by a factor of 27. In the strong-coupling regime (Fig. 21B) one can observe the effects of sideband cooling, which heavily reduces the variance of the unconditional mechanical state. In addition, the strong optomechanical interaction introduces a notable asymmetry, which shows that the state is not thermal anymore and equipartition is violated. The marginals drawn along the phase space axes nicely demonstrate the Gaussian distribution of the optomechanical state. Fig. 21C and D shows correlations between the mechanical position and the phase quadrature of the resonant beam. Analogous to the mechanical phase space, the uncertainty reduction due to sideband cooling is clearly visible in the shown correlations.

Finally we analyze the obtained innovation sequence to verify the consistency of our assumed model with the measured data. As discussed in Section 5.1, in the ideal situation the innovation sequence should be white, and should therefore posses a flat noise power spectrum with a certain variance. Fig. 22A shows the noise power spectrum of the normalized innovation sequence for both homodyne measurements in the weak and the strong-coupling regime. The dashed lines indicate the confidence interval that 95% of the datapoints should fall into. The fractions of datapoints effectively contained in this region are given in the plots. We can see that we have a fairly good agreement with the hypothetical values, especially in the region around the mechanical frequency (indicated by the vertical gray-shaded areas). The deviations from the theoretical flat spectrum are mainly due to low-frequency noise at around 200 kHz, caused by the spectral response of the photodetectors, and sharp piezo-resonances between 2 and 5 MHz. Neither of these features is taken into account in the state-space model. In Fig. 22B we compare the empirical cumulative distribution function (CDF) (Bar-Shalom et al., 2001, p. 33) and the corresponding probability

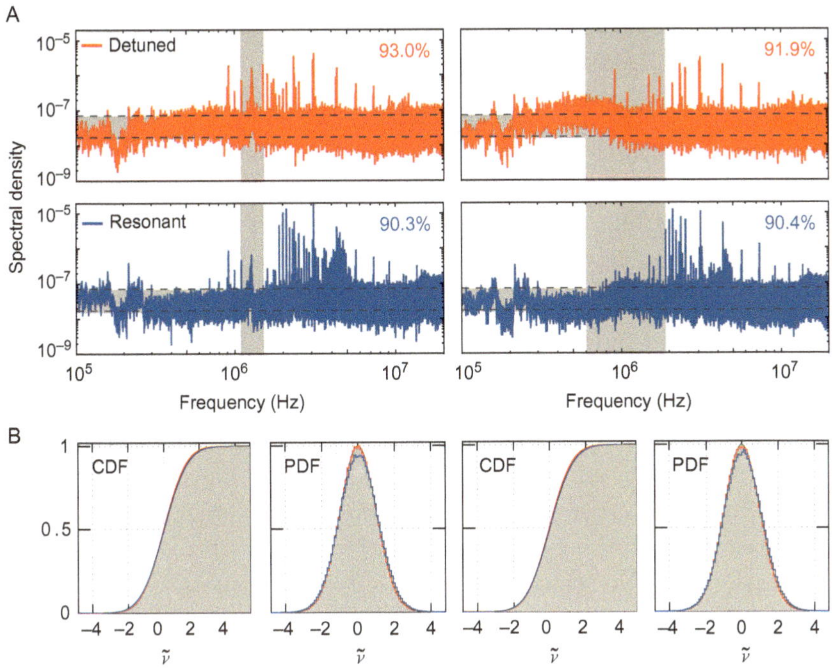


**Fig. 22** (A) Noise power spectrum of the normalized innovation sequence for the detuned beam (*upper row*) and the resonant beam (*lower row*). The *left column* shows the results for weak coupling, the *right column* for strong coupling. The 95% confidence region is shown by the *dashed horizontal lines*. The effective percentages of points in the confidence region are given in the plots. The vertical *gray-shaded area* marks the fundamental mechanical mode. (B) Comparison of the empirical cumulative distribution function (CDF) and probability density function (PDF) (*red* and *blue lines*) with the theoretical form (*gray-shaded areas*) for the weak (*left*) and strong (*right*) coupling regime. *Adapted from Wieczorek, W., Hofer, S.G., Hoelscher-Obermaier, J., Riedinger, R., Hammerer, K., Aspelmeyer, M., 2015. Optimal state estimation for cavity optomechanical systems. Phys. Rev. Lett. 114 (22), 223601. doi:10.1103/PhysRevLett.114. 223601.*

density function (PDF) to the hypothetical Gaussian distribution. We again find good overall agreement of our model with the measured data.

Although this experiment is conducted in a classical regime, it demonstrates the ability to evaluate the conditional quantum state of optomechanical systems via Kalman filtering. The estimation procedure was applied in postprocessing, but can in principle be operated in real time. Combining the presented techniques with continuous measurements in the strong cooperativity regime (as demonstrated in Wilson et al., 2015) enables measurement-based feedback cooling of the mechanical motion into the ground state, as well as the realization of the more sophisticated optomechanical quantum control protocols laid out in this review.

# 6. CONCLUSION

In this chapter we reviewed in detail the generation of optomechanical entanglement in continuous-wave and pulsed regimes, and explored various ways to make use of this entanglement to implement control protocols for (opto-)mechanical quantum systems. This led us to discuss both pulsed and time-continuous teleportation protocols allowing us to transfer the state of the light field onto the mechanical system, a time-continuous entanglement swapping protocol to prepare two mechanical oscillators in an bipartite entangled state, and optimal optomechanical feedback cooling. Additionally we presented a first step toward an experimental implementation of these schemes in the form of a Kalman filter for a real-world optomechanical system.

To describe the presented protocols we heavily draw from the well-developed and elegant frameworks of quantum stochastic calculus, continuous measurement theory, and quantum control theory, which—in our eyes—are not widely enough known in the optomechanics community. It was thus our goal to make our text accessible and enjoyable to researchers from both communities, and ultimately bring them closer together.

# APPENDICES

# APPENDIX A. QUANTUM STOCHASTIC CALCULUS

## A.1 Quantum-Optical Model

In this section we briefly review the quantum-optical model of an open quantum system coupling to the electromagnetic field, following the presentation in Gardiner and Collett (1985). A detailed discussion of this topic can for example be found in Gardiner and Zoller (2004), Zoller and Gardiner (1997), and Barchielli (2006).

We consider a quantum system $S$ described by some Hamiltonian $H_{\rm sys}$ which couples to a 1d electromagnetic field $A$, consisting of a continuum of harmonic oscillators. The environment (or bath) is described in the Schrödinger picture by the creation and annihilation operators $a_0^\dagger(\omega)$ and $a_0(\omega)$, which fulfill the canonical commutation relations

$$[a_0(\omega), a_0^\dagger(\omega')] = \delta(\omega - \omega'). \tag{A.1}$$

The bath is thus described by the free Hamiltonian

$$H_{\text{bath}} = \int_{-\infty}^{\infty} \mathrm{d}\omega\omega a_0^\dagger(\omega) a_0(\omega), \tag{A.2}$$

and in a RWA we assume the coupling to the system to be described by

$$H_{\text{int}} = \int_{-\infty}^{\infty} \mathrm{d}\omega k(\omega)[s a_0^\dagger(\omega) + s^\dagger a_0(\omega)] \tag{A.3}$$

with a frequency-dependent coupling constant $k(\omega)$. The system operator $s$ we often call *jump operator*. Note that both operators $s$ and $a(t)$ have a dimension $[s] = [a(t)] = \sqrt{\text{Hz}}$. Typically, the system will evolve on a time scale $1/\omega_0$ set by the large optical frequency $\omega_0$ (e.g., the cavity resonance frequency), which we eliminate by moving to the corresponding rotating frame; this amounts to the replacement $s \rightarrow s\mathrm{e}^{-\mathrm{i}\omega_0 t}$. Similarly, we move into an interaction picture with the free bath Hamiltonian $H_{\text{bath}}$ generated by the unitary

$$\Theta(t) = \exp(-\mathrm{i}H_{\text{bath}}t), \tag{A.4}$$

and thus substitute $a_0(\omega) \rightarrow a_0(\omega)\mathrm{e}^{-\mathrm{i}\omega t}$. In a Markov approximation, which consists of taking the limit of a flat coupling, i.e., $k(\omega) \equiv 1$, we can then write

$$H_{\text{int}} = s a_{\text{in}}^\dagger(t) + s^\dagger a_{\text{in}}(t) \tag{A.5}$$

where we introduced the field operators

$$a_{\text{in}}(t) = \frac{1}{\sqrt{2\pi}} \int_{-\infty}^{\infty} \mathrm{d}\omega a_0(\omega) \mathrm{e}^{-\mathrm{i}(\omega-\omega_0)t}. \tag{A.6}$$

Due to the Markov approximation they obey the singular commutation relation

$$[a_{\text{in}}(t), a_{\text{in}}^\dagger(t')] = \delta(t-t'). \tag{A.7}$$

The system dynamics are then typically described in terms of QLEs in the Stratonovich form for an arbitrary system operator $X$

$$\begin{aligned} (\mathbf{S}) \quad \dot{X} &= -\mathrm{i}[X, H_{\text{sys}}] - \frac{1}{2}([X, s^\dagger]s - s^\dagger[X, s]) \\ &\quad - [X, s^\dagger] a_{\text{in}}(t) + [X, s] a_{\text{in}}^\dagger(t) \end{aligned} \tag{A.8}$$

The field operator $a_{\text{in}}$ acts as a driving term for the equations of motion of the system $S$ and thus be regarded as an input channel. Note that $a_{\text{in}}$ is defined in terms of Schrödinger bath operators at the initial time. Its time argument thus does not denote a time dependence in the sense of the Heisenberg picture, but should rather be interpreted as the time the initial input interacts with the system. Depending on the state of the bath at this initial time, this input can either be a noise term (e.g., for a thermal state of the bath), a classical drive (for a coherent state), or something more general such as a single-photon input. In order to interpret $a_{\text{in}}$ as quantum white noise, we need to assume it to be $\delta$-correlated, e.g.,

$$\langle a_{\text{in}}^{\dagger}(t)a_{\text{in}}(t')\rangle := \text{tr}(a_{\text{in}}^{\dagger}(t)a_{\text{in}}(t')\sigma_0) = N\delta(t-t'), \tag{A.9a}$$

$$\langle a_{\text{in}}(t)a_{\text{in}}^{\dagger}(t')\rangle := \text{tr}(a_{\text{in}}(t)a_{\text{in}}^{\dagger}(t')\sigma_0) = (N+1)\delta(t-t'), \tag{A.9b}$$

where $\sigma_0$ is the initial state of the system + bath (here in a thermal state with mean occupation number $N$). This does not correspond to a thermal state where the photon number follows the Bose–Einstein distribution $n_{\text{B}}(\omega)$, but rather an ensemble with a constant number of photons per unit bandwidth. In this sense quantum white noise is an approximation that is valid in a certain (small) bandwidth only. In principle quantum stochastic differential equation (QSDE) in Stratonovich form can also be formulated for nonwhite noise terms, as is the case for mechanical oscillators (see Section 2.3.1).

Similar to the input field $a_{\text{in}}$ we can define the output field $a_{\text{out}}$ of the system $S$, which is related to the input field via the *input–output* relations (Gardiner and Collett, 1985)

$$a_{\text{out}}(t) = s(t) + a_{\text{in}}(t). \tag{A.10}$$

Note that $s(t)$ denotes the Heisenberg operator given by the corresponding QLE, see Appendix A.3. The outgoing field at time $t$ thus consists of a reflected part of the input field and a contribution from the internal system modes (which in turn contain contributions from the input field at earlier times).

## A.2 The Hudson–Parthasarathy Equation

In the white noise limit of the Markov approximation (including an appropriate state of the bath) and in the interaction picture introduced above, one can develop a noncommutative quantum analog of the classical stochastic Itō-calculus to describe the joint unitary evolution of the system $S$ and the white noise field $A$ (Hudson and Parthasarathy, 1984). This leads to the so-called *Hudson–Parthasarathy equation*, an Itō QSDE for the evolution operator $U$ given by

$$\mathrm{d}U(t) = \left[-\mathrm{i}\left(H_{\mathrm{sys}} - \mathrm{i}\frac{1}{2}s^{\dagger}s\right)\mathrm{d}t + s\mathrm{d}A^{\dagger}(t) - s^{\dagger}\mathrm{d}A(t)\right]U(t) \tag{A.11}$$

with the initial condition $U(0) = \mathbb{1}$. Here we introduced the *quantum Wiener process* $A(t)$ (analog to a classical complex Wiener process) and its (forward-pointing) Itō-increments

$$\mathrm{d}A(t) = A(t+\mathrm{d}t) - A(t), \tag{A.12a}$$

$$\mathrm{d}A^{\dagger}(t) = A^{\dagger}(t+\mathrm{d}t) - A^{\dagger}(t). \tag{A.12b}$$

Formally the quantum Wiener process (A.12) is related to the field operators from the previous section by

$$A(t) = \int_0^t \mathrm{d}s\, a_{\mathrm{in}}(s), \quad A^{\dagger}(t) = \int_0^t \mathrm{d}s\, a_{\mathrm{in}}^{\dagger}(s). \tag{A.13}$$

For the electromagnetic field initially in the vacuum state the Itō-increments obey the following multiplication rules:

| $\times$ | $\mathrm{d}A$ | $\mathrm{d}A^{\dagger}$ | $\mathrm{d}t$ |
|---|---|---|---|
| $\mathrm{d}A$ | 0 | $\mathrm{d}t$ | 0 |
| $\mathrm{d}A^{\dagger}$ | 0 | 0 | 0 |
| $\mathrm{d}t$ | 0 | 0 | 0 |

(A.14)

More generally two quantum stochastic processes $X(t)$, $Y(t)$ obey the Itō product rule

$$\mathrm{d}[X(t)Y(t)] = [\mathrm{d}X(t)]Y(t) + X(t)\mathrm{d}Y(t) + \mathrm{d}X(t)\mathrm{d}Y(t), \tag{A.15}$$

again with $\mathrm{d}X(t) := X(t+\mathrm{d}t) - X(t)$, etc. The standard chain rule is modified in a similar way: for a differentiable function $f$, we have

$$\mathrm{d}f(X(t)) = f'(X(t))\mathrm{d}X(t) + \frac{1}{2}f''(X(t))\mathrm{d}X(t)^2, \tag{A.16}$$

which in particular leads to

$$f(X(t+\mathrm{d}t)) = f(X(t)) + f'(X(t))\mathrm{d}X(t) + \frac{1}{2}f''(X(t))\mathrm{d}X(t)^2. \tag{A.17}$$

Due to the forward-pointing definition of the Itō-increments non-anticipating functions such as $X$ commute with the increments at equal times, i.e.,

$$[X(t), \mathrm{d}A(t)] = [X(t), \mathrm{d}A^\dagger(t)] = 0. \tag{A.18}$$

In particular this property also holds for the evolution operator $U$. It will be convenient to introduce the non-Hermitian, effective Hamiltonian

$$H_{\text{eff}} = H_{\text{sys}} - \mathrm{i}\frac{1}{2}s^\dagger s, \tag{A.19}$$

which contains the so-called Itō-correction $-\mathrm{i}s^\dagger s/2$.

Using Itō-calculus one can show that $\mathrm{d}[U(t)U^\dagger(t)] = \mathrm{d}[U^\dagger(t)U(t)] = 0$, i.e., $U(t)$ is unitary for all $t$ as expected for the solution of a Schrödinger equation. $U(t)$ does not form a group, however, but rather fulfills the *cocycle* property (Barchielli, 2006)

$$U(t+t') = \Theta^\dagger(t')U(t)\Theta(t')U(t') \quad \forall t, t' > 0 \tag{A.20}$$

with respect to the free evolution of the electromagnetic bath $\Theta$. Moving out of the interaction picture we can define

$$V(t) = \begin{cases} \Theta(t)U(t) & t \geq 0 \\ U^\dagger(|t|)\Theta(t) & t \leq 0 \end{cases}, \tag{A.21}$$

which form a unitary group, and thus $V^\dagger(t) = V(-t)$ and $V(t+t') = V(t)V(t')$, etc.

## A.3 Heisenberg–Langevin Equations

For a given QSDE defining $U$, the Heisenberg equations of motion of an arbitrary system operator $X$ can be found by applying the formalism introduced above. We first transform to the Heisenberg picture using the map (also called quantum flow)

$$j_t(X) = U^\dagger(t)(X \otimes \mathbb{1}_{\text{field}})U(t), \tag{A.22}$$

and then use Itō-calculus to deduce the equation for $\mathrm{d}j_t(X)$. For a vacuum field we find (introducing $\bar{X} = X \otimes \mathbb{1}_{\text{field}}$ for notational convenience)

$$\begin{aligned}
\mathrm{d}j_t(X) &= \mathrm{d}U^\dagger(t)\bar{X}U(t) + U^\dagger(t)\bar{X}\mathrm{d}U(t) + \mathrm{d}U^\dagger(t)\bar{X}\mathrm{d}U(t)\\
&= \mathrm{i}U^\dagger(t)(H_{\mathrm{eff}}^\dagger\bar{X} - \bar{X}H_{\mathrm{eff}})U(t)\mathrm{d}t\\
&\quad + U^\dagger(t)\bar{X}[s\mathrm{d}A^\dagger(t) - s^\dagger\mathrm{d}A(t)]U(t)\\
&\quad + U^\dagger(t)[s^\dagger\mathrm{d}A(t) - s\mathrm{d}A^\dagger(t)]\bar{X}U(t)\\
&\quad + U^\dagger(t)[s^\dagger\mathrm{d}A(t) - s\mathrm{d}A^\dagger(t)]\bar{X}[s\mathrm{d}A^\dagger(t) - s^\dagger\mathrm{d}A(t)]U(t)\\
&= U^\dagger(t)[\mathrm{i}H_{\mathrm{eff}}^\dagger\bar{X} - \mathrm{i}\bar{X}H_{\mathrm{eff}} + s^\dagger\bar{X}s]U(t)\mathrm{d}t\\
&\quad + U^\dagger(t)[\bar{X}, s\mathrm{d}A^\dagger(t)]U(t) - U^\dagger(t)[s^\dagger\mathrm{d}A(t),\bar{X}]U(t)\\
&= U^\dagger(t)\left\{\mathrm{i}[H,\bar{X}] + [s^\dagger\bar{X}s - \frac{1}{2}(s^\dagger s\bar{X} + \bar{X}s^\dagger s)]\right\}U(t)\mathrm{d}t\\
&\quad + U^\dagger(t)[\bar{X},s]U(t)\mathrm{d}A^\dagger(t) - U^\dagger(t)[s^\dagger,\bar{X}]U(t)\mathrm{d}A(t).
\end{aligned}$$

We thus obtain the Itō QLE (Barchielli, 2006)

$$\mathrm{d}j_t(X) = j_t(\mathcal{L}^\dagger X)\mathrm{d}t + j_t([X,s])\mathrm{d}A^\dagger(t) + j_t([s^\dagger,X])\mathrm{d}A(t), \tag{A.23}$$

where $\mathcal{L}^\dagger$ is the adjoint of the Liouvillian introduced in (A.31) given by

$$\mathcal{L}^\dagger X = -\mathrm{i}[X, H_{\mathrm{sys}}] + \mathcal{D}^\dagger[s]X, \tag{A.24a}$$

$$\mathcal{D}^\dagger[s]X = s^\dagger Xs - \frac{1}{2}(s^\dagger sX + Xs^\dagger s). \tag{A.24b}$$

Following the usual notation used in quantum optics we often denote the Heisenberg-picture operators by $X(t) := j_t(X)$.

The evolution of the output field is obtained in a similar way. We define $A_{\mathrm{out}}(t)$ by moving to the Heisenberg picture, i.e.,

$$A_{\mathrm{out}}(t) = U^\dagger(t)[\mathbb{1}_{\mathrm{sys}} \otimes A(t)]U(t). \tag{A.25}$$

We can easily obtain the corresponding QSDE (Barchielli, 1986)

$$\mathrm{d}A_{\mathrm{out}}(t) = j_t(s)\mathrm{d}t + \mathrm{d}A(t), \tag{A.26}$$

which corresponds to the input–output relations (A.10).

## A.4 Stochastic Evolution of the Quantum State

The time evolution of the state vector $|\Psi\rangle$ (describing system + field) is just given by $|\Psi(t)\rangle = U(t)|\Psi_0\rangle$ with an initial state $|\Psi_0\rangle = |\psi_0\rangle\otimes|\mathrm{vac}\rangle$. The vacuum state $|\mathrm{vac}\rangle$ of the field obeys $\mathrm{d}A(t)|\mathrm{vac}\rangle = 0$. As we have established that $[U(t), \mathrm{d}A(t)] = 0$ we can write the resulting equation in the form

$$\mathrm{d}|\Psi(t)\rangle = \left[-\mathrm{i}H_{\mathrm{eff}}\mathrm{d}t + s\mathrm{d}A^\dagger(t)\right]|\Psi(t)\rangle. \tag{A.27}$$

If we define the stochastic density operator $\sigma(t) = |\Psi(t)\rangle\langle\Psi(t)|$ we readily find the generalization of the von-Neumann equation (Gardiner and Zoller, 2004, p. 349)

$$\begin{aligned} d\sigma(t) &= -i[H_{\text{eff}}\sigma(t) - \sigma(t)H_{\text{eff}}^\dagger]dt \\ &\quad + dA^\dagger(t)s\sigma(t)s^\dagger dA(t) + dA^\dagger(t)s\sigma(t) + \sigma(t)dA(t)s^\dagger. \end{aligned} \tag{A.28}$$

## A.5 Master Equation

If we are exclusively interested in the evolution of the system we can trace-out the bath variables and define the system state by

$$\rho(t) = \text{tr}_{\text{field}}\big(U(t)\rho_0 \otimes |\text{vac}\rangle\langle\text{vac}|U^\dagger(t)\big). \tag{A.29}$$

We then find (by evaluating $d\rho$ in Itō-calculus and dividing by $dt$ afterward)

$$\dot{\rho}(t) = \mathcal{L}\rho(t) := -i[H_{\text{eff}}\rho(t) - \rho(t)H_{\text{eff}}^\dagger] + s\rho(t)s^\dagger, \tag{A.30}$$

which shows a clear similarity to (A.28). The Liouvillian $\mathcal{L}$ in its expanded form reads

$$\mathcal{L}\rho = -i[H_{\text{sys}}, \rho] + \mathcal{D}[s]\rho, \tag{A.31a}$$

$$\mathcal{D}[s]\rho = s\rho s^\dagger - \frac{1}{2}(s^\dagger s\rho + \rho s^\dagger s). \tag{A.31b}$$

Note that we have the identity

$$\text{tr}(X\mathcal{L}\rho) = \text{tr}(\rho\mathcal{L}^\dagger X) \tag{A.32}$$

with the adjoint Liouvillian $\mathcal{L}^\dagger$ defined in (A.24).

## A.6 Stochastic Equations in Stratonovich Form

To convert between the Itō and Stratonovich formulation of quantum-stochastic processes we can for *linear* equations use the following approach (Gardiner and Zoller, 2004, p. 346f). Consider the Stratonovich stochastic differential equation

$$(\mathbf{S}) \quad dX(t) = [\mathcal{A}dt + \mathcal{B}dA(t) + \mathcal{C}dA^\dagger(t)]X(t), \tag{A.33}$$

with linear operators $\mathcal{A}$, $\mathcal{B}$, $\mathcal{C}$ and a given initial condition $X(0)$. This equation has the formal solution

$$X(t) = \mathcal{T}_+\exp\left\{\int_0^t [\mathcal{A}dt + \mathcal{B}dA(t) + \mathcal{C}dA^\dagger(t)]\right\}X(0), \tag{A.34}$$

where $\mathcal{T}_+$ denotes the time-ordered product. We can now calculate the Itō increment $dX(t) = X(t + dt) - X(t)$ by using the rules of Itō-calculus outlined in the previous section and find

$$\begin{aligned} dX(t) &= \left\{ \exp\left[\mathcal{A}dt + \mathcal{B}dA(t) + \mathcal{C}dA^\dagger(t)\right] - 1 \right\} X(t) \\ &= \left\{ [\mathcal{A} + \frac{1}{2}\mathcal{B}\mathcal{C}]dt + \mathcal{B}dA(t) + \mathcal{C}dA^\dagger(t) \right\} X(t), \end{aligned} \tag{A.35}$$

where we assumed the field to be in vacuum and thus $dA\, dA^\dagger = dt$ to be the only nonvanishing quadratic term. Applying this to the Hudson–Parthasarathy equation (A.11) we find the corresponding Stratonovich equation to be

$$(\mathbf{S}) \quad dU(t) = \left[-iH_{sys}dt + s dA^\dagger(t) - s^\dagger dA(t)\right] U(t). \tag{A.36}$$

Using Eq. (A.34) this has the formal solution

$$U(t) = \mathcal{T}_+ \exp\left\{ \int_0^t \left[-iH_{sys}dt + s dA^\dagger(t) - s^\dagger dA(t)\right] \right\} \mathbb{1}. \tag{A.37}$$

As it is conventional in quantum optics, in this review stochastic differential equations started by a fluxion of the form $\dot{X}$ are written in Stratonovich form without explicit denotation by **(S)**. Equations starting with an increment $dX$ are to be read as of Itō-type, if not stated otherwise.

## A.7 Nonvacuum Field States

For a general (squeezed) Gaussian initial state of the electromagnetic field the field operators obey the correlation function

$$\langle a_{in}^\dagger(t) a_{in}(t') \rangle_{(N,M)} = N\,\delta(t - t'), \tag{A.38a}$$

$$\langle a_{in}(t) a_{in}(t') \rangle_{(N,M)} = M\,\delta(t - t'), \tag{A.38b}$$

with parameters $M \in \mathbb{C}$, $N \geq 0$, and $|M|^2 \leq N(N + 1)$, where equality is attained for a pure state. The vacuum multiplication table (A.14) is thus modified to

| × | $dA$ | $dA^\dagger$ | $dt$ |
|---|---|---|---|
| $dA$ | $M\,dt$ | $(N+1)\,dt$ | 0 |
| $dA^\dagger$ | $N\,dt$ | $M^*\,dt$ | 0 |
| $dt$ | 0 | 0 | 0 |

(A.39)

Note that while the Stratonovich form of the corresponding QSDEs does not change, the Itō-form has to be modified accordingly (Barchielli, 1986). We find instead of (A.11) (Gardiner and Zoller, 2004, p. 350f)

$$\begin{aligned} dU(t) &= -iH_{sys}U(t)dt \\ &\quad -\frac{1}{2}\left[ss^\dagger N + s^\dagger s(N+1) - ssM^* - s^\dagger s^\dagger M\right]U(t)dt \\ &\quad + \left[s dA^\dagger(t) - s^\dagger dA(t)\right]U(t). \end{aligned} \tag{A.40}$$

Similar modification are found for the corresponding Langevin and master equation.

# APPENDIX B. QUANTUM FILTERING AND CONTROL

In this section we briefly review the most important results from the theory of quantum filtering and (linear–quadratic–Gaussian) control. Quantum filtering—the extraction of information about the quantum state from noisy measurements—was developed in the mathematical physics community (Belavkin, 1980, 1992) and later in the context of quantum optics (Carmichael, 1993).

## B.1 The Belavkin Equation

### *B.1.1 Quantum Optical Derivation*

Here we discuss a derivation of the stochastic Schrödinger equation and the corresponding stochastic master equation for homodyne detection from a quantum-optics perspective (see Carmichael, 1993 and Gardiner and Zoller, 2004 for an in-depth discussion). We consider the situation depicted in Fig. B1, where a system $S$ couples to the one-dimensional electromagnetic field $A$, which is initially in vacuum and is subject to homodyne detection. Homodyne detection is realized by a strong local oscillator (LO) which is mixed with the field $A$ on a (balanced) beam-splitter and subsequent detection by photodetectors. We first assume unit detection efficiency, but will discuss the case of inefficient detection at the end of the section.

We start from the Schrödinger equation (A.27) which describes the full system ($S + A$) and we assume an initial state $|\Psi_0\rangle = |\psi_0\rangle_S \otimes |\text{vac}\rangle_A$. Eq. (A.27) can then be written in the form

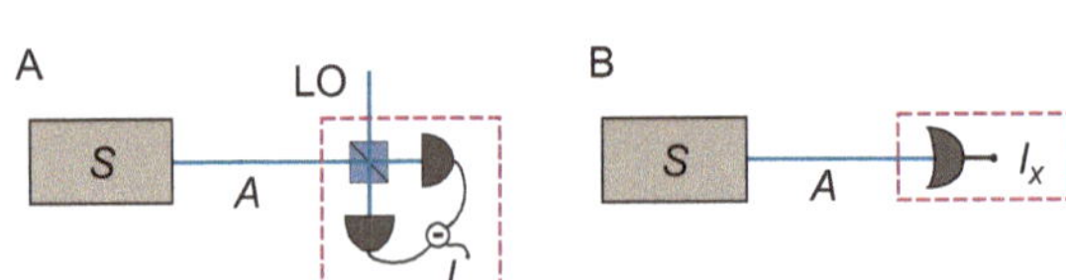


**Fig. B1** (A) Physical realization of homodyne detection and (B) its equivalent depiction as it is used throughout the review.

$$d|\Psi(t)\rangle = \{-iH_{\text{eff}}dt + s[dA^{\dagger}(t) + dA(t)]\}|\Psi(t)\rangle, \tag{B.1}$$

where we used the fact that $dA(t)|\Psi(t)\rangle = dA(t)|\Psi_0\rangle = dA(t)|\text{vac}\rangle = 0$ (Gardiner and Zoller, 2004, p. 375). A homodyne measurement of an electromagnetic quadrature $x = a + a^{\dagger}$ with a result $I_x \in \mathbb{R}$ effectively projects the state of the light field onto the eigenstate $|I_x\rangle_A$ of $x$, where $x|I_x\rangle_A = I_x|I_x\rangle_A$. Projecting (B.1) onto $|I_x\rangle$ results in Goetsch and Graham (1994)

$$\langle I_x|\Psi(t+dt)\rangle \otimes |I_x\rangle = [\mathbb{1} - iH_{\text{eff}}dt + s\,I_x dt]\langle I_x|\Psi(t)\rangle \otimes |I_x\rangle.$$

Defining $|\tilde{\psi}_c(t+dt)\rangle := \langle I_x|\Psi(t+dt)\rangle$ as the state *after the measurement* and factoring off the bath state thus leads to the linear stochastic Schrödinger equation (Carmichael, 1993)

$$d|\tilde{\psi}_c(t)\rangle = [-iH_{\text{eff}}dt + s\,I_x(t)dt]|\psi_c(t)\rangle, \tag{B.2}$$

with the forward pointing Itō-increment $d|\tilde{\psi}_c(t)\rangle = |\tilde{\psi}_c(t+dt)\rangle - |\psi_c(t)\rangle$. $|\tilde{\psi}_c\rangle$ is the unnormalized system state which is conditioned on the homodyne photo-current $I_x$, which is a Gaussian random variable with a probability distribution $\Upsilon_t(I_x) = |\langle I_x|\Psi(t+dt)\rangle|^2$. One obtains (Barchielli and Paganoni, 1996; Goetsch and Graham, 1994; Wiseman and Milburn, 2009)

$$\Upsilon_t(I_x)dI_x = \sqrt{\frac{dt}{2\pi}}\exp\left(-\frac{dt}{2}[I_x - \langle s + s^{\dagger}\rangle_c]^2\right)dI_x, \tag{B.3}$$

where the conditional expectation value is defined with respect to the normalized conditional state, i.e., $\langle s\rangle_c = \langle\psi_c(t)|s|\psi_c(t)\rangle$. Using this, $I_x$ can be written as

$$I_x dt = \langle s + s^{\dagger}\rangle_c dt + dW, \tag{B.4}$$

where $W$ is a classical Wiener process with $dW(t)^2 = dt$ and zero mean. We can introduce the classical stochastic process $\tilde{X}$ defined by $d\tilde{X}(t) = I_x(t)dt$, which is statistically equivalent to $dX(t) = dA_{\text{out}}(t) + dA^{\dagger}_{\text{out}}(t)$. This is due to nondemolition properties of the measurement operator, see Appendix B.1.2 and Bouten et al. (2007). It obeys $d\tilde{X}(t)^2 = dt$.

From the stochastic Schrödinger equation (B.2) we can obtain a stochastic master equation for the conditional density operator in the following way. We first find the equation of motion for the unnormalized conditional state $\tilde{\rho}_c = |\tilde{\psi}_c\rangle\langle\tilde{\psi}_c|$ by multiplying out (B.2). We obtain

$$\begin{aligned} d\tilde{\rho}_c(t) &= \tilde{\rho}_c(t+dt) - \rho_c(t) \\ &= |\psi_c(t)\rangle\left[d\langle\tilde{\psi}_c(t)|\right] + \left[d|\tilde{\psi}_c(t)\rangle\right]\langle\psi_c(t)| + \left[d|\tilde{\psi}_c(t)\rangle\right]\left[d\langle\tilde{\psi}_c(t)|\right] \\ &= \mathcal{L}\rho_c(t)dt + \left[s\rho_c(t) + \rho_c(t)s^\dagger\right]d\tilde{X}(t). \end{aligned} \tag{B.5}$$

Eq. (B.5) is the quantum analog to the classical Zakai equation (Xiong, 2008). Note that although the Liouvillian is trace-preserving [i.e., $\mathrm{tr}(\mathcal{L}\rho)=0$], the second term in (B.5) does not possess this property. The equation for the normalized state $\rho_c(t) = \tilde{\rho}_c(t)/\mathrm{tr}(\tilde{\rho}_c(t))$ is then found by noting that

$$\begin{aligned} \mathrm{tr}(\tilde{\rho}_c(t+dt)) &= \mathrm{tr}(\rho_c(t) + d\tilde{\rho}_c(t)) \\ &= 1 + \langle s+s^\dagger\rangle_c d\tilde{X} - i\langle H_{\mathrm{eff}} - H_{\mathrm{eff}}^\dagger\rangle_c dt + \langle ss^\dagger\rangle_c dt \\ &= 1 + \langle s+s^\dagger\rangle_c d\tilde{X}, \end{aligned} \tag{B.6}$$

where now $\langle A\rangle_c = \mathrm{tr}(\rho_c(t)A)$ and we used $\mathrm{tr}(\rho_c(t)) = 1$. Thus we find

$$\mathrm{tr}(\tilde{\rho}_c(t+dt))^{-1} = 1 - \langle s+s^\dagger\rangle_c d\tilde{X} + \langle s+s^\dagger\rangle_c^2 dt, \tag{B.7}$$

which is obtained by expanding $\mathrm{tr}(\tilde{\rho}_c(t+dt))^{-1}$ to *second order* in $d\tilde{X}$ (leading to a first-order expansion in d$t$). Using Itō multiplication rules this leads to

$$\begin{aligned} d\rho_c(t) &= \frac{\tilde{\rho}_c(t+dt)}{\mathrm{tr}(\tilde{\rho}_c(t+dt))} - \rho_c(t) \\ &= \mathcal{L}\rho_c(t)dt + \mathcal{H}[s]\rho_c(t)dW(t), \end{aligned} \tag{B.8}$$

which is the desired result (see Gardiner and Zoller, 2004, p. 375 and Wiseman and Milburn, 2009, p. 161). The (nonlinear) measurement term is given by

$$\mathcal{H}[s]\rho_c = [s - \mathrm{tr}(\rho_c s)]\rho_c + \rho_c[s - \mathrm{tr}(\rho_c s)]^\dagger. \tag{B.9}$$

It is clear from the derivation that under made assumptions the SSE (B.2) is equivalent to the SME (B.8). The stochastic master equation is more general, however, as it can accommodate for additional, unobserved decay channels (such as photon losses/inefficient detection or coupling of a mechanical oscillator to a heat bath), as well as for mixed initial states.

We can generalize the homodyne master equation in several ways: above we assumed a measurement of a specific light quadrature $x = a + a^\dagger$. To measure a rotated quadrature $x^\phi = ae^{-i\phi} + a^\dagger e^{+i\phi}$ we have to make the

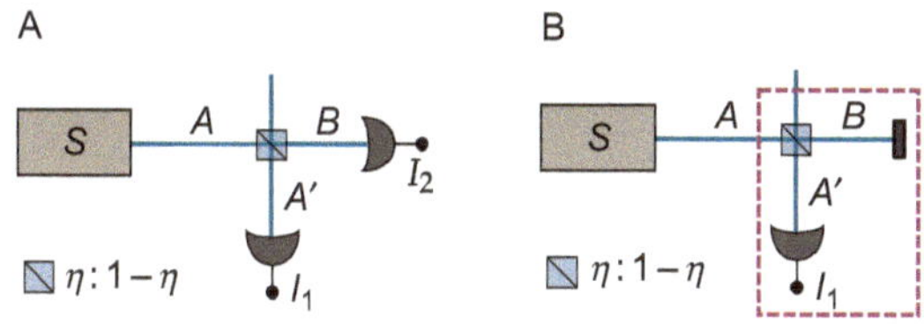


**Fig. B2** (A) Dual homodyne detection. (B) Effective description of inefficient homodyne detection.

replacement $s \rightarrow e^{i\phi}s$, which simply follows from replacing $a \rightarrow e^{-i\phi}a$. We can as easily obtain the SME corresponding to heterodyne detection at a LO frequency $\omega_{lo} \neq \omega_0$ by substituting $s \rightarrow e^{i\Delta_{lo}t}s$, where $\Delta_{lo} = \omega_{lo} - \omega_0$.

Below we will discuss the situation where we split up the field with a beam-splitter (with a splitting ration $\eta : 1 - \eta$, $0 \leq \eta \leq 1$) and perform two simultaneous homodyne measurements on its outputs. The measured modes $A'$, $B$ after the beam-splitter (see Fig. B2) are related to $A$ before the beam-splitter via (Gough and James, 2009) $a(t) = \sqrt{\eta}\, a'(t) + \sqrt{1-\eta}\, b(t)$, where $A'$ and $B$ are both initially in vacuum and are uncorrelated such that $dA(t)dB^{\dagger}(t) = 0$, etc. Plugging this relation into (B.1) and projecting onto the quadratures $e^{-i\phi_1}a' + e^{i\phi_1}a'^{\dagger}$, $e^{-i\phi_2}b + e^{i\phi_2}b^{\dagger}$ one can repeat above steps and find

$$d\rho_c = \mathcal{L}\rho_c dt + \sqrt{\eta}\mathcal{H}[e^{i\phi_1}s]\rho_c dW_1 + \sqrt{1-\eta}\mathcal{H}[e^{i\phi_2}s]\rho_c dW_2, \tag{B.10}$$

with uncorrelated Wiener processes $W_i$, i.e., $dW_i dW_j = \delta_{ij}dt$. To model photon losses or inefficient photodetectors, we average over, say, the second measurement process, and thus discard all information obtained from it. Due to the fact that $W_2$ is a Wiener process and thus has zero mean, the equation of motion for the resulting conditional state—which is now conditioned on the measurement results of the first channel only—is obtained by averaging, and thus dropping, the last term in (B.10). The beam-splitter transmissivity is then identified with the efficiency of the photo-detection. Formally we can obtain the same result from (B.8) by replacing $s \rightarrow \sqrt{\eta}s$ in the measurement term, while keeping the Liouvillian unchanged. The general SME for heterodyne detection at a LO frequency $\omega_{lo} = \omega_0 + \Delta_{lo}$, at a phase $\phi$, and with a detection efficiency $\eta$ thus is

$$d\rho_c = \mathcal{L}\rho_c dt + \sqrt{\eta}\mathcal{H}[e^{i(\Delta_{lo}t+\phi)}s]\rho_c dW. \tag{B.11}$$

Often we will be faced with the situation where the LO detuning is large on the typical frequencies of the system. In this case, a time coarse-grained description of the heterodyne measurement is appropriate (Wiseman and Milburn, 2009, p. 166f). To this end we choose a time interval $\delta t$ which is large compared to $\Delta_{\mathrm{lo}}^{-1}$ but short on all system timescales, and define the time-averaged density matrix $\overline{\rho}$ by

$$\int_t^{t+\delta t} \mathrm{d}\rho = \rho(t+\delta t) - \rho(t) =: \mathrm{d}\overline{\rho}(t). \tag{B.12}$$

At the same time we have to assume that $\rho(t)$ is slowly varying on the timescale $\delta t$ and we can consequently make the approximation

$$\int_t^{t+\delta t} \mathrm{d}\tau\rho(\tau) \approx \rho(t)\delta t =: \overline{\rho}(t)\mathrm{d}t. \tag{B.13}$$

This leads us to define the Fourier cosine and sine components of the Wiener process $W(t)$ by

$$\sqrt{2}\int_t^{t+\delta t} \mathrm{d}W(\tau)\mathrm{e}^{-\mathrm{i}\Delta_{\mathrm{lo}}\tau} =: \mathrm{d}W_{\mathrm{c}} + \mathrm{i}\mathrm{d}W_{\mathrm{s}}, \tag{B.14}$$

where $\mathrm{d}W_{\mathrm{s}}$ and $\mathrm{d}W_{\mathrm{s}}$ are both real. Calculating their variances we find

$$\begin{aligned}\mathrm{d}W_{\mathrm{c}}(t)^2 &= 2\int_t^{t+\delta t}\int_t^{t+\delta t} \mathrm{d}W(\tau)\mathrm{d}W(\tau')\cos(\Delta_{\mathrm{lo}}\tau)\cos(\Delta_{\mathrm{lo}}\tau')\\ &= \int_t^{t+\delta t} \mathrm{d}W(\tau)[1+\cos(2\Delta_{\mathrm{lo}}\tau)] = [1+\mathcal{O}(\Delta_{\mathrm{lo}}\delta t)^{-1}]\delta t,\end{aligned}$$

and in a similar way $\mathrm{d}W_{\mathrm{s}}(t)^2 = [1+\mathcal{O}(\Delta_{\mathrm{lo}}\delta t)^{-1}]\delta t$. For the cross-correlations we have

$$\begin{aligned}\mathrm{d}W_{\mathrm{s}}(t)\mathrm{d}W_{\mathrm{c}}(t) &= 2\int_t^{t+\delta t}\int_t^{t+\delta t} \mathrm{d}W(\tau)\mathrm{d}W(\tau')\cos(\Delta_{\mathrm{lo}}\tau)\sin(\Delta_{\mathrm{lo}}\tau')\\ &= \int_t^{t+\delta t} \mathrm{d}W(\tau)\sin(2\Delta_{\mathrm{lo}}\tau) = \mathcal{O}(\Delta_{\mathrm{lo}}\delta t)^{-1}\delta t.\end{aligned}$$

We thus see that $W_{\mathrm{c}}$, $W_{\mathrm{s}}$ are Wiener processes which are (approximately) uncorrelated. On timescales much longer than $\delta t$ we can make the replacement $\delta t \rightarrow \mathrm{d}t$ and can thus use the Itō table

$$\mathrm{d}W_{\mathrm{c}}^2 = \mathrm{d}W_{\mathrm{s}}^2 = \mathrm{d}t, \tag{B.15a}$$

$$\mathrm{d}W_{\mathrm{c}}\mathrm{d}W_{\mathrm{s}} = 0. \tag{B.15b}$$

We therefore find an effective stochastic master equation for heterodyne detection, reading

$$d\overline{\rho}_c = \mathcal{L}\overline{\rho}_c dt + \sqrt{\frac{\eta}{2}}\mathcal{H}[e^{i\phi}s]\overline{\rho}_c dW_c + \sqrt{\frac{\eta}{2}}\mathcal{H}[-ie^{i\phi}s]\overline{\rho}_c dW_s. \tag{B.16}$$

We thus see that in this limit a heterodyne measurement is equivalent to simultaneous homodyne measurements of two conjugate quadratures, each with half efficiency (as is depicted in Fig. B2).

More generally we can describe an open quantum system coupling to $m$ vacuum field channels, $m' \leq m$ of which are subject to homodyne detection, by the equation

$$d\rho_c = -i[H,\rho_c]dt + \sum_{i=1}^{m}\mathcal{D}[s_i]\rho_c dt + \sum_{i=1}^{m'}\mathcal{H}[s_i]\rho_c dW_i. \tag{B.17}$$

The case $m' < m$ can be used to describe inefficient photo-detection, as we just demonstrated in the discussion above, or decoherence channels which cannot be observed at all, e.g., phonon losses of a mechanical oscillator. Equations of this form are often called Belavkin equations (Belavkin, 1992).

### B.1.2 Derivation by Nonlinear Filtering

In this section we derive the stochastic master equation for homodyne detection from the perspective of classical nonlinear filtering. In our derivation we will follow the technique laid out in Gough and Kostler (2010). First we briefly review the most important properties of the conditional expectation value (see, e.g., Bouten et al., 2007). In the following we denote by $\mathbb{E}[A]$ the expectation value of a random variable $A$ with respect to some probability space. For a quantum observable $A$ and a quantum state $\rho$ we identify $\mathbb{E}[A] = \text{tr}(A\rho)$. The following construction works independently of this specific choice, however.

#### B.1.2.1 Conditional Expectation Value

Let $\mathcal{A}$ be a (von-Neumann) algebra of operators and $\mathcal{A}' = \{B : [B, A] = 0, \forall A \in \mathcal{A}\}$ its *commutant*. We then call the map $\mathbb{E}[\cdot|\mathcal{A}] : \mathcal{A}' \to \mathcal{A}$ the *conditional expectation* from $\mathcal{A}$ onto $\mathcal{A}'$ if

$$\mathbb{E}[\mathbb{E}[B|\mathcal{A}]g(A)] = \mathbb{E}[Bg(A)] \tag{B.18}$$

for all $A \in \mathcal{A}$, $B \in \mathcal{A}'$ and some measureable function $g$ (in particular $g = \text{id}$ and $g(A) \equiv 1$). The conditional expectation has the following useful properties:

$$\mathbb{E}[B|\mathcal{A}] = B \text{ for all } B \in \mathcal{A} \tag{B.19a}$$

$$\mathbb{E}[\mathbb{E}[B|\mathcal{A}]|\mathcal{C}] = \mathbb{E}[B|\mathcal{C}] \text{ for } \mathcal{C} \subset \mathcal{A} \tag{B.19b}$$

$$\mathbb{E}[AB|\mathcal{C}] = A\mathbb{E}[B|\mathcal{C}] \text{ for } A \in \mathcal{C} \tag{B.19c}$$

The expectation $\mathbb{E}$ defines a scalar product $\langle A, B\rangle_{\mathbb{E}} := \mathbb{E}[A^\dagger B]$, which induces the norm $\| A\|_{\mathbb{E}} := \langle A, A\rangle_{\mathbb{E}}^{1/2}$. From the definition (B.18) then directly follows the orthogonality relation

$$\langle \mathbb{E}[B|A] - B, g(A)\rangle_{\mathbb{E}} = 0 \tag{B.20}$$

for all $A \in \mathcal{A}$, $B \in \mathcal{A}'$. The conditional expectation thus also satisfies the least-squares property (Bouten et al., 2007)

$$\| B - \mathbb{E}[B|\mathcal{A}]\|_{\mathbb{E}} \leq \| B - A\|_{\mathbb{E}}. \tag{B.21}$$

This means that $\mathbb{E}[\cdot|\mathcal{A}]$ is the orthogonal projection from $\mathcal{A}'$ onto $\mathcal{A}$.

#### B.1.2.2 Characteristic Function Technique

Suppose we continuously monitor the output of a quantum system whose evolution is described by Eq. (A.11) via homodyne detection, yielding a measurement of the field quadrature $Y(t) = A_{\text{out}}(t) + A_{\text{out}}^\dagger(t)$ [where $A_{\text{out}}$ is defined in (A.25)]. We denote by $\mathcal{Y}_t = \{Y(t') : 0 \leq t' \leq t\}$ the algebra generated by the observation process $Y$ and by $\mathcal{Y}_t'$ its commutant, $\mathcal{Y}_t' = \{A : [A, Y] = 0, \forall Y \in \mathcal{Y}_t\}$. For $0 \leq t' \leq t$ we then have the *nondemolition* property (Bouten et al., 2007)

$$[j_t(X), Y(t')] = 0 \tag{B.22}$$

[such that $j_t(X) \in \mathcal{Y}_t$] and the *self-nondemolition* property

$$[Y(t), Y(t')] = 0. \tag{B.23}$$

This allows us to map the quantum filtering problem to a classical filtering problem which we can tackle by methods from classical nonlinear filtering theory.

We define the quantum state $\mathbb{E}[\cdot] = \text{tr}(\rho_0 \otimes \rho_{\text{vac}} \cdot)$ and seek to find an estimate $\pi_t(X) \in \mathcal{Y}_t$ for the Heisenberg operator $X(t) = j_t(X)$ which minimizes the quadratic error $\mathbb{E}[(\pi_t(X) - j_t(X))^2]$ based on the observations $Y(t)$. As this requirement is equivalent to Eq. (B.21) we directly find

$$\pi_t(X) = \mathbb{E}[j_t(X)|\mathcal{Y}_t]. \tag{B.24}$$

We will now show how to calculate $\pi_t(X)$ explicitly using the characteristic function technique (Gough and Kostler, 2010), for the case of a single vacuum input. Due to the orthogonality property (B.20) of the conditional expectation we have the identity $\mathbb{E}[(\pi_t(X) - j_t(X))C_t] = 0$ for all $C_t \in \mathcal{Y}_t$. We can thus make the following Ansatz

$$\mathrm{d}C_t = g_t C_t \mathrm{d}Y(t), \tag{B.25a}$$

$$\mathrm{d}\pi_t(X) = \alpha_t \mathrm{d}t + \beta_t \mathrm{d}Y(t), \tag{B.25b}$$

for adapted stochastic processes $\alpha_t, \ \beta_t \in \mathcal{Y}_t$. This implies $C_t, \pi_t(X) \in \mathcal{Y}_t$ [while on the other hand $j_t(X) \notin \mathcal{Y}_t$]. We can then use $\mathrm{d}\mathbb{E}[(\pi_t(X) - j_t(X))C_t] = 0$, which, due to Itō-rules consists of three terms. We find for the first term

$$\begin{aligned} \mathrm{I} &= \mathbb{E}[(\mathrm{d}\pi_t(X) - \mathrm{d}j_t(X))C_t] \\ &= \mathbb{E}[\alpha_t C_t]\mathrm{d}t + \mathbb{E}[\beta_t \mathrm{d}Y(t) C_t] - \mathbb{E}[\mathrm{d}j_t(X) C_t] \\ &= \mathbb{E}[\alpha_t C_t]\mathrm{d}t + \mathbb{E}[\beta_t j_t(s + s^\dagger) C_t]\mathrm{d}t - \mathbb{E}[j_t(\mathcal{L}^* X) C_t]\mathrm{d}t \\ &= \mathbb{E}[\alpha_t C_t]\mathrm{d}t + \mathbb{E}[\beta_t \pi_t(s + s^\dagger) C_t]\mathrm{d}t - \mathbb{E}[\pi_t(\mathcal{L}^* X) C_t]\mathrm{d}t, \end{aligned}$$

where to go from the second to the third line we used (i) the fact that $Y$ is generated by a Wiener process and therefore the increments $\mathrm{d}Y(t)$ are independent of $Y(t)$ at equal times, i.e., $\mathbb{E}[\mathrm{d}Y(t) f(Y(t))] = \mathbb{E}[\mathbb{E}[\mathrm{d}Y(t)] f(Y(t))]$ for a measureable function $f$, (ii) $\mathbb{E}[\mathrm{d}Y(t)] = \mathbb{E}[j_t(s + s^\dagger)]\mathrm{d}t$, and (iii) for the field initially in vacuum we have $\mathbb{E}[\mathrm{d}A(t)] = 0$ and $\mathbb{E}[\mathrm{d}A^\dagger(t)] = 0$. In particular this means that $\mathbb{E}[\beta_t \mathrm{d}A(t) C_t] = \mathbb{E}[\beta_t \mathrm{d}A^\dagger(t) C_t] = 0$. To go to the fourth line we used the orthogonality property, thus $\mathbb{E}[j_t(X) C_t] = \mathbb{E}[\pi_t(X) C_t]$. For the second term we find

$$\begin{aligned} \mathrm{II} &= \mathbb{E}[(\pi_t(X) - j_t(X))\mathrm{d}C_t] \\ &= \mathbb{E}[(\pi_t(X) - j_t(X)) g_t C_t j_t(s + s^\dagger)]\mathrm{d}t \\ &= \mathbb{E}[\pi_t(X)\pi_t(s + s^\dagger) g_t C_t]\mathrm{d}t - \mathbb{E}[\pi_t(X(s + s^\dagger)) g_t C_t]\mathrm{d}t \end{aligned}$$

by the same reasoning as above. Going from the second to the third line we used (B.19c), which implies $\pi_t(\pi_t(X) j_t(s)) = \pi_t(X)\pi_t(s)$, and $j_t(XY) = j_t(X) j_t(Y)$. For the third term we obtain by using the Itō table (A.14)

$$\begin{aligned} \mathrm{III} &= \mathbb{E}[(\mathrm{d}\pi_t(X) - \mathrm{d}j_t(X))\mathrm{d}C_t] \\ &= \mathbb{E}[(\beta_t(X)\mathrm{d}Y(t) - \mathrm{d}j_t(X)) g_t C_t \mathrm{d}Y(t)] \\ &= \mathbb{E}[\beta_t(X) g_t C_t]\mathrm{d}t - \mathbb{E}[j_t([s^\dagger, X]) g_t C_t]\mathrm{d}t \\ &= \mathbb{E}[\beta_t(X) g_t C_t]\mathrm{d}t - \mathbb{E}[\pi_t([s^\dagger, X]) g_t C_t]\mathrm{d}t. \end{aligned}$$

As $g_t$ is an arbitrary function we can deduce from I + II + III = 0 by equating the coefficients of $g_t$ and $g_t C_t$ that

$$\begin{aligned}0 &= \alpha_t + \beta_t \pi_t(s+s^\dagger) - \pi_t(\mathcal{L}^* X),\\ 0 &= \pi_t(X)\pi_t(s+s^\dagger) - \pi_t(X(s+s^\dagger)) + \beta_t - \pi_t([s^\dagger, X])\\ &= \pi_t(X)\pi_t(s+s^\dagger) - \pi_t(Xs + s^\dagger X) + \beta_t,\end{aligned}$$

where both equalities hold almost surely with respect to $\mathbb{E}$. Solving this for $\alpha_t$ and $\beta_t$, and using the ansatz (B.25b) we find for the quantum filter

$$d\pi_t(X) = \pi_t(\mathcal{L}^* X)dt + [\pi_t(Xs + s^\dagger X) - \pi_t(X)\pi_t(s+s^\dagger)]dW, \tag{B.26}$$

where

$$dW(t) = dY(t) - \pi_t(s+s^\dagger)dt \tag{B.27}$$

is the *innovations process*. $W(t)$ is a Wiener process with $\mathbb{E}[dW] = 0$, $dW^2 = dt$ and therefore $\mathbb{E}[\pi_s(X)dW(t)] = 0$ for $0 \le s \le t$. The innovations process can be interpreted as the difference between the observed changed $dY(t)$ and the expected change $\pi_t(s + s^\dagger)dt$ of the field. We can obtain the adjoint equation for the conditional density matrix $\rho_c$ by identifying $\pi_t(X) = \mathrm{tr}(\rho_c(t)X)$. This leads to the SME (B.8).

## B.2 Markovian Homodyne Feedback

In this section we discuss how to describe Markovian (i.e., instantaneous) homodyne feedback in a framework developed in Wiseman (1994). The setup is basically the one from Fig. 12, where the photo-current $I_x(t)$ is fed back to the system. Markovian feedback dynamics can be described by the expression

$$(\mathbf{S}) \quad [d\rho_c]_{\mathrm{fb}} = \mathcal{K}[F]\rho_c d\tilde{X}, \tag{B.28}$$

the strength of the feedback being proportional to $d\tilde{X} = I_x(t)dt$. The feedback is assumed to be described by a Hamiltonian evolution and we thus write the linear operator $\mathcal{K}$ as

$$\mathcal{K}[F]\rho = -i[F, \rho], \tag{B.29}$$

where the Hermitian operator $F = F^\dagger$ describes a generalized force. As explicitly stated, Eq. (B.28) has to be interpreted as a Stratonovich equation (Wiseman, 1994). In order to reconcile this with a master equation in the Itō-formalism [e.g., Eq. (B.8)] we first have to convert it to Itō form. For a general master equation (with generic linear operators $\mathcal{L}$ and $\mathcal{K}$, and a nonlinear operator $\mathcal{H}$) this procedure yields the following:

- We start from a conditional Itō master equation

$$d\rho_c = \mathcal{L}\rho_c dt + \mathcal{H}\rho_c dW$$

- We convert Eq. (B.28) to the Itō form by applying the relations from Appendix A.6,

$$(\mathbf{I}) \quad [d\rho_c]_{\text{fb}} = \frac{1}{2}\mathcal{K}[F]^2\rho_c dt + \mathcal{K}[F]\rho_c d\tilde{X}$$

- As analyzed in Wiseman (1994), Markovian feedback can be described by alternately applying the generators of the conditional and the feedback evolution, applying feedback *after* the measurement, i.e.,

$$\rho_c(t+dt) = \exp\left(\frac{1}{2}\mathcal{K}^2 dt + \mathcal{K}d\tilde{X}\right)\exp(\mathcal{L}dt + \mathcal{H}dW)\rho_c(t).$$

Expanding each exponential to first order in $dt$ yields

$$\begin{aligned}\rho_c(t+dt) &= \left(1 + \frac{1}{2}\mathcal{K}^2 dt + \mathcal{K}d\tilde{X}\right)(1 + \mathcal{L}dt + \mathcal{H}dW)\rho_c(t)\\ &= \left(1 + \mathcal{L} + \frac{1}{2}\mathcal{K}^2 + \mathcal{K}\mathcal{H}\right)\rho_c dt + \left(\mathcal{H}dW + \mathcal{K}d\tilde{X}\right)\rho_c.\end{aligned}$$

Using the definitions of $\mathcal{K} := \mathcal{K}[F]$ and $\mathcal{H} := \mathcal{H}[s]$ [with Eqs. (B.9) and (B.29)], and the expression of the homodyne current $d\tilde{X}$ [Eq. (B.4)] we can show

$$\begin{aligned}\mathcal{K}\mathcal{H}\rho dt + \mathcal{K}\rho d\tilde{X} &= \mathcal{K}(\mathcal{H} + \langle s + s^\dagger\rangle_c)\rho dt + \mathcal{K}\rho dW\\ &= \mathcal{K}(s\rho + \rho s^\dagger)\rho dt + \mathcal{K}\rho dW.\end{aligned}$$

Additionally we note that

$$\frac{1}{2}\mathcal{K}^2\rho = -\frac{1}{2}[F,[F,\rho]] = \mathcal{D}[F]\rho. \tag{B.30}$$

Plugging in these two relations we can then take the classical average over the measurement results (dropping all terms proportional to $dW$) to obtain an unconditional master equation,

$$\dot{\rho} = \mathcal{L}\rho + \mathcal{D}[F]\rho + \mathcal{K}(s\rho + \rho s^\dagger).$$

Finally, using the identity

$$\mathcal{D}[s-\mathrm{i}F]\rho=\mathcal{D}[s]\rho+\mathcal{D}[F]\rho-\mathrm{i}[F,s\rho+\rho s^{\dagger}]-\frac{\mathrm{i}}{2}[\rho,Fs+s^{\dagger}F], \tag{B.31}$$

this can be transformed to

$$\dot{\rho}=\mathcal{L}\rho-\mathcal{D}[s]\rho+\mathcal{D}[s-\mathrm{i}F]\rho-\mathrm{i}[\frac{1}{2}(Fs+s^{\dagger}F),\rho]. \tag{B.32}$$

If we now use the form (A.31) of the Lindblad operator we end up with

$$\dot{\rho}=-\mathrm{i}[H_{\mathrm{sys}},\rho]+\mathcal{D}[s-\mathrm{i}F]\rho-\mathrm{i}[\frac{1}{2}(Fs+s^{\dagger}F),\rho]. \tag{B.33}$$

Let us discuss a two-fold generalization of this equation: first, we consider $m$ independent measurement and feedback channels that are described by the operators $s_i$ and $F_i$ ($i=1, \ldots, m$), respectively. This generalization is immediate in the light of Eq. (B.17). Second, we take into account a nonunit detection efficiency $0<\eta<1$. We know from Appendix B.1.1 that we can obtain the corresponding SME by rescaling the jump operator $s\longrightarrow\sqrt{\eta}s$, but *only in the measurement terms* $\mathcal{H}[s]$, while the decoherence terms $\mathcal{D}[s]$ are unchanged. We use Eq. (B.32) as our starting point and rescale $s\longrightarrow\sqrt{\eta}s$ in *the feedback terms only* and at the same time adjust the feedback gain to $F\longrightarrow F/\sqrt{\eta}$. This yields

$$\begin{aligned}\dot{\rho}&=-\mathrm{i}[H,\rho]+\sum_i\left\{\mathcal{D}[s_i]\rho+\frac{1}{\eta}\mathcal{D}[F_i]\rho-\mathrm{i}[F_i,s_i\rho+\rho s_i^{\dagger}]\right\}\\&=-\mathrm{i}[H,\rho]+\sum_i\left\{\mathcal{D}[s_i-\mathrm{i}F_i]\rho+\frac{1-\eta}{\eta}\mathcal{D}[F_i]\rho-\mathrm{i}\frac{1}{2}[s_i^{\dagger}F_i+F_is_i,\rho]\right\},\end{aligned} \tag{B.34}$$

which agrees with (Wiseman and Milburn, 2009, p. 246f).

## B.3 Quantum Filtering in Linear Systems

In this section we briefly review the most important equations of quantum LQG control, following closely the presentation in Edwards and Belavkin (2005). We start by deriving the Gaussian quantum filter—the so-called Kalman–Bucy filter—and afterward discuss how its output can be combined with optimal control techniques.

### *B.3.1 Kalman Filtering*

Consider a Gaussian $n$-dimensional open quantum system coupling to $m$ vacuum field channels, $m'\leq m$ of which are subject to homodyne detection, yielding the measurement outputs $Y_i=A_{i,\mathrm{out}}+A_{i,\mathrm{out}}$. The system's state conditioned on the outcomes of the homodyne measurements is described by the Belavkin equation (B.17) or equivalently by a slightly generalized version of the quantum filter (B.26)

$$\begin{aligned} d\pi_t(X) &= -i\pi_t([X,H])dt + \sum_{i=1}^{m} \pi_t(\mathcal{D}^\dagger[s_i]X)dt \\ &+ \sum_{i=1}^{m'} [\pi_t(Xs_i + s_i^\dagger X) - \pi_t(X)\pi_t(s_i + s_i^\dagger)]dW_i. \end{aligned} \tag{B.35}$$

$W_i$ are Wiener processes with $dW_i dW_j = \delta_{ij}\,dt$ and the Hamiltonian is at most quadratic in the system's quadratures, which we collect into a column vector $\boldsymbol{X} = (X_1, \ldots, X_{2n})^{\mathrm{T}}$. The canonical commutation relations can then be written as $[X_i, X_j] = iJ_{ij}$, where $\mathbf{J}$ is a skew-symmetric real matrix. Similarly we define $\boldsymbol{Y} = (Y_i, \ldots, Y_{m'})$. We can parametrize the jump operators $\boldsymbol{s} = (s_1, \ldots, s_m)^{\mathrm{T}}$, and the Hamiltonian $H$ as

$$\boldsymbol{s} = \boldsymbol{\Lambda X}, \tag{B.36}$$

$$H = \frac{1}{2}\mathbf{X}^T\mathbf{R}\mathbf{X} + \left[\mathbf{X}^{\mathrm{T}}\tilde{\mathbf{R}}\boldsymbol{u}(t) + \mathrm{H.c.}\right], \tag{B.37}$$

where $\mathbf{R} \in \mathbb{R}^{2n\times 2n}$ is symmetric, $\tilde{\mathbf{R}} \in \mathbb{C}^{2n\times m}$, and $\boldsymbol{u}(t)$ is a $m$-dimensional input signal, which will later be used as a control input. The system can then be described in terms of a linear vector quantum Langevin equation and an output equation (see the subsection below for a derivation)

$$d\boldsymbol{X}(t) = [\mathbf{F}\boldsymbol{X}(t) + \mathbf{G}\boldsymbol{u}(t)]dt + d\boldsymbol{V}(t), \tag{B.38a}$$

$$d\boldsymbol{Y}(t) = \mathbf{H}\boldsymbol{X}(t)dt + \left[d\boldsymbol{A}(t) + d\boldsymbol{A}(t)^\dagger\right], \tag{B.38b}$$

with the real matrices

$$\mathbf{F} = \mathbf{J}[\mathbf{R} + \mathrm{Im}(\boldsymbol{\Lambda}^\dagger\boldsymbol{\Lambda})], \tag{B.39a}$$

$$\mathbf{H} = 2\mathrm{Re}(\boldsymbol{\Lambda}), \tag{B.39b}$$

$$\mathbf{G} = \mathbf{J}(\tilde{\mathbf{R}} + \tilde{\mathbf{R}}^*), \tag{B.39c}$$

and the real, vector-valued noise process

$$d\boldsymbol{V} = i\mathbf{J}(\boldsymbol{\Lambda}^{\mathrm{T}}d\boldsymbol{A}^\dagger - \boldsymbol{\Lambda}^\dagger d\boldsymbol{A}), \tag{B.40}$$

where $d\boldsymbol{A} = (dA_1, \ldots, dA_m)^{\mathrm{T}}$. We assume the field is in the vacuum state $\rho_{\mathrm{vac}}$, such that $dA_i(t)dA_j(t) = \delta_{ij}dt$. The (co)variances of $d\boldsymbol{V}$ are thus determined by

$$\mathbf{N}dt = \mathrm{Re}\left(d\boldsymbol{V}d\boldsymbol{V}^{\mathrm{T}}\right) = \mathbf{J}\mathrm{Re}(\boldsymbol{\Lambda}^\dagger\boldsymbol{\Lambda})\mathbf{J}^{\mathrm{T}}dt, \tag{B.41a}$$

$$\mathbf{M}\mathrm{d}t = \mathrm{Re}\big(\mathrm{d}\boldsymbol{V}[\mathrm{d}\boldsymbol{A} + \mathrm{d}\boldsymbol{A}^\dagger]^\mathrm{T}\big) = -\mathbf{J}\,\mathrm{Im}(\boldsymbol{\Lambda}^\mathrm{T})\mathrm{d}t. \tag{B.41b}$$

From the quantum filter equations we can deduce the equations of motion for the conditional mean values $\hat{X}_i = \pi_t(X_i)$ and symmetric covariance matrix $\hat{\Sigma}_{ij} = \frac{1}{2}\pi_t(X_iX_j + X_jX_i) - \hat{X}_i\hat{X}_j$. We find

$$\mathrm{d}\hat{\boldsymbol{X}}(t) = \big[\mathbf{F}\hat{\boldsymbol{X}}(t) + \mathbf{G}\boldsymbol{u}(t)\big]\mathrm{d}t + \mathbf{K}(t)\big[\mathrm{d}\boldsymbol{Y}(t) - \mathbf{H}\hat{\boldsymbol{X}}(t)\mathrm{d}t\big], \tag{B.42a}$$

$$\frac{\mathrm{d}}{\mathrm{d}t}\hat{\boldsymbol{\Sigma}}(t) = \mathbf{F}\hat{\boldsymbol{\Sigma}}(t) + \hat{\boldsymbol{\Sigma}}(t)\mathbf{F}^\mathrm{T} + \mathbf{N} - \big[\hat{\boldsymbol{\Sigma}}(t)\mathbf{H}^\mathrm{T} + \mathbf{M}\big]\big[\hat{\boldsymbol{\Sigma}}(t)\mathbf{H}^\mathrm{T} + \mathbf{M}\big]^\mathrm{T}, \tag{B.42b}$$

where

$$\mathbf{K}(t) = \hat{\boldsymbol{\Sigma}}(t)\mathbf{H}^\mathrm{T} + \mathbf{M}. \tag{B.43}$$

Eq. (B.42) together with (B.43) are known as the *Kalman–Bucy* filter in classical estimation theory (Kalman and Bucy, 1961). Presuming certain stability conditions (Wiseman and Milburn, 2009, p. 292f), the steady-state solution of the conditional covariance matrix $\hat{\Sigma}$ can be found by setting the right-hand side of (B.42b) to zero, and solving the resulting algebraic Riccati equation. If instead we are interested in the properties of the unconditional state, we can solve the Lyapunov equation obtained from (B.42b) by dropping the last term. [The resulting equation can also be obtained from (B.38a) by application of Itō calculus.]

#### B.3.1.1 Derivation of the Linear Langevin Equations

We can derive the linear vector Langevin equation (B.38a) from the general form (A.23) by using the definition of the Liouville operator (A.24) together with Eqs. (B.36) and (B.37). In the following we are using a sum convention where repeated indices are summed over. For the unitary evolution we find

$$\begin{aligned}
-\mathrm{i}[X_k, H] &= -\frac{\mathrm{i}}{2}R_{ij}[X_k, X_iX_j] - \mathrm{i}(\tilde{R}_{ij}u_j[X_k, X_i] - \mathrm{H.c.}) \\
&= -\frac{\mathrm{i}}{2}R_{ij}([X_k, X_i]X_j + X_i[X_k, X_j]) + (\tilde{R}_{ij} + \tilde{R}^*_{ij})u_jJ_{ki} \\
&= \frac{1}{2}R_{ij}(J_{ki}X_j + X_iJ_{kj}) + (\tilde{R}_{ij} + \tilde{R}^*_{ij})u_jJ_{ki} \\
&= J_{ki}[R_{ij}X_j + (\tilde{R}_{ij} + \tilde{R}^*_{ij})u_j],
\end{aligned}$$

where in the second to last line we used $\mathbf{R} = \mathbf{R}^{\mathrm{T}}$. The Lindblad term yields

$$\begin{aligned}\sum_l \mathcal{D}^\dagger[s_l]X_k &= s_l^\dagger X_k s_l - \frac{1}{2}[s_l^\dagger s_l X_k + X_k s_l^\dagger s_l] \\ &= \Lambda_{li}^* \Lambda_{lj}\left( X_i^\dagger X_k X_j - \frac{1}{2}[X_i^\dagger X_j X_k + X_k X_i^\dagger X_j]\right) \\ &= \frac{1}{2}\Lambda_{li}^* \Lambda_{lj}(X_i[X_k, X_j] + [X_i, X_k]X_j) \\ &= \frac{\mathrm{i}}{2}\Lambda_{li}^* \Lambda_{lj}[J_{kj}X_i - J_{ki}X_j] \\ &= \frac{\mathrm{i}}{2}J_{kj}(\Lambda_{li}^*\Lambda_{lj} - \Lambda_{lj}^*\Lambda_{li})X_i = -J_{kj}\,\mathrm{Im}(\Lambda_{lj}\Lambda_{li}^*)X_i,\end{aligned}$$

while the noise terms result in

$$\begin{aligned}[X_k, s_l]\mathrm{d}A_l^\dagger &= \Lambda_{li}[X_k, X_i]\mathrm{d}A_l^\dagger = \mathrm{i}\Lambda_{li}J_{ki}\mathrm{d}A_l^\dagger, \\ [s_l^\dagger, X_k]\mathrm{d}A_l &= \Lambda_{li}^*[X_i, X_k]\mathrm{d}A_l = \mathrm{i}\Lambda_{li}^* J_{ik}\mathrm{d}A_l.\end{aligned}$$

Putting this in vector form we immediately find Eq. (B.38a). The output Eq. (B.38b) follows directly from the input–output relations (A.26) and the definition of the jump operators.

#### B.3.1.2 Derivation of the Kalman Filter Equations

The Kalman filter equations can be derived starting from the quantum filter equation (B.35) by using Itō-calculus. The derivation is very similar to the one of the classical Kalman filter, see, e.g., Xiong (2008). We start with the conditional mean and find an explicit expression for $\mathrm{d}\hat{\mathbf{X}}(t) = \mathrm{d}\pi_t(\mathbf{X})$. The result contains the three terms: the terms generated by the Hamiltonian evolution and the diffusive dynamics we can copy from above, thus

$$\begin{aligned}-\mathrm{i}\pi_t([X_k, H]) &= J_{ki}[R_{ij}\pi_t(X_j) + (\widetilde{R}_{ij} + \widetilde{R}_{ij}^*)u_j], \\ \sum_l \pi_t(\mathcal{D}^\dagger[s_l]X_k) &= -J_{kj}\,\mathrm{Im}(\Lambda_{lj}\Lambda_{li}^*)\pi_t(X_i).\end{aligned}$$

Introducing the shorthand notation $\tilde{s}_i = s_i - \pi_t(s_i)$ and $\widetilde{X}_i = X_i - \pi_t(X_i)$, the measurement term (without the noise increment) can be written as

$$\begin{aligned}\pi_t(X_k\tilde{s}_l + \tilde{s}_l X_k) &= \Lambda_{li}\pi_t(X_k\widetilde{X}_i) + \Lambda_{li}^*\pi_t(\widetilde{X}_i X_k) \\ &= \Lambda_{li}\pi_t(\widetilde{X}_k\widetilde{X}_i) + \Lambda_{li}^*\pi_t(\widetilde{X}_i\widetilde{X}_k) + (\Lambda_{li} + \Lambda_{li}^*)\pi_t(X_k)\pi_t(\widetilde{X}_i) \\ &= \mathrm{Re}(\Lambda_{li})[\pi_t(\widetilde{X}_k\widetilde{X}_i + \widetilde{X}_i\widetilde{X}_k)] + \mathrm{i}\,\mathrm{Im}(\Lambda_{li})\pi_t([\widetilde{X}_k, \widetilde{X}_i]) \\ &= 2\,\mathrm{Re}(\Lambda_{li})\hat{\Sigma}_{ki} - \mathrm{Im}(\Lambda_{li})J_{ki},\end{aligned}$$

where we used $\pi_t(\tilde{s}_i) = \pi_t(\tilde{X}) = 0$ and $\hat{\Sigma}_{ij} = \pi_t(\tilde{X}_i\tilde{X}_j + \tilde{X}_j\tilde{X}_i)$.

Adding up the resulting expressions and casting them into vector form leaves us with

$$\begin{aligned} d\pi_t(\mathbf{X}) =& \mathbf{J}[\mathbf{R} + \mathrm{Im}(\mathbf{\Lambda}^\dagger\mathbf{\Lambda})]\pi_t(\mathbf{X})dt + \mathbf{J}(\mathbf{R} + \mathbf{R}^*)\boldsymbol{u}(t)dt \\ &+ [2\hat{\mathbf{\Sigma}}(t)\mathrm{Re}(\mathbf{\Lambda})^\mathrm{T} - \mathbf{J}\,\mathrm{Im}(\mathbf{\Lambda})^\mathrm{T}]d\boldsymbol{W}(t). \end{aligned} \tag{B.44}$$

It remains to show that

$$dW_i = dY_i - \pi_t(s_i + s_i^\dagger)dt = dY_i - (\Lambda_{ij} + \Lambda_{ij}^*)\pi_t(X_j)dt. \tag{B.45}$$

This leads to Eq. (B.42a). The evolution of the corresponding covariance is determined by

$$d\hat{\Sigma}_{lk} = \frac{1}{2}d\pi_t(X_lX_k + X_kX_l) - d[\pi_t(X_l)\pi_t(X_k)]. \tag{B.46}$$

For simplicity we here neglect the deterministic input and thus set $\tilde{\mathbf{R}} = 0$; it is, however, straightforward to extend the presented derivation to the more general case $\tilde{\mathbf{R}} \neq 0$. The second term in Eq. (B.46) can be directly obtained from Eq. (B.44) by using Itō-rules. It is made up of the terms

$$\pi_t(X_k)d\pi_t(X_l) = [F_{ki}\pi_t(X_i)dt + K_{ki}dW_i]\pi_t(X_l), \tag{B.47a}$$

$$d\pi_t(X_k)d\pi_t(X_l) = K_{ki}K_{il}dt, \tag{B.47b}$$

plus the corresponding expression for $\pi_t(X_l)d\pi_t(X_k)$. The expressions for the first term in Eq. (B.46) are more involved. For the Hamiltonian evolution we find

$$\begin{aligned} -i\pi_t([X_lX_k, H]) &= -\frac{i}{2}R_{mn}\pi_t([X_lX_k, X_mX_n]) \\ &= \frac{1}{2}R_{mn}\pi_t(X_lX_mJ_{ln} + X_lX_nJ_{km} + X_mX_kJ_{ln} + X_nX_kJ_{lm}) \\ &= R_{mn}\pi_t(X_lX_nJ_{km} + X_nX_kJ_{lm}), \end{aligned}$$

and after symmetrization

$$\begin{aligned} -i\pi_t([X_kX_l + X_lX_k, H]) =& J_{lm}R_{mn}\pi_t([X_n, X_k]_+) + J_{km}R_{mn}\pi_t([X_n, X_l]_+) \\ =& 2J_{lm}R_{mn}[\hat{\Sigma}_{nk} + \pi_t(X_n)\pi_t(X_k)] \\ &+ 2J_{km}R_{mn}[\hat{\Sigma}_{nl} + \pi_t(X_n)\pi_t(X_l)]. \end{aligned}$$

We now note that we can write the Lindblad term $\mathcal{D}^\dagger[s]X = \frac{1}{2}(s^\dagger[X,s] + [s^\dagger, X]s)$. Together with the definition $\lambda_{nm} = (\boldsymbol{\Lambda}^\dagger\boldsymbol{\Lambda})_{nm}$ this leads to

$$
\begin{aligned}
\sum_j \pi_t(\mathcal{D}^\dagger[s_j]X_lX_k) &= \frac{1}{2}\Lambda_{jn}^*\Lambda_{jm}\pi_t(X_n[X_lX_k, X_m] + [X_n, X_lX_k]X_m) \\
&= \frac{\mathrm{i}}{2}\Lambda_{jn}^*\Lambda_{jm}\pi_t(X_n[J_{km}X_l + J_{lm}X_k] + [J_{nl}X_k + J_{nk}X_l]X_m) \\
&= \frac{\mathrm{i}}{4}\lambda_{nm}\pi_t(J_{km}X_nX_l + J_{lm}X_nX_k + J_{nl}X_kX_m + J_{nk}X_lX_m) \\
&\quad - \frac{\mathrm{i}}{4}\lambda_{nm}^*\pi_t(J_{km}X_lX_n + J_{lm}X_kX_n + J_{nl}X_mX_k + J_{nk}X_mX_l).
\end{aligned}
$$

We now decompose $\lambda_{nm} = \mathrm{Re}(\lambda_{nm}) + \mathrm{i}\,\mathrm{Im}(\lambda_{nm})$ into its real and imaginary part, and observe that $\mathrm{Re}(\lambda_{nm}) = \mathrm{Re}(\lambda_{mn})$ and $\mathrm{Im}(\lambda_{nm}) = -\mathrm{Im}(\lambda_{mn})$. Above expression can thus be written as the sum of the two terms

$$
\begin{aligned}
&\frac{\mathrm{i}}{4}\mathrm{Re}(\lambda_{nm})\pi_t(J_{km}[X_n, X_l] + J_{lm}[X_n, X_k]) \\
&\quad + \frac{\mathrm{i}}{4}\mathrm{Re}(\lambda_{nm})\pi_t(J_{nl}[X_k, X_m] + J_{nk}[X_l, X_m]) \\
&= \frac{\mathrm{i}}{2}\mathrm{Re}(\lambda_{nm})\pi_t(J_{km}[X_n, X_l] + J_{lm}[X_n, X_k]) \\
&= -\frac{1}{2}\mathrm{Re}(\lambda_{nm})(J_{km}J_{nl} + J_{lm}J_{nk}) = -J_{ln}\mathrm{Re}(\lambda_{nm})J_{mk}
\end{aligned}
$$

and

$$
\begin{aligned}
&-\frac{1}{4}\mathrm{Im}(\lambda_{nm})\pi_t(J_{km}[X_l, X_n]_+ + J_{nk}[X_m, X_l]_+) \\
&\quad - \frac{1}{4}\mathrm{Im}(\lambda_{nm})\pi_t(J_{nl}[X_m, X_k]_+ + J_{lm}[X_k, X_n]_+) \\
&= -\frac{1}{2}\mathrm{Im}(\lambda_{nm})\pi_t(J_{km}[X_l, X_n]_+ + J_{nl}[X_m, X_k]_+) \\
&= [\hat{\Sigma}_{ln} + \pi_t(X_l)\pi_t(X_n)]\,\mathrm{Im}(\lambda_{nm})J_{mk} \\
&\quad + J_{ln}\,\mathrm{Im}(\lambda_{nm})[\hat{\Sigma}_{mk} + \pi_t(X_m)\pi_t(X_k)].
\end{aligned}
$$

Note that both expressions are symmetric under the substitution $k \Leftrightarrow l$. For the measurement term we obtain

$$
\begin{aligned}
\pi_t(X_l X_k \tilde{s}_i + \tilde{s}_i^\dagger X_l X_k) &= \Lambda_{ij}\pi_t(X_l X_k \tilde{X}_j) + \Lambda_{ij}^*\pi_t(\tilde{X}_j X_l X_k)\\
&= \Lambda_{ij}\pi_t([\tilde{X}_l + \pi_t(X_l)][\tilde{X}_k + \pi_t(X_k)]\tilde{X}_j)\\
&\quad + \Lambda_{ij}^*\pi_t(\tilde{X}_j[\tilde{X}_l + \pi_t(X_l)][\tilde{X}_k + \pi_t(X_k)])\\
&= \Lambda_{ij}[\pi_t(X_k)\pi_t(\tilde{X}_l\tilde{X}_j) + \pi_t(X_l)\pi_t(\tilde{X}_k\tilde{X}_j)]\\
&\quad + \Lambda_{ij}^*[\pi_t(X_k)\pi_t(\tilde{X}_j\tilde{X}_l) + \pi_t(X_l)\pi_t(\tilde{X}_j\tilde{X}_k)]\\
&= \mathrm{Re}(\Lambda_{ij})[\pi_t(X_k)\pi_t([\tilde{X}_l, \tilde{X}_j]_+) + \pi_t(X_l)\pi_t([\tilde{X}_k, \tilde{X}_j]_+)]\\
&\quad - \mathrm{i}\,\mathrm{Im}(\Lambda_{ij})[\pi_t(X_k)\pi_t([\tilde{X}_j, \tilde{X}_l]) + \pi_t(X_l)\pi_t([\tilde{X}_j, \tilde{X}_k])]\\
&= 2\mathrm{Re}(\Lambda_{ij})[\pi_t(X_k)\hat{\Sigma}_{lj} + \pi_t(X_l)\hat{\Sigma}_{kj}]\\
&\quad + \mathrm{Im}(\Lambda_{ij})[\pi_t(X_k)J_{jl} + \pi_t(X_l)J_{jk}],
\end{aligned}
$$

where we used that for Gaussian states central moments of odd order must vanish, i.e., $\pi_t(\tilde{X}_i\tilde{X}_j\tilde{X}_k) = 0$ for all $i$, $j$, $k$. Combining these expressions yields

$$
\begin{aligned}
\frac{1}{2}\mathrm{d}\pi_t(X_l X_k + X_k X_l) &= F_{lj}\hat{\Sigma}_{jk} + \hat{\Sigma}_{lj}F_{kj} + N_{lk}\\
&\quad + F_{lj}\hat{X}_j\hat{X}_k + \hat{X}_l\hat{X}_j F_{kj} + K_{lj}\mathrm{d}W_j\hat{X}_k + \hat{X}_l\mathrm{d}W_j K_{kj}.
\end{aligned}
\tag{B.48}
$$

The first line corresponds to the standard Lyapunov equation. Plugging Eqs. (B.47) and (B.48) into (B.46), the second line of (B.48) cancels with Eq. (B.47a) and the corresponding term for $\pi_t(X_l)\mathrm{d}\pi_t(X_k)$. We eventually obtain the Riccati equation (B.42b).

### *B.3.2 Quantum LQG control*

Linear–quadratic–Gaussian control concerns the control of linear systems subject to additive Gaussian noise such that a quadratic cost function is minimized. The LQG controller turns out to be a combination of the system's Kalman filter plus a quadratic regulator, where the control signal $\boldsymbol{u}(t)$ depends on a set of observations $\{\boldsymbol{Y}(s) : 0 \leq s \leq t\}$. In this review we only deal with the asymptotic control problem for $t \to \infty$ as we are interested in the steady state of our systems. We therefore want to find a feedback strategy which minimizes the total cost (Bouten and van Handel, 2008)

$$
\int_0^\infty \mathbb{E}[h(\boldsymbol{X}(t), \boldsymbol{u}(t))]\mathrm{d}t. \tag{B.49}
$$

We choose $h$ to be of the form

$$h(\mathbf{X},\mathbf{u}) = \mathbf{X}^{\mathrm{T}}\mathbf{P}\mathbf{X} + \mathbf{u}^{\mathrm{T}}\mathbf{Q}\mathbf{u}, \tag{B.50}$$

where $\mathbf{P} \geq 0$ and $\mathbf{Q} > 0$ are both real, symmetric matrices of appropriate dimensions. Under the assumption of certain stability conditions (Wiseman and Milburn, 2009, p. 298) the optimal feedback signal is given by (Edwards and Belavkin, 2005)

$$\mathbf{u}(t) = -\mathbf{C}(t)\hat{\mathbf{X}}(t), \tag{B.51}$$

$$\mathbf{C} = \mathbf{Q}^{-1}\mathbf{G}^{\mathrm{T}}\boldsymbol{\Omega}^{\mathrm{ss}}, \tag{B.52}$$

with $\boldsymbol{\Omega}^{\mathrm{ss}}$ the solution of the algebraic Riccati equation

$$\mathbf{F}^{\mathrm{T}}\boldsymbol{\Omega}^{\mathrm{ss}} + \boldsymbol{\Omega}^{\mathrm{ss}}\mathbf{F} + \mathbf{P} - \boldsymbol{\Omega}^{\mathrm{ss}}\mathbf{G}\mathbf{Q}^{-1}\mathbf{G}^{\mathrm{T}}\boldsymbol{\Omega}^{\mathrm{ss}} = 0. \tag{B.53}$$

In Section 4.1 we need to calculate the steady-state covariance matrix of a linear system including optimal feedback. This can be achieved by first noting that (due to the separation principle (Bouten and van Handel, 2008)) we can write (B.42a) as

$$\mathrm{d}\hat{\mathbf{X}}(t) = (\mathbf{F} - \mathbf{G}\mathbf{C})\hat{\mathbf{X}}(t)\mathrm{d}t + \mathbf{K}\mathrm{d}\mathbf{W}, \tag{B.54}$$

where $\mathbf{W}$ is a Wiener process with $\mathrm{d}\mathbf{W}(t)\mathrm{d}\mathbf{W}(t)^{\mathrm{T}} = \mathbb{1}_m \mathrm{d}t$ (the innovations process). We also need that (Bouten et al., 2007; Edwards and Belavkin, 2003)

$$\mathrm{Re}\left(\mathbb{E}[(\mathbf{X}(t) - \hat{\mathbf{X}}(t))(\mathbf{X}(t) - \hat{\mathbf{X}}(t))^{\mathrm{T}}]\right) = \hat{\boldsymbol{\Sigma}}(t), \tag{B.55}$$

$$\mathbb{E}[(\mathbf{X}(t) - \hat{\mathbf{X}}(t))\hat{\mathbf{X}}(t)^{\mathrm{T}}] = 0, \tag{B.56}$$

where the first relation follows from the definition of $\hat{\mathbf{X}}$ and $\hat{\Sigma}$, and the second from the orthogonality principle (B.20). We therefore find

$$\mathrm{Re}\left(\mathbb{E}[\mathbf{X}(t)\mathbf{X}(t)^{\mathrm{T}}]\right) = \hat{\boldsymbol{\Sigma}}(t) + \mathbb{E}[\hat{\mathbf{X}}(t)\hat{\mathbf{X}}(t)^{\mathrm{T}}], \tag{B.57}$$

where the equation of motion for the last term on the right-hand side $\boldsymbol{\Xi}(t) = \mathbb{E}[\hat{\mathbf{X}}(t)\hat{\mathbf{X}}(t)^{\mathrm{T}}]$ can be deduced from (B.54). The steady-state solution $\boldsymbol{\Xi}^{\mathrm{ss}}$ which fulfills

$$(\mathbf{F} - \mathbf{G}\mathbf{C})\boldsymbol{\Xi}^{\mathrm{ss}} + \boldsymbol{\Xi}^{\mathrm{ss}}(\mathbf{F} - \mathbf{G}\mathbf{C})^{\mathrm{T}} + \mathbf{K}\mathbf{K}^{\mathrm{T}} = 0. \tag{B.58}$$

The steady-state solution of the symmetric covariance matrix of the controlled quantum system is thus given by

$$\lim_{t\to\infty} \mathrm{Re}\left(\mathbb{E}[\boldsymbol{X}(t)\boldsymbol{X}(t)^{\mathrm{T}}]\right) = \hat{\boldsymbol{\Sigma}}^{\mathrm{ss}} + \boldsymbol{\Xi}^{\mathrm{ss}}. \tag{B.59}$$

Finally, we want to estimate the magnitude of the expected feedback signal. We quantify this by $\mathbb{E}[\boldsymbol{u}^{\mathrm{T}}(t)\boldsymbol{u}(t)]$. In the steady state we find

$$\mathbb{E}[\boldsymbol{u}^{\mathrm{T}}(t)\boldsymbol{u}(t)] = \mathbb{E}[\hat{\boldsymbol{X}}^{\mathrm{T}}\mathbf{C}^{\mathrm{T}}\mathbf{C}\hat{\boldsymbol{X}}] = \mathrm{tr}(\mathbf{C}\boldsymbol{\Xi}^{\mathrm{ss}}\mathbf{C}^{\mathrm{T}}). \tag{B.60}$$

## REFERENCES

Abdi, M., Barzanjeh, S., Tombesi, P., Vitali, D., 2011. Effect of phase noise on the generation of stationary entanglement in cavity optomechanics. Phys. Rev. A 84 (3), 032325. http://dx.doi.org/10.1103/PhysRevA.84.032325.

Ang, S.Z., Harris, G.I., Bowen, W.P., Tsang, M., 2013. Optomechanical parameter estimation. New J. Phys. 15 (10), 103028. ISSN: 1367–2630. http://dx.doi.org/10.1088/1367-2630/15/10/103028. http://iopscience.iop.org/1367-2630/15/10/103028.

Aspelmeyer, M., Gröblacher, S., Hammerer, K., Kiesel, N., 2010. Quantum optomechanics—throwing a glance. J. Opt. Soc. Am. B 27 (6), A189–A197. http://dx.doi.org/10.1364/JOSAB.27.00A189. http://josab.osa.org/abstract.cfm?URI=josab-27-6-A189.

Aspelmeyer, M., Meystre, P., Schwab, K., 2012. Quantum optomechanics. Phys. Today 65 (7), 29–35. http://dx.doi.org/10.1063/PT.3.1640. http://link.aip.org/link/?PTO/65/29/1.

Aspelmeyer, M., Kippenberg, T., Marquardt, F., 2014a. Cavity Optomechanics. Springer, Berlin Heidelberg.

Aspelmeyer, M., Kippenberg, T.J., Marquardt, F., 2014b. Cavity optomechanics. Rev. Mod. Phys. 86 (4), 1391–1452. http://dx.doi.org/10.1103/RevModPhys.86.1391.

Bar-Shalom, Y., Li, X.R., Kirubarajan, T., 2001. Estimation With Applications to Tracking and Navigation: Theory Algorithms and Software. John Wiley & Sons, Hoboken, NJ. ISBN 978-0-471-41655-5.

Barchielli, A., 1986. Measurement theory and stochastic differential equations in quantum mechanics. Phys. Rev. A 34 (3), 1642–1649. http://dx.doi.org/10.1103/PhysRevA.34.1642.

Barchielli, A., 2006. Continual measurements in quantum mechanics and quantum stochastic calculus. In: Attal, S., Joye, A., Pillet, C.-A. (Eds.), Open Quantum Systems III, In: vol. 1882. Springer-Verlag, Berlin/Heidelberg, pp. 207–292, http://link.springer.com/chapter/10.1007%2F3-540-33967-1_5?LI=true.

Barchielli, A., Paganoni, A.M., 1996. Detection theory in quantum optics: stochastic representation. Quantum Semiclass. Opt. 8 (1), 133–156. ISSN 1355-5111, 1361-6625. http://dx.doi.org/10.1088/1355-5111/8/1/011. http://iopscience.iop.org/1355-5111/8/1/011.

Barker, P.F., Shneider, M.N., 2010. Cavity cooling of an optically trapped nanoparticle. Phys. Rev. A 81 (2), 023826. http://dx.doi.org/10.1103/PhysRevA.81.023826.

Barrett, M.D., Chiaverini, J., Schaetz, T., Britton, J., Itano, W.M., Jost, J.D., Knill, E., Langer, C., Leibfried, D., Ozeri, R., Wineland, D.J., 2004. Deterministic quantum teleportation of atomic qubits. Nature 429 (6993), 737–739. ISSN 0028-0836, 1476-4679. http://dx.doi.org/10.1038/nature02608.

Belavkin, V., 1980. Optimal filtering of Markov signals with quantum white noise. Radio Eng. Electron Phys. 25, 1445–1453.

Belavkin, V.P., 1992. Quantum stochastic calculus and quantum nonlinear filtering. J. Multivar. Anal. 42 (2), 171–201. ISSN 0047259X. http://dx.doi.org/10.1016/0047-259X(92)90042-E. http://linkinghub.elsevier.com/retrieve/pii/0047259X9290042E.

Belavkin, V., 1999. Measurement, filtering and control in quantum open dynamical systems. Rep. Math. Phys. 43 (3), A405–A425. ISSN 0034-4877. http://dx.doi.org/10.1016/S0034-4877(00)86386-7. http://www.sciencedirect.com/science/article/pii/S0034487700863867.

Belavkin, V.P., Guta, M., 2008. Quantum Stochastics and Information: Statistics, Filtering and Control. World Scientific, Singapore. ISBN 978-981-283-295-5.

Benguria, R., Kac, M., 1981. Quantum Langevin Equation. Phys. Rev. Lett. 46 (1), 1–4. http://dx.doi.org/10.1103/PhysRevLett.46.1.

Bennett, C.H., Brassard, G., Crépeau, C., Jozsa, R., Peres, A., Wootters, W.K., 1993. Teleporting an unknown quantum state via dual classical and Einstein-Podolsky-Rosen channels. Phys. Rev. Lett. 70 (13), 1895. http://dx.doi.org/10.1103/PhysRevLett.70.1895.

Bergquist, J.C., Jefferts, S.R., Wineland, D.J., 2001. Time measurement at the millennium. Phys. Today 54 (3), 37–42. ISSN 0031-9228, 1945-0699. http://dx.doi.org/10.1063/1.1366066.

Bouten, L., van Handel, R., 2008. On the separation principle of quantum control. In: V.P. Belavkin, M.I. Guta, (Eds.), Quantum Stochastics and Information: Statistics, Filtering and Control. World Scientific, Singapore. http://www.worldscientific.com/worldscibooks/10.1142/6942.

Bouten, L., Van Handel, R., James, M.R., 2007. An introduction to quantum filtering. SIAM J. Control Optim. 46 (6), 2199–2241. ISSN 0363-0129, 1095-7138. http://dx.doi.org/10.1137/060651239.

Bouwmeester, D., Pan, J.-W., Mattle, K., Eibl, M., Weinfurter, H., Zeilinger, A., 1997. Experimental quantum teleportation. Nature 390 (6660), 575–579. ISSN 0028-0836. http://dx.doi.org/10.1038/37539.

Bowen, W.P., Milburn, G.J., 2016. Quantum Optomechanics. CRC Press, Boca Raton, FL.

Braginskii, V.B., Manukin, A.B., 1967. Ponderomotive effects of electromagnetic radiation. Sov. Phys. JETP 25, 653. ISSN 1063-7761. http://adsabs.harvard.edu/abs/1967JETP..25.653B.

Braginskii, V.B., Manukin, A.B., Tikhonov, M.Y., 1970. Investigation of dissipative ponderomotive effects of electromagnetic radiation. Sov. Phys. JETP 31, 829. ISSN 1063-7761. http://adsabs.harvard.edu/abs/1970JETP..31.829B.

Braginsky, V.B., Vyatchanin, S.P., 2002. Low quantum noise tranquilizer for Fabry–Perot interferometer. Phys. Lett. A 293 (5–6), 228–234. ISSN 0375-9601. http://dx.doi.org/10.1016/S0375-9601(02)00020-8. http://www.sciencedirect.com/science/article/pii/S0375960102000208.

Braginsky, V.B., Vorontsov, Y.I., Thorne, K.S., 1980. Quantum Nondemolition Measurements. Science 209 (4456), 547–557. ISSN 0036-8075, 1095-9203. http://dx.doi.org/10.1126/science.209.4456.547. http://www.sciencemag.org/content/209/4456/547.

Braginsky, V.B., Gorodetsky, M.L., Khalili, F.Y., 1997. Optical bars in gravitational wave antennas. Phys. Lett. A 232 (5), 340–348. ISSN 0375-9601. http://dx.doi.org/10.1016/S0375-9601(97)00413-1. http://www.sciencedirect.com/science/article/pii/S0375960197004131.

Braunstein, S.L., Kimble, H.J., 1998. Teleportation of continuous quantum variables. Phys. Rev. Lett. 80 (4), 869. http://dx.doi.org/10.1103/PhysRevLett.80.869. http://link.aps.org/abstract/PRL/v80/p869.

Braunstein, S.L., van Loock, P., 2005. Quantum information with continuous variables. Rev. Mod. Phys. 77 (2), 513. http://dx.doi.org/10.1103/RevModPhys.77.513.

Brooks, D.W.C., Botter, T., Schreppler, S., Purdy, T.P., Brahms, N., Stamper-Kurn, D.M., 2012. Non-classical light generated by quantum-noise-driven cavity optomechanics. Nature 488 (7412), 476–480. ISSN 0028-0836, 1476-4687. http://dx.doi.org/10.1038/nature11325.

Caldeira, A.O., Leggett, A.J., 1983. Path integral approach to quantum Brownian motion. Physica A 121 (3), 587–616. ISSN 0378-4371. http://dx.doi.org/10.1016/0378-4371(83)90013-4. http://www.sciencedirect.com/science/article/pii/0378437183900134.

Carmichael, H., 1993. An Open Systems Approach to Quantum Optics. Springer, Berlin. ISBN 978-0-387-56634-4.

Cerrillo, J., Machnes, S., Aspelmeyer, M., Wieczorek, W., Plenio, M.B., Retzker, A., 2011. Pulsed laser cooling for cavity-optomechanical resonators. arXiv:1104.5448. http://arxiv.org/abs/1104.5448.

Chan, J., Alegre, T.P.M., Safavi-Naeini, A.H., Hill, J.T., Krause, A., Groblacher, S., Aspelmeyer, M., Painter, O., 2011. Laser cooling of a nanomechanical oscillator into its quantum ground state. Nature 478 (7367), 89–92. ISSN 0028-0836. http://dx.doi.org/10.1038/nature10461.

Chang, D.E., Regal, C.A., Papp, S.B., Wilson, D.J., Ye, J., Painter, O., Kimble, H.J., Zoller, P., 2010. Cavity opto-mechanics using an optically levitated nanosphere. Proc. Natl. Acad. Sci. 107 (3), 1005–1010. ISSN 0027-8424, 1091-6490. http://dx.doi.org/10.1073/pnas.0912969107. http://www.pnas.org/content/107/3/1005.

Cirac, J.I., Blatt, R., Zoller, P., Phillips, W.D., 1992. Laser cooling of trapped ions in a standing wave. Phys. Rev. A 46 (5), 2668–2681. http://dx.doi.org/10.1103/PhysRevA.46.2668.

Clerk, A.A., Marquardt, F., Jacobs, K., 2008. Back-action evasion and squeezing of a mechanical resonator using a cavity detector. New J. Phys. 10 (9), 095010. ISSN 1367-2630. http://dx.doi.org/10.1088/1367-2630/10/9/095010. http://iopscience.iop.org/1367-2630/10/9/095010.

Cohadon, P.F., Heidmann, A., Pinard, M., 1999. Cooling of a mirror by radiation pressure. Phys. Rev. Lett. 83 (16), 3174–3177. http://dx.doi.org/10.1103/PhysRevLett.83.3174.

Cole, G.D., Wilson-Rae, I., Werbach, K., Vanner, M.R., Aspelmeyer, M., 2011. Phonon-tunnelling dissipation in mechanical resonators. Nat. Commun. 2, 231. http://dx.doi.org/10.1038/ncomms1212.

Corbitt, T., Chen, Y., Innerhofer, E., Müller-Ebhardt, H., Ottaway, D., Rehbein, H., Sigg, D., Whitcomb, S., Wipf, C., Mavalvala, N., 2007. An all-optical trap for a gram-scale mirror. Phys. Rev. Lett. 98 (15), 150802. http://dx.doi.org/10.1103/PhysRevLett.98.150802.

Courty, J.-M., Heidmann, A., Pinard, M., 2001. Quantum limits of cold damping with optomechanical coupling. Eur. Phys. J. D 17 (3), 399–408. ISSN 1434-6060, 1434-6079. http://dx.doi.org/10.1007/s100530170014.

Courty, J.-M., Heidmann, A., Pinard, M., 2003. Quantum locking of mirrors in interferometers. Phys. Rev. Lett. 90 (8), 083601. http://dx.doi.org/10.1103/PhysRevLett.90.083601.

Cubitt, T.S., Verstraete, F., Dür, W., Cirac, J.I., 2003. Separable states can be used to distribute entanglement. Phys. Rev. Lett. 91 (3), 037902. http://dx.doi.org/10.1103/PhysRevLett.91.037902.

Dąbrowska, A., Gough, J., 2016. Belavkin filtering with squeezed light sources. Russ. J. Math. Phys. 23 (2), 172–184. ISSN 1061-9208, 1555-6638. http://dx.doi.org/10.1134/S1061920816020035.

Doherty, A.C., Jacobs, K., 1999. Feedback control of quantum systems using continuous state estimation. Phys. Rev. A 60 (4), 2700–2711. http://dx.doi.org/10.1103/PhysRevA.60.2700.

Dolde, F., Bergholm, V., Wang, Y., Jakobi, I., Naydenov, B., Pezzagna, S., Meijer, J., Jelezko, F., Neumann, P., Schulte-Herbrüggen, T., Biamonte, J., Wrachtrup, J., 2014. High-fidelity spin entanglement using optimal control. Nat. Commun. 5, 3371. http://dx.doi.org/10.1038/ncomms4371. http://www.nature.com/ncomms/2014/140228/ncomms4371/abs/ncomms4371.html.

Dong, D., Petersen, I.R., 2010a. Quantum control theory and applications: a survey. IET Control Theory Appl. 4 (12), 2651–2671. ISSN 1751-8644. http://dx.doi.org/10.1049/iet-cta.2009.0508.

Dong, D., Petersen, I.R., 2010b. Quantum control theory and applications: a survey. IET Control Theory Appl. 4 (20), 2651–2671. ISSN 1751-8644. http://digital-library.theiet.org/content/journals/10.1049/iet-cta.2009.0508.

Dorsel, A., McCullen, J.D., Meystre, P., Vignes, E., Walther, H., 1983. Optical bistability and mirror confinement induced by radiation pressure. Phys. Rev. Lett. 51 (17), 1550. http://dx.doi.org/10.1103/PhysRevLett.51.1550.

Duan, L.M., Giedke, G., Cirac, J.I., Zoller, P., 2000. Inseparability criterion for continuous variable systems. Phys. Rev. Lett. 84 (12), 2722. http://dx.doi.org/10.1103/PhysRevLett.84.2722. http://link.aps.org/abstract/PRL/v84/p2722.

Edwards, S., Belavkin, V., 2003. On the duality of quantum filtering and optimal feedback control in quantum open linear dynamical systems. In: Physics and Control, 2003. Proceedings. 2003 International Conference, vol. 3, pp. 768–772. http://dx.doi.org/10.1109/PHYCON.2003.1237001.

Edwards, S.C., Belavkin, V.P., 2005. Optimal Quantum Filtering and Quantum Feedback Control. ArXiv:quant-ph/0506018. http://arxiv.org/abs/quant-ph/0506018.

Einstein, A., 1909. Zum gegenwärtigen Stande des Strahlungsproblems. Phys. Z. (10), 185–193.

Elste, F., Girvin, S.M., Clerk, A.A., 2009. Quantum noise interference and backaction cooling in cavity nanomechanics. Phys. Rev. Lett. 102 (20), 207209. http://dx.doi.org/10.1103/PhysRevLett.102.207209.

Favero, I., Karrai, K., 2009. Optomechanics of deformable optical cavities. Nat. Photon 3 (4), 201–205. ISSN 1749-4885. http://dx.doi.org/10.1038/nphoton.2009.42.

Favero, I., Sankey, J., Weig, E.M., 2014. Mechanical resonators in the middle of an optical cavity. In: Aspelmeyer, M., Kippenberg, T.J., Marquardt, F. (Eds.), Cavity Optomechanics. Quantum Science and Technology. Springer, Berlin/Heidelberg, pp. 83–119. ISBN 978-3-642-55311-0 978-3-642-55312-7. http://link.springer.com/chapter/10.1007/978-3-642-55312-7_5.

Furusawa, A., Sorensen, J.L., Braunstein, S.L., Fuchs, C.A., Kimble, H.J., Polzik, E.S., 1998. Unconditional quantum teleportation. Science 282 (5389), 706–709. http://dx.doi.org/10.1126/science.282.5389.706. http://www.sciencemag.org/cgi/content/abstract/282/5389/706.

Galve, F., Pachón, L.A., Zueco, D., 2010. Bringing entanglement to the high temperature limit. Phys. Rev. Lett. 105 (18), 180501. http://dx.doi.org/10.1103/PhysRevLett.105.180501.

Gammelmark, S., Julsgaard, B., Mølmer, K., 2013. Past quantum states of a monitored system. Phys. Rev. Lett. 111 (16), 160401. http://dx.doi.org/10.1103/PhysRevLett.111.160401.

Gardiner, C.W., Collett, M.J., 1985. Input and output in damped quantum systems: quantum stochastic differential equations and the master equation. Phys. Rev. A 31 (6), 3761. http://dx.doi.org/10.1103/PhysRevA.31.3761. http://link.aps.org/abstract/PRA/v31/p3761.

Gardiner, C.W., Zoller, P., 2004. Quantum Noise, third ed. Springer, Berlin. ISBN 978-3-540-22301-6.

Genes, C., Mari, A., Tombesi, P., Vitali, D., 2008a. Robust entanglement of a micromechanical resonator with output optical fields. Phys. Rev. A 78 (3), 032316.

http://dx.doi.org/10.1103/PhysRevA.78.032316. http://link.aps.org/abstract/PRA/v78/e032316.

Genes, C., Vitali, D., Tombesi, P., Gigan, S., Aspelmeyer, M., 2008b. Ground-state cooling of a micromechanical oscillator: comparing cold damping and cavity-assisted cooling schemes. Phys. Rev. A 77 (3), 033804. http://dx.doi.org/10.1103/PhysRevA.77.033804. http://link.aps.org/abstract/PRA/v77/e033804.

Genes, C., Mari, A., Vitali, D., Tombesi, P., 2009. Quantum effects in optomechanical systems. In: Advances in Atomic Molecular and Optical Physics, vol. 57. Academic Press, Cambridge, MA, pp. 33–86 (Chapter 2) ISBN 1049-250X. http://www.sciencedirect.com/science/article/pii/S1049250X09570024.

Geremia, J., Stockton, J.K., Doherty, A.C., Mabuchi, H., 2003. Quantum Kalman liltering and the Heisenberg limit in atomic magnetometry. Phys. Rev. Lett. 91 (25), 250801. http://dx.doi.org/10.1103/PhysRevLett.91.250801.

Ghobadi, R., Bahrampour, A.R., Simon, C., 2011a. Optomechanical entanglement in the presence of laser phase noise. Phys. Rev. A 84 (6), 063827. http://dx.doi.org/10.1103/PhysRevA.84.063827.

Ghobadi, R., Bahrampour, A.R., Simon, C., 2011b. Quantum optomechanics in the bistable regime. Phys. Rev. A 84 (3), 033846. http://dx.doi.org/10.1103/PhysRevA.84.033846.

Giovannetti, V., Vitali, D., 2001. Phase-noise measurement in a cavity with a movable mirror undergoing quantum Brownian motion. Phys. Rev. A 63 (2), 023812. http://dx.doi.org/10.1103/PhysRevA.63.023812. http://link.aps.org/abstract/PRA/v63/e023812.

Goetsch, P., Graham, R., 1994. Linear stochastic wave equations for continuously measured quantum systems. Phys. Rev. A 50 (6), 5242–5255. http://dx.doi.org/10.1103/PhysRevA.50.5242.

Gough, J.E., 2012. Principles and applications of quantum control engineering. Philos. Trans. R. Soc. A Math. Phys. Eng. Sci. 370 (1979), 5239–5240. ISSN 1364-503X, 1471-2962. http://dx.doi.org/10.1098/rsta.2012.0331. http://rsta.royalsocietypublishing.org/content/370/1979/5239.

Gough, J., James, M., 2009. The series product and its application to quantum feedforward and feedback networks. IEEE Trans. Autom. Control 54 (11), 2530–2544. ISSN 0018-9286. http://dx.doi.org/10.1109/TAC.2009.2031205.

Gough, J., Kostler, C., 2010. Quantum filtering in coherent states. Commun. Stoch. Anal. 4 (4), 505–521. https://www.math.lsu.edu/cosa/contents.htm.

Gozzini, A., Maccarrone, F., Mango, F., Longo, I., Barbarino, S., 1985. Light-pressure bistability at microwave frequencies. J. Opt. Soc. Am. B 2 (11), 1841–1845. http://dx.doi.org/10.1364/JOSAB.2.001841. http://josab.osa.org/abstract.cfm?URI=josab-2-11-1841.

Gröblacher, S., Hertzberg, J.B., Vanner, M.R., Cole, G.D., Gigan, S., Schwab, K.C., Aspelmeyer, M., 2009. Demonstration of an ultracold micro-optomechanical oscillator in a cryogenic cavity. Nat. Phys. 5 (7), 485–488. ISSN 1745-2473. http://dx.doi.org/10.1038/nphys1301.

Gröblacher, S., Trubarov, A., Prigge, N., Cole, G.D., Aspelmeyer, M., Eisert, J., 2015. Observation of non-Markovian micromechanical Brownian motion. Nat. Commun. 6. http://dx.doi.org/10.1038/ncomms8606. http://www.nature.com/ncomms/2015/150728/ncomms8606/full/ncomms8606.html.

Hamerly, R., Mabuchi, H., 2013. Coherent controllers for optical-feedback cooling of quantum oscillators. Phys. Rev. A 87 (1), 013815. http://dx.doi.org/10.1103/PhysRevA.87.013815.

Hammerer, K., Polzik, E.S., Cirac, J.I., 2005. Teleportation and spin squeezing utilizing multimode entanglement of light with atoms. Phys. Rev. A 72 (5), 052313. http://dx.doi.org/10.1103/PhysRevA.72.052313.

Hartmann, M.J., Plenio, M.B., 2008. Steady state entanglement in the mechanical vibrations of two dielectric membranes. Phys. Rev. Lett. 101 (20), 200503. http://dx.doi.org/10.1103/PhysRevLett.101.200503.

Heijden, F.v.d., Duin, R., Ridder, D.d., Tax, D.M.J., 2005. Classification, Parameter Estimation and State Estimation: An Engineering Approach Using MATLAB. John Wiley & Sons, Hoboken, NJ. ISBN 978-0-470-09014-5.

Helstrom, C.W., 1969. Quantum detection and estimation theory. J. Stat. Phys. 1 (2), 231–252. ISSN 0022-4715, 1572-9613. http://dx.doi.org/10.1007/BF01007479.

Hofer, S.G., Hammerer, K., 2015. Entanglement-enhanced time-continuous quantum control in optomechanics. Phys. Rev. A 91 (3), 033822. http://dx.doi.org/10.1103/PhysRevA.91.033822.

Hofer, S.G., Wieczorek, W., Aspelmeyer, M., Hammerer, K., 2011. Quantum entanglement and teleportation in pulsed cavity optomechanics. Phys. Rev. A 84 (5), 052327. http://dx.doi.org/10.1103/PhysRevA.84.052327.

Hofer, S.G., Vasilyev, D.V., Aspelmeyer, M., Hammerer, K., 2013. Time-continuous bell measurements. Phys. Rev. Lett. 111 (17), 170404. http://dx.doi.org/10.1103/PhysRevLett.111.170404.

Hofer, S.G., 2015. Quantum Control of Optomechanical Systems. Doctoral Thesis, University of Vienna, Vienna. http://othes.univie.ac.at/38975/.

Huang, S., Agarwal, G.S., 2009. Entangling nanomechanical oscillators in a ring cavity by feeding squeezed light. New J. Phys. 11 (10), 103044. ISSN 1367-2630. http://dx.doi.org/10.1088/1367-2630/11/10/103044. http://iopscience.iop.org/1367-2630/11/10/103044/.

Hudson, R.L., Parthasarathy, K.R., 1984. Quantum Ito's formula and stochastic evolutions. Commun. Math. Phys. 93 (3), 301–323. ISSN 0010-3616. http://dx.doi.org/10.1007/BF01258530. http://www.springerlink.com/content/r475342gnu684346/abstract/.

Gradshteyn, I.S., Ryshik, I.M., 2007. Table of Integrals, Series, and Products, seventh ed. Academic Press, Amsterdam. ISBN 978-0-12-373637-6.

Iwasawa, K., Makino, K., Yonezawa, H., Tsang, M., Davidovic, A., Huntington, E., Furusawa, A., 2013. Quantum-limited mirror-motion estimation. Phys. Rev. Lett. 111 (16), 163602. http://dx.doi.org/10.1103/PhysRevLett.111.163602.

Jacobs, K., 2014. Quantum Measurement Theory and Its Applications. Cambridge University Press, Cambridge.

Kalman, R.E., 1960. A new approach to linear filtering and prediction problems. J. Basic Eng. 82 (1), 35–45. ISSN 0098-2202. http://dx.doi.org/10.1115/1.3662552.

Kalman, R.E., Bucy, R.S., 1961. New results in linear filtering and prediction theory. J. Fluids Eng. 83 (1), 95–108. ISSN 0098-2202. http://dx.doi.org/10.1115/1.3658902.

Kippenberg, T., Vahala, K., 2007. Cavity opto-mechanics. Opt. Express 15 (25), 17172–17205. http://dx.doi.org/10.1364/OE.15.017172. http://www.opticsexpress.org/abstract.cfm?URI=oe-15-25-17172.

Kippenberg, T.J., Vahala, K.J., 2008. Cavity Optomechanics: Back-Action at the Mesoscale. Science 321 (5893), 1172–1176. http://dx.doi.org/10.1126/science.1156032. http://www.sciencemag.org/cgi/content/abstract/321/5893/1172.

Kloeden, P.E., 1994. Numerical Solution of Sde Through Computer Experiments. Springer. ISBN 978-3-540-57074-5.

Krauter, H., Muschik, C.A., Jensen, K., Wasilewski, W., Petersen, J.M., Cirac, J.I., Polzik, E.S., 2011. Entanglement generated by dissipation and steady state entanglement of two macroscopic objects. Phys. Rev. Lett. 107 (8), 080503. http://dx.doi.org/10.1103/PhysRevLett.107.080503.

Kuzmich, A., Mandel, L., Bigelow, N.P., 2000. Generation of spin squeezing via continuous quantum nondemolition measurement. Phys. Rev. Lett. 85 (8), 1594–1597. http://dx.doi.org/10.1103/PhysRevLett.85.1594.

Law, C.K., 1995. Interaction between a moving mirror and radiation pressure: a Hamiltonian formulation. Phys. Rev. A 51 (3), 2537. http://dx.doi.org/10.1103/PhysRevA.51.2537. http://link.aps.org/abstract/PRA/v51/p2537.

Lecocq, F., Clark, J.B., Simmonds, R.W., Aumentado, J., Teufel, J.D., 2015. Quantum non-demolition measurement of a nonclassical state of a massive object. Phys. Rev. X 5, 041037. http://dx.doi.org/10.1103/PhysRevX.5.041037.

Lindblad, G., 1976. On the generators of quantum dynamical semigroups. Commun. Math. Phys. 48 (2), 119–130. http://dx.doi.org/10.1007/BF01608499.

Ludwig, M., Hammerer, K., Marquardt, F., 2010. Entanglement of mechanical oscillators coupled to a nonequilibrium environment. Phys. Rev. A 82 (1), 012333. http://dx.doi.org/10.1103/PhysRevA.82.012333.

Møller, C.B., Thomas, R.A., Vasilakis, G., Zeuthen, E., Tsaturyan, Y., Jensen, K., Schliesser, A., Hammerer, K., Polzik, E.S., 2016. Back Action Evading Quantum Measurement of Motion in a Negative Mass Reference Frame. arxiv:1608.03613. http://arxiv.org/abs/1608.03613.

Mancini, S., Tombesi, P., 1994. Quantum noise reduction by radiation pressure. Phys. Rev. A 49 (5), 4055. http://dx.doi.org/10.1103/PhysRevA.49.4055. http://link.aps.org/abstract/PRA/v49/p4055.

Mancini, S., Vitali, D., Tombesi, P., 1998. Optomechanical cooling of a macroscopic oscillator by homodyne feedback. Phys. Rev. Lett. 80 (4), 688–691. http://dx.doi.org/10.1103/PhysRevLett.80.688.

Mancini, S., Giovannetti, V., Vitali, D., Tombesi, P., 2002. Entangling macroscopic oscillators exploiting radiation pressure. Phys. Rev. Lett. 88 (12), 120401. http://dx.doi.org/10.1103/PhysRevLett.88.120401.

Mari, A., Eisert, J., 2009. Gently modulating optomechanical systems. Phys. Rev. Lett. 103 (21), 213603. http://dx.doi.org/10.1103/PhysRevLett.103.213603.

Marquardt, F., Girvin, S., 2009. Optomechanics. Physics 2, 40. ISSN 1943-2879. http://dx.doi.org/10.1103/Physics.2.40.

Marquardt, F., Chen, J.P., Clerk, A.A., Girvin, S.M., 2007. Quantum theory of cavity-assisted sideband cooling of mechanical motion. Phys. Rev. Lett. 99 (9), 093902. http://dx.doi.org/10.1103/PhysRevLett.99.093902.

Meystre, P., 2013. A short walk through quantum optomechanics. Ann. Phys. 525 (3), 215–233. ISSN 1521-3889. http://dx.doi.org/10.1002/andp.201200226.

Miao, H., Danilishin, S., Chen, Y., 2010. Universal quantum entanglement between an oscillator and continuous fields. Phys. Rev. A 81 (5), 052307. http://dx.doi.org/10.1103/ PhysRevA.81.052307.

Mišta, L., Korolkova, N., 2009. Improving continuous-variable entanglement distribution by separable states. Phys. Rev. A 80 (3), 032310. http://dx.doi.org/10.1103/PhysRevA.80.032310.

Nielsen, W.H.P., Tsaturyan, Y., Møller, C.B., Polzik, E.S., Schliesser, A., 2017. Multimode optomechanical system in the quantum regime. Proc. Natl. Acad. Sci. 114 (1), 62–66. http://dx.doi.org/10.1073/pnas.1608412114. http://www.pnas.org/content/114/1/62.abstract.

Ockeloen-Korppi, C.F., Damskägg, E., Pirkkalainen, J.-M., Clerk, A.A., Woolley, M.J., Sillanpää, M.A., 2016. Quantum backaction evading measurement of collective mechanical modes. Phys. Rev. Lett. 117, 140401. http://dx.doi.org/10.1103/PhysRevLett.117.140401.

O'Connell, A.D., Hofheinz, M., Ansmann, M., Bialczak, R.C., Lenander, M., Lucero, E., Neeley, M., Sank, D., Wang, H., Weides, M., Wenner, J., Martinis, J.M., Cleland, A.N., 2010. Quantum ground state and single-phonon control of a mechanical resonator. Nature 464 (7289), 697–703. ISSN 0028-0836. http://dx.doi.org/10.1038/nature08967.

Palomaki, T.A., Harlow, J.W., Teufel, J.D., Simmonds, R.W., Lehnert, K.W., 2013a. Coherent state transfer between itinerant microwave fields and a mechanical oscillator. Nature 495 (7440), 210–214. ISSN: 0028-0836. http://dx.doi.org/10.1038/nature11915.

Palomaki, T.A., Teufel, J.D., Simmonds, R.W., Lehnert, K.W., 2013b. Entangling mechanical motion with microwave fields. Science 342 (6159), 710–713. ISSN 0036-8075, 1095-9203. http://dx.doi.org/10.1126/science.1244563. http://www.sciencemag.org/content/342/6159/710.

Paris, M.G.A., 2009. Quantum estimation for quantum technology. Int. J. Quant. Inf. 7 (Suppl. 1), 125–137. ISSN 0219-7499. http://dx.doi.org/10.1142/S0219749909004839.

Paternostro, M., Vitali, D., Gigan, S., Kim, M.S., Brukner, C., Eisert, J., Aspelmeyer, M., 2007. Creating and probing multipartite macroscopic entanglement with light. Phys. Rev. Lett. 99 (25), 250401. http://dx.doi.org/10.1103/PhysRevLett.99.250401.

Pinard, M., Hadjar, Y., Heidmann, A., 1999. Effective mass in quantum effects of radiation pressure. Eur. Phys. J. D 7 (1), 107–116. http://dx.doi.org/10.1007/s100530050354.

Pinard, M., Dantan, A., Vitali, D., Arcizet, O., Briant, T., Heidmann, A., 2005. Entangling movable mirrors in a double-cavity system. Europhys. Lett. 72 (5), 747. http://dx.doi.org/10.1209/epl/i2005-10317-6.

Pirandola, S., Vitali, D., Tombesi, P., Lloyd, S., 2006. Macroscopic Entanglement by Entanglement Swapping. Phys. Rev. Lett. 97 (15), 150403. http://dx.doi.org/10.1103/PhysRevLett.97.150403.

Pirkkalainen, J.-M., Damskägg, E., Brandt, M., Massel, F., Sillanpää, M.A., 2015. Squeezing of quantum noise of motion in a micromechanical resonator. Phys. Rev. Lett. 115, 243601. http://dx.doi.org/10.1103/PhysRevLett.115.243601.

Plenio, M.B., 2005. Logarithmic negativity: a full entanglement monotone that is not convex. Phys. Rev. Lett. 95 (9), 090503. http://dx.doi.org/10.1103/PhysRevLett.95.090503.

Poggio, M., Degen, C.L., Mamin, H.J., Rugar, D., 2007. Feedback cooling of a cantilever's fundamental mode below 5 mK. Phys. Rev. Lett. 99 (1), 017201. http://dx.doi.org/10.1103/PhysRevLett.99.017201.

Purdy, T.P., Peterson, R.W., Regal, C.A., 2013. Observation of radiation pressure shot noise on a macroscopic object. Science 339 (6121), 801–804. ISSN: 0036-8075, 1095-9203. http://dx.doi.org/10.1126/science.1231282. http://www.sciencemag.org/content/339/6121/801.

Rabl, P., Genes, C., Hammerer, K., Aspelmeyer, M., 2009. Phase-noise induced limitations on cooling and coherent evolution in optomechanical systems. Phys. Rev. A 80 (6), 063819. http://dx.doi.org/10.1103/PhysRevA.80.063819.

Riebe, M., Häffner, H., Roos, C.F., Hänsel, W., Benhelm, J., Lancaster, G.P.T., Körber, T.W., Becher, C., Schmidt-Kaler, F., James, D.F.V., Blatt, R., 2004. Deterministic quantum teleportation with atoms. Nature 429 (6993), 734–737. ISSN 0028-0836. http://dx.doi.org/10.1038/nature02570. http://www.nature.com/nature/journal/v429/n6993/full/nature02570.html#B2.

Romero-Isart, O., Juan, M.L., Quidant, R., Cirac, J.I., 2010. Toward quantum superposition of living organisms. New J. Phys. 12 (3), 033015. ISSN 1367-2630. http://dx.doi.org/10.1088/1367-2630/12/3/033015. http://iopscience.iop.org/1367-2630/12/3/033015.

Romero-Isart, O., Pflanzer, A.C., Juan, M.L., Quidant, R., Kiesel, N., Aspelmeyer, M., Cirac, J.I., 2011. Optically levitating dielectrics in the quantum regime: theory and protocols. Phys. Rev. A 83 (1), 013803. http://dx.doi.org/10.1103/PhysRevA.83.013803.

Safavi-Naeini, A.H., Gröblacher, S., Hill, J.T., Chan, J., Aspelmeyer, M., Painter, O., 2013. Squeezed light from a silicon micromechanical resonator. Nature 500 (7461), 185–189.

ISSN 0028-0836. http://dx.doi.org/10.1038/nature12307. http://www.nature.com/nature/journal/v500/n7461/full/nature12307.html.

Saulson, P.R., 1990. Thermal noise in mechanical experiments. Phys. Rev. D 42 (8), 2437–2445. http://dx.doi.org/10.1103/PhysRevD.42.2437.

Sawadsky, A., Kaufer, H., Nia, R.M., Tarabrin, S.P., Khalili, F.Y., Hammerer, K., Schnabel, R., 2015. Observation of generalized optomechanical coupling and cooling on cavity resonance. Phys. Rev. Lett. 114 (4), 043601. http://dx.doi.org/10.1103/PhysRevLett.114.043601.

Schliesser, A., Kippenberg, T.J., 2010. Cavity optomechanics with whispering-gallery mode optical micro-resonators. In: Paul Berman, E.A., Lin, C. (Eds.), Advances in Atomic, Molecular, and Optical Physics, vol. 58. Academic Press, London, Oxford, Boston, New York, and San Diego, pp. 207–323. http://www.sciencedirect.com/science/article/pii/S1049250X10058106.

Sherson, J.F., Krauter, H., Olsson, R.K., Julsgaard, B., Hammerer, K., Cirac, I., Polzik, E.S., 2006. Quantum teleportation between light and matter. Nature 443 (7111), 557–560. ISSN 0028-0836. http://dx.doi.org/10.1038/nature05136.

Simon, R., 2000. Peres-Horodecki separability criterion for continuous variable systems. Phys. Rev. Lett. 84 (12), 2726–2729. http://dx.doi.org/10.1103/PhysRevLett.84.2726.

Stengel, R.F., 1994. Optimal Control and Estimation, Reissue ed. Dover Publications, New York. ISBN 978-0-486-68200-6.

Szorkovszky, A., Doherty, A.C., Harris, G.I., Bowen, W.P., 2012. Position estimation of a parametrically driven optomechanical system. New J. Phys. 14 (9), 095026. ISSN 1367-2630. http://dx.doi.org/10.1088/1367-2630/14/9/095026. http://iopscience.iop.org/1367-2630/14/9/095026.

Teufel, J.D., Donner, T., Li, D., Harlow, J.W., Allman, M.S., Cicak, K., Sirois, A.J., Whittaker, J.D., Lehnert, K.W., Simmonds, R.W., 2011. Sideband cooling of micromechanical motion to the quantum ground state. Nature 475 (7356), 359–363. ISSN 0028-0836. http://dx.doi.org/10.1038/nature10261.

Thompson, J.D., Zwickl, B.M., Jayich, A.M., Marquardt, F., Girvin, S.M., Harris, J.G.E., 2008. Strong dispersive coupling of a high-finesse cavity to a micromechanical membrane. Nature 452 (7183), 72–75. ISSN: 0028-0836. http://dx.doi.org/10.1038/nature06715.

Thorne, K.S., Drever, R.W.P., Caves, C.M., Zimmermann, M., Sandberg, V.D., 1978. Quantum nondemolition measurements of harmonic oscillators. Phys. Rev. Lett. 40 (11), 667–671. http://dx.doi.org/10.1103/PhysRevLett.40.667.

Tsang, M., 2009. Time-symmetric quantum theory of smoothing. Phys. Rev. Lett. 102 (25), 250403. http://dx.doi.org/10.1103/PhysRevLett.102.250403.

Tsang, M., Wiseman, H.M., Caves, C.M., 2011. Fundamental quantum limit to waveform estimation. Phys. Rev. Lett. 106 (9), 090401. http://dx.doi.org/10.1103/PhysRevLett.106. 090401.

Vacanti, G., Paternostro, M., Massimo Palma, G., Vedral, V., 2008. Optomechanical to mechanical entanglement transformation. New J. Phys. 10 (9), 095014. ISSN 1367-2630. http://dx.doi.org/10.1088/1367-2630/10/9/095014. http://iopscience.iop.org/1367-2630/10/9/095014/.

Vacchini, B., 2000. Completely positive quantum dissipation. Phys. Rev. Lett. 84 (7), 1374–1377. http://dx.doi.org/10.1103/PhysRevLett.84.1374.

Vaidman, L., 1994. Teleportation of quantum states. Phys. Rev. A 49 (2), 1473. http://dx.doi.org/10.1103/PhysRevA.49.1473.

Vanner, M.R., Pikovski, I., Cole, G.D., Kim, M.S., Brukner, C., Hammerer, K., Milburn, G.J., Aspelmeyer, M., 2011. Pulsed quantum optomechanics. Proc. Natl. Acad. Sci. 108 (39), 16182–16187. http://dx.doi.org/10.1073/pnas.1105098108. http://www.pnas.org/content/108/39/16182.abstract.

Vanner, M.R., Hofer, J., Cole, G.D., Aspelmeyer, M., 2013. Cooling-by-measurement and mechanical state tomography via pulsed optomechanics. Nat. Commun. 4, 2295. http://dx.doi.org/10.1038/ncomms3295. http://www.nature.com/ncomms/2013/130815/ncomms3295/full/ncomms3295.html.

Vasilyev, D.V., Muschik, C.A., Hammerer, K., 2013. Dissipative versus conditional generation of Gaussian entanglement and spin squeezing. Phys. Rev. A 87 (5), 053820. http://dx.doi.org/10.1103/PhysRevA.87.053820.

Verhagen, E., Deléglise, S., Weis, S., Schliesser, A., Kippenberg, T.J., 2012. Quantum-coherent coupling of a mechanical oscillator to an optical cavity mode. Nature 482 (7383), 63–67. ISSN 0028-0836. http://dx.doi.org/10.1038/nature10787. http://www.nature.com/nature/journal/v482/n7383/full/nature10787.html.

Vidal, G., Werner, R.F., 2002. Computable measure of entanglement. Phys. Rev. A 65 (3), 032314. http://dx.doi.org/10.1103/PhysRevA.65.032314.

Vitali, D., Mancini, S., Ribichini, L., Tombesi, P., 2002. Mirror quiescence and high-sensitivity position measurements with feedback. Phys. Rev. A 65 (6), 063803. http://dx.doi.org/10.1103/PhysRevA.65.063803.

Vitali, D., Gigan, S., Ferreira, A., Böhm, H.R., Tombesi, P., Guerreiro, A., Vedral, V., Zeilinger, A., Aspelmeyer, M., 2007a. Optomechanical entanglement between a movable mirror and a cavity field. Phys. Rev. Lett. 98 (3), 030405. http://dx.doi.org/10.1103/PhysRevLett.98.030405.

Vitali, D., Mancini, S., Tombesi, P., 2007b. Stationary entanglement between two movable mirrors in a classically driven Fabry–Perot cavity. J. Phys. A Math. Theor. 40 (28), 8055–8068. ISSN 1751-8113. http://dx.doi.org/10.1088/1751-8113/40/28/S14. http://iopscience.iop.org/1751-8121/40/28/S14?fromSearchPage=true.

Vitali, D., Tombesi, P., Woolley, M.J., Doherty, A.C., Milburn, G.J., 2007c. Entangling a nanomechanical resonator and a superconducting microwave cavity. Phys. Rev. A 76 (4), 042336. http://dx.doi.org/10.1103/PhysRevA.76.042336.

Vogel, W., Welsch, D.-G., 2006. Quantum Optics. John Wiley & Sons, Hoboken, NJ. ISBN 978-3-527-60845-4.

Walls, D., Milburn, G.J., 2008. Quantum Optics. Springer, Berlin. ISBN: 978-3-540-28573-1, 978-3-540-28573-3; 978-3-540-28574-8, 978-3-540-28574-1.

Wieczorek, W., Hofer, S.G., Hoelscher-Obermaier, J., Riedinger, R., Hammerer, K., Aspelmeyer, M., 2015. Optimal state estimation for cavity optomechanical systems. Phys. Rev. Lett. 114 (22), 223601. http://dx.doi.org/10.1103/PhysRevLett.114.223601.

Wilson, D.J., Sudhir, V., Piro, N., Schilling, R., Ghadimi, A., Kippenberg, T.J., 2015. Measurement-based control of a mechanical oscillator at its thermal decoherence rate. Nature 524 (7565), 325–329. ISSN 0028-0836. http://dx.doi.org/10.1038/nature14672. http://www.nature.com/nature/journal/v524/n7565/full/nature14672.html.

Wilson-Rae, I., Nooshi, N., Zwerger, W., Kippenberg, T.J., 2007. Theory of ground state cooling of a mechanical oscillator using dynamical backaction. Phys. Rev. Lett. 99 (9), 093901. http://dx.doi.org/10.1103/PhysRevLett.99.093901.

Wiseman, H.M., 1994. Quantum theory of continuous feedback. Phys. Rev. A 49 (3), 2133–2150. http://dx.doi.org/10.1103/PhysRevA.49.2133.

Wiseman, H.M., 1995. Using feedback to eliminate back-action in quantum measurements. Phys. Rev. A 51 (3), 2459–2468. http://dx.doi.org/10.1103/PhysRevA.51.2459.

Wiseman, H.M., Milburn, G.J., 1993. Quantum theory of field-quadrature measurements. Phys. Rev. A 47 (1), 642–662. http://dx.doi.org/10.1103/PhysRevA.47.642.

Wiseman, H.M., Milburn, G.J., 2009. Quantum Measurement and Control. Cambridge University Press, Cambridge. ISBN 978-0-521-80442-4.

Wollman, E.E., Lei, C.U., Weinstein, A.J., Suh, J., Kronwald, A., Marquardt, F., Clerk, A.A., Schwab, K.C., 2015. Quantum squeezing of motion in a mechanical resonator. Science 349 (6251), 952–955. ISSN 0036-8075. http://dx.doi.org/10.1126/science.aac5138. http://science.sciencemag.org/content/349/6251/952.

Woolley, M.J., Clerk, A.A., 2013. Two-mode back-action-evading measurements in cavity optomechanics. Phys. Rev. A 87 (6), 063846. http://dx.doi.org/10.1103/PhysRevA.87.063846.

Woolley, M.J., Doherty, A.C., Milburn, G.J., Schwab, K.C., 2008. Nanomechanical squeezing with detection via a microwave cavity. Phys. Rev. A 78 (6), 062303. http://dx.doi.org/10.1103/PhysRevA.78.062303.

Xiong, J., 2008. An Introduction to Stochastic Filtering Theory. Oxford University Press, Oxford. ISBN 978-0-19-155139-0.

Xuereb, A., Schnabel, R., Hammerer, K., 2011. Dissipative optomechanics in a Michelson-Sagnac interferometer. Phys. Rev. Lett. 107 (21), 213604. http://dx.doi.org/10.1103/PhysRevLett.107.213604.

Zhang, J., Peng, K., Braunstein, S.L., 2003. Quantum-state transfer from light to macroscopic oscillators. Phys. Rev. A 68 (1), 013808. http://dx.doi.org/10.1103/PhysRevA.68.013808.

Zoller, P., Gardiner, C.W., 1997. Quantum noise in quantum optics: the stochastic Schrödinger equation. In: Reynaud, S., Giacobino, E., David, F. (Eds.), Quantum Fluctuations, vol. 63. North Holland, Amsterdam. ISBN 0-444-82593-2.

CHAPTER SIX

# Efficient Description of Bose–Einstein Condensates in Time-Dependent Rotating Traps

**Matthias Meister*[,1], Stefan Arnold[†], Daniela Moll*, Michael Eckart*, Endre Kajari*, Maxim A. Efremov*, Reinhold Walser[‡], Wolfgang P. Schleich*[,§]**

*Institut für Quantenphysik and Center for Integrated Quantum Science and Technology (IQ$^{ST}$), Universität Ulm, Ulm, Germany
[†]Institut für Theoretische Informatik, Universität Ulm, Ulm, Germany
[‡]Institut für Angewandte Physik, Technische Universität Darmstadt, Darmstadt, Germany
[§]Hagler Institute for Advanced Study at Texas A&M University, Institute for Quantum Science and Engineering (IQSE), Texas A&M University, TX, United States
[1]Corresponding author: e-mail address: matthias.meister@uni-ulm.de

## Contents



*Advances in Atomic, Molecular, and Optical Physics*, Volume 66
ISSN 1049-250X
http://dx.doi.org/10.1016/bs.aamop.2017.03.006

## Abstract

Quantum sensors based on matter-wave interferometry are promising candidates for high-precision gravimetry and inertial sensing in space. The favorable sources for the coherent matter waves in these devices are Bose–Einstein condensates. A reliable prediction of their dynamics, which is governed by the Gross–Pitaevskii equation, requires suitable analytical and numerical methods, which take into account the center-of-mass motion of the condensate, its rotation, and its spatial expansion by many orders of magnitude.

In this chapter, we present an efficient way to study their dynamics in time-dependent rotating traps that meet this objective. Both an approximate analytical solution for condensates in the Thomas–Fermi regime and dedicated numerical simulations on a variable adapted grid are discussed. We contrast and relate our approach to previous alternative methods and provide further results, such as analytical expressions for the one- and two-dimensional spatial density distributions and the momentum distribution in the long-time limit that are of immediate interest to experimentalists working in this field of research.

## 1. INTRODUCTION

Since the first creation of a Bose–Einstein condensate (BEC) in the mid-1990s (Anderson et al., 1995; Davis et al., 1995), the field of ultracold quantum gases (Dalfovo et al., 1999; Giorgini et al., 2008) has enormously developed and BECs are nowadays commonly used in a broad variety of applications. They include the generation of vortices (Fetter, 2009), the exploration of different quantum phase transitions (Bloch et al., 2008), and inertial sensors that are realized with the help of matter-wave interferometry (Berman, 1997; Cronin et al., 2009; Tino and Kasevich, 2014). In many of these phenomena the Gross–Pitaevskii (GP) equation (Gross, 1961, 1963; Pitaevskii, 1961) provides a reliable theoretical description of the BEC dynamics. Analytic solutions of the GP equation can be derived within the Thomas–Fermi (TF) approximation when combined with the so-called scaling approach (Castin and Dum, 1996; Kagan et al., 1996, 1997), where most of the BEC dynamics is described by an appropriate time-dependent coordinate transformation. In this chapter, we generalize this scaling approach and apply it directly to the GP equation to obtain an efficient analytic description of the dynamics of a BEC in time-dependent rotating traps.

Our approach was motivated and used by seminal experiments realized within the QUANTUS collaboration (Müntinga et al., 2013; van Zoest et al., 2010), which has successfully performed matter-wave interferometry with BECs in microgravity at the drop tower in Bremen (ZARM). These pioneering experiments study and manipulate BECs that expand freely for several seconds and hence evolve into matter waves of macroscopic dimensions. In order to reliably predict and describe the outcome of such experiments, it is essential to have dedicated analytical and numerical tools at hand, which take into account translational and rotational motions of the BEC as well as the fact that the spatial size of the condensate changes by many orders of magnitude during its free expansion. To account for these effects, the scaling approach has been generalized within the hydrodynamic framework (Edwards et al., 2002; Storey and Olshanii, 2000) by employing an affine transformation, where a translation first eliminates the center-of-mass motion and a subsequent linear map absorbs most of the remaining dynamics of the BEC. Here, we apply this affine transformation directly to the GP equation, resulting in an efficient description of the time evolution of a BEC in a time-dependent quadratic potential including slow rotations. Our approach has a straightforward application to matter-wave interferometry (Roura et al., 2014), facilitates an efficient numerical computation of the condensate wave function, and provides valuable analytical insights into the dynamics of BECs.

Our chapter is organized as follows. In Section 2, we first present the affine transformation of the GP equation, where we displace the wave function to eliminate the center-of-mass motion and then apply a linear transformation of the coordinates to account for the inner dynamics of the BEC. The combination of the affine approach and the time-dependent TF approximation results in an approximate, but astonishingly accurate solution for the dynamics of a BEC subject to a time-dependent rotating trap. Moreover, we derive expressions for the one- and two-dimensional integrated density distributions, which are experimentally accessible through time-of-flight pictures.

To verify the accuracy of our efficient description of the BEC dynamics, we perform in Section 3 full numerical simulations of the GP equation for a purely rotating trap as well as for the free expansion of an initially rotating BEC. We find an excellent agreement with our approximate analytical solution and show that the affine transformation can be used to improve the performance of numerical simulations by solving the transformed GP equation rather than the original one. Indeed, since the affine transformation itself does not contain any approximation, our technique does not alter the accuracy of the numerical simulations but speeds up the computation.

In Section 4, we establish a connection between the dynamics of the time-dependent affine transformation and the corresponding Hamiltonian formalism triggered by Kagan et al. (1997). In this context, we discuss two constants of motion of the affine transformation matrix, which we relate to the conservation of the total energy and the angular momentum of a BEC.

Since almost every experiment dealing with BECs makes use of time-of-flight pictures, it is essential to have a thorough understanding of the dynamics of a BEC during its free expansion. Indeed, due to their mean-field interaction, BECs possess a more complex time-of-flight dynamics compared with the simple ballistic expansion known from noninteracting quantum gases. Therefore, in Section 5, we compare the free time evolution of the GP equation with the Schrödinger equation and derive a relation between the long-time behavior of the momentum distribution of a freely expanding BEC and its initial spatial density distribution.

In order to keep our consideration self-contained but focused on the central ideas, we have included further details and calculations in three appendices. In Appendix A, we sketch the derivation of the affinely transformed GP equation and show how to calculate the integrated density distributions of a BEC within the TF regime. The important steps to determine the total energy of a BEC and its angular momentum within the time-dependent TF approximation are presented in Appendix B. Finally, in Appendix C, we discuss the behavior of the affine transformation matrix for a freely expanding BEC that is released from an isotropic harmonic trap.

## 2. EFFICIENT DESCRIPTION OF THE TIME EVOLUTION OF A BEC

This section introduces a natural generalization of the scaling approach (Castin and Dum, 1996; Kagan et al., 1996, 1997) to the case of *rotating harmonic traps*. In contrast to the hydrodynamical approach (Storey and Olshanii, 2000), we carry out this generalization directly to the GP equation. We apply these results to a BEC within the TF regime and obtain an approximate analytical description of its time evolution, the so-called time-dependent TF approximation. Moreover, we then derive simple expressions for the integrated density distributions of a BEC within the TF regime, which establish a direct link to experimental time-of-flight observations. We conclude this section with a brief discussion of an alternative but equivalent description (Edwards et al., 2002) of the internal dynamics of a BEC. Further details on the results presented in this section are collected in Appendix A.

## 2.1 Affine Transformation of the GP Equation

The affine transformation of the GP equation is realized in two steps (Eckart, 2008): namely (i) we eliminate the center-of-mass motion using Kohn's theorem (Bialynicki-Birula and Bialynicka-Birula, 2002; Dobson, 1994; Kohn, 1961), and (ii) we introduce a linear mapping of the coordinates in order to account for the main contributions to the internal dynamics of a BEC. Moreover, we show how this linear mapping is reduced to the well-known scaling approach (Castin and Dum, 1996; Kagan et al., 1996, 1997) in the case of a nonrotating harmonic potential.

### *2.1.1 Basic Setting*

We describe the dynamics of a BEC in an inertial frame of reference by the macroscopic wave function $\psi(t, \boldsymbol{x})$, which satisfies the GP equation

$$i\hbar\frac{\partial\psi}{\partial t}=\left[-\frac{\hbar^2}{2m}\nabla_{\boldsymbol{x}}^2+V(t,\boldsymbol{x})+g\left|\psi(t,\boldsymbol{x})\right|^2\right]\psi(t,\boldsymbol{x}), \tag{1}$$

where the position vector $\boldsymbol{x}$ is considered here as an element of the $d$-dimensional vector space $\mathbb{R}^d$ with $d = 1, 2,$ or $3$. In this way, our results are applicable to the general three-dimensional case (Clancy et al., 2007; Dalfovo et al., 1999; Edwards et al., 2002; Hechenblaikner et al., 2002; Jin et al., 1996; Ketterle et al., 1999; Mewes et al., 1996; Stamper-Kurn et al., 1998) as well as to the cases of BECs confined in one or two dimensions (Bongs et al., 2001; Görlitz et al., 2001; Petrov et al., 2000b; Plaja and Santos, 2002). Throughout this chapter the macroscopic wave function $\psi(t, \boldsymbol{x})$ is normalized to the number of particles $N$ in the condensate, that is,

$$\int_{\mathbb{R}^d}|\psi(t,\boldsymbol{x})|^2\,\mathrm{d}^d x=N. \tag{2}$$

The general potential $V(t, \boldsymbol{x})$ in Eq. (1) describes the interaction of an atom of mass $m$ with the external fields that correspond, for example, to a magneto-optical trap and Earth's gravity. Moreover, we assume that the atoms interact with each other via a repulsive contact interaction, leading to the nonlinear term $g\,|\psi(t,\,\boldsymbol{x})|^2$ in Eq. (1) with the positive coupling constant $g$.

In many experimental situations a semiclassical treatment of the underlying quantum mechanical dynamics allows an accurate and transparent interpretation of the measurement results. Here, we apply this approach to study the BEC dynamics around a classical trajectory $\boldsymbol{\rho}(t)$ and expand the potential $V(t,\,\boldsymbol{x})$ into a Taylor series up to second order around this trajectory

$$V(t,\boldsymbol{x}) = V(t,\boldsymbol{\rho}(t)) - \boldsymbol{F}(t)[\boldsymbol{x}-\boldsymbol{\rho}(t)] + \frac{m}{2}[\boldsymbol{x}-\boldsymbol{\rho}(t)]^{\mathrm{T}}\Omega^2(t)\,[\boldsymbol{x}-\boldsymbol{\rho}(t)]. \quad (3)$$

Any anharmonicity of the potential can be neglected, as long as the size of the BEC remains sufficiently small within the vicinity of the trajectory $\boldsymbol{\rho}(t)$. The latter can be associated with the center-of-mass motion of the atomic cloud or the minimum of the external potential $V(t,\,\boldsymbol{x})$, as discussed in more detail in Section 2.1.6.

Each term in Eq. (3) has a clear meaning. Indeed, the zeroth-order term $V(t,\,\boldsymbol{\rho}(t))$ represents the value of the potential along the trajectory $\boldsymbol{\rho}(t)$. The second term $-\boldsymbol{F}(t)[\boldsymbol{x}-\boldsymbol{\rho}(t)]$ corresponds to the force $\boldsymbol{F}(t)$ acting on the atoms at the point $\boldsymbol{x} = \boldsymbol{\rho}(t)$. The third term $\frac{1}{2}m[\boldsymbol{x}-\boldsymbol{\rho}(t)]^{\mathrm{T}}\Omega^2(t)[\boldsymbol{x}-\boldsymbol{\rho}(t)]$ is the purely quadratic trapping potential represented in terms of the symmetric positive definite matrix $\Omega^2(t)$. Its eigenvalues coincide with the squared trap frequencies $\omega_i^2(t)$ along the principal axes of the harmonic trap.

### *2.1.2 Elimination of the Center-of-Mass Motion*

We incorporate the center-of-mass motion of the BEC in a straightforward way and thereby eliminate all $\boldsymbol{\rho}$-dependent terms in the potential (3) by making use of the transformation

$$\psi(t,\boldsymbol{x}) = \mathrm{e}^{\frac{\mathrm{i}}{\hbar}S_1(t)}\,\mathrm{e}^{\frac{\mathrm{i}}{\hbar}[\boldsymbol{\mathcal{P}}(t)\,\boldsymbol{x}-\boldsymbol{\mathcal{R}}(t)\,\hat{\boldsymbol{p}}]}\,\psi_{\mathrm{D}}(t,\boldsymbol{x}) \quad (4)$$

from the original wave function $\psi(t,\,\boldsymbol{x})$ to the new one $\psi_{\mathrm{D}}(t,\,\boldsymbol{x})$. Note that $\hat{\boldsymbol{p}} \equiv -\mathrm{i}\hbar\boldsymbol{\nabla}_{\boldsymbol{x}}$ represents the momentum operator in the position representation.

The time-dependent vectors $\boldsymbol{\mathcal{R}}(t)$ and $\boldsymbol{\mathcal{P}}(t)$ in Eq. (4) describe the time evolution of the center of mass of the condensate (see Section 4.2.2 for more details) and obey the classical equations of motion

$$\begin{aligned}\frac{\mathrm{d}\mathcal{R}}{\mathrm{d}t}&=\frac{\mathcal{P}(t)}{m},\\ \frac{\mathrm{d}\mathcal{P}}{\mathrm{d}t}&=-m\Omega^2(t)[\mathcal{R}(t)-\boldsymbol{\rho}(t)]+\boldsymbol{F}(t),\end{aligned}\tag{5}$$

which are the Hamilton equations corresponding to the Lagrangian function

$$\mathcal{L}(\mathcal{R},\dot{\mathcal{R}},t)=\frac{m}{2}\dot{\mathcal{R}}^2-V(t,\boldsymbol{\rho})+\boldsymbol{F}(\mathcal{R}-\boldsymbol{\rho})-\frac{m}{2}(\mathcal{R}-\boldsymbol{\rho})^{\mathrm{T}}\Omega^2(\mathcal{R}-\boldsymbol{\rho}).\tag{6}$$

The global phase $\mathcal{S}_1(t)$ in Eq. (4) depends on the classical action via the generalized definition

$$\mathcal{S}_k(t)=\int_0^t\mathcal{L}(\mathcal{R},\dot{\mathcal{R}},t')\mathrm{d}t'-\frac{k}{2}[\mathcal{R}(t)\mathcal{P}(t)-\mathcal{R}(0)\mathcal{P}(0)],\tag{7}$$

where the additional integer $k\in\mathbb{Z}$ has been introduced for later purposes. As outlined in Appendix A.1, we arrive at the GP equation

$$\mathrm{i}\hbar\frac{\partial\psi_{\mathrm{D}}}{\partial t}=\left[-\frac{\hbar^2}{2m}\nabla_{\boldsymbol{x}}^2+\frac{m}{2}\boldsymbol{x}^{\mathrm{T}}\Omega^2(t)\,\boldsymbol{x}+g\,|\psi_{\mathrm{D}}(t,\boldsymbol{x})|^2\right]\psi_{\mathrm{D}}(t,\boldsymbol{x})\tag{8}$$

for the transformed wave function $\psi_{\mathrm{D}}(t,\boldsymbol{x})$ by inserting Eq. (4) into Eq. (1) and taking advantage of Eqs. (5)–(7) for $k=1$.

The decoupling of the center-of-mass motion of a BEC is possible as long as the potential is at most quadratic (Bialynicki-Birula and Bialynicka-Birula, 2002; Dobson, 1994; Nandi et al., 2007). It no longer holds true for anharmonic potentials for which a nontrivial coupling of the center-of-mass motion and the inner dynamics of the condensate exists (Dum et al., 1998).

### *2.1.3 Linear Transformation as a Natural Generalization of the Scaling Approach*

After having eliminated the center-of-mass motion, we perform a time-dependent linear transformation to account for the main internal dynamics of the BEC. For this purpose, we also introduce a new time coordinate $\tau$ such that the linear mapping between the "original coordinates" $(t,\boldsymbol{x})$ and the new ones $(\tau,\boldsymbol{\xi})$ reads

$$\begin{aligned}t&=\tau,\\ \boldsymbol{x}&=\Lambda(\tau)\,\boldsymbol{\xi},\end{aligned}\tag{9}$$

with $\Lambda(\tau)$ being an arbitrary, time-dependent matrix. In what follows, we refer to $\Lambda(\tau)$ as "adaptive matrix" and to $(\tau, \boldsymbol{\xi})$ as "adapted coordinates."

In analogy to Eq. (4), the coordinate transformation (9) goes hand in hand with the transformation

$$\psi_{\mathrm{D}}(t,\boldsymbol{x}) = \frac{1}{\sqrt{\det\Lambda(\tau)}}\, \mathrm{e}^{\frac{\mathrm{i}}{\hbar}\left[\boldsymbol{\xi}^{\mathrm{T}} A(\tau)\boldsymbol{\xi} - \beta(\tau)\right]}\, \psi_{\Lambda}(\tau,\boldsymbol{\xi}) \tag{10}$$

to the so-called affinely transformed wave function $\psi_{\Lambda}(t, \boldsymbol{x})$. The scalar phase $\beta(\tau)$ and the symmetric matrix $A(\tau)$ introduced in this transformation depend on the adaptive matrix $\Lambda(\tau)$ via

$$\beta(\tau) \equiv \int_0^{\tau} \frac{\mu}{\det\Lambda(\tau')}\, \mathrm{d}\tau' \tag{11}$$

and

$$A(\tau) \equiv \frac{m}{2}\, \Lambda^{\mathrm{T}}(\tau)\, \frac{\mathrm{d}\Lambda}{\mathrm{d}\tau}. \tag{12}$$

The constant $\mu$ that appears in the definition (11) represents the chemical potential associated with the ground state of the initial BEC at the time $\tau = 0$. In Appendix A.2, we outline the derivation of the affinely transformed GP equation

$$\begin{aligned} \mathrm{i}\hbar\, \frac{\partial \psi_{\Lambda}}{\partial \tau} = &-\frac{\hbar^2}{2m}\left[\Lambda^{-\mathrm{T}}(\tau)\boldsymbol{\nabla}_{\boldsymbol{\xi}}\right]^2 \psi_{\Lambda}(\tau,\boldsymbol{\xi}) \\ &+ \frac{1}{\det\Lambda(\tau)}\left[\frac{m}{2}\,\boldsymbol{\xi}^{\mathrm{T}}\Omega^2(0)\,\boldsymbol{\xi} + g\,|\psi_{\Lambda}(\tau,\boldsymbol{\xi})|^2 - \mu\right]\psi_{\Lambda}(\tau,\boldsymbol{\xi}) \end{aligned} \tag{13}$$

for the wave function $\psi_{\Lambda}(\tau, \boldsymbol{\xi})$, where we have made use of the shorthand notation $\Lambda^{-\mathrm{T}} \equiv (\Lambda^{-1})^{\mathrm{T}}$ and the requirement that the matrix $\Lambda(\tau)$ obeys the nonlinear matrix differential equation

$$\Lambda^{\mathrm{T}}(\tau)\left[\frac{\mathrm{d}^2\Lambda}{\mathrm{d}\tau^2} + \Omega^2(\tau)\,\Lambda(\tau)\right] = \frac{\Omega^2(0)}{\det\Lambda(\tau)}. \tag{14}$$

The additional assumption that at $\boldsymbol{x} = \boldsymbol{\xi}$ and $t = 0$ we have $\psi_{\mathrm{D}}(0, \boldsymbol{x}) = \psi_{\Lambda}(0, \boldsymbol{x})$ yields the initial conditions

$$\Lambda(0) = \mathbb{1} \quad \text{and} \quad \left.\frac{\mathrm{d}\Lambda}{\mathrm{d}\tau}\right|_{\tau=0} = 0 \tag{15}$$

for the nonlinear matrix differential equation (14). As discussed in Appendix A.3, the symmetry of the matrix $A(\tau)$, Eq. (12), is connected to the so-called irrotationality condition

$$\Lambda^{\mathrm{T}}(\tau)\,\frac{\mathrm{d}\Lambda}{\mathrm{d}\tau}=\frac{\mathrm{d}\Lambda^{\mathrm{T}}}{\mathrm{d}\tau}\,\Lambda(\tau), \tag{16}$$

that gives rise to $d(d-1)/2$ constants of motion of the matrix differential equation (14).

So far, we have made no approximation in deriving the affinely transformed GP equation (13). Although it looks much more complicated than Eq. (8), the solutions of Eq. (13) show almost no time dependence within the TF regime. This observation enables us to establish an efficient description of the BEC dynamics, as discussed in Section 2.2.

### 2.1.4 The Affinely Transformed Wave Function

By combining the two transformations given by Eqs. (4), (9), and (10), we obtain the following relation between the original and the affinely transformed wave function

$$\psi(t,\boldsymbol{x})=\frac{1}{\sqrt{\det\Lambda(t)}}\,\mathrm{e}^{\mathrm{i}\Phi(t,\boldsymbol{x})}\psi_{\Lambda}\left(t,\Lambda^{-1}(t)[\boldsymbol{x}-\mathcal{R}(t)]\right), \tag{17}$$

where we have introduced the local phase

$$\Phi(t,\boldsymbol{x})\equiv\frac{1}{\hbar}\left\{\mathcal{S}_1(t)-\beta(t)+\mathcal{P}(t)\left[\boldsymbol{x}-\frac{1}{2}\mathcal{R}(t)\right]+\frac{m}{2}[\boldsymbol{x}-\mathcal{R}(t)]^{\mathrm{T}}C(t)[\boldsymbol{x}-\mathcal{R}(t)]\right\} \tag{18}$$

and the time-dependent symmetric matrix

$$C(t)\equiv\frac{\mathrm{d}\Lambda}{\mathrm{d}t}\,\Lambda^{-1}(t). \tag{19}$$

Since the matrix $C(t)$ determines the time evolution of the quadratic phase term, we refer to it as the "quadratic phase matrix." For later purposes, we also note that by inverting Eq. (17), we find the affinely transformed wave function $\psi_{\Lambda}(t, \boldsymbol{x})$ in terms of $\psi(t, \boldsymbol{x})$ via

$$\psi_{\Lambda}(t,\boldsymbol{x})=\sqrt{\det\Lambda(t)}\;\mathrm{e}^{-\mathrm{i}\Phi(t,\Lambda(t)\boldsymbol{x}+\mathcal{R}(t))}\psi(t,\Lambda(t)\boldsymbol{x}+\mathcal{R}(t)). \tag{20}$$

In summary, the affine transformation of the macroscopic wave function is realized by Eq. (17) and leads to the affinely transformed GP equation (13).

In order to determine $\psi(t, \boldsymbol{x})$, one has to solve the time-dependent partial differential equation (13) together with the ordinary differential equations (5) and (14) for the center-of-mass variables $\mathcal{R}(t)$, $\mathcal{P}(t)$, and the adaptive matrix $\Lambda(t)$, respectively. Throughout this chapter, we refer to this generalization of the standard scaling method (Castin and Dum, 1996; Kagan et al., 1996, 1997) as the "affine approach."

### 2.1.5 Nonrotating Trap: Connection to the Scaling Approach

We continue by pointing out the connection between our affine approach and the scaling method introduced by Castin and Dum (1996) and Kagan et al. (1996, 1997). We consider a nonrotating trap and assume for simplicity that the principal axes of the harmonic potential coincide with the coordinate axes. This can always be achieved by an appropriate choice of coordinates for which the trap matrix $\Omega^2(t)$ in Eq. (3) possesses the diagonal form

$$\Omega^2(t) \equiv \mathrm{diag}\left[\omega_1^2(t), \ldots, \omega_d^2(t)\right], \tag{21}$$

with $\omega_i(t)$ being the trapping frequency along the $x_i$ direction. Accordingly, we assume that the adaptive matrix $\Lambda(\tau)$ is of diagonal form and substitute $\Lambda(\tau) \equiv \mathrm{diag}[\lambda_1(\tau), \ldots, \lambda_d(\tau)]$ into Eq. (14), which yields the coupled nonlinear differential equations

$$\frac{\mathrm{d}^2\lambda_i}{\mathrm{d}\tau^2} + \omega_i^2(\tau)\lambda_i(\tau) = \frac{\omega_i^2(0)}{\lambda_i(\tau)\, \Pi_{k=1}^d \lambda_k(\tau)}, \tag{22}$$

where $i \in 1, \ldots, d$. The functions $\lambda_i(\tau)$ characterize the time evolution of the condensate in terms of an individual scaling along the three principal axes of the potential. From Eq. (15), we obtain for the corresponding initial conditions

$$\lambda_i(0) = 1 \quad \text{and} \quad \left.\frac{\mathrm{d}\lambda_i}{\mathrm{d}\tau}\right|_{\tau=0} = 0. \tag{23}$$

Clearly, the irrotationality condition, Eq. (16), is automatically satisfied for a diagonal adaptive matrix $\Lambda(\tau)$. For the case $d = 3$, Eqs. (22) and (23) precisely rephrase the well-known equations derived for the scaling factors in Castin and Dum (1996) and Kagan et al. (1997).

### 2.1.6 Interpretation of the Classical Trajectory

During the preparation of the BEC by laser and evaporative cooling at $t \leq 0$, we denote by $\boldsymbol{\rho}(t)$ the position of the minimum of the full external potential $V(t, \boldsymbol{x})$, which acts on the atomic cloud. For this reason, the classical trajectory $\boldsymbol{\rho}(t)$ is found as a solution of the equation

$$\nabla_{\boldsymbol{x}} V(t, \boldsymbol{x})|_{\boldsymbol{x}=\boldsymbol{\rho}(t)} = 0 \tag{24}$$

for all times $t \leq 0$. Since the trapping potential $V(t, \boldsymbol{x})$ is located in an Earth-bound laboratory, a capsule freely falling in a drop tower, or a satellite in space, inertial effects due to the local acceleration and rotation of the comoving frame of reference attached to the trapping potential $V(t, \boldsymbol{x})$ along $\boldsymbol{\rho}(t)$ do occur in general. However, here we assume that these inertial effects can be neglected for all times $t \leq 0$ due to the dominating influence of the trapping potential and the repulsive interaction of the atoms.

As pointed out in Section 2.1.2, the time-dependent vectors $\mathcal{R}(t)$ and $\mathcal{P}(t)$ are associated with the center-of-mass motion of the atomic cloud. This interpretation implies that the initial conditions $\mathcal{R}(0)$ and $\mathcal{P}(0)$ for the classical equations of motion (5) are directly linked to the preparation of the BEC in the external potential $V(t, \boldsymbol{x})$ according to

$$\begin{aligned} \mathcal{R}(0) &= \boldsymbol{\rho}(0), \\ \mathcal{P}(0) &= m\dot{\boldsymbol{\rho}}(0). \end{aligned} \tag{25}$$

In other words, the center of mass of the initial BEC at $t = 0$ is supposed to be at rest within the comoving frame of reference attached to the external potential $V(t, \boldsymbol{x})$ at the position $\boldsymbol{\rho}(0)$.

After the initial preparation phase, the dynamical evolution of the BEC for times $t > 0$ can be analyzed based on two different interpretations of the classical trajectory $\boldsymbol{\rho}(t)$. In case one is interested in the relative motion of the center of mass of the BEC with respect to the trap minimum, the trajectory $\boldsymbol{\rho}(t)$ can still be associated with the minimum of the external potential $V(t, \boldsymbol{x})$ and Eq. (24) has to be satisfied by $\boldsymbol{\rho}(t)$ also for $t > 0$.

In terms of a semiclassical approach, one can likewise associate $\boldsymbol{\rho}(t)$ for all $t > 0$ with the center-of-mass motion of the atomic cloud itself. In this case, the trajectory $\boldsymbol{\rho}(t)$ is determined as the solution of the classical equation of motion

$$m\ddot{\boldsymbol{\rho}}(t) = -\nabla_{\boldsymbol{x}} V(t, \boldsymbol{x})|_{\boldsymbol{x}=\boldsymbol{\rho}(t)} \tag{26}$$

for $t > 0$ and describes the semiclassical motion of the center of mass of the BEC in the full external potential $V(t, \boldsymbol{x})$. Thus, the time-dependent vectors $\mathcal{R}(t)$ and $\mathcal{P}(t)$ follow directly from $\boldsymbol{\rho}(t)$ via

$$\begin{aligned} \mathcal{R}(t) &= \boldsymbol{\rho}(t), \\ \mathcal{P}(t) &= m\dot{\boldsymbol{\rho}}(t), \end{aligned} \tag{27}$$

whereas the classical equations of motion, Eq. (5), mathematically coincide with Eq. (26).

## 2.2 The Macroscopic Wave Function in the Time-Dependent TF Approximation

We start this section by recalling the TF approximation (Fetter and Feder, 1998; Pethick and Smith, 2002) for the ground state of a BEC in a harmonic trap. We then apply this method also to the case of a time-dependent rotating trap in order to provide an efficient description of the dynamics of the macroscopic wave function of a BEC within the TF regime (Castin and Dum, 1996; Dalfovo et al., 1999; Kagan et al., 1996, 1997; Storey and Olshanii, 2000).

In addition, we present a simple relationship between the spatial density distribution of a BEC in three dimensions and its corresponding integrated density distributions in one and two dimensions, which allows us to fully characterize the BEC dynamics within the TF regime by a sequence of three mutually orthogonal time-of-flight pictures. We conclude this section by briefly discussing an alternative description of the internal dynamics of a rotating BEC within the TF regime (Edwards et al., 2002).

### *2.2.1 TF Approximation for the Initial Ground State*

Without loss of generality, we assume that $V(0, \boldsymbol{\rho}(0)) = 0$ at the initial time $t = 0$. Furthermore, we recall that $\boldsymbol{F}(0) = 0$ for the harmonic potential (3) due to the validity of Eq. (24). Taking also Eq. (25) into account, the ground state $\phi(\boldsymbol{x})$ of a BEC at $t = 0$ is defined as the solution of the stationary GP equation

$$\mu\,\phi(\boldsymbol{x}) = \left\{ -\frac{\hbar^2}{2m}\nabla_{\boldsymbol{x}}^2 + \frac{m}{2}[\boldsymbol{x} - \mathcal{R}(0)]^{\mathrm{T}}\Omega^2(0)[\boldsymbol{x} - \mathcal{R}(0)] + g|\phi(\boldsymbol{x})|^2 \right\}\phi(\boldsymbol{x}), \tag{28}$$

with $\mu$ being the chemical potential of the ground state.

The TF regime is characterized by a dominant contribution of the atomic interactions to the total energy of the BEC (Fetter and Feder, 1998; Pethick and Smith, 2002). In this case, the kinetic energy term in the stationary GP equation (28) only accounts for a negligible contribution to the total energy in comparison with those arising from the harmonic potential and the atom–atom interactions. For this reason, within the TF approximation one simply neglects the kinetic energy term, which gives rise to the approximate ground state of the BEC

$$\phi_{\mathrm{TF}}(\boldsymbol{x}) = \frac{1}{\sqrt{g}}\left\{\mu_{\mathrm{TF}} - \frac{m}{2}[\boldsymbol{x} - \mathcal{R}(0)]^{\mathrm{T}}\Omega^2(0)[\boldsymbol{x} - \mathcal{R}(0)]\right\}_+^{\frac{1}{2}} \mathrm{e}^{\frac{\mathrm{i}}{\hbar}\mathcal{P}(0)\,\boldsymbol{x}}. \tag{29}$$

Here, the chemical potential $\mu_{\mathrm{TF}}$ within the TF approximation

$$\mu_{\mathrm{TF}} = \frac{m}{2}\left(\frac{2\,\Gamma\left(2+\frac{d}{2}\right)}{\pi^{d/2}}\frac{Ng}{m}\sqrt{\det\Omega^2(0)}\right)^{\frac{2}{d+2}} \tag{30}$$

is determined by the normalization condition (2) for the wave function $\phi_{\mathrm{TF}}(\boldsymbol{x})$, $\Gamma(x)$ denotes the Gamma function, and

$$\{x\}_+^{\alpha} \equiv x^{\alpha}\Theta(x) \tag{31}$$

stands for the positive part of $x$ to the power of $\alpha$, $\alpha > 0$, with $\Theta(x)$ being the Heaviside function. The additional phase $\mathrm{i}\mathcal{P}(0)\,\boldsymbol{x}/\hbar$ in Eq. (29) is due to the nonvanishing momentum $\mathcal{P}(0)$, Eq. (25), of the initial BEC as discussed in Section 2.1.6.

The ground state within the TF approximation $\phi_{\Lambda,\mathrm{TF}}(\boldsymbol{\xi})$ rewritten in terms of the adapted coordinates $\boldsymbol{\xi}$ follows directly from Eqs. (20) and (29) according to

$$\begin{aligned}\phi_{\Lambda,\mathrm{TF}}(\boldsymbol{\xi}) &= \mathrm{e}^{-\mathrm{i}\Phi(0,\boldsymbol{\xi}+\mathcal{R}(0))}\phi_{\mathrm{TF}}(\boldsymbol{\xi}+\mathcal{R}(0)) \\ &= \frac{1}{\sqrt{g}}\left\{\mu_{\mathrm{TF}} - \frac{m}{2}\boldsymbol{\xi}^{\mathrm{T}}\Omega^2(0)\,\boldsymbol{\xi}\right\}_+^{\frac{1}{2}} \mathrm{e}^{\frac{\mathrm{i}}{2\hbar}\mathcal{R}(0)\mathcal{P}(0)}.\end{aligned} \tag{32}$$

Besides the initial conditions for the adaptive matrix $\Lambda(t)$, Eq. (15), we have used the defining Eqs. (7) for $k = 1$, (11) and (19) to find $\Phi(0,\boldsymbol{x}) = \mathcal{P}(0)\left[\boldsymbol{x} - \frac{1}{2}\mathcal{R}(0)\right]/\hbar$ for the local phase (18) at $t = 0$ in the derivation of Eq. (32).

### 2.2.2 Time-Dependent TF Approximation for the Dynamical Evolution

Within the time-dependent TF approximation (Castin and Dum, 1996; Kagan et al., 1997), one neglects the kinetic energy term in the affinely transformed GP equation (13) for all times $\tau \geq 0$ in order to arrive at the approximate, ordinary differential equation

$$i\hbar \frac{\partial \psi_\Lambda}{\partial \tau} \approx \frac{1}{\det \Lambda(\tau)} \left[ \frac{m}{2} \boldsymbol{\xi}^{\mathrm{T}} \Omega^2(0) \boldsymbol{\xi} + g |\psi_\Lambda(\tau, \boldsymbol{\xi})|^2 - \mu \right] \psi_\Lambda(\tau, \boldsymbol{\xi}). \tag{33}$$

As we will show below, the solution $\psi_\Lambda(\tau, \boldsymbol{\xi})$ of Eq. (33) is time independent and therefore agrees with the approximate ground state $\phi_{\Lambda,\,\mathrm{TF}}(\boldsymbol{\xi})$, Eq. (32), for all times $\tau \geq 0$

$$\psi_\Lambda(\tau, \boldsymbol{\xi}) = \psi_\Lambda(0, \boldsymbol{\xi}) \approx \phi_{\Lambda,\mathrm{TF}}(\boldsymbol{\xi}). \tag{34}$$

This observation is the most important result of the time-dependent TF approximation and it manifests its full strength by providing an efficient and accurate description of the BEC dynamics within the TF regime.

In order to prove the validity of Eq. (34), we first multiply the ordinary differential equation (33) by $\psi_\Lambda^*(\tau, \boldsymbol{\xi})$ and subtract the imaginary part of the resulting equation to find

$$\frac{\partial}{\partial \tau}\left(|\psi_\Lambda(\tau, \boldsymbol{\xi})|^2\right) \approx 0. \tag{35}$$

Thus, the absolute value of the affinely transformed wave function does not change in time and we conclude that

$$|\psi_\Lambda(\tau, \boldsymbol{\xi})|^2 = |\psi_\Lambda(0, \boldsymbol{\xi})|^2 \approx |\phi_{\Lambda,\mathrm{TF}}(\boldsymbol{\xi})|^2 \quad \text{and} \quad \mu \approx \mu_{\mathrm{TF}}. \tag{36}$$

We infer from Eqs. (32) and (36) that the square bracket on the right-hand side of the ordinary differential equation (33) vanishes, giving rise to $\partial \psi_\Lambda / \partial \tau = 0$, which finally justifies the validity of Eq. (34).

To find the above-mentioned efficient description of the BEC dynamics, we take the advantage of Eq. (17) with the function $\psi_\Lambda(\tau, \boldsymbol{\xi})$ being determined by combining Eqs. (34) and (32). As a result, we obtain the central expression

$$\psi(t, \boldsymbol{x}) \approx \psi_{\mathrm{TF}}(t, \boldsymbol{x}) = \frac{e^{i\Phi_{\mathrm{TF}}(t, \boldsymbol{x})}}{\sqrt{g \det \Lambda(t)}} \left\{ \mu_{\mathrm{TF}} - \frac{m}{2} [\boldsymbol{x} - \mathcal{R}(t)]^{\mathrm{T}} \Sigma^{-1}(t) [\boldsymbol{x} - \mathcal{R}(t)] \right\}_+^{\frac{1}{2}}, \tag{37}$$

which characterizes the time evolution of a BEC within the time-dependent TF approximation. In Eq. (37), we have introduced the new local phase

$$\Phi_{\mathrm{TF}}(t,\boldsymbol{x}) \equiv \frac{1}{\hbar}\left\{\mathcal{S}_2(t) - \beta(t) + \boldsymbol{\mathcal{P}}(t)\,\boldsymbol{x} + \frac{m}{2}[\boldsymbol{x} - \boldsymbol{\mathcal{R}}(t)]^{\mathrm{T}} C(t)[\boldsymbol{x} - \boldsymbol{\mathcal{R}}(t)]\right\}, \tag{38}$$

where the generalized action $\mathcal{S}_2(t)$ is given by Eq. (7) for $k = 2$. Moreover, we have defined the so-called TF matrix

$$\Sigma(t) = \Lambda(t)\left[\Omega^2(0)\right]^{-1}\Lambda^{\mathrm{T}}(t), \tag{39}$$

which is positive definite.

According to Eq. (37), the boundary at which the spatial density distribution of the BEC vanishes corresponds to an ellipse if $d = 2$ or to an ellipsoid if $d = 3$. The TF matrix (39) defines the orientation and semiprincipal axes of this ellipse or ellipsoid whereby its points $\boldsymbol{x} \in \mathbb{R}^d$ satisfy the condition $[\boldsymbol{x} - \boldsymbol{\mathcal{R}}(t)]^{\mathrm{T}}\Sigma^{-1}(t)[\boldsymbol{x} - \boldsymbol{\mathcal{R}}(t)] = 2\mu_{\mathrm{TF}}/m$. The eigenvalues $\sigma_i$ of the TF matrix $\Sigma(t)$ determine the lengths of the semiprincipal axes, which coincide with the individual TF radii $r_i$ of the BEC, via $r_i = \sqrt{2\mu_{\mathrm{TF}}\,\sigma_i/m}$.

Fig. 1 illustrates the time evolution of such an ellipsoid in three dimensions. The initial density distribution of the ground state is centered around $\boldsymbol{\mathcal{R}}(0)$ with its principal axes being parallel to the coordinate axes, as shown in Fig. 1A. Changes in the orientation and position of the quadratic potential (3) give rise to a center-of-mass motion $\boldsymbol{\mathcal{R}}(t)$ and a transformation of the elliptic contour of the BEC density distribution via the time evolution of the TF matrix $\Sigma(t)$, as depicted in Fig. 1B.

We summarize this section by pointing out once more that the expression given by Eq. (37) provides a valuable description of the time evolution of a rotating BEC within the TF regime. Having the time-dependent TF

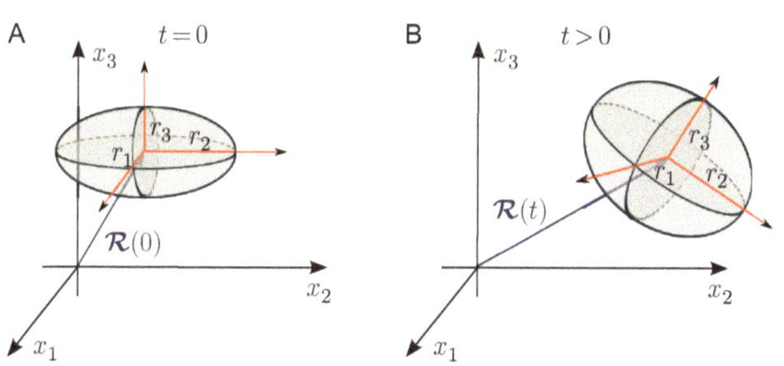


**Fig. 1** Boundaries of an ellipsoidal density distribution of a BEC within the time-dependent TF approximation. At $t = 0$ the BEC is centered around $\boldsymbol{\mathcal{R}}(0)$ and the principal axes of its boundary are parallel to the coordinate axes (A). Modifications in the quadratic trapping potential at $t > 0$ result in a center-of-mass motion $\boldsymbol{\mathcal{R}}(t)$ and changes in the ellipsoidal shape and the TF radii $r_i$ of the BEC (B).

approximation in mind, one might be tempted to believe that Eq. (37) is only valid for times shortly after $t = 0$. However, Eq. (37) yields very accurate results also for freely expanding BECs, since in this case the adaptive matrix $\Lambda(\tau)$ grows linearly in time, as discussed in Section 5.1, and therefore strongly reduces the influence of the kinetic energy term in Eq. (13), the so-called quantum pressure.

### 2.2.3 Integrated Density Distributions for Time-of-Flight Pictures

In practice, absorption imaging techniques (Reinaudi et al., 2007) are the prevalent method to study the dynamics of a three-dimensional BEC. Identically prepared BECs are illuminated by a laser field at different times of flight and their shadows are detected by a CCD camera, resulting in a sequence of two-dimensional pictures of the spatial density distribution of a BEC. For this reason, these techniques provide only partial information about the orientation and size of the original three-dimensional density distribution.

In this section, we discuss the one- and two-dimensional density distributions that result from an originally three-dimensional density profile of a BEC within the TF regime. In addition, we sketch the determination of the underlying three-dimensional density distribution from three mutually orthogonal time-of-flight pictures of a single BEC. The high efficiency of our approach is due to the knowledge of the density profile of the macroscopic wave function (Eq. 37). In case one does not have any a priori knowledge about the wave function under consideration, a reconstruction is still possible with the help of the Radon transformation (Gindikin and Michor, 1994; Schleich, 2001).

We start with the density distribution of a three-dimensional BEC within the TF regime

$$|\psi_{\mathrm{TF}}(t,\boldsymbol{x})|^2 = \frac{1}{g\,\det\Lambda(t)}\left\{\mu_{\mathrm{TF}} - \frac{m}{2}[\boldsymbol{x} - \boldsymbol{\mathcal{R}}(t)]^{\mathrm{T}}\Sigma^{-1}(t)[\boldsymbol{x} - \boldsymbol{\mathcal{R}}(t)]\right\}_{+}, \tag{40}$$

which directly follows from Eq. (37). The two-dimensional integrated density distribution in the $x$–$y$ plane

$$n_{\mathrm{TF}}^{(2\mathrm{D})}(t,x,y) \equiv \int_{\mathbb{R}} |\psi_{\mathrm{TF}}(t,\boldsymbol{x})|^2 \,\mathrm{d}z \tag{41}$$

can be evaluated as discussed in Appendix A.4 and reads

$$n_{\mathrm{TF}}^{(2\mathrm{D})}(t,x,y)=\frac{4\sqrt{2}}{3g\sqrt{m\det\overline{\Sigma}(t)\det\Omega^2(0)}}\left\{\mu_{\mathrm{TF}}-\frac{m}{2}\left[\bar{x}-\overline{\mathcal{R}}(t)\right]^{\mathrm{T}}\overline{\Sigma}^{-1}(t)\left[\bar{x}-\overline{\mathcal{R}}(t)\right]\right\}_{+}^{\frac{3}{2}}. \tag{42}$$

Here, we have introduced the reduced vectors $\bar{\boldsymbol{x}}\equiv(x,y)^{\mathrm{T}}$ and $\overline{\mathcal{R}}\equiv(\mathcal{R}_1,\mathcal{R}_2)^{\mathrm{T}}$ together with the symmetric submatrix

$$\overline{\Sigma}\equiv\begin{pmatrix}\Sigma_{11} & \Sigma_{12}\\ \Sigma_{12} & \Sigma_{22}\end{pmatrix}\in\mathbb{R}^{2\times 2}, \tag{43}$$

which is obtained from the TF matrix $\Sigma$, Eq. (39), by removing the row $\Sigma_{3i}$ and the column $\Sigma_{i3}$ that correspond to the $z$-coordinate. In a similar way, the two-dimensional integrated density distributions $n_{\mathrm{TF}}^{2\mathrm{D}}(t,y,z)$ and $n_{\mathrm{TF}}^{2\mathrm{D}}(t,x,z)$ depend on the two-dimensional submatrices in which the rows and columns corresponding to the $x$- and $y$-coordinates have been eliminated, respectively.

For the one-dimensional integrated density distribution along the $x$-axis

$$n_{\mathrm{TF}}^{(1\mathrm{D})}(t,x)=\int_{\mathbb{R}}\int_{\mathbb{R}}|\psi_{\mathrm{TF}}(t,\boldsymbol{x})|^2\,\mathrm{d}y\,\mathrm{d}z,$$

we obtain

$$n_{\mathrm{TF}}^{(1\mathrm{D})}(t,x)=\frac{\pi}{mg\sqrt{\Sigma_{11}(t)\det\Omega^2(0)}}\left\{\mu_{\mathrm{TF}}-\frac{m}{2\Sigma_{11}(t)}\left[x-\mathcal{R}_1(t)\right]^2\right\}_{+}^{2}. \tag{44}$$

The one-dimensional integrated density distributions $n_{\mathrm{TF}}^{(1\mathrm{D})}(t,y)$ and $n_{\mathrm{TF}}^{(1\mathrm{D})}(t,z)$ possess a similar structure. A general derivation of the previous results for arbitrary dimensions is presented in Appendix A.4.

To reconstruct the unknown three-dimensional density distribution of a BEC within the TF regime, Eq. (40), from three mutually orthogonal time-of-flight pictures, one needs to fit the two-dimensional integrated density distributions $n_{\mathrm{TF}}^{(2\mathrm{D})}(t,x,y)$, $n_{\mathrm{TF}}^{(2\mathrm{D})}(t,y,z)$, and $n_{\mathrm{TF}}^{(2\mathrm{D})}(t,x,z)$ to the corresponding experimental data and thereby determine the submatrix (43) and its two other counterparts. These three submatrices provide us with all elements of the TF matrix $\Sigma$, where each diagonal element of $\Sigma$ is found twice. This redundancy can be used to check the quality of the data obtained from the three mutually orthogonal time-of-flight pictures.

### 2.2.4 Alternative Description of Rotating and Vortex-Free Condensates

Our description of the internal dynamics of a BEC within the TF regime rests upon the solution $\Lambda(t)$ of the nonlinear matrix differential equation (14) with the corresponding initial conditions, Eq. (15). Based on the adaptive matrix $\Lambda(t)$, we then determine the TF matrix $\Sigma(t)$ and the quadratic phase matrix $C(t)$ defined by Eqs. (39) and (19), respectively. The two symmetric matrices $\Sigma(t)$ and $C(t)$ characterize the time evolution of the absolute value and the phase of the macroscopic wave function $\psi(t, \boldsymbol{x})$ (Eq. 37). Whereas the TF matrix $\Sigma(t)$ plays a crucial role in the context of time-of-flight pictures, the quadratic phase matrix $C(t)$ is of great importance for BEC-based interferometry (Altin et al., 2013; Debs et al., 2011; Müntinga et al., 2013; Simsarian et al., 2000; Torii et al., 2000).

One might wonder whether there is a way to study the BEC dynamics directly in terms of the experimentally more accessible matrices $\Sigma^{-1}(t)$ and $C(t)$ without a link to the adaptive matrix $\Lambda(t)$ at all. In fact, such an alternative description is possible and involves the solution of the two coupled first-order nonlinear matrix differential equations

$$\frac{\mathrm{d}\Sigma^{-1}}{\mathrm{d}t} = -C(t)\,\Sigma^{-1}(t) - \Sigma^{-1}(t)\,C(t), \tag{45}$$

$$\frac{\mathrm{d}C}{\mathrm{d}t} = -C^2(t) - \Omega^2(t) + \Sigma^{-1}(t)\sqrt{\frac{\det\Sigma^{-1}(t)}{\det\Omega^2(0)}} \tag{46}$$

with the corresponding initial conditions

$$\begin{aligned} \Sigma^{-1}(0) &= \Omega^2(0), \\ C(0) &= 0. \end{aligned} \tag{47}$$

These first-order differential equations can be derived by taking the time derivative on both sides of Eqs. (19) and (39) and using the matrix differential equation (14), the two identities $(d\Sigma^{-1}/dt) = -\Sigma^{-1}(d\Sigma/dt)\Sigma^{-1}$ and $\det\Lambda(t) = \sqrt{\det\Omega^2(0)/\det\Sigma^{-1}(t)}$ as well as the symmetry of the matrix $C(t)$.

At present, only a simplified version (Edwards et al., 2002) of the system of Eqs. (45) and (46) has been used to study the dynamics of a freely expanding BEC that was initially prepared in a rotating anisotropic harmonic trap (Hechenblaikner et al., 2002). The trap was instantly turned off at the time $t = T_{\mathrm{off}}$. In fact, by inserting the trap matrix $\Omega^2(t) = 0$ for all $t \geq T_{\mathrm{off}}$ into both Eqs. (45) and (46), it can be shown that the resulting matrix differential equations are mathematically equivalent to the equations of motion (12) and (13) derived in Edwards et al. (2002). However, the approach presented in this chapter facilitates an analysis of the time evolution of BECs within the TF regime that goes far beyond the scope considered in Edwards et al. (2002) and Hechenblaikner et al. (2002).

Next, we briefly outline how to obtain $\Lambda(t)$ from the quadratic phase matrix $C(t)$ and the TF matrix $\Sigma(t)$. When the time evolution of $C(t)$ is known, the adaptive matrix $\Lambda(t)$ follows from Eq. (19) by simple integration of the first-order matrix differential equation

$$\frac{d\Lambda}{dt} = C(t)\Lambda(t)$$

with the initial condition $\Lambda(0) = 1$. In contrast, the knowledge of the TF matrix $\Sigma(t)$ does not suffice to fully determine the adaptive matrix $\Lambda(t)$. From Eq. (39), the latter can only be identified up to an arbitrary orthogonal matrix $U(t)$ via the relation

$$\Lambda(t) = \Sigma^{\frac{1}{2}}(t)\, U(t)\, \Omega(0),$$

where the square root of $\Sigma(t)$ is defined in terms of its spectral decomposition. The specification of the orthogonal matrix $U(t)$ involves again the quadratic phase matrix $C(t)$.

There are two main reasons that make our approach based on the adaptive matrix $\Lambda(t)$ more suitable for the characterization of BEC dynamics within the TF regime than the alternative description just presented. First, the second-order matrix differential equation (14) for $\Lambda(t)$ can be recast in the form of Hamilton's equations, as shown in Section 4.1. This fact allows us to apply the full mathematical machinery available for Hamiltonian mechanics to study the time evolution of $\Lambda(t)$. Second, for studying the density distribution of a BEC near its surface at the scale of the corresponding healing length, the TF ground state given by Eq. (29) provides no longer an adequate solution of Eq. (28) and the macroscopic wave function Eq. (37) does not suffice for this purpose. Hence, the TF matrix $\Sigma(t)$, which rests upon the validity of Eq. (37), loses its immediate physical significance. In contrast, the affine approach based on the adaptive matrix $\Lambda(t)$ provides a valuable tool to study scenarios that are beyond the scope of the TF approximation with the help of dedicated numerical simulations, as discussed in the next section.

# 3. APPLICATION OF THE AFFINE APPROACH TO NUMERICAL SIMULATIONS

In this section, we show how numerical simulations of the BEC dynamics can benefit from the affine approach and we likewise apply this approach to quantify the accuracy of the time-dependent TF approximation for several different scenarios. For simplicity, we present here the study of the dynamics of rotating BECs on a 2D grid. We emphasize that full 3D simulations of BECs in the quasi-2D regime (Mateo and Delgado, 2008; Petrov et al., 2000a; Salasnich et al., 2002) have also been performed, which verify the validity of our 2D results.

## 3.1 Efficient Simulation of the Time Evolution of a BEC

Solving numerically the time-dependent GP equation (1) in the case of a freely expanding BEC in the original coordinates $(t, \boldsymbol{x})$, one immediately faces the problem that the size of the condensate grows by several orders of magnitude after switching off the trap. To deal with this problem, a large and well-resolved grid is required, leading to a huge increase of the computational costs of the numerical simulations, especially for two- or three-dimensional cases.

Here, we present an alternative approach to overcome this problem. Namely we are solving the affinely transformed GP equation (13) in the adapted coordinates $(\tau, \boldsymbol{\xi})$, rather than the GP equation (1) in the original coordinates $(t, \boldsymbol{x})$. Since the external and internal dynamics of the BEC are handled by the time dependence of the center-of-mass position $\mathcal{R}(t)$ and the adaptive matrix $\Lambda(\tau)$, respectively, only the time evolution of the affinely transformed wave function $\psi_\Lambda(\tau, \boldsymbol{\xi})$ is left and can be computed very efficiently (Eckart, 2008). As a result, in order to obtain the macroscopic wave function $\psi(t, \boldsymbol{x})$ in the original coordinates, it is sufficient to apply the transformation equation (17) once at the end of the simulation. Moreover, our method is not limited to the case of free expansion, but can enhance the efficiency of numerical simulations for various experimental scenarios such as rotating traps and delta-kick collimation (Ammann and Christensen, 1997).

## 3.2 Quantifying the Accuracy of the Time-Dependent TF Approximation

Since the time-dependent TF approximation plays a key role in deriving the analytic expression, Eq. (37), for the wave function $\psi(t, \boldsymbol{x})$ of a BEC, it is worthwhile to have a closer look at its accuracy. The time-dependent TF approximation is based on assumption (34), meaning that the wave function in adapted coordinates $\psi_\Lambda(\tau, \boldsymbol{\xi})$ remains approximately in its initial value $\psi_\Lambda(0, \boldsymbol{\xi})$ and does not undergo any time evolution at all. With our numerical simulations, we have quantified the validity of this assumption for two different scenarios involving rotating traps.

We take advantage of the Bures distance to quantify how much the numerically obtained state $\psi_\Lambda(\tau, \boldsymbol{\xi})$ differs from the initial state $\psi_\Lambda(0, \boldsymbol{\xi})$. In the following, we first recall the definition of the Bures distance and its properties and then describe the 2D model of a rotating BEC. Finally, we discuss the results of the simulations and characterize the parameter regime where it is safe to apply the TF approximation.

### 3.2.1 Bures Distance

The Bures distance (Bures, 1969) of two macroscopic wave functions $\psi_1$ and $\psi_2$ is given by

$$B(\psi_1,\psi_2) \equiv \left(2 - \frac{2|\langle\psi_1|\psi_2\rangle|}{\sqrt{\langle\psi_1|\psi_1\rangle\langle\psi_2|\psi_2\rangle}}\right)^{\frac{1}{2}} = \left(2 - \frac{2}{N}|\langle\psi_1|\psi_2\rangle|\right)^{\frac{1}{2}}, \tag{48}$$

where we have used the normalized condition given by Eq. (2). First, we note that one obtains $B(\psi_1, \psi_2) = 0$ if $\psi_1 = \psi_2$ and $B(\psi_1,\psi_2) = \sqrt{2}$ if $\psi_1$ and $\psi_2$ are orthogonal. Second, for two $d$-dimensional spherically symmetric Gaussian wave packets

$$\psi_{\mathrm{G}}(\boldsymbol{x},\sigma_{1,2}) = \frac{\sqrt{N}}{\left(2\pi\sigma_{1,2}^2\right)^{d/4}}\, \mathrm{e}^{-\frac{\boldsymbol{x}^2}{4\sigma_{1,2}^2}}$$

with different widths $\sigma_1$ and $\sigma_2$, the Bures distance (Eq. 48) reads

$$B_{\mathrm{G}}^{(d)}(\psi_{\mathrm{G}}(\boldsymbol{x},\sigma_1),\psi_{\mathrm{G}}(\boldsymbol{x},\sigma_2)) = \sqrt{2\left[1 - \left(\frac{2\sigma_1\sigma_2}{\sigma_1^2+\sigma_2^2}\right)^{d/2}\right]}. \tag{49}$$

In the case of two-dimensional Gaussian wave packets with the relative difference $\delta_\sigma \equiv (\sigma_2 - \sigma_1)/\sigma_1$ in their widths, Eq. (49) gives rise to

$$B_{\mathrm{G}}^{(2)}(\delta_\sigma) \equiv B_{\mathrm{G}}^{(2)}(\psi_{\mathrm{G}}(\boldsymbol{x},\sigma_1),\psi_{\mathrm{G}}(\boldsymbol{x},(1+\delta_\sigma)\sigma_1)) = \frac{\delta_\sigma}{\sqrt{1+\delta_\sigma+\frac{1}{2}\delta_\sigma^2}}. \tag{50}$$

Thus, a small relative change $\delta_\sigma$ in the width of a two-dimensional Gaussian wave packet results in a small Bures distance, that is, $B_G^{(2)}(\delta_\sigma) \cong \delta_\sigma \ll 1$. Despite the fact that the macroscopic wave function of a BEC is not typically Gaussian, this estimate can help us to evaluate the time dependence of the remaining inner dynamics, which is not included in the affine approach.

### 3.2.2 2D Model of a BEC in a Rotating Trap

The numerical simulations discussed in this section have all been performed with a 2D model based on realistic values that are accessible in state-of-the-art experiments. Our system consists of atoms of mass $m$, which are condensed and harmonically trapped along the $x$- and $y$-directions with the frequencies $\omega_x$ and $\omega_y$, respectively. The time evolution of this system is described by the affinely transformed GP equation (13) with $d = 2$. The number of discretization steps in time and the physical length of the grid are chosen as a trade-off between accuracy and computational time.

The free parameters of our system are (i) the anisotropy factor $\epsilon \equiv \omega_y/\omega_x$, (ii) the strength $\bar{g}N$ of the zero-range interaction between the atoms, and (iii) the final rotation rate $\dot{\varphi}_{\text{end}}$ of the trap. All quantities are measured with respect to the chosen timescale $1/\omega_x$ and the length scale $a_x \equiv \sqrt{\hbar/(m\omega_x)}$, respectively. Table 1 displays the parameters for the numerical simulations as well as the relations between the physical quantities and the free parameters. For convenience, we refer to the dimensionless quantity $\bar{g}N = g_{2D}N/(\hbar\omega_x a_x^2)$ as the interaction strength as it combines the 2D coupling constant $g_{2D}$ with the number of particles $N$.

It is worth emphasizing that our 2D model can be experimentally realized as a 3D disk-shaped BEC in a highly anisotropic trap (Burger et al., 2002; Görlitz et al., 2001; Rychtarik et al., 2004), with the

**Table 1** List of the Grid Settings, Physical Trap Frequencies, and 2D-Interaction Strength in Terms of the Simulation Parameters

| | |
|---|---|
| Number of discretization steps | $N_x = N_y = 2^7$ |
| Physical length of the grid | $L_x = L_y = 20\ a_x$ |
| Trap frequency along the $x$-axis | $\omega_x$ |
| Trap frequency along the $y$-axis | $\omega_y = \epsilon\ \omega_x$ |
| 2D-interaction strength | $g_{2D}N = \bar{g}N\ \hbar\omega_x a_x^2$ |

confinement length $a_z \equiv \sqrt{\hbar/(m\omega_z)}$ along the $z$-direction being much smaller than those along the $x$- and $y$-directions (Mateo and Delgado, 2008; Petrov et al., 2000a; Salasnich et al., 2002), where $\omega_z$ is the trap frequency along the $z$-axis. As a result, the 2D coupling constant $g_{2D}$ is determined by the 3D one $g_{3D}$ as $g_{2D} = g_{3D}/(\sqrt{2\pi}a_z)$, where $g_{3D} = 4\pi\hbar^2 a_s/m$ and $a_s$ is the $s$-wave scattering length. Using these relations, the results based on the full 3D simulations have been shown to be in good agreement with the results of the 2D simulations.

### 3.2.3 Time Evolution Within a Rotating Trap

The starting point for all simulations is the ground-state wave function $\psi_\Lambda(0, \boldsymbol{\xi})$ of the BEC obtained by imaginary time propagation of Eq. (13) with $\Lambda(\tau) = \mathbb{1}$ in combination with the Newton method. For the time evolution itself, the affinely transformed GP equation (13) is solved with an implicit Adams–Bashforth–Moulton multistep algorithm (Press et al., 2007).

The first scenario in which the accuracy of the time-dependent TF approximation is checked numerically is the rotating trap. In this case the time-dependent trap matrix $\Omega^2(\tau) = O(\tau)D\,O^T(\tau)$ contains the diagonal matrix

$$D = \begin{pmatrix} \omega_x^2 & 0 \\ 0 & \epsilon^2\omega_x^2 \end{pmatrix}$$

with the time-independent trap frequencies and an orthogonal matrix

$$O(\tau) = \begin{pmatrix} \cos\varphi(\tau) & \sin\varphi(\tau) \\ -\sin\varphi(\tau) & \cos\varphi(\tau) \end{pmatrix}$$

describing a rotation around the $z$-axis with the time-dependent angle $\varphi(\tau)$.

Triggered by experiments (Hechenblaikner et al., 2002; Müntinga et al., 2013), the rotation rate $\dot{\varphi}(\tau)$ of the trap has been chosen to have a sigmoidal ramp for $0 \leq \tau \leq T_{\text{end}}$, as depicted in Fig. 2, and to keep equal to the rotation rate $\dot{\varphi}_{\text{end}}$ for $\tau \geq T_{\text{end}}$. We assume that the characteristic time $T_{\text{end}}$ is large enough, that is, $\omega_x T_{\text{end}} \gg 1$, to have an adiabatically slow increase of the rotation rate in order to avoid collective excitations of the BEC. In addition, the final rotation rate $\dot{\varphi}_{\text{end}}$ is kept significantly smaller than the trap frequency $\omega_x$ in order to avoid the generation of vortices. For the rest of this section, we introduce the dimensionless characteristic time $\bar{T}_{\text{end}} \equiv \omega_x T_{\text{end}}/(2\pi)$ and use $\bar{T}_{\text{end}} = 15$ for all numerical simulations.

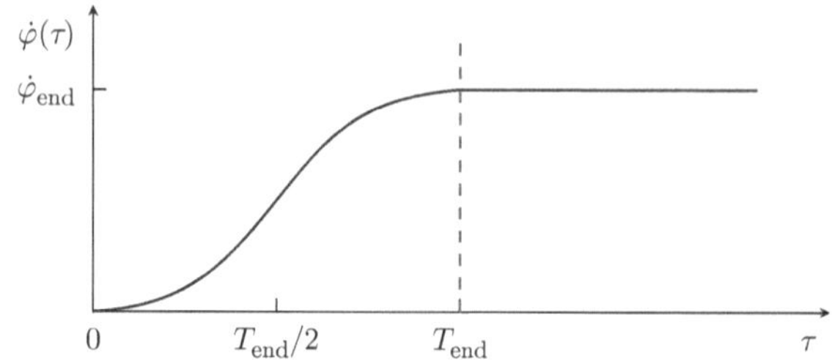


**Fig. 2** The dependence of the rotation rate $\dot{\varphi}(\tau)$ of the harmonic trap on time $\tau$. After a smooth ramp within the characteristic time $T_{\text{end}}$, the rotation rate $\dot{\varphi}(\tau)$ reaches its maximum value $\dot{\varphi}_{\text{end}}$ and is kept constant for all $\tau \geq T_{\text{end}}$.

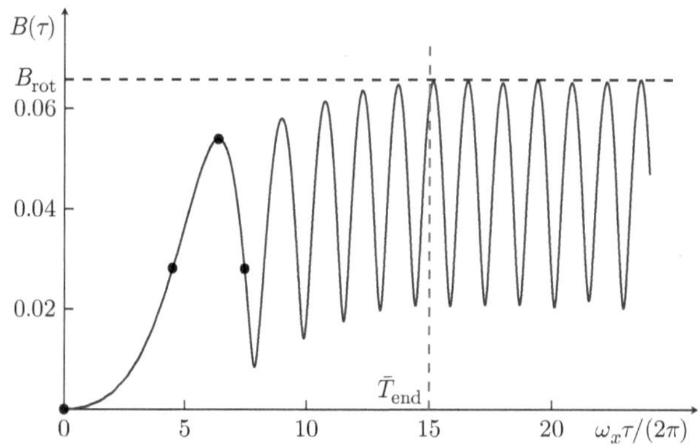


**Fig. 3** Time dependence of the Bures distance $B(\tau) = B(\psi_\Lambda(\tau, \boldsymbol{\xi}), \psi_\Lambda(0, \boldsymbol{\xi}))$, Eq. (48), between the time-evolved state $\psi_\Lambda(\tau, \boldsymbol{\xi})$ and the initial state $\psi_\Lambda(0, \boldsymbol{\xi})$ in adapted coordinates for a rotating trap with the anisotropy factor $\epsilon = 1.5$, the final rotation rate $\dot{\varphi}_{\text{end}} = 0.4\,\omega_x$, and the interaction strength $\bar{g}N = 100$. The Bures distance $B(\tau)$ exhibits oscillations during the entire simulation, while the maxima of this oscillation grow until the rotation rate $\dot{\varphi}(\tau)$, shown in Fig. 2, reaches its maximum value $\dot{\varphi}_{\text{end}}$ at the characteristic time $\bar{T}_{\text{end}} = 15$, after which they remain constant at the value $B_{\text{rot}}$. The *black circles* indicate the times for which the corresponding density distributions are displayed in Fig. 4.

Fig. 3 shows the time dependence of the Bures distance $B(\tau) = B(\psi_\Lambda(\tau, \boldsymbol{\xi}), \psi_\Lambda(0, \boldsymbol{\xi}))$, Eq. (48), between the time-evolved state $\psi_\Lambda(\tau, \boldsymbol{\xi})$, being the numerical solution of Eq. (13), and the initial state $\psi_\Lambda(0, \boldsymbol{\xi})$ in adapted coordinates. The Bures distance $B(\tau)$ exhibits oscillations over the whole simulation time $\tau$ and the local maxima of this oscillation grow as long as the rotation rate increases until they reach their maximal value $B_{\text{rot}}$ at the characteristic time $\bar{T}_{\text{end}}$ and stay constant from that point on. Since the magnitude of the Bures distance is very small, $B_{\text{rot}} < 0.07$, the time-evolved state does not differ significantly from the initial state.

For the parameters used in Fig. 3 the Bures distance oscillates with the frequency $\omega_{\text{Bures}} = 1.8\,\dot{\varphi}_{\text{end}}$ for $\tau > T_{\text{end}}$. Since the Bures distance, Eq. (48), only measures the absolute value of the overlap between two wave functions, the time-evolved state $\psi_\Lambda(\tau, \boldsymbol{\xi})$ actually undergoes collective oscillations with the frequency $0.5\ \omega_{\text{Bures}}$. In general, the frequency of the collective oscillations caused by the rotation of the trap depends on the rotation rate $\dot{\varphi}(\tau)$ as well as the anisotropy factor $\epsilon$ of the trap.

In Fig. 4, we contrast the dynamics of a rotating BEC in original and adapted coordinates, while the whole time evolution is shown in Movie 1 (http://dx.doi.org/10.1016/bs.aamop.2017.03.006), which is available in the online supplementary material. Fig. 4A depicts the two-dimensional (nonintegrated) density distributions $|\psi(t, \boldsymbol{x})|^2$ of a BEC in the original coordinates at different evolution times, whereas the first column of Fig. 4B displays the corresponding density distributions $|\psi_\Lambda(\tau, \boldsymbol{\xi})|^2$ in the adapted coordinates. In contrast to the density distribution $|\psi(t, \boldsymbol{x})|^2$, which follows the clockwise rotation induced by the rotating trap, the density distribution $|\psi_\Lambda(\tau, \boldsymbol{\xi})|^2$ in adapted coordinates does not show any visible changes. To highlight the residual dynamics, we present in the second column of Fig. 4B the difference $||\psi_\Lambda(\tau, \boldsymbol{\xi})| - |\psi_\Lambda(0, \boldsymbol{\xi})||^2$ of the absolute values of the wave functions, which depends only on local effects of the density distributions and not on their phases. This difference has nonzero values only at the edge of the BEC forming a ring-like structure, which rotates clockwise in accordance with the density distribution $|\psi(t, \boldsymbol{x})|^2$ presented in Fig. 4A. However, the magnitude of this residual dynamics is very small compared to the magnitude of the density distribution $|\psi_\Lambda(\tau, \boldsymbol{\xi})|^2$ and it oscillates in accordance with the Bures distance shown in Fig. 3.

In summary, we have proven that the time-dependent TF approximation applied to the affinely transformed GP equation (13) provides an accurate description of the dynamics of a rotating BEC. Only at the very edge of the BEC a minor amount of residual dynamics occurs that goes beyond the time-dependent TF approximation. In Section 3.2.5, we study how the magnitude of this residual dynamics depends on the different setup parameters.

### *3.2.4 Free Expansion After Switching Off a Rotating Trap*

As the second scenario to verify the time-dependent TF approximation, we have studied the subsequent free expansion of an initially rotating BEC.

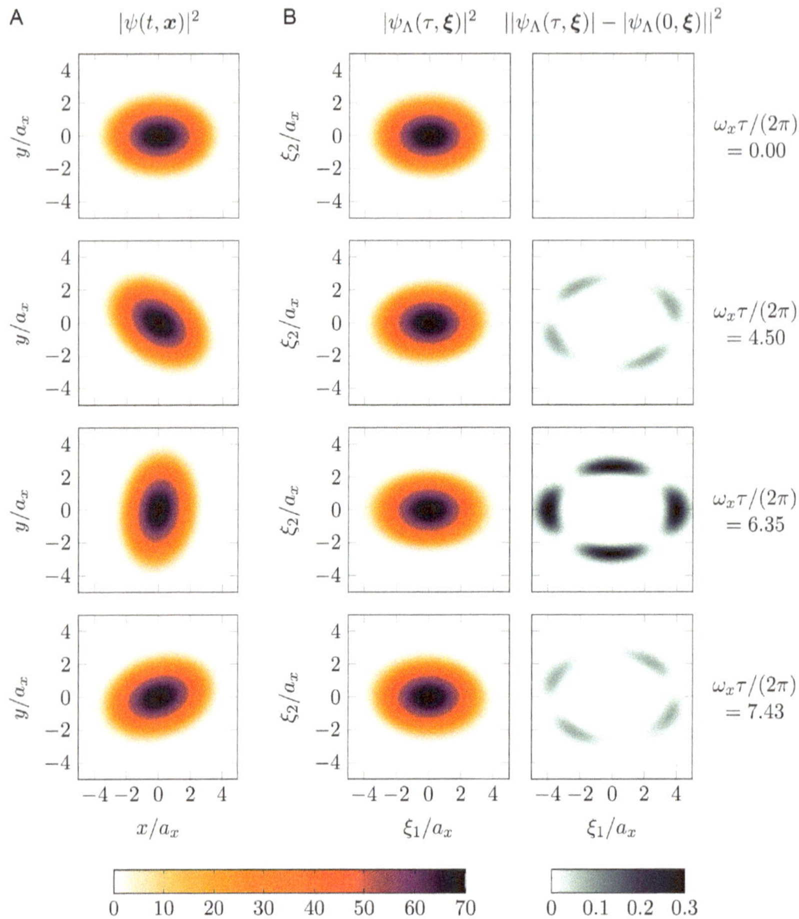


**Fig. 4** Density distributions $|\psi(t,\boldsymbol{x})|^2$ and $|\psi_\Lambda(\tau,\boldsymbol{\xi})|^2$ as well as the difference $||\psi_\Lambda(\tau,\boldsymbol{\xi})| - |\psi_\Lambda(0,\boldsymbol{\xi})||^2$ of a rotating BEC plotted for different times as indicated in Fig. 3. In original coordinates, (A) the density distribution undergoes a clockwise rotation, while it stays almost constant in adapted coordinates (B). The residual dynamics in adapted coordinates are made visible in the third column by considering the difference $||\psi_\Lambda(\tau,\boldsymbol{\xi})| - |\psi_\Lambda(0,\boldsymbol{\xi})||^2$ between the time-evolved state $\psi_\Lambda(\tau, \boldsymbol{\xi})$ and the initial state $\psi_\Lambda(0, \boldsymbol{\xi})$. Only at the edge of the condensate, this difference obtains nonzero values, while its magnitude corresponds well with the oscillations of the Bures distance $B(\tau)$ shown in Fig. 3. Movie 1 (http://dx.doi.org/10.1016/bs.aamop.2017.03.006) displaying the whole time evolution is available in the online supplementary material.

Here, the setup is the same as in the first scenario with the exception that the trap is switched off at $\tau = T_{\text{off}} > T_{\text{end}}$, that is, shortly after the final rotation rate $\dot{\varphi}_{\text{end}}$ is reached. Since the timing of the switch off has a strong influence on the final value $B_{\text{free}}$ of the Bures distance, we have varied $T_{\text{off}}$ within the first period of the Bures distance after $T_{\text{end}}$ for each set of parameters to find the maximal value of $B_{\text{free}}$.

The time dependence of the Bures distance for this case is depicted in Fig. 5. During the slow initial ramp, shown in Fig. 2, the Bures distance displays the collective oscillations discussed in Section 3.2.3. After switching off the trap at $\tau = T_{\text{off}}$, these oscillations stop and the Bures distance approaches its final value $B_{\text{free}}$. This behavior is a consequence of the fact that for a freely expanding BEC the adaptive matrix grows linearly for large times, as discussed in Section 5.1, and thus the right-hand side of the affinely transformed GP equation (13) vanishes asymptotically.

The two-dimensional (nonintegrated) density distributions in the original coordinates $|\psi(t, \boldsymbol{x})|^2$, Fig. 6A, as well as in the adapted coordinates $|\psi_\Lambda(\tau, \boldsymbol{\xi})|^2$, Fig. 6B, are displayed at different times indicated by the black dots in Fig. 5. The entire time evolution of these density distributions is presented in Movie 2 (http://dx.doi.org/10.1016/bs.aamop.2017.03.006), which is available in the online supplementary material. Whereas the size of the BEC in the original coordinates quickly grows after the trap is switched off, the rotation of the BEC stops completely after a further rotation of about $\pi/2$, in accordance with the irrotationality condition, as well as the energy and momentum conservation. It is worth mentioning that our

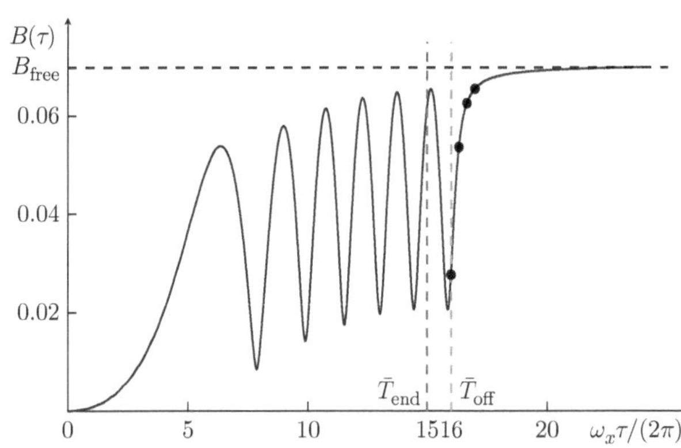


**Fig. 5** Bures distance $B(\tau) = B(\psi_\Lambda(\tau, \boldsymbol{\xi}), \psi_\Lambda(0, \boldsymbol{\xi}))$ between the time-evolved state and the initial state in adapted coordinates for free expansion after switching off a rotating trap with the anisotropy factor $\epsilon = 1.5$, the final rotation rate $\dot{\varphi}_{\text{end}} = 0.4\,\omega_x$, and the interaction strength $\bar{g}N = 100$. The trap is switched off at $\bar{T}_{\text{off}} = 16$ within the first period of the Bures distance $B(\tau)$ after the characteristic time $\bar{T}_{\text{end}} = 15$. The oscillation of the Bures distance stops after the trap is switched off and it approaches its final value $B_{\text{free}}$. The *black circles* indicate the times for which the corresponding density distributions are displayed in Fig. 6.

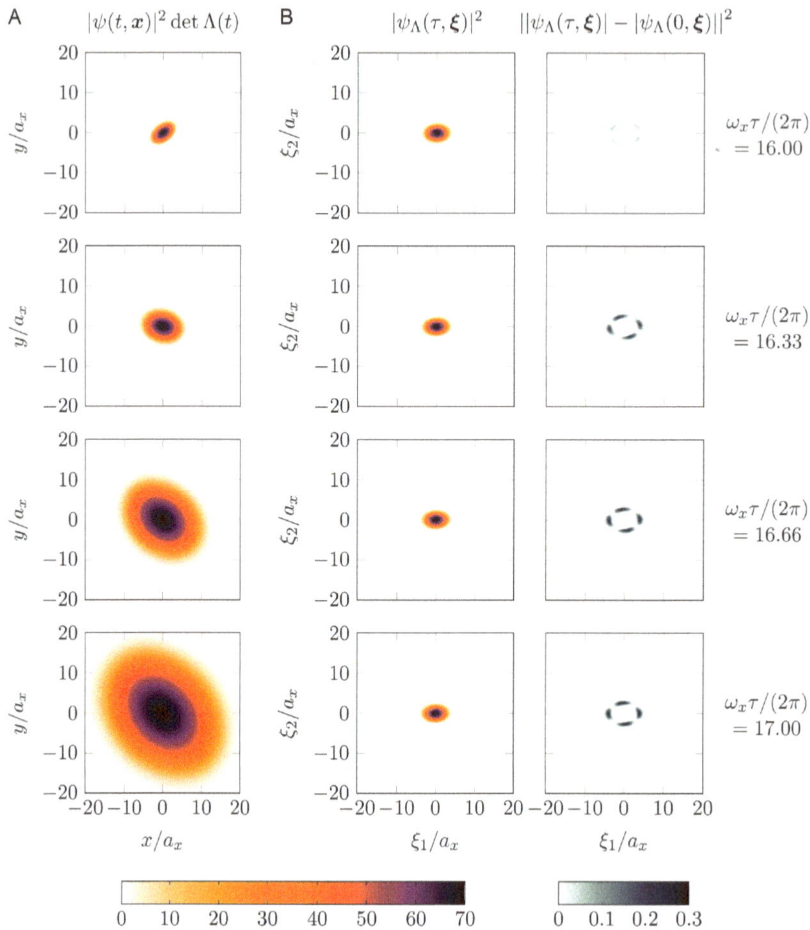


**Fig. 6** Density distributions $|\psi(t,\boldsymbol{x})|^2$ and $|\psi_\Lambda(\tau,\boldsymbol{\xi})|^2$ together with the difference $||\psi_\Lambda(\tau,\boldsymbol{\xi})| - |\psi_\Lambda(0,\boldsymbol{\xi})||^2$ of a free expanding BEC after release from a rotating trap plotted for different times as indicated in Fig. 5. In original coordinates, (A) the rotation of the condensate comes to an end after an angle of about $\pi/2$, while the size of the condensate grows continuously. In adapted coordinates, (B) no such effects are visible and the density distribution stays almost constant. In the third column the difference $||\psi_\Lambda(\tau,\boldsymbol{\xi})| - |\psi_\Lambda(0,\boldsymbol{\xi})||^2$ between the time-evolved state $\psi_\Lambda(\tau, \boldsymbol{\xi})$ and the initial state $\psi_\Lambda(0, \boldsymbol{\xi})$ illustrates the residual dynamics in adapted coordinates. This difference only obtains nonzero values at the edge of the condensate, while its magnitude grows together with the Bures distance $B(\tau)$ shown in Fig. 5. Movie 2 (http://dx.doi.org/10.1016/bs.aamop.2017.03.006) displaying the whole time evolution is available in the online supplementary material.

results are in good agreement with the theoretical consideration (Edwards et al., 2002) and the experimental data (Hechenblaikner et al., 2002).

However, in adapted coordinates, Fig. 6B, neither an increase of the size of the BEC nor a significant rotation of the density distribution can be observed. Only by looking at the residual dynamics visualized again by the difference $||\psi_\Lambda(\tau, \boldsymbol{\xi})|-|\psi_\Lambda(0, \boldsymbol{\xi})||^2$, minor changes of $\psi_\Lambda(\tau, \boldsymbol{\xi})$ become visible at the edge of the condensate. Thus, the time-dependent TF approximation used for the affine approach is an excellent tool to predict the free expansion of an initially rotating BEC as well.

### 3.2.5 Dependence of the Bures Distance on the Setup Parameters

In extension of the two previous scenarios, we now provide a broader study of the accuracy of the time-dependent TF approximation. In particular, we have solved numerically Eq. (13) for a rotating BEC with and without subsequent free expansion for many different combinations of the interaction strength $\bar{g}N$, the anisotropy factor $\epsilon$, and the final rotation rate $\dot{\varphi}_{\text{end}}$, thereby using $\bar{T}_{\text{end}} = 15$ as before. The results for the Bures distances $B_{\text{rot}}$ and $B_{\text{free}}$ are presented in Fig. 7.

As displayed in Fig. 7A the Bures distance $B_{\text{rot}}$ decreases with an increase of the interaction strength $\bar{g}N$ and increases for larger values of the anisotropy factor $\epsilon$ and the maximal rotation rate $\dot{\varphi}_{\text{end}}$. These observations can be explained as follows:

- **(i)** With an increase of $\bar{g}N$, the interaction energy term in Eq. (13) becomes more dominant compared to the kinetic energy term, resulting in an improved accuracy of the time-dependent TF approximation. For this reason, the Bures distance $B_{\text{rot}}$ decreases as the interaction strength $\bar{g}N$ grows.
- **(ii)** Since a spherically symmetric trap with $\epsilon = 1$ cannot transfer any angular momentum to a BEC, the wave function $\psi_\Lambda(\tau, \boldsymbol{\xi})$ will remain in the initial state $\psi_\Lambda(0, \boldsymbol{\xi})$, giving rise to $B_{\text{rot}} = 0$. However, for $\epsilon > 1$ the rotation of the trap affects the dynamics of the BEC. The Bures distance $B_{\text{rot}}$ increases with an increase of $\epsilon$, since the kinetic energy term in Eq. (13) depends on the spatial derivatives of the wave function and therefore increases for a stronger confinement along one axis. Thus, the time-dependent TF approximation works better for a system with small anisotropies.
- **(iii)** For larger values of the maximal rotation rate $\dot{\varphi}_{\text{end}}$, the total energy transferred from the trap to the BEC increases, raising the chance to

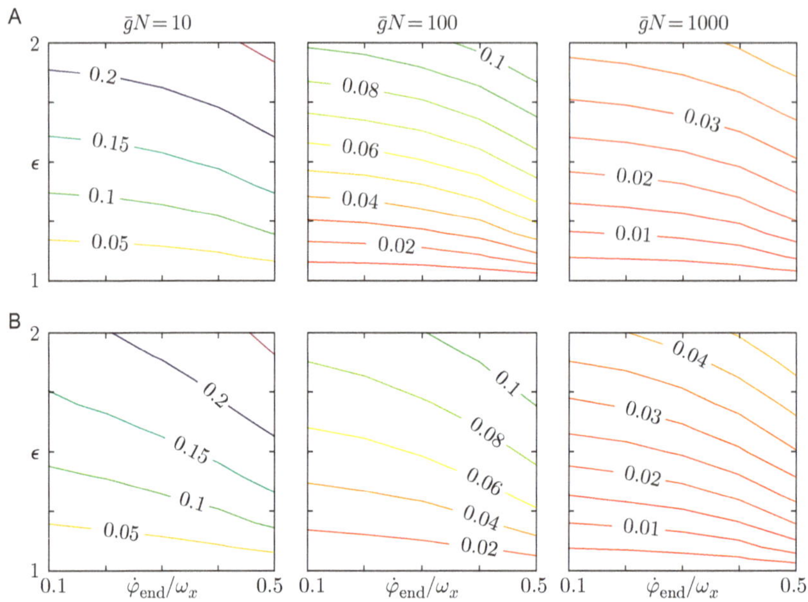


**Fig. 7** Contour plot of (A) the maximal Bures distance $B_{\rm rot}$ for a purely rotating trap with a time-dependent rotation rate as shown in Fig. 2 and (B) the final Bures distance $B_{\rm free}$ for a subsequent free expansion of an initially rotating trap as discussed in Section 3.2.4. For both scenarios the interaction strength $\bar{g}N$ increases from *left* to *right*, while the ranges of the anisotropy factor $\epsilon$ and the final rotation rate $\dot{\varphi}_{\rm end}$ are the same for all plots, respectively. The differences between the case of a purely rotating trap (A) and a free expansion after the rotation (B) are minor. In general, an increase in the interaction strength $\bar{g}N$ leads to smaller values of the Bures distance, while in contrast the Bures distance takes on larger values as the anisotropy factor $\epsilon$ and the final rotation rate $\dot{\varphi}_{\rm end}$ increase.

excite collective modes of the condensate. However, these modes are not taken into account within the framework of the TF approximation. As a result, the amount of residual dynamics grows with the increase of the maximal rotation rate $\dot{\varphi}_{\rm end}$. The quadrupole mode is well known to be excited when $\dot{\varphi}_{\rm end}$ approaches the critical value $\omega_c = 0.71\ \omega_x$(Hodby et al., 2001; Madison et al., 2001) and the time-dependent TF approximation breaks down completely in this regime. We have verified this fact with our numerical simulations, and for the results presented here, we have always made sure to stay well below the threshold for vortex creation.

Since the dependencies of $B_{\rm rot}$ and $B_{\rm free}$ on the simulation parameters are very similar to each other, as shown in Fig. 7, we conclude here that the subsequent free expansion of an initially rotating BEC does not substantially amplify the residual time evolution in the adapted coordinates.

In summary, we find that the accuracy of the time-dependent TF approximation depends on various parameters with the most important one being the interaction strength $\bar{g}N$. For values $\bar{g}N \geq 100$, the transformed wave function $\psi_\Lambda(\tau, \boldsymbol{\xi})$ is practically constant in time and the time-dependent TF approximation holds true. In most experimental setups, one typically aims at a large number of particles in a BEC to improve the signal-to-noise ratio. For this reason, the TF approximation provides a valuable tool to describe the dynamics of a large class of rotating BECs.

# 4. CONSTANTS OF MOTION

In the preceding section, we have confirmed the accuracy of our method based on the time-dependent TF approximation by dedicated numerical simulations of the affinely transformed GP equation (13). However, the reliability of such numerical simulations themselves must also be guaranteed. One possible way to test it is based on simulations of suitable scenarios that possess certain underlying symmetries, which by Noether's theorem give rise to specific constants of motion. In this chapter, we focus on the time-translational invariance and the rotational symmetry and discuss the corresponding energy and angular momentum conservation for both the matrix differential equation (14) and the GP equation (1).

In this section, we first show how the matrix differential equation (14) can be rewritten in terms of a Hamiltonian formalism. Second, we analyze the center-of-mass motion of the BEC with regard to a generalized version of the Ehrenfest theorem (Bodurov, 1998), and third, we present expressions for specific constants of motion of the GP equation (1) valid within the time-dependent TF approximation.

## 4.1 Hamiltonian Formalism for the Matrix Differential Equation

Here, we establish a connection between the matrix differential equation (14) and the corresponding Hamiltonian formalism. For this purpose, we first perform a transformation of the adaptive matrix and the time coordinate to bring the matrix differential equation (14) into a specific form that does no longer contain the initial trap matrix $\boldsymbol{\Omega}^2(0)$. We call the resulting equation the canonical form of the matrix differential equation. We then introduce the Hamiltonian and verify that the corresponding Hamilton equations of motion are equivalent to the canonical form of the matrix differential equation. With the help of this Hamiltonian, we finally obtain two constants of motion for the matrix differential equation, which we later on relate to the conservation of the total energy and the angular momentum of the BEC.

### 4.1.1 Canonical Form of the Matrix Differential Equation

We start by defining the orthogonal matrix $O$ and the diagonal matrix $D$ via the diagonalization of the initial trap matrix $\Omega^2(0)$ according to

$$O\,D\,O^{\mathrm{T}} \equiv \Omega^2(0). \tag{51}$$

Moreover, we introduce a new timescale with the help of the quantity

$$\alpha \equiv \left[\det \Omega^2(0)\right]^{\frac{1}{2d}}. \tag{52}$$

Based on these new quantities, we now consider the transformation

$$\begin{aligned} \Lambda &= \frac{1}{\alpha}\, O\,\tilde{\Lambda}\, D^{\frac{1}{2}}\, O^{\mathrm{T}}, \\ t &= \frac{1}{\alpha}\, \tilde{t} \end{aligned} \tag{53}$$

to the new adaptive matrix $\tilde{\Lambda}$ and the dimensionless time $\tilde{t}$. The canonical form of the matrix differential equation is simply obtained by inserting the transformation (53) into the matrix differential equation (14), thereby taking into account the relation $\det(\Lambda) = \det(\tilde{\Lambda})$. It describes the time evolution of the new adaptive matrix $\tilde{\Lambda}(\tilde{t})$ and reads

$$\tilde{\Lambda}^{\mathrm{T}}(\tilde{t})\left[\frac{\mathrm{d}^2\tilde{\Lambda}}{\mathrm{d}\tilde{t}^2} + \tilde{\Omega}^2(\tilde{t})\tilde{\Lambda}(\tilde{t})\right] = \frac{\mathbb{1}}{\det \tilde{\Lambda}(\tilde{t})}, \tag{54}$$

where we have introduced the transformed trap matrix

$$\tilde{\Omega}^2(\tilde{t}) \equiv \frac{1}{\alpha^2}\, O^{\mathrm{T}}\Omega^2(\tilde{t}/\alpha)\, O. \tag{55}$$

The initial conditions (15) accordingly transform to

$$\tilde{\Lambda}(0) = \alpha\, D^{-\frac{1}{2}} \ \text{ and } \ \left.\frac{\mathrm{d}\tilde{\Lambda}}{\mathrm{d}\tilde{t}}\right|_{\tilde{t}=0} = 0, \tag{56}$$

whereas the irrotationality condition (16) preserves its form under the transformation (53)

$$\tilde{\Lambda}^{\mathrm{T}}(\tilde{t})\frac{\mathrm{d}\tilde{\Lambda}}{\mathrm{d}\tilde{t}} = \frac{\mathrm{d}\tilde{\Lambda}^{\mathrm{T}}}{\mathrm{d}\tilde{t}}\tilde{\Lambda}(\tilde{t}). \tag{57}$$

Next we show that the canonical form of the matrix differential equation (54) can be embedded in a Hamiltonian formalism.

### 4.1.2 Hamiltonian and Equations of Motion

We start by introducing the momenta

$$\tilde{\Pi}_{\alpha\beta} = \frac{\mathrm{d}\tilde{\Lambda}_{\alpha\beta}}{\mathrm{d}\tilde{t}} \tag{58}$$

as conjugate variables to the matrix elements $\tilde{\Lambda}_{\alpha\beta}$ together with the Hamiltonian

$$H(\tilde{\Lambda},\tilde{\Pi},\tilde{t}) = \frac{1}{2}\,\mathrm{Tr}\left\{\tilde{\Pi}^{\mathrm{T}}\tilde{\Pi} + \tilde{\Lambda}^{\mathrm{T}}\tilde{\Omega}^2(\tilde{t})\tilde{\Lambda}\right\} + \frac{1}{\det\tilde{\Lambda}}. \tag{59}$$

The Hamilton equations of motion

$$\begin{aligned}\frac{\mathrm{d}\tilde{\Lambda}_{\alpha\beta}}{\mathrm{d}\tilde{t}} &= \{H,\tilde{\Lambda}_{\alpha\beta}\}_{(\tilde{\Lambda},\tilde{\Pi})} = \frac{\partial H}{\partial\tilde{\Pi}_{\alpha\beta}} = \tilde{\Pi}_{\alpha\beta},\\ \frac{\mathrm{d}\tilde{\Pi}_{\alpha\beta}}{\mathrm{d}\tilde{t}} &= \{H,\tilde{\Pi}_{\alpha\beta}\}_{(\tilde{\Lambda},\tilde{\Pi})} = -\frac{\partial H}{\partial\tilde{\Lambda}_{\alpha\beta}} = -\tilde{\Omega}^2\tilde{\Lambda} + \frac{1}{\det\tilde{\Lambda}}\,\tilde{\Lambda}^{-\mathrm{T}}\end{aligned} \tag{60}$$

are mathematically equivalent to the canonical form of the matrix differential equation (54). Here, we have introduced the corresponding Poisson bracket

$$\{f,g\}_{(\tilde{\Lambda},\tilde{\Pi})} \equiv -\sum_{\alpha,\beta}\left(\frac{\partial f}{\partial\tilde{\Lambda}_{\alpha\beta}}\frac{\partial g}{\partial\tilde{\Pi}_{\alpha\beta}} - \frac{\partial g}{\partial\tilde{\Lambda}_{\alpha\beta}}\frac{\partial f}{\partial\tilde{\Pi}_{\alpha\beta}}\right) \tag{61}$$

of two functions $f$ and $g$ that depend on the matrix elements $\tilde{\Lambda}_{\alpha\beta}$ and $\tilde{\Pi}_{\alpha\beta}$ and made use of the relation

$$\frac{\mathrm{d}}{\mathrm{d}\tilde{\Lambda}_{\alpha\beta}}\left(\det\tilde{\Lambda}\right) = \det\tilde{\Lambda}\cdot\mathrm{Tr}\left\{\tilde{\Lambda}^{-1}\frac{\partial\tilde{\Lambda}}{\partial\tilde{\Lambda}_{\alpha\beta}}\right\} = \tilde{\Lambda}_{\alpha\beta}^{-\mathrm{T}}\,\det\tilde{\Lambda}. \tag{62}$$

When we accordingly define the "momenta" $\Pi \equiv \mathrm{d}\Lambda/\mathrm{d}t$, Eq. (53) implies the following transformation of the conjugate variables:

$$\begin{aligned}\tilde{\Lambda} &= \alpha\, O^{\mathrm{T}}\,\Lambda\, O\, D^{-\frac{1}{2}},\\ \tilde{\Pi} &= O^{\mathrm{T}}\,\Pi\, O\, D^{-\frac{1}{2}}.\end{aligned} \tag{63}$$

We emphasize that Eq. (63) is not a canonical transformation and that a corresponding Hamiltonian formalism for the original matrices $\Lambda$ and $\Pi$ does not exist, as can be seen by inserting the transformation (63) into the expression for the Poisson bracket (61). As a result, we arrive at

$$\{f,g\}_{(\tilde{\Lambda},\tilde{\Pi})} = -\frac{1}{\alpha}\sum_{\mu,\nu,\lambda}\left(\frac{\partial f}{\partial\Lambda_{\mu\nu}}\Omega^2_{\nu\lambda}(0)\frac{\partial g}{\partial\Pi_{\mu\lambda}} - \frac{\partial g}{\partial\Lambda_{\mu\nu}}\Omega^2_{\nu\lambda}(0)\frac{\partial f}{\partial\Pi_{\mu\lambda}}\right), \quad (64)$$

which clearly shows that the structure of the Poisson bracket (61) is not preserved by the transformation (63). Thus, there exist no corresponding Hamilton equations for the original matrix differential equation (14).

### *4.1.3 Energy Conservation*

When we consider a time-independent trapping potential with the constant trap matrix $\tilde{\Omega}^2(\tilde{t}) = \tilde{\Omega}^2_*$, the energy is a constant of motion given by the Hamiltonian $H(\tilde{\Lambda},\tilde{\Pi})$, Eq. (59). In this spirit, we call the quantity

$$\tilde{E}(\tilde{\Lambda},\tilde{\Pi};\tilde{t}) \equiv \frac{1}{2}\,\mathrm{Tr}\{\tilde{\Pi}^{\mathrm{T}}\,\tilde{\Pi} + \tilde{\Lambda}^{\mathrm{T}}\,\tilde{\Omega}^2(\tilde{t})\,\tilde{\Lambda}\} + \frac{1}{\det\tilde{\Lambda}} \quad (65)$$

the total energy associated with the matrix differential equation in canonical form also for a time-dependent trap matrix $\tilde{\Omega}^2(\tilde{t})$.

Using Eq. (58) together with the inverse of the transformation (53) in Eq. (65), the total energy associated with the matrix differential equation $E_\Lambda(t) = \tilde{E}(\tilde{\Lambda},\tilde{\Pi};\tilde{t})$ reads in terms of the original adaptive matrix

$$E_\Lambda(t) = \frac{1}{2}\,\mathrm{Tr}\left\{\left(\frac{\mathrm{d}\Lambda^{\mathrm{T}}}{\mathrm{d}t}\frac{\mathrm{d}\Lambda}{\mathrm{d}t} + \Lambda^{\mathrm{T}}(t)\,\Omega^2(t)\,\Lambda(t)\right)\Omega^{-2}(0)\right\} + \frac{1}{\det\Lambda(t)}, \quad (66)$$

where $\Omega^2(t) = \alpha^2\,O\,\tilde{\Omega}^2(\tilde{t})\,O^{\mathrm{T}}$ and $\Omega^{-2}(0) = O\,D^{-1}O^{\mathrm{T}}$. The fact that the total energy (66) is indeed a constant of motion if $\Omega^2(t) = \Omega^2_*$ can be easily verified by taking the time derivative of Eq. (66) on both sides and inserting the matrix differential equation (14) on the right-hand side of the resulting equation.

### *4.1.4 Angular Momentum Conservation*

For an isotropic harmonic potential with corresponding trap matrix $\tilde{\Omega}^2(\tilde{t}) = \tilde{\omega}^2(\tilde{t})\,\mathbb{1}$, the angular momentum of a BEC is preserved. In order to see how this conservation law is related to the matrix differential equation (14), we study the quantity

$$\tilde{L}(\tilde{\Lambda},\tilde{\Pi}) \equiv \tilde{\Lambda}\,\tilde{\Pi}^{\mathrm{T}} - \tilde{\Pi}\,\tilde{\Lambda}^{\mathrm{T}}, \quad (67)$$

which we denote as angular momentum matrix associated with the matrix differential equation in the canonical form. The angular momentum matrix (67) can be found with the help of Noether's theorem and the irrotationality condition (57), which reads in terms of both conjugate variables $\tilde{\Lambda}^{\mathrm{T}}\,\tilde{\Pi} - \tilde{\Pi}^{\mathrm{T}}\,\tilde{\Lambda} = 0$.

We now prove that this angular momentum matrix is preserved for an isotropic trapping potential. For this purpose, we take the total derivative of $\tilde{L}$ with respect to the time $\tilde{t}$ and obtain

$$\frac{\mathrm{d}\tilde{L}}{\mathrm{d}\tilde{t}} = \frac{\partial \tilde{L}}{\partial \tilde{t}} + \{H, \tilde{L}\}_{(\tilde{\Lambda},\tilde{\Pi})}, \tag{68}$$

where the Poisson bracket $\{H, \tilde{L}\}_{(\tilde{\Lambda},\tilde{\Pi})}$ is defined by Eq. (61). Since the angular momentum matrix $\tilde{L}$ does not explicitly depend on the time $\tilde{t}$, the partial derivative $\partial\tilde{L}/\partial\tilde{t}$ vanishes. With the Hamiltonian (59), the evaluation of the Poisson bracket finally yields

$$\frac{\mathrm{d}\tilde{L}}{\mathrm{d}\tilde{t}} = \tilde{\Lambda}\,\tilde{\Lambda}^{\mathrm{T}}\,\tilde{\Omega}^2(\tilde{t}) - \tilde{\Omega}^2(\tilde{t})\,\tilde{\Lambda}\,\tilde{\Lambda}^{\mathrm{T}}. \tag{69}$$

Thus, the angular momentum matrix $\tilde{L}$ is a constant of motion if the matrix $\tilde{\Lambda}\,\tilde{\Lambda}^{\mathrm{T}}\,\tilde{\Omega}^2(\tilde{t})$ is symmetric, which is indeed the case for an isotropic trap with $\tilde{\Omega}^2(\tilde{t}) = \tilde{\omega}^2(\tilde{t})\,\mathbb{1}$.

When we insert the transformation (63) into the definition (67) and use the transformation law $L_\Lambda(t) = O\,\tilde{L}(\tilde{\Lambda},\tilde{\Pi})\,O^{\mathrm{T}}$, we obtain the angular momentum matrix in terms of the original adaptive matrix

$$L_\Lambda(t) = \alpha\left[\Lambda(t)\,\Omega^{-2}(0)\,\frac{\mathrm{d}\Lambda^{\mathrm{T}}}{\mathrm{d}t} - \frac{\mathrm{d}\Lambda}{\mathrm{d}t}\Omega^{-2}(0)\Lambda^{\mathrm{T}}(t)\right]. \tag{70}$$

We emphasize that the angular momentum matrix is antisymmetric and explicitly depends on the initial trapping potential.

## 4.2 Constants of Motion of the GP Equation

Our efficient description of the time evolution of a rotating BEC within the TF regime, Eq. (37), rests upon two different, but related, approximations: (i) the initial macroscopic wave function $\psi(0, \boldsymbol{x})$ is assumed to be given by the ground state in the TF approximation, Eq. (29), and (ii) the kinetic energy term in the affinely transformed GP equation (13) is supposed to be negligible in comparison to the potential and interaction energies involved.

According to this second assumption known as the time-dependent TF approximation, the affinely transformed wave function, being the solution of Eq. (13), does not display any essential dynamics

$$\psi_\Lambda(\tau,\boldsymbol{\xi}) \approx \psi_\Lambda(0,\boldsymbol{\xi}). \tag{71}$$

In Section 2.2.2, we used a combination of both approximations, (i) and (ii), to derive the efficient description of the BEC dynamics within the TF regime given by Eq. (37).

However, for BECs that are not deep within the TF regime and whose ground state differs significantly from Eq. (29), the assumption (71) can still provide very accurate results for some scenarios such as freely expanding BECs. For this reason, we first seek an efficient description that does not involve approximation (i), but includes the exact ground state $\psi(0, \boldsymbol{x})$ of a BEC or its numerically determined approximation. Based on this slightly more general description, we derive approximate expressions for the total energy and the angular momentum of a BEC that allow further tests of assumption (71) by analyzing possible time dependencies of the constants of motion that should not occur due to the chosen symmetry of the GP equation. Eventually, we also include approximation (i) and relate the resulting expressions for the total energy and the angular momentum based on the TF ground state (29) to the corresponding constants of motion associated with the matrix differential equation.

### 4.2.1 Effective Description of the Dynamics Based Solely on the Time-Dependent TF Approximation

In order to determine the slightly generalized efficient description of the BEC dynamics, we first relate the initial state in adapted coordinates $\psi_\Lambda(0, \boldsymbol{\xi})$ with the original ground state $\psi(0, \boldsymbol{x})$ by evaluating Eq. (20) at $t = 0$ giving rise to

$$\psi_\Lambda(0,\boldsymbol{\xi}) = \mathrm{e}^{-\frac{\mathrm{i}}{\hbar}\mathcal{P}(0)\left[\boldsymbol{\xi}+\frac{1}{2}\mathcal{R}(0)\right]}\,\psi(0,\boldsymbol{\xi}+\mathcal{R}(0)). \tag{72}$$

Next, we obtain $\psi_\Lambda(\tau, \boldsymbol{\xi})$ simply by combining Eqs. (71) and (72). We insert the result into Eq. (17) and thereby use the substitution $\boldsymbol{\xi} = \Lambda^{-1}(t)\left[\boldsymbol{x} - \mathcal{R}(t)\right]$. Hence, we find as efficient description of the BEC dynamics that is solely based on the time-dependent TF approximation

$$\psi(t,\boldsymbol{x}) \approx \frac{1}{\sqrt{\det\Lambda(t)}}\,\mathrm{e}^{\mathrm{i}\,\chi(t,\boldsymbol{x})}\,\psi\big(0,\Lambda^{-1}(t)\left[\boldsymbol{x}-\mathcal{R}(t)\right]+\mathcal{R}(0)\big), \tag{73}$$

where we have introduced the phase

$$\chi(t,\boldsymbol{x}) = \frac{1}{\hbar}\Big\{\mathcal{S}_2(t) - \beta(t) + \mathcal{P}(t)\,\boldsymbol{x} + \frac{m}{2}[\boldsymbol{x} - \mathcal{R}(t)]^{\mathrm{T}}\, C(t)[\boldsymbol{x} - \mathcal{R}(t)] \\ -\mathcal{P}(0)\big(\Lambda^{-1}(t)[\boldsymbol{x} - \mathcal{R}(t)] + \mathcal{R}(0)\big)\Big\}. \tag{74}$$

Note that by insertion of $\psi(0, \boldsymbol{x}) = \phi_{\mathrm{TF}}(\boldsymbol{x})$ together with Eq. (29) into Eq. (73), we consequentially arrive at the expression (37).

### 4.2.2 Generalized Ehrenfest Theorem for a BEC in a Harmonic Trap

According to the generalized Ehrenfest theorem (Bodurov, 1998), the expectation values

$$\langle \hat{\boldsymbol{x}} \rangle_{\psi(t)} \equiv \frac{1}{N}\int_{\mathbb{R}^d} \psi^*(t,\boldsymbol{x})\,\boldsymbol{x}\,\psi(t,\boldsymbol{x})\,\mathrm{d}^d x, \tag{75}$$

$$\langle \hat{\boldsymbol{p}} \rangle_{\psi(t)} \equiv \frac{1}{N}\int_{\mathbb{R}^d} \psi^*(t,\boldsymbol{x})\,\hat{\boldsymbol{p}}\,\psi(t,\boldsymbol{x})\,\mathrm{d}^d x \tag{76}$$

of the position $\hat{\boldsymbol{x}} = \boldsymbol{x}$ and the momentum $\hat{\boldsymbol{p}} \equiv -\mathrm{i}\hbar\boldsymbol{\nabla}_{\boldsymbol{x}}$ operators satisfy the classical equations of motion

$$\frac{\mathrm{d}}{\mathrm{d}t}\langle \hat{\boldsymbol{x}} \rangle_{\psi(t)} = \frac{1}{m}\langle \hat{\boldsymbol{p}} \rangle_{\psi(t)}, \\ \frac{\mathrm{d}}{\mathrm{d}t}\langle \hat{\boldsymbol{p}} \rangle_{\psi(t)} = -m\,\Omega^2(t)\Big[\langle \hat{\boldsymbol{x}} \rangle_{\psi(t)} - \boldsymbol{\rho}(t)\Big] + \boldsymbol{F}(t). \tag{77}$$

The expectation values are evaluated with the full solution $\psi(t, \boldsymbol{x})$ of the GP equation (1) for the harmonic potential (Eq. 3). Since Eqs. (77) coincide with Eqs. (5) for the time-dependent vectors $\mathcal{R}(t)$ and $\mathcal{P}(t)$, we can identify the latter ones with the center-of-mass position $\langle \hat{\boldsymbol{x}} \rangle_{\psi(t)}$ and the average momentum $\langle \hat{\boldsymbol{p}} \rangle_{\psi(t)}$ by simply choosing the initial conditions according to

$$\mathcal{R}(0) = \langle \hat{\boldsymbol{x}} \rangle_{\psi_0} \quad \text{and} \quad \mathcal{P}(0) = \langle \hat{\boldsymbol{p}} \rangle_{\psi_0}. \tag{78}$$

Here, we have introduced the shorthand notation of the exact initial state $\psi_0 \equiv \psi(0, \boldsymbol{x})$. We emphasize that the parameters $\mathcal{R}(t)$ and $\mathcal{P}(t)$ introduced in Section 2.1.2 have strictly speaking no physical meaning unless their initial conditions are fixed in accordance with Eq. (78).

Next, we discuss if the above assignment

$$\mathcal{R}(t) = \langle \hat{\boldsymbol{x}} \rangle_{\psi(t)} \quad \text{and} \quad \mathcal{P}(t) = \langle \hat{\boldsymbol{p}} \rangle_{\psi(t)} \tag{79}$$

is also consistent with our efficient description, Eq. (73), based on the time-dependent TF approximation. For this purpose, we insert Eq. (73) into the Eqs. (75) and (76) and make use of the substitution

$$\boldsymbol{x}' = \Lambda^{-1}(t)[\boldsymbol{x} - \mathcal{R}(t)] + \mathcal{R}(0) \tag{80}$$

to arrive at the expectation values

$$\langle \hat{\boldsymbol{x}} \rangle_{\psi(t)} \approx \mathcal{R}(t) + \Lambda(t)\left[\langle \hat{\boldsymbol{x}} \rangle_{\psi_0} - \mathcal{R}(0)\right], \tag{81}$$

$$\langle \hat{\boldsymbol{p}} \rangle_{\psi(t)} \approx \mathcal{P}(t) + \Lambda^{-\mathrm{T}}(t)\left[\langle \hat{\boldsymbol{p}} \rangle_{\psi_0} - \mathcal{P}(0)\right] + m\frac{\mathrm{d}\Lambda}{\mathrm{d}t}\left[\langle \hat{\boldsymbol{x}} \rangle_{\psi_0} - \mathcal{R}(0)\right]. \tag{82}$$

Thus, using $\mathcal{R}(0)$ and $\mathcal{P}(0)$ according to Eq. (78), the last two expressions simply reduce to Eq. (79), which proves that our efficient description based on the time-dependent TF approximation, Eq. (73), satisfies the generalized Ehrenfest theorem.

### 4.2.3 Energy of a BEC

Here, we use the approximate BEC wave function $\psi(t, \boldsymbol{x})$ given by Eq. (73) to derive an explicit expression for the total energy

$$E(t) = E_{\mathrm{kin}}(t) + E_{\mathrm{pot}}(t) + E_{\mathrm{int}}(t) \tag{83}$$

of the BEC within the time-dependent TF approximation, where the kinetic, the potential, and the interaction energy are defined by

$$E_{\mathrm{kin}}(t) = \frac{\hbar^2}{2m}\int_{\mathbb{R}^d} |\boldsymbol{\nabla}_{\boldsymbol{x}}\psi(t,\boldsymbol{x})|^2\,\mathrm{d}^d x = \frac{N}{2m}\langle \hat{\boldsymbol{p}}^2 \rangle_{\psi(t)}, \tag{84}$$

$$E_{\mathrm{pot}}(t) = \int_{\mathbb{R}^d} V(t,\boldsymbol{x})\,|\psi(t,\boldsymbol{x})|^2\,\mathrm{d}^d x = N\langle V(t,\boldsymbol{x}) \rangle_{\psi(t)}, \tag{85}$$

$$E_{\mathrm{int}}(t) = \frac{g}{2}\int_{\mathbb{R}^d} |\psi(t,\boldsymbol{x})|^4\,\mathrm{d}^d x = \frac{Ng}{2}\langle |\psi(t,\boldsymbol{x})|^2 \rangle_{\psi(t)}. \tag{86}$$

In Appendix B.1, we evaluate each of these energy terms and obtain as the energy per particle

$$\frac{E(t)}{N}=\frac{\mathcal{P}^2(t)}{2m}+V(t,\mathcal{R}(t))+\frac{1}{2m}\,\mathrm{Tr}\left[\Lambda^{-1}(t)\,\Lambda^{-\mathrm{T}}(t)\,\langle\hat{\boldsymbol{p}}\otimes\hat{\boldsymbol{p}}^{\mathrm{T}}\rangle_{|\psi_0|}\right]+\frac{E_{\mathrm{int}}(0)}{N\det\Lambda(t)}$$
$$+\frac{m}{2}\,\mathrm{Tr}\left\{\left[\frac{\mathrm{d}\Lambda^{\mathrm{T}}}{\mathrm{d}t}\frac{\mathrm{d}\Lambda}{\mathrm{d}t}+\Lambda^{\mathrm{T}}(t)\Omega^2(t)\Lambda(t)\right]\langle[\hat{\boldsymbol{x}}-\mathcal{R}(0)]\otimes[\hat{\boldsymbol{x}}-\mathcal{R}(0)]^{\mathrm{T}}\rangle_{\psi_0}\right\}, \tag{87}$$

where we have assumed that the initial wave function $\psi(0, \boldsymbol{x})$ is given by

$$\psi(0,\boldsymbol{x})=\phi_{\mathrm{D}}(\boldsymbol{x}-\mathcal{R}(0))\,\mathrm{e}^{\frac{\mathrm{i}}{\hbar}\mathcal{P}(0)\,\boldsymbol{x}}. \tag{88}$$

Here, $\phi_{\mathrm{D}}(\boldsymbol{x})$ is the real-valued ground state of the BEC in the comoving frame of the trap as discussed in Section 2.1.6 and the phase $\mathrm{i}\mathcal{P}(0)\,\boldsymbol{x}/\hbar$ accounts for the initial momentum $\mathcal{P}(0)$ of the BEC in the inertial frame of reference $\boldsymbol{x}$, analogous to the TF ground state (29). The first two terms in Eq. (87) display the kinetic and the potential energy of the center of mass of the BEC, respectively, whereby we have used $V(t,\mathcal{R}(t))$ as shorthand notation for the harmonic potential (3). The third term corresponds to the quantum pressure and its expectation value is evaluated with respect to the absolute value $|\psi(0, \boldsymbol{x})|$ of the initial wave function. The forth term describes the mean-field interaction energy. The last term is determined by the explicit form of the initial wave function $\psi(0, \boldsymbol{x})$ and represents the inner dynamics of the condensate associated with the deformation of the BEC, which corresponds to the evolution of the adaptive matrix $\Lambda(t)$.

When the initial wave function $\psi(0, \boldsymbol{x})$ is given by the TF ground state $\phi_{\mathrm{TF}}(\boldsymbol{x})$, Eq. (29), the energy per particle reads

$$\frac{E_{\mathrm{TF}}(t)}{N}=\frac{\mathcal{P}^2(t)}{2m}+V(t,\mathcal{R}(t))+\frac{2}{d+4}\,\mu_{\mathrm{TF}}\,E_{\Lambda}(t) \tag{89}$$

as shown in Appendix B.2. Here, $E_{\Lambda}(t)$ denotes the total energy associated with the matrix differential equation given by Eq. (66), whereas $\mu_{\mathrm{TF}}$ is the chemical potential within the TF approximation (Eq. 30).

By applying Eq. (15) to the energy per particle (89) and assuming without loss of generality that $V(0,\mathcal{R}(0))=0$, the initial energy per particle for a BEC in the TF ground state $\phi_{\mathrm{TF}}(\boldsymbol{x})$ reads

$$\frac{E_{\mathrm{TF}}(0)}{N}=\frac{\mathcal{P}^2(0)}{2m}+\frac{d+2}{d+4}\,\mu_{\mathrm{TF}} \tag{90}$$

and is determined solely by the initial momentum $\mathcal{P}(0)$ and the chemical potential $\mu_{\mathrm{TF}}$.

### 4.2.4 Angular Momentum of a BEC

In analogy to the previous section, we now discuss the angular momentum of a BEC within the time-dependent TF approximation. Indeed, as derived in Appendix B.3 the angular momentum operator with respect to the center-of-mass coordinates $\mathcal{R}(t)$ and $\mathcal{P}(t)$ is defined by

$$\hat{\boldsymbol{L}} = (\hat{\boldsymbol{x}} - \mathcal{R}(t)) \times (\hat{\boldsymbol{p}} - \mathcal{P}(t)). \tag{91}$$

In Appendix B.4, we show that the expectation value of the angular momentum operator, Eq. (91), evaluated with respect to the approximate wave function (73), is given by the relation

$$\langle \hat{\boldsymbol{L}} \rangle_{\psi(t)} = \left\langle \{\Lambda(t)[\hat{\boldsymbol{x}} - \mathcal{R}(0)]\} \times \left\{ m \frac{\mathrm{d}\Lambda}{\mathrm{d}t} [\hat{\boldsymbol{x}} - \mathcal{R}(0)] + \Lambda^{-\mathrm{T}} \hat{\boldsymbol{p}} \right\} \right\rangle_{\psi_0}, \tag{92}$$

where we have used the initial conditions (78). Here, the first vector of the cross product represents the position, while the second vector corresponds to the momentum associated with the inner dynamics of the BEC. For the sake of a concise mathematical notation, we map the angular momentum operator to the so-called angular momentum matrix operator via $\hat{L}_{kl} = \epsilon_{jkl} \hat{L}_j$, where $\varepsilon_{jkl}$ denotes the Levi-Civita symbol. Applied to Eq. (92), this mapping yields the following expression for the matrix operator:

$$\begin{aligned} \langle \hat{L} \rangle_{\psi(t)} = m \bigg[ & \Lambda(t) \langle [\hat{\boldsymbol{x}} - \mathcal{R}(0)] \otimes [\hat{\boldsymbol{x}} - \mathcal{R}(0)]^{\mathrm{T}} \rangle_{\psi_0} \frac{\mathrm{d}\Lambda^{\mathrm{T}}}{\mathrm{d}t} \\ & - \frac{\mathrm{d}\Lambda}{\mathrm{d}t} \langle [\hat{\boldsymbol{x}} - \mathcal{R}(0)] \otimes [\hat{\boldsymbol{x}} - \mathcal{R}(0)]^{\mathrm{T}} \rangle_{\psi_0} \Lambda^{\mathrm{T}}(t) \bigg] \\ & + \Lambda(t) \langle [\hat{\boldsymbol{x}} - \mathcal{R}(0)] \otimes \hat{\boldsymbol{p}}^{\mathrm{T}} \rangle_{\psi_0} \Lambda^{-1}(t) - \Lambda^{-\mathrm{T}}(t) \langle \hat{\boldsymbol{p}} \otimes [\hat{\boldsymbol{x}} - \mathcal{R}(0)]^{\mathrm{T}} \rangle_{\psi_0} \Lambda^{\mathrm{T}}(t). \end{aligned} \tag{93}$$

This matrix notation of the angular momentum is especially useful if the initial wave function $\psi(0, \boldsymbol{x})$ is approximated by the TF ground state $\phi_{\mathrm{TF}}(\boldsymbol{x})$, Eq. (29), since in that case Eq. (93) reduces to

$$\langle \hat{L} \rangle_{\mathrm{TF}} = \frac{2}{d+4} \frac{\mu_{\mathrm{TF}}}{\left[\det \boldsymbol{\Omega}^2(0)\right]^{\frac{1}{2d}}} L_\Lambda(t) \tag{94}$$

as discussed in Appendix B.5 with the angular momentum matrix $L_\Lambda(t)$ given by Eq. (70).

In conclusion, the deep connection between the constants of motion of the GP equation, Eq. (1), and the matrix differential equation,

Eq. (14), illustrated by Eqs. (89) and (94), once more highlights the benefit of the affine approach for the description of the BEC dynamics. Not only grants the solution of Eq. (14) valuable insights into the time evolution of a BEC, but it is also relevant for numerical applications.

# 5. SPECIAL PROPERTIES OF FREE EXPANDING BECs

Experiments dealing with quantum gases usually involve time-of-flight pictures to deduce the state of the system under study. It is well known, and we will briefly recall this statement later, that in the long-time limit the spatial density distribution obtained by time-of-flight measurements is determined by the initial momentum distribution of the quantum gas if the free time evolution is described by the Schrödinger equation. In other words, a noninteracting quantum gas undergoes a ballistic expansion if it is released from the trapping potential. However, it is not obvious that this statement holds true for free expanding BECs governed by the nonlinear GP equation. In fact, for BECs a reverse relation applies (Eckart, 2008), namely that the momentum distribution in the long-time limit is given by the initial spatial density distribution. This relation is not only relevant for the interpretation of time-of-flight pictures of BECs, but it also plays a role for matter-wave interferometry (Kleinert et al., 2015), where the visibility of the phase oscillations quickly decreases when the momentum distribution of the atomic clouds exceeds the corresponding line width of the laser pulses.

This section starts with the discussion of the analytic solutions of the matrix differential equation (14) for freely expanding BECs in the long-time limit that start from isotropic initial traps. We then recall the long-time behavior for the Schrödinger equation before we turn to the GP equation.

## 5.1 Analytic Solutions of the Matrix Differential Equation

For a pure free expansion, that is, $\Omega^2(t) = 0$ for all times $t > 0$, the matrix differential equation (14) possesses analytical solutions that can be used to determine the macroscopic wave function of a BEC in the long-time limit. In order to derive these solutions, we assume without loss of generality that the coordinate system $\boldsymbol{x}$ is aligned with the principal axes of the trap at the initial time $t = 0$. As a consequence, the adaptive matrix $\Lambda(t)$ is diagonal during the free expansion, that is, $\Lambda(t) = \mathrm{diag}[\lambda_1(t), \ldots, \lambda_d(t)]$ for $t > 0$, and the scaling parameters $\lambda_i$ with $i = 1, 2, \ldots, d$ are determined by the system of nonlinear differential equations

$$\frac{d^2\lambda_i}{dt^2} = \frac{\omega_i^2(0)}{\lambda_i \prod_{j=1}^{d} \lambda_j}, \tag{95}$$

which are a special case of Eq. (14) or (22) for free evolution. Here, the frequencies $\omega_i(0)$ denote the initial trap frequencies along the principal axes. The initial conditions (15) reduce to

$$\lambda_i(0) = 1 \quad \text{and} \quad \left.\frac{d\lambda_i}{dt}\right|_{t=0} = 0, \tag{96}$$

accordingly.

### 5.1.1 Long-Time Limit

Since the scaling parameters $\lambda_i(t)$ determine the size of the condensate, they grow as the condensate expands freely. Hence, the right-hand side of Eq. (95) vanishes in the long-time limit and the scaling parameters approach the linear dependence

$$\lambda_i(t) \cong a_i + b_i\, t \tag{97}$$

as $t \to \infty$. In general, the constants $a_i$ and $b_i$ are determined numerically. However, in the case of an isotropic initial trap, Eq. (95) can be solved analytically also yielding the values for $a_i$ and $b_i$ as shown in the next section.

### 5.1.2 Isotropic Case

For an isotropic initial trap, that is, $\omega_i(0) = \omega_0$, the scaling parameters $\lambda_i(t)$ are given by the single scaling parameter $\lambda(t)$ for all $i = 1, 2, \ldots, d$ and the system of nonlinear differential equations, Eq. (95), decouples. We show in Appendix C that the function $\lambda(t)$ is a solution of the integral equation

$$\int_1^{\lambda(t)} d\zeta \, \frac{\zeta^{d/2}}{\sqrt{\zeta^d - 1}} = \sqrt{\frac{2}{d}}\, \omega_0 t. \tag{98}$$

Depending on the dimensionality $d$, Eq. (98) gives rise to explicit, $d = 2$, or implicit, $d = 1$ and $d = 3$, solutions for the scaling parameter $\lambda(t)$, which are discussed in the next paragraphs.

#### *The case* $d = 1$

The left-hand side of Eq. (98) gives rise to the transcendental equation

$$\sqrt{\lambda(\lambda-1)} + \ln\left(\sqrt{\lambda}+\sqrt{\lambda-1}\right) = \sqrt{2}\,\omega_0 t,$$

which can also be written as

$$\lambda = \frac{\sqrt{2}\,\omega_0 t}{\sqrt{1-\frac{1}{\lambda}+\frac{1}{\lambda}\ln\left(\sqrt{\lambda}+\sqrt{\lambda-1}\right)}}. \tag{99}$$

It is clear that for long times, that is, $\omega_0 t \gg 1$, the function $\lambda \cong \sqrt{2}\,\omega_0 t$, since the denominator in the right-hand side of Eq. (99) approaches unity as $\lambda \to \infty$. Taking into account the terms in the denominator scaling as $\lambda^{-1}$ and $\ln(\lambda)\lambda^{-1}$, we finally arrive at the approximate solution

$$\lambda_{1\mathrm{D}}(t) \cong \sqrt{2}\,\omega_0 t - \frac{1}{2}\left[\ln\left(4\sqrt{2}\,\omega_0 t\right) - 1\right] \tag{100}$$

for $\omega_0 t \gg 1$. Here, we have neglected all terms that vanish in the long-time limit.

***The case*** $d = 2$

In the case of two dimensions, the integral in the left-hand side of Eq. (98) yields $\sqrt{\lambda^2(t)-1}$, leading to the exact solution

$$\lambda_{2\mathrm{D}}(t) = \sqrt{1+\omega_0^2 t^2} \tag{101}$$

valid for all $t \geq 0$.

***The case*** $d = 3$

For a three-dimensional BEC the integral in the left-hand side of Eq. (98) can be expressed in terms of the Gaussian hypergeometric function ${}_2F_1(a, b, c; z)$, leading to the implicit equation

$${}_2F_1\left(-\frac{1}{3}, \frac{1}{2}, \frac{2}{3}; \frac{1}{\lambda^3}\right)\lambda - \frac{\sqrt{\pi}\,\Gamma\left(\frac{2}{3}\right)}{\Gamma\left(\frac{1}{6}\right)} = \sqrt{\frac{2}{3}}\,\omega_0 t \tag{102}$$

for $\lambda(t)$, where $\Gamma(x)$ denotes the Gamma function. Since the hypergeometric function ${}_2F_1(-1/3, 1/2, 2/3; 1/\lambda^3)$ approaches unity as $\lambda \to \infty$, the solution $\lambda(t)$ of Eq. (102) in the long-time limit, that is, $\omega_0 t \gg 1$, is given by

$$\lambda_{3\mathrm{D}}(t) \cong \frac{\sqrt{\pi}\,\Gamma\left(\frac{2}{3}\right)}{\Gamma\left(\frac{1}{6}\right)} + \sqrt{\frac{2}{3}}\,\omega_0 t. \tag{103}$$

When we compare the relations (100), (101), and (103) with Eq. (97), we find that the coefficient $b_i = \sqrt{2/d}\,\omega_0$.

## 5.2 Connection Between the Long-Time Behavior of the Wave Function and the Initial State

Here, we show how the long-time behavior of the macroscopic wave function of a BEC can be estimated in the case of free expansion, although the dynamics is governed by the nonlinear GP equation. Since the approximate analytic result for BECs contrasts the long-time solution of noninteracting wave functions, we first have a look at the Schrödinger equation before discussing the GP equation.

### 5.2.1 *Schrödinger Equation*

The Schrödinger equation describing the free evolution of the noninteracting wave function $\psi(t, \boldsymbol{x})$ in a $d$-dimensional space reads

$$\mathrm{i}\hbar\,\frac{\partial\psi(t,\boldsymbol{x})}{\partial t} = \frac{\hat{\boldsymbol{p}}^2}{2m}\,\psi(t,\boldsymbol{x}) \tag{104}$$

and has the formal solution

$$\psi(t,\boldsymbol{x}) = \mathrm{e}^{-\frac{\mathrm{i}}{\hbar}\frac{\hat{\boldsymbol{p}}^2}{2m}t}\psi(0,\boldsymbol{x}) \tag{105}$$

in terms of the initial wave function $\psi(0, \boldsymbol{x})$. When we use the Fourier transformation

$$\tilde{\psi}(0,\boldsymbol{p}) \equiv \frac{1}{(2\pi\hbar)^{d/2}}\int_{\mathbb{R}^d} \mathrm{e}^{-\frac{\mathrm{i}}{\hbar}\boldsymbol{p}\boldsymbol{x}}\,\psi(0,\boldsymbol{x})\,\mathrm{d}^d x \tag{106}$$

of the wave function $\psi(0, \boldsymbol{x})$, we can cast Eq. (105) into the form

$$\psi(t,\boldsymbol{x}) = \frac{1}{(2\pi\hbar)^{d/2}}\,\mathrm{e}^{-\frac{\mathrm{i}}{\hbar}\frac{m\boldsymbol{x}^2}{2t}}\int_{\mathbb{R}^d} \mathrm{e}^{-\frac{\mathrm{i}}{\hbar}\frac{t}{2m}\left(\boldsymbol{p}-\frac{m\boldsymbol{x}}{t}\right)^2}\,\tilde{\psi}(0,\boldsymbol{p})\,\mathrm{d}^d p. \tag{107}$$

For long times, that is, in the limit $t \longrightarrow \infty$, the exponential function inside the integral on the right-hand side of Eq. (107) oscillates very rapidly and only the values of $\boldsymbol{p}$ in a narrow region around $\boldsymbol{p} = m\boldsymbol{x}/t$ contribute to the integral. Hence, when we assume that the momentum distribution $\tilde{\psi}(0,\boldsymbol{p})$ is a smooth function of $\boldsymbol{p}$ in this region, we can evaluate $\tilde{\psi}(0,\boldsymbol{p})$ at $\boldsymbol{p} = m\boldsymbol{x}/t$ and perform the remaining integral, leading to the relation

$$\psi(t,\boldsymbol{x}) \cong \left(\frac{\mathrm{i}m}{t}\right)^{d/2} \mathrm{e}^{-\frac{\mathrm{i}}{\hbar}\frac{m}{2t}\boldsymbol{x}^2}\,\tilde{\psi}\left(0,\frac{m\boldsymbol{x}}{t}\right). \tag{108}$$

Consequently, as is well known, the spatial density distribution in the long-time limit

$$|\psi(t,\boldsymbol{x})|^2 \cong \left(\frac{m}{t}\right)^d \left|\tilde{\psi}\left(0,\frac{m\boldsymbol{x}}{t}\right)\right|^2 \tag{109}$$

is determined by the shape of the initial momentum distribution $\tilde{\psi}(0,\boldsymbol{p})$.

### 5.2.2 GP Equation

For free expanding BECs a similar relation can be established based on the affine approach. In order to compare the results derived for the Schrödinger equation and the GP equation, respectively, it is suitable to consider the wave function $\psi_{\mathrm{D}}(t, \boldsymbol{x})$ that describes the inner dynamics of the BEC where the center-of-mass motion has already been eliminated. As shown in Section 2.1, the time evolution of $\psi_{\mathrm{D}}(t, \boldsymbol{x})$ is governed by the transformed GP equation (8), which is related to the affinely transformed wave function $\psi_{\Lambda}(\tau, \boldsymbol{\xi})$ by Eq. (10). If the initial wave function $\psi_{\mathrm{D}}(0, \boldsymbol{x})$ coincides with the TF ground state, Eq. (32), the time-dependent TF approximation (34) holds true during the free expansion and we arrive at the solution

$$\psi_{\mathrm{D}}(t,\boldsymbol{x}) \cong \frac{1}{\sqrt{\det\Lambda(t)}}\mathrm{e}^{\frac{\mathrm{i}}{\hbar}\left[\frac{m}{2}\boldsymbol{x}^{\mathrm{T}}\frac{\mathrm{d}\Lambda}{\mathrm{d}t}\Lambda^{-1}\boldsymbol{x}-\beta(t)\right]}\,\psi_{\mathrm{D}}\left(0,\Lambda^{-1}(t)\,\boldsymbol{x}\right), \tag{110}$$

where we have used the definition (12) of the symmetric matrix $A(t)$.

In the long-time limit and for free expansion the adaptive matrix $\Lambda(t)$ grows linearly, Eq. (97), that is, $\Lambda(t) \cong B\,t$ as $t \to \infty$ with the diagonal matrix $B = \mathrm{diag}(b_1, \ldots, b_d)$. When we apply this limit to expression (110), we obtain

$$\psi_{\mathrm{D}}(t,\boldsymbol{x}) \cong \frac{1}{\sqrt{t^d \det B}}\,\mathrm{e}^{\frac{\mathrm{i}}{\hbar}\left[\frac{m\boldsymbol{x}^2}{2t}-\beta(t)\right]}\,\psi_{\mathrm{D}}\left(0,B^{-1}\frac{\boldsymbol{x}}{t}\right). \tag{111}$$

As a result, the spatial density distribution in the long-time limit

$$|\psi_{\mathrm{D}}(t,\boldsymbol{x})|^2 \cong \frac{1}{t^d \det B}\left|\psi_{\mathrm{D}}\left(0,B^{-1}\frac{\boldsymbol{x}}{t}\right)\right|^2 \tag{112}$$

is determined by the initial spatial density distribution and the time-dependent scaling. Obviously, Eq. (112) is a direct consequence of the time-dependent TF approximation on which this derivation is based and stands in contrast to relation (109) derived for the Schrödinger equation.

In the next step, we obtain a similar relation for the momentum distribution of a free expanding BEC in the long-time limit. In order to do so, we insert Eq. (111) into Eq. (106) for the Fourier transform and arrive at the expression

$$\tilde{\psi}_{\mathrm{D}}(t,\boldsymbol{p}) \cong \frac{\mathrm{e}^{-\frac{\mathrm{i}}{\hbar}\beta(t)-\frac{\mathrm{i}}{\hbar}\frac{\boldsymbol{p}^2 t}{2m}}}{(2\pi\hbar\, t)^{d/2}\sqrt{\det B}} \int_{\mathbb{R}^d} \mathrm{e}^{\frac{\mathrm{i}}{\hbar}\frac{m}{2t}\boldsymbol{x}'^2}\, \psi_{\mathrm{D}}\left(0, B^{-1}\frac{\boldsymbol{x}'}{t} + B^{-1}\frac{\boldsymbol{p}}{m}\right) \mathrm{d}^d x', \quad (113)$$

where we have introduced the new integration variable $\boldsymbol{x}' \equiv \boldsymbol{x} - \boldsymbol{p}t/m$. For long times $t \to \infty$, the argument of the initial wave function $\psi_{\mathrm{D}}$ gets practically independent of $\boldsymbol{x}'$, which allows to perform the remaining integration and to obtain

$$\tilde{\psi}_{\mathrm{D}}(t,\boldsymbol{p}) \cong \frac{\mathrm{e}^{-\frac{\mathrm{i}}{\hbar}\beta(t)-\frac{\mathrm{i}}{\hbar}\frac{\boldsymbol{p}^2 t}{2m}}}{(\mathrm{i}\, m)^{d/2}\sqrt{\det B}} \psi_{\mathrm{D}}\left(0, B^{-1}\frac{\boldsymbol{p}}{m}\right). \quad (114)$$

Hence, in the case of the GP equation for free expansion, the momentum distribution in the long-time limit

$$|\tilde{\psi}_{\mathrm{D}}(t,\boldsymbol{p})|^2 \cong \frac{1}{m^d \det B}\left|\psi_{\mathrm{D}}\left(0, B^{-1}\frac{\boldsymbol{p}}{m}\right)\right|^2 \quad (115)$$

is given by the initial spatial density distribution, while for the Schrödinger equation it is vice versa, (Eq. 109). Moreover, according to Eq. (115) the momentum distribution $|\tilde{\psi}_{\mathrm{D}}(t,\boldsymbol{p})|^2$ of a free expanding BEC becomes time independent in the long-time limit. However, we emphasize that for the derivation of Eq. (115), we have required the validity of the time-dependent TF approximation during the free expansion. As a consequence, for a proper interpretation of time-of-flight pictures of BECs the interactions play an important role, especially deep within the TF regime.

In addition, we point out that the methods presented in this section have been successfully applied by Roura et al. (2014) to study the influence of gravity gradients on the properties of an atom interferometer.

## 6. CONCLUSION AND OUTLOOK

In this chapter, we have introduced a natural generalization of the scaling approach for time-dependent rotating traps, the so-called affine approach. In contrast to the hydrodynamical approach, we have carried out this generalization directly to the GP equation by first eliminating the center-of-mass motion and by subsequently applying a linear mapping to account for the main part of the remaining internal dynamics of a BEC. With the help of this affine approach, we have established an efficient analytic description of the BEC dynamics based on two main assumptions: (i) the external potential in the vicinity of the BEC is at most quadratic and (ii) the time-dependent TF approximation holds true during the whole time evolution, which manifests itself in the fact that the density distribution of the condensate preserves its parabolic shape.

We have verified the accuracy of this approach by performing full numerical simulations of the GP equation for two scenarios: (i) a pure rotation of the trap, and (ii) a free expansion of an initially rotating BEC. In both cases, we have found an excellent agreement of the numerical results with our approximate analytical solution for many different combinations of the interaction strength, the anisotropy factor, and the final rotation rate of the trap. Moreover, the affine transformation can be successfully used to improve the performance of numerical simulations for a variety of experimentally relevant scenarios by solving the transformed GP equation rather than the original one.

By employing an additional transformation, we have introduced the canonical form of the matrix differential equation that governs the time evolution of the adaptive matrix and presented its corresponding Hamiltonian formalism. We have furthermore analyzed the relation between the constants of motion of this matrix differential equation and the constants of motion of the GP equation in connection with the conservation of the total energy and the angular momentum of a BEC. For both cases, we have derived explicit expressions valid within the TF regime.

In agreement with our affine approach, the *spatial* density distribution of a freely expanding BEC in the long-time limit *does not* display the initial *momentum* distribution in general, unlike a freely expanding atomic cloud that evolves according to the Schrödinger equation. Instead, it resembles the affinely transformed initial spatial density distribution. However, there exists a converse relation between the spatial and momentum distributions for a BEC within the TF regime; that is the *momentum* distribution of a free expanding BEC in the long-time limit is indeed given by its initial *spatial* density distribution. These relations are a direct consequence of the nonlinearity of the GP equation and stand in contrast to the well-known ballistic expansion that wave functions governed by the free Schrödinger equation show. Hence, in order to properly interpret time-of-flight pictures of BECs, the mean-field interaction has to be taken into account. This observation is of special importance for the TF regime where the interactions play a dominant role.

In summary, scaling solutions like the affine approach presented in this chapter are valuable tools for studying the dynamics of strongly interacting quantum gases. Not only do they provide efficient ways to accurately describe their time evolution, but they also allow to unravel and separate the different layers of their dynamics. Similar approaches have been successfully used in a multitude of interesting physical phenomena, such as (i) the Gaussian wave packet, which keeps its form in an at most harmonic potential, (ii) the Airy wave packet (Berry and Balazs, 1979; Kajari et al., 2010), and (iii) the Bessel beam (Durnin, 1987; Durnin et al., 1987), which freely propagate without dispersion.

We emphasize that there are a few scenarios in which one of the two above-mentioned assumptions for our efficient analytic description of the BEC dynamics is not fulfilled. Indeed, the external potential could be of a form incompatible with the harmonic expansion, Eq. (3), as is the case for (i) realistic trapping potentials created by optical dipole traps, optical lattices, and magnetic chip traps, where the anharmonic terms play a crucial role especially in the context of delta-kick collimation techniques (Ammann and Christensen, 1997; Zeller, 2016), and (ii) trapping potentials that are not differentiable due to sharp corners like linear potential wells (the irrigation canal) (Bestle et al., 1995; Gallas et al., 1995) or gravitational traps (Aminoff et al., 1993; Nesvizhevsky et al., 2002). Moreover, the TF approximation is well known to break down when the interaction strength approaches very small values, which can be caused by a low number of atoms, or a decreasing coupling constant. In order to accurately describe BECs in this regime, one can either expand the theoretical model beyond the TF approximation (Jamison et al., 2011) or resort to numerical methods based on the affinely transformed GP equation. For all these scenarios the affine approach can still offer valuable insights by dedicated numerical simulations that take advantage of the affine transformation to substantially reduce the computational costs of the simulations (Eckart, 2008).

Finally, it is noteworthy that the affine approach can even be applied to multispecies BECs to study their expansion dynamics both from an analytical and numerical perspective. However, a detailed discussion of this method is beyond the scope of this chapter and will be addressed in a future publication (Meister et al., 2017).

## ACKNOWLEDGMENTS

We would like to thank H. Ahlers, M. Buser, A. Friedrich, N. Gaaloul, E. Giese, N. Harshman, W. Herr, J. Jenewein, S. Kleinert, M. Krutzik, W. Lewoczko-Adamczyk, A. Roura, E. Sadurni, S.T. Seidel, E.M. Rasel, C. Ufrecht, V. Yakovlev, W. Zeller, and T. van Zoest for many fruitful discussions and helpful suggestions. M.A.E. thanks the Alexander von Humboldt Stiftung and W.P.S. is grateful to the Hagler Institute for Advanced Study at Texas A&M University for a Faculty Fellowship. This project was generously supported by the German Space Agency (DLR) with funds provided by the Federal Ministry for Economic Affairs and Energy (BMWi) under the Grant Nos. 50WM0346, 50WM0837, 50WM1136, and 50WM1556.

# APPENDICES

## APPENDIX A. AFFINE TRANSFORMATION

In this appendix, we outline the derivation of the affinely transformed GP equation (13). This is done by first eliminating the center-of-mass motion of the condensate and second, by performing a linear transformation of the coordinates together with a unitary transformation of the wave function. Moreover, we show how to determine the integrated density distributions of a BEC within the TF regime.

### A.1 Center-of-Mass Motion

Our starting point is the GP equation (1)

$$\mathrm{i}\hbar\frac{\partial\psi}{\partial t}=\left[-\frac{\hbar^2}{2m}\nabla_x^2+V(t,\boldsymbol{\rho})-\boldsymbol{F}(\boldsymbol{x}-\boldsymbol{\rho})+\frac{m}{2}(\boldsymbol{x}-\boldsymbol{\rho})^{\mathrm{T}}\Omega^2(\boldsymbol{x}-\boldsymbol{\rho})+g|\psi|^2\right]\psi \tag{A.1}$$

with the explicit time-dependent harmonic potential given by Eq. (3). We then rewrite the transformation (4) by making use of the displacement operator in position representation

$$\hat{D}(\mathcal{R},\mathcal{P})\equiv e^{\frac{\mathrm{i}}{\hbar}(\mathcal{P}\,\boldsymbol{x}-\mathcal{R}\,\hat{\boldsymbol{p}})} \tag{A.2}$$

and obtain

$$\psi(t,\boldsymbol{x})=e^{\frac{i}{\hbar}S_1(t)}\hat{D}(\mathcal{R},\mathcal{P})\psi_{\mathrm{D}}(t,\boldsymbol{x}). \tag{A.3}$$

The displacement operator satisfies the relations

$$\hat{D}(\mathcal{R},\mathcal{P})\,\psi_{\rm D}(t,\boldsymbol{x}) = {\rm e}^{-\frac{\rm i}{2\hbar}\mathcal{R}\mathcal{P}}{\rm e}^{\frac{\rm i}{\hbar}\mathcal{P}\boldsymbol{x}}\psi_{\rm D}(t,\boldsymbol{x}-\mathcal{R}), \tag{A.4}$$

$$\hat{D}^{\dagger}(\mathcal{R},\mathcal{P})f(\boldsymbol{x},\hat{\boldsymbol{p}})\,\hat{D}(\mathcal{R},\mathcal{P}) = f(\boldsymbol{x}+\mathcal{R},\hat{\boldsymbol{p}}+\mathcal{P}) \tag{A.5}$$

for any smooth wave function $\psi_{\rm D}(t,\,\boldsymbol{x})$ and phase-space operator $f(\boldsymbol{x},\hat{\boldsymbol{p}})$. Taking advantage of Eqs. (A.3)–(A.5), as well as the fact that

$$\hat{D}^{\dagger}\,|\psi(t,\boldsymbol{x})|^2\,\hat{D} = \hat{D}^{\dagger}\,|\psi_{\rm D}(t,\boldsymbol{x}-\mathcal{R})|^2\,\hat{D} = |\psi_{\rm D}(t,\boldsymbol{x})|^2,$$

we arrive at the transformed GP equation

$$\begin{aligned}{\rm i}\hbar\frac{\partial\psi_{\rm D}}{\partial t} = &\left\{\frac{\hat{\boldsymbol{p}}^2}{2m}+\frac{m}{2}\boldsymbol{x}^{\rm T}\Omega^2\boldsymbol{x}+g\,|\psi_{\rm D}|^2+\left[-\boldsymbol{F}+m\Omega^2(\mathcal{R}-\boldsymbol{\rho})+\frac{{\rm d}\mathcal{P}}{{\rm d}t}\right]\boldsymbol{x}\right.\\ &+\left[\frac{\mathcal{P}}{m}-\frac{{\rm d}\mathcal{R}}{{\rm d}t}\right]\hat{\boldsymbol{p}}+\frac{{\rm d}\mathcal{S}_1}{{\rm d}t}+\frac{\rm d}{{\rm d}t}\left(\frac{\mathcal{R}\mathcal{P}}{2}\right)-\frac{{\rm d}\mathcal{R}}{{\rm d}t}\mathcal{P}\\ &\left.+\frac{\mathcal{P}^2}{2m}+V(t,\boldsymbol{\rho})-\boldsymbol{F}(\mathcal{R}-\boldsymbol{\rho})+\frac{m}{2}(\mathcal{R}-\boldsymbol{\rho})^{\rm T}\Omega^2(\mathcal{R}-\boldsymbol{\rho})\right\}\psi_{\rm D}\end{aligned} \tag{A.6}$$

for the wave function $\psi_{\rm D}$.

In order to eliminate all terms in Eq. (A.6) that depend linearly on $\boldsymbol{x}$ and $\hat{\boldsymbol{p}}$, as well as arbitrarily on $\boldsymbol{\rho}$, we require the parameters $\mathcal{R}(t)$ and $\mathcal{P}(t)$ to satisfy the classical equations of motion (5) and the phase $\mathcal{S}_1(t)$ to be a solution of the first-order differential equation

$$\frac{{\rm d}\mathcal{S}_1}{{\rm d}t} = \mathcal{L}(\mathcal{R},\dot{\mathcal{R}},t)-\frac{1}{2}\frac{\rm d}{{\rm d}t}(\mathcal{R}\mathcal{P}). \tag{A.7}$$

Here, we have used the first equation in Eq. (5) and the definition of the Lagrange function $\mathcal{L}(\mathcal{R},\dot{\mathcal{R}},t)$ (Eq. 6). As a result, we have simplified the transformed GP equation (A.6) to Eq. (8).

## A.2 Linear Transformation

We continue the derivation of the affinely transformed GP equation by recalling Eq. (9) for the linear transformation of the spatial coordinates and the substitution of the time variable. This mapping implies that the corresponding partial derivatives in Eq. (8) transform as

$$\frac{\partial}{\partial t}=\frac{\partial}{\partial \tau}-\left(\Lambda^{-1}(\tau)\,\frac{\partial \Lambda}{\partial \tau}\,\boldsymbol{\xi}\right)^{\mathrm{T}}\boldsymbol{\nabla}_{\xi},$$
$$\boldsymbol{\nabla}_{x}=\Lambda^{-\mathrm{T}}(\tau)\,\boldsymbol{\nabla}_{\xi}. \tag{A.8}$$

As mentioned in Section 2.1.3, the linear mapping (9) is accompanied by the unitary transformation of the wave function (10), which guarantees that the affinely transformed wave function $\psi_{\Lambda}(\tau,\,\boldsymbol{\xi})$ is again normalized according to Eq. (2). Thus, by using Eqs. (9), (10), and (A.8), as well as the identity

$$\frac{\mathrm{d}}{\mathrm{d}\tau}(\det\Lambda)=\det\Lambda\cdot\mathrm{Tr}\left\{\Lambda^{-1}\,\frac{\mathrm{d}\Lambda}{\mathrm{d}\tau}\right\}, \tag{A.9}$$

we arrive at the GP equation for the affinely transformed wave function $\psi_{\Lambda}(\tau,\,\boldsymbol{\xi})$

$$\begin{aligned}\mathrm{i}\hbar\,\frac{\partial \psi_{\Lambda}}{\partial t}= &\left(-\frac{\hbar^{2}}{2m}\boldsymbol{\nabla}_{\xi}^{\mathrm{T}}\Lambda^{-1}\,\Lambda^{-\mathrm{T}}\boldsymbol{\nabla}_{\xi}+\frac{g}{\det\Lambda}|\psi_{\Lambda}|^{2}\right)\psi_{\Lambda}\\ &+\mathrm{i}\hbar\,\psi_{\Lambda}\left[\mathrm{Tr}\left\{\frac{\Lambda^{-1}}{2}\left(\frac{\mathrm{d}\Lambda}{\mathrm{d}\tau}-\frac{2}{m}\Lambda^{-\mathrm{T}}A\right)\right\}+\frac{\mathrm{i}}{\hbar}\frac{\mathrm{d}\beta}{\mathrm{d}\tau}\right]\\ &+\psi_{\Lambda}\,\boldsymbol{\xi}^{\mathrm{T}}\left[\frac{\mathrm{d}A}{\mathrm{d}\tau}-2A\,\Lambda^{-1}\left(\frac{\mathrm{d}\Lambda}{\mathrm{d}\tau}-\frac{1}{m}\Lambda^{-\mathrm{T}}A\right)+\frac{m}{2}\Lambda^{\mathrm{T}}\Omega^{2}\Lambda\right]\boldsymbol{\xi}\\ &+\mathrm{i}\hbar\,\left(\boldsymbol{\nabla}_{\xi}\psi_{\Lambda}\right)^{\mathrm{T}}\Lambda^{-1}\left[\frac{\mathrm{d}\Lambda}{\mathrm{d}\tau}-\frac{2}{m}\Lambda^{-\mathrm{T}}A\right]\boldsymbol{\xi}.\end{aligned} \tag{A.10}$$

Now we first choose $A(\tau)$ to be given by Eq. (12), which serves the purpose of eliminating the last term in Eq. (A.10) as well as the part that contains the trace in the second term on the right-hand side. Next we insert $\beta(\tau)$ as defined by Eq. (11). In order to simplify the third term on the right-hand side of Eq. (A.10), we therein replace $A(\tau)$ according to Eq. (12) and require the adaptive matrix $\Lambda(\tau)$ to satisfy the matrix differential equation (14). Finally, by taking advantage of the irrotationality condition (16) whose validity is a direct consequence of Eqs. (14) and (15) as discussed in the next section, we achieve our goal to simplify Eq. (A.10) to the affinely transformed GP equation (13).

## A.3 Irrotationality Condition

Within our approach the adaptive matrix $\Lambda(\tau)$ automatically satisfies the irrotationality condition, Eq. (16), since the time derivative of the auxiliary matrix function

$$Z(\tau) = \Lambda^{\mathrm{T}}(\tau)\frac{\mathrm{d}\Lambda}{\mathrm{d}\tau} - \frac{\mathrm{d}\Lambda^{\mathrm{T}}}{\mathrm{d}\tau}\Lambda(\tau)$$

vanishes for all $\tau \geq 0$ provided $\Lambda(\tau)$ is a solution of the matrix differential equation (14) with the initial conditions (15). For this reason, the antisymmetric matrix $Z(\tau) = Z(0) = 0$ represents $d(d-1)/2$ constants of motion of the second-order matrix differential equation (14) and the matrix $A(\tau)$, Eq. (12), is indeed symmetric.

However, as pointed out by Storey and Olshanii (2000), the irrotationality condition (16) can likewise be derived by using the notion of classical particle trajectories that constitute a velocity field $\boldsymbol{v}(t, \boldsymbol{x})$, which is irrotational, namely $\boldsymbol{\nabla}\times\boldsymbol{v}(t, \boldsymbol{x}) = 0$ at any point in space. We briefly sketch this derivation within our approach. Let $\boldsymbol{x} = \boldsymbol{x}(\boldsymbol{x}_0, t)$ be the classical trajectory of a particle within a BEC, which depends on the initial position $\boldsymbol{x}_0$ and the time $t$. As we have shown in Section 2.2.2 a BEC does not show any time evolution in the adapted coordinates if the time-dependent TF approximation holds true. Hence, the trajectory in adapted coordinates is given by $\boldsymbol{\xi}(t) = \boldsymbol{\xi}_0 = \boldsymbol{x}_0$, keeping in mind that the original and the adapted coordinates coincide at $t = 0$. Since the transformation between the two coordinate systems is defined by the linear mapping $\boldsymbol{\xi} = \Lambda^{-1}(t)[\boldsymbol{x} - \boldsymbol{\mathcal{R}}(t)]$ with the center-of-mass position $\boldsymbol{\mathcal{R}}(t)$ and the adaptive matrix $\Lambda(t)$, the classical trajectory is given by

$$\boldsymbol{x}_0 = \Lambda^{-1}(t)[\boldsymbol{x} - \boldsymbol{\mathcal{R}}(t)] \tag{A.11}$$

or rather

$$\boldsymbol{x}(\boldsymbol{x}_0, t) = \boldsymbol{\mathcal{R}}(t) + \Lambda(t)\,\boldsymbol{x}_0. \tag{A.12}$$

Thus, the velocity field

$$\boldsymbol{v}(\boldsymbol{x}, t) \equiv \frac{\mathrm{d}\boldsymbol{x}}{\mathrm{d}t} = \frac{\mathrm{d}\boldsymbol{\mathcal{R}}}{\mathrm{d}t} + \frac{\mathrm{d}\Lambda}{\mathrm{d}t}\Lambda^{-1}(\boldsymbol{x} - \boldsymbol{\mathcal{R}}) \tag{A.13}$$

is irrotational, that is, $\boldsymbol{\nabla}\times\boldsymbol{v}(t, \boldsymbol{x}) = 0$, if the matrix $(\mathrm{d}\Lambda/\mathrm{d}t)\Lambda^{-1}$ is symmetric

$$\Lambda^{-\mathrm{T}}\frac{\mathrm{d}\Lambda^{\mathrm{T}}}{\mathrm{d}t} = \frac{\mathrm{d}\Lambda}{\mathrm{d}t}\Lambda^{-1}, \tag{A.14}$$

which is equivalent to the irrotationality condition (16).

## A.4 Integration of the Density Distribution of a BEC Within the TF Regime

In this appendix, we discuss the integration of the density distribution $|\psi_{\mathrm{TF}}(t,\boldsymbol{x})|^2$ of a $d$-dimensional BEC within the TF regime over $n$ dimensions with $n < d$. We denote by $\boldsymbol{x}_{[2]} \in \mathbb{R}^n$ the vector consisting of the components of $\boldsymbol{x} \in \mathbb{R}^d$ over which the integration is performed. Accordingly, the vector $\boldsymbol{x}_{[1]} \in \mathbb{R}^{d-n}$ is composed of the remaining coordinates. Thus, the integrated density distribution is defined as

$$n_{\mathrm{TF}}(t,\boldsymbol{x}_{[1]}) = \int_{\mathbb{R}^n} |\psi_{\mathrm{TF}}(t,\boldsymbol{x})|^2 \, \mathrm{d}^n x_{[2]}. \tag{A.15}$$

Without loss of generality, we can rearrange the coordinates in the original density distribution, Eq. (40), such that $\boldsymbol{x}^{\mathrm{T}} = (\boldsymbol{x}_{[1]}^{\mathrm{T}}, \boldsymbol{x}_{[2]}^{\mathrm{T}})$, which furthermore translates into $\mathcal{R}^{\mathrm{T}} = (\mathcal{R}_{[1]}^{\mathrm{T}}, \mathcal{R}_{[2]}^{\mathrm{T}})$ for the center-of-mass position and

$$\Sigma = \begin{pmatrix} \Sigma_{[11]} & \Sigma_{[12]} \\ \Sigma_{[12]}^{\mathrm{T}} & \Sigma_{[22]} \end{pmatrix}$$

for the TF matrix defined by Eq. (39). By using the decomposition of the matrix $\Sigma$ into the Schur complements (Horn and Johnson, 1990)

$$\Sigma = \begin{pmatrix} \mathbb{1}_{d-n} & 0 \\ \Sigma_{[12]}^{\mathrm{T}}\Sigma_{[11]}^{-1} & \mathbb{1}_n \end{pmatrix} \begin{pmatrix} \Sigma_{[11]} & 0 \\ 0 & \Sigma_{[22]} - \Sigma_{[12]}^{\mathrm{T}}\Sigma_{[11]}^{-1}\Sigma_{[12]} \end{pmatrix} \begin{pmatrix} \mathbb{1}_{d-n} & 0 \\ \Sigma_{[12]}^{\mathrm{T}}\Sigma_{[11]}^{-1} & \mathbb{1}_n \end{pmatrix}^{\mathrm{T}}, \tag{A.16}$$

the inverse of the TF matrix can be written as

$$\Sigma^{-1} = \begin{pmatrix} \mathbb{1}_{d-n} & 0 \\ -\Sigma_{[12]}^{\mathrm{T}}\Sigma_{[11]}^{-1} & \mathbb{1}_n \end{pmatrix}^{\mathrm{T}} \begin{pmatrix} \Sigma_{[11]}^{-1} & 0 \\ 0 & \left(\Sigma_{[22]} - \Sigma_{[12]}^{\mathrm{T}}\Sigma_{[11]}^{-1}\Sigma_{[12]}\right)^{-1} \end{pmatrix} \begin{pmatrix} \mathbb{1}_{d-n} & 0 \\ -\Sigma_{[12]}^{\mathrm{T}}\Sigma_{[11]}^{-1} & \mathbb{1}_n \end{pmatrix}.$$

By inserting this expression into Eq. (40), the quadratic form $(\boldsymbol{x} - \mathcal{R})^{\mathrm{T}}\Sigma^{-1}(\boldsymbol{x} - \mathcal{R})$ splits into two terms: (i) a term involving only the free coordinates $\boldsymbol{x}_{[1]}$ and (ii) a term depending on both the coordinates $\boldsymbol{x}_{[2]}$ over which the integration is performed and the free coordinates $\boldsymbol{x}_{[1]}$. In view of the integration in Eq. (A.15), we perform a change of the integration variables from $\boldsymbol{x}_{[2]}$ to $\boldsymbol{y}$ via

$$\boldsymbol{y}=\sqrt{\frac{m}{2}}\left(\Sigma_{[22]}-\Sigma_{[12]}^{\mathrm{T}}\Sigma_{[11]}^{-1}\Sigma_{[12]}\right)^{-\frac{1}{2}}\left[\boldsymbol{x}_{[2]}-\boldsymbol{\mathcal{R}}_{[2]}-\Sigma_{[12]}^{\mathrm{T}}\Sigma_{[11]}^{-1}\left(\boldsymbol{x}_{[1]}-\boldsymbol{\mathcal{R}}_{[1]}\right)\right].$$

Note that the inverse square root of the matrix $(\Sigma_{[22]}-\Sigma_{[12]}^{\mathrm{T}}\Sigma_{[11]}^{-1}\Sigma_{[12]})$ is defined in terms of its spectral decomposition. As a result, we obtain for the integrated density distribution (A.15)

$$n_{\mathrm{TF}}(t,\boldsymbol{x}_{[1]})=\frac{1}{g\det\Lambda}\sqrt{\frac{2^n}{m^n}\det\left(\Sigma_{[22]}-\Sigma_{[12]}^{\mathrm{T}}\Sigma_{[11]}^{-1}\Sigma_{[12]}\right)}\int_{\mathbb{R}^n}\left\{R^2(\boldsymbol{x}_{[1]})-\boldsymbol{y}^2\right\}_+\mathrm{d}^n y, \tag{A.17}$$

where we have introduced the auxiliary function

$$R(\boldsymbol{x}_{[1]})\equiv\left\{\mu_{\mathrm{TF}}-\frac{m}{2}\left(\boldsymbol{x}_{[1]}-\boldsymbol{\mathcal{R}}_{[1]}\right)^{\mathrm{T}}\Sigma_{[11]}^{-1}\left(\boldsymbol{x}_{[1]}-\boldsymbol{\mathcal{R}}_{[1]}\right)\right\}_+^{\frac{1}{2}}.$$

We perform the integration in Eq. (A.17) with the help of the $n$-dimensional spherical coordinates

$$\begin{aligned}\int_{\mathbb{R}^n}\{R^2-\boldsymbol{y}^2\}_+\mathrm{d}^n y&=\int_{S_{n-1}}\mathrm{d}^{n-1}\Omega\int_0^R\left(R^2-r^2\right)r^{n-1}\,\mathrm{d}r\\&=\frac{4\pi^{\frac{n}{2}}}{n(n+2)\Gamma\left(\frac{n}{2}\right)}R^{n+2},\end{aligned}\tag{A.18}$$

where $S_{n-1}$ denotes the $(n-1)$-dimensional spherical surface. Moreover, by applying the determinant on both sides of the decomposition (A.16), we find with the help of Eq. (39) the useful relation

$$\det\left(\Sigma_{[22]}-\Sigma_{[12]}^{\mathrm{T}}\,\Sigma_{[11]}^{-1}\,\Sigma_{[12]}\right)=\frac{\det\Sigma}{\det\Sigma_{[11]}}=\frac{(\det\Lambda)^2}{\det\Omega^2(0)\det\Sigma_{[11]}}.$$

Taking advantage of the latter, the integrated density distribution equation (A.17) finally reads

$$\begin{aligned}n_{\mathrm{TF}}(t,\boldsymbol{x}_{[1]})=&\frac{1}{\Gamma\left(\frac{n}{2}+2\right)g}\sqrt{\frac{(2\pi/m)^n}{\det\Omega^2(0)\det\Sigma_{[11]}}}\\&\left\{\mu_{\mathrm{TF}}-\frac{m}{2}\left(\boldsymbol{x}_{[1]}-\boldsymbol{\mathcal{R}}_{[1]}\right)^{\mathrm{T}}\Sigma_{[11]}^{-1}\left(\boldsymbol{x}_{[1]}-\boldsymbol{\mathcal{R}}_{[1]}\right)\right\}_+^{\frac{n}{2}+1}.\end{aligned}$$

For $n=1$ and $n=2$, this result reduces to Eqs. (42) and (44), respectively.

# APPENDIX B. ENERGY AND ANGULAR MOMENTUM OF A BEC

This appendix presents the important steps to determine the different contributions to the total energy of a BEC and its angular momentum within the time-dependent TF approximation.

## B.1 Energy Terms in the Time-Dependent TF Approximation

We now evaluate the energy terms, Eqs. (84)–(86), by using the approximate wave function, Eq. (73), valid in the time-dependent TF approximation. With the help of Eq. (80), we can rewrite the expectation values in Eqs. (84)–(86) in terms of the expectation value with respect to the initial state $\psi(0, \boldsymbol{x})$. Thus, for the kinetic energy, Eq. (84), we obtain

$$\begin{aligned}\frac{E_{\text{kin}}(t)}{N} =& \frac{1}{2m}\left[\mathcal{P}(t) - \Lambda^{-\mathrm{T}}(t)\,\mathcal{P}(0)\right]^2 \\ &+\frac{1}{m}\left[\mathcal{P}(t) - \Lambda^{-\mathrm{T}}(t)\mathcal{P}(0)\right]^{\mathrm{T}}\Lambda^{-\mathrm{T}}(t)\langle\hat{\boldsymbol{p}}\rangle_{\psi_0} \\ &+\frac{1}{2m}\langle\hat{\boldsymbol{p}}^{\mathrm{T}}\Lambda^{-1}(t)\Lambda^{-\mathrm{T}}(t)\,\hat{\boldsymbol{p}}\rangle_{\psi_0} + \frac{m}{2}\left\langle[\hat{\boldsymbol{x}} - \mathcal{R}(0)]^{\mathrm{T}}\frac{\mathrm{d}\Lambda^{\mathrm{T}}}{\mathrm{d}t}\frac{\mathrm{d}\Lambda}{\mathrm{d}t}[\hat{\boldsymbol{x}} - \mathcal{R}(0)]\right\rangle_{\psi_0} \\ &+\left[\mathcal{P}(t) - \Lambda^{-\mathrm{T}}(t)\mathcal{P}(0)\right]^{\mathrm{T}}\frac{\mathrm{d}\Lambda}{\mathrm{d}t}\left[\langle\hat{\boldsymbol{x}}\rangle_{\psi_0} - \mathcal{R}(0)\right] \\ &+\mathrm{Re}\left\{\left\langle[\hat{\boldsymbol{x}} - \mathcal{R}(0)]^{\mathrm{T}}\frac{\mathrm{d}\Lambda^{\mathrm{T}}}{\mathrm{d}t}\Lambda^{-\mathrm{T}}(t)\,\hat{\boldsymbol{p}}\right\rangle_{\psi_0}\right\}.\end{aligned} \tag{B.1}$$

In the same manner the potential energy given by Eqs. (85) and (3) reads

$$\begin{aligned}\frac{E_{\text{pot}}(t)}{N} =& V(t,\boldsymbol{\rho}(t)) + \frac{m}{2}\langle[\hat{\boldsymbol{x}} - \mathcal{R}(0)]^{\mathrm{T}}\Lambda^{\mathrm{T}}(t)\Omega^2(t)\Lambda(t)[\hat{\boldsymbol{x}} - \mathcal{R}(0)]\rangle_{\psi_0} \\ &- \boldsymbol{F}(t)\Lambda(t)\left[\langle\hat{\boldsymbol{x}}\rangle_{\psi_0} - \mathcal{R}(0)\right] - \boldsymbol{F}(t)[\mathcal{R}(t) - \boldsymbol{\rho}(t)] \\ &+m[\mathcal{R}(t) - \boldsymbol{\rho}(t)]^{\mathrm{T}}\Omega^2(t)\Lambda(t)\left[\langle\hat{\boldsymbol{x}}\rangle_{\psi_0} - \mathcal{R}(0)\right] \\ &+\frac{m}{2}[\mathcal{R}(t) - \boldsymbol{\rho}(t)]^{\mathrm{T}}\Omega^2(t)[\mathcal{R}(t) - \boldsymbol{\rho}(t)].\end{aligned} \tag{B.2}$$

Likewise, the dependence on time of the interaction energy equation (86) is given by

$$\frac{E_{\text{int}}(t)}{N} = \frac{E_{\text{int}}(0)}{N \det \Lambda(t)}. \tag{B.3}$$

We can further simplify Eqs. (B.1) and (B.2) for the kinetic and the potential energies by taking into account the initial conditions (78) and by using the identity

$$\langle \hat{\boldsymbol{f}}^{\mathrm{T}} M \hat{\boldsymbol{f}} \rangle = \sum\nolimits_{j,k} M_{jk} \langle \hat{f}_j \hat{f}_k \rangle = Tr\left[ M \langle \hat{\boldsymbol{f}} \otimes \hat{\boldsymbol{f}}^{\mathrm{T}} \rangle \right], \tag{B.4}$$

which is valid for any symmetric matrix $M$. Hence, Eqs. (B.1) and (B.2) reduce to

$$\begin{aligned} \frac{E_{\text{kin}}(t)}{N} = & \frac{\boldsymbol{\mathcal{P}}^2(t)}{2m} + \frac{1}{2m} \mathrm{Tr}\left[ \Lambda^{-1}(t)\, \Lambda^{-\mathrm{T}}(t)\, \langle \hat{\boldsymbol{p}} \otimes \hat{\boldsymbol{p}}^{\mathrm{T}} \rangle_{\psi_0} \right] \\ & - \frac{1}{2m} \left[ \Lambda^{-\mathrm{T}}(t)\, \boldsymbol{\mathcal{P}}(0) \right]^2 + \frac{m}{2} \mathrm{Tr}\left\{ \frac{\mathrm{d}\Lambda^{\mathrm{T}}}{\mathrm{d}t} \frac{\mathrm{d}\Lambda}{\mathrm{d}t} \langle [\hat{\boldsymbol{x}} - \boldsymbol{\mathcal{R}}(0)] \otimes [\hat{\boldsymbol{x}} - \boldsymbol{\mathcal{R}}(0)]^{\mathrm{T}} \rangle_{\psi_0} \right\} \\ & + \mathrm{Re}\left\{ \left\langle [\hat{\boldsymbol{x}} - \boldsymbol{\mathcal{R}}(0)]^{\mathrm{T}} \frac{\mathrm{d}\Lambda^{\mathrm{T}}}{\mathrm{d}t} \Lambda^{-\mathrm{T}}(t)\, \hat{\boldsymbol{p}} \right\rangle_{\psi_0} \right\} \end{aligned} \tag{B.5}$$

and

$$\begin{aligned} \frac{E_{\text{pot}}(t)}{N} = & V(t, \boldsymbol{\mathcal{R}}(t)) \\ & + \frac{m}{2} \mathrm{Tr}\left\{ \Lambda^{\mathrm{T}}(t) \Omega^2(t) \Lambda(t) \langle [\hat{\boldsymbol{x}} - \boldsymbol{\mathcal{R}}(0)] \otimes [\hat{\boldsymbol{x}} - \boldsymbol{\mathcal{R}}(0)]^{\mathrm{T}} \rangle_{\psi_0} \right\}, \end{aligned} \tag{B.6}$$

respectively, where $V(t, \boldsymbol{\mathcal{R}}(t))$ denotes the harmonic potential (3).

In addition, by making use of the explicit form of the initial wave function, Eq. (88), we can evaluate the second and last terms of Eq. (B.5). We start by inserting Eq. (88) into the expectation value of the second term of Eq. (B.5) and arrive at the relation

$$\begin{aligned} \langle \hat{\boldsymbol{p}} \otimes \hat{\boldsymbol{p}}^{\mathrm{T}} \rangle_{\psi_0} = & \langle \hat{\boldsymbol{p}} \otimes \hat{\boldsymbol{p}}^{\mathrm{T}} \rangle_{|\psi_0|} + \boldsymbol{\mathcal{P}}(0) \otimes \boldsymbol{\mathcal{P}}^{\mathrm{T}}(0) \\ & + \langle \hat{\boldsymbol{p}} \rangle_{|\psi_0|} \otimes \boldsymbol{\mathcal{P}}^{\mathrm{T}}(0) + \boldsymbol{\mathcal{P}}(0) \otimes \langle \hat{\boldsymbol{p}} \rangle^{\mathrm{T}}_{|\psi_0|}, \end{aligned} \tag{B.7}$$

where the expectation values are now evaluated with respect to the absolute value of the initial state. Since the BEC is initially at rest in the comoving frame of the trap, we obtain $\langle \hat{\boldsymbol{p}} \rangle_{|\psi_0|} = 0$ and thus the second term of Eq. (B.5) fully reads

$$\frac{1}{2m}\mathrm{Tr}\ \left[\Lambda^{-1}(t)\,\Lambda^{-\mathrm{T}}(t)\,\langle\hat{\boldsymbol{p}}\otimes\hat{\boldsymbol{p}}^{\mathrm{T}}\rangle_{\psi_0}\right] \\ =\frac{1}{2m}\mathrm{Tr}\left[\Lambda^{-1}(t)\,\Lambda^{-\mathrm{T}}(t)\,\langle\hat{\boldsymbol{p}}\otimes\hat{\boldsymbol{p}}^{\mathrm{T}}\rangle_{|\psi_0|}\right]+\frac{1}{2m}\left[\Lambda^{-\mathrm{T}}(t)\,\boldsymbol{\mathcal{P}}(0)\right]^2. \tag{B.8}$$

Next, we apply Eq. (88) to the last term of Eq. (B.5) and arrive at the expression

$$\mathrm{Re}\ \left\{\left\langle[\hat{\boldsymbol{x}}-\boldsymbol{\mathcal{R}}(0)]^{\mathrm{T}}\frac{\mathrm{d}\Lambda^{\mathrm{T}}}{\mathrm{d}t}\Lambda^{-\mathrm{T}}(t)\,\hat{\boldsymbol{p}}\right\rangle_{\psi_0}\right\} \\ =\left[\langle\hat{\boldsymbol{x}}\rangle_{\psi_0}-\boldsymbol{\mathcal{R}}(0)\right]^{\mathrm{T}}\frac{\mathrm{d}\Lambda^{\mathrm{T}}}{\mathrm{d}t}\Lambda^{-\mathrm{T}}(t)\,\boldsymbol{\mathcal{P}}(0)=0, \tag{B.9}$$

which vanishes if we take into account the initial conditions, Eq. (78), for the center-of-mass position $\boldsymbol{\mathcal{R}}(0)$ of the BEC. Thus, by inserting Eqs. (B.8) and (B.9) into Eq. (B.5), the kinetic energy is finally given by

$$\frac{E_{\mathrm{kin}}(t)}{N}=\frac{\boldsymbol{\mathcal{P}}^2(t)}{2m}+\frac{1}{2m}\mathrm{Tr}\left[\Lambda^{-1}(t)\,\Lambda^{-\mathrm{T}}(t)\,\langle\hat{\boldsymbol{p}}\otimes\hat{\boldsymbol{p}}^{\mathrm{T}}\rangle_{|\psi_0|}\right] \\ +\frac{m}{2}\mathrm{Tr}\left\{\frac{\mathrm{d}\Lambda^{\mathrm{T}}}{\mathrm{d}t}\frac{\mathrm{d}\Lambda}{\mathrm{d}t}\langle[\hat{\boldsymbol{x}}-\boldsymbol{\mathcal{R}}(0)]\otimes[\hat{\boldsymbol{x}}-\boldsymbol{\mathcal{R}}(0)]^{\mathrm{T}}\rangle_{\psi_0}\right\}. \tag{B.10}$$

In conclusion, by summing up the contributions given by Eqs. (B.3), (B.6), and (B.10), we arrive at Eq. (87) for the energy per particle of a BEC within the time-dependent TF approximation.

## B.2 Energy Terms for the TF Ground State

In this appendix, we use the TF ground state $\phi_{\mathrm{TF}}(\boldsymbol{x})$, Eq. (29), to derive explicit expressions for the energy of a BEC given by Eq. (87). Since we have neglected the quantum pressure term in the derivation of the TF ground state, Eq. (29), we consequently neglect it also for the calculation of the energy within the TF approximation, that is,

$$\langle\hat{\boldsymbol{p}}\otimes\hat{\boldsymbol{p}}^{\mathrm{T}}\rangle_{|\phi_{\mathrm{TF}}|}\approx 0. \tag{B.11}$$

Next we use Eq. (29), the transformation $\boldsymbol{\xi}=\boldsymbol{x}-\boldsymbol{\mathcal{R}}(0)$, and the relation (30) to rewrite the expectation value

$$\left\langle [\hat{\boldsymbol{x}} - \boldsymbol{\mathcal{R}}(0)] \otimes [\hat{\boldsymbol{x}} - \boldsymbol{\mathcal{R}}(0)]^{\mathrm{T}} \right\rangle_{\phi_{\mathrm{TF}}} = \frac{1}{Ng} \int_{\mathbb{R}^d} \boldsymbol{\xi} \otimes \boldsymbol{\xi}^{\mathrm{T}} \left[ \mu_{\mathrm{TF}} - \frac{m}{2} \boldsymbol{\xi}^{\mathrm{T}} \Omega^2(0)\, \boldsymbol{\xi} \right]_+ \mathrm{d}^d \xi$$

in the form

$$\left\langle [\hat{\boldsymbol{x}} - \boldsymbol{\mathcal{R}}(0)] \otimes [\hat{\boldsymbol{x}} - \boldsymbol{\mathcal{R}}(0)]^{\mathrm{T}} \right\rangle_{\phi_{\mathrm{TF}}} = \frac{2}{d+4} \frac{\mu_{\mathrm{TF}}}{m} \Omega^{-2}(0), \tag{B.12}$$

which is valid in $d$ dimensions. Similarly, by applying Eq. (29) to Eq. (86), we obtain

$$E_{\mathrm{int}}(0) = \frac{2}{d+4} N \mu_{\mathrm{TF}}. \tag{B.13}$$

Finally, when we insert Eqs. (B.11)–(B.13) into Eq. (87), we arrive at the expression for the total energy of a BEC, Eq. (89), valid within the TF regime.

## B.3 Angular Momentum Operator in Original Coordinates

Here, we establish a relation for the angular momentum operator with respect to the center of mass in original coordinates. We start with the displaced wave function $\psi_{\mathrm{D}}(t, \boldsymbol{x})$, which is the solution of the GP equation (8) in the center-of-mass coordinates. In these coordinates the angular momentum operator $\hat{\boldsymbol{L}}_{\mathrm{D}}$ is defined in the usual way by

$$\hat{\boldsymbol{L}}_{\mathrm{D}} = \hat{\boldsymbol{x}} \times \hat{\boldsymbol{p}}. \tag{B.14}$$

With the help of the inverse of transformation (4), we can rewrite the expectation value of the angular momentum operator (B.14) as follows:

$$\begin{aligned} \left\langle \hat{\boldsymbol{L}}_{\mathrm{D}} \right\rangle_{\psi_{\mathrm{D}}} &\equiv \frac{1}{N} \int_{\mathbb{R}^3} \mathrm{d}^3 x\, \psi_{\mathrm{D}}^*(t, \boldsymbol{x})\, \hat{\boldsymbol{L}}_{\mathrm{D}}\, \psi_{\mathrm{D}}(t, \boldsymbol{x}) \\ &= \frac{1}{N} \int_{\mathbb{R}^3} \mathrm{d}^3 x\, \psi^*(t, \boldsymbol{x})\, \hat{D}(\boldsymbol{\mathcal{R}}(t), \boldsymbol{\mathcal{P}}(t))\, \hat{\boldsymbol{L}}_{\mathrm{D}}\, \hat{D}^\dagger(\boldsymbol{\mathcal{R}}(t), \boldsymbol{\mathcal{P}}(t))\, \psi(t, \boldsymbol{x}) \\ &= \frac{1}{N} \int_{\mathbb{R}^3} \mathrm{d}^3 x\, \psi^*(t, \boldsymbol{x})\, \hat{\boldsymbol{L}}\, \psi(t, \boldsymbol{x}) \equiv \left\langle \hat{\boldsymbol{L}} \right\rangle_{\psi(t)}, \end{aligned} \tag{B.15}$$

where $\hat{D}(\mathcal{R}(t), \mathcal{P}(t))$ is the displacement operator, which fulfills the relations (A.4) and (A.5). Hence, the angular momentum operator in original coordinates reads

$$\hat{\boldsymbol{L}} = (\hat{\boldsymbol{x}} - \mathcal{R}(t)) \times (\hat{\boldsymbol{p}} - \mathcal{P}(t)). \tag{B.16}$$

## B.4 Angular Momentum in the Time-Dependent TF Approximation

We now evaluate the expectation value, Eq. (B.15), of the angular momentum operator defined by Eq. (B.16) with respect to the approximate wave function, Eq. (73), valid in the time-dependent TF approximation. By making use of the coordinate transformation (80), we obtain the relation

$$\begin{aligned}\langle \hat{\boldsymbol{L}} \rangle_{\psi(t)} = & \left\langle \{\Lambda(t)[\hat{\boldsymbol{x}} - \mathcal{R}(0)]\} \times \left\{ m \frac{\mathrm{d}\Lambda}{\mathrm{d}t} [\hat{\boldsymbol{x}} - \mathcal{R}(0)] + \Lambda^{-\mathrm{T}}(t)\, \hat{\boldsymbol{p}} \right\} \right\rangle_{\psi_0} \\ & - \left\{ \Lambda(t) \left[ \langle \hat{\boldsymbol{x}} \rangle_{\psi_0} - \mathcal{R}(0) \right] \right\} \times \left\{ \Lambda^{-\mathrm{T}}(t)\, \mathcal{P}(0) \right\}.\end{aligned} \tag{B.17}$$

Taking into account the initial condition, Eq. (78), for the center-of-mass position $\mathcal{R}(0)$, the second term in Eq. (B.17) vanishes and we arrive at Eq. (92).

## B.5 Angular Momentum for the TF Ground State

When we consider the TF ground state $\phi_{\mathrm{TF}}(\boldsymbol{x})$, Eq. (29), as the initial state $\psi(0, \boldsymbol{x})$ for the expectation value of the angular momentum matrix operator, Eq. (93), the last two terms vanish, that is,

$$\Lambda(t) \langle [\hat{\boldsymbol{x}} - \mathcal{R}(0)] \otimes \hat{\boldsymbol{p}}^{\mathrm{T}} \rangle_{\phi_{\mathrm{TF}}} \Lambda^{-1}(t) \; - \; \Lambda^{-\mathrm{T}}(t) \langle \hat{\boldsymbol{p}} \otimes [\hat{\boldsymbol{x}} - \mathcal{R}(0)]^{\mathrm{T}} \rangle_{\phi_{\mathrm{TF}}} \Lambda^{\mathrm{T}}(t) = 0. \tag{B.18}$$

This results from the fact that the contributions of these two terms cancel each other when the momentum operator is applied to $|\phi_{\mathrm{TF}}(\boldsymbol{x})|$. In addition, the contributions due to the phase $\mathrm{i}\mathcal{P}(0)\, \boldsymbol{x}/\hbar$ of the TF ground state, Eq. (29), vanish in consequence of the initial condition, Eq. (78), for $\mathcal{R}(0)$. What remains is the contribution of the first term in Eq. (93), which can be evaluated with the help of Eq. (B.12). Hence, we obtain the expression

$$\langle \hat{L} \rangle_{\mathrm{TF}} = \frac{2\mu_{\mathrm{TF}}}{d+4} \left[ \Lambda(t) \Omega^{-2}(0) \frac{\mathrm{d}\Lambda^{\mathrm{T}}}{\mathrm{d}t} - \frac{\mathrm{d}\Lambda}{\mathrm{d}t} \Omega^{-2}(0) \Lambda^{\mathrm{T}}(t) \right] \tag{B.19}$$

for the expectation value of the angular momentum matrix operator within the TF regime. By applying the angular momentum matrix associated with the matrix differential equation as given by Eq. (70) to (B.19), we obtain Eq. (94).

# APPENDIX C. SOLUTIONS OF THE MATRIX DIFFERENTIAL EQUATION FOR ISOTROPIC TRAPS

In this appendix, we find the solution of the nonlinear differential equation (95) in the implicit form given by Eq. (98) valid for isotropic initial traps. In the case of such a trap, that is, $\omega_i(0) = \omega_0$, all scaling parameters $\lambda_i(t)$ are identical, $\lambda_i(t) = \lambda(t)$, and Eq. (95) then reads

$$\frac{\mathrm{d}^2\lambda}{\mathrm{d}t^2} = \frac{\omega_0^2}{\lambda^{d+1}} \tag{C.1}$$

with the initial conditions

$$\lambda(0) = 1 \quad \left.\frac{\mathrm{d}\lambda}{\mathrm{d}t}\right|_{t=0} = 0, \tag{C.2}$$

resulting from Eq. (96).

By considering Eq. (C.1) as the classical Newton equation for the "coordinate" $\lambda$, we can integrate this equation with the integral of motion

$$\frac{1}{2}\left(\frac{\mathrm{d}\lambda}{\mathrm{d}t}\right)^2 + \frac{\omega_0^2}{d}\frac{1}{\lambda^d} = \frac{\omega_0^2}{d}, \tag{C.3}$$

which plays the role of "energy" corresponding to Eq. (C.1). Here, we have used the initial conditions given by Eq. (C.2). After solving Eq. (C.3) with respect to $d\lambda/dt$ and using again Eq. (C.2), we obtain the implicit solution

$$\int_1^{\lambda(t)} \mathrm{d}\tilde{\lambda}\, \frac{\tilde{\lambda}^{\frac{d}{2}}}{\sqrt{\tilde{\lambda}^d - 1}} = \sqrt{\frac{2}{d}}\,\omega_0\, t \tag{C.4}$$

of Eq. (C.1) for the function $\lambda(t)$.

# REFERENCES

Altin, P.A., Johnsson, M.T., Negnevitsky, V., Dennis, G.R., Anderson, R.P., Debs, J.E., Szigeti, S.S., Hardman, K.S., Bennetts, S., McDonald, G.D., Turner, L.D., Close, J.D., Robins, N.P., 2013. Precision atomic gravimeter based on Bragg diffraction. New J. Phys. 15 (2), 023009. http://dx.doi.org/10.1088/1367-2630/15/2/023009.

Aminoff, C.G., Steane, A.M., Bouyer, P., Desbiolles, P., Dalibard, J., Cohen-Tannoudji, C., 1993. Cesium atoms bouncing in a stable gravitational cavity. Phys. Rev. Lett. 71, 3083–3086. http://dx.doi.org/10.1103/PhysRevLett.71.3083.

Ammann, H., Christensen, N., 1997. Delta kick cooling: a new method for cooling atoms. Phys. Rev. Lett. 78, 2088–2091. http://dx.doi.org/10.1103/PhysRevLett.78.2088.

Anderson, M.H., Ensher, J.R., Matthews, M.R., Wieman, C.E., Cornell, E.A., 1995. Observation of Bose-Einstein condensation in a dilute atomic vapor. Science 269 (5221), 198–201. http://dx.doi.org/10.1126/science.269.5221.198.

Berman, P.R. (Ed.), 1997. Atom Interferometry. Academic Press, San Diego

Berry, M.V., Balazs, N.L., 1979. Nonspreading wave packets. Am. J. Phys. 47 (3), 264–267. http://dx.doi.org/10.1119/1.11855.

Bestle, J., Schleich, W.P., Wheeler, J.A., 1995. Anti-stealth: WKB grapples with a corner. Appl. Phys. B. 60 (2), 289–299. http://dx.doi.org/10.1007/BF01135876.

Bialynicki-Birula, I., Bialynicka-Birula, Z., 2002. Center-of-mass motion in the many-body theory of Bose-Einstein condensates. Phys. Rev. A 65, 063606. http://dx.doi.org/10.1103/PhysRevA.65.063606.

Bloch, I., Dalibard, J., Zwerger, W., 2008. Many-body physics with ultracold gases. Rev. Mod. Phys. 80, 885–964. http://dx.doi.org/10.1103/RevModPhys.80.885.

Bodurov, T.G., 1998. Generalized Ehrenfest theorem for nonlinear Schrodinger equations. Int. J. Theor. Phys. 37, 1299–1306. http://dx.doi.org/10.1023/A:1026632006040.

Bongs, K., Burger, S., Dettmer, S., Hellweg, D., Arlt, J., Ertmer, W., Sengstock, K., 2001. Waveguide for Bose-Einstein condensates. Phys. Rev. A 63 (3), 031602. http://dx.doi.org/10.1103/PhysRevA.63.031602.

Bures, D., 1969. An extension of Kakutani's theorem on infinite product measures to the tensor product of semidefinite *w* *-algebras. Trans. Am. Math. Soc. 135, 199–212. http://dx.doi.org/10.2307/1995012.

Burger, S., Cataliotti, F.S., Fort, C., Maddaloni, P., Minardi, F., Inguscio, M., 2002. Quasi-2D Bose-Einstein condensation in an optical lattice. Europhys. Lett. 57 (1), 1. http://dx.doi.org/10.1209/epl/i2002-00532-1.

Castin, Y., Dum, R., 1996. Bose-Einstein condensates in time dependent traps. Phys. Rev. Lett. 77, 5315–5319. http://dx.doi.org/10.1103/PhysRevLett.77.5315.

Clancy, B., Luo, L., Thomas, J.E., 2007. Observation of nearly perfect irrotational flow in normal and superfluid strongly interacting Fermi gases. Phys. Rev. Lett. 99, 140401. http://dx.doi.org/10.1103/PhysRevLett.99.140401.

Cronin, A.D., Schmiedmayer, J., Pritchard, D.E., 2009. Optics and interferometry with atoms and molecules. Rev. Mod. Phys. 81, 1051–1129. http://dx.doi.org/10.1103/RevModPhys.81.1051.

Dalfovo, F., Giorgini, S., Pitaevskii, L.P., Stringari, S., 1999. Theory of Bose-Einstein condensation in trapped gases. Rev. Mod. Phys. 71, 463–512. http://dx.doi.org/10.1103/RevModPhys.71.463.

Davis, K.B., Mewes, M.O., Andrews, M.R., van Druten, N.J., Durfee, D.S., Kurn, D.M., Ketterle, W., 1995. Bose-Einstein condensation in a gas of sodium atoms. Phys. Rev. Lett. 75, 3969–3973. http://dx.doi.org/10.1103/PhysRevLett.75.3969.

Debs, J.E., Altin, P.A., Barter, T.H., Döring, D., Dennis, G.R., McDonald, G., Anderson, R.P., Close, J.D., Robins, N.P., 2011. Cold-atom gravimetry with a Bose-Einstein condensate. Phys. Rev. A 84, 033610. http://dx.doi.org/10.1103/PhysRevA.84.033610.

Dobson, J.F., 1994. Harmonic-potential theorem: implications for approximate many-body-theories. Phys. Rev. Lett. 73 (16), 2244–2247. http://dx.doi.org/10.1103/PhysRevLett.73.2244.

Dum, R., Sanpera, A., Suominen, K.A., Brewczyk, M., Kuś, M., Rzążewski, K., Lewenstein, M., 1998. Wave packet dynamics with Bose-Einstein condensates. Phys. Rev. Lett. 80, 3899–3902. http://dx.doi.org/10.1103/PhysRevLett.80.3899.

Durnin, J., 1987. Exact solutions for nondiffracting beams. I. The scalar theory. J. Opt. Soc. Am. A 4 (4), 651–654. http://dx.doi.org/10.1364/JOSAA.4.000651.
Durnin, J., Miceli, J.J., Eberly, J.H., 1987. Diffraction-free beams. Phys. Rev. Lett. 58, 1499–1501. http://dx.doi.org/10.1103/PhysRevLett.58.1499.
Eckart, M., 2008. Non-equilibrium dynamics of trapped gases in controlled geometries. PhD thesis, Universität Ulm.
Edwards, M., Clark, C.W., Pedri, P., Pitaevskii, L., Stringari, S., 2002. Consequence of superfluidity on the expansion of a rotating Bose-Einstein condensate. Phys. Rev. Lett. 88 (7), 070405. http://dx.doi.org/10.1103/PhysRevLett.88.070405.
Fetter, A.L., 2009. Rotating trapped Bose-Einstein condensates. Rev. Mod. Phys. 81, 647–691. http://dx.doi.org/10.1103/RevModPhys.81.647.
Fetter, A.L., Feder, D.L., 1998. Beyond the Thomas-Fermi approximation for a trapped condensed Bose-Einstein gas. Phys. Rev. A 58 (4), 3185–3194. http://dx.doi.org/10.1103/PhysRevA.58.3185.
Gallas, J.A.C., Schleich, W.P., Wheeler, J.A., 1995. Waves at walls, corners, heights: looking for simplicity. Appl. Phys. B . 60 (2), 279–287. http://dx.doi.org/10.1007/BF01135875.
Gindikin, S., Michor, P. (Eds.). 1994. 75 Years of Radon Transform. Proceedings of the Conference Held at the Erwin Schrödinger International Institute for Mathematical Physics in Vienna, Austria, August 31 - September 4, 1992. International Press, Cambridge, MA.
Giorgini, S., Pitaevskii, L.P., Stringari, S., 2008. Theory of ultracold atomic Fermi gases. Rev. Mod. Phys. 80, 1215–1274. http://dx.doi.org/10.1103/RevModPhys.80.1215.
Görlitz, A., Vogels, J.M., Leanhardt, A.E., Raman, C., Gustavson, T.L., Abo-Shaeer, J.R., Chikkatur, A.P., Gupta, S., Inouye, S., Rosenband, T., Ketterle, W., 2001. Realization of Bose-Einstein condensates in lower dimensions. Phys. Rev. Lett. 87 (13), 130402. http://dx.doi.org/10.1103/PhysRevLett.87.130402.
Gross, E.P., 1961. Structure of a quantized vortex in boson systems. Nuovo Cimento 20, 454–477.
Gross, E.P., 1963. Hydrodynamics of a superfluid condensate. J. Math. Phys. 4, 195–207.
Hechenblaikner, G., Hodby, E., Hopkins, S.A., Maragò, O.M., Foot, C.J., 2002. Direct observation of irrotational flow and evidence of superfluidity in a rotating Bose-Einstein condensate. Phys. Rev. Lett. 88 (7), 070406. http://dx.doi.org/10.1103/PhysRevLett.88.070406.
Hodby, E., Hechenblaikner, G., Hopkins, S.A., Maragò, O.M., Foot, C.J., 2001. Vortex nucleation in Bose-Einstein condensates in an oblate, purely magnetic potential. Phys. Rev. Lett. 88, 010405. http://dx.doi.org/10.1103/PhysRevLett.88.010405.
Horn, R.A., Johnson, C.R., 1990. Matrix Analysis. Cambridge University Press, Cambridge.
Jamison, A.O., Kutz, J.N., Gupta, S., 2011. Atomic interactions in precision interferometry using Bose-Einstein condensates. Phys. Rev. A 84, 043643. http://dx.doi.org/10.1103/PhysRevA.84.043643.
Jin, D.S., Ensher, J.R., Matthews, M.R., Wieman, C.E., Cornell, E.A., 1996. Collective excitations of a Bose-Einstein condensate in a dilute gas. Phys. Rev. Lett. 77, 420–423. http://dx.doi.org/10.1103/PhysRevLett.77.420.
Kagan, Y., Surkov, E.L., Shlyapnikov, G.V., 1996. Evolution of a Bose-condensed gas under variations of the confining potential. Phys. Rev. A 54, R1753–R1756. http://dx.doi.org/10.1103/PhysRevA.54.R1753.
Kagan, Y., Surkov, E.L., Shlyapnikov, G.V., 1997. Evolution of a Bose gas in anisotropic time-dependent traps. Phys. Rev. A 55, R18–R21. http://dx.doi.org/10.1103/PhysRevA.55.R18.
Kajari, E., Harshman, N.L., Rasel, E.M., Stenholm, S., Süßmann, G., Schleich, W.P., 2010. Inertial and gravitational mass in quantum mechanics. Appl. Phys. B. 100 (1), 43–60. http://dx.doi.org/10.1007/s00340-010-4085-8.

Ketterle, W., Durfee, D., Stamper-Kurn, D., 1999. Making, probing and understanding Bose-Einstein condensates. In: Inguscio, M., Stringari, S., Wieman, C.E. (Eds.), Bose-Einstein Condensation in Atomic Gases. Proceedings of the International School of Physics "Enrico Fermi", Course CXL. IOS Press, Amsterdam, pp. 67–176.

Kleinert, S., Kajari, E., Roura, A., Schleich, W.P., 2015. Representation-free description of light-pulse atom interferometry including non-inertial effects. Phys. Rep. 605, 1–50. http://dx.doi.org/10.1016/j.physrep.2015.09.004 (and references therein).

Kohn, W., 1961. Cyclotron resonance and de Haas-van Alphen oscillations of an interacting electron gas. Phys. Rev. 123 (4), 1242–1244. http://dx.doi.org/10.1103/PhysRev.123.1242.

Madison, K.W., Chevy, F., Bretin, V., Dalibard, J., 2001. Stationary states of a rotating Bose-Einstein condensate: routes to vortex nucleation. Phys. Rev. Lett. 86, 4443–4446. http://dx.doi.org/10.1103/PhysRevLett.86.4443.

Mateo, A.M., Delgado, V., 2008. Effective mean-field equations for cigar-shaped and disk-shaped Bose-Einstein condensates. Phys. Rev. A 77, 013617. http://dx.doi.org/10.1103/PhysRevA.77.013617.

Meister, M., Roura, A., Schleich, W.P., 2017. Expansion dynamics of dual-species Bose-Einstein condensates. (in preparation).

Mewes, M.-O., Andrews, M.R., van Druten, N.J., Kurn, D.M., Durfee, D.S., Townsend, C.G., Ketterle, W., 1996. Collective excitations of a Bose-Einstein condensate in a magnetic trap. Phys. Rev. Lett. 77, 988–991. http://dx.doi.org/10.1103/PhysRevLett.77.988.

Müntinga, H., Ahlers, H., Krutzik, M., Wenzlawski, A., Arnold, S., Becker, D., Bongs, K., Dittus, H., Duncker, H., Gaaloul, N., Gherasim, C., Giese, E., Grzeschik, C., Hänsch, T.W., Hellmig, O., Herr, W., Herrmann, S., Kajari, E., Kleinert, S., Lämmerzahl, C., Lewoczko-Adamczyk, W., Malcolm, J., Meyer, N., Nolte, R., Peters, A., Popp, M., Reichel, J., Roura, A., Rudolph, J., Schiemangk, M., Schneider, M., Seidel, S.T., Sengstock, K., Tamma, V., Valenzuela, T., Vogel, A., Walser, R., Wendrich, T., Windpassinger, P., Zeller, W., van Zoest, T., Ertmer, W., Schleich, W.P., Rasel, E.M., 2013. Interferometry with Bose-Einstein condensates in microgravity. Phys. Rev. Lett. 110, 093602. http://dx.doi.org/10.1103/PhysRevLett.110.093602.

Nandi, G., Walser, R., Kajari, E., Schleich, W.P., 2007. Dropping cold quantum gases on Earth over long times and large distances. Phys. Rev. A 76, 063617. http://dx.doi.org/10.1103/PhysRevA.76.063617.

Nesvizhevsky, V.V., Börner, H.G., Petukhov, A.K., Abele, H., Baessler, S., Ruess, F.J., Stöferle, T., Westphal, A., Gagarski, A.M., Petrov, G.A., Strelkov, A.V., 2002. Quantum states of neutrons in the Earth's gravitational field. Nature 415, 297–299. http://dx.doi.org/10.1038/415297a.

Pethick, C., Smith, H., 2002. Bose-Einstein Condensation in Dilute Gases. Cambridge University Press, Cambridge.

Petrov, D.S., Holzmann, M., Shlyapnikov, G.V., 2000. Bose-Einstein condensation in quasi-2D trapped gases. Phys. Rev. Lett. 84, 2551–2555. http://dx.doi.org/10.1103/PhysRevLett.84.2551.

Petrov, D.S., Shlyapnikov, G.V., Walraven, J.T.M., 2000. Regimes of quantum degeneracy in trapped 1D gases. Phys. Rev. Lett. 85 (18), 3745–3749. http://dx.doi.org/10.1103/PhysRevLett.85.3745.

Pitaevskii, L.P., 1961. Vortex lines in an imperfect Bose gas. Sov. Phys. JETP 13, 451.

Plaja, L., Santos, L., 2002. Expansion of a Bose-Einstein condensate in an atomic waveguide. Phys. Rev. A 65 (3), 035602. http://dx.doi.org/10.1103/PhysRevA.65.035602.

Press, W.H., Teukolsky, S.A., Vetterling, W.T., Flannery, B.P., 2007. Numerical Recipes—The Art of Scientific Computing, third ed. Cambridge University Press, Cambridge.

Reinaudi, G., Lahaye, T., Wang, Z., Guéry-Odelin, D., 2007. Strong saturation absorption imaging of dense clouds of ultracold atoms. Opt. Lett. 32 (21), 3143–3145. http://dx.doi.org/10.1364/OL.32.003143.

Roura, A., Zeller, W., Schleich, W.P., 2014. Overcoming loss of contrast in atom interferometry due to gravity gradients. New J. Phys. 16 (12), 123012. http://dx.doi.org/10.1088/1367-2630/16/12/123012.

Rychtarik, D., Engeser, B., Nägerl, H.-C., Grimm, R., 2004. Two-dimensional Bose-Einstein condensate in an optical surface trap. Phys. Rev. Lett. 92, 173003. http://dx.doi.org/10.1103/PhysRevLett.92.173003.

Salasnich, L., Parola, A., Reatto, L., 2002. Effective wave equations for the dynamics of cigar-shaped and disk-shaped Bose condensates. Phys. Rev. A 65, 043614. http://dx.doi.org/10.1103/PhysRevA.65.043614.

Schleich, W.P., 2001. Quantum Optics in Phase Space. Wiley-VCH Verlag, Berlin.

Simsarian, J.E., Denschlag, J., Edwards, M., Clark, C.W., Deng, L., Hagley, E.W., Helmerson, K., Rolston, S.L., Phillips, W.D., 2000. Imaging the phase of an evolving Bose-Einstein condensate wave function. Phys. Rev. Lett. 85, 2040–2043. http://dx.doi.org/10.1103/PhysRevLett.85.2040.

Stamper-Kurn, D.M., Miesner, H.-J., Inouye, S., Andrews, M.R., Ketterle, W., 1998. Collisionless and hydrodynamic excitations of a Bose-Einstein condensate. Phys. Rev. Lett. 81 (3), 500–503. http://dx.doi.org/10.1103/PhysRevLett.81.500.

Storey, P., Olshanii, M., 2000. Closed class of hydrodynamical solutions for the collective excitations of a Bose-Einstein condensate. Phys. Rev. A 62, 033604. http://dx.doi.org/10.1103/PhysRevA.62.033604.

Tino, G.M., Kasevich, M.A. (Eds.), 2014. Proceedings of the International School of Physics "Enrico Fermi" Course 188 "Atom Interferometry". IOS Press, Amsterdam.

Torii, Y., Suzuki, Y., Kozuma, M., Sugiura, T., Kuga, T., Deng, L., Hagley, E.W., 2000. Mach-Zehnder Bragg interferometer for a Bose-Einstein condensate. Phys. Rev. A 61, 041602(R). http://dx.doi.org/10.1103/PhysRevA.61.041602.

van Zoest, T., Gaaloul, N., Singh, Y., Ahlers, H., Herr, W., Seidel, S.T., Ertmer, W., Rasel, E., Eckart, M., Kajari, E., Arnold, S., Nandi, G., Schleich, W.P., Walser, R., Vogel, A., Sengstock, K., Bongs, K., Lewoczko-Adamczyk, W., Schiemangk, M., Schuldt, T., Peters, A., Könemann, T., Müntinga, H., Lämmerzahl, C., Dittus, H., Steinmetz, T., Hänsch, T.W., Reichel, J., 2010. Bose-Einstein condensation in microgravity. Science 328 (5985), 1540–1543. http://dx.doi.org/10.1126/science.1189164.

Zeller, W., 2016. The impact of wave-packet dynamics in long-time atom interferometry. PhD thesis, Universität Ulm.

CHAPTER SEVEN

# Optical Nanofibers: A New Platform for Quantum Optics

**Pablo Solano*, Jeffrey A. Grover*, Jonathan E. Hoffman*, Sylvain Ravets*,†, Fredrik K. Fatemi‡, Luis A. Orozco*,[1], Steven L. Rolston***

*Joint Quantum Institute, University of Maryland and NIST, College Park, MD, United States
†Laboratoire Charles Fabry, Institut d'Optique, CNRS, Univ. Paris Sud, Palaiseau Cedex, France
‡Army Research Laboratory, Adelphi, MD, United States
[1]Corresponding author: e-mail address: lorozco@umd.edu

## Contents



*Advances in Atomic, Molecular, and Optical Physics*, Volume 66
ISSN 1049-250X
http://dx.doi.org/10.1016/bs.aamop.2017.02.003

## Abstract

The development of optical nanofibers (ONFs) and the study and control of their optical properties when coupling atoms to their electromagnetic modes has opened new possibilities for their use in quantum optics and quantum information science. These ONFs offer tight optical mode confinement (less than the wavelength of light) and diffraction-free propagation. The small cross section of the transverse field allows probing of linear and nonlinear spectroscopic features of atoms with exquisitely low power. The cooperativity—the figure of merit in many quantum optics and quantum information systems—tends to be large even for a single atom in the mode of an ONF, as it is proportional to the ratio of the atomic cross section to the electromagnetic mode cross section. ONFs offer a natural bus for information and for inter-atomic coupling through the tightly confined modes, which opens the possibility of one-dimensional many-body physics and interesting quantum interconnection applications. The presence of the ONF modifies the vacuum field, affecting the spontaneous emission rates of atoms in its vicinity. The high gradients in the radial intensity naturally provide the potential for trapping atoms around the ONF, allowing the creation of one-dimensional arrays of atoms. The same radial gradient in the transverse direction of the field is responsible for the existence of a large longitudinal component that introduces the possibility of spin–orbit coupling of the light and the atom, enabling the exploration of chiral quantum optics.

# 1. INTRODUCTION

Quantum optics has followed a path to achieve the ideal limit of one single quantum of light, a photon, interacting with one single quantum of matter, an atom. The interest in this realization has theoretical and experimental implications that have illuminated and guided much of the contemporary discussion on quantum information.

In the case of photon-mediated interactions, the advent of cavity quantum electrodynamics (QED) (Haroche and Raimond, 2006), whereby cavities formed by mirrors or other structures modify the vacuum modes of the electromagnetic field while providing a preferential mode for the atom–light coupling, marked a transformative milestone. Cavity QED ushered in the ability to sufficiently isolate a quantum system from its environment and control nearly all of its degrees of freedom (Kimble, 1998). This has led, for instance, to the demonstration of the Purcell effect—increased or inhibited spontaneous emission rates (Hinds, 1994)—and to generation of highly nonclassical photon states (Haroche, 2013), among other phenomena. Remarkably, these studies have been realized within many different regions of the electromagnetic spectrum, e.g., from the microwave (Haroche, 2013; Wallraff et al., 2004) to optical domains (Kimble, 1994; Reiserer and Rempe, 2015). The emerging field of waveguide QED, which applies the machinery of cavity QED to propagating modes in electromagnetic structures, is growing and the topic of this review falls into this subject (Calajó et al., 2016; Fang and Baranger, 2015; Lodahl et al., 2015; Paulisch et al., 2016).

Our study of atom–light interactions is framed within this context. We motivate the use of evanescent waveguides helped by the notion of cooperativity, which we define in the next paragraph.

## 1.1 Cooperativity and Optical Depth

Consider a two-level atom with dipole moment $\vec{d}$ interacting with an electric field $\vec{E}$ whose average energy is equivalent to one photon ($\hbar\omega$). The parameter, $g$, gives the strength of the coupling in frequency units, numerically equivalent to half of the vacuum Rabi splitting (Sanchez-Mondragon et al., 1983),

$$g = \frac{\vec{d} \cdot \vec{E}}{\hbar}. \tag{1}$$

For an atom with decay rate, $\gamma$, and a field with decay rate, $\kappa$, we define the single-atom cooperativity to be (Kimble, 1998)

$$C_1 = \frac{g^2}{\kappa\gamma}. \tag{2}$$

A cooperativity of $C_1 > 1$ means that the rate that governs the interaction between the atom and the field mode is larger than the atomic and field reservoir coupling rates. This places the system in the so-called strong coupling regime, which was a longstanding goal within the quantum optics community and that has been achieved in several systems, such as ions (Wineland, 2013), Rydberg atoms in microwave cavities (Haroche, 2013), neutral atoms in optical cavities (Kimble, 1994), excitons in semiconductor microcavities (Gibbs et al., 2011), optomechanical systems (Groblacher et al., 2009), and superconducting circuits in planar waveguides (Blais et al., 2004).

To better understand how one can coerce a system into the strong coupling regime where a single mode of the electromagnetic field and one atom preferentially exchange an excitation, it can be useful to relate the cooperativity to the optical depth (OD). This way of thinking comes from the predecessor of cavity QED, optical bistability (Lugiato, 1984), which is part of the more general area of dissipative systems in quantum optics as treated by Bonifacio and Lugiato (1982). We illustrate the argument by considering a high-finesse Fabry–Perot cavity with mirror transmission $T$ and mirror separation length $L$ so that its full width at half maximum (FWHM) is $2\kappa$, with $\kappa = cT/2L$ as a frequency half-width. The electric field amplitude for a field with an average energy of a single photon within this mode is given by

$$E = \sqrt{\frac{\hbar\omega}{2\varepsilon_0 V}}, \tag{3}$$

defining the mode volume to be $V = A_{\text{mode}} \times L$. The free-space decay rate of the atom from Fermi's golden rule is

$$\gamma_0 = \frac{4\omega^3}{3c^2}\frac{d^2}{4\pi\varepsilon_0\hbar c}, \tag{4}$$

where $d$ is the magnitude of the dipole moment of the atom and $\omega = 2\pi c/\lambda$ is the resonant angular frequency of the decay transition associated with the wavelength $\lambda$. Typical values of $d$ for the D2 line of alkali atoms are about

$5a_0e$, where $a_0$ is the Bohr radius and $e$ the electron charge (Safronova et al., 2004). Combining Eq. (3) and Eq. (4) into Eq. (2) yields a single-atom cooperativity of (Tanji-Suzuki et al., 2011)

$$C_1 = \frac{A_{\text{atom}}}{A_{\text{mode}}} \frac{1}{T}. \tag{5}$$

Here we have defined the "area" of the atom, $A_{\text{atom}}$, to be the resonant scattering cross section $\sigma_0 = 3\lambda^2/2\pi$. This motivates a geometric framework to think about the cooperativity by realizing that the OD for a dipole transition is

$$\text{OD} = \frac{A_{\text{atom}}}{A_{\text{mode}}}. \tag{6}$$

We conclude from Eq. (5) that $C_1$ is just the product of the optical density of a single atom times the cavity enhancement factor $1/T$.

Eq. (5) states that the cooperativity is independent of the atomic dipole moment and the cavity length $L$, but depends on the overlap of the two "areas." See Bonifacio and Lugiato (1982) for a discussion not only of resonance fluorescence, but also optical bistability and superradiance and their connection through the idea of cooperativity. Efforts to increase this figure of merit have followed a few different paths.

For certain processes with $N$ atoms, the total cooperativity, $C = C_1N$, is important. Increasing $N$ achieves an appropriate threshold of the system, e.g., vapor cells with high atomic densities facilitate the observation of coherent processes such as electromagnetically induced transparency (EIT) (Fleischhauer et al., 2005).

The Nobel Prize-worthy efforts of Serge Haroche focused on decreasing $T$ with microwave cavities possessing finesses greater than $10^9$ while making sure that the cavity-mode cross section significantly overlaps with the properly aligned Rydberg atom cross section. With this system, his group created highly nonclassical states and performed quantum nondemolition measurements of photon jumps (Haroche, 2013).

Recent advances in superconducting technology have allowed physicists to create nonlinear quantum circuits that behave like "artificial atoms" (Makhlin et al., 2001). By coupling these so-called qubits to a high-quality-factor coplanar resonator, scientists have engineered an analog of cavity QED, dubbed circuit QED, that achieves couplings far beyond what have been realized in optical systems (Blais et al., 2004; Wallraff et al., 2004). This architecture not only relies on the high finesse of the cavity to increase $C$, but the area of the artificial atoms (antennae, qubits) has also been increased significantly beyond that of the mode. This limit can not yet be realized with atoms in free-space, but may be achievable for atoms near photonic and plasmonic structures, where the field can be confined beyond the diffraction limit.

Finally, we mention recent efforts made by some groups which have moved away from the use of a traditional cavity altogether, trying to increase the cooperativity of an atom in free-space, making the enhancement factor $1/T = 1$ as in Eq. (5). One possibility is to use high-NA optical systems to focus light to a small spot and achieve high coupling in free-space (Hétet et al., 2011; Streed et al., 2012; Tey et al., 2008). Another possibility is to use a parabolic mirror that focuses a laser such that the focused beam has the same structure as the dipole radiation pattern of a single atom, thereby increasing the ratio of the atomic area to the mode area (Golla et al., 2012; Heugel et al., 2012; Stobińska et al., 2009).

We will consider, in this review, the coupling of atoms to one kind of nanophotonic waveguide: an optical nanofiber (ONF). Nanophotonic waveguides are not like the traditional optical cavities with high finesses as discussed earlier, but they do modify the vacuum mode structure in a nontrivial way. As a result of this modification and the appreciable overlap between the atomic and optical areas, one can couple an atom to the electric field of a photon in a single pass. In fact, there is an active area of research studying waveguides constructed via nanofabrication techniques, whose mode produces a large OD for a single atom (Goban et al.,2014, 2015; Hood et al., 2016; Thompson et al., 2013). Important advances have happened, for example, with hollow-core fibers: encasing an atomic vapor into the hollow core of a photonic crystal fiber to confine atoms and photons in the waveguide increases $C$, but the manipulation of the atoms is not as straightforward as if they are outside the photonic structure as in Ghosh et al. (2006), Bajcsy et al. (2009), Venkataraman et al. (2011), and Sprague et al. (2014). ONFs formed by thinning single-mode optical fibers to subwavelength diameters, as shown in Fig. 1 (not to scale), are another example of this kind of structure. It has been demonstrated that ONFs provide an excellent platform to interface trapped atoms to the evanescent field of the mode around a nanometer-size waist region (Béguin et al., 2014; Corzo et al., 2016; Goban et al., 2012; Kato and Aoki, 2015; Lee et al., 2015; Vetsch et al., 2010). We review the platform of ONFs with atoms and their implications and applications for quantum physics.

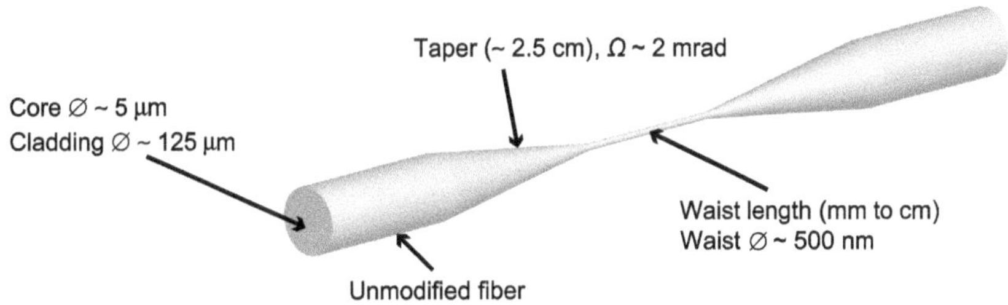


**Fig. 1** Schematic of an optical nanofiber where the transverse dimension has been greatly expanded compared to the longitudinal. Atoms are around the waist region either free or trapped. There are three distinctive sections of the ONF and the typical values are those used in our experiments (Grover et al., 2015).

## 1.2 Nanofiber Platform

Before embarking on a thorough discussion of the nanofiber platform it is important to point out the advantages over other nanophotonic structures that we see.

Nanofibers can be produced in-house, using a heat-and-pull method (Hoffman et al., 2014; Ward et al., 2014). The glass malleability ensures low surface roughness. The smoothness of the surface is a great asset since it leads to ultra-high transmission structures that can withstand high optical powers (almost one Watt in vacuum (Hoffman et al., 2014)) without permanent damage to the fiber or degradation of the transmission.

Optical fibers also show great versatility in terms of connectivity to other systems. The advanced state of fiber optic technology is an enormous advantage to pursuing quantum information devices on this platform as they facilitate the interaction and communication among different modes and modular devices as stated by Kimble (2008). Kurizki et al. (2015) and Xiang et al. (2013) review the growing area of hybrid quantum systems that are increasing in importance in quantum optics and quantum information. They combine different kinds of systems—e.g., atomic, Rydberg, ionic, photonic, condensed matter—to utilize the best coherence available for different tasks: processing or memory. Hoffman et al. (2011) and Kim et al. (2011) propose the use of a trapped atoms around an ONF to couple them through their magnetic dipole them to a superconducting circuit in a cryogenic environment. Hafezi et al. (2012) study the atomic interface between microwave and optical photons in such a system.

One of the most fascinating developments is the use of ONFs in quantum optics for the study of chiral quantum optics (Lodahl et al., 2017) and its connections with many-body physics. ONFs indeed provide a unique platform to study this nascent area.

One difficulty with nanofibers comes from the polarization structure of the modes and its limited control along the waist length. However, they have not been a major drawback for experiments. Also, the values of atom coupling to the nanofiber currently do not reach those recently seen in nanophotonic devices (Goban et al., 2015; Hood et al., 2016; Yu et al., 2014), but the entire parameter space for traps has yet to be explored, and improvements may be possible.

## 1.3 Cooperativity in a Nanofiber

We now present, following closely the discussion in Quan et al. (2009) related to waveguide QED with atoms, the connections between emission enhancement, coupling efficiency, cooperativity, and the Purcell effect. Two quantities characterize the coupling between the nanofiber mode and the atomic dipoles around the fiber but inside the mode.

The emission enhancement parameter $\alpha$

$$\alpha = \frac{\gamma_{1D}}{\gamma_0} \tag{7}$$

is the ratio between the emission rate into the quasi one-dimensional (1D) mode $\gamma_{1D}$ of the nanofiber in both possible directions combined and the intrinsic spontaneous emission rate of an atom in free-space $\gamma_0$ (see Eq. 4).

The waveguide coupling efficiency $\beta$

$$\beta = \frac{\gamma_{1D}}{\gamma_{Tot}} \tag{8}$$

is the ratio between the emission rate into the waveguide mode in both directions combined and the total emission rate into all radiative channels $\gamma_{Tot}$.

The parameter $\alpha$ is proportional to the interaction rate of the atom and the single mode, while $\beta$ quantifies the fraction of the total rate that couples to the single mode. As an example to illustrate the two concepts consider an atom in a cavity with dimensions less than $\lambda/2$. The system has $\alpha = 0$ and $\beta = 1$, implying that there will be no decay signal despite $\beta = 1$.

The single-atom cooperativity in terms of these parameters is

$$C_1 = \frac{\beta}{(1-\beta)} = \frac{\gamma_{1D}}{\gamma_{Tot} - \gamma_{1D}}. \tag{9}$$

This expression relates the parameter $\beta$, often used in the microcavity literature (Chang and Campillo, 1996), to $C_1$ and thus to what we have presented above. The meaning of Eq. (9) in terms of rates is equivalent to Section 1.1 in the sense that $C$ is the ratio of the rate of interaction between atom and field mode to the interaction to other reservoirs. The Purcell factor, the total enhancement of spontaneous emission, is the ratio $\alpha/\beta = \gamma_{Tot}/\gamma_0$. It is noteworthy that, in the presence of a nanofiber with index of refraction $n$, $\alpha$ (connected to the energy density) is directly related to the OD (connected to the energy flux) via the index of refraction, as we learn from classical electricity and magnetism (Blow et al., 1990). Furthermore, $C$ is related to the OD through an enhancement factor $1/T$ in cavity QED and for the nanofibers $[\gamma_0/(\gamma_{Tot} - \gamma_{1D})]$.

## 1.4 ONFs as Enablers of High Cooperativity and Optical Depth

Fiber optics in present day telecommunications have been a game-changing technology allowing enormous bandwidth for current and future uses. It began with the pioneering observation of the low-loss properties of glass fibers in Kao and Hockham (1966). This was recognized with the 2009 Nobel Prize awarded to Charles K. Kao (see Kao, 2010).

More recently, ONFs have seen widespread use in science and engineering (Brambilla, 2010; Chen et al., 2013; Morrissey et al., 2013). The tight confinement of light around ONFs (Le Kien et al., 2004), unique geometries provided by the fiber modes (Le Kien et al., 2004; Reitz and Rauschenbeutel., 2012; Sagué et al., 2008), low loss, and promise of improved atom–light interaction (Alton et al., 2011; Goban et al., 2012; Le Kien et al., 2005b; Vetsch et al., 2010; Wuttke et al., 2012) have led to increased interest in the physics community. Optical micro- or nanofibers are used for sensing and detection (Knight et al., 1997; Nayak et al., 2007), and coupling light to resonators (Fujiwara et al., 2012; Kakarantzas et al., 2002; Knight et al., 1997; Louyer et al., 2005; Morrissey et al., 2009; Spillane et al., 2003), NV centers (Schröder et al., 2012), or photonic crystals (Sadgrove et al., 2013; Thompson et al., 2013).

Reducing the thickness of an optical fiber to subwavelength diameters (Tong et al., 2003) modifies the boundary conditions of the field so that a significant fraction of the light propagates in an evanescent field around the fiber waist. Nanofibers thus provide an excellent platform to interface light with atoms. To confine atoms along the nanofiber, one can couple laser beams into the fiber to create an optical dipole potential around the waist

(Le Kien et al., 2004). Optical dipole trapping of atoms is a well-developed technique applied to numerous atomic species. The reader is referred to the excellent review of Grimm et al. (2000) for details on the requirements depending on specific magnetic substates and polarizations used.

Typical trapping schemes allow trap depths of fractions of a millikelvin located a few hundred nanometers from the fiber (Béguin et al., 2014; Corzo et al., 2016; Goban et al., 2012; Kato and Aoki, 2015; Lee et al., 2015; Vetsch et al., 2010). Trapping lifetimes of tens of milliseconds and coherence times of $\sim$ 600 μs Reitz et al. (2013) have been obtained and formation of a one-dimensional lattice along the nanofiber waist has also been demonstrated. In this regime, the OD per atom is as large as a few percent, so that a modest atom number can achieve a large optical depth. This confirms that ONFs are a viable platform for studying the physics of light-matter interactions, as they enable high optical depth and cooperativity.

### 1.5 Outline of Review

We start the review in Section 2 with a treatment of the nanofiber electromagnetic modes and show intensity profiles with particular attention to the polarization properties of these modes. Section 3 details the fabrication and characterization of the nanofibers. We then in Section 4 concentrate on recent studies with untrapped atoms around nanofibers, including spectroscopy. Current implementations of atom trapping around a nanofiber and some interesting experiments with such 1D arrays are presented in Section 5. Experiments in quantum optics and quantum information come in Section 6 including realizations of EIT, optical memories, and cavity QED. Section 7, the last section, discusses how nanofibers are currently contributing to the emergence of the field of chiral quantum optics. The review ends with concluding remarks.

## 2. NANOFIBER ELECTROMAGNETIC MODES

A standard step-index optical fiber is a cylindrical waveguide having a light-guiding core of typical radius of 2–3 μm and refractive index $n_1$, surrounded by a cylindrical cladding with index $n_2$. In commercial, single-mode fibers, the refractive index difference is small, with $0.001 \lesssim n_1 - n_2 \lesssim 0.02$, and the waveguide is considered to have "weakly guiding" core-cladding guidance. As shown schematically in Fig. 1, as the fiber is tapered down this geometry adiabatically transforms to a step-index waveguide in which the light is entirely guided with air (or vacuum) as the cladding ("cladding-air" guidance), having an index difference close to 0.5. Detailed vector-mode solutions are required for such "strongly guiding" waveguides and are described in a number of treatments (Marcuse, 1981, 1991; Sagué, 2008; Snyder and Love, 1983; Vetsch, 2010; Yariv, 1990), while in the weakly guiding limit, a scalar treatment is sufficient. Here we present a summary of the vector-mode solutions from Hoffman (2014), and discuss some of the relevant aspects of ONFs.

## 2.1 Field Equations

The electric and magnetic fields, **E** and **H**, are shown below in cylindrical coordinates for a single-step index geometry with core radius, $a$, of index $n_1$, and an infinite cladding of index $n_2$. The solution of the Maxwell Equations in cylindrical coordinates leads to the following expressions for the field components along the radial ($r$), azimuthal ($\phi$), and longitudinal ($z$, propagation) directions inside ($r < a$) and outside ($r > a$) the core.

For $r < a$:

$$E_{r,\pm} = \frac{-i\beta}{h^2}\left[\pm\frac{i\mu_0\omega l}{\beta r}BJ_l(hr) + AhJ_l'(hr)\right]\mathrm{e}^{i(\omega t-\beta z\pm l\varphi)} \tag{10}$$

$$E_{\phi,\pm} = \frac{-i\beta}{h^2}\left[\pm\frac{il}{r}AJ_l(hr) - \frac{\mu_0\omega h}{\beta}BJ_l'(hr)\right]\mathrm{e}^{i(\omega t-\beta z\pm l\varphi)} \tag{11}$$

$$E_{z,\pm} = AJ_l(hr)\mathrm{e}^{i(\omega t-\beta z\pm l\varphi)} \tag{12}$$

$$H_{r,\pm} = \frac{-i\beta}{h^2}\left[\mp\frac{i\varepsilon_1\omega l}{\beta r}AJ_l(hr) + BhJ_l'(hr)\right]\mathrm{e}^{i(\omega t-\beta z\pm l\varphi)} \tag{13}$$

$$H_{\phi,\pm} = \frac{-i\beta}{h^2}\left[\frac{\varepsilon_1\omega}{\beta}AhJ_l'(hr) \pm \frac{il}{r}BJ_l(hr)\right]\mathrm{e}^{i(\omega t-\beta z\pm l\varphi)} \tag{14}$$

$$H_{z,\pm} = BJ_l(hr)\mathrm{e}^{i(\omega t-\beta z\pm l\varphi)}, \tag{15}$$

and $r > a$,

$$E_{r,\pm} = \frac{i\beta}{q^2}\left[\pm\frac{i\mu_0\omega l}{\beta r}DK_l(qr) + ChK_l'(qr)\right]\mathrm{e}^{i(\omega t-\beta z\pm l\varphi)} \tag{16}$$

$$E_{\phi,\pm} = \frac{i\beta}{q^2}\left[\pm\frac{il}{r}CK_l(qr) - \frac{\mu_0\omega h}{\beta}DK_l'(qr)\right]\mathrm{e}^{i(\omega t-\beta z\pm l\varphi)} \tag{17}$$

$$E_{z,\pm} = CK_l(qr)\mathrm{e}^{i(\omega t-\beta z\pm l\varphi)} \tag{18}$$

$$H_{r,\pm} = \frac{i\beta}{q^2}\left[\mp\frac{i\varepsilon_2\omega l}{\beta r}CK_l(qr) + DqK_l'(qr)\right]\mathrm{e}^{i(\omega t-\beta z\pm l\varphi)} \tag{19}$$

$$H_{\phi,\pm} = \frac{i\beta}{q^2}\left[\frac{\varepsilon_2\omega}{\beta}CqJ_l'(qr) \pm \frac{il}{r}DK_l(qr)\right]\mathrm{e}^{i(\omega t-\beta z\pm l\varphi)} \tag{20}$$

$$H_{z,\pm} = DK_l(qr)\mathrm{e}^{i(\omega t-\beta z\pm l\varphi)}, \tag{21}$$

where $\beta$ is the mode propagation constant, $h = \sqrt{k^2-\beta^2}$, $q = \sqrt{\beta^2-k^2}$, $k = 2\pi/\lambda$ is the wavenumber, and $\varepsilon_i$ gives the dielectric constant in regions $i = 1, 2$. The parameter $l$ is a nonnegative integer that gives the order of the guided mode and its angular momentum. We also use the notation $J_l'(hr) = \partial J_l(hr)/\partial(hr)$, $K_l'(qr) = \partial K_l(qr)/\partial(qr)$ for derivatives of Bessel functions $J_l$ and modified Bessel functions of the second kind $K_l$ of order $l$.

Boundary conditions impose the following relations for the interior constants $A$ and $B$, and for the exterior constants $C$ and $D$:

$$\frac{B}{A} = \pm\left[\left(\frac{1}{ha}\right)^2 + \left(\frac{1}{qa}\right)^2\right]\left[\frac{J_l'(ha)}{haJ_l(ha)} + \frac{K_l'(qa)}{qaK_l(qa)}\right]^{-1} \tag{22}$$

$$\frac{C}{A} = \frac{J_l(ha)}{K_l(qa)} \tag{23}$$

$$\frac{D}{A} = \frac{B}{A}\frac{J_l(ha)}{K_l(qa)} - \frac{l^2\beta^2}{k_0^2}\left[\left(\frac{1}{ha}\right)^2 + \left(\frac{1}{qa}\right)^2\right]^2, \tag{24}$$

so that the knowledge of the propagation constant $\beta$ and amplitude normalization constant $A$ (Section 2.3) completely define the system.

## 2.2 Propagation Constant

An eigenvalue equation determines the propagation constant:

$$\frac{J_{l-1}(ha)}{haJ_l(ha)} = \frac{\left(n_1^2+n_2^2\right)}{4n_1^2}\left[\frac{K_{l-1}(qa)+K_{l+1}(qa)}{qaK_l(qa)}\right] + \frac{l}{(ha)^2} \pm R, \tag{25}$$

where

$$R = \sqrt{\frac{\left(n_1^2-n_2^2\right)^2}{\left(4n_1^2\right)^2}\left[\frac{K_{l-1}(qa)+K_{l+1}(qa)}{qaK_l(qa)}\right]^2 + \frac{l^2\beta^2}{n_1^2k_0^2}\left[\left(\frac{1}{ha}\right)^2 + \left(\frac{1}{qa}\right)^2\right]^2}, \tag{26}$$

and the $\pm$ $R$ solutions correspond to $EH$ and $HE$ modes, respectively. A normalized frequency called the $V$-number is defined by the relation

$$V = \frac{2\pi}{\lambda} a \sqrt{n_1^2 - n_2^2}, \tag{27}$$

which scales the optical frequency by the fiber radius and its numerical aperture ($\sqrt{n_1^2 - n_2^2}$). We can numerically solve Eq. (25) for a particular $V$-number (Eq. 27) and $l$ by finding the points of intersection of its left hand side and right hand side. Modes are labeled with subscripts $lm$, e.g., $HE_{lm}$, where for a given $l$, the successive points of intersection signify increasing $m$. Fig. 2 plots the result of this calculation (where $n_{\text{eff}} = \beta/k$) as a function of $V$ for various families of modes. Note that the cutoff occurs at $V = 2.405$ and that the fundamental $HE_{11}$ mode propagates for any $V > 0$. In the weak-guiding limit, these vector-mode solutions, $HE_{lm}$, become degenerate in the linearly polarized (LP) basis, meaning that they have the same propagation constant.

## 2.3 Normalization

The last parameter to determine is $A$, which is calculated using energy conservation. We normalize the time-averaged Poynting vector in the $z$-direction relative to the input power,

$$P = \langle S_z \rangle_t = A^2 \pi (D_{\text{in}} + D_{\text{out}}), \tag{28}$$

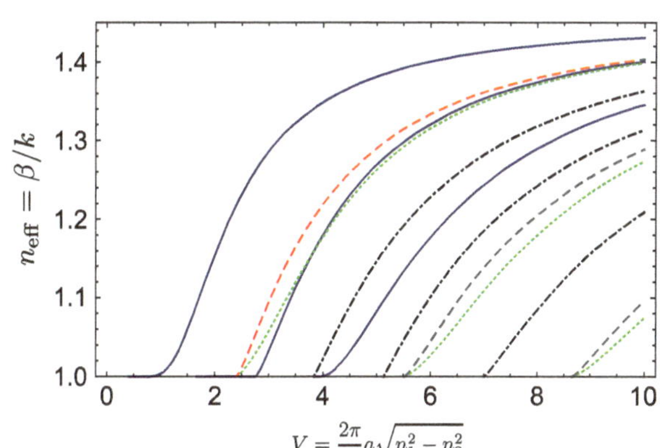


**Fig. 2** Effective index of refraction as a function of V-number. The families of modes and their *colors* are *HE* (*solid blue*), *EH* (*dashed-dotted black*), *TE* (*dashed red*), and *TM* (*dotted green*). *Reprinted from fig. 4.1 with permission from Hoffman, J.E., 2014. Optical Nanofiber Fabrication and Analysis Towards Coupling Atoms to Superconducting Qubits (Ph.D. thesis). University of Maryland College Park. Copyright (2014).*

where $D_{\text{in}}$ and $D_{\text{out}}$ will be found analytically. For the $HE_{lm}$ and $EH_{lm}$ modes, these parameters are

$$D_{\text{in}} = \frac{\pi a \beta^2}{4\mu_0 \omega} \frac{\beta}{h^2} [(1+sl)(N_1^2+sl)[J_{l+1}^2(ha) - J_l(ha)J_{l+2}(ha)] + (1-sl)(N_1^2-sl)[J_{l-1}^2(ha) - J_l(ha)J_{l-2}(ha)]] \tag{29}$$

$$D_{\text{out}} = \frac{-\pi a \beta^2}{4\mu_0 \omega} \frac{\beta}{q^2} \left(\frac{J_l(ha)}{K_l(qa)}\right)^2 [(1+sl)(N_2^2+sl)[K_{l+1}^2(qa) - K_l(qa)K_{l+2}(ha)] + (1-sl)(N_2^2-sl)[K_{l-1}^2(qa) - K_l(qa)K_{l-2}(qa)]], \tag{30}$$

where $N_i = n_i k/\beta$ and $s = B\mu_0\omega/(il\beta)$ (with $B$ given by Eq. (22)).

Fig. 3 offers a summary by plotting the mode structure of the $HE_{11}$ mode, showing the intensity (Fig. 3A) as well as the norm of each electric

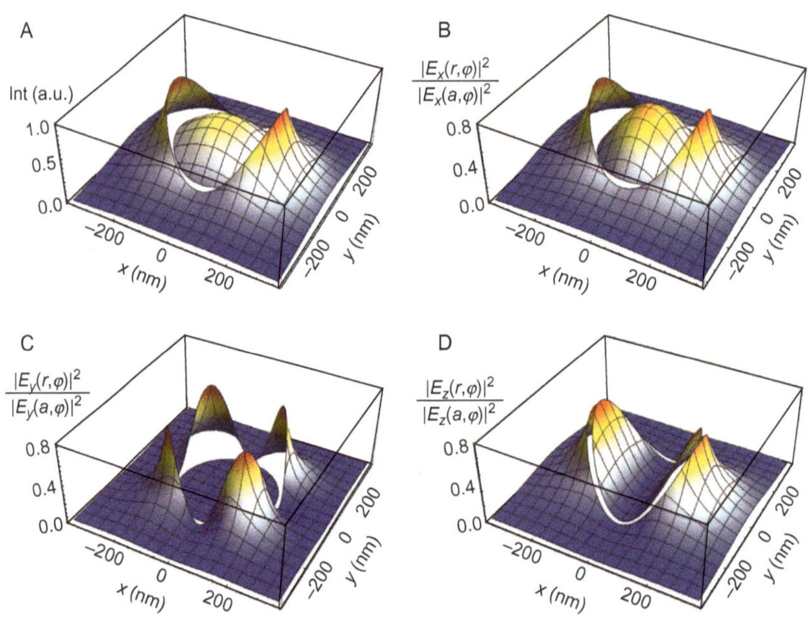


**Fig. 3** Fundamental ($HE_{11}$) mode structure of 360-nm diameter nanofiber. (A) Total intensity normalized to its maximum value. (B) Intensity of the $x$ component normalized to its maximum value at the nanofiber radius ($a$). (C) Intensity of the $y$ component normalized to its maximum value at the nanofiber radius ($a$). (D) Intensity of the $z$ component normalized to its maximum value at the nanofiber radius ($a$). *Reprinted from fig. A.7 with permission from Hoffman, J.E., 2014. Optical Nanofiber Fabrication and Analysis Towards Coupling Atoms to Superconducting Qubits (Ph.D. thesis). University of Maryland College Park.*

field component normalized to their value at the fiber surface (Fig. 3B–D). These values correspond to a 360-nm diameter fiber with index of refraction $n_1 = 1.45367$ and propagating wavelength of 780 nm. Note that the intensity has a sharp discontinuity at the fiber surface. The two largest components, $E_x$ and $E_z$, are comparable. The presence of a sizable longitudinal component $E_z$ plays a critical role in nanophotonic atom–photon interactions (Le Kien et al., 2004).

The radial decay of the evanescent field amplitude is not a simple exponential, but a complicated sum of modified Bessel functions of the second kind $K_l(qr)$ (see Eqs. 16–18). Since $r$ is the distance from the center of the ONF and we are interested in the field outside the dielectric media, the asymptotic expansion for large argument $K_l(qr) \approx \sqrt{\pi/2qr}e^{-qr}$ is a good approximation, for any order $l$. Considering this, the radial dependence of the evanescent electric field is

$$E_i(qr) \approx c_i r^{-\frac{1}{2}} e^{-qr}, \tag{31}$$

where $E_i$ is the $i$-th component of the electric field with amplitude proportional to $c_i$. This shows that the evanescent field decays at a shorter distance than an exponential decay. We have tested this approximation against exact numerical calculation showing excellent agreement, providing a simpler and more intuitive mathematical expression for the evanescent field.

## 2.4 Quasilinear Polarization

When we launch linearly polarized light into a fiber it will excite both the $\pm l\phi$ solutions. We can represent the mode in the quasilinear basis, as represented below.

$$\mathbf{E_{lin}} = \frac{1}{\sqrt{2}}[\mathbf{E}_+ \pm \mathbf{E}_-] \tag{32}$$

$$\mathbf{H_{lin}} = \frac{1}{\sqrt{2}}[\mathbf{H}_+ \pm \mathbf{H}_-]. \tag{33}$$

It is useful to think of this as analogous to representing linearly polarized light as a superposition of left and right circularly polarized.

## 2.5 A Note About the Polarization

The polarization of the propagating field can be characterized at the ONF waist with polarimetric measurement of the scattered light. Because Rayleigh scattering is dipolar, the scattered field preserves the polarization of the incoming one. This allows us to measure the polarization of the guided field at each point along the nanofiber. We can control the polarization at the ONF waist by modifying the input polarization. For more details the reader is referred to Vetsch et al. (2012).

The calculations in Section 2.1 show that $E_z$ can be significant in an ONF (Le Kien et al., 2004). This is generally true for tightly focused laser beams and waveguides in the strong-guiding regime. The existence of longitudinal polarizations is intimately related to the first of Maxwell's equations (Gauss' law). While in everyday paraxial optics rays and transverse polarizations suffice to characterize optical phenomena, when there are significant gradients on the transverse field they have to be accompanied by a corresponding longitudinal component of the electric field. Lax et al. (1975) carries out a systematic (perturbational) approach starting with Gauss' law as a function of a small parameter $\lambda/w$ where $\lambda$ is the wavelength of the light and $w$ the characteristic transverse width of the field: the transverse size of the beam. As $w$ decreases the ray optics approximation with transverse fields fails and longitudinal components appear. This is a serious issue when focusing laser beams to small sizes and their polarization properties have been studied and measured in Erikson and Singh (1994).

Gauss' law ($\nabla \cdot \mathbf{E} = 0$) establishes that a focused light beam of wavelength $\lambda$ and angular frequency $\omega$ that propagates along $\pm z$ with a slowly varying amplitude has a longitudinal component

$$E_z = \pm \frac{i}{k} \nabla \cdot \mathbf{E}_\perp \quad (34)$$

where the $\pm$ corresponds to the two possible directions of propagation and $\mathbf{E}_\perp$ is the transverse field. The presence of $i$ in the expression for $E_z$ means that $E_z$ is $\pm \pi/2$ out of phase with the transverse components, depending on the propagation direction. The existence of this longitudinal field and resulting transverse component of elliptical polarization is at the source of the chirality observed in nanophotonic systems as stated in Lodahl et al. (2017).

## 2.6 Higher-Order Modes

As seen in Fig. 2, when $V \leq 2.405$, the nanofiber supports only the two $HE_{11}$ degenerate modes. For a typical fused silica fiber, this requires $a \lesssim 0.35\lambda$. Nanofibers supporting higher-order modes (HOMs) provide additional degrees of freedom to enable complex evanescent field profiles for interaction with the surrounding medium. A few works have explored the use for HOM in nanofibers for atom trapping (Phelan et al., 2013; Sagué et al., 2008), interactions with atoms (Kumar et al., 2015a), and high-resolution fiber profilometry (Fatemi et al., 2017; Hoffman et al., 2015).

The treatment above provides modal solutions for these higher orders when $a$ increases. Referring to Fig. 2, the next grouping or family of modes is comprised of the $TM_{01}$, $TE_{01}$, and two degenerate $HE_{21}$ modes, with transverse profiles inside the fiber shown in Fig. 4 for an ONF with $a = 390$ nm and $\lambda = 795$ nm; in the weak-guiding limit, these are in the $LP_{11}$ basis. For $TM_{0m}$ and $TE_{0m}$, only three of the six components of $\mathbf{E}$ and $\mathbf{H}$ are nonzero. In particular, the $TE_{01}$ mode is purely azimuthally polarized, having $E_z = E_r = 0$ everywhere; while the $TM_{01}$ mode is quasi-radially polarized, with $E_\phi = 0$. These two modes are cylindrically symmetric and achieve cutoff at the same $V = 2.405$. The degenerate $HE_{21}$ modes have azimuthally varying polarization, with transverse field components that alternate between radial and tangential polarization, and mode cutoff near $V = 2.8$. These are labeled $HE^e_{21}$ and $HE^o_{21}$ with orthogonal transverse polarizations at every point. Direct observation of a nanofiber radius reaching cutoff condition during the tapering process is shown in Frawley et al. (2012b) and Ravets et al. (2013b).

To control the HOM composition on the nanofiber waist, the field at the input of the standard optical fiber (Fig. 1) must also be controlled. At the fiber input, the waveguide is in the weakly guiding regime, described by

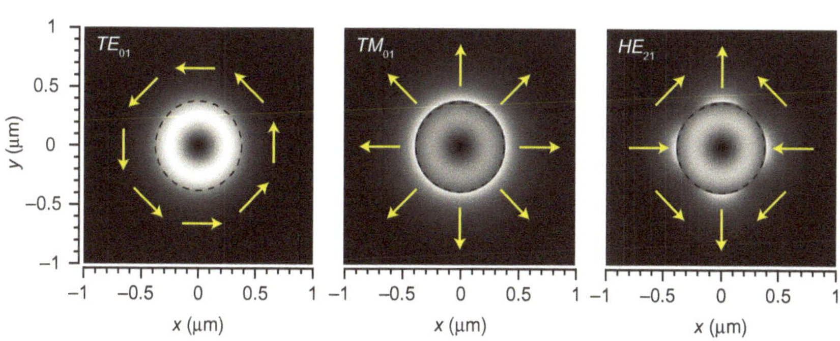


**Fig. 4** Transverse intensity profiles of modes in the $LP_{11}$ family, along with the $\phi$-dependent local polarization direction qualitatively indicated by the *yellow arrows*. The ONF outline of the $a = 390$ nm ONF is indicated by the *dashed black line*.

the LP-basis, with negligible longitudinal field and near unity overlap with superposition of the free-space propagation Hermite–Gauss (HG) or Laguerre–Gauss (LG) modes. By tailoring an incident Gaussian beam with spatial light modulators (Frawley et al., 2012b) or appropriate phase plates and mode conversion (Fatemi, 2011; Ravets et al., 2013b), we can efficiently and selectively excite and control specific modes (or their superpositions) in the nanofiber waist (Hoffman et al., 2015). Diagnostics of light propagation within the nanofiber using Rayleigh scattering (Hoffman et al., 2015) or near-field detection (Fatemi et al., 2017) have shown that mode discrimination is possible even in the subwavelength region of the waist.

## 2.7 Light Propagation in Nanofibers

As indicated at the beginning of this section, the nanofiber system consists of three distinct propagation regions: the fiber input, made of standard optical fiber in the weakly guiding regime; the nanofiber waist in the strongly guiding regime; and the tapered region connecting both. Computational analysis of light propagation from the input to the waist can be achieved with finite-difference time-domain (FDTD) methods. The analytical treatment above is sufficient for the two-layer input and waist regions, having only two $n_i$, but within the taper region, the effect of finite cladding radius becomes strong and calculations must consider the third medium (usually air or vacuum).

The evolution of the light propagation within the nanofiber from core-cladding guidance to cladding-air guidance can be measured by imaging Rayleigh elastic scattering (Hoffman et al., 2015), shown in Fig. 5. The figure shows propagation of the $LP_{01}$ mode family through the waveguide. For the Fibercore SM1500 fiber used, the field leaves the core near $2a = 27$ μm and enters the cladding volume. The sudden large change from guiding with small $V$-number ($V \approx 2$) to large $V$-number ($V \approx 100$) represents a mode volume change of more than 1000. It is this region that places the most stringent requirements on adiabatic propagation, as described in analytical treatments (Frawley et al., 2012b; Nagai and Aoki, 2014; Ravets et al., 2013a).

With multiple allowed modes, beating due to different modal propagation constants is easily detected. Because the fields in cylindrically symmetric waveguides are exactly solvable, and because the $\beta_i$ are strongly dependent on the nanofiber radius, the measured beat frequencies can be used as a non-destructive, in situ method for evaluating the fiber radius with subangstrom

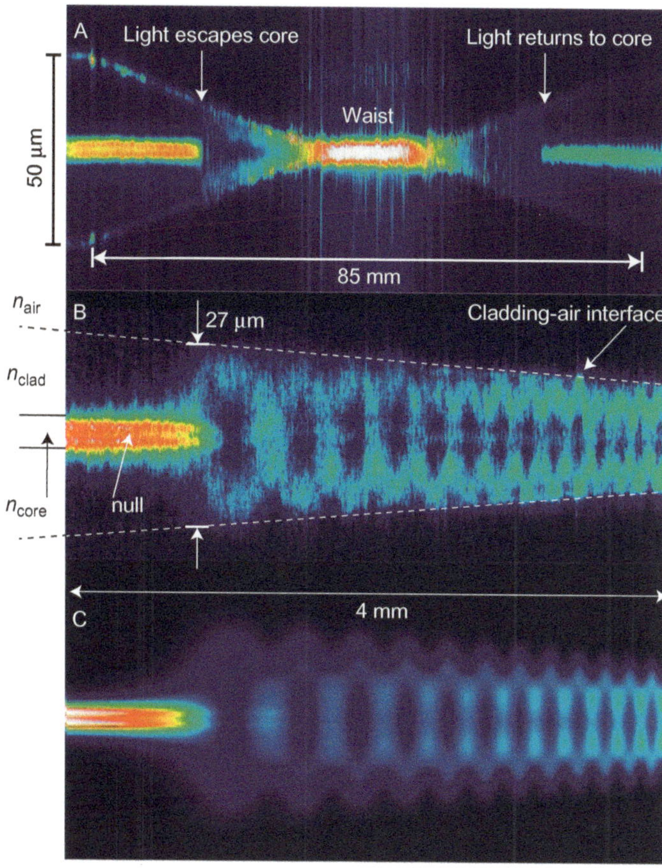


**Fig. 5** Propagation of the $LP_{11}$ mode family through an ONF by imaging Rayleigh scattered light. (A) The entire tapered fiber waveguide. Most notable is the transition from core-cladding guidance to cladding-air guidance, magnified in (B). The transition is sudden, occurring over a radius change of only $\approx$20 nm. The intensity null observed in the core is due to the "donut"-like intensity profile of this mode family (see Fig. 4). (C) Simulated propagation using FIMMPROP. The excellent agreement between simulation and experiment, and high dependence of spatial intensity beating on radius, enables a high-precision measure of fiber radius. *Reprinted from fig. 4 with permission from Hoffman, J.E., Fatemi, F.K., Beadie, G., Rolston, S.L., Orozco, L.A., 2015. Rayleigh scattering in an optical nanofiber as a probe of higher-order mode propagation. Optica 2, 416. Copyright (2015) by the Optical Society.*

radius sensitivity (Fatemi et al., 2017). Near-field methods based on whispering gallery modes (Sumetsky and Dulashko, 2010; Sumetsky et al., 2006) or effects on transmitted light (Madsen et al., 2016) can also provide subnanometer radius precision.

## 3. FABRICATION AND CHARACTERIZATION

Experiments in quantum optics and quantum information with atom traps around ONFs can benefit from high-transmission nanofibers. This reduces laser power requirements and unwanted stray light from both scattering and nonadiabatic mode excitation. The review by Ward et al. (2014) analyzes a variety of ONF fabrication techniques.

Common techniques for ONF production make use of micro-furnaces, fusion splicers, chemical etchants, and $CO_2$ lasers (Dimmick et al., 1999; Ding et al., 2010; Kbashi, 2012; Lambelet et al., 1998; Sørensen et al., 2014; Ward et al.,2006, 2014; Yokota et al., 1997). Chemical etching generally produces lower transmission than other heat and pull methods but offers more variety over the shape of the taper and the length of the waist. A $CO_2$ laser produces high-transmission ONFs but the final diameter is limited by the power and focus of the laser. Our work Hoffman et al. (2014) follows closely that of Warken (2007) and we summarize it in this section.

Our pulling technique is based on an existing methodology (Brambilla, 2010). It requires two pulling motors and a stationary oxyhydrogen heat source. This flame-brushing method allows us to reliably produce ONFs with controllable taper geometries and uniform waists. With our setup the waist can vary in length from 1 to 100 mm, and we can achieve radii as small as 150 nm (Bilodeau et al., 1988; Birks and Li, 1992; Garcia-Fernandez et al., 2011; Warken, 2007; Warken et al., 2007). Rather than sweep the flame back and forth over the fiber, we keep the flame stationary while sweeping the motors; this action reduces air currents, which could lead to nonuniformities on the fiber waist. This approach is applicable to other pulling techniques as well, so there would be no need to scan a heat source.

We focus on the critical prepull steps necessary to achieve an ultra-high transmission, before handling the known environmental effects (Fujiwara et al., 2011). Following the protocols described in this section we have produced fibers with 99.95% transmission when launching the fundamental mode (Hoffman et al., 2014). This high transmission corresponds to a loss of $2.6 \times 10^{-5}$ dB/mm for the fundamental mode (Brambilla, 2010; Stiebeiner et al., 2010), with controllable taper geometries and long fiber waists. We have also launched HOMs (Ravets et al., 2013a,b) and achieved transmissions of greater than 97% for the first family of excited modes. This level of transmission requires a thorough optimization of the pulling algorithm and fiber cleaning procedure. However, we can routinely achieve transmissions above 99% with root-mean-square (rms) fractional uniformities on the waist region smaller than 0.02.

## 3.1 The Fiber-Pulling Apparatus

The entire pulling apparatus is inside a nominally specified ISO Class 100 clean room. The apparatus consists of a heat source that brings the glass to a temperature greater than its softening point (1585°C for fused silica) and two motors that pull the fiber from both ends. We use two computer-controlled motors, mounted to a precision-ground granite slab. The granite slab serves two purposes: its weight suppresses vibrations and it provides a flat surface. The weight of the granite slab damps the recoil from the fiber motors as they change direction at the end of every pull step. Without a flat surface the motors will not work to specification, leading to distortion of the ONF during the pull: the pitch or yaw of the motor can vary the distance between the fiber and the flame, changing the effective size of the flame and pulling the fiber in unintended directions (negating any prepull alignment). The granite is mounted on an optical table at three points so that surface imperfections of the optical table do not distort the granite slab.

We attach flexure stages with a fiber clamp to the moving platforms of the motors. We position the v-grooves of the fiber clamp on each stage at the minimum separation allowed by the parameters of a given pull. Separating the fiber clamps at the closest allowed distance minimizes the fiber sag during the pull, which can result in the ONF breaking during the pulling process. The v-grooves of the fiber clamps on the left and right fiber motors must be aligned within a few micrometers to achieve a high transmission. We align the v-grooves using micrometers, attached to the flexure stages to allow for three axis translation assisted by an in situ optical microscope. The optical microscope is also used to check the prepull cleanliness of the fiber and for postpull characterization. We use an oxyhydrogen flame to heat the fibers, in a stoichiometric mixture of hydrogen and oxygen to ensure that water vapor is the only byproduct. Stainless steel gas lines introduce the hydrogen and oxygen through flow meters. The gases mix in a tee and pass through a high-quality, 3-nm filter. Finally the hydrogen–oxygen mixture exits through a custom-made nozzle with 29 holes of 200-μm diameter in a $1 \times 2$ mm$^2$ area. The nozzle serves as a flame arrestor, while still allowing for the gas flow to be in the laminar regime. The nozzle is clamped to a computer-controlled motorized stage that translates the flame to about 0.5 mm in front of the fiber for the duration of the pulling process. If any fiber buffer remains or dust lands on the fiber at any time the transmission will degrade.

## 3.2 Algorithm

To choose the parameters for a pull we developed an algorithm,[1] based on Warken et al. (2007) and Warken (2007), that calculates the trajectories of the motors needed to produce a fiber with the desired final radius, length of uniform waist, and taper geometry. The tapers are formed by a series of small fiber sections that are well-approximated by lines, allowing us to form a linear taper with a given angle down to a radius of 6 μm. This connects to an exponential section that smoothly reduces and connects to a uniform section with a submicron radius, typically 250 nm. The slope of the linear taper section generally varies between 0.3 and 5 mrad. The algorithm divides the pull into steps defined by their pulling velocity and the traveling length of the flame. It recursively calculates the parameters, starting from the desired radius, until reaching the initial one.

## 3.3 Cleaning and Alignment Procedure

Obtaining high transmission through an ONF requires careful attention to the prepull cleanliness of the fiber. If any particulate remains from the fiber buffer or if dust arrives on the fiber before being introduced to the flame, the particulate will burn and greatly diminish the final transmission, see Hoffman et al. (2014). Residue from solvents can also decrease transmission.

Our cleaning procedure starts by mechanically removing the protective plastic buffer to expose the glass of the fiber to the flame. Then we use isopropyl alcohol on lens tissue to remove larger particulates. A few wipes of acetone are then applied with class 10a clean room wipes, in order to dissolve smaller remnants of the buffer. A final cleaning with methanol using class 10a clean room wipes removes residues left from the previous solvents. We carefully lay the fiber into the grooves of the fiber clamps on the pulling apparatus and image the entire length of cleaned fiber using the in situ optical system. If there is any visible dust, particulate, or evaporate, within the 2-μm resolution of the optical microscope, we repeat the cleaning procedure over. If the fiber is clean, we proceed to align it.

[1] The program is available at the Digital Repository of the University of Maryland (DRUM) at http://hdl.handle.net/1903/15069.

We begin the alignment procedure by properly tensioning the fiber. Moving the pulling motors apart in 200-μm increments until the fiber slides through produces a uniform tension. We align the sections of fiber directly next to the left clamp and right clamp equidistant from the in situ optical microscope and at the same height. There are more critical details about untensioning the fiber before starting the pull that the interested reader can find in Hoffman et al. (2014). The fiber alignment is on the order of micrometers over a length of centimeters.

## 3.4 Characterization and Microscopy Validation

We validate the accuracy of our algorithm expected fiber profile using both an in situ optical microscope and a scanning electron microscope (SEM). The measurements of the radius with SEM agree with our targeted ONF radius to better than 5%, while optical measurements across the whole fiber including the taper region give rms fractional variations around 2%. We can also test the profile with exquisite (nanometric) sensitivity when using the beating spatial frequencies among higher order modes as seen through Rayleigh scattering (Hoffman et al., 2015) and by evanescent coupling to a second fiber (subangstrom) (Fatemi et al., 2017).

## 3.5 Transmission

During the pull we monitor the transmission of a few nW of light through the fiber using a laser of 780-nm wavelength. We record data for the duration of the pull on a digital oscilloscope in high-resolution mode (16-bit digitization). We normalize the output signal to the laser power drift throughout the pull. Fig. 6 shows the transmission as a function of time during the pull for an ONF with a 2-mrad-angle taper to a radius of 6 μm and exponential profile to reach a final waist radius of 250 nm, with a fiber waist length of 5 mm. It achieves a transmission of $99.95 \pm 0.02\%$, corresponding to a loss of $2.6 \times 10^{-5}$ dB/mm when taken over the entire stretch.

Using the numerical Maxwell's equations solver FIMMPROP, we calculate the expected transmission through a fiber with the same profile as in our pulls. We find the expected transmission to be consistent with our experimental result that measures the transmission through the entire ONF. Furthermore, when launching the next family of modes ($LP_{11}$) through the fiber the FIMMPROP simulations are well matched to the achieved transmissions in Ravets et al. (2013a).

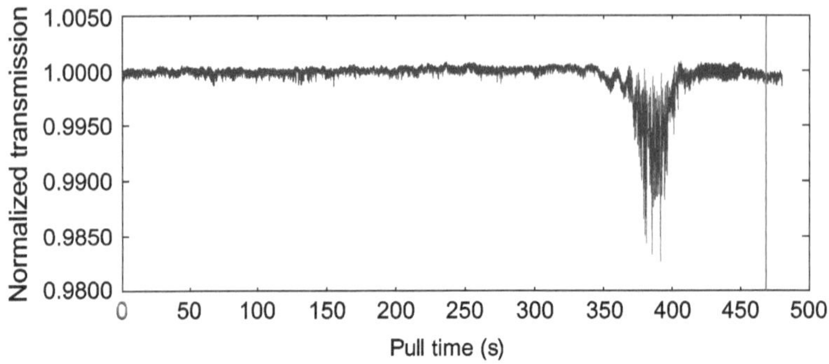


**Fig. 6** The normalized transmission as a function of time during the pull through an ONF with an angle of 2 mrad to a radius of 6 μm and exponential profile to a final waist radius of 250 nm. The length of the waist is 5 mm. The fiber has a final transmission of 99.95 ± 0.02% or equivalently a loss of $2.6 \times 10^{-5}$ dB/mm. The *red line* is the point where we evaluate the final transmission. *Reprinted with permission from Hoffman, J.E., Ravets, S., Grover, J.A., Solano, P., Kordell, P.R., Wong-Campos, J.D., Orozco, L.A., Rolston, S.L., 2014. Ultrahigh transmission optical nanobers. AIP Adv. 4 (6), 067124.*

## 3.6 Spectrogram Analysis

We analyze the quality of the ONF using a spectrogram, a short-time Fourier transform also sometimes referred to as the Gabor transform, of the transmission data during the pull. The spectrogram allows us to extract the modal change in the ONF during the pull. Each curve in the spectrogram corresponds to the evolution of the spatial beat frequency between the fundamental mode and excited modes propagating in the fiber, while the contrast corresponds to the energy transferred from the fundamental mode to the HOMs. We use simulations to identify all modes that are excited during the pull. Ravets et al. (2013a) has a detailed description and the full theoretical background.

Fig. 7 is a spectrogram of the transmission data from Fig. 6. For a successful 2-mrad pull with SM800 fiber we expect to observe few HOM excitations. If modes asymmetric to the fundamental are excited, we know the cylindrical symmetry of the fiber was broken during the pulling process (Ravets et al., 2013a), which can aid in identifying and fixing the error in the pulling apparatus. We have seen that such coupling to asymmetric modes can occur for multiple-angle fiber pulls (Ravets et al., 2013b) and uncleaned fibers (Hoffman et al., 2014).

## 3.7 ONF Radius Extraction

We have developed a nondestructive tool to accurately and precisely measure the ONF radius and characterize the propagation of HOMs. These

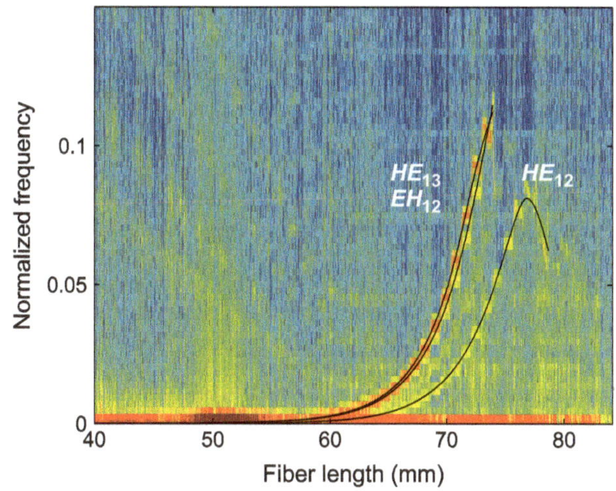


**Fig. 7** Spectrogram of the transmission data from Fig. 6 as a function of the stretch of the fiber, illustrating the chirp of the beating frequency and the abrupt end of the beating. The *colormap* corresponds to the power spectral density. The *curves* are calculations of the higher-order mode excitations of the same symmetry as the $HE_{11}$ fundamental mode: $EH_{11}$, $HE_{12}$, and $HE_{13}$. *Reprinted fig. 12 with permission from Ravets, S., Hoffman, J.E., Kordell, P.R., Wong-Campos, J.D., Rolston, S.L., Orozco, L.A., 2013. Intermodal energy transfer in a tapered optical fiber: optimizing transmission. JOSA A 30, 2361–2371. Copyright (2013) Optical Society.*

properties are crucial for optimizing both fundamental and applied uses of ONFs. The ONF radius, $a$, governs the propagation of the HOMs through Eq. (27) and the coupling between a nearby atom (trapped or free) to the allowed nanofiber modes (see Le Kien et al., 2005b; Sections 4 and 5). As stated in Section 2 these modes have distinct effective refractive indices that depend strongly on the radius.

Extraction of beating frequencies from Rayleigh scattering has the limitations of far-field imaging (Hoffman et al., 2015; Szczurowski et al., 2011). Contact techniques between an ONF and a probe microfiber avoid these limitations. They have been used to measure the local radius variations of an ONF with subangstrom/subnanometer precision by propagating light through the probe microfiber and observing the mode spectrum of a whispering gallery or a composite photonic crystal cavity (Birks et al., 2000; Keloth et al., 2015; Semenova et al., 2015; Sumetsky and Dulashko, 2010). Other contact techniques rely on changes in the amplitude of the transmitted light, becoming sensitive to polarization and van der Waals forces, as in the recent work by Madsen et al. (2016), but do not directly sample the modal decomposition of the local field.

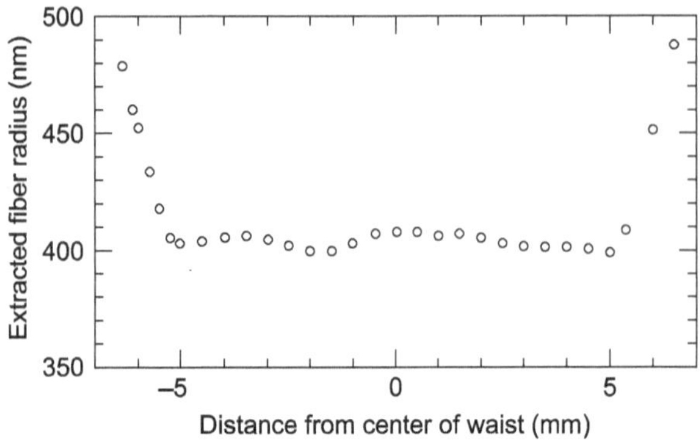


**Fig. 8** Radius extraction of a 10-mm nanofiber using the evanescent coupling (Fatemi et al., 2017). The radius comes from the $HE_{11}$:$TM_{01}$ beat frequency, using FFT windows every 600 μm. The error bars are smaller than the circles.

We measure the beat lengths between propagating nanofiber modes in the near-field over the ONF length through evanescent coupling to a probe microfiber in Fatemi et al. (2017). This near-field approach provides high ($\approx$1μm) longitudinal resolution by taking the position-dependent spectrogram of the coupled signal. The spatial frequencies map the mean of the ONF radius to within 0.4 Å over 600-μm measurement windows along the ONF taper and waist. This technique has allowed us to measure how the 380 nm radius waist changes over its 10 mm length finding it to be better than $\pm$5 nm (the measured radius is closer to 400 nm), as Fig. 8 shows with a resolution smaller than the size of the data points (0.4 Å). Such control on the fabrication process is beneficial for cold atom experiments in the vicinity of a nanofiber as detailed in Sections 4 and 5.

## 4. ATOMS AROUND THE NANOFIBER

This section presents studies of unconfined atoms and their interactions with ONFs. Some experiments investigate the atom–surface interactions, others use an ONF as a passive probe to measure properties of atomic clouds of atoms, while others perform linear and nonlinear spectroscopy using the nanofiber for light collection and/or directly probing the atoms.

The experiments in this section show the versatility of ONFs, and how they are making inroads on a wide variety of fields (Brambilla, 2010). We foresee many other uses, such as studies of molecules and molecular formation, and other forms of spectroscopy.

## 4.1 Atom–Surface Interactions

The interaction of atoms with a dielectric nanofiber has been studied theoretically in a number of papers (Frawley et al., 2012a; Klimov and Ducloy, 2004; Le Kien et al., 2005b, 2006; Minogin and Nic Chormaic, 2010; Russell et al., 2009). They demonstrate that the nanofiber modifies the boundary conditions and density of modes, affecting decay properties of the atom. These modifications can be measured in the frequency domain, through fluorescence and absorption spectroscopy, by observing lineshape asymmetries, or in the time domain through direct measurement of the spontaneous emission rate.

The van der Waals (vdW) and Casimir–Polder (CP) potentials will produce a red-shifted lineshape. This has been measured in two experiments using cesium atoms. The first observes the fluorescence spectrum, collected through the ONF, with a long red tail attributed to the vdW interaction (Nayak et al., 2007). They use the distance-dependence of the vdW potential to relate spectroscopic features to atoms passing at particular distances from the fiber surface.

Nayak et al. (2012) employ a similar technique with the addition of a violet laser to control the surface conditions of atoms adsorbed to the nanofiber. After turning off the violet laser, they measure sequential spectra over time as more and more atoms adhere to the fiber, as displayed in Fig. 9. The long tail on the red side of the spectrum, due to vdW interactions, becomes visible in Fig. 9C and D.

Sagué et al. (2007) instead use a low-power, through-fiber probe laser to study atom–surface interactions in a cloud of cesium atoms. Because of the evanescent decay of the field on the fiber waist, the probe excites only atoms close to the nanofiber, and only on the order of nanowatts of laser power are needed to saturate the atoms. At low powers, they see an asymmetry and broadening of the absorption profile due to light-induced dipole forces, the vdW potential, and emission enhancement. As they increase the probe power, they observe a saturation of the linewidth, which is unexpected given power broadening. This is attributed to a reduced atomic density near the fiber because of the probe beam, which lowers the absorbance and narrows the line. Simulations using atomic trajectories agree closely with their data in both the low- and high-power regimes.

Our group has also measured the transmission spectra of cold $^{87}$Rb atoms close to an ONF. We produce near-surface (<50 nm), thermally desorbed atoms with a far-detuned, variable-intensity laser beam coupled to the ONF.

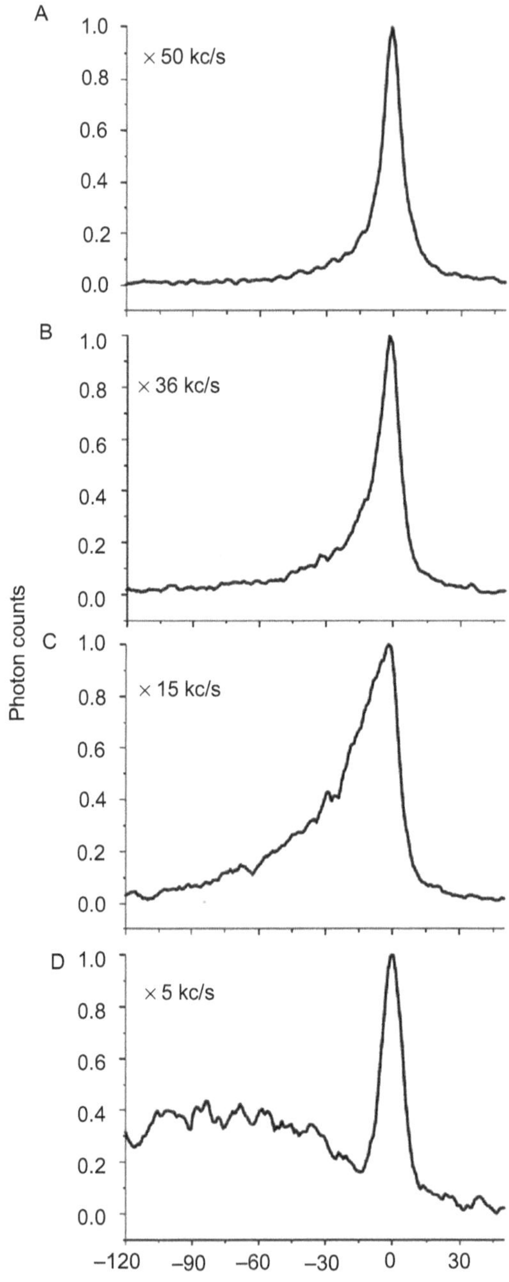


**Fig. 9** Measured fluorescence excitation spectra. The spectra (A)–(D) are measured one after another consecutively and the measurement time for each spectrum is around 6.5 min. The scaling factors are shown in the figure. The spectra are background corrected. *Reprinted fig. 2 with permission from Nayak, K., Das, M., Le Kien, F., Hakuta, K., 2012. Spectroscopy of near-surface atoms using an optical nanober. Opt. Commun. 285, 4698. Copyright (2012) Elsevier.*

The spectra show asymmetries on the red side of the transmission spectrum due to the van der Waals interaction of the atoms with the ONF (Patterson, 2016).

## 4.2 Atom–Cloud Characteristics

Fluorescence from cold atom clouds is a key tool for determining properties of those clouds, e.g., temperature or atom number. Typically the emitted photons are collected using free-space optics. Another possible path is to employ an ONF to collect the photons emitted into the guided mode to discern atom–cloud characteristics.

Several groups have used nanofibers immersed in cold atom clouds in multiple experiments to both replicate and improve upon these standard techniques. Morrissey et al. (2009) uses the nanofiber to measure several characteristics of a magneto-optical trap (MOT): cloud shape, cloud size, atom loading rate, and atom number. In another experiment, Russell et al. (2012) measure sub-Doppler temperatures of an atom cloud by dithering the magnetic trapping field to extract the trap frequency. The method is simultaneously corroborated with the standard free-space collection method. Finally, Russell et al. (2013) perform release and recapture with an ONF to measure atomic temperature, as well as diagnose optical misalignment in their MOT setups.

Our photon-correlation measurements of Grover et al. (2015) and Grover (2015) of a cloud of atoms around an ONF allow us to extract the atomic-cloud temperature from the correlations, by the knowledge of the evanescent field and the coupling of the moving atom fluorescence to the nanofiber mode. We validate the method against temperature measurements from time-of-flight images of the expanding atomic cloud. We measure the temporal width of the intensity–intensity correlation function due to atomic transit time, as presented in Fig. 10, and use it to determine the most probable atomic velocity, hence the temperature. The results are further validated with atomic trajectory simulations.

## 4.3 Linear and Nonlinear Spectroscopy

The small mode profile of light propagating along the nanofiber waist can result in high intensities for even modest laser power. As demonstrated in Sagué et al. (2007), only nanowatts of resonant probe laser power are enough to saturate atoms near the fiber waist. This makes nanofibers a nice platform for studying nonlinear optical effects that require high intensities. Various

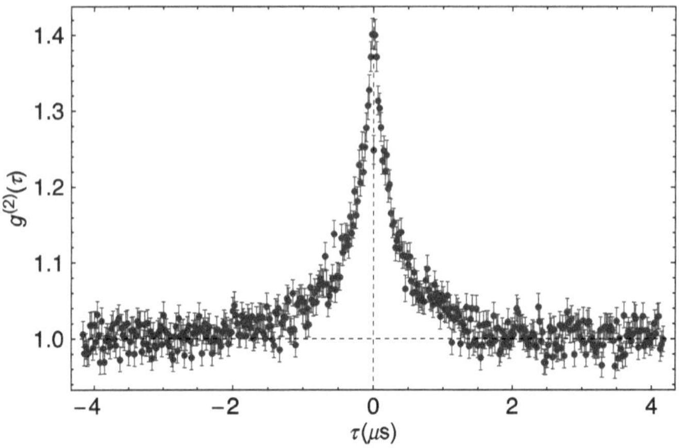


**Fig. 10** Second-order correlation function $g_\tau^{(2)}$ as a function of delay time $\tau$ for a cold atom cloud that still shows an antibunching feature at $\tau = 0$.

low-power, nonlinear optical experiments have been conducted with nanofibers immersed in warm atomic vapors.

Using powers lower than 105 nW, Hendrickson et al. (2010) observe two-photon absorption in room temperature atoms around a nanofiber. They attribute the neither Gaussian nor Lorentzian profiles of the two-photon absorption peaks to the transit time of atoms passing through the nanofiber mode. In a deviation from typical alkali-atom experiments, Pittman et al. (2013) measure nonlinear transmission and low-power saturation through a nanofiber surrounded by metastable xenon atoms. This is a promising avenue for future nonlinear optics experiments since the inert xenon atoms will not accumulate on the fiber surface. Jones et al. (2014) make a direct comparison between saturation effects in free-space vs through an ONF. They present an empirical nonlinear absorption model that accurately fits the ONF data but fails when applied to the free-space data.

Another experiment measured the Autler–Townes splitting, in $^{87}$Rb, generating up-converted photons at 420 nm (Kumar et al., 2015b).

Das et al. (2010) use quantum correlations to measure the fluorescence emission spectrum of a few strongly driven atoms around an ONF. Crucial to their success is the fact that they collect light on the nanofiber, then perform a heterodyne mixing before measuring the intensity–intensity correlation function, on which they perform a fast Fourier Transform to obtain the fluorescence spectrum. They observe that the fluorescence spectrum of the cold atoms near the nanofiber shows negligible effects of the atom–surface interaction, meaning they are far from the surface and the coupling $\beta$ is small, and agrees well with the Mollow triplet spectrum of free-space atoms at high excitation intensity.

## 5. TRAPPING ATOMS AROUND AN ONF

The evanescent field of the guided modes of an ONF provides an intensity gradient suitable for generating an optical dipole trap (Grimm et al., 2000). All current realizations use two lasers in the fundamental $HE_{11}$ guided mode (Vetsch et al., 2010). A laser red-detuned from resonance creates an attractive potential, while a blue-detuned one creates a repulsive potential preventing the atoms from sticking to the nanofiber surface. The characteristic decay length of both evanescent fields is proportional to their wavelengths (see Eq. 31), producing a longer (shorter) range attractive (repulsive) potential for the red- (blue-) detuned beam. By adjusting their relative intensities it is possible to find a configuration where there is a minimum of the dipole potential away from the ONF surface, achieving trap depths of a few hundreds of microkelvins. A longitudinal confinement of the trap atoms can be implemented by counter-propagating two red-detuned beams, creating a standing wave. Stable traps can be created for different trapping beam polarizations and intensities, generating different azimuthal confinements.

The gradient of the transversal component of the guided field is always on the order of $2\pi/\lambda$, creating a nonnegligible longitudinal component (as explained in Section 2.5), which is generally not the case for trapping beams propagating in free-space, unless they are tightly focused. The phase relation between the transversal and the longitudinal component of the guided field creates elliptically polarized light, introducing significant vector light shifts in the trapped atoms (Le Kien et al., 2004). Differential light shifts induced among atomic sublevels can be suppressed with a proper choice of trapping wavelengths and polarizations (Lacroûte et al., 2012) or they can be used as a tool for atomic state preparation (Albrecht et al., 2016).

These types of trapping schemes have shown to be a promising platform for integrating laser-cooled atomic ensembles with optical fibers, opening a route towards applications in the context of quantum information and quantum technologies.

### 5.1 An Atomic Dipole Trap Around an ONF

We now summarize how our own ONF trap operates (Lee et al., 2015; Solano et al., 2017c) (see Fig. 11). The ONF has a waist diameter of 500 nm and length of 7 mm, with tapering regions of 28 mm in length and 2 mrad angle.

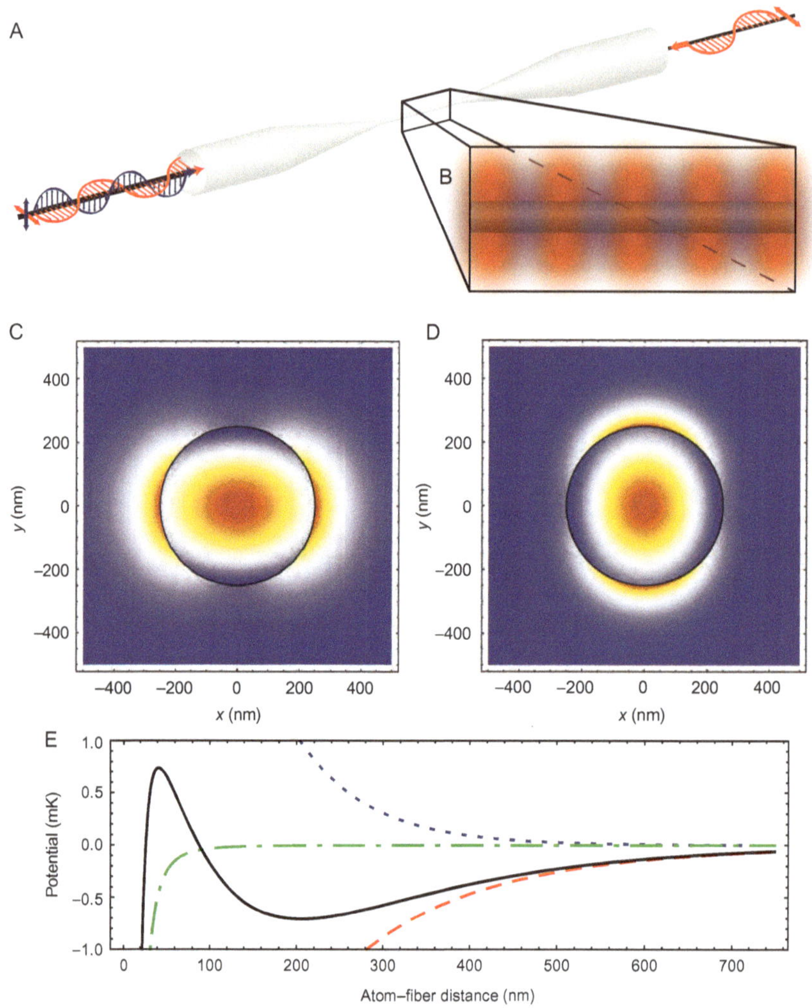


**Fig. 11** (A) Schematic of an ONF with counter-propagating 1064-nm beams and an orthogonally polarized 750-nm beam. (B) Illustration of the intensity of the fields at the ONF waist with lattice formed by 1064-nm beams. (C) Intensity plot of quasilinearly polarized, 1064-nm light in an ONF with diameter 500 nm. The *color scale* indicates increasing intensity from *blue* to *red*. (D) Intensity profile of vertically polarized 750-nm light through the same ONF. (E) Total trapping potential (*black solid*) for a 500-nm diameter ONF with contributions from 3 mW in each 1064-nm beam (*red dashed*), 6.5 mW of 750-nm power (*blue dotted*), and van der Waals (*green dashed* and *dotted*). The potentials are calculated from only the scalar shifts. *Reprinted from fig. 1.4 with permission from Hoffman, J.E., 2014. Optical Nanofiber Fabrication and Analysis Towards Coupling Atoms to Superconducting Qubits (Ph.D. thesis). University of Maryland College Park. Copyright (2014).*

A MOT loaded from a background vapor of $^{87}$Rb produces a cloud of $\sim 10^8$ atoms. We overlap the cloud with the ONF waist using magnetic field shim coils and an ultra-high vacuum mechanical manipulator. Two orthogonal imaging systems ensure the overlap between the two of them. Atoms fall into the ONF trap (the red- and blue-detuned trap beams are on throughout the experiment) after 90 ms of increased MOT detuning and a 1-ms-duration optical molasses stage, cooling the atoms to $\sim$15 μK.

An ONF trap requires light that is tuned to the red of the resonant frequency (with respect to the $^{87}$Rb D2 line) to provide an attractive potential and light tuned blue of resonance to prevent atoms from striking the ONF surface (see Fig. 11). A 1064-nm wavelength laser beam provides the (red) attractive potential and a 750-nm wavelength beam the (blue) repulsive force. A potential minimum of a few hundred μK in depth is formed $\sim$200 nm from the fiber surface, as calculated with a simple two-level atom and only scalar shifts (see Fig. 11E).

Counter-propagating the red-detuned attractive beams, generating a standing wave, confine the atoms along the nanofiber. This configuration creates a one-dimensional lattice on each side of the ONF. The presence of a longitudinal component of the propagating electric field prevents us from having a standing wave with perfect contrast. The contrast varies for different wavelength and ONF radius, and in our case is $\sim$65%. However, this is not an impediment for trapping the atoms in a lattice configuration.

The azimuthal confinement of the trapped atoms is achieved by using quasi-linearly polarized trapping beams in the $HE_{11}$ mode (see Fig. 11C and D). There is no particular requirement for the relative polarization of the blue- and red-detuned beams. A trapping minimum can be generated for polarizations varying from parallel (Goban et al., 2012) to perpendicular (Vetsch et al., 2012), provided the proper optical powers of the trapping beams are used. To verify the polarization of each beam on the nanofiber waist, we take polarization-sensitive measurements of the Rayleigh scattering of the propagating beams at the ONF waist.

Fig. 12 shows results from the pioneering experiment by Vetsch et al. (2010). The authors demonstrate trapping and optical interfacing of neutral Cs atoms around an ONF. The atoms are localized in two one-dimensional optical lattice arrays about 200 nm above and below the nanofiber surface and can be efficiently interrogated with a resonant light field sent through the nanofiber. Their operation parameters allow only zero or one atoms on each independent well because of collisional blockade (Schlosser et al., 2002), creating on average a filling factor of 50%.

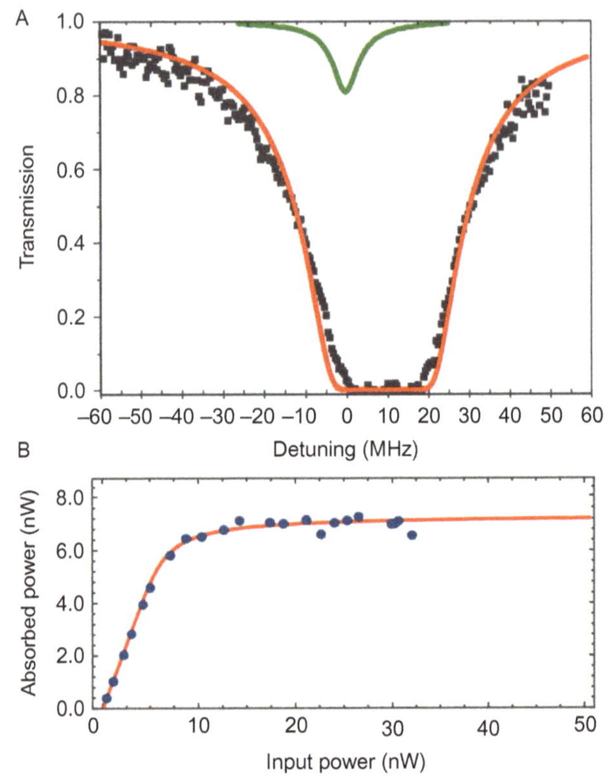


**Fig. 12** (A) Transmission spectrum of the probe beam through the nanofiber after loading the trap (*black squares*). For reference, the spectrum of the MOT cloud (*green line*) is plotted. The *red line* is a theoretical fit, see Vetsch et al. (2010). (B) Saturation measurement yielding the number of trapped atoms. The *red line* is a theoretical fit. *Reprinted fig. 3 with permission from Vetsch, E., Reitz, D., Sagué, G., Schmidt, R., Dawkins, S. T., Rauschenbeutel, A., 2010. Optical interface created by laser-cooled atoms trapped in the evanescent field surrounding an optical nanofiber. Phys. Rev. Lett. 104 (20), 203603, with Copyright (2010) by the American Physical Society.*

The number of trapped atoms in an ONF two-color dipole trap can be measured by propagating a probe beam through the nanofiber. Possible strategies are based on transmission spectroscopy and off-resonance dispersive measurements.

## 5.2 Transmission Spectroscopy

Transmission spectroscopy is based on the Beer–Lambert–Bouguer law for optical absorption in a thick medium and the knowledge of the atomic spectral lineshape. By scanning the frequency of a weak near-resonance probe beam and measuring its transmission as a function of frequency, we can obtain the optical density of the sample (Foot, 2005). If the atom–light coupling strength is known, the OD can be translated to a number of trapped atoms. When implemented in an ONF the probe power is typically of the order pW, to avoid saturation of the atoms and power broadening. The transmitted photons are counted with avalanche photodiodes and appropriate electronics. A careful bandwidth filtering is necessary to keep the background from the blue-detuned laser and what appears to be fluorescence of the fiber. Fig. 12A shows an example of this measurement. Similar results can be found in Goban et al. (2012).

Nonetheless, the number of trapped atoms cannot always be simply determined by a Lorentzian line shape of the probe beam absorption. This depends on the trapping potentials and the atomic species. Ref. Lee et al. (2015) shows that distinct asymmetries are observed in the spectrum due to the effects of the vector light shifts associated with the optical trapping fields. Although Rb and Cs are nominally atoms with similar atomic structure, the light shifts can in fact be quite different, with larger effects in Rb than Cs, leading to a modified absorption profile.

This method of measuring the number of trapped atoms is a rather destructive one. The absorbed photon transfers a momentum kick to the atoms. After repeated scattering events, the atoms heat up, and eventually leave the trap.

## 5.3 Dispersive Measurements

A less destructive atom number measurement is to send an off-resonance probe beam through the ONF. The dispersive atom–light interaction creates an effective modification of the refractive index experienced by the propagating field, but minimally altering the atomic state. The tight mode confinement provides a large atom–light nonlinear interaction and efficient readout. The main features of this technique are explored theoretically in detail by Qi et al. (2016), in the context of quantum nondemolition measurements. The dispersive atom–light interaction produces a phase shift in the propagating field, proportional to the number of atoms and the atom–light coupling strength. This phase shift can be read out with polarimetric (Dawkins et al., 2011; Solano et al., 2017c) and interferometric measurements (Béguin et al., 2014).

The polarimetric measurement relies on the azimuthally asymmetric geometry of the trap. An ensemble of trapped atoms at each side of the

ONF defines a preferential propagation axis for the evanescent field of an off-resonant quasi-linearly polarized probe beam. The atoms create an effective birefringence for the probe. In Dawkins et al. (2011), 1000 cesium atoms are trapped around an ONF, producing a phase shift per atom of ≈1 mrad at a detuning of six times the natural linewidth. Fig. 13 shows the measured phase shift for different probe detuning.

Béguin et al. (2014) realize an interferometric measurement of the probe beam. They send two probe beams detuned by the same frequency above and below resonance. This eliminates any possibility of modification of the potential landscape experienced by the trapped atoms. Because the sign of the phase shift depends on which side of the resonant frequency the probe is, both probe beams experience equal-magnitude but opposite-sign phase shifts. The differential phase shift between the two probes is measured using optical homodyne interferometry, allowing them to extract the number of trapped atoms.

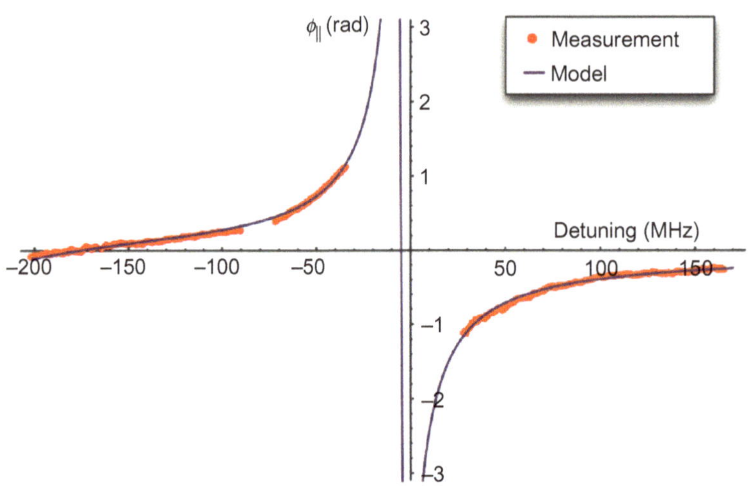


**Fig. 13** Phase shift $\phi_{||}$ of the eigenmode $e_{||}$ induced by 1000 atoms and measured as a function of the detuning from the $F=4 \rightarrow F'=5$ free-space transition frequency, $(\omega - \omega_5)/2\pi$, with 128 averages (*red dots*). The *blue line* is a fit taking into account the two nearest transitions $F=4 \rightarrow F'=4$ and $F=4 \rightarrow F'=5$. The former gives rise to the zero-crossing of the signal near-180 MHz. *Reprinted fig. 3 with permission from Dawkins, S., Mitsch, R., Reitz, D., Vetsch, E., Rauschenbeutel, A., 2011. Dispersive optical interface based on nanober-trapped atoms. Phys. Rev. Lett. 107, 243601, with Copyright (2011) by the American Physical Society.*

## 5.4 State-Sensitive and State-Insensitive Traps

When a two-level atom interacts with an oscillating electric field, e.g., a dipole trap, the two energy levels shift in opposite directions, an effect known as the ac-Stark shift. If the field is red-detuned from the atomic resonance, the ground state shifts down in energy and the excited state shifts up; the opposite case is true for a blue-detuned optical beam. When the shift only depends on the total amplitude of the electric field it is called scalar light shift, and when it depends on field polarization it is called vector light shift. In a dipole trap produced by circularly polarized light the polarization of the electric field couples to the components of the angular momentum of the real (multilevel) atom, shifting also the hyperfine sublevels, similar to what happens to the atoms in the presence of a "fictitious" external magnetic field. This is the case for ONF-based dipole traps, due to the significant longitudinal component of the propagating field that produces elliptically polarized light along the nanofiber. The differential shifts of the atomic energy levels cause different electronic states to experience different trapping potentials. State-dependent potentials create difficulties for coherent control of atoms, because internal states will couple to the noisier center-of-mass motion.

We can solve this issue using the multilevel structure of real atoms. In a multilevel-atom the scalar shift that the first excited atomic state experiences due to its coupling with the ground states can be counteracted by an opposite shift due to its coupling with further-excited states. The wavelength of light that creates the same shift for the ground and excited atomic states is called a magic wavelength. The idea of magic wavelengths, originally stated by Cho (1997) in the context of a possible systematic effect in precision measurements, shows that in a dipole trap the ground and excited electronic states of an atom can experience the same trapping potential, permitting coherent control of electronic transitions independent of the atomic center-of-mass motion (Arora and Sahoo, 2012; Ye et al., 2008). On the other hand, a vector light shift is not present when using only linearly polarized light. In the case of an ONF dipole trap, this can be achieved by canceling the longitudinal component of the electric field by using equal intensities and polarization for the counter-propagating trapping beams. A frequency difference between blue-detuned counter-propagating beams creates a traveling wave that averages to produce a uniform repulsive potential. These methods for implementing a state-insensitive dipole trap in an ONF are proposed in Le Kien et al. (2005a) and first implemented in Goban et al. (2012), Lacroûte et al. (2012), and Ding et al. (2012), where they trap Cs atoms

215 nm from the surface of a nanofiber, and suppress the differential scalar and vector light shifts by a factor of 250.

State-sensitive traps, which remove the degeneracy of the atomic hyperfine-sublevel, can be used for state preparation and interrogation of trapped atoms. For these applications, the vector light shift in ONF-based dipole trap can be particularly useful (Albrecht et al., 2016; Le Kien et al., 2013). Applying additional real or fictitious magnetic fields, the state dependence of the trapping potential can be controlled, providing a means to probe and to manipulate the motional state of the atoms in the trap by driving transitions between Zeeman sublevels. In particular, microwave sideband cooling can be implemented to bring the trapped atoms to the motional ground state (Albrecht et al., 2016). Another possible application is to use the fictitious magnetic field induced by a nanofiber-guided light field in conjunction with an external magnetic bias field to create an effective trapping potential for atoms around the ONF (Schneeweiss et al., 2014).

## 5.5 Ground State Coherence

Despite the difficulties that state-dependent trapping potentials impose in the coherent control of arbitrary atomic states, ground states can keep coherences for times suitable for some quantum information applications. Reitz et al. (2013) study the ground state coherence properties of Cs atoms in a nanofiber-based two-color dipole trap. Using microwave radiation to coherently drive the clock transition, they record Ramsey fringes as well as spin echo signals and infer a reversible dephasing time of $T_2^* = 0.6$ ms and an irreversible dephasing time of $T_2' = 3.7$ ms. Both time constants are limited by the finite initial temperature of the atomic ensemble and the heating rate, respectively.

## 5.6 Reflectivity

The regime of collective coupling with high OD provided by ONF-based dipole traps opens a door to the study of long-range interactions and, for example, the formation of so-called atomic mirrors (Chang et al., 2012) and the possible observation of self-crystallization (Chang et al., 2013; Grießer and Ritsch, 2013).

A high reflectivity due to an array of atoms requires the separation between the atoms to be commensurate with the resonance wavelength (Deutsch et al., 1995). In general, this condition is not satisfied for atoms trapped around an ONF. The counter-propagating red-detuned beams that provide the lattice configuration for trapping the atoms is usually far-detuned from resonance, helping to avoid losses from spontaneous emission. This imposes an extra difficulty for the implementation of optical reflectors that has to be circumvented.

Two groups have recently reported coherent backscattering of light (reflectivity) off trapped Cs atoms around a nanofiber. Sørensen et al. (2016) shows coherent Bragg scattering in a one-dimensional system realized by trapping atoms in the evanescent field of an ONF. They report nearly 12% power reflection of a probe light sent through the nanofiber interacting with only about 1000 cesium atoms, an enhancement of two orders of magnitude compared to the reflection from randomly positioned atoms. This is achieved by optically pumping the atoms to the appropriate state with a standing wave that imprints the necessary periodicity. The main mechanism that lowers the reflectivity seems to be the dephasing from the oscillations of the atoms in each individual trap that start at different places in the potentials. Fig. 14 shows the reflected intensity and its decay. There is a

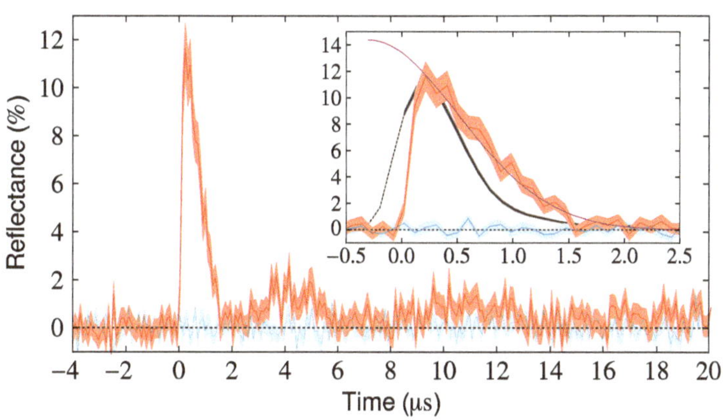


**Fig. 14** Reflectance off the atomic strings within a 2.8 MHz detection bandwidth (3 dB). The probe is tuned 8 MHz above atomic resonance and turned on at $t = 0$ with an independently measured rise time of 80 ns. *Blue curve*: unstructured atomic ensemble, average of 250 experiments. *Red curve*: structured ensemble, average of 200 experiments. The *shaded* regions signify the $1\sigma$ uncertainty interval with contributions from statistical averaging and a 5% input probe power fluctuation during the measurement. *Inset*: zoom on the first reflection peak with a fit to a Gaussian decay (*purple*) and the theoretical prediction (*black*) with uncertainty band given by the statistical averaging. *Reprinted fig. 2 with permission from Sørensen, H.L., Béguin J.-B., Kluge, K.W., Iakoupov, I., Sørensen, A.S., Müller, J.H., Polzik, E.S., Appel, J., 2016. Coherent backscattering of light o one-dimensional atomic strings. Phys. Rev. Lett. 117, 133604. Copyright (2016) by the American Physical Society.*

revival of the reflectivity about 4 μs after the structure pulse, showing a brief rephasing of some of the pumped atoms oscillating in the trap.

Corzo et al. (2016) also report Bragg reflection from atoms trapped on a nanofiber. By using an optical lattice with a period nearly commensurate with the resonant wavelength, realized by a pair of close-to-resonance red-detuned counter-propagating beams, they observe a reflectance of up to 75%. Each atom behaves as a partially reflecting mirror and an ordered chain of about 2000 atoms is sufficient to realize an efficient Bragg mirror. They report measurements of the reflection spectra as a function of the lattice period and the probe polarization (see Fig. 15).

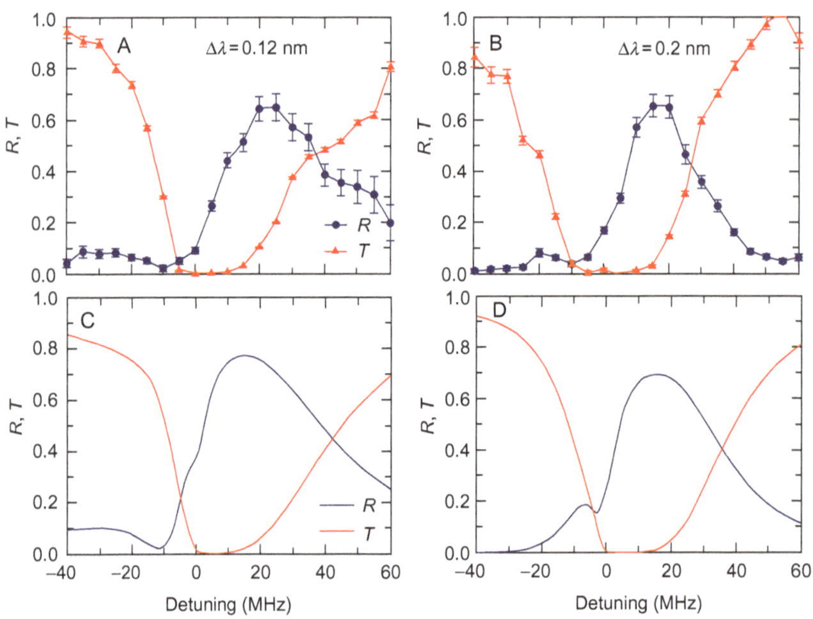


**Fig. 15** Reflection and transmission spectra for a probe quasi-linearly polarized along the *y*-direction. (A) and (B) Experimental results for $\Delta\lambda = 0.12$ nm and $\Delta\lambda = 0.2$ nm. (C) and (D) Simulated spectra for $N = 2000$ atoms, $\gamma_{1D}/\gamma_0 = 0.007$, and a filling factor of the lattice sites is 0.3. The coupling value and filling factor have been adjusted to fit the measured spectra reported in (A). *Reprinted fig. 3 with permission from Corzo, N.V., Gouraud, B., Chandra, A., Goban, A., Sheremet, A.S., Kupriyanov, D., Laurat, J., 2016. Large Bragg reflection from one-dimensional chains of trapped atoms near a nanoscale waveguide. Phys. Rev. Lett. 117, 133603. Copyright (2016) by the American Physical Society.*

## 5.7 Trapping With Higher-Order Modes

Allowing higher-order modes to propagate through an ONF enables new optical trapping geometries. These modes have a transversal intensity profile and a field polarization richer than the fundamental mode (see Section 2 and Fig. 4), facilitating modified transversal confinements of the atoms, usually further away from the ONF surface. Different modes also have different propagation constants, creating an interference pattern along the nanofiber as they propagate. This enables different possibilities for trapping patterns in the longitudinal direction (Sagué et al., 2008).

Phelan et al. (2013) investigate trapping geometries that combine the fundamental mode with one of the next lowest possible modes, namely the $HE_{21}$ mode. Counter-propagating, red-detuned $HE_{21}$ modes are combined with a blue-detuned $HE_{11}$ fundamental mode to form a potential in the shape of four intertwined spirals. By changing the polarization from circular to linear in each of the two counter-propagating $HE_{21}$ modes simultaneously, the 4-helix configuration can be transformed into a lattice configuration. Helical trapping geometries are also proposed for combinations of different polarizations of beams only in the fundamental $HE_{11}$ mode (Reitz and Rauschenbeutel., 2012).

The interaction between atoms and higher-order propagating modes can produce mode-conversion, the redistribution of power of an incoming field from one mode to another (Kumar et al., 2015a). The different propagation constant of the modes implies different momentum for each propagating mode. When a process of mode-conversion happens the differential change in momentum is compensated by a momentum transfer to the nearby atom. Based on this effect, recent papers have explored the possibility of using higher-order modes to control center-of-mass motion of trapped atoms (Ebongue et al., 2016) and trapped microparticles (Maimaiti et al., 2015) along the nanofiber.

## 5.8 Possible Heating Mechanisms

Atoms trapped around an ONF eventually leave the trap. The heating mechanisms that leads the atoms to escape have not been fully characterized. One contribution comes from elastic scattering events of the far-detuned trapping beams (Grimm et al., 2000), a well-known effect in dipole traps. A less familiar contribution comes from torsional modes of the ONF (Wuttke et al., 2013). The torsional modes couple to the polarization of all the guided fields. Since the trapping potential is sensitive to the polarization of the trapping beam, the torsional modes create a time-dependent perturbation of the potential. The frequencies of these modes are densely populated, and they are close enough to the trapping frequencies ($\sim$200 kHz) to consider parametric heating of the atoms as particularly strong mechanisms for losses. Torsional modes have been measured and characterized (Wuttke et al., 2013), and even optically excited (Fatemi et al., 2016; Fenton et al., 2016). However, it is necessary to suppress them or increase their frequencies to prevent direct and parametric heating of the atoms.

# 6. QUANTUM OPTICS AND QUANTUM INFORMATION WITH ONFS

Section 1 has motivated the importance of ONFs for quantum optics and quantum information. They provide a unique platform for many reasons, one being that the presence of an ONF modifies the structure of the vacuum electromagnetic field experienced by a nearby atom. This can modify atomic properties such as coherence or spontaneous emission rate, the latter attributed to the Purcell effect (Klimov and Ducloy, 2004; Le Kien and Hakuta, 2008, 2012; Le Kien et al., 2005b; Verhart et al., 2014).

Second, ONF guided modes create a preferential channel for the atomic radiation field. This is a virtually lossless channel that mediates interaction between distant atoms along the nanofiber, even when they are not trapped. When the interaction time is within the coherence of the system, they present a novel realization of many-body states in one-dimension with long-range interaction.

Third, atom–light interactions mediated by ONFs can be further increased by modifying the geometry of the platform. For example, through the introduction of Bragg mirrors in the optical fiber, the single atom cooperativity can reach the strong coupling regime.

Interfacing guided light with atomic ensembles has been actively pursued. Atoms around ONFs have been used as quantum information storage for memories and repeaters using the coherence of atoms similarly to EIT schemes.

## 6.1 Geometric Dependence of the Atom Light Coupling

We start by presenting calculations related to atom–light coupling (Section 1) relevant for quantum optics experiments. We look first to the emission enhancement coefficient, $\alpha$, of Eq. (7). The proximity of the nanofiber changes the spontaneous emission rate of the atom, and this change is reflected in this parameter (Klimov and Ducloy, 2004). It will serve as a proxy by which we can maximize atom–light coupling through nanofiber and optical trap design.

Fig. 16 shows a calculation based on the mode structure in Section 2 for a single Rb atom operating on the D2 line in the $HE_{11}$ evanescent mode of the ONF, following the work of Le Kien et al. (2005b). The four relevant parameters—emission enhancement $\alpha$, coupling coefficient $\beta$, cooperativity $C_1$, and Purcell factor—are plotted as a function of fiber radius and atom–surface distance.

The emission enhancement can be up to 20% larger than that of free-space near to the fiber surface according to Fig. 16A. As expected, it decreases to the

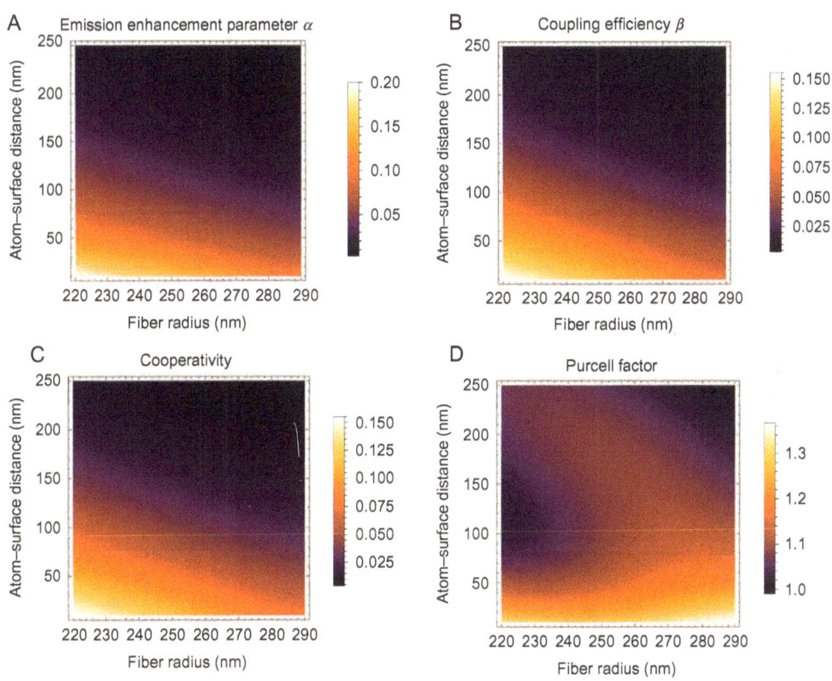


**Fig. 16** Atom–mode coupling parameters as a function of nanofiber radius and atom–surface distance for our ONF configuration from Grover et al. (2015). (A) Emission enhancement $\alpha$, (B) coupling efficiency, $\beta$ (C) single atom cooperativity $C_1$, and (D) Purcell factor $\alpha/\beta$. The Purcell factor, $C_1$, and $\beta$ depends strongly on the atomic dipole alignment relative to the ONF surface. Here we assume that the atomic dipole moment points in the same direction than the electric field of the circularly polarized fundamental mode of the ONF.

free-space value as the atom–fiber distance increases. For a typical nanofiber $\beta \ll 1$, it follows from Eq. (9) that $C_1 \approx \beta$, as a simple comparison between of the plots in Fig. 16B and C shows. Both the coupling efficiency and cooperativity display similar spatial dependence as the emission enhancement. Performing a global search we find that for a given mode there is an optimal ONF radius; reducing the radius decreases the coupling. The modification of the total spontaneous emission due to the presence of the nanofiber is given by the Purcell factor $(\alpha/\beta)$, plotted in Fig. 16D. Its nontrivial dependence on the geometrical parameters comes from the atomic dipole alignment and is discussed in more detail in Section 6.2.

## 6.2 Purcell Effect of Atoms Around an ONF

The Purcell effect is the change of the atomic spontaneous emission due to the modification of the vacuum electromagnetic field in the presence of an object (Purcell, 1946). As explained in Section 1.1, it tells us about the properties of atoms coupled to a given structure of modes, in our case radiated (reservoir) and guided modes. When only the coupling into the preferential guided mode is known, the measurement of a Purcell factor helps obtain information about the coupling of the system to the reservoir. This provides us with a tool to characterize coherence properties of the system.

Solano et al. (2017a) presents measurements of the radiative lifetime changes in the vicinity of an ONF due to the presence of a dielectric surface and to atomic dipole alignment effects. A dielectric surface modifies the dipole moment of a proximal atom, in turn changing its rate of spontaneous emission in a manner analogous to the Purcell effect. The alignment of the induced atomic dipoles from the excitation relative to the ONF surface determines if there is an enhancement or inhibition of the atomic decay rate. The difference can be as large as ~80% enhancement to ~20% inhibition for our current ONF parameters. Fig. 17 shows this dependence, suggesting that for a particular alignment atomic dipoles could couple differently to the environment (particularly nonguided modes). We use a sample of free cold atoms around the nanofiber. In this configuration, it is possible to probe closer distances than with trapped atoms and still avoid the complications of the van der Waals interaction. We measure enhancements and inhibitions of the spontaneous emission rate of ~10% and ~5%, respectively, depending on the alignment of the induced dipoles and averaged over the given atomic distribution. FDTD simulations are in agreement with the measurements.

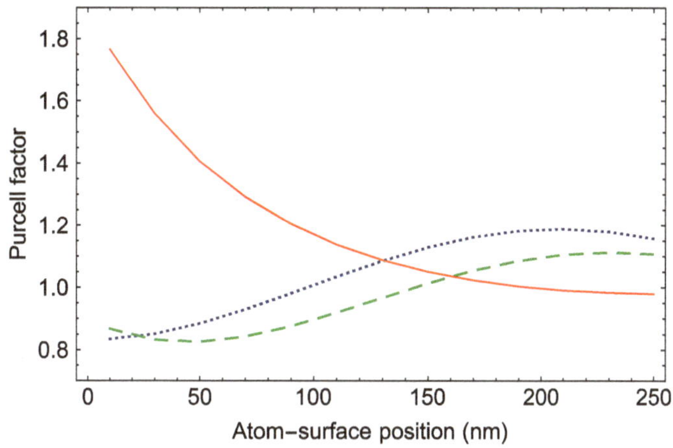


**Fig. 17** Purcell factor of an ONF of 237-nm radius as a function of the atom-to-surface distance for an atomic dipole aligned perpendicular to (*solid red*), parallel to (*dotted blue*), and along (*dashed green*) the ONF. The *curves* are calculated from finite-difference time-domain calculations.

## 6.3 ONF-Mediated Coherent Interactions

Chang et al. (2012) look at the ONF platform and find a way to convert it into a cavity QED system with atomic mirrors where it is possible to reach strong coupling. The reader is referred to Section 5 for the recent results on the reflectivity of atoms trapped around an ONF.

Grießer and Ritsch (2013) study the collective off-resonant scattering of coherent light by a cold gas in a 1D configuration such as that of atoms trapped around an ONF. This scattering induces long-range interactions via interference of light scattered by different particles that can propagate through the ONF. They study the possible phase transitions of the atomic spatial order as a function of the external pump intensity which can show self-sustained crystallization, similar to Domokos and Ritsch (2002), but the parameter space is rich and there can coexist multiple ordered states with distinct appearances. Similar studies have been carried for slightly different nanophotonic platforms (Chang et al., 2013; Eldredge et al., 2016; Holzmann et al., 2014). For a detailed discussion about the role of long-range interaction in self-organized systems the reader is referred to Ritsch et al. (2013).

Collective effects become readily visible for a strong enough coupling efficiency. By using optical pumping to positionally select the atoms such that their average distance is 30 nm from the ONF surface, Solano et al. (2017b) reach an efficiency of $\beta \sim 0.13$. This allows coherent interactions of atomic ensembles mediated by the nanofiber. Collective effects are evidenced by measuring the dependence of the atomic decay rate on the number of atoms, observing subradiant and superradiant behavior. In particular, superradiant decay times are observed for atoms separated by hundreds of resonant wavelengths, providing evidence of long-range coherent interactions.

## 6.4 Antibunching With Atoms Around an ONF

Intensity–intensity correlation measurements are another class of experiments that can be applied to untrapped atoms around nanofibers. Widely used throughout quantum optics, the intensity autocorrelation function, $g^{(2)}(\tau)$, measures correlations in the fluctuations of light intensity, e.g., the photon statistics, and can reveal both classical and quantum aspects of the light and its sources.

Using these methods, Nayak and Hakuta (2008) observe the presence of a single atom emitting photons into the guided mode of the fiber. A second paper enhances these results by varying the atom number in the vicinity of the nanofiber and by considering the direction of emission of the photons (Nayak et al., 2009). For photons emitted into the same direction, they observe a clear transition from antibunching to bunching at zero time delay as the number of atoms increases, as expected from the theory. When measuring the correlations of photons emitted into opposite ends of the nanofiber, however, they find that antibunching persists regardless of atom number, attributable to a geometric effect depending on the phase of the excitation field. An example in our work is in Fig. 10.

## 6.5 EIT and Optical Memories

Experiments have recently demonstrated the relevance of ONFs for quantum information and for general quantum optics. These experiments use the tight confinement of the field to induce nonlinear interactions of atoms. The experiments use atoms that are trapped, cold, or even at room temperature to observe EIT (Fleischhauer et al., 2005) and exploit these properties together with well-established quantum memory protocols to store information and retrieve it (Duan et al., 2001; Lvovsky et al., 2009).

The theoretical treatment of Le Kien and Rauschenbeutel (2015) provides a framework to study the propagation of guided light along an array of three-level atoms in the vicinity of an ONF under the condition of EIT. They study different polarization arrangements and find that when the input guided light is quasilinearly polarized along the major principal axis, the transmissivity and the group delay of the transmitted field substantially depend on the propagation direction of the probe field. Under the Bragg resonance condition, an array of atoms prepared in an appropriate internal state can transmit guided light polarized along the major principal axis in one specific direction. They state the importance of the longitudinal polarization component for these effects.

An experiment with warm ($\approx$373 K) atoms by Spillane et al. (2008) reports the observation of low-light level optical interactions through an ONF. They show V-type EIT with continuous wave powers below 10 nW in the D1 manifold of $^{85}$Rb. They were sensitive to the transit time dephasing given the small mode cross section of the ONF.

Two recent experiments with Rb use the transitions between the $5S_{1/2} \rightarrow 5P_{3/2} \rightarrow 5D_{5/2}$ to establish a ladder configuration for EIT. Jones et al. (2015) use an ONF (300 nm diameter) suspended in a warm rubidium vapor with the natural abundance of isotopes. Their signal and control fields propagate through the nanofiber. Transit-time broadening, given the room temperature velocities, is a significant EIT decoherence mechanism in this tightly confined waveguiding geometry. Their setup includes a way to keep the nanofiber warm to avoid deposits of Rb metal on the surface that prevent the propagation of light. They observe significant EIT and controlled polarization rotation using control-field powers of only a few nanowatts in this warm-atom nanofiber system.

Kumar et al. (2015c) use cold $^{87}$Rb atoms at about 200 μK from a MOT to realize EIT also in the ladder configuration with the two fields (probe at 780 nm) for the $5S_{1/2} \rightarrow 5P_{3/2}$ transition (coupling strong field at 556 nm) for the $5P_{3/2} \rightarrow 5D_{5/2}$. The fields propagate in opposite directions but overlap in the waist of the 350 nm diameter ONF. They are able to see the multilevel cascaded EIT. They observe and analyze the multipeak transparency spectra of the probe beam. They demonstrate all-optical-switching in the all-fiber system modulating the 80 nW coupling beam in the EIT region by observing the probe beam ($\approx$5 pW) that inputs with a constant intensity and shows modulation at the output.

Gouraud et al. (2015) work with cold Cs atoms around their 400-nm diameter ONF in a well-compensated magnetic environment. They use a $\Lambda$ system in the D2 line and involve the two hyperfine ground states. A narrow transparency window is observed for one of the transitions (signal), as Fig. 18 shows. The polarization control is important, and they use

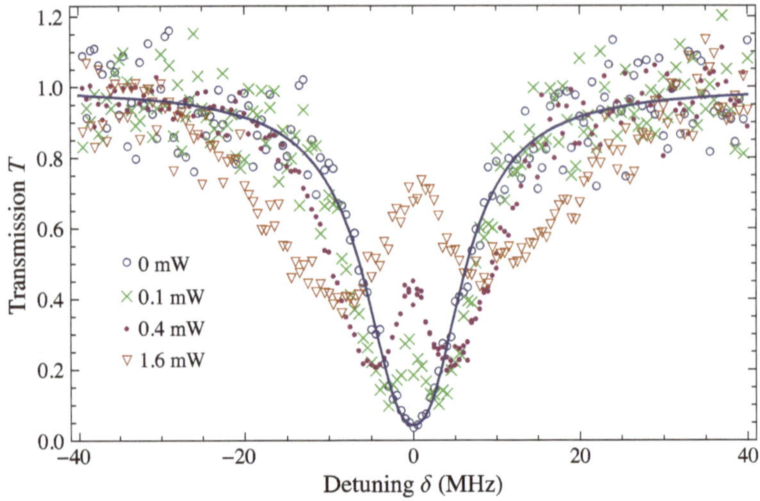


**Fig. 18** EIT for the guided light. The control is on resonance on the $|s\rangle \rightarrow |e\rangle$ transition while the signal is detuned by $\delta$ from the $|g\rangle \rightarrow |e\rangle$ resonance ($|s\rangle$, $|g\rangle$, and $|e\rangle$ are the two ground and the excited states, respectively, in $\Lambda$ configuration). Profiles are displayed as a function of $\delta$, for four values of the control power. *Reprinted fig. 2 with permission from Gouraud, B., Maxein, D., Nicolas, A., Morin, O., Laurat, J., 2015. Demonstration of a memory for tightly guided light in an optical nanober. Phys. Rev. Lett. 114, 180503, with Copyright (2015) by the American Physical Society.*

Rayleigh scattering to ensure its properties (see Section 2.5). The control field is a free-space beam propagating through the atomic cloud, while the signal from the upper hyperfine ground state ($F=4$) is coupled through the ONF. They interact with about 2000 atoms and achieve coherent storage of a pulse of light with intensities low enough to be single photons. They obtain an overall efficiency of retrieval of $10 \pm 0.5\%$ despite the limited delay available in the system from the EIT, the pulse cannot be contained entirely in the ensemble, and a leakage is observed before the control is switched off.

Sayrin et al. (2015a) use an ONF interfaced to an ensemble of trapped neutral Cs atoms to store fiber-guided light. They demonstrate EIT, slow light, and the storage of fiber-guided optical pulses in an ensemble of trapped cold atoms. They measure light group velocities of 50 m/s and store optical pulses at the single-photon level and retrieve them on demand in the fiber after 2 μs with an overall efficiency of $3.0 \pm 0.4\%$. They work on the D2 line of Cs with their probe between the low ($F = 3$, $m_F = 3$) hyperfine state detuned from the excited $6P_{3/2}, F' = 4$ state, while they use as control the transition that starts on the $6S_{1/2}$, $F=4$, $m_F=4$ in the presence of a magnetic field perpendicular to the propagation direction of the light on the ONF.

The control and probe beams copropagate through the ONF. The spectral transparency window can be as narrow as 30 kHz with a transparency of 60%. They store and retrieve classical light pulses that contained fewer than one photon on average and should be able to show storage of quantum states with nonclassical correlations in the future.

## 6.6 Cavity QED With Modified ONFs

One of the advantages of using ONFs for atom–light interaction is the high single atom cooperativity achieved due to the high mode confinement. However, there are methods to further increase the single atom cooperativity as stated in Section 1. A particular approach is to increase the number of times the light in the guided field interacts with the atomic dipole, as Eq. (2) describes. This can be achieved by an optical cavity that recycles the field, effectively increasing the atom–light interaction time. It is possible to fabricate in-fiber cavities by modifying the optical fiber geometry to create Bragg mirrors that reflect the guided mode. Modified-ONFs are promising platforms for interfacing atoms and guided light in the strong coupling regime.

### 6.6.1 *Photonic Crystal Cavities*

The evanescent field of a guided mode can be used not only for interacting light with single atoms but also with microscopic structures. A particular example is a photonic crystal in a Bragg mirror configuration. Sadgrove et al. (2013) and Yalla et al. (2014) implement an in-nanofiber Fabry–Perot-type cavity, coupling the evanescent field of the ONF guided mode to a photonic crystal structure by bringing them in contact. Fig. 19 shows an example of the narrow-bandwidth transmission achieved with this technique by Yalla et al. (2014). They observe quality factors better than 2000, offering a noninvasive method to create in-fiber cavities to enhance atom–light coupling.

### 6.6.2 *Bragg Resonators*

The refractive index of an optical fiber can be modified using nanosecond pulses of UV-light in a highly germanium doped photosensitive fiber. A spatially dependent refractive index can be imprinted in the fiber by modulating the intensity pattern of the writing laser. Lindner et al. (2009) use a Talbot interferometer to create a periodic modulation of the refractive index in a Bragg mirror configuration. Two Bragg reflectors produce an in-fiber Fabry–Perot-type optical cavity. Wuttke et al. (2012) and Kato and Aoki (2015)

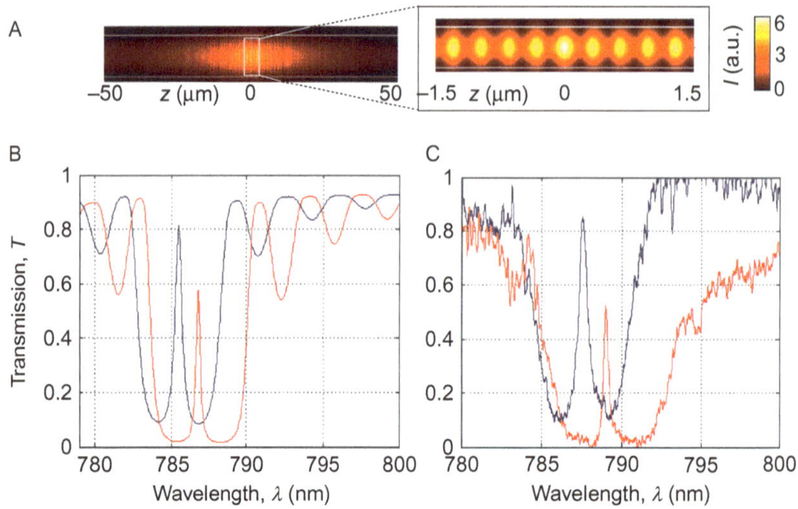


**Fig. 19** (A) Simulated electric field intensity inside the CPCC for the *y*-mode. The *inset* shows an expanded view of the central region of the cavity. *White horizontal lines* mark the position of the edges of the nanofiber. (B and C) Simulated and measured transmission spectra, respectively, for the *x*- (*blue line*) and *y*-modes (*red line*). *Reprinted fig. 2 with permission from Yalla, R., Sadgrove, M., Nayak, K.P., Hakuta, K., 2014. Cavity quantum electrodynamics on a nanofiber using a composite photonic crystal cavity. Phys. Rev. Lett. 113 (14), 143601, with Copyright (2014) by the American Physical Society.*

use an optical fiber with printed Bragg reflectors and taper it with the flame-brushing technique to integrate an ONF inside an in-fiber optical cavity. Fig. 20 shows the observation of vacuum Rabi splitting in this platform using a single trapped Cs atom in a state-insensitive trap. This transmission measurement is evidence of atom–light interaction in the strong coupling regime where the coherent coupling is larger than the incoherent losses through spontaneous emission or cavity loss.

Bragg reflectors can also be implemented at the ONF waist by making a periodic array of holes in the fused silica nanofiber, like photonic crystal cavities. This has been achieved by photon ablation using femtosecond pulses of UV-light (Nayak and Hakuta, 2012; Nayak et al., 2014). Using this configuration, Nayak et al. (2014) report the fabrication of in-fiber cavities with a finesse up to 500. Another approach has been to use a focused ion beam milling technique for drilling periodic nano-grooves in the ONF (Nayak et al., 2011). This is the same method used in Daly et al. (2014, 2016) to create a slot at the ONF waist to enhance the interaction of the guided mode with particles inside the carved space.

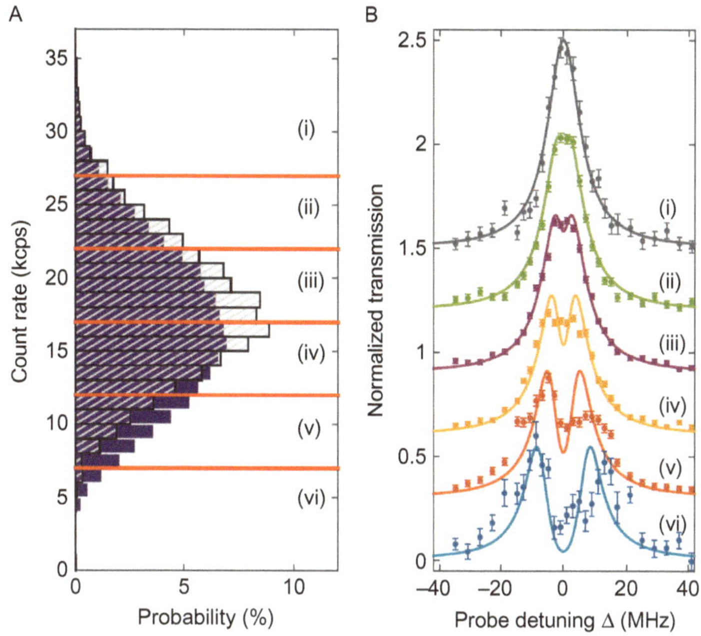


**Fig. 20** (A) Histogram of the transmission intensity of the detection probe with (*blue*) and without (*gray*) the optical molasses. The dark count of the detector is $1.4 \times 10^3$ counts per second (cps), and the measurement background noise including the dark count is about 10 kcps. (B) Transmission spectra as function of the probe detuning $\Delta$. Data sets and fits are normalized to the empty-cavity transmission and are vertically offset for clarity. The observed asymmetry in the spectra is presumably due to the effect of the probe pulse on the center-of-mass motion of the atom. Error bars are the standard error of the mean. The atom–cavity coupling rates $g$ for the fits are $2\pi \times (1.3; 1.9; 2.9; 4.3; 7.8)$ MHz for (ii) to (vi), respectively. *Reprinted fig. 2 with permission from Kato, S., Aoki, T., 2015. Strong coupling between a trapped single atom and an all-fiber cavity. Phys. Rev. Lett. 115, 093603, with Copyright (2015) by the American Physical Society.*

## 7. CHIRALITY IN NANOPHOTONIC SYSTEMS

The existence of longitudinal components of the electromagnetic field is well-established in ONFs (see Section 2 for a discussion of the polarization). Such polarizations also exist in the microwave regime and have been thoroughly used to make directional couplers and other important technological devices (Baden Fuller, 1986). There, the waveguides are accompanied by materials with the appropriate electro-magnetic properties, that together with the geometry of the guides ensure the directionality. This enables microwave isolators that protect delicate hardware pieces, but also direct information and even quantum information to sensitive detectors.

Atoms are excellent materials for manipulating the directionality of electro-magnetic waves through the chiral coupling of their momentum to the spin angular momentum of the light. ONFs with trapped atoms are becoming an important platform for chirality. Now it is possible to talk about chiral quantum optics, as the title of the recent review by Lodahl et al. (2017) is called (see Fig. 21 for an ONF setup). We point the interested reader to that reference as it covers this area in more detail and makes the important connection with many-body physics, as it may become possible to realize interesting Hamiltonians as suggested by Ramos et al. (2014) and many others. The mode structure of the ONF also opens the platform to the rich spin–orbit interactions of light (Bliokh et al., 2015). The intrinsic asymmetry of the problem is a new important tool in the harnessing of the atom–light interaction for further studies of quantum optics and quantum information.

## 7.1 Optical Control of Directionality in Photonic Systems

Realization of directionality through use of polarization selection rules, internal structure of the atoms, waveguide structures that spin–orbit couple the light, and chiral effects in nanophotonic systems has happened in different platforms: quantum dots in a cross waveguide, atoms in

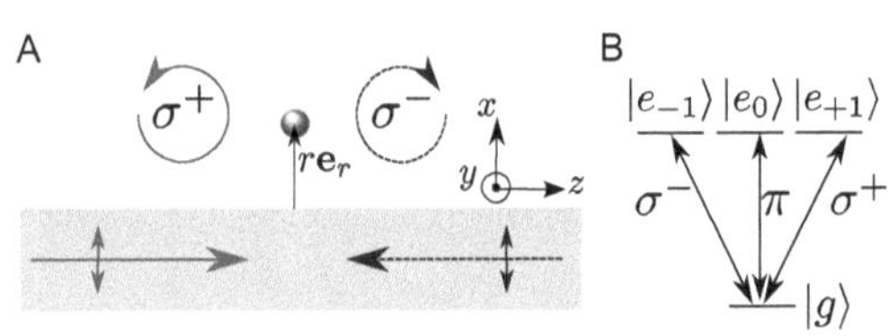


**Fig. 21** Chiral photons in evanescent fields coupled to spin-polarized atoms. (A) Polarization properties of the evanescent light field that surrounds an optical nanofiber (*gray*). A light field that propagates in the (+*z*) direction and whose main polarization axis (*double arrow*) is along the *x*-axis is almost fully $\sigma^+$ polarized (*green solid arrows*) in the ($y = 0$) plane. If it propagates in the ($-z$)direction, it is almost fully $\sigma^-$ polarized (*blue dashed arrows*). The quantization axis is chosen along *y*, i.e., orthogonal to the propagation direction. An atom (*light blue sphere*) placed at a distance *r* to the nanofiber surface couples to the evanescent field. (B) Relevant energy levels of the atom. The ground state $|g\rangle$ is coupled to the excited states $|e_{-1}\rangle$, $|e_0\rangle$, and $|e_1\rangle$ via $\sigma^-$, $\pi$, and $\sigma^+$ transitions, respectively. *Reprinted fig. 1 with permission from Sayrin, C., Junge, C., Mitsch, R., Albrecht, B., O'Shea, D., Schneeweiss, P., Volz, J., Rauschenbeutel, A., 2015b. Nanophotonic optical isolator controlled by the internal state of cold atoms. Phys. Rev. X 5, 041036.*

whispering-gallery-mode microresonator (WGM) and gold nanoparticles and atoms on an ONF. They are an important starting point for what promises to be an active area of research.

Crucial to the quantum dot work is first the highly polarized emission of its excitons using optical pumping, and second the emission properties of the quantum dots in a crossed-waveguide device with broken inversion symmetry. Together they enable optical control of the emission direction in Luxmoore et al. (2013).

Junge et al. (2013) realize in a WGM microresonator strong coupling between single atoms and nontransversal photons. They profit from the fundamental change that the longitudinal polarization brings to the interaction between light and matter in this resonator. Contrary to a macroscopic Fabry Perot, here the two propagation directions of the photons are correlated with two nearly orthogonal polarization states. As a consequence, counter-propagating photons are distinguishable by their polarization and cannot interfere destructively.

Shomroni et al. (2014) and Scheucher et al. (2016) use the directional coupling and its dependence on the internal atomic state to create a single atom switch and router. Helped by an optical circulator (a WGM) evanescently coupled to an ONF and to a single atom, they are able to change the transmission and reflection of a single photon propagating through the nanofiber by preparing the state of the atom with a single triggering photon. Applications like these boost the role of ONF in photonic-based quantum technologies.

## 7.2 Chirality in ONF Systems

Le Kien et al. (2004) states that under certain circumstances if the transverse component of the field is linearly polarized then the phase of the longitudinal component differs from that of the transverse components by $\pm\ \pi/2$ depending on the direction of propagation, as explained in Eq. (34) and Fig. 21. Due to this phase difference, the total electric field **E** rotates elliptically with time, in a plane parallel to the fiber axis $z$. This observation invites extending the tools used in the atom–light interaction in ONFs to include the longitudinal component of the field into the chiral domain. Mitsch et al. (2014) demonstrate the directional spontaneous emission of photons by trapped Cs atoms into an ONF controlling their asymmetric propagation direction with the excited state of the atomic emitters.

Petersen et al. (2014) control the flow of light with nanophotonic waveguides with the help of the spin–orbit coupling of light, and gold nanoparticles attached to an ONF. The nanoparticle breaks the mirror symmetry of the scattering of light realizing a chiral waveguide coupler in which the handedness of the incident light determines the propagation direction in the ONF. They achieve directionality of the scattering process up to 94% of the in-coupled light into a given direction.

Sayrin et al. (2015b) realize a nanophotonic optical isolator with high optical isolation at the single-photon level. The direction of the resulting optical isolator is controlled by the spin state of cold Cs atoms. The basic photonic and atomic structure is in Fig. 21. They perform measurements with an ensemble of trapped cold atoms where each atom is weakly coupled (a few percent for the reasons explained in Sections 4 and 5) to the ONF. To significantly enhance the coupling and be able to work with a single atom, they strongly couple the atom to a WGM bottle microresonator similar to what they use in Junge et al. (2013). They observe simultaneously high isolation and high forward transmission.

We expect in the near future increasing use for technology such as isolators, but also the system is set to address questions in cascaded quantum open systems (Carmichael, 1993; Gardiner, 1993) that have been open for more than two decades. Early work as that attempted by some of us (Gripp et al., 1995) could not reach the important couplings nor the directionality that the recent experiments have advanced, inviting exploration of the cascaded open systems and other questions related to chiral reservoirs for atoms around an ONF (Eldredge et al., 2016; Guimond et al., 2016; Pichler et al., 2015; Ramos et al., 2014).

## 8. CONCLUSIONS

ONF platforms with atoms are becoming an important player in the development of quantum optics and quantum information at the beginning of the 21st century. The structure has a simplicity and elegance that attracts many users. The large community that has grown using them has been able to control many of its technical challenges. Atoms are now routinely trapped around the nanofibers and provide robust platforms for quantum optics experiments. This would not have been possible without the many studies theoretical and experimental with atoms, both hot and cold, around the nanofibers. These arrays of atoms around the ONF not only enable the exploration of the many-body physics of condensed matter system but given the peculiarities and interactions that are available, may allow for the study of Hamiltonians that do not exist in current materials. The atom–atom interaction could in principle be infinite range, but it remains to be seen if the coherence times of the different ONF modes permit long-range interactions. This may be possible as atoms around an ONF could have a phase imprinted without concern of any boundary conditions. The guided mode of the nanofiber manages to propagate without diffraction, maintaining a tight cross-sectional area that can interact with many atoms along the way. Most work to date is with the fundamental mode, but there are many other modes available that also do not diffract and could be used. There are still some challenges and difficulties to be solved in the future, but the versatility of the three field polarizations available and the possibility to use the spin–orbit coupling of the light together with single atoms is an open avenue that will lead to many surprises and discoveries. We expect that parallel to the quantum optics and quantum information advances there will continue to be many technological applications that could have the potential to affect future areas of telecommunications. ONFs are a beautiful, versatile, and fun platform for future experiments in quantum optics.

## ACKNOWLEDGMENTS

This work has been supported by National Science Foundation of the United States (NSF) (PHY-1307416) and NSF Physics Frontier Center at the Joint quantum Institute (PHY-1430094), S.R. acknowledges support from Fulbright Foundation.

## REFERENCES

Albrecht, B., Meng, Y., Clausen, C., Dareau, A., Schneeweiss, P., Rauschenbeutel, A., 2016. Fictitious magnetic-field gradients in optical microtraps as an experimental tool for interrogating and manipulating cold atoms. Phys. Rev. A 94, 061401. http://dx.doi.org/10.1103/PhysRevA.94.061401.

Alton, D.J., Stern, N.P., Aoki, T., Lee, H., Ostby, E., Vahala, K.J., Kimble, H.J., 2011. Strong interactions of single atoms and photons near a dielectric boundary. Nat. Phys. 7 (2), 159–165. ISSN 1745-2473. http://dx.doi.org/10.1038/nphys1837.

Arora, B., Sahoo, B.K., 2012. State-insensitive trapping of Rb atoms: linearly versus circularly polarized lights. Phys. Rev. A 86 (3), 033416. ISSN 1050-2947. http://dx.doi.org/10.1103/PhysRevA.86.033416.

Baden Fuller, A.J., 1986. Ferrites at Microwave Frequencies. The Institution of Engineering and Technology, London.

Bajcsy, M., Hofferberth, S., Balic, V., Peyronel, T., Hafezi, M., Zibrov, A.S., Vuletic, V., Lukin, M.D., 2009. Efficient all-optical switching using slow light within a hollow fiber. Phys. Rev. Lett. 102, 203902. http://dx.doi.org/10.1103/PhysRevLett.102.203902.

Béguin, J.-B., Bookjans, E., Christensen, S., Sørensen, H., Müller, J., Polzik, E., Appel, J., 2014. Generation and detection of a sub-Poissonian atom number distribution in a one-dimensional optical lattice. Phys. Rev. Lett. 113 (26), 263603. ISSN 0031-9007. http://dx.doi.org/10.1103/PhysRevLett.113.263603.

Bilodeau, F., Hill, K.O., Faucher, S., Johnson, D.C., 1988. Low-loss highly overcoupled: fabrication and sensitivity to external pressure. J. Lightwave Technol. 6, 1476–1482. ieeexplore.ieee.org/iel1/50/417/00007904.pdf.

Birks, T., Li, Y., 1992. The shape of fiber tapers. J. Lightwave Technol. 10 (4), 432–438. ISSN 0733-8724. http://dx.doi.org/10.1109/50.134196.

Birks, T.A., Knight, J.C., Dimmick, T.E., 2000. High-resolution measurement of the fiber diameter variations using whispering gallery modes and no optical alignment. IEEE Photon. Technol. Lett. 12 (2), 182–183. ISSN 1041-1135. http://dx.doi.org/10.1109/68.823510.

Blais, A., Huang, R., Wallraff, A., Girvin, S., Schoelkopf, R., 2004. Cavity quantum electrodynamics for superconducting electrical circuits: an architecture for quantum computation. Phys. Rev. A 69 (6), 62320. http://journals.aps.org/pra/abstract/10.1103/PhysRevA.69.062320.

Bliokh, K.Y., Rodriguez-Fortuno, F.J., Nori, F., Zayats, A.V., 2015. Spin-orbit interactions of light. Nat. Photon 9 (12), 796–808. http://dx.doi.org/10.1038/nphoton.2015.201.

Blow, K.J., Loudon, R., Phoenix, S.J.D., Shepherd, T.J., 1990. Continuum fields in quantum optics. Phys. Rev. A 42, 4102–4114. http://dx.doi.org/10.1103/PhysRevA.42.4102.

Bonifacio, R., Lugiato, L.A. (Eds.) 1982. Dissipative Systems in Quantum Optics. Springer Series Topics in Current Physics, vol. 27. Springer-Verlag, Berlin, Heidelberg, New York.

Brambilla, G., 2010. Optical fibre nanowires and microwires: a review. J. Opt. 12 (4), 043001. ISSN 2040-8978. http://dx.doi.org/10.1088/2040-8978/12/4/043001.

Calajó, G., Ciccarello, F., Chang, D., Rabl, P., 2016. Atom-field dressed states in slow-light waveguide QED. Phys. Rev. A 93, 033833. http://dx.doi.org/10.1103/PhysRevA.93.033833.

Carmichael, H.J., 1993. Quantum trajectory theory for cascaded open systems. Phys. Rev. Lett. 70, 2273–2276. http://dx.doi.org/10.1103/PhysRevLett.70.2273.

Chang, R.K., Campillo, A.J. (Eds.) 1996. Optical Processes in Microcavities. Advanced Series in Applied Physics, vol. 3. World Scientific, Singapore. http://www.worldscientific.com/worldscibooks/10.1142/2828?t=aboutBook.

Chang, D.E., Jiang, L., Gorshkov, A.V., Kimble, H.J., 2012. Cavity QED with atomic mirrors. New J. Phys. 14, 063003. http://iopscience.iop.org/1367-2630/14/6/063003.

Chang, D.E., Cirac, J.I., Kimble, H.J., 2013. Self-organization of atoms along a nanophotonic waveguide. Phys. Rev. Lett. 110, 113606. http://journals.aps.org/prl/abstract/10.1103/PhysRevLett.110.113606.

Chen, G.Y., Ding, M., Newson, T.P., Brambilla, G., 2013. A review of microfiber and nanofiber based optical sensors. Open Opt. J. 7 (Suppl. 1, M3), 32–57. https://benthamopen.com/ABSTRACT/TOOPTSJ-7-32.

Cho, D., 1997. Analogous Zeeman effect from the tensor polarizability in alkali atoms. J. Korean Phys. Soc. 30, 373. http://old.kps.or.kr/jkps/abstract_view.asp?articleuid=A7132FDE-8732-42EA-9C20-907094FA6BEC&globalmenu=3&localmenu=10.

Corzo, N.V., Gouraud, B., Chandra, A., Goban, A., Sheremet, A.S., Kupriyanov, D., Laurat, J., 2016. Large Bragg reflection from one-dimensional chains of trapped atoms near a nanoscale waveguide. Phys. Rev. Lett. 117, 133603. http://dx.doi.org/10.1103/PhysRevLett.117.133603.

Daly, M., Truong, V.G., Phelan, C.F., Deasy, K., Nic Chormaic, S., 2014. Nanostructured optical nanofibres for atom trapping. New J. Phys. 16 (5), 053052. http://stacks.iop.org/1367-2630/16/i=5/a=053052.

Daly, M., Truong, V., Nic Chormaic, S., 2016. Evanescent field trapping of nanoparticles using nanostructured ultrathin optical fibers. Opt. Express 24 (13), 14470–14482. http://dx.doi.org/10.1364/OE.24.014470.

Das, M., Shirasaki, A., Nayak, K., Morinaga, M., Le Kien, F., Hakuta, K., 2010. Measurement of fluorescence emission spectrum of few strongly driven atoms using an optical nanofiber. Opt. Express 18 (16), 17154–17164. https://www.osapublishing.org/oe/abstract.cfm?uri=oe-18-16-17154.

Dawkins, S., Mitsch, R., Reitz, D., Vetsch, E., Rauschenbeutel, A., 2011. Dispersive optical interface based on nanofiber-trapped atoms. Phys. Rev. Lett. 107, 243601. http://link.aps.org/doi/10.1103/PhysRevLett.107.243601.

Deutsch, I.H., Spreeuw, R.J.C., Rolston, S.L., Phillips, W.D., 1995. Photonic band gaps in optical lattices. Phys. Rev. A 52, 1394–1410. http://dx.doi.org/10.1103/PhysRevA.52.1394.

Dimmick, T.E., Kakarantzas, G., Birks, T.A., Russell, P.S., 1999. Carbon dioxide laser fabrication of fused-fiber couplers and tapers. Appl. Opt. 38 (33), 6845–6848. ISSN 0003-6935. http://www.ncbi.nlm.nih.gov/pubmed/18324225.

Ding, L., Belacel, C., Ducci, S., Leo, G., Favero, I., 2010. Ultralow loss single-mode silica tapers manufactured by a microheater. Appl. Opt. 49 (13), 2441–2445. http://dx.doi.org/10.1364/AO.49.002441. http://ao.osa.org/abstract.cfm?URI=ao-49-13-2441.

Ding, D., Goban, A., Choi, K.S., Kimble, H.J., 2012. Corrections to our results for optical nanofiber traps in Cesium. ArXiv:1212.4941 [physics, physics:quant-ph], arXiv: 1212.4941, http://arxiv.org/abs/1212.4941.

Domokos, P., Ritsch, H., 2002. Collective cooling and self-organization of atoms in a cavity. Phys. Rev. Lett. 89, 253003. http://dx.doi.org/10.1103/PhysRevLett.89.253003.

Duan, L.M., Lukin, M.D., Cirac, J.I., Zoller, P., 2001. Long-distance quantum communication with atomic ensembles and linear optics. Nature 414 (6862), 413–418. ISSN 0028-0836. http://dx.doi.org/10.1038/35106500.

Ebongue, C.A., Holzmann, D., Ostermann, S., Ritsch, H., 2016. Generating a stationary infinite range tractor force via a multimode optical fiber. ArXiv:1611.05207 [physics, physics:quant-ph], arXiv: 1611.05207, http://arxiv.org/abs/1611.05207.

Eldredge, Z., Solano, P., Chang, D., Gorshkov, A.V., 2016. Self-organization of atoms coupled to a chiral reservoir. Phys. Rev. A 94, 053855. http://dx.doi.org/10.1103/PhysRevA.94.053855.

Erikson, W.L., Singh, S., 1994. Polarization properties of Maxwell-Gaussian laser beams. Phys. Rev. E 49, 5778–5786. http://dx.doi.org/10.1103/PhysRevE.49.5778.

Fang, Y.-L.L., Baranger, H.U., 2015. Waveguide QED: power spectra and correlations of two photons scattered off multiple distant qubits and a mirror. Phys. Rev. A 91, 053845. http://dx.doi.org/10.1103/PhysRevA.91.053845.

Fatemi, F.K., 2011. Cylindrical vector beams for rapid polarization-dependent measurements in atomic systems. Opt. Express 19 (25), 25143–25150. http://dx.doi.org/10.1364/OE.19.025143.

Fatemi, F.K., Solano, P., Orozco, L.A., Rolston, S.L., 2016. Polarimetric measurement of the motion of trapped atoms around a nanofiber. In: APS (Ed.), Bull. Am. Phys. Soc. Vol. 61 of 47th Annual Meeting of the APS Division of Atomic, Molecular and Optical Physics, vol. 61, p. G6.00009, http://meetings.aps.org/Meeting/DAMOP16/Session/G6.9.

Fatemi, F.K., Hoffman, J.E., Solano, P., Fenton, E.F., Beadie, G., Rolston, S.L., Orozco, L.A., 2017. Near-field modal interference in optical nano-fibers for sub-angstrom radius sensitivity. Optica 4, 157. https://doi.org/10.1364/OPTICA.4.000157.

Fenton, E., Khan, A., Patterson, B., Solano, P., Rolston, S.L., Orozco, L.A., Fatemi, F.K., 2016. Torsional modes of a nanofiber: polarimetric excitation and read out. In: Front. Opt. 2016. Optical Society of America, p. FTu3I.6, http://www.osapublishing.org/abstract.cfm?URI=FiO-2016-FTu3I.6.

Fleischhauer, M., Imamoglu, A., Marangos, J.P., 2005. Electromagnetically induced transparency: optics in coherent media. Rev. Mod. Phys. 77, 633–673. http://dx.doi.org/10.1103/RevModPhys.77.633.

Foot, C.J., 2005. Atomic Physics. Oxford University Press, Oxford.
Frawley, M.C., Nic Chormaic, S., Minogin, V.G., 2012a. The van der Waals interaction of an atom with the convex surface of a nanocylinder. Phys. Scripta 85 (5), 058103. http://dx.doi.org/10.1088/0031-8949/85/05/058103.
Frawley, M.C., Petcu-Colan, A., Truong, V.G., Nic Chormaic, S., 2012b. Higher order mode propagation in an optical nanofiber. Opt. Commun. 285 (23), 4648–4654. ISSN 00304018. http://dx.doi.org/10.1016/j.optcom.2012.05.016.
Fujiwara, M., Toubaru, K., Takeuchi, S., 2011. Optical transmittance degradation in tapered fibers. Opt. Express 19 (9), 8596–8601. ISSN 1094-4087. http://www.ncbi.nlm.nih.gov/pubmed/21643110.
Fujiwara, M., Noda, T., Tanaka, A., Toubaru, K., Zhao, H.Q., Takeuchi, S., 2012. Coupling of ultrathin tapered fibers with high-Q microsphere resonators at cryogenic temperatures and observation of phase-shift transition from undercoupling to overcoupling. Opt. Express 20, 19545–19553. https://www.osapublishing.org/oe/abstract.cfm?URI=oe-20-17-19545.
Garcia-Fernandez, R., Alt, W., Bruse, F., Dan, C., Karapetyan, K., Rehband, O., Stiebeiner, A., Wiedemann, U., Meschede, D., Rauschenbeutel, A., 2011. Optical nanofibers and spectroscopy. Appl. Phys. B 105 (1), 3–15. ISSN 1432-0649. http://link.springer.com/article/10.1007%2Fs00340-011-4730-x.
Gardiner, C.W., 1993. Driving a quantum system with the output field from another driven quantum system. Phys. Rev. Lett. 70, 2269–2272. http://dx.doi.org/10.1103/PhysRevLett.70.2269.
Ghosh, S., Bhagwat, A.R., Renshaw, C.K., Goh, S., Gaeta, A.L., Kirby, B.J., 2006. Low-light-level optical interactions with rubidium vapor in a photonic band-gap fiber. Phys. Rev. Lett. 97, 023603. http://dx.doi.org/10.1103/PhysRevLett.97.023603.
Gibbs, H.M., Khitrova, G., Koch, S.W., 2011. Exciton-polariton light-semiconductor coupling effects. Nat. Photon. 5 (5). http://dx.doi.org/10.1038/nphoton.2011.15273–273.
Goban, A., Choi, K.S., Alton, D.J., Ding, D., Lacroûte, C., Pototschnig, M., Thiele, T., Stern, N.P., Kimble, H.J., 2012. Demonstration of a state-insensitive, compensated nanofiber trap. Phys. Rev. Lett. 109, 033603. http://dx.doi.org/10.1103/PhysRevLett.109.033603.
Goban, A., Hung, C.-L., Yu, S.-P., Hood, J., Muniz, J., Lee, J., Martin, M., McClung, A., Choi, K., Chang, D., Painter, O., Kimble, H., 2014. Atom–light interactions in photonic crystals. Nat. Commun. 5. http://dx.doi.org/10.1038/ncomms4808.
Goban, A., Hung, C.-L., Hood, J.D., Yu, S.-P., Muniz, J.A., Painter, O., Kimble, H.J., 2015. Superradiance for atoms trapped along a photonic crystal waveguide. Phys. Rev. Lett. 115, 063601. http://dx.doi.org/10.1103/PhysRevLett.115.063601.
Golla, A., Chalopin, B., Bader, M., Harder, I., Mantel, K., Maiwald, R., Lindlein, N., Sondermann, M., Leuchs, G., 2012. Generation of a wave packet tailored to efficient free space excitation of a single atom. Eur. Phys. J. D 66 (7), 190. ISSN 1434-6060. http://dx.doi.org/10.1140/epjd/e2012-30293-y.
Gouraud, B., Maxein, D., Nicolas, A., Morin, O., Laurat, J., 2015. Demonstration of a memory for tightly guided light in an optical nanofiber. Phys. Rev. Lett. 114, 180503. http://dx.doi.org/10.1103/PhysRevLett.114.180503.
Grießer, T., Ritsch, H., 2013. Light-induced crystallization of cold atoms in a 1D optical trap. Phys. Rev. Lett. 111, 055702. http://dx.doi.org/10.1103/PhysRevLett.111.055702.
Grimm, R., Weidemuller, M., Ovchinnikov, Y.B., 2000. Optical dipole traps for neutral atoms. In: Bederson, B., Walther, H. (Eds.), Advances in Atomic, Molecular, and Optical Physics, vol. 42. Academic Press, Boston, pp. 95–170. http://www.sciencedirect.com/science/article/pii/S1049250X0860186X.
Gripp, J., Mielke, S.L., Orozco, L.A., 1995. Cascaded optical cavities with two-level atoms: Steady state. Phys. Rev. A 51, 4974–4981. http://dx.doi.org/10.1103/PhysRevA.51.4974.

Groblacher, S., Hammerer, K., Vanner, M.R., Aspelmeyer, M., 2009. Observation of strong coupling between a micromechanical resonator and an optical cavity field. Nature 460 (7256), 724–727. ISSN 0028-0836. http://dx.doi.org/10.1038/nature08171.

Grover, J.A., 2015. Atom-Trapping And Photon-Counting Experiments With Optical Nanofibers (Ph.D. thesis). University of Maryland College Park. http://drum.lib.umd.edu/handle/1903/16638.

Grover, J.A., Solano, P., Orozco, L.A., Rolston, S.L., 2015. Photon-correlation measurements of atomic-cloud temperature using an optical nanofiber. Phys. Rev. A 92, 013850. http://dx.doi.org/10.1103/PhysRevA.92.013850.

Guimond, P.-O., Pichler, H., Rauschenbeutel, A., Zoller, P., 2016. Chiral quantum optics with v-level atoms and coherent quantum feedback. Phys. Rev. A 94, 033829. http://dx.doi.org/10.1103/PhysRevA.94.033829.

Hafezi, M., Kim, Z., Rolston, S.L., Orozco, L.A., Lev, B.L., Taylor, J.M., 2012. Atomic interface between microwave and optical photons. Phys. Rev. A 85 (2), 020302. ISSN 1050-2947. http://dx.doi.org/10.1103/PhysRevA.85.020302. http://link.aps.org/doi/10.1103/PhysRevA.85.020302.

Haroche, S., 2013. Nobel lecture: controlling photons in a box and exploring the quantum to classical boundary. Rev. Mod. Phys. 85 (3), 1083–1102. ISSN 0034-6861. http://dx.doi.org/10.1103/RevModPhys.85.1083.

Haroche, S., Raimond, J.-M., 2006. Exploring the Quantum: Atoms, Cavities, and Photons. Oxford University Press, Oxford.

Hendrickson, S.M., Lai, M.M., Pittman, T.B., Franson, J.D., 2010. Observation of two-photon absorption at low power levels using tapered optical fibers in rubidium vapor. Phys. Rev. Lett. 105, 173602. http://dx.doi.org/10.1103/PhysRevLett.105.173602.

Hétet, G., Slodička, L., Hennrich, M., Blatt, R., 2011. Single atom as a mirror of an optical cavity. Phys. Rev. Lett. 107 (13), 133002. ISSN 0031-9007. http://dx.doi.org/10.1103/PhysRevLett.107.133002.

Heugel, S., Chalopin, B., Sondermann, M., Leuchs, G., 2012. Collecting more than half the fluorescence photons from a single ion. Phys. Rev. A 86, 043431. ISSN 1094-1622. http://pra.aps.org/abstract/PRA/v86/i4/e043431.

Hinds, E.A., 1994. Perturbative cavity quantum electrodynamics. In: Berman, P.R. (Ed.), Advances in Atomic, Molecular, and Optical Physics, Supplement 2, Cavity Quantum Electrodynamics. Academic Press, Boston, pp. 1–56.

Hoffman, J.E., 2014. Optical Nanofiber Fabrication and Analysis Towards Coupling Atoms to Superconducting Qubits (Ph.D. thesis). University of Maryland College Park.

Hoffman, J.E., Grover, J.A., Kim, Z., Wood, A.K., Anderson, J.R., Dragt, A.J., Hafezi, M., Lobb, C.J., Orozco, L.A., Rolston, S.L., Taylor, J.M., Vlahacos, C.P., Wellstood, F.C., 2011. Atoms Talking to SQUIDs. Rev. Mex. Fís. S 57, 1. http://arxiv.org/abs/1108.4153v2.

Hoffman, J.E., Ravets, S., Grover, J.A., Solano, P., Kordell, P.R., Wong-Campos, J.D., Orozco, L.A., Rolston, S.L., 2014. Ultrahigh transmission optical nanofibers. AIP Adv. 4 (6), 067124. ISSN 2158-3226. http://dx.doi.org/10.1063/1.4879799.

Hoffman, J.E., Fatemi, F.K., Beadie, G., Rolston, S.L., Orozco, L.A., 2015. Rayleigh scattering in an optical nanofiber as a probe of higher-order mode propagation. Optica 2 (5), 416–423. http://dx.doi.org/10.1364/OPTICA.2.000416.

Holzmann, D., Sonnleitner, M., Ritsch, H., 2014. Self-ordering and collective dynamics of transversely illuminated point-scatterers in a 1D trap. Eur. Phys. J. D 68, 352. http://dx.doi.org/10.1140/epjd/e2014-50692-2.

Hood, J.D., Goban, A., Asenjo-Garcia, A., Lu, M., Yu, S.-P., Chang, D.E., Kimble, H.J., 2016. Atom–atom interactions around the band edge of a photonic crystal waveguide. Proc. Natl. Acad. Sci. 113 (38), 10507–10512. http://dx.doi.org/10.1073/pnas.1603788113.

Jones, D.E., Franson, J.D., Pittman, T.B., 2014. Saturation of atomic transitions using subwavelength diameter tapered optical fibers in rubidium vapor. J. Opt. Soc. Am. B 31 (8), 1997–2001. http://dx.doi.org/10.1364/JOSAB.31.001997.

Jones, D.E., Franson, J.D., Pittman, T.B., 2015. Ladder-type electromagnetically induced transparency using nanofiber-guided light in a warm atomic vapor. Phys. Rev. A 92, 043806. http://dx.doi.org/10.1103/PhysRevA.92.043806.

Junge, C., O'Shea, D., Volz, J., Rauschenbeutel, A., 2013. Strong coupling between single atoms and nontransversal photons. Phys. Rev. Lett. 110, 213604. http://dx.doi.org/10.1103/PhysRevLett.110.213604.

Kakarantzas, G., Birks, T.A., Russell, P.S.J., 2002. Structural long-period gratings in photonic crystal fibers. Opt. Lett. 27, 1013–1015. https://doi.org/10.1364/OL.27.001013.

Kao, C.K., 2010. Nobel lecture: sand from centuries past: send future voices fast. Rev. Mod. Phys. 82, 2299–2303. http://dx.doi.org/10.1103/RevModPhys.82.2299.

Kao, K.C., Hockham, G.A., 1966. Dielectric-fibre surface waveguides for optical frequencies. Proc. IET 113 (7), 1151–1158. https://doi.org/10.1049/piee.1966.0189.

Kato, S., Aoki, T., 2015. Strong coupling between a trapped single atom and an all-fiber cavity. Phys. Rev. Lett. 115, 093603. http://dx.doi.org/10.1103/PhysRevLett.115.093603.

Kbashi, H.J., 2012. Fabrication of submicron-diameter and taper fibers using chemical etching. J. Mater. Sci. Technol. 28, 308–312. www.sciencedirect.com/science/article/pii/S1005030212600590.

Keloth, J., Sadgrove, M., Yalla, R., Hakuta, K., 2015. Diameter measurement of optical nanofibers using a composite photonic crystal cavity. Opt. Lett. 40 (17), 4122–4125. http://dx.doi.org/10.1364/OL.40.004122.

Kim, Z., Vlahacos, C.P., Hoffman, J.E., Grover, J.A., Voigt, K.D., Cooper, B.K., Ballard, C.J., Palmer, B.S., Hafezi, M., Taylor, J.M., Anderson, J.R., Dragt, A.J., Lobb, C.J., Orozco, L.A., Rolston, S.L., Wellstood, F.C., 2011. Thin-film superconducting resonator tunable to the ground-state hyperfine splitting of 87Rb. AIP Adv. 1 (4), 042107. ISSN 21583226. http://dx.doi.org/10.1063/1.3651466.

Kimble, H.J., 1994. Structure and dynamics in cavity quantum electrodynamics. In: Berman, P.R. (Ed.), Advances in Atomic, Molecular, and Optical Physics, Supplement 2, Cavity Quantum Electrodynamics. Academic Press, Boston, pp. 203–267.

Kimble, H.J., 1998. Strong interactions of single atoms and photons in cavity QED. Phys. Scripta T76 (1), 127. ISSN 0031-8949. http://dx.doi.org/10.1238/Physica.Topical.076a00127.

Kimble, H.J., 2008. The quantum internet. Nature 453 (7198), 1023–1030. ISSN 1476-4687. http://dx.doi.org/10.1038/nature07127.

Klimov, V.V., Ducloy, M., 2004. Spontaneous emission rate of an excited atom placed near a nanofiber. Phys. Rev. A 69, 013812. http://dx.doi.org/10.1103/PhysRevA.69.013812.

Knight, J.C., Cheung, G., Jacques, F., Birks, T.A., 1997. Phase-matched excitation of whispering-gallery-mode resonances by a fiber taper. Opt. Lett. 22, 1129–1131. https://doi.org/10.1364/OL.22.001129.

Kumar, R., Gokhroo, V., Deasy, K., Maimaiti, A., Frawley, M.C., Phelan, C., Nic Chormaic, S., 2015a. Interaction of laser-cooled $^{87}$Rb atoms with higher order modes of an optical nanofibre. New J. Phys. 17 (1), 013026. ISSN 1367-2630. http://dx.doi.org/10.1088/1367-2630/17/1/013026.

Kumar, R., Gokhroo, V., Deasy, K., Nic Chormaic, S., 2015b. Autler-Townes splitting via frequency up-conversion at ultralow-power levels in cold $^{87}$Rb atoms using an optical nanofiber. Phys. Rev. A 91, 053842. http://dx.doi.org/10.1103/PhysRevA.91.053842.

Kumar, R., Gokhroo, V., Nic Chormaic, S., 2015c. Multi-level cascaded electromagnetically induced transparency in cold atoms using an optical nanofibre interface. New J. Phys. 17 (12), 123012. http://stacks.iop.org/1367-2630/17/i=12/a=123012.

Kurizki, G., Bertet, P., Kubo, Y., Mølmer, K., Petrosyan, D., Rabl, P., Schmiedmayer, J., 2015. Quantum technologies with hybrid systems. Proc. Natl. Acad. Sci. 112 (13), 3866–3873. http://dx.doi.org/10.1073/pnas.1419326112.

Lacroûte, C., Choi, K.S., Goban, A., Alton, D.J., Ding, D., Stern, N.P., Kimble, H.J., 2012. A state-insensitive, compensated nanofiber trap. New J. Phys. 14, 023056. http://iopscience.iop.org/article/10.1088/1367-2630/14/2/023056/pdf.

Lambelet, P., Sayah, A., Pfeffer, M., Philipona, C., Marquis-Weible, F., 1998. Chemically etched fiber tips for near-field optical microscopy: a process for smoother tips. Appl. Opt. 37 (31), 7289–7292. http://dx.doi.org/10.1364/AO.37.007289.

Lax, M., Louisell, W.H., McKnight, W.B., 1975. From Maxwell to paraxial wave optics. Phys. Rev. A 11, 1365–1370. http://dx.doi.org/10.1103/PhysRevA.11.1365.

Le Kien, F., Hakuta, K., 2008. Effect of the orientation of a weak magnetic field on the radiative decay of a cesium atom near a nanofiber. Phys. Rev. A 78, 063803. http://dx.doi.org/10.1103/PhysRevA.78.063803.

Le Kien, F., Hakuta, K., 2012. Cooperative enhancement of channeling of emission from atoms into a nanofiber. Adv. Nat. Sci.: Nanosci. Nanotechnol. 3, 035001. http://iopscience.iop.org/article/10.1088/2043-6262/3/3/035001/pdf.

Le Kien, F., Rauschenbeutel, A., 2015. Electromagnetically induced transparency for guided light in an atomic array outside an optical nanofiber. Phys. Rev. A 91, 053847. http://dx.doi.org/10.1103/PhysRevA.91.053847.

Le Kien, F., Balykin, V.I., Hakuta, K., 2004. Atom trap and waveguide using a two-color evanescent light field around a subwavelength-diameter optical fiber. Phys. Rev. A 70, 063403. http://dx.doi.org/10.1103/PhysRevA.70.063403.

Le Kien, F., Liang, J., Hakuta, K., Balykin, V., 2004. Field intensity distributions and polarization orientations in a vacuum-clad subwavelength-diameter optical fiber. Opt. Commun. 242 (4–6), 445–455. ISSN 0030-4018. http://dx.doi.org/10.1016/j.optcom.2004.08.044.

Le Kien, F., Balykin, V., Hakuta, K., 2005a. State-insensitive trapping and guiding of cesium atoms using a two-color evanescent field around a subwavelength-diameter fiber. J. Phys. Soc. Jpn 74 (3), 910–917. ISSN 0031-9015. http://dx.doi.org/10.1143/JPSJ.74.910.

Le Kien, F., Dutta Gupta, S., Balykin, V., Hakuta, K., 2005b. Spontaneous emission of a cesium atom near a nanofiber: efficient coupling of light to guided modes. Phys. Rev. A 72 (3), 32509. ISSN 1094-1622. http://journals.aps.org/pra/abstract/10.1103/PhysRevA.72.032509.

Le Kien, F., Balykin, V., Hakuta, K., 2006. Scattering of an evanescent light field by a single cesium atom near a nanofiber. Phys. Rev. A 73 (1), 13819. ISSN 1094-1622. http://journals.aps.org/pra/abstract/10.1103/PhysRevA.73.013819.

Le Kien, F., Schneeweiss, P., Rauschenbeutel, A., 2013. Dynamical polarizability of atoms in arbitrary light fields: general theory and application to cesium. Eur. Phys. J. D 67 (5), 92. ISSN 1434-6079. http://dx.doi.org/10.1140/epjd/e2013-30729-x.

Lee, J., Grover, J.A., Hoffman, J.E., Orozco, L.A., Rolston, S.L., 2015. Inhomogeneous broadening of optical transitions of 87 Rb atoms in an optical nanofiber trap. J. Phys. B: At. Mol. Opt. Phys. 48 (16), 165004. http://stacks.iop.org/0953-4075/48/i=16/a=165004.

Lindner, E., Chojetzki, C., Brückner, S., Becker, M., Rothhardt, M., Bartelt, H., 2009. Thermal regeneration of fiber Bragg gratings in photosensitive fibers. Opt. Express 17 (15), 12523–12531. http://dx.doi.org/10.1364/OE.17.012523.

Lodahl, P., Mahmoodian, S., Stobbe, S., 2015. Interfacing single photons and single quantum dots with photonic nanostructures. Rev. Mod. Phys. 87, 347–400. http://dx.doi.org/10.1103/RevModPhys.87.347.

Lodahl, P., Mahmoodian, S., Stobbe, S., Schneeweiss, P., Volz, J., Rauschenbeutel, A., Pichler, H., Zoller, P., 2017. Chiral quantum optics. Nature 541, 473–480. http://dx.doi.org/10.1038/nature21037.

Louyer, Y., Meschede, D., Rauschenbeutel., A., 2005. Tunable whispering-gallery-mode resonators for cavity quantum electrodynamics. Phys. Rev. A 72, 031801. http://journals.aps.org/pra/abstract/10.1103/PhysRevA.72.031801.

Lugiato, L.A., 1984. Theory of optical bistability. In: Wolf, E. (Ed.), Progress in Optics, vol. 21. North-Holland, Amsterdam, pp. 69–216.

Luxmoore, I.J., Wasley, N.A., Ramsay, A.J., Thijssen, A.C.T., Oulton, R., Hugues, M., Fox, A.M., Skolnick, M.S., 2013. Optical control of the emission direction of a quantum dot. Appl. Phys. Lett. 103 (24), 241102. http://dx.doi.org/10.1063/1.4845975.

Lvovsky, A.I., Sanders, B.C., Tittel, W., 2009. Optical quantum memory. Nat. Photon. 3 (12), 706–714. http://dx.doi.org/10.1038/nphoton.2009.231.

Madsen, L.S., Baker, C., Rubinsztein-Dunlop, H., Bowen, W.P., 2016. Nondestructive profilometry of optical nanofibers. Nano Lett. 16 (12), 7333–7337. http://dx.doi.org/10.1021/acs.nanolett.6b02460.

Maimaiti, A., Truong, V.G., Sergides, M., Gusachenko, I., Nic Chormaic, S., 2015. Higher order microfibre modes for dielectric particle trapping and propulsion. Sci. Rep. 5, 9077. http://dx.doi.org/10.1038/srep09077.

Makhlin, Y., Schön, G., Shnirman, A., 2001. Quantum-state engineering with Josephson-junction devices. Rev. Mod. Phys. 73 (2), 357–400. ISSN 0034-6861. http://dx.doi.org/10.1103/RevModPhys.73.357.

Marcuse, D., 1981. Principles of Optical Fiber Measurement. Academic Press, Boston.

Marcuse, D., 1991. Theory of Dielectric Optical Waveguides, second ed. Academic Press, Boston.

Minogin, V.G., Nic Chormaic, S., 2010. Manifestation of the van der Waals surface interaction in the spontaneous emission of atoms into an optical nanofiber. Laser Phys. 20 (1), 32–37. http://dx.doi.org/10.1134/S1054660X09170137.

Mitsch, R., Sayrin, C., Albrecht, B., Schneeweiss, P., Rauschenbeutel, A., 2014. Exploiting the local polarization of strongly confined light for sub-micrometer-resolution internal state preparation and manipulation of cold atoms. Phys. Rev. A 89 (6), 063829. ISSN 1050-2947. http://dx.doi.org/10.1103/PhysRevA.89.063829.

Morrissey, M.J., Deasy, K., Chakrabarti, S., Nic Chormaic, S., 2009. Tapered optical fibers as tools for probing magneto-optical trap characteristics. Rev. Sci. Instrum. 80 (5), 053102. http://dx.doi.org/10.1063/1.3117201.

Morrissey, M.J., Deasy, K., Frawley, M., Kumar, R., Prel, E., Russell, L., Truong, V.G., Nic Chormaic, S., 2013. Spectroscopy, manipulation and trapping of neutral atoms, molecules, and other particles using optical nanofibers: a review. Sensors 13 (8), 10449–10481. ISSN 1424-8220. http://dx.doi.org/10.3390/s130810449.

Nagai, R., Aoki, T., 2014. Ultra-low-loss tapered optical fibers with minimal lengths. Opt. Express 22 (23), 28427. ISSN 1094-4087. http://dx.doi.org/10.1364/OE.22.028427.

Nayak, K.P., Hakuta, K., 2008. Single atoms on an optical nanofibre. New J. Phys. 10 (5), 053003. http://stacks.iop.org/1367-2630/10/i=5/a=053003.

Nayak, K.P., Hakuta, K., 2012. Photonic crystal formation on optical nanofibers using femtosecond laser ablation technique. http://arxiv.org/abs/1210.7592v1. ArXiv physics.op.

Nayak, K.P., Melentiev, P.N., Morinaga, M., Kien, F.L., Balykin, V.I., Hakuta, K., 2007. Optical nanofiber as an efficient tool for manipulating and probing atomic fluorescence. Opt. Express 15 (9), 5431–5438. http://dx.doi.org/10.1364/OE.15.005431.

Nayak, K., Le Kien, F., Morinaga, M., Hakuta, K., 2009. Antibunching and bunching of photons in resonance fluorescence from a few atoms into guided modes of an optical nanofiber. Phys. Rev. A 79, 021801. http://dx.doi.org/10.1103/PhysRevA.79.021801.

Nayak, K.P., Kien, F.L., Kawai, Y., Hakuta, K., Nakajima, K., Miyazaki, H.T., Sugimoto, Y., 2011. Cavity formation on an optical nanofiber using focused ion beam milling technique. Opt. Express 19 (15), 14040–14050. http://dx.doi.org/10.1364/OE.19.014040.

Nayak, K., Das, M., Le Kien, F., Hakuta, K., 2012. Spectroscopy of near-surface atoms using an optical nanofiber. Opt. Commun. 285, 4698. ISSN 0030-4018. http://www.sciencedirect.com/science/article/pii/S0030401812003653.

Nayak, K.P., Zhang, P., Hakuta, K., 2014. Optical nanofiber-based photonic crystal cavity. Opt. Lett. 39 (2), 232–235. http://dx.doi.org/10.1364/OL.39.000232. http://ol.osa.org/abstract.cfm?URI=ol-39-2-232.

Patterson, B.D., 2016. Transmission Spectra of Rb 87 Atoms Near an Optical Nanofiber (Ph.D. thesis). University of Maryland College Park. http://drum.lib.umd.edu/handle/1903/18780.

Paulisch, V., Kimble, H.J., González-Tudela, A., 2016. Universal quantum computation in waveguide QED using decoherence free subspaces. New J. Phys. 18 (4), 043041. http://stacks.iop.org/1367-2630/18/i=4/a=043041.

Petersen, J., Volz, J., Rauschenbeutel, A., 2014. Chiral nanophotonic waveguide interface based on spin-orbit interaction of light. Science 346 (6205), 67–71. ISSN 1095-9203. http://dx.doi.org/10.1126/science.1257671.

Phelan, C., Hennessy, T., Busch, T., 2013. Shaping the evanescent field of optical nanofibers for cold atom trapping. Opt. Express 21 (22), 27093–27101. http://dx.doi.org/10.1364/OE.21.027093.

Pichler, H., Ramos, T., Daley, A.J., Zoller, P., 2015. Quantum optics of chiral spin networks. Phys. Rev. A 91, 042116. http://dx.doi.org/10.1103/PhysRevA.91.042116.

Pittman, T.B., Jones, D.E., Franson, J.D., 2013. Ultralow-power nonlinear optics using tapered optical fibers in metastable xenon. Phys. Rev. A 88, 053804. http://dx.doi.org/10.1103/PhysRevA.88.053804.

Purcell, E.M., 1946. Spontaneous emission probabilities at radio frequencies. Phys. Rev. 69, 681.

Qi, X., Baragiola, B.Q., Jessen, P.S., Deutsch, I.H., 2016. Dispersive response of atoms trapped near the surface of an optical nanofiber with applications to quantum non-demolition measurement and spin squeezing. Phys. Rev. A 93, 023817. http://dx.doi.org/10.1103/PhysRevA.93.023817.

Quan, Q., Bulu, I., Lončar, M., 2009. Broadband waveguide QED system on a chip. Phys. Rev. A 80, 011810. http://dx.doi.org/10.1103/PhysRevA.80.011810.

Ramos, T., Pichler, H., Daley, A.J., Zoller, P., 2014. Quantum spin dimers from chiral dissipation in cold-atom chains. Phys. Rev. Lett. 113, 237203. http://dx.doi.org/10.1103/PhysRevLett.113.237203.

Ravets, S., Hoffman, J.E., Kordell, P.R., Wong-Campos, J.D., Rolston, S.L., Orozco, L.A., 2013a. Intermodal energy transfer in a tapered optical fiber: optimizing transmission. JOSA A 30 (11), 2361–2371. ISSN 1520-8532. http://dx.doi.org/10.1364/JOSAA.30.002361.

Ravets, S., Hoffman, J.E., Orozco, L.A., Rolston, S.L., Beadie, G., Fatemi, F.K., 2013b. A low-loss photonic silica nanofiber for higher-order modes. Opt. Express 21 (15), 18325–18335. ISSN 1094-4087. http://dx.doi.org/10.1364/OE.21.018325.

Reiserer, A., Rempe, G., 2015. Cavity-based quantum networks with single atoms and optical photons. Rev. Mod. Phys. 87, 1379–1418. http://dx.doi.org/10.1103/RevModPhys.87.1379.

Reitz, D., Rauschenbeutel., A., 2012. Nanofiber-based double-helix dipole trap for cold neutral atoms. Opt. Commun. 285, 4705–4708. http://dx.doi.org/10.1016/j.optcom.2012.06.034.

Reitz, D., Sayrin, C., Mitsch, R., Schneeweiss, P., Rauschenbeutel, A., 2013. Coherence properties of nanofiber-trapped cesium atoms. Phys. Rev. Lett. 110 (24), 243603. ISSN 0031-9007. http://dx.doi.org/10.1103/PhysRevLett.110.243603.

Ritsch, H., Domokos, P., Brennecke, F., Esslinger, T., 2013. Cold atoms in cavity-generated dynamical optical potentials. Rev. Mod. Phys. 85, 553–601. http://dx.doi.org/10.1103/RevModPhys.85.553.

Russell, L., Gleeson, D.A., Minogin, V.G., Nic Chormaic, S., 2009. Spectral distribution of atomic fluorescence coupled into an optical nanofibre. J. Phys. B: At. Mol. Opt. Phys. 42 (18), 185006. http://stacks.iop.org/0953-4075/42/i=18/a=185006.

Russell, L., Deasy, K., Daly, M.J., Morrissey, M.J., Nic Chormaic, S., 2012. Sub-Doppler temperature measurements of laser-cooled atoms using optical nanofibres. Meas. Sci. Technol. 23 (1), 015201. ISSN 0957-0233. http://dx.doi.org/10.1088/0957-0233/23/1/015201.

Russell, L., Kumar, R., Tiwari, V., Nic Chormaic, S., 2013. Measurements on release–recapture of cold 85Rb atoms using an optical nanofibre in a magneto-optical trap. Opt. Commun. 309, 313–317. ISSN 00304018. http://dx.doi.org/10.1016/j.optcom.2013.07.080.

Sadgrove, M., Yalla, R., Nayak, K.P., Hakuta, K., 2013. Photonic crystal nanofiber using an external grating. Opt. Lett. 38 (14), 2542–2545. https://doi.org/10.1364/OL.38.002542.

Safronova, M.S., Williams, C.J., Clark, C.W., 2004. Relativistic many-body calculations of electric-dipole matrix elements, lifetimes, and polarizabilities in rubidium. Phys. Rev. A 69, 022509. http://dx.doi.org/10.1103/PhysRevA.69.022509.

Sagué, G., 2008. Cold Atom Physics Using Ultra-Thin Optical Fibres (Ph.D. thesis). University of Mainz.

Sagué, G., Vetsch, E., Alt, W., Meschede, D., Rauschenbeutel, A., 2007. Cold-atom physics using ultrathin optical fibers: light-induced dipole forces and surface interactions. Phys. Rev. Lett. 99 (16), 163602. http://dx.doi.org/10.1103/PhysRevLett.99.163602.

Sagué, G., Baade, A., Rauschenbeutel, A., 2008. Blue-detuned evanescent field surface traps for neutral atoms based on mode interference in ultrathin optical fibres. New J. Phys. 10 (11), 113008. http://stacks.iop.org/1367-2630/10/i=11/a=113008.

Sanchez-Mondragon, J.J., Narozhny, N.B., Eberly, J.H., 1983. Theory of spontaneous-emission line shape in an ideal cavity. Phys. Rev. Lett. 51, 550–553. http://dx.doi.org/10.1103/PhysRevLett.51.550.

Sayrin, C., Clausen, C., Albrecht, B., Schneeweiss, P., Rauschenbeutel, A., 2015a. Storage of fiber-guided light in a nanofiber-trapped ensemble of cold atoms. Optica 2 (4), 353–356. http://dx.doi.org/10.1364/OPTICA.2.000353.

Sayrin, C., Junge, C., Mitsch, R., Albrecht, B., O'Shea, D., Schneeweiss, P., Volz, J., Rauschenbeutel, A., 2015b. Nanophotonic optical isolator controlled by the internal state of cold atoms. Phys. Rev. X 5, 041036. http://dx.doi.org/10.1103/PhysRevX.5.041036.

Scheucher, M., Hilico, A., Will, E., Volz, J., Rauschenbeutel, A., 2016. Quantum optical circulator controlled by a single chirally coupled atom. Science 354 (6319), 1577–1580. http://dx.doi.org/10.1126/science.aaj2118.

Schlosser, N., Reymond, G., Grangier, P., 2002. Collisional blockade in microscopic optical dipole traps. Phys. Rev. Lett. 89 (2), 023005. http://dx.doi.org/10.1103/PhysRevLett.89.023005.

Schneeweiss, P., Kien, F.L., Rauschenbeutel, A., 2014. Nanofiber-based atom trap created by combining fictitious and real magnetic fields. New J. Phys. 16 (1), 013014. http://dx.doi.org/10.1088/1367-2630/16/1/013014.

Schröder, T., Fujiwara, M., Noda, T., Zhao, H.-Q., Benson, O., Takeuchi, S., 2012. A nanodiamond-tapered fiber system with high single-mode coupling efficiency. Opt. Express 20 (10), 10490–10497. http://dx.doi.org/10.1364/OE.20.010490.

Semenova, Y., Kavungal, V., Wu, Q., Farrell, G., 2015. Submicron accuracy fiber taper profiling using whispering gallery modes in a cylindrical fiber micro-resonator. Proc. SPIE 9634, 24th International Conference on Optical Fibre Sensors, 96343F; http://dx.doi.org/10.1117/12.2194986.

Shomroni, I., Rosenblum, S., Lovsky, Y., Bechler, O., Guendelman, G., Dayan, B., 2014. All-optical routing of single photons by a one-atom switch controlled by a single photon. Science 345 (6199), 903–906. http://dx.doi.org/10.1126/science.1254699.

Snyder, A.W., Love, J.D., 1983. Optical Waveguide Theory. Chapman and Hall, London.

Solano, P., Grover, J.A., Xu, Y., Barberis-Blostein, P., Munday, J.N., Orozco, L.A., Phillips, W.D., Rolston, S.L., 2017a. Alignment-dependent decay rate of an atomic dipole near an optical nanofiber. (to be published).

Solano, P., Barberis-Blostein, P., Fatemi, F.K., Orozco, L.A., Rolston, S.L., 2017b. Super- and sub-radiance reveal infinite-range interactions through a nanofiber. (to be published).

Solano, P., Fatemi, F.K., Orozco, L.A., Rolston, S.L., 2017c. Dynamics of trapped atoms around an optical nanofiber probed through polarimetry. arXiv:1703.09122. https://arxiv.org/abs/1703.09122.

Sørensen, H.L., Polzik, E.S., Appel, J., 2014. Heater self-calibration technique for shape prediction of fiber tapers. J. Lightwave Technol. 32 (10), 1886–1891. http://dx.doi.org/10.1109/JLT.2014.2314319.

Sørensen, H.L., Béguin, J.-B., Kluge, K.W., Iakoupov, I., Sørensen, A.S., Müller, J.H., Polzik, E.S., Appel, J., 2016. Coherent backscattering of light off one-dimensional atomic strings. Phys. Rev. Lett. 117, 133604. http://dx.doi.org/10.1103/PhysRevLett.117.133604.

Spillane, S.M., Kippenberg, T.J., Painter, O.J., Vahala, K.J., 2003. Ideality in a fiber-taper-coupled microresonator system for application to cavity quantum electrodynamics. Phys. Rev. Lett. 91, 043902. http://dx.doi.org/10.1103/PhysRevLett.91.043902.

Spillane, S.M., Pati, G.S., Salit, K., Hall, M., Kumar, P., Beausoleil, R.G., Shahriar, M.S., 2008. Observation of nonlinear optical interactions of ultralow levels of light in a tapered optical nanofiber embedded in a hot rubidium vapor. Phys. Rev. Lett. 100, 233602. http://dx.doi.org/10.1103/PhysRevLett.100.233602.

Sprague, M.R., Michelberger, P.S., Champion, T.F.M., England, D.G., Nunn, J., Jin, X.M., Kolthammer, W.S., Abdolvand, A., Russell, P.S.J., Walmsley, I.A., 2014. Broadband single-photon-level memory in a hollow-core photonic crystal fibre. Nat. Photon 8 (4), 287–291. http://dx.doi.org/10.1038/nphoton.2014.45.

Stiebeiner, A., Garcia-Fernandez, R., Rauschenbeutel, A., 2010. Design and optimization of broadband tapered optical fibers with a nanofiber waist. Opt. Express 18 (22), 22677–22685. http://dx.doi.org/10.1364/OE.18.022677.

Stobińska, M., Alber, G., Leuchs, G., 2009. Perfect excitation of a matter qubit by a single photon in free space. Europhys. Lett. 86, 14007. http://iopscience.iop.org/0295-5075/86/1/14007.

Streed, E.W., Jechow, A., Norton, B.G., Kielpinski, D., 2012. Absorption imaging of a single atom. Nat. Commun. 3, 933. http://dx.doi.org/10.1038/ncomms1944.

Sumetsky, M., Dulashko, Y., 2010. Radius variation of optical fibers with angstrom accuracy. Opt. Lett. 35 (23), 4006–4008. http://dx.doi.org/10.1364/OL.35.004006.

Sumetsky, M., Dulashko, Y., Fini, J.M., Hale, A., Nicholson, J.W., 2006. Probing optical microfiber nonuniformities at nanoscale. Opt. Lett. 31 (16), 2393–2395. http://dx.doi.org/10.1364/OL.31.002393.

Szczurowski, M., Urbanczyk, W., Napiorkowski, M., Hlubina, P., Hollenbach, U., Sieber, H., Mohr, J., 2011. Differential Rayleigh scattering method for measurement of polarization and intermodal beat length in optical waveguides and fibers. Appl. Opt. 50, 2594–2600. http://dx.doi.org/10.1364/AO.50.002594. http://ao.osa.org/abstract.cfm?URI=ao-50-17-2594.

Tanji-Suzuki, H., Leroux, I.D., Schleier-Smith, M.H., Cetina, M., Grier, A.T., Simon, J., Vuletić, V., 2011. Interaction between atomic ensembles and optical resonators: classical description. In: Adv. At. Mol. Opt. Phys., vol. 60. Elsevier, pp. 201–237. http://www.sciencedirect.com/science/article/pii/B9780123855084000048.

Tey, M.K., Chen, Z., Aljunid, S.A., Chng, B., Huber, F., Maslennikov, G., Kurtsiefer, C., 2008. Strong interaction between light and a single trapped atom without the need for a cavity. Nat. Phys. 4 (12), 924–927. http://dx.doi.org/10.1038/nphys1096.

Thompson, J.D., Tiecke, T.G., de Leon, N.P., Feist, J., Akimov, A.V., Gullans, M., Zibrov, A.S., Vuletić, V., Lukin, M.D., 2013. Coupling a single trapped atom to a nanoscale optical cavity. Science 340 (6137), 1202. http://dx.doi.org/10.1126/science.1237125.

Tong, L., Gattass, R.R., Ashcom, J.B., He, S., Lou, J., Shen, M., Maxwell, I., Mazur, E., 2003. Subwavelength-diameter silica wires for low-loss optical wave guiding. Nature 426 (6968), 816–819. http://dx.doi.org/10.1038/nature02193.

Venkataraman, V., Saha, K., Londero, P., Gaeta, A.L., 2011. Few-photon all-optical modulation in a photonic band-gap fiber. Phys. Rev. Lett. 107, 193902. http://dx.doi.org/10.1103/PhysRevLett.107.193902.

Verhart, N.R., Lepert, G., Billing, A.L., Hwang, J., Hinds, E.A., 2014. Single dipole evanescently coupled to a multimode waveguide. Opt. Express 22 (16), 19633. http://dx.doi.org/10.1364/OE.22.019633.

Vetsch, E., 2010. Optical Interface Based on a Nanofiber (Ph.D. thesis). University of Mainz.

Vetsch, E., Reitz, D., Sagué, G., Schmidt, R., Dawkins, S.T., Rauschenbeutel, A., 2010. Optical interface created by laser-cooled atoms trapped in the evanescent field surrounding an optical nanofiber. Phys. Rev. Lett. 104 (20), 203603. http://dx.doi.org/10.1103/PhysRevLett.104.203603.

Vetsch, E., Dawkins, S.T., Mitsch, R., Reitz, D., Schneeweiss, P., Rauschenbeutel, A., 2012. Nanofiber-based optical trapping of cold neutral atoms. IEEE J. Sel. Top. Quantum Electron. 18 (6), 1763–1770. http://dx.doi.org/10.1109/JSTQE.2012.2196025.

Wallraff, A., Schuster, D.I., Blais, A., Frunzio, L., Huang, R.-S., Majer, J., Kumar, S., Girvin, S.M., Schoelkopf, R.J., 2004. Strong coupling of a single photon to a superconducting qubit using circuit quantum electrodynamics. Nature 431 (7005), 162–167. http://dx.doi.org/10.1038/nature02851.

Ward, J., O'shea, D., Shortt, B., Morrissey, M., Deasy, K., Nic Chormaic, S., 2006. Heat-and-pull rig for fiber taper fabrication. Rev. Sci. Instrum. 77 (8), 83105. papers://fba91f4c-b378-44ee-a4a1-924eac936bf5/Paper/p39.

Ward, J.M., Maimaiti, A., Le, V.H., Nic Chormaic, S., 2014. Contributed review: optical micro- and nanofiber pulling rig. Rev. Sci. Instrum. 85 (11), 111501. http://dx.doi.org/10.1063/1.4901098.

Warken, F., 2007. Ultra Thin Glass Fibers as a Tool for Coupling Light and Matter (Ph.D. thesis). Rheinische Friedrich-Wilhelms Universitat, Mainz, Germany.

Warken, F., Vetsch, E., Meschede, D., Sokolowski, M., Rauschenbeutel, A., 2007. Ultra-sensitive surface absorption spectroscopy using sub-wavelength diameter optical fibers. Opt. Express 15 (19), 11952–11958. http://dx.doi.org/10.1364/OE.15.011952.

Wineland, D.J., 2013. Nobel lecture: superposition, entanglement, and raising Schrödinger's cat. Rev. Mod. Phys. 85 (3), 1103–1114. http://dx.doi.org/10.1103/RevModPhys.85.1103.

Wuttke, C., Becker, M., Brückner, S., Rothhardt, M., Rauschenbeutel, A., 2012. Nanofiber Fabry–Perot microresonator for nonlinear optics and cavity quantum electrodynamics. Opt. Lett. 37 (11), 1949–1951. http://dx.doi.org/10.1364/OL.37.001949.

Wuttke, C., Cole, G., Rauschenbeutel, A., 2013. Optically active mechanical modes of tapered optical fibers. Phys. Rev. A 88 (6), 061801. http://dx.doi.org/10.1103/PhysRevA.88.061801.

Xiang, Z.-L., Ashhab, S., You, J.Q., Nori, F., 2013. Hybrid quantum circuits: superconducting circuits interacting with other quantum systems. Rev. Mod. Phys. 85, 623–653. http://dx.doi.org/10.1103/RevModPhys.85.623.

Yalla, R., Sadgrove, M., Nayak, K.P., Hakuta, K., 2014. Cavity quantum electrodynamics on a nanofiber using a composite photonic crystal cavity. Phys. Rev. Lett. 113 (14), 143601. http://dx.doi.org/10.1103/PhysRevLett.113.143601.

Yariv, A., 1990. Optical Electronics. Oxford University Press, Oxford.

Ye, J., Kimble, H.J., Katori, H., 2008. Quantum state engineering and precision metrology using state-insensitive light traps. Science 320 (5884), 1734–1738. http://dx.doi.org/10.1126/science.1148259.

Yokota, H., Sugai, E., Sasaki, Y., 1997. Optical irradiation method for fiber coupler fabrications. Opt. Rev. 4 (1), A104. http://dx.doi.org/10.1007/BF02936004.

Yu, S., Hood, J.D., Muniz, J.A., Martin, M.J., Norte, R., Hung, C.-L., Meenehan, S.M., Cohen, J.D., Painter, O., Kimble, H.J., 2014. Nanowire photonic crystal waveguides for single-atom trapping and strong light-matter interactions. Appl. Phys. Lett. 104 (11), 111103. http://dx.doi.org/10.1063/1.4868975.

CHAPTER EIGHT

# The Role of Projectile Coherence in the Few-Body Dynamics of Simple Atomic Systems

**Michael Schulz**[1]
Missouri University of Science and Technology, Rolla, MO, United States
[1]Corresponding author: e-mail address: schulz@mst.edu

## Contents



## Abstract

The reaction dynamics in simple atomic systems have been studied extensively in order to address the fundamentally important, but yet unsolved few-body problem. For the description of electron impact-induced reactions theory has made remarkable progress over the last decade. However, describing the reaction dynamics for the same processes induced by ion impact proved to be much more challenging. Surprising discrepancies between experiment and theory were found even for cases which were considered to be "easy" for theory. Only a few years ago an important step toward a resolution of this puzzle was made when strong experimental indications were reported that measured cross sections can sensitively depend on the projectile coherence properties, while theory usually assumes that the projectile beam is always coherent. Owing to the much larger de Broglie wavelength electron projectiles tend to be much more coherent relative to the target dimension, which results in much better agreement between experiment and theory. In this chapter, recent studies on the effect of the projectile coherence properties on the reaction dynamics are reviewed.

*Advances in Atomic, Molecular, and Optical Physics*, Volume 66
ISSN 1049-250X
http://dx.doi.org/10.1016/bs.aamop.2017.01.001

## 1. INTRODUCTION

Since atoms are the basic building blocks of all matter, atomic physics is not only of fundamental importance from a basic science point of view, but it also has significant practical impacts in areas well beyond the boundaries of physics like material science, chemistry, and even biology. Our understanding of the properties of matter on an atomic level has advanced tremendously since the advent of quantum mechanics. Chemical bonds, for example, are to a large extent understood and even complex systems can often be described with satisfactory accuracy.

In contrast, our understanding of dynamic processes, like e.g., fragmentation processes, is still surprisingly incomplete even for simple atomic systems. The basic problem is that the Schrödinger equation is not analytically solvable for more than two mutually interacting particles even if the underlying forces are precisely known (e.g., Rescigno et al., 1999; Schulz et al., 2003a). This is known as the few-body problem (FBP). As a result, theory has to resort heavy numerical modeling efforts and these models have to be tested by detailed experimental data. Especially kinematically complete experiments, in which the complete momentum vectors of all final-state products are determined, are crucially important as they offer the most sensitive tests of theory.

For target ionization by electron impact kinematically complete experiments have been performed (e.g., de Harak et al., 2008; Dorn et al., 1997; Duguet et al., 1987; Dürr et al., 2006, 2008; Murray et al., 1992; Ren et al., 2010; Röder et al., 1997; Rosel et al., 1992; Stefani et al., 1990) since the pioneering work of Ehrhardt et al. (1986). Furthermore, remarkable theoretical progress has been achieved over the last 15–20 years (e.g., Bartlett and Stelbovics, 2004; Bray, 2002; Colgan et al., 2002; Jones and Madison, 2000; Rescigno et al., 1999). As a result, very good agreement between theory and experiment is now routinely achieved for ionization of simple one- or two-electron targets (e.g., Dürr et al., 2006, 2008; Ren et al., 2010) even close to threshold, which was considered a particularly challenging regime. More recently, the research efforts have shifted to studying many-electron targets (e.g., Childers et al., 2001; Naja et al., 2008; Stevenson and Lohmann, 2006; Ren et al., 2012, 2015b) including systems as complex as biomolecules (e.g., Builth-Williams et al., 2012, 2013; da Silva et al., 2014; Ren et al., 2014). There, agreement between theory and experiment is much less satisfactory, although impressive progress was made recently for atomic many-electron targets (Ren et al., 2015a).

For ion impact, on the other hand, the status of research on the few-body dynamics in ionizing collisions is not as advanced as for electron impact. The basic difficulties stem from the much larger projectile mass of ions compared to electrons. From an experimental point of view, this results in very small (for fast and heavy ions immeasurably small) scattering angles and projectile energy losses relative to the initial energy. Therefore, a direct measurement of the complete momentum vector of the scattered projectile is only feasible for light ions at small or intermediate speeds (Alexander et al., 2008; Egodapitiya et al., 2011a; Htwe et al., 1994; Laforge et al., 2009; Park et al., 1977, 1978; Vajnai et al., 1995). From a theoretical point of view, one major challenge introduced by the large projectile mass is that an enormous number of angular momentum states contribute to the scattered projectile state. This makes the implementation of nonperturbative approaches rather challenging.

Owing to these difficulties kinematically complete experiments on ionization by ion impact only became feasible with the development of cold target recoil-ion momentum spectroscopy (COLTRIMS) (for review, see Dörner et al., 2000; Ullrich et al., 2003). For fast and heavy ion impact, fully differential cross sections (FDCS) were measured using this method by detecting the momentum-analyzed recoil ions and ejected electrons in coincidence (e.g., Fischer et al., 2003; Hubele et al., 2013; Schulz et al., 2001, 2002, 2003a, 2005, 2013; Voitkiv et al., 2004) (for review, see Schulz, 2009; Schulz and Madison, 2006). For light ions at intermediate collision energies coincidence setups between the momentum-analyzed scattered projectiles and the recoil ions were also used to measure FDCS (Arthanayaka et al., 2015a, 2016; Hasan et al., 2014a; Maydanyuk et al., 2005; Sharma et al., 2014a).

Given the challenges theory is facing for ion impact it was not necessarily expected that the agreement between calculated and measured FDCS would be as good as for electron impact ionization of simple atoms. However, these problems were not believed to represent a serious limitation on calculations for fast collisions with small perturbation parameters $\eta$ (i.e., the projectile charge to speed ratio). In fact, in this regime, it was anticipated that even the first Born approximation (FBA) would provide an adequate description of the collision process. It was therefore very surprising when for certain kinematic conditions rather poor agreement between state-of-the-art calculations, including nonperturbative and time-dependent approaches, and measured data were found (Fiol et al., 2006; Kouzakov et al., 2012; Madison et al., 2002; McGovern et al., 2010a; Voitkiv et al., 2003a). With increasing $\eta$, these discrepancies increased and were severe for $\eta > 1$ (Fiol and Olson, 2004; McGovern et al., 2010b; Pedlow et al., 2005).

These discrepancies, especially for small $\eta$, have led to vivid debates over the last decade (e.g., Fiol et al., 2006; Kouzakov et al., 2012; Madison et al., 2002; McGovern et al., 2010a; Voitkiv et al., 2003a). Already in the early stages of these discussions, there were indications that the explanation for the unsatisfactory agreement between theory and experiment may have to involve the role of the interaction between the nuclei of the collision partners (Madison et al., 2003; Schulz et al., 2003a,b). On the other hand, in several models this interaction was incorporated in a rather sophisticated manner and it was difficult to see how these calculations could go so severely wrong (e.g., Kouzakov et al., 2012; Madison et al., 2003; McGovern et al., 2010a; Voitkiv et al., 2003a).

Finally, a major breakthrough was achieved when ionization of $H_2$ by intermediate energy proton impact was studied by Egodapitiya et al. (2011b). In that work, it was suggested that measured cross sections can sensitively depend on the coherence properties of the projectiles. Theory usually assumes that the projectile has a sharp momentum, i.e., that it is coherent relative to the dimension of the target atom. However, in reality there is an inherent momentum uncertainty associated with the projectiles, which can be quite large especially for fast and heavy ions. As a result, the dimension of the target which is coherently illuminated, the transverse coherence length $\Delta x$, is accordingly small, i.e., the projectiles are often incoherent relative to the target dimension. In collisions with $H_2$ the projectile can be diffracted from both atomic centers of the molecule simultaneously if $\Delta x$ is larger than the internuclear separation $D$ (coherent beam). In this case, interference structures are observable in the scattering angle dependence of the cross sections (Alexander et al., 2008). However, if $\Delta x < D$ only one atomic center is coherently illuminated at a time and no interference is expected. This was indeed observed by Egodapitiya et al. (2011b). Furthermore, in that work it was pointed out that a similar coherence effect may also be present in the FDCS data for ionization of atomic targets by fast heavy ion impact. Egodapitiya et al. argued that there interference may occur between first- and higher-order transition amplitudes, the latter involving the nucleus–nucleus interaction. However, higher-order transitions tend to occur at smaller impact parameters than first-order transitions. Therefore, an interference structure between both contributions is only observable if $\Delta x$ is larger than a typical impact parameter separation between the corresponding amplitudes. But for fast heavy ions $\Delta x$ is often smaller than the target dimension by several orders of magnitude, while theory usually assumes $\Delta x = \infty$.

Since the work on projectile coherence effects in atomic collisions by Egodapitiya et al. numerous experiments have been performed and the initial findings were to a large extent confirmed. In this chapter, these studies will be reviewed. The basic ideas underlying these studies did not remain completely unchallenged and critical views will also be analyzed.

## 2. EXPERIMENTAL METHODS

The experiments which will be discussed in this chapter share two major features: almost all of them (with one exception) are kinematically complete and all of them were carried out for varying transverse projectile coherence lengths relative to the dimension of the diffracting object. First, we will describe the methods used to perform a kinematically complete experiment and then explain how the projectile coherence properties can be varied.

As mentioned in Section 1, a kinematically complete experiment on single target ionization requires determining the momentum vectors of the three collision fragments, i.e., of the scattered projectile, the recoiling target ion, and the ejected electron. Two of these momentum vectors need to be measured directly; the third one can then be deduced from momentum conservation (in fact, since in addition to momentum energy is also conserved, for one of the fragments it is sufficient to directly measure two momentum components). Therefore, there are three possibilities to perform a kinematically complete experiment: one can directly measure the momentum vectors of (a) the scattered projectiles and the ejected electrons, (b) the scattered projectiles and the recoil ions, or (c) the recoil ions and the ejected electrons. Method (a) was used in most electron impact studies while so far only (b) and (c) were applied to ion impact studies. Owing to the difficulties, mentioned in Section 1, introduced by the large masses of heavy and fast ionic projectiles, methods (a) and (b) are only feasible for light ions at small or intermediate energies.

In Fig. 1, a sketch of the setup of an experiment directly measuring the scattered projectile and recoil-ion momenta is shown. The projectile beam intersects a neutral target beam from a supersonic jet. Typical scattering

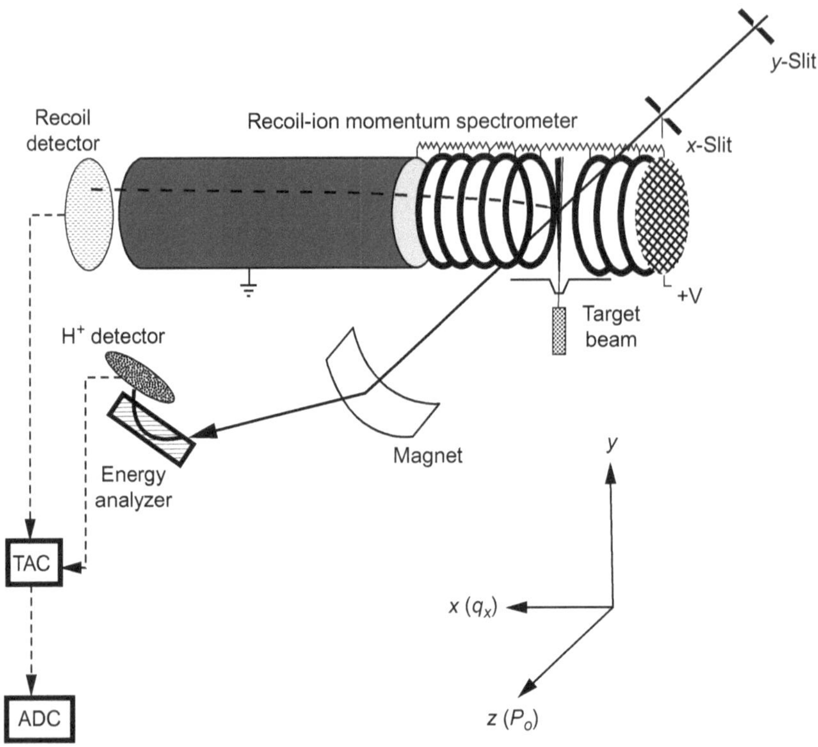


**Fig. 1** Setup of an experiment directly measuring the momentum vectors of the scattered projectile and the recoil ions. The ejected electron momentum is deduced from momentum conservation. The various components of the apparatus are not shown to scale.

angles in ionizing collisions are of the order of a few mrad or less so that the projectiles traverse the collision region nearly undeflected. The projectiles are then sent through a switching magnet to clean the beam from the component which was charge exchanged in the collision. The charge-selected beam is energy analyzed by an electrostatic parallel-plate analyzer (Gaus et al., 1994) yielding the magnitude of the projectile momentum. Finally, the projectiles are detected by a two-dimensional position-sensitive channel-plate detector. From the position information, the direction of the projectile momentum can be determined. The orientation of the analyzer slit defines the $x$-axis and the initial projectile beam direction the $z$-axis of the coordinate system in which the momentum components are calculated. Because the slit is very narrow in the $y$-direction the $y$-component of the projectile is fixed at zero within the resolution.

The recoil ions are extracted from the collision region by a weak electric field (a few V/cm) which is generated by applying a voltage to a set of ring-shaped electrodes which are connected to each other by a resistor chain. The ions then traverse a field-free region, provided by a grounded cylinder, before hitting another two-dimensional position-sensitive channel-plate detector. From the position information, the $y$- and $z$-components (where the $y$-direction is defined by the target beam direction) are obtained. The two detectors are set in coincidence and the coincidence time is, apart from a constant offset, identical to the time of flight of the recoil ions from the collision region to the detector. From the time of flight, the $x$-component of the recoil-ion momentum can be calculated.

In Fig. 2, a sketch of the setup of an experiment directly measuring the recoil-ion and electron momenta (Ullrich et al., 2003) is shown. Here, the measurement of the recoil-ion momentum is conceptually the same as in the setup shown in Fig. 1. The most important difference is that the recoil ions are not extracted in a direction perpendicular to the projectile beam axis, but nearly parallel to it (longitudinal extraction). However, spectrometers measuring the electron and recoil-ion momenta directly by extracting the recoil ions perpendicular to the projectile beam (transverse extraction) are also in operation (Dörner et al., 2000). The extraction field guiding the recoil ions to the left in Fig. 2 simultaneously guides the ejected electrons in the opposite direction, where a second two-dimensional position-sensitive detector is located. However, as the electrons have similar momenta as the recoil ions,

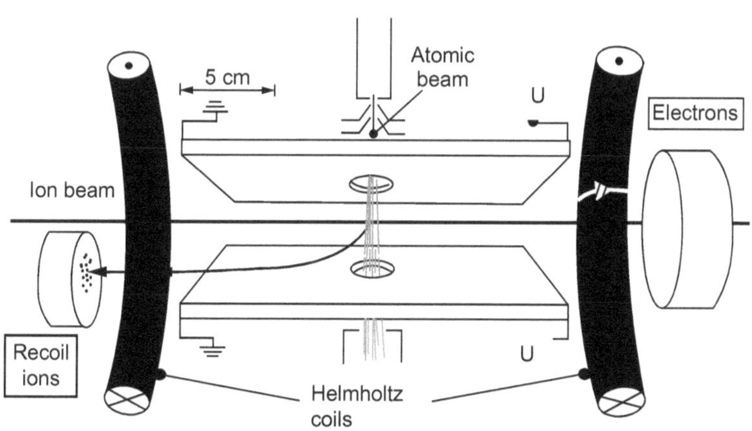


**Fig. 2** Setup of an experiment directly measuring the momentum vectors of the recoil ions and the ejected electrons. The projectile momentum transfer to the target atom is obtained from momentum conservation.

but a much smaller mass, the electric field is not sufficient to guide all electrons onto the detector. Therefore a uniform magnetic field of about 10–20 G, generated by a set of Helmholtz coils, is superimposed on the electric field forcing the electrons into cyclotron motion with a radius proportional to the transverse electron momentum. Therefore, all electrons with transverse momenta smaller than a few a.u. will hit the detector. With the position and timing information and with the knowledge of the electric and magnetic field strengths, it is possible to disentangle the cyclotron motion and to calculate the electron momentum resulting from the collision (Moshammer et al., 1996).

In analogy to classical optics an initially incoherent beam can be made coherent at the target by placing a collimating slit before the collision region. The transverse coherence length is then given by

$$\Delta x = (L/2a)\lambda \tag{1}$$

where $L$ is the distance of the slit to the target, $a$ is its width, and $\lambda$ is the de Broglie wavelength of the projectiles. The coherence length can thus be varied by changing either $L$ or $a$. However, changing the distance has the advantage that this has a much smaller effect on the instrumental resolution than changing $a$ as long as the focusing properties of the beam are not changed. If $\Delta x$ is larger than the dimension of the diffracting object the projectile is considered coherent and incoherent otherwise.

For fast heavy ions, $\lambda$ is so small that it is very difficult to make the beam coherent by means of a collimating slit. For example, if the dimension of the diffracting object was of the order of 1 a.u. and $a = 1$ mm, for a 25 MeV Li beam the slit would have to be placed at a distance of at least 50 m from the target. As an alternative, the experiments can be performed at a storage ring where very small transverse and longitudinal ion momentum spreads can be achieved, e.g., through electron cooling (Steck et al., 1990). In this case, a beam size throughout the storage ring of less than 1 mm can be obtained without using a collimation slit. The coherence length is then determined by an "effective distance" of the source from the target, which is of the order of the circumference of the ring. The disadvantage of doing the experiments at a storage ring is that the investigator has only limited control over the coherence length. Increasing the momentum spread in an attempt to make the projectiles more incoherent means that the beam can be lost after only a few revolutions. Therefore, the data for an incoherent beam may have to be collected at a "conventional" linear accelerator or in single pass mode.

# 3. COHERENCE EFFECTS IN COLLISIONS WITH $H_2$

## 3.1 Single and Double Differential Studies

A well suited case to investigate whether the projectile coherence properties can have an observable impact on measured cross sections is a collision system and a process for which interference structures were observed already. One such example is ionization of $H_2$ by 75 keV proton impact, for which interference effects were reported in the scattering angle dependence of double differential cross sections (DDCS) for fixed projectile energy losses $\varepsilon$ (Alexander et al., 2008). However, in that study data were only obtained for a coherent beam. Later, the experiment was repeated for coherent and incoherent beams (Egodapitiya et al., 2011b). The results are shown in the left panel of Fig. 3 for $\varepsilon = 30$ eV. Throughout this chapter closed and open symbols represent data taken for a coherent and incoherent beam, respectively. Clear differences between both datasets are seen suggesting that indeed the cross sections are affected by the projectile coherence properties. A calculation based on the molecular 3-body distorted wave (M3DW) approach (Chowdhury et al., 2011), which assumes a coherent projectile beam, is in very good agreement with the coherent data, but in poor agreement with the incoherent data. In analogy to classical optics, the interference term is given by the ratio $R$ between the coherent and incoherent cross sections. This ratio is plotted in the right panel of Fig. 3 as a function of

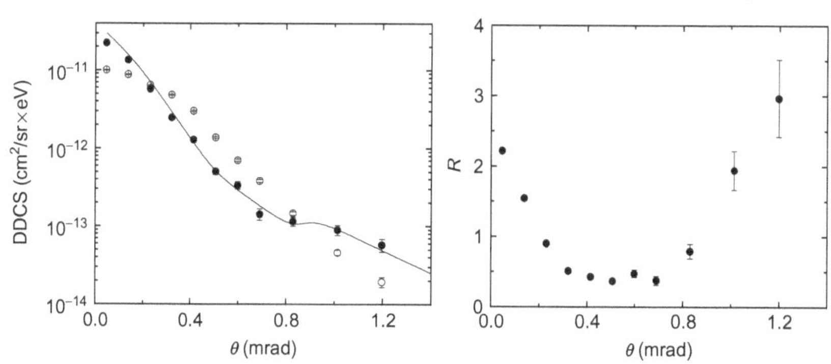


**Fig. 3** Double differential ionization cross sections (*left panel*) in 75 keV $p + H_2$ collisions for $\varepsilon = 30$ eV as a function of scattering angle. The *closed symbols* show the data taken for a coherent beam and the *open symbols* those taken for an incoherent beam. The *solid curve* shows a M3DW calculation, which assumes a coherent beam. The *right panel* shows the ratios between the coherent and incoherent cross sections.

scattering angle. In $R$ a structure resembling a typical interference pattern is more prominent because the very steep scattering angle dependence of the DDCS, which is quite characteristic to atomic collisions, cancels to a large extent in the ratios.

In Fig. 4, the same double differential $R$ is plotted for $\varepsilon = 57$ eV as a function of scattering angle (Arthanayaka et al., 2015b). This energy loss represents a special case because it corresponds to an ejected electron speed which is very close to the projectile speed. In this regime, higher-order contributions in the projectile–electron interaction, known as postcollision interaction (PCI), are important. In a first-order process, the electrons are predominantly emitted in the direction of the momentum transfer $q$ (defined as the difference between the initial and final projectile momenta $p_o - p_f$). But PCI has a tendency of diverting a significant fraction of the electron flux in the forward direction. In Alexander et al. (2008), it was found that the interference structures, which were quite pronounced for $\varepsilon < 53$ eV and for $\varepsilon > 65$ eV, were strongly suppressed at $\varepsilon = 57$ eV. It was vaguely argued that this may be due to a loss of coherence caused by PCI, but the specific mechanism that could lead to such a loss of coherence was not known at the time. The variations in the scattering angle dependence of the ratios shown in Fig. 4 are much weaker than for $\varepsilon = 30$ eV (Fig. 3) and are thus consistent with the observations of Alexander et al. (2008). Furthermore, the minimum in $R(\theta)$ has moved from 0.5 mrad at $\varepsilon = 30$ eV to 0.3 mrad at

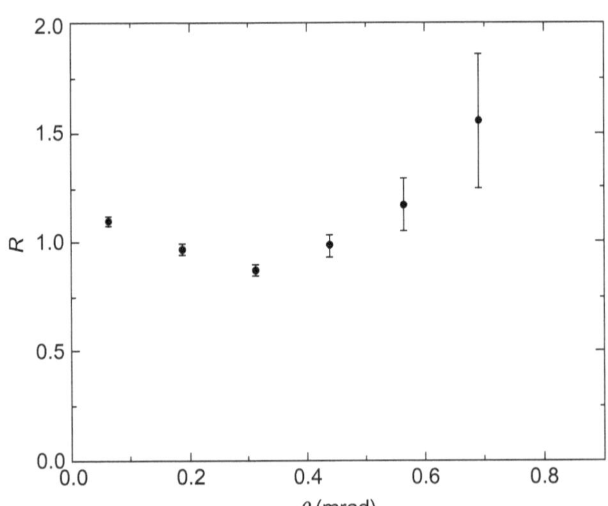


**Fig. 4** Same as *right panel* of Fig. 3 for $\varepsilon = 57$ eV.

$\varepsilon = 57$ eV. A fully differential analysis for the same energy loss, which will be discussed in Section 3.2, provided an explanation for the loss of coherence suggested in Alexander et al. (2008).

Later, a very similar experiment was performed for the same collision system but studying electron capture from the target to the projectile (Sharma et al., 2012). The single differential coherent to incoherent cross section ratio for this process is shown in Fig. 5 as a function of scattering angle. The structures observed in the capture data look very similar to those observed for ionization, except that, if any minimum is present at all, it occurs at significantly larger scattering angle than in ionization.

While several datasets taken for two different processes seem to consistently illustrate projectile coherence effects this interpretation of the data was disputed by Feagin and Hargreaves (2013). They asserted that the differences observed between the cross sections taken for different slit distances are not due to different projectile coherence properties, but merely due to different beam divergences depending on the slit distance. Indeed, by convoluting theoretical cross sections (which assume a fully coherent beam) over what Feagin and Hargreaves assumed to be the beam divergence they could simultaneously reasonably reproduce the measured cross sections for the coherent and incoherent beams. However, they assumed that the beam divergence was identical to the local collimation angle, an assumption which is not justified. The local collimation angle, which determines the transverse

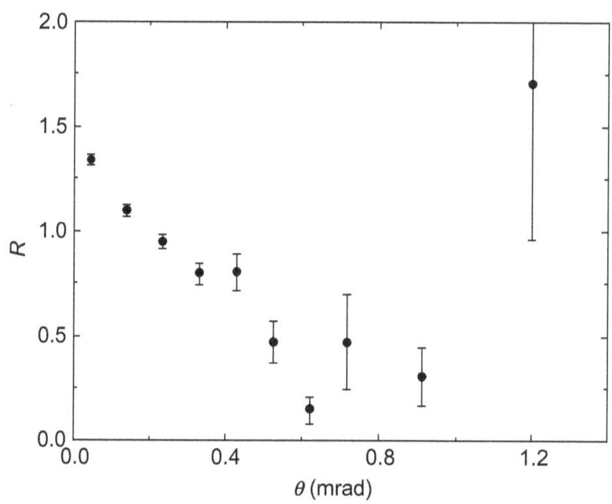


**Fig. 5** Coherent to incoherent differential capture cross section ratio for 75 keV $p+H_2$ collisions as a function of scattering angle.

coherence length (see Eq. 1), is defined as the angle subtended by the collimating slit at the target location. In contrast, the beam divergence is determined by the focusing properties of the beam. Only if the focal point is located at the target is the local collimation angle and the beam divergence equal to each other. However, in measurements of cross sections as a function of scattering angle, it is important to move the focal point as far away as possible from the target location (i.e., to minimize the beam divergence) in order to optimize the angular resolution.

Another theoretical study, analyzing the double differential data of Egodapitiya et al. (2011b), was performed by Sarkadi and Barrachina (2016). They followed the method of Karlovets et al. (2015) who earlier had shown that differential cross sections for an (partly) incoherent beam can be expressed as a convolution of the differential cross sections for a fully coherent beam with a Gaussian wave packet describing the projectile, where the width of the wave packet reflects the coherence length. As shown in Fig. 6, Sarkadi et al. were also able to simultaneously reproduce the shape of the scattering angle dependence of the coherent and incoherent DDCS measured by Egodapitiya et al., although in the incoherent case there was a

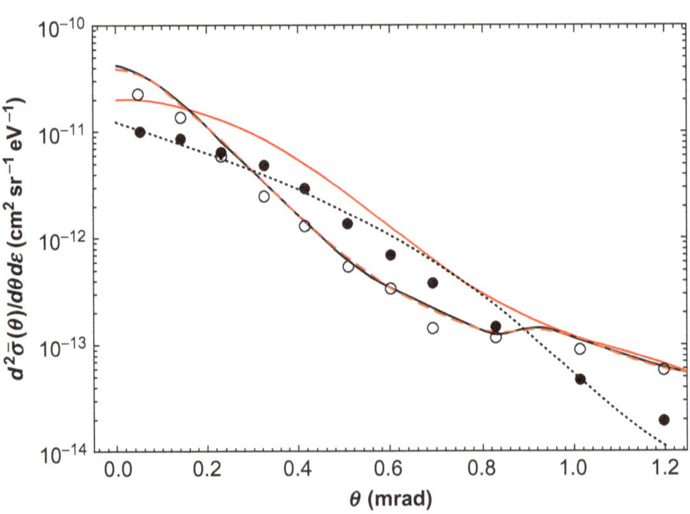


**Fig. 6** Same as *right panel* of Fig. 3. The *solid black curve* shows a calculation based on the M3DW model assuming a fully coherent beam (Chowdhury et al., 2011). The *dotted curve* represents a convolution of the M3DW with a beam divergence of 2.3 mrad for the small slit distance performed by Feagin and Hargreaves (2013). The *red dashed (solid) curves* show the calculations of Sarkadi and Barrachina (2016) for coherence lengths corresponding to the small (large) slit distances.

discrepancy in absolute magnitude by about a factor of two. However, they obtained the agreement in shape without convoluting the theoretical cross sections with any beam divergence. They therefore concluded that the differences between the DDCS measured with coherent and incoherent projectile beams are indeed due to a true, resolution-independent coherence effect and not merely due to a beam divergence effect, as suggested by Feagin and Hargreaves.

Nevertheless, one cannot rule out the possibility that the beam divergences for the coherent and incoherent beams in the experiment of Egodapitiya et al. may indeed have differed (i.e., that the focus of the beam was not located far enough from the target). In that case, the beam divergence could contribute to the differences in the cross sections measured for both slit distances in addition to coherence effects. Therefore, a potential influence of the beam divergence was investigated by performing a test which was actually proposed by Feagin and Hargreaves (2013). They pointed out that the projectile scattering angle can be obtained using two independent methods: it can be obtained from (a) the momentum information of the detected projectiles and (b) from the sum of the $x$-components of the electron and recoil-ion momenta. This sum is equal to the transverse component of $q$, which in turn is related to the scattering angle by

$$p_{elx} + p_{recx} = q_x = p_o \sin\theta \tag{2}$$

If the experimental resolution was infinitely good the scattering angles determined from both methods would have to be identical, i.e., the difference between them $\Delta\theta$ would be a $\delta$-function. Therefore, the width of the actual distribution of $\Delta\theta$ reflects the overall experimental resolution, i.e., it provides an upper limit for the beam divergence.

With the setup shown in Fig. 1, the electron momentum cannot be measured. However, the experiment is kinematically overdetermined because a total of six momentum components are measured and only five of them are independent (because energy is conserved in addition to momentum). The redundant information can be used to obtain $p_{elx}$ independently of $q_x$. An experimental spectrum of $\Delta\theta$ was recently reported (Sharma et al., 2014b) and is shown in Fig. 7. No differences in the resolution between the coherent and incoherent beam are found. Furthermore, the width of the peak is less than 0.3 mrad, which stems mostly from the resolution in the recoil-ion momentum. The resolution in the scattering angle from the direct projectile measurement is about a factor of two better. In contrast,

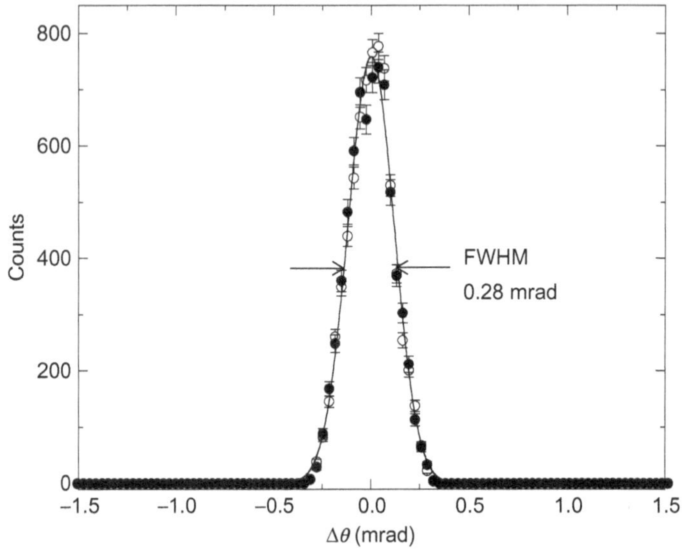


**Fig. 7** Difference spectrum between the scattering angles determined from the projectiles directly and from the target fragments (see text). *Closed symbols*: data taken for a coherent beam; *open symbols*: data taken for an incoherent beam. The *solid curve* shows a Gaussian fit.

Feagin and Hargreaves had to assume a beam divergence of more than 2 mrad in order to reproduce the experimental DDCS taken for an incoherent beam. Therefore, the data of Fig. 7 do not support their analysis.

## 3.2 Fully Differential Studies

After potential differences in the beam divergence can be ruled out as a source for the differences in the DDCS taken for varying slit distances the data shown in Figs. 3–5 strongly suggest that indeed the collision dynamics can be significantly affected by the projectile coherence properties. However, in the DDCS not all kinematic parameters are determined, which makes it difficult to extract detailed information about the interference from $R$ and, more specifically, about the phase factor. For example, the phase angle in molecular two-center interference is expected to be determined by the recoil-ion momentum (Corchs et al., 1994). But in the DDCS shown in Fig. 3, a distribution of recoil-ion momenta contributes to each scattering angle, which means that the phase angle is not fixed. In order to study the phase factor in more detail a series of experiments measuring FDCS was performed (Arthanayaka et al., 2015a,b; Hasan et al., 2014b; Sharma et al., 2014c).

Fig. 8 illustrates the geometry and coordinate system that is used to present the FDCS. The positive $x$-direction is defined by the transverse component of the momentum transfer, i.e., by choice of coordinate system $q_x$ is always positive and $q_y = 0$. The positive $z$-axis is defined by the initial projectile beam direction ($p_o$). For a given electron momentum, $p_{el}$ the polar angle $\theta_{el}$ is measured relative to the $z$-axis and the azimuthal angle $\varphi_{el}$ is (somewhat unconventionally) the angle between the positive $y$-axis and the projection of $p_{el}$ onto the $xy$-plane. The FDCS are presented for fixed energy loss $\varepsilon$ (or equivalently fixed electron energy $E_{el} = \varepsilon - I$, where $I$ is the ionization potential), $\theta_{el}$, and either $q_x$ or $p_{recx}$. Fixing $\theta_{el}$ means that only electrons are analyzed for which $p_{el}$ lies on the surface of a cone centered on the $z$-axis indicated in Fig. 8 by the shaded area. The data are then plotted as a function of $\varphi_{el}$. With the coordinate system introduced in Fig. 8 and using Eq. (2), $\varphi_{el}$ is given by

$$\sin\varphi_{el} = (q_x - p_{recx})/(p_{el}\sin\theta_{el}) \tag{3}$$

If $p_{recx}$ is fixed in addition to $p_{el}$ and $\theta_{el}$ then $q_x$ is the only variable on the right-hand side of Eq. (3) so that any variation of a spectrum in $\varphi_{el}$ corresponds to a variation in $q_x$. Likewise, if $q_x$ is fixed any variation in $\varphi_{el}$ corresponds to a variation in $p_{recx}$.

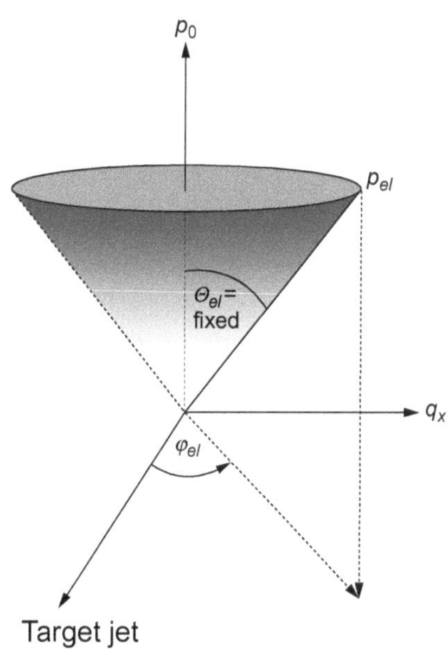


**Fig. 8** Coordinate system and geometry in which FDCS are presented.

In Fig. 9, the ratio $R$ between the FDCS measured by Sharma et al. (2014c) for a coherent and an incoherent beam, i.e., the fully differential interference term, is plotted for $\varepsilon = 30$ eV, $\theta_{el} = 35$ degree, and $p_{recx} = 0.2$ a.u. as a function of $\varphi_{el}$. Once again, the pronounced structure in these data shows the important role of projectile coherence and the presence of interference. However, this structure is not consistent with what is expected for molecular two-center interference, for which the interference term was thought to be given by (Corchs et al., 1994):

$$I_2 = 1 + \sin(p_{rec}D)/(p_{rec}D) \tag{4}$$

This expression, which is plotted as the dashed curve in Fig. 9, is not even in qualitative agreement with the data. Only a rather weak structure is present in $I_2$ because $p_{recx}$ and $p_{recz}$ are fixed ($p_{recz}$ only depends on $\varepsilon$ and $p_{elz}$, which is determined by the fixed values of $\varepsilon$ and $\theta_{el}$) so that a variation of the phase angle can only occur through $p_{recy}$. Furthermore, a maximum occurs where the data exhibit a minimum.

Based on the poor agreement between $I_2$ and the measured $R$ of Fig. 9, Sharma et al. (2014b,c) concluded that the variation in $R$ reflects a

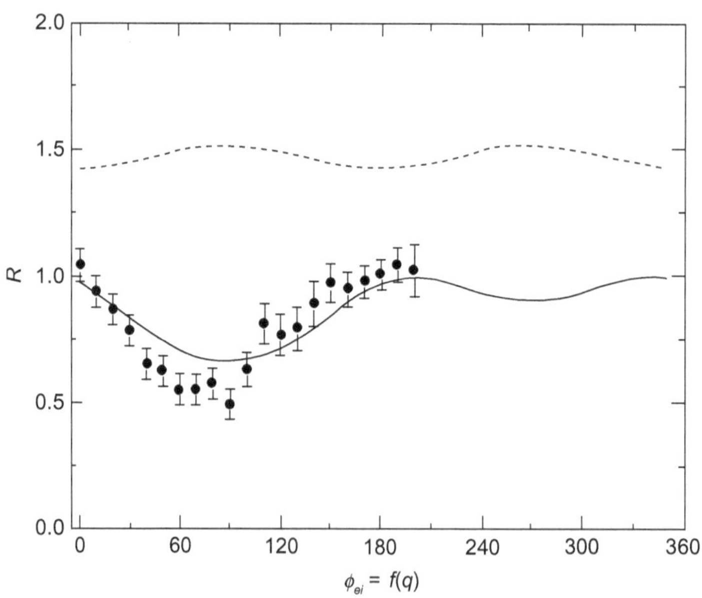


**Fig. 9** Fully differential coherent to incoherent cross section ratios for ionization in 75 keV $p+H_2$ collisions for $\varepsilon = 30$ eV, $\theta_{el} = 35$ degree, and $p_{recx} = 0.2$ a.u. as a function of $\varphi_{el}$. *Dashed curve*: two-center interference term based on Eq. (4); *solid curve*: single-center interference term based on Eq. (5).

different type of interference. They interpreted the structure as due to interference between different impact parameters $b$ leading to the same scattering angle. Such interference is particularly important when both first- and higher-order transition amplitudes are large because the former tends to contribute at relatively large $b$ and the latter at small $b$. Since this type of interference does not require multiple scattering centers it is referred to as single-center interference. Using a simple geometric model, Sharma et al. found the following expression for the single-center interference term

$$I_1 = 1 + \alpha \cos(q_x \Delta b) \quad (5)$$

Here, $\Delta b$ is the impact parameter separation between the interfering amplitudes and $\alpha$ accounts for the damping of the interference structure due to incomplete coherence (incoherence) at the large (small) slit distance and due to the experimental resolution. The solid curve in Fig. 9 shows a fit of $I_1$ to the measured ratios using $\Delta b = 2$ a.u. and $\alpha = 0.5$. The good agreement with the data lends some support to the interpretation that the structure in $R$ is due to single-center interference.

Similar observations were made for $\varepsilon = 57$ eV (Arthanayaka et al., 2015b), for which the same ratio using the same kinematic parameters as for $\varepsilon = 30$ eV are shown in Fig. 10. A structure very similar to the one found for the smaller energy loss is observed. Here, too, the data are very well reproduced by Eq. (5) using $\Delta b = 2$ a.u. However, in this case a best fit is obtained for $\alpha = 0.3$ indicating that single-center interference is not as pronounced as for $\varepsilon = 30$ eV.

The observation that the structures in Figs. 9 and 10 exhibit the characteristic features of single-center interference does not imply that molecular two-center interference is insignificant. For these plots, the recoil-ion momentum (except for $p_{recy}$), and therefore the phase angle, is nearly constant. Therefore, this presentation of the FDCS ratios is not ideally suited to identify two-center interference. Instead, these ratios should be analyzed for fixed momentum transfers. In this case, any variation in $\varphi_{el}$ corresponds to a variation in $p_{rec}$ so that the ratios become sensitive to the phase angle of the molecular two-center interference term. Such a plot is shown in Fig. 11 for $\varepsilon = 57$ eV, $\theta_{el} = 35$ degree, and $q_x = 0.64$ a.u. as a function of $\varphi_{el}$. Again, a structure is found, which, however, is not as pronounced as in the plots for constant $p_{recx}$.

The dependence of $R$ on $\varphi_{el}$ for fixed $q_x$ cannot be fitted by the single-center interference term of Eq. (5) if $\Delta b$ is constant because that

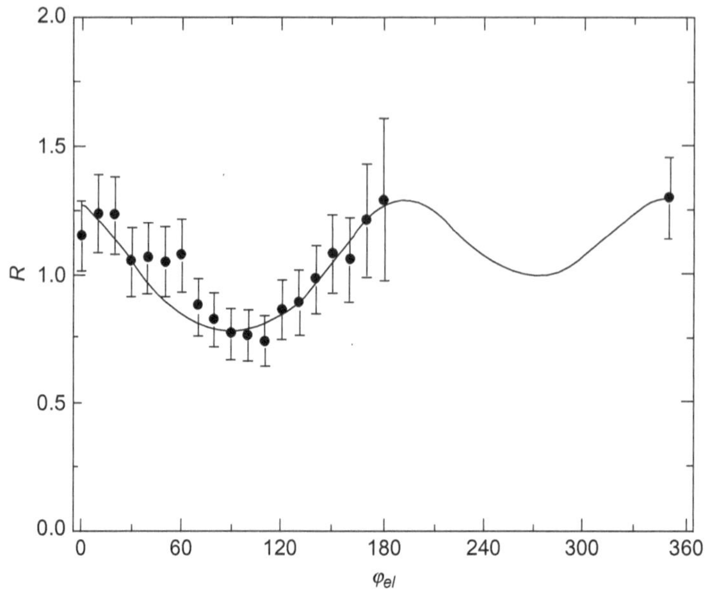


**Fig. 10** Fully differential coherent to incoherent cross section ratios for ionization in 75 keV $p+H_2$ collisions for $\varepsilon=57$ eV, $\theta_{el}=35$ degree, and $p_{recx}=0.2$ a.u. as a function of $\varphi_{el}$. *Solid curve*: $I_1$ calculated with Eq. (5).

would make $I_1$ independent of $\varphi_{el}$. On the other hand, it cannot be taken for granted that $\Delta b$ is independent of $\varphi_{el}$. Therefore, the observation that $R$ is not constant does not necessarily mean that the structure cannot be due to single-center interference. However, the success of reproducing the ratios for fixed $p_{recx}$ by the single-center interference term with constant $\Delta b$ suggest that for this collision system it may be a reasonable approximation to treat $\Delta b$ as constant. Furthermore, the data of Fig. 11 can very well be reproduced by the molecular two-center interference term of Eq. (4) using a damping factor of $\alpha=0.7$ (solid curve in Fig. 11). Therefore, it seems likely that the structure in Fig. 11 is to a large extent, perhaps even predominantly, due to molecular two-center interference.

The analysis of the coherent and incoherent FDCS for $\varepsilon=57$ eV also provides an explanation for the weaker interference structure observed in Alexander et al. (2008) in the coherent DDCS for the same energy loss. Since these cross sections are presented as a function of scattering angle (or equivalently transverse momentum transfer) they are mostly affected by single-center interference, for which the phase angle depends on $q_x$. Since the phase angle of two-center interference depends on the recoil-ion

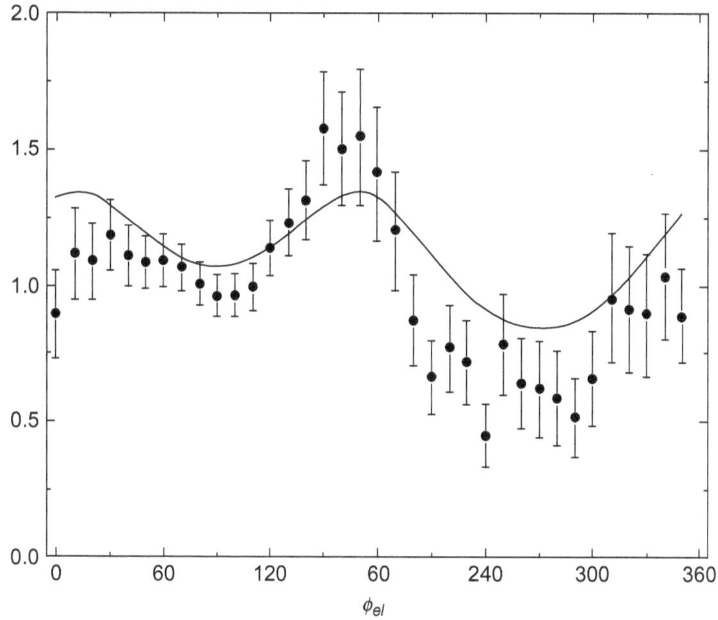


**Fig. 11** Fully differential coherent to incoherent cross section ratios for ionization in 75 keV $p+H_2$ collisions for $\varepsilon=57$ eV, $\theta_{el}=35$ degree, and $q_x=0.64$ a.u. as a function of $\varphi_{el}$. *Solid curve*: $I_2$ calculated with Eq. (4).

momentum, which is not determined in the DDCS reported in Alexander et al. (2008), any structure due to this type of interference is to a large extent "washed out." On the other hand, the measured ratios plotted in Figs. 9 and 10 show that single-center interference is weaker at $\varepsilon=57$ eV than at $\varepsilon=30$ eV (as indicated by the smaller fitted value of $\alpha$). This supports the hypothesis expressed in Alexander et al. (2008) that for some reason the coherence may be suppressed at 57 eV compared to 30 eV.

Recently, Arthanayaka et al. (2015a) have provided a possible explanation for a loss of coherence near the electron–projectile velocity matching (i.e., for $\varepsilon=57$ eV). They analyzed DDCS with an additional condition on the longitudinal recoil-ion momentum to either enhance or suppress PCI. If the electron is ejected in the forward direction due to momentum conservation the recoil ion must have a large momentum opposite to the beam direction. Therefore, setting a condition on large negative longitudinal recoil-ion momenta (to which we refer as the "PCI on" condition) enhances PCI. Likewise, a condition on positive longitudinal recoil-ion momenta ("PCI off" condition) suppresses PCI. Arthanyaka et al. found a more pronounced single-center interference structure in the data with the PCI off condition than in those with the PCI on condition. This observation was interpreted by analyzing a classical trajectory Monte Carlo calculation (for details of the model, see Sarkadi, 2010). There, it was found that for $\varepsilon = 57$ eV typical impact parameters contributing to ionization are much larger for the PCI on condition than for the PCI off condition. The authors argued that that would likely make $\Delta b$, entering in the phase angle of the single-center interference term, larger as well. Therefore, the coherence condition $\Delta x > \Delta b$ is more difficult to satisfy for the PCI on condition, which results in the loss of coherence suggested in Alexander et al. (2008). Larger impact parameters are favored by the PCI on condition for $\varepsilon = 30$ eV as well, however, the effect is not nearly as large as for $\varepsilon = 57$ eV.

## 4. COHERENCE EFFECTS IN COLLISIONS WITH HELIUM

While the studies reviewed in Section 3 demonstrate the important role of projectile coherence for molecular targets, the discussion on this topic was actually triggered by the unsatisfactory agreement between theory and experiment for ionization of atomic targets by fast heavy ions. Therefore, several experiments were performed studying coherence effects in the FDCS for single ionization and transfer ionization (TI) of helium (Arthanayaka et al., 2016; Gassert et al., 2016; Schneider et al., 2013; Wang et al., 2012).

Arthanayaka et al. measured FDCS for ionization of helium by 75 keV proton impact for the same two slit distances (i.e., for the same $\Delta x$) as in the experiments on 75 keV $p + H_2$ collisions (Arthanayaka et al., 2016). The data were analyzed using the same geometry and coordinate system as was used for the $H_2$ target and which is illustrated in Fig. 6. The FDCS are shown in Fig. 12 as a function of $\varphi_{el}$ for $p_{recx} = 0.2$ a.u. and $\theta_{el} = 25$ degree (left panel), $p_{recx} = 0.7$ a.u. and $\theta_{el} = 65$ degree (center panel), and $p_{recx} = 1.25$ a.u. and $\theta_{el} = 45$ degree (right panel). The closed (open) symbols represent the data measured with a coherent (incoherent) beam. The shape of the angular dependence of the FDCS is quite sensitive to the kinematic parameters. While for $p_{recx} = 0.2$ a.u. a broad maximum is seen at $\varphi_{el} = 90$ degree, the largest structure is centered at $\varphi_{el} = 270$ degree for the larger values of $p_{recx}$. Furthermore, for $p_{recx} = 0.7$ a.u. that structure is split into a double maximum, resulting in a shallow minimum at $\varphi_{el} = 270$ degree, at least for the coherent case. Generally, the structures become increasingly narrow with increasing $p_{recx}$.

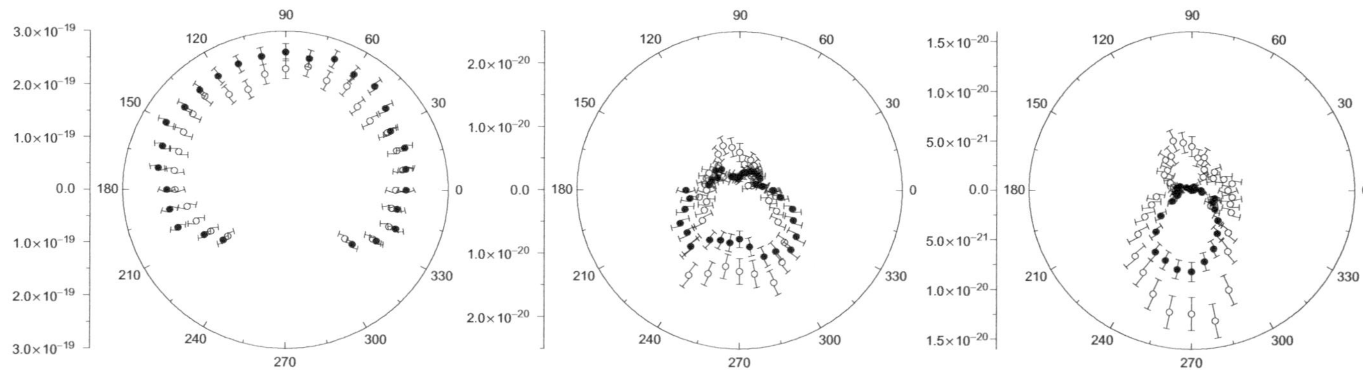


**Fig. 12** Fully differential cross sections for ionization in 75 keV $p+\text{He}$ for incoherent (*open symbols*) and coherent (*closed symbols*) projectile beams as a function of $\varphi_{el}$. The energy loss was fixed at 30 eV and $p_{recx}$ and $\theta_{el}$ at 0.2 a.u. and $\theta_{el}=25$ degree (*left panel*), 0.7 a.u. and 65 degree (*center panel*), and 1.25 a.u. and 45 degree (*right panel*).

The comparison between the FDCS for coherent and incoherent projectiles is quite sensitive to the kinematic parameters as well. At $p_{recx}=0.2$ a.u. basically no differences between both datasets can be discerned, except that the coherent FDCS are systematically slightly larger. In contrast, significant differences in shape are quite apparent for $p_{recx}=0.7$ and 1.25 a.u. Most notably, in the incoherent data a second pronounced maximum at $\varphi_{el}=90$ degree is observed. On the other hand, in the coherent case this maximum is much weaker and, like the structure at 270 degree, split into two maxima for $p_{recx}=0.7$ a.u. and it is not present all for $p_{recx}=1.25$ a.u.

The differences between the coherent and incoherent FDCS are once again more prominent in the ratios, which are plotted for the same kinematic settings as the FDCS in Fig. 13. The solid curves show nonperturbative calculations (Arthanayaka et al., 2016). In this model, which builds on previous theoretical studies of coherence effects (Jarai-Szabo and Nagy, 2015), an impact parameter-dependent transition amplitude $a(b)$ is computed by solving the time-dependent Schrödinger equation numerically (Borbély et al., 2014). The interaction between the two nuclei of the collision system is accounted for in terms of an eikonal phase. $a(b)$ is then multiplied by a Gaussian wave packet describing the incoming projectile wave, where the width of the wave packet reflects the coherence length. Finally, this product is Fourier-transformed to obtain a momentum transfer-dependent transition amplitude $R(q)$. The ratio is then obtained from calculations using widths for the projectile wave packet which reflect the two coherence lengths realized in the experiment.

There are significant quantitative discrepancies between the experimental and calculated data. These are believed to originate mostly from the omission of projectile states in the basis set. This means that the capture channel is not accounted for in the calculation. Rather, capture events are erroneously counted as ionization events. However, the $\varphi_{el}$ dependence of the ratios is qualitatively well reproduced by the calculations. More specifically, experiment and theory agree that the ratio is nearly constant for $p_{recx}=0.2$ a.u., that there are minima at $\varphi_{el}=90$ and 270 degree for $p_{recx}=0.7$ a.u., that the minimum at 90 degree approaches 0 with increasing $p_{recx}$, and that the minimum at 270 degree is turned into a maximum for $p_{recx}=1.25$ a.u. Apart from the observation that the beam divergence does not differ for the two slit distances (see Fig. 7), it is also difficult to see how beam divergence effects could mimic in so much detail the coherence effects predicted by theory. Rather, one would expect that, if the beam divergence was larger for the

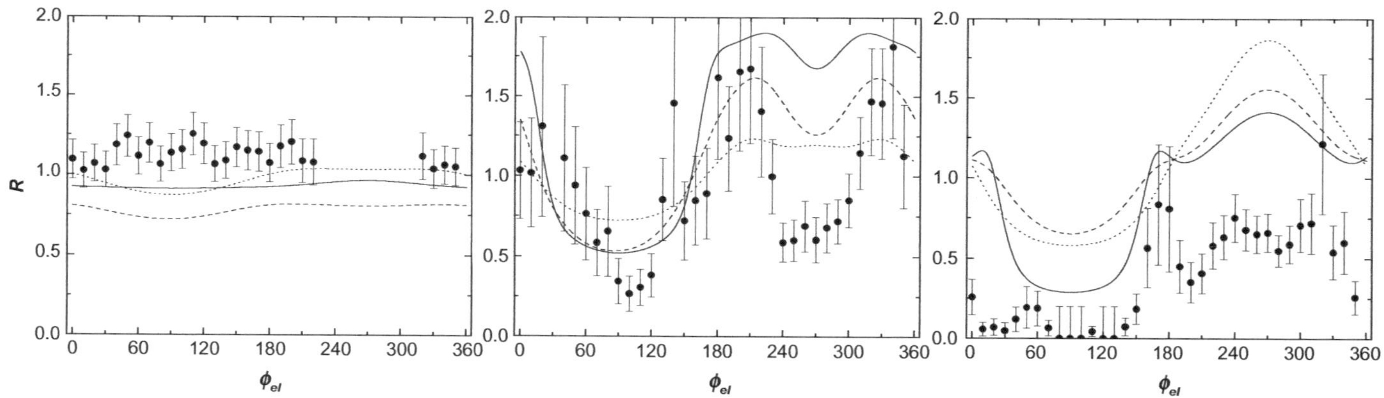


**Fig. 13** Ratio between the coherent and incoherent FDCS from Fig. 12. The *curves* show calculations based on an impact parameter formulation in which the projectile is described by a wave packet, where the *width* reflects the coherence length. *Dotted curves*: first-order treatment of the transition amplitude; *dashed curve*: transition amplitude includes higher-order contributions in the projectile–electron interaction; *solid curves*: transition amplitude is based on numeric solutions of the time-dependent Schrödinger equation and includes all higher-order contributions.

small slit distance, it would lead to a systematic broadening of the FDCS. Instead, the structures in the FDCS of Fig. 12 appear to be of equal width and even narrower for the small slit distance for $p_{recx} = 0.7$ a.u. Therefore, the comparison between the experimental and theoretical ratios strongly suggests that true, resolution-independent coherence effects are present in the data.

In addition to confirming coherence effects theory also provides valuable information about the single-center interference term. The dotted and dashed curves in Fig. 13 show different variations of the calculations using the same basic model as the one underlying the solid curves in that figure. In case of the dotted curve, all higher-order contributions to the transition amplitude are omitted in the calculation. Therefore, the collision proceeds under the influence of only a single interaction between the projectile and the active electron. In case of the dashed curve, multiple interactions between the projectile and the active electron (i.e., PCI) are accounted for, but higher-order contributions involving the projectile–target nucleus interaction are not. The solid curves represent the full calculation, including all higher-order contributions.

In all three calculations pronounced structures in the ratios are found. This shows that single-center interference is not necessarily due to a superposition of first- and higher-order amplitudes, as suggested by Egodapitiya et al. (2011b). Rather, it should be interpreted as interference between different impact parameters contributing to the same scattering angle. Regardless of how important higher-order contributions may be, there is always a broad spectrum of impact parameters contributing to the same scattering angle since the impact parameter is not an observable, i.e., the projectiles do not follow well-defined trajectories. In theory, the interference between different impact parameters is reflected by the Fourier transform from $a(b)$ to $R(q)$. However, multiplying $a(b)$ by a narrow wave packet suppresses the contributions of a large range of impact parameters to the Fourier integral and asymptotically, for the width approaching 0, the classical (completely incoherent) limit of a single well-defined trajectory is reached. Likewise, if the wave packet has infinite width then all impact parameters are weighted equally, which corresponds to the fully coherent case. By varying the width of the wave packet all degrees of coherence between these extreme scenarios can be realized in the calculation.

The three sets of calculations plotted in Fig. 13 differ substantially from each other. Therefore, although higher-order contributions are not essential for single-center interference, they are nevertheless important for this specific collision system and for the specific kinematic parameters for which the data were analyzed. Furthermore, the comparison between these calculations sensitively shows the effects of the various types of higher-order contributions. The differences between the dashed curves, accounting only for PCI, and the dotted curves, not accounting for any higher-order effects, illustrates how the interference term is affected by PCI. Likewise, the important role of the projectile–target nucleus interaction is reflected by the differences between the solid curves (all higher-order contributions included) and the dashed curves. Furthermore, the differences between the full calculation and the dashed curves seem to increase with increasing $p_{recx}$ showing that the projectile–target nucleus interaction becomes increasingly important with increasing $p_{recx}$, as expected. This sensitivity of the interference term on how higher-order contributions are treated in the theoretical model opens the possibility to extract detailed information about the few-body dynamics from the analysis of the interference term.

The aforementioned discrepancies between experimental and theoretical FDCS for ionization were particularly surprising for 100 MeV/amu $C^{6+}$ + He collisions (Schulz et al., 2003a). The FDCS for this collision system are plotted as open symbols in Fig. 14 for $q = 0.75$ a.u., $E_{el} = 6.5$ eV, and $\theta_{el} = 90$ degree as a function of $\varphi_{el}$. The solid curve shows a fully quantum–mechanical calculation based on the 3-body distorted wave (3DW) approach (Madison et al., 2002). While in the calculation two pronounced peak structures (known as the binary and recoil peaks) are visible, which are separated from each other by a deep minimum, in the data that minimum is almost completely filled up.

Since the transverse coherence length of the $C^{6+}$ projectiles was only about $10^{-3}$ a.u., and coherence effects had been demonstrated for molecular targets already, it seemed sensible to investigate whether similar effects could also be responsible for these discrepancies. However, for the very small de Broglie wavelength of such a fast and massive projectile ($\lambda \approx 5 \times 10^{-6}$ a.u.), it is extremely difficult to generate a coherent beam even when taking advantage of electron cooling at a storage ring. Therefore, a follow-up experiment was performed for 3 MeV $p$ projectiles ($\lambda \approx 3 \times 10^{-4}$ a.u.) at the storage ring in Heidelberg (Wang et al., 2012). Owing to a combination of the much smaller momentum compared to a 100 MeV/amu $C^{6+}$ ion and the narrow momentum spread obtained by electron cooling (Steck et al., 1990), an increase in coherence length by 3–4 orders of magnitude could be achieved. Furthermore, the perturbation parameters $\eta$ (projectile charge to speed ratio) for 3 MeV $p$ and 100 MeV/amu $C^{6+}$ are identical. For the

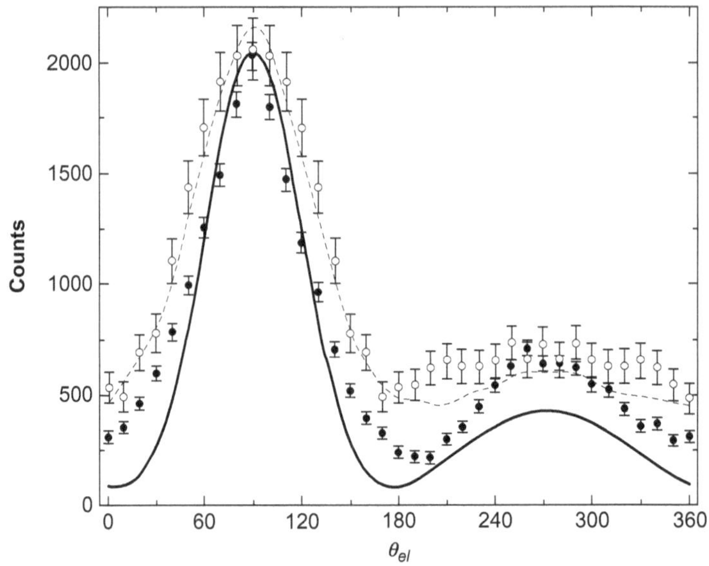


**Fig. 14** Fully differential cross sections for ionization in 100 MeV/amu $C^{6+}$ + He (*open symbols*, incoherent beam) and 3 MeV $p$ + He collisions (*closed symbols*, coherent beam) as a function of $\varphi_{el}$. In both cases $E_{el} = 6.5$ eV, $\theta_{el} = 90$ degree, and $q = 0.75$ a.u. The *solid curve* shows a coherent calculation for the $p$ projectiles based on the 3DW model and the *dashed curve* a simulation of a calculation for an incoherent beam (see text).

same perturbation of such a small value ($\eta = 0.1$), theories assuming a coherent beam predict that the FDCS are practically identical.

The FDCS for 3 MeV $p$ + He are plotted as closed symbols in Fig. 14 for the same kinematic conditions as for the $C^{6+}$ projectiles. Indeed, both datasets differ significantly and qualitatively from each other. More specifically, a pronounced minimum separating the binary and recoil peaks is visible in the FDCS for proton impact, as predicted by theory. Furthermore, both peak structures are much narrower than in the $C^{6+}$ case, although the resolution was better in the latter case. Overall, the agreement with theory is significantly improved. The remaining discrepancies are probably due to the incomplete coherence even for the $p$ projectiles (i.e., $\Delta x \neq \infty$) and due to the experimental resolution. At the same time, the $C^{6+}$ data are very well reproduced by a calculation simulating an incoherent projectile beam (dashed curve in Fig. 14). Here, the results of a first Born calculation were convoluted with classical elastic scattering between the projectile and the target nucleus (Schulz et al., 2007). In this convolution, a classical relation between the momentum transfer and the impact parameter, based on the Rutherford

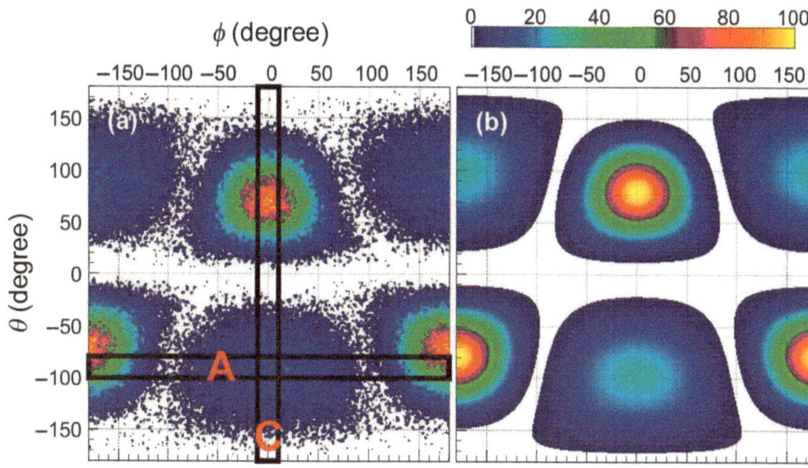


**Fig. 15** Fully differential cross sections for ionization in 1 MeV $p$+He collisions for $E_{el}$=6.5 eV and $q$=0.75 a.u. as a function of $\varphi_{el}$ and $\theta_{el}$. *Left panel*: experimental data; *right panel*, calculation based on the first Born approximation assuming a fully coherent beam.

model, was assumed, which implies $\Delta x = 0$. The comparison between the theory and the experiment suggests that the FDCS contain significant contributions from single-center interference, which is, however, not observable for 100 MeV/amu $C^{6+}$ projectiles due to the tiny coherence length.

More recently, FDCS for ionization were measured for 1 MeV $p$+He collisions at the accelerator facility of the University of Frankfurt (Gassert et al., 2016). The widths of the slits defining the local collimation angle were different in the vertical and horizontal directions resulting in coherence lengths of $\Delta x = 1.3$ and 2.6 a.u., respectively.[1] In the left panel of Fig. 15, the FDCS measured for $q = 0.75$ a.u. and electron energy $E_{el} = 6.5$ eV are plotted as a function of the azimuthal (horizontal axis) and polar (vertical axis) electron ejection angles. Here, no selection was made for projectiles scattering in the vertical or horizontal direction so that the effective coherence length is given by the average ($\Delta x \approx 2.0$ a.u.) of the values for the vertical and horizontal collimating slit orientations. The right panel of that figure shows a calculation based on the FBA for the same kinematic parameters, which treats the collision fully coherently. The experimental data are reasonably well reproduced by theory. Furthermore, no significant

[1] The coherence lengths reported in Gassert et al. (2016) were calculated incorrectly (Schöffler and Dörner, 2016) and are too small by nearly a factor of two for the smaller coherence length.

differences between the FDCS for $\Delta x = 1.3$ and 2.6 a.u. were observed. The authors of that study therefore concluded that no projectile coherence effects were observed in their data. However, even the smaller value of $\Delta x$ is three orders of magnitude larger than for the 100 MeV/amu $C^{6+}$ data and larger than the size of the He atom. Furthermore, for this collision system, corresponding to a small perturbation ($\eta = 0.16$), the impact parameters contributing to ionization are much smaller than for 75 keV $p+$He. It is thus not clear whether $\Delta x = 1.3$ a.u. is small enough to observe a significant departure of measured FDCS from fully coherent calculations. Regardless of what conclusions may be drawn from these data regarding a potential role of projectile coherence effects this study is very valuable as it demonstrates that single ionization of simple atoms by coherent projectiles corresponding to a small perturbation parameter is essentially understood.

Kouzakov et al. argued that in experiments momentum-analyzing only the recoil ions and the ejected electrons (and not the scattered projectiles), the FDCS are not affected by the projectile coherence properties (Kouzakov et al., 2013). However, in their analysis, a crucial approximation is used and it is not clear to what extent this approximation is justified. The cross section for wave packet scattering is proportional to $\int d^3p|T_{fi}(p,p')\varphi(p)|^2$, where $T_{fi}(p,p')$ is the transition amplitude, $p'$ is the final projectile momentum, and $\varphi(p)$ is the wave packet describing the incident projectile (see, e.g., Taylor, 2006). Therefore, the initial inherent momentum spread of the projectile wave packet directly alters the transition amplitude, regardless of whether or not the projectile is actually momentum analyzed in the experiment. Kouzakov et al. made the approximation that the variation in $T_{fi}(p,p')$ can be neglected in the region where the wave packet differs significantly from 0 (Kouzakov et al., 2013). This condition does not appear to be satisfied for small coherence lengths. For example, the very small value of $\Delta x$ for the $C^{6+}$ case corresponds to a transverse width of the wave packet in momentum space of about $10^3$ a.u. Over such an extremely large transverse momentum range the transition amplitudes varies over many orders of magnitude. Even within just a few a.u. the cross sections typically drop by several orders of magnitude (e.g., Voitkiv et al., 2003b). By assuming that the spread of the wave packet in momentum space is small compared to the variation in $T_{fi}(p,p')$. Kouzakov et al. implicitly assume right from the onset that the projectile is coherent. Therefore, their analysis does not allow for any conclusions regarding potential variations of the cross sections with the coherence length.

Another theoretical analysis of coherence effects in ionization of He by 100 MeV/amu $C^{6+}$ and 1 MeV $p$ impact was recently reported by Navarrete et al. (2016). They used the same approach that was already successfully applied to reproduce coherence effects observed in the DDCS for $p+H_2$ collisions. Although this method is similar to the one used by Kouzakov et al., Navarrete et al. reach very different conclusions. They were able to simultaneously reproduce the experimental data for both 100 MeV/amu $C^{6+}$ and 1 MeV $p$ impact with amazing accuracy using an incoherent calculation. Furthermore, they found significant differences between the incoherent and a fully coherent calculation for 100 MeV/amu $C^{6+}$ projectiles, but only very small differences for 1 MeV $p$ impact. Their results therefore support the interpretation that the discrepancies between experiment and theory for 100 MeV/amu $C^{6+}+He$ collisions are due to the projectile coherence properties. Furthermore, they concluded that the data reported in Gassert et al. (2016) are not inconsistent with that interpretation.

Overall, the literature on studies of coherence effects in single ionization by fast ion impact is not as extensive as it is for slow collisions. Therefore, a final conclusion as to whether or not the discrepancies between measured and calculated FDCS for 100 MeV/amu $C^{6+}+He$ collisions are indeed mainly due to the projectile coherence properties has to await further experimental and theoretical studies.

Another strong indication for the importance of projectile coherence effects was obtained by the simultaneous measurement of FDCS for single ionization and TI in 16 MeV $O^{7+}+He$ collisions (Schneider et al., 2013). In TI one target electron is ejected to the continuum and the second is captured by the projectile. In single-center interference, the coherence requirement is $\Delta x > \Delta b$ where $\Delta b$ is the separation between the interfering impact parameters. The range of impact parameters contributing to such interference is not well known so that it is not easy to estimate $\Delta b$. However, it is well established that capture processes for fast ion impact are very selective on very small impact parameters. One would therefore expect that $\Delta b$ for TI is much smaller than for single ionization. This offers another possibility of varying the projectile coherence properties: instead of varying $\Delta x$ through the collimating slit geometry one can vary $\Delta b$ by simultaneously studying different processes in the same experiment, thereby keeping $\Delta x$ fixed. Such an experiment was performed by Schneider et al. studying single ionization and TI (Schneider et al., 2013).

In Fig. 16, fourfold differential cross sections (4DCS) for single ionization (open symbols) and TI (closed symbols) in 16 MeV $O^{7+}+He$ collisions

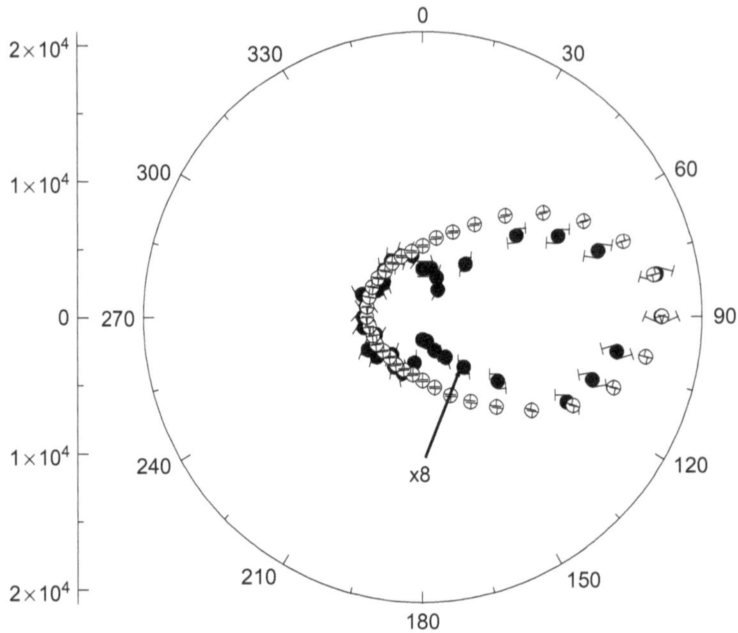


**Fig. 16** Fourfold differential cross sections for ionization (*open symbols*) and transfer ionization (*closed symbols*) in 16 MeV $O^{7+}$+He collisions as a function of $\varphi_{el}$ for $E_{el}$=8 eV eV and $\theta_{el}$=90 degree. The data are integrated overall scattering angles.

are shown as a function of $\varphi_{el}$. The electron energy $E_{el}$ is fixed at 8 eV and $\theta_{el}$ at 90 degree, while the data are integrated over all scattering angles. The transverse coherence length was estimated to be 0.5 a.u. For the large perturbation parameter for this collision system ($\eta \approx 1.1$) typical impact parameters (and therefore $\Delta b$) contributing to single ionization are expected to be larger than the size of the target atom, while for TI it should be much smaller. Therefore, Schneider et al. argued that the projectile beam was coherent with respect to TI and simultaneously incoherent with respect to single ionization. Indeed, the shape of the single ionization data closely resembles the one of the FDCS for ionization by 100 MeV/amu $C^{6+}$ impact (incoherent), while for TI the shape is very similar to the FDCS for 3 MeV $p$+He collisions (coherent).

The disadvantage of the experiment of Schneider et al. is that the angular electron ejection pattern is compared for two different processes. It cannot be entirely ruled out that the pronounced minimum separating the binary and recoil peaks in the TI data are caused by factors which do not affect single ionization, like e.g., electron–electron correlation effects. However, for this large perturbation parameter the capture and ionization steps are expected to be uncorrelated, i.e., they occur via two independent interactions between the projectile and the respective electron. Therefore, the shape of the 4DCS should be unaffected by the capture step and thus be very similar to single ionization if coherence effects were unimportant. The decisive advantage of this experiment is that the data for the supposedly coherent and incoherent cases are recorded simultaneously under identical experimental conditions. Any differences between features observed for single ionization and TI can thus not be explained by experimental artifacts, like e.g., the instrumental resolution or beam divergence effects. Therefore, the results of Schneider et al. have to be regarded as providing strong support for the important role of true, resolution-independent projectile coherence effects.

## 5. CONCLUSIONS

Since the beginning of quantum–mechanical scattering theory, it has been assumed that the projectile can generally be treated as being coherent. This forms the basis of stationary scattering theory. For decades, this assumption seemed to be supported by experimental data. The reason is that initially multiple differential data as a function of projectile parameters, especially the scattering angle, were mostly obtained for electron impact. Because of their small mass the de Broglie wavelength of electrons is usually so large that it is difficult to make the beam incoherent relative to atomic dimensions even at large energies.

For ion impact scattering angles and energy losses are much more difficult to measure because of the much larger mass. Therefore, experiments measuring these parameters were for a long time restricted to light and/or slow projectiles. Furthermore, only single or, at best, DDCS were measured. Therefore, the data were often not sensitive enough to observe coherence effects. In addition, in this regime too, the momenta were often still sufficiently small (i.e., the de Broglie sufficiently large) to make the beam at least partly coherent (for a typical collimating slit geometry) relative to the size of the diffracting object. This explains why the important role of the projectile coherence properties could go unnoticed for several decades.

With the advent of COLTRIMS FDCS measurements became feasible practically for any kind of projectile ion. Surprising discrepancies between experiment and theory seemed to suggest that there was some qualitative difference between the few-body dynamics for electron and for ion impact. But it took another decade before a possible explanation of this apparent puzzle emerged. A series of experiments, which were reviewed in this chapter, strongly suggest that cross sections can be significantly affected by the projectile coherence properties, especially for heavy and/or fast ions. This represents a major advancement in our understanding of atomic collision dynamics in general and the difference between electron and ion impact in particular. As far as basic one-electron processes occurring in collision systems with simple one- or two-electron targets are concerned, this may well represent the last chapter (apart from quantitative detail) that needed to be closed regarding the few-body dynamics of atomic fragmentation processes. If this optimistic view is further supported by future studies, the remaining major challenges in atomic collision physics are presented by processes involving multiple active electrons, where electron–electron correlation effects can be important, and by systems involving complex targets.

## ACKNOWLEDGMENTS

This work was supported by the National Science Foundation under grant no. PHY-1401586. Fruitful discussions with many friends and colleagues are gratefully acknowledged.

## REFERENCES

Alexander, J.S., Laforge, A.C., Hasan, A., Machavariani, Z.S., Ciappina, M.F., Rivarola, R.D., Madison, D.H., Schulz, M., 2008. Interference effects due to projectile target nucleus scattering in single ionization of $H_2$ by 75-keV proton impact. Phys. Rev. A 78, 060701.

Arthanayaka, T.P., Sharma, S., Lamichhane, B.R., Hasan, A., Remolina, J., Gurung, S., Sarkadi, L., Schulz, M., 2015a. Influence of the post-collision interaction on interference effects in ionization of $H_2$ by proton impact. J. Phys. B 48, 175204.

Arthanayaka, T.P., Sharma, S., Lamichhane, B.R., Hasan, A., Remolina, J., Gurung, S., Schulz, M., 2015b. Separation of single- and two-center interference in ionization of $H_2$ by proton impact. J. Phys. B 48, 071001.

Arthanayaka, T.P., Lamichhane, B.R., Hasan, A., Gurung, S., Remolina, J., Borbély, S., Járai-Szabó, F., Nagy, L., Schulz, M., 2016. Fully differential study of wave packet scattering in ionization of helium by proton impact. J. Phys. B 49, 13LT02.

Bartlett, P.L., Stelbovics, A.T., 2004. Threshold behavior of e-H ionizing collisions. Phys. Rev. Lett. 93, 233201.

Borbély, S., Feist, J., Tőkési, K., Nagele, S., Nagy, L., Burgdörfer, J., 2014. Ionization of helium by slow antiproton impact: total and differential cross sections. Phys. Rev. A 90, 052706.

Bray, I., 2002. Close-coupling approach to coulomb three-body problems. Phys. Rev. Lett. 89, 273201.

Builth-Williams, J.D., Bellm, S.M., Jones, D.B., Chaluvadi, H., Madison, D.H., Ning, C.G., Lohmann, B., Brunger, M.J., 2012. Experimental and theoretical investigation of the triple differential cross section for electron impact ionization of pyrimidine molecules. J. Chem. Phys. 136, 024304.

Builth-Williams, J.D., Bellm, S.M., Chiari, L., Thorn, P.A., Jones, D.B., Chaluvadi, H., Madison, D.H., Ning, C.G., Lohmann, B., da Silva, G.B., Brunger, M.J., 2013. A dynamical (e,2e) investigation of the structurally related cyclic ethers tetrahydrofuran, tetrahydropyran, and 1,4-dioxane. J. Chem. Phys. 139, 034306.

Childers, J.G., Thompson, D.B., Martin, N.L.S., 2001. (e,2e) experiments on the autoionizing levels of Xe between the $2P_{3/2}$ and $2P_{1/2}$ ionic limits. Phys. Rev. A 64, 062703.

Chowdhury, U., Schulz, M., Madison, D.H., 2011. Differential cross sections for single ionization of $H_2$ by 75-keV proton impact. Phys. Rev. A 83, 032712.

Colgan, J., Pindzola, M.S., Robicheaux, F.J., Griffin, D.C., Baertschy, M., 2002. Time-dependent close-coupling calculations of the triple-differential cross section for electron-impact ionization of hydrogen. Phys. Rev. A 65, 042721.

Corchs, S.E., Rivarola, R.D., McGuire, J.H., Wang, Y.D., 1994. Electron capture from $H_2$ targets by $H^+$ and $He^{2+}$ ions. Dependence of the cross sections on the orientation of the molecular axis. Phys. Scr. 50, 469.

da Silva, G.B., Neves, R.F.C., Chiari, L., Jones, D.B., Ali, E., Madison, D.H., Ning, C.G., Nixon, K.L., Lopes, M.C.A., Brunger, M.J., 2014. Triply differential (e,2e) studies of phenol. J. Chem. Phys. 141, 124307.

de Harak, B.A., Bartschat, K., Martin, N.L.S., 2008. Out-of-plane (e, 2e) experiments on helium $L=0$, 1, 2 autoionizing levels. Phys. Rev. Lett. 100, 063201.

Dorn, A., Elliot, A., Guo, X., Hurn, J., Lower, J., Mazevet, S., McCarthy, I.E., Shen, Y., Weigold, E., 1997. (e, 2e) collisions on xenon with spin-polarized electrons. J. Phys. B 30, 4097.

Dörner, R., Mergel, V., Jagutzski, O., Spielberger, L., Ullrich, J., Moshammer, R., Schmidt-Böcking, H., 2000. Cold target recoil Ion momentum spectroscopy: a 'Momentum Microscope' to view atomic collision dynamics. Phys. Rep. 330, 95.

Duguet, A., Cherid, M., Lahmam-Bennani, A., Franz, A., Klar, H., 1987. High-accuracy (e,2e) cross sections for helium: reference data in the first Born approximation. J. Phys. B 20, 6145.

Dürr, M., Dimopoulou, C., Najjari, B., Dorn, A., Bray, I., Fursa, D.V., Chen, Z.J., Madison, D.H., Bartschat, K., Ullrich, J., 2006. Single ionization of helium by 102 eV electron impact: three-dimensional images for electron emission. J. Phys. B 39, 4097.

Dürr, M., Dimopoulou, C., Najjari, B., Dorn, A., Bartschat, K., Bray, I., Fursa, D.V., Chen, Z.J., Madison, D.H., Ullrich, J., 2008. Higher-order contributions observed in three-dimensional (e,2e) cross-section measurements at 1-keV impact energy. Phys. Rev. A 77, 032717.

Egodapitiya, K.N., Sharma, S., Laforge, A.C., Schulz, M., 2011a. Scattering-angle dependence of doubly differential cross sections for fragmentation of $H_2$ by proton impact. Phys. Rev. A 83, 012709.

Egodapitiya, K.N., Sharma, S., Hasan, A., Laforge, A.C., Madison, D.H., Moshammer, R., Schulz, M., 2011b. Manipulating atomic fragmentation processes by controlling the projectile coherence. Phys. Rev. Lett. 106, 153202.

Ehrhardt, H., Jung, K., Knoth, G., Schlemmer, P., 1986. Differential cross sections of direct single electron impact ionization. Z. Phys. D 1, 3.

Feagin, J.M., Hargreaves, L., 2013. Loss of wave-packet coherence in stationary scattering experiments. Phys. Rev. A 88, 032705.

Fiol, J., Olson, R.E., 2004. Three- and four-body dynamics in fast heavy ion–atom ionization. J. Phys. B 37, 3947.

Fiol, J., Otranto, S., Olson, R.E., 2006. Critical comparison between theory and experiment for $C^{6+}$ + He fully differential ionization cross sections. J. Phys. B 39, L285.

Fischer, D., Moshammer, R., Schulz, M., Voitkiv, A., Ullrich, J., 2003. Fully differential cross sections for the single ionization of helium by ion impact. J. Phys. B 36, 3555.

Gassert, H., Chuluunbaatar, O., Waitz, M., Trinter, F., Kim, H.-K., Bauer, T., Laucke, A., Müller, C., Voigtsberger, J., Weller, M., Rist, J., Pitzer, M., Zeller, S., Jahnke, T., Schmidt, L.P.H., Williams, J.B., Zaytsev, S.A., Bulychev, A.A., Kouzakov, K.A., Schmidt-Böcking, H., Dörner, R., Popov, Y.V., Schöffler, M.S., 2016. Agreement of experiment and theory on the single ionization of helium by fast proton impact. Phys. Rev. Lett. 116, 073201.

Gaus, A.D., Htwe, W., Brand, J.A., Gay, T.J., Schulz, M., 1994. Energy-spread measurements of various ion sources. Rev. Sci. Instrum. 65, 3739.

Hasan, A., Sharma, S., Arthanayaka, T.P., Lamichhane, B.R., Remolina, J., Akula, S., Madison, D.H., Schulz, M., 2014a. Triple differential study of ionization of $H_2$ by proton impact for varying electron ejection geometries. J. Phys. B 47, 215201.

Hasan, A., Sharma, S., Arthanayaka, T.P., Lamichhane, B.R., Remolina, J., Akula, S., Madison, D.H., Schulz, M., 2014b. Triple differential study of ionization of $H_2$ by proton impact for varying electron ejection geometries. J. Phys. B 47, 215201.

Htwe, W., Vajnai, T., Barnhart, M., Gaus, A.D., Schulz, M., 1994. Differential double excitation cross sections in proton-helium collisions studied by energy-loss spectroscopy. Phys. Rev. Lett. 73, 1348.

Hubele, R., Laforge, A.C., Schulz, M., Goullon, J., Wang, X., Najjari, B., Ferreira, N., Grieser, M., de Jesus, V.L.B., Moshammer, R., Schneider, K., Voitkiv, A.B., Fischer, D., 2013. Polarization and interference effects in ionization of Li by ion impact. Phys. Rev. Lett. 110, 133201.

Jarai-Szabo, F., Nagy, L., 2015. Theoretical investigations on the projectile coherence effects in fully differential ionization cross sections. Eur. Phys. J. D 69, 4.

Jones, S., Madison, D.H., 2000. Ionization of hydrogen atoms by fast electrons. Phys. Rev. A 62, 042701.

Karlovets, D.V., Kotkin, G.L., Serbo, V.G., 2015. Scattering of wave packets on atoms in the Born approximation. Phys. Rev. A 92, 052703.

Kouzakov, K.A., Zaytsev, S.A., Popov, Y.V., Takahashi, M., 2012. Singly ionizing 100-MeV/amu C6++He collisions with small momentum transfer. Phys. Rev. A 86, 032710.

Kouzakov, K.A., Zaytsev, S.A., Popov, Y.V., Takahashi, M., 2013. Reply to "Comment on 'Singly ionizing 100-MeV/amu C6++He collisions with small momentum transfer'" Phys. Rev. A 87, 046702.

Laforge, A.C., Egodapitiya, K.N., Alexander, J.S., Hasan, A., Ciappina, M.F., Khakoo, M.A., Schulz, M., 2009. Three-body dynamics in single ionization of atomic hydrogen by 75 keV proton impact. Phys. Rev. Lett. 103, 053201.

Madison, D., Schulz, M., Jones, S., Foster, M., Moshammer, R., Ullrich, J., 2002. Comparison of theoretical and absolute experimental fully differential cross sections for ion-atom impact ionization. J. Phys. B 35, 3297.

Madison, D.H., Fischer, D., Foster, M., Schulz, M., Moshammer, R., Jones, S., Ullrich, J., 2003. Probing scattering wave functions close to the nucleus. Phys. Rev. Lett. 91, 253201.

Maydanyuk, N.V., Hasan, A., Foster, M., Tooke, B., Nanni, E., Madison, D.H., Schulz, M., 2005. Fully differential studies on single ionization of helium by slow proton impact. Phys. Rev. Lett. 94, 243201.

McGovern, M., Whelan, C.T., Walters, H.R.J., 2010a. $C^{6+}$-impact ionization of helium in the perpendicular plane: ionization to the ground state, excitation-ionization, and relativistic effects. Phys. Rev. A 82, 032702.

McGovern, M., Assafrao, D., Mohallem, J.R., Whelan, C.T., Walters, H.R.J., 2010b. Coincidence studies of He ionized by $C^{6+}$, $Au^{24+}$, and $Au^{53+}$. Phys. Rev. A 81, 042704.

Moshammer, R., Unverzagt, M., Schmitt, W., Ullrich, J., Schmidt-Böcking, H., 1996. A 4π recoil-ion electron momentum analyzer: a high-resolution 'Microscope' for the investigation of the dynamics of atomic, molecular and nuclear reactions. Nucl. Instrum. Meth. B 108, 425.

Murray, A.J., Woolf, M.B.J., Read, F.H., 1992. Results from symmetric and non-symmetric energy sharing (e, 2e) experiments in the perpendicular plane. J. Phys. B 25, 3021.

Naja, A., Casagrande, E.M.S., Lahmam-Bennani, A., Stevenson, M., Lohmann, B., Dal Cappello, C., Bartschat, K., Kheifets, A., Bray, I., Fursa, D.V., 2008. (e, 2e) triple differential cross-sections for ionization beyond helium: the neon case at large energy transfer. J. Phys. B 41, 085205.

Navarrete, F., Ciappina, M.F., Sarkadi, L., Barrachina, R.O., 2016. The role of wave packet coherence on the ionization cross section of He by $p^+$ and $C^{6+}$ projectiles. Nucl. Instrum. Meth. B. (submitted for publication, see original manuscript).

Park, J.T., Aldag, J.E., George, J.M., Peacher, J.L., McGuire, J.H., 1977. Differential energy-loss cross sections for ionization of atomic hydrogen by 25-200-keV protons. Phys. Rev. A 15, 508.

Park, J.T., Aldag, J.E., Peacher, J.L., George, J.M., 1978. Angular differential cross sections for excitation of atomic hydrogen by 25-, 50-, and 100-keV protons. Phys. Rev. Lett. 40, 1646.

Pedlow, R.T., O'Rourke, S.F.C., Crothers, D.S.F., 2005. Fully differential cross sections for $3.6MeVu^{-1}$ $Au^{ZP+}$ + He collisions. Phys. Rev. A 72, 062719.

Ren, X., Senftleben, A., Pflüger, T., Dorn, A., Colgan, J., Pindzola, M.S., Al-Hagan, O., Madison, D.H., Bray, I., Fursa, D.V., Ullrich, J., 2010. Tracing multiple scattering patterns in absolute (e,2e) cross sections for $H_2$ and He over a 4π solid angle. Phys. Rev. A 82, 032712.

Ren, X., Pflüger, T., Ullrich, J., Zatsarinny, O., Bartschat, K., Madison, D.H., Dorn, A., 2012. Low-energy electron-impact ionization of argon: three-dimensional cross section. Phys. Rev. A 85, 032702.

Ren, X.G., Pflüger, T., Weyland, M., Baek, W.Y., Rabus, H., Ullrich, J., Dorn, A., 2014. An (e, 2e + ion) study of low-energy electron-impact ionization and fragmentation of tetrahydrofuran with high mass and energy resolutions. J. Chem. Phys. 141, 134314.

Ren, X.G., Amami, S., Zatsarinny, O., Pflüger, T., Weyland, M., Baek, W.Y., Rabus, H., Bartschat, K., Madison, D., Dorn, A., 2015a. Kinematically complete study of low-energy electron-impact ionization of neon: internormalized cross sections in three-dimensional kinematics. Phys. Rev. A 91, 032707.

Ren, X.G., Pflüger, T., Weyland, M., Baek, W.Y., Rabus, H., Ullrich, J., Dorn, A., 2015b. High-resolution (e, 2e + ion) study of electron-impact ionization and fragmentation of methane. J. Chem. Phys. 142, 174313.

Rescigno, T., Baertschy, M., Isaacs, W.A., McCurdy, C.E., 1999. R-matrix theory of atomic collisions: application to atomic, molecular and optical processes. Science 286, 2474.

Röder, J., Ehrhardt, H., Pan, C., Starace, A.F., Bray, I., Fursa, D.V., 1997. Absolute triply differential (e, 2e) cross section measurements for H with comparison to theory. Phys. Rev. Lett. 79, 1666.

Rosel, T., Röder, J., Frost, L., Jung, K., Ehrhardt, H., Jones, S., Madison, D.H., 1992. Absolute triple differential cross section for ionization of helium near threshold. Phys. Rev. A 46, 2539.

Sarkadi, L., 2010. Classical trajectory Monte Carlo model calculations for ionization of atomic hydrogen by 75-keV proton impact. Phys. Rev. A 82, 052710.

Sarkadi, L., Barrachina, R., 2016. Loss of wave-packet coherence in ion-atom collisions. Phys. Rev. A 93, 032702.

Schneider, K., Schulz, M., Wang, X., Kelkar, A., Grieser, M., Krantz, C., Ullrich, J., Moshammer, R., Fischer, D., 2013. Role of projectile coherence in close heavy ion-atom collisions. Phys. Rev. Lett. 110, 113201.

Schöffler, M., Dörner, R., 2016. Private communication.
Schulz, M., 2009. The role of elastic scattering in inelastic atomic fragmentation processes. comments on atomic, molecular, and optical physics. Phys. Scr. 80, 068101.
Schulz, M., Madison, D.H., 2006. Studies of the few-body problem in atomic break-up processes. Int. J. Mod. Phys. A 21, 3649.
Schulz, M., Moshammer, R., Madison, D.H., Olson, R.E., Marchalant, P., Whelan, C.T., Jones, S., Foster, M., Kollmus, H., Cassimi, A., Ullrich, J., 2001. Triply differential single ionization cross sections in coplanar and non-coplanar geometry for fast heavy ion-atom collisions. J. Phys. B 34, L305.
Schulz, M., Moshammer, R., Perumal, A.N., Ullrich, J., 2002. Triply differential single-ionization cross sections in fast ion-atom collisions at large perturbation. J. Phys. B 35, L161.
Schulz, M., Moshammer, R., Fischer, D., Kollmus, H., Madison, D.H., Jones, S., Ullrich, J., 2003a. Three-dimensional imaging of atomic four-body processes. Nature 422, 48.
Schulz, M., Moshammer, R., Fischer, D., Ullrich, J., 2003b. Evidence for significant projectile–target nucleus scattering in single ionization of helium. J. Phys. B 36, L311.
Schulz, M., Moshammer, R., Voitkiv, A., Najjari, B., Ullrich, J., 2005. Fully differential single ionization cross sections of helium by relativistic heavy-ion impact. Nucl. Instrum. Meth. B 235, 296.
Schulz, M., Dürr, M., Najjari, B., Moshammer, R., Ullrich, J., 2007. Reconciliation of measured fully differential single ionization data with the first Born approximation convoluted with elastic scattering. Phys. Rev. A 76, 032712.
Schulz, M., Najjari, B., Voitkiv, A.B., Schneider, K., Wang, X., Laforge, A.C., Hubele, R., Goullon, J., Ferreira, N., Kelkar, A., Grieser, M., Moshammer, R., Ullrich, J., Fischer, D., 2013. Postcollision effects in target ionization by ion impact at large momentum transfer. Phys. Rev. A 88, 022704.
Sharma, S., Hasan, A., Egodapitiya, K.N., Arthanayaka, T.P., Schulz, M., 2012. Projectile coherence effects in electron capture by protons colliding with $H_2$ and He. Phys. Rev. A 86, 022706.
Sharma, S., Arthanayaka, T.P., Hasan, A., Lamichhane, B.R., Remolina, J., Smith, A., Schulz, M., 2014a. Fully differential study of interference effects in the ionization of $H_2$ by proton impact. Phys. Rev. A 90, 052710.
Sharma, S., Arthanayaka, T.P., Hasan, A., Lamichhane, B.R., Remolina, J., Smith, A., Schulz, M., 2014b. Complete momentum balance in ionization of $H_2$ by 75-keV-proton impact for varying projectile coherence. Phys. Rev. A 89, 052703.
Sharma, S., Arthanayaka, T.P., Hasan, A., Lamichhane, B.R., Remolina, J., Smith, A., Schulz, M., 2014c. Fully differential study of interference effects in the ionization of $H_2$ by proton impact. Phys. Rev. A 90, 052710.
Steck, M., Bisoffi, G., Blum, M., Friedrich, A., Geyer, C., Grieser, M., Holzer, B., Jaeschke, E., Jung, M., Krämer, D., Matl, K., Ott, W., Repnow, R., 1990. Electron cooling of heavy ions. Nucl. Instrum. Methods Phys. Res., Sect. A 287, 324.
Stefani, G., Avaldi, L., Camilloni, R., 1990. On the relevance of the final ionic state to the (e, 2e) TDCS. J. Phys. B 23, L227.
Stevenson, M.A., Lohmann, B., 2006. Triple-differential cross-section measurements of electron-impact ionization of argon 3s in the backward scattering direction. Phys. Rev. A 73, 020701.
Taylor, J.R., 2006. Scattering Theory: The Quantum Theory of Nonrelativistic Collisions. Dover, New York.
Ullrich, J., Moshammer, R., Dorn, A., Dörner, R., Schmidt, L., Schmidt-Böcking, H., 2003. Recoil-ion and electron momentum spectroscopy: reaction-microscopes. Rep. Prog. Phys. 66, 1463.

Vajnai, T., Gaus, A.D., Brand, J.A., Htwe, W., Madison, D.H., Olson, R.E., Peacher, J.L., Schulz, M., 1995. Observation of postcollision effects in the scattered projectile spectra for ionizing proton-helium collisions. Phys. Rev. Lett. 74, 3588.
Voitkiv, A.B., Najjari, B., Ullrich, J., 2003a. On the higher-order effects in target single ionization by bare ions in the perturbative regime. J. Phys. B 36, 2591.
Voitkiv, A.B., Najjari, B., Ullrich, J., 2003b. On the higher-order effects in target single ionization by bare ions in the perturbative regime. J. Phys. B 36, 2591.
Voitkiv, A.B., Najjari, B., Moshammer, R., Schulz, M., Ullrich, J., 2004. Three-body quantum dynamics of helium single ionization by 1 GeV/u $U^{92+}$ impact. J. Phys. B 37, L365.
Wang, X., Schneider, K., LaForge, A., Kelkar, A., Grieser, M., Moshammer, R., Ullrich, J., Schulz, M., Fischer, D., 2012. Projectile coherence effects in single ionization of helium. J. Phys. B 45, 211001.

CHAPTER NINE

# Recent Progress in Dissociative Electron Attachment: From Diatomics to Biomolecules

**Ilya I. Fabrikant*[,1], Samuel Eden[†], Nigel J. Mason[†], Juraj Fedor[‡]**
*University of Nebraska, Lincoln, Nebraska, United States
[†]The Open University, Milton Keynes, United Kingdom
[‡]J. Heyrovský Institute of Physical Chemistry v.v.i., Academy of Sciences of the Czech Republic, Prague, Czech Republic
[1]Corresponding author: e-mail address: ifabrikant@unl.edu

## Contents



*Advances in Atomic, Molecular, and Optical Physics*, Volume 66
ISSN 1049-250X
http://dx.doi.org/10.1016/bs.aamop.2017.02.002

## Abstract

Dissociative electron attachment (DEA) processes occur in many important applied contexts, particularly gas discharges, plasmas, biological systems, and astrophysical environments. In this review, we survey the basic physics of DEA and the progress that has been made during past 14 years since the last important review on DEA (Hotop et al., Adv. At. Mol. Opt. Phys. 49, 86). This progress includes studies of DEA to simple diatomic and polyatomic molecules with high energy resolution revealing vibrational Feshbach resonances and threshold structures, studies of angular distribution of the fragmentation products allowing analysis of the symmetries of the resonances involved, and theoretical developments in investigating the dynamics of nuclear motion in DEA processes. Particular attention is paid to recent advances in DEA to biological molecules as the process is important for understanding radiation damage. Recent progress in understanding electron attachment to van der Waals clusters and the influence of cluster environments on DEA is also reviewed. The review concludes with a forward look and suggestions for new research directions.

# 1. INTRODUCTION

## 1.1 Introducing DEA

It is just over a century since the first experiment was performed that clearly defined the products of a single well-defined collision between electrons and atoms, the Franck–Hertz experiment (Franck and Hertz, 1914). The Franck–Hertz experiment is widely regarded as an experiment that provided validation of the Bohr theory of atomic structure, but it should also be viewed as the birth of scientific study of atomic and molecular phenomena by collisions. Today, we recognize that electron–atom and electron–molecule collisions are prevalent across nature, for example, as the fundamental collisions in the aurora both on Earth and now observed in atmospheres of other planets in our solar system. The exploitation of such collisions also underpins many of the technologies upon which modern society relies, for example, the use of plasmas in device fabrication, medicine, and pollution abatement.

The most important electron collisions are those that are energy transfer processes, leading to excitation (rotational, vibrational, and electronic), ionization, and dissociation as these change the inherent "reactivity" of the local medium in which the collision occurs. Ionization processes dominate the creation of plasmas, the fourth (and most common) state of matter in the universe, while excitation (and deexcitation) processes underpin most of the chemical processes used in modern technologies. More recently, the development of bright well-defined electron sources has led to new instrumentation such as the scanning tunnel microscope that is now the analytical tool of choice to the developing science of nanotechnology.

The majority of electron collisions either in nature or in technology involve low-energy (<20 eV) electron interactions with molecules. An important phenomenon of low-energy electron collisions, first observed in helium (Schulz, 1963) but subsequently found to be general in atomic and molecular gases, on surfaces, and in liquids and solids, is the formation of temporary negative ions (TNIs on the timescale $10^{-12}$–$10^{-15}$ s). TNIs are observed at specific energies and reveal themselves as discontinuities (sharp peaks, troughs) or increases in the interaction cross section, hence their designation as "electron scattering resonances" (Schulz, 1973). The resonance may decay by autodetachment

$$e + AB \rightarrow AB^- \rightarrow AB + e$$

or, in the case of molecules, by complete fragmentation of the TNI to form an anionic product

$$AB^- \rightarrow A + B^-$$

a process entitled dissociative electron attachment (DEA). First revealed in the 1960s, DEA is a process common to most molecules though the probability (cross section) varies over many orders of magnitude (Christophorou, 1984). The fact that the anionic product can be readily detected (e.g., by conventional mass spectrometry) has enabled extensive experimental studies to be carried out. DEA may be explained by an understanding of the potential energy surfaces of the anion and the neutral parent (Fig. 1). The magnitude of the cross section is defined by the overlap (Franck–Condon

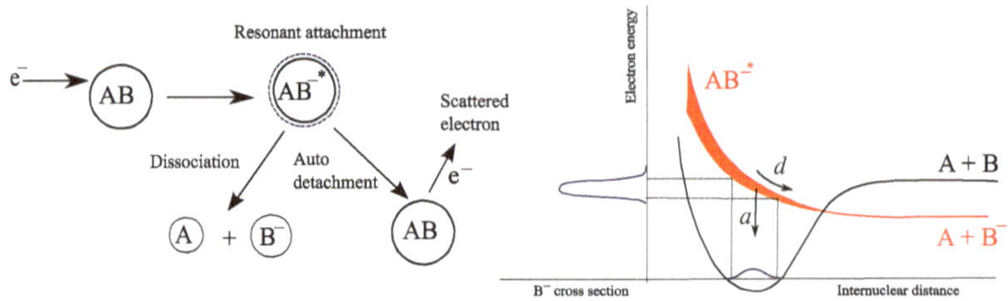


**Fig. 1** A schematic of the dissociative electron attachment process (DEA) with an illustration of the potential energy surfaces of the parent neutral and temporary negative-ionic states.

factor) of the wave functions for the nuclear motion in the neutral and anion states and by the survival probability of the TNI state. The residual kinetic energies of the fragments are defined by the asymptotic behavior of the potential energy curves and the internal energies of the products.

DEA dynamics involves a strong interaction between the electron motion and nuclear motion, and this makes the theoretical description of DEA processes particularly challenging. In fact a complete theory exists only for diatomic molecules, and calculations involving polyatomics typically involve several approximations, which will be discussed in Section 2. Because of the complexity of DEA processes, their studies help us to understand a lot about electronic and nuclear dynamics and their interactions. Typically the electron motion should be treated quantum-mechanically, whereas the nuclear motion is basically classical because of the relatively large masses of nuclei. Therefore, theoretical studies of DEA processes allow us to explore a variety of semiclassical and quasiclassical approaches.

DEA is now known to be an important dissociation channel for many molecules and has important applications in industry as well as occurring naturally, for example, in the Earth's ionosphere where $O^-$ anions are formed by DEA to molecular oxygen. Recently, it has been shown to play a key role in electron-induced fragmentation of biomolecules and hence radiation damage induced within radiotherapy (see below).

As discussed earlier, DEA is a "resonant process" with each decay channel being open only over a specific energy range and it is strongly dependent on the environment within which the target is placed. Accordingly, changes in operational conditions (e.g., pressure and temperature) and electron energy can be used to control DEA and selectively fragment molecules to produce chemically reactive species that will drive chemical reactions to enhance the product of key processes. Radicals are intermediate species found in many chemical reactions in a diversity of environments from planetary atmospheres, fusion reactors, combustion processes, and chemical synthesis, through to biological environments. They are prevalent in different media from single molecules to more complex stages of aggregation, such as clusters, liquids, and condensed phases. The production of radicals as neutral products of DEA therefore provides a means to control the production of such species and thence control the local chemistry. DEA thus offers the opportunity to control chemical reaction pathways that has long been a goal of physical chemistry. DEA-induced chemical control is complementary to the methodology of laser-induced coherent control, which has been demonstrated and explored for more than two decades (Potter et al., 1992) but, due to its expense and technological challenges, has not yet proven to be a viable commercial methodology. In contrast, the ease of production of low-energy electrons and the established methods by which electron energy can be varied (Section 3.1) makes DEA chemical control both technologically and commercially feasible and thus "electron-induced chemistry" is one of the most rapidly growing fields of atomic and molecular physics. Fig. 2 illustrates this electron-induced chemical control. In Fig. 2, very low energy electrons (~0 eV) incident upon gas-phase $CFCl_3$ are able to cleave the C–Cl bond exclusively (producing a $Cl^-$ anion and $CFCl_2$ neutral), while electrons at ~3.2 eV cleave the C–F bond with the electron being shared by the F and $CCl_3$ fragments (Illenberger et al., 1979). Hence by changing the energy of the incident electron, it is possible to select the fragmentation pathway and the production of specific reactive species for subsequent local chemistry.

In this review, we will summarize the current state of knowledge of the dynamics of DEA determined using increasingly refined experiments to probe the fragmentation pathways and cross sections. We will review DEA in several common molecules (from simple diatomics to large biomolecules) and how DEA is influenced by the local medium, e.g., when the molecule is in a cluster or in a surface adsorbate. The review will end with a forward look describing future work. Before starting this review with the description of theoretical approaches, let us look at some of the research topics in which DEA plays a major role.

## 1.2 Plasma Technologies

Underpinning all plasma technologies are the fundamental collision processes that initiate and maintain the plasma and the species generated therein that

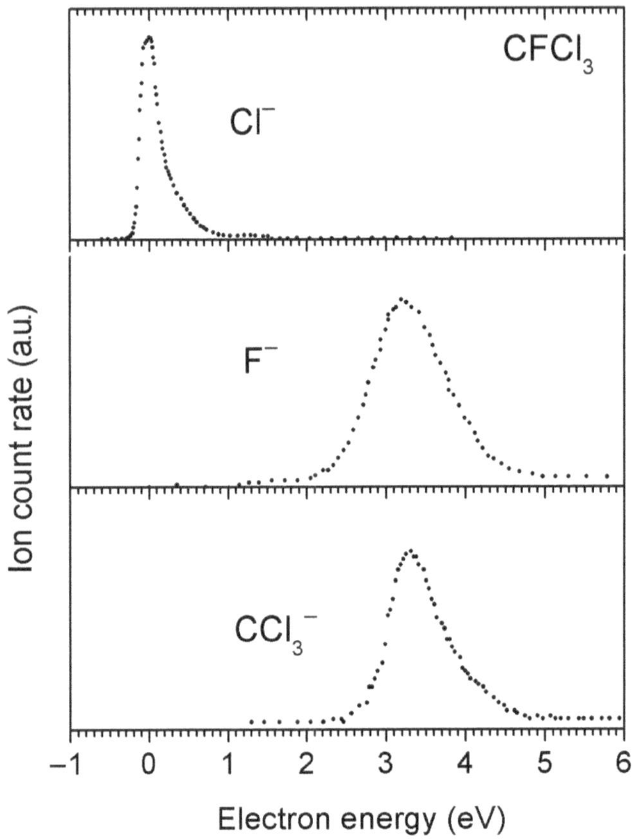


**Fig. 2** Demonstration of the ability of low-energy electrons to selectivity cleaves molecular bonds. At low energies (≈0 eV) electrons cleave the C–Cl bond in $CFCl_3$, while at circa 3 eV they cleave the C–Cl bond (Illenberger et al., 1979).

subsequently control the chemistry induced by the plasma in its local environment. The most important processes in a plasma are those that produce the ions that characterize the plasma. The most fundamental electron processes in a plasma are therefore those that lead to the formation of long-lived ions. These include electron-impact (dissociative) ionization $e + AB \rightarrow e + AB^+$ with the parent cation often subsequently decaying $AB^+ \rightarrow A^+ + B$, and, in those plasmas that are designated "electronegative," DEA. The most common electronegative plasmas containing the highest concentrations of anions (and thus within which DEA is a dominant process) are oxygen plasmas and fluorocarbon plasmas.

Fluorocarbon plasmas have dominated the study of technological plasmas for more than three decades since they are the principle plasma source used in plasma etching to fabricate microstructures in "chip" technology (Posseme, 2015; Samukawa et al., 2012). Much of the technological progress in modern electronics and computational power has been due to advances in our understanding and exploitation of fluorocarbon plasmas. Today, some of the most detailed and realistic simulations of any technological process involve the simulation of plasma etching building microstructures using fluorocarbon plasmas. Such models have been entitled "virtual factories" since they are sufficiently robust that they may be used to test new methodologies for chip manufacture before expensive technical plants are constructed and since they may test dozens of different scenarios they provide a high cost benefit to the industry (Makabe and Petrovic, 2014).

Typical fluorocarbons used in plasma technology include $CF_4$ $C_2F_6$, $C_3F_8$, $SiF_4$, $C_5F_{12}$, $CHF_3$, $CCl_2F_2$, $SF_6$, HF, and variants thereof. Accordingly, the electron interactions with these species have been extensively studied and a detailed (but still not complete) database of such interactions has been compiled (LXcat, https://fr.lxcat.net/home/). It should be noted that all halogenated species have strong cross sections for DEA (see below) and hence fluorocarbon plasmas are "electronegative plasmas" with the anions playing a strong role in subsequent plasma chemistry. Indeed, since DEA is a resonant process, changing the electron temperature may alter the relevant DEA pathways allowing the chemistry in the plasma to be "tuned." Unfortunately, all of these commonly used fluorocarbon feed gases have high "Global Warming Potentials" (GWPs relative to $CO_2$) (Christophorou and Olthoff, 2000). Indeed, $SF_6$ has the highest GWP (=23,000) of any known compound, i.e., one molecule of $SF_6$ is equivalent to 23,000 $CO_2$ molecules. Accordingly, with increasing awareness of the role of potential effect of global warming, protocols have been put in place to "ban" the use of such compounds within the decade. Such protocols provided the semiconductor industry with its greatest challenge since it attacked the very basis of the industry. Major research programs were established to explore alternatives, but to date the alternatives (e.g., $CF_3I$, $NF_3$) are not as efficient as those gases they replace, or may have other drawbacks (e.g., toxicity or reactivity). An alternative is to use nonfluorocarbon species such as $NH_3$, but the reproducibility and stability of the microstructures such plasmas create remain uncertain, in part due to the lack of a comprehensive database to develop a "virtual factory" using such feed gases against which to test fabrication conditions. In all such cases, absolute data on DEA processes are vital, but to date, such studies are at best incomplete and often such studies as do exist are in conflict.

The plasma industry has been the first to exploit electron-induced chemistry through DEA since the electron energy (temperature) in a plasma can be controlled by the voltages applied to the plasma generating electrodes. Samukawa (2006) demonstrated that, by pulsing the electrode voltages generating the plasma in an industrial etching reactor, the electron temperature may be lowered and thus DEA channels opened (Figs. 3 and 4). These anions may be drawn out of the plasma and, by passing through a "carbon filter," converted to neutrals. Resultant neutral radicals may then be used to etch the surface without the disadvantages of charging the surface or UV damage from the plasma. Such neutral beam etching has been shown to produce well-defined 20–30 nm structures and is being developed to produce even smaller (10 nm) structures.

## 1.3 Radiation Chemistry

The second field in which the study of electron collisions was found to be critical is radiation chemistry and the development of future radiotherapy technologies. In 2000, Sanche and coworkers (Boudaïffa et al., 2000) published an astonishing result in electron-induced damage on strands of DNA. Among the various types of structural damage in the DNA helix, single-strand breaks (ssbs) and double-strand breaks (dsbs) are particularly important

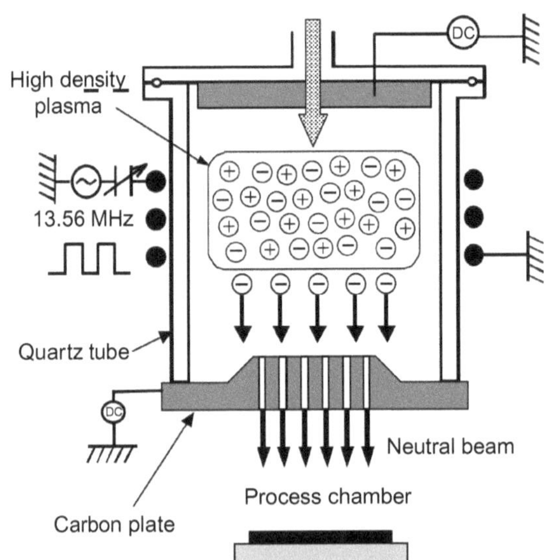


**Fig. 3** Schematic of a modulated plasma reactor that exploits electron-induced chemistry to produce anions, which are withdrawn from the plasma, charge exchanged with a carbon surface and then used to make a neutral beam for surface etching (Samukawa, 2006).

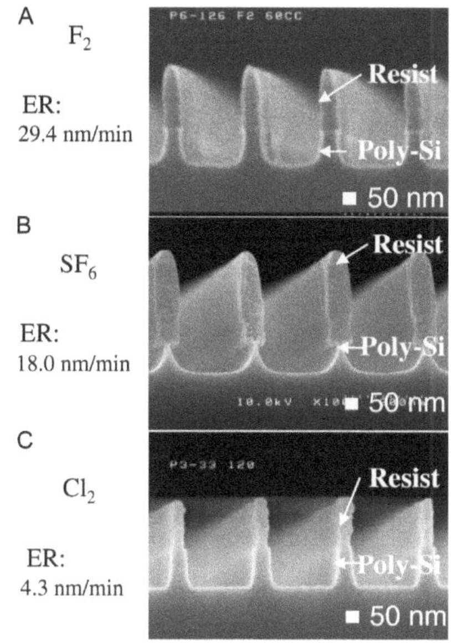


**Fig. 4** Nanostructures in polysilicon created using the neutral beam etching technique with (A) $F_2$, (B) $SF_6$, and (C) $Cl_2$ as plasma feed gases (all of which have strong DEA cross sections) (Samukawa, 2006).

in the initiation of lasting radiation damage. Sanche and coworkers showed that, contrary to expectation, electrons with <10 eV could produce both ssbs and dsbs and their dependences on the incident electron energy (Fig. 5) are reminiscent of "resonances" seen in DEA (notably in DNA bases, as shown in Fig. 6). These results challenged many assumptions of the radiation chemistry community, and our understanding of how DNA, the most important molecule in the cell nucleus, is damaged when irradiated by ionizing radiation. In particular, as low-energy electrons are most abundant *secondary* products of ionizing radiation (notably the X-rays, ion beams, electron beams, and UV light used in radiotherapy), the processes that they can initiate need to be fully understood and taken into account. Thus radiation chemistry and the models that underpin protocols for radiotherapy and environmental exposure should include atomic and molecular interactions, operating on the nanoscale rather than solely on the traditional macroscale. This led to an explosion of experimental studies that explored electron interactions with biomolecular building blocks such as nucleobases and provided theory with enormous challenges to model electron interactions, including DEA, with such complex molecular targets. This will be discussed further in Section 8.

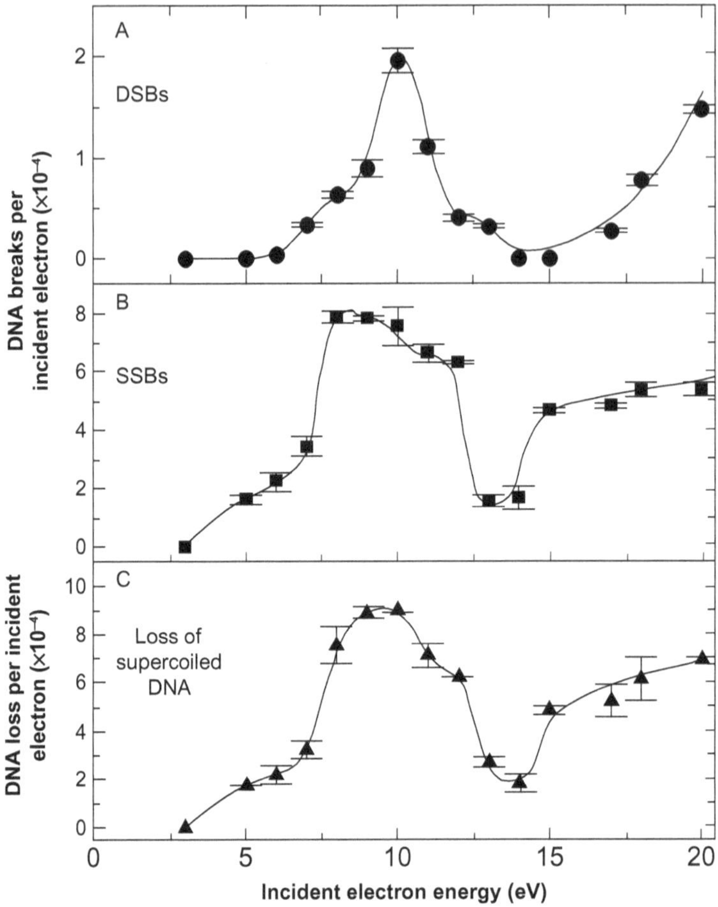


**Fig. 5** The now famous result of Sanche and coworkers showing resonance structure in electron interactions with DNA (Boudaïffa et al., 2000).

In radiation chemistry, the recognition that DEA may play an important role in damage induced by ionizing radiation led to the proposal that molecules that have high DEA cross sections may be used as "radiosensitizers" to enhance targeted damage in cancer cells (Kopyra et al., 2014). Indeed, many of the radiosensitizers currently used in the clinic (such as bromouracil and cisplatin ($Cl_2H_6N_2Pt$)) contain halogenated species and thus have high DEA cross sections (Abdoul-Carime et al., 2001; Chen et al., 2014)—does this play a critical role in their effect? New radiosensitizers have been proposed based on their DEA properties and ability to bind to DNA with higher probability of inducing ssb or dsb rupture when interacting with secondary

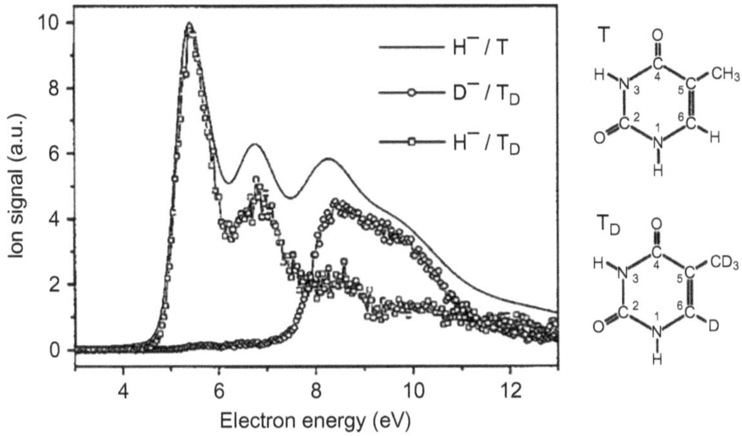


**Fig. 6** H⁻ production via DEA to thymine (T, *solid line*), a constituent nucleobase of DNA illustrating resonance structure comparable with that observed in DNA strand breaks. The figure also shows H⁻ production (*open squares*) and D⁻ production (*open circles*) from partially deuterated thymine ($T_D$). Comparing the three profiles demonstrates selective bond cleavage using low-energy electrons; H atoms are liberated from different sites at distinct energies leading to fine structure in the DEA cross section. *Adapted from Ptasińska, S., Denifl, S., Mróz, B., Probst, M., Grill, V., Illenberger, E., Scheier, P., Märk, T. D., 2005c. Bond selective dissociative electron attachment to thymine. J. Chem. Phys. 123, 124302 (1–8).*

electrons, e.g., 5-bromouridine and 5-nitrouracil (Abdoul-Carime et al., 2004b; Ptasinska et al., 2008). More recently, the introduction of metallic nanoparticles (normally gold or platinum) into cancer cells (ideally into the cell nucleus) has been suggested and tested with the expectation that such particles will be a ready source of secondary electrons which may then rupture DNA by DEA-driven processes (Coulter et al., 2013; Haume et al., 2016; Porcel et al., 2010). Indeed, electron-induced processes (including DEA) are now being included in radiation damage modeling codes (e.g., LEPTS, GEANT4-DNA), which is a testimony to how far electron-induced processes have been recognized within the radiation chemistry community in the 17 years since the pioneering work of Sanche and coworkers.

## 1.4 DEA: A Universal Phenomenon

This review does not provide a comprehensive survey of all fields and phenomena in which DEA plays a major role nor does it present a complete database review of all molecules demonstrating DEA. Rather we focus on fundamental advances and challenges in the field and discuss specific molecules and clusters that are important in plasma technology and radiation damage. Nevertheless, we would note that the role of electrons in astrophysics and astrochemistry is increasingly apparent and the discovery of anions in space ($C_xH^-$ and $C_xN^-$ (McCarthy et al., 2006)) and their major role in the upper atmosphere of Titan (Vuitton et al., 2009) shows that DEA is a truly "universal" phenomenon. Electron collisions and DEA play important roles in space craft reentry phenomena (Celiberto et al., 2016). With the development of commercialization of space (e.g., in asteroid mining and space tourism), such modeling will become increasingly relevant. It is also becoming recognized that detailed understanding of electron-induced processes is important in combustion studies (Bisettia and Morsli, 2014); indeed the correlation between compounds used as fire retardants and their large DEA cross sections has been noted for some time (Mills, 1968). Therefore, one of the future areas ripe for development is exploration of electron chemistry in combustion.

## 2. THEORY

### 2.1 Introduction

Theoretical description of DEA processes is a challenging task. Unlike the case of elastic and inelastic electron scattering by molecules, there are no routine methods and no routine computer packages for solving DEA problems. Each particular process requires a separate theoretical analysis including identification of the resonance responsible for the process. Most of the calculations performed so far were based on the projection-operator theory (Feshbach, 1962) and the Fano (1961) configuration-interaction resonance theory developed by O'Malley (1966) and Bardsley (1968) in the late 1960s. These theories lead to description of nuclear motion by an effective Schroedinger equation with a complex nonlocal potential. The alternative resonance *R*-matrix method (Lane and Thomas, 1958) leads to an equivalent description of the nuclear motion (Fabrikant, 1990; Schneider et al., 1979). The direct application of these theories presents very challenging computational problems. Some practical computational schemes were developed in the early 1990s (Domcke, 1991; Fabrikant, 1991; Kalin and Kazansky, 1990), but they can be applied only to simple diatomic molecules. Model assumptions need to be introduced for more complicated polyatomic cases. In particular, a DEA process leading to breaking of a certain bond can be treated with the assumption that all other bonds are kept frozen. A more advanced method (Chourou and Orel, 2008; Haxton et al., 2007a,b; McCurdy et al., 2003) based on multidimensional potential surfaces includes motion of all atoms during the dissociation process. However, so far only the local version of this theory has been developed. Generally, in local treatment the resonance width depends only on the nuclear configuration (nuclear distance in the case of diatomic molecules), whereas in the nonlocal theories, it is also a function of the electron energy.

## 2.2 Long-Range Effects

For many molecules, long-range electron–molecule interactions, namely dipolar and polarization interactions, play an important role in the DEA process. First, the long-range interaction changes the resonance decay width and, accordingly, the capture amplitude. Second, the long-range interaction can give rise to a weakly bound or a virtual anion state. This state can be formed by electron capture by the molecule with its simultaneous vibrational excitation. This process leads to formation of vibrational Feshbach resonance (VFR) (Gauyacq and Herzenberg, 1982).

A natural choice for description of these effects is the resonance *R*-matrix theory (Lane and Thomas, 1958), which can be considered as an extension of the modified effective range theory (MERT) to the resonance processes. The resonance *R*-matrix theory uses one-pole approximation for the *R*-matrix expansion that allows description of resonance scattering in terms of the Breit–Wigner formula (Breit and Wigner, 1936). On the other hand, by including the long-range electron–molecule interaction outside the *R*-matrix sphere, we can study how this interaction affects the cross sections for resonance processes. The projection-operator approach (Feshbach, 1962; O'Malley, 1966) leads to description of the nuclear motion in a nonlocal energy-dependent complex potential (Bardsley, 1968). The long-range electron–molecule interaction is present in this potential in the form of the threshold energy dependence of the nonlocal resonance width and resonance shift (Domcke, 1991). If the long-range electron–molecule interaction is properly incorporated in the nonlocal potential, then the nonlocal complex potential theory is completely equivalent to the resonance *R*-matrix theory with the inclusion of nuclear motion (Fabrikant, 1990). Due to the computational difficulties in implementing nonlocal effects, quite often the local approximation, or the boomerang model (Birtwistle and Herzenberg, 1971), of the nonlocal theory is used.

## 2.3 Low-Energy Behavior

The local approximation is not able to describe properly threshold behavior of DEA cross sections. The well-known example is DEA to halogen molecules at low energies, which is dominated by the $\Sigma_u$ resonance (a discussion of experimental analysis of DEA to halogens is presented in Section 5.5). According to the Wigner threshold law (Wigner, 1948), the DEA cross section should behave as $E^{1/2}$ at low energies (Bardsley and Wadehra, 1983; Barsotti et al., 2002a; Braun et al., 2007a; Fabrikant et al., 2000). However, local version of the resonance theory gives the wrong energy behavior exhibiting infinite cross section for $E \to 0$ (Hall, 1978).

Due to the high electron affinity of the halogen atoms, DEA processes involving formation of the corresponding anions are quite often exothermic and therefore exhibit very high cross sections in the low-energy region. The low-energy behavior typically is dominated by the s-wave capture, unless such capture is forbidden by the resonance symmetry (as in halogen molecules, for example). According to the Wigner threshold law (Wigner, 1948), the cross section for this process behaves as $E^{-1/2}$ at low energies. However, the dipolar interaction modifies this behavior (Fabrikant, 1977). If the electron energy exceeds the rotational spacing, and the molecular dipole moment $D$ is less than $D_{cr}=0.6395$ a.u., the cross section behaves as $E^{\lambda-1/2}$, where $\lambda$ is a parameter dependent on $D$. When $D$ grows from 0 to $D_{cr}$, $\lambda$ decreases from 0 to $-1/2$. For $D>D_{cr}$, in addition to the $1/E$ divergence, the cross section exhibits oscillations as a function of ln $E$. However, these oscillations cannot be observed in practice. If $D$ is not large (of the order of 1 a.u.) the period of oscillations is too large to fit even one oscillation within the range of validity of the threshold theory limited by the rotational spacing from below. For large $D$ the oscillation amplitude is strongly suppressed by the exponential factor $\exp(-\pi\mu)$, where $\mu=\mathrm{Im}(\lambda)$. Oscillations of this kind were found first by Gailitis and Damburg (1963) and are more relevant to electron scattering by excited hydrogen atoms and to positronium scattering by charged particles (Gailitis, 1982).

When the electron energy is comparable to the rotational spacing, the dipole threshold behavior gradually switches to the Wigner threshold law. However, the presence of a weakly bound or a virtual state leads to significant changes in the energy dependence of the DEA cross section, and deviations from the Wigner threshold law are still observed for energies comparable to the energy of the bound or virtual state (Fabrikant, 1996; Frey et al., 1994; Hill et al., 1996). For symmetric top molecules, the situation is even more complicated because the remnant dipole moment does not average out to 0 due to rotations (Engelking, 1982). The threshold exponent is determined by the average dipole moment in this case.

## 2.4 VFRs and Threshold Cusps in Polyatomic Molecules

If the long-range electron–molecule interaction is strong enough to support a weakly bound state, resonances may appear below the vibrational excitation threshold. These multichannel resonances are called VFR (Gauyacq and Herzenberg, 1982, although these authors called them "nuclear-excited Feshbach resonances"). They were first observed in VE of hydrogen halides (Knoth et al., 1989), but the most pronounced feature appears in the $CH_3I$ molecule below the first excitation threshold of the C–I stretch vibrations (Schramm et al., 1999; Wilde et al., 2000). The dipole moment of $CH_3I$ at the equilibrium internuclear separation, $D=0.638$ a.u., is *subcritical.* However, due to the additional long-range polarization interaction, the $CH_3I^-$ anion becomes stable in the fixed-nuclei approximation. The binding energy at the equilibrium internuclear separation is very small, smaller than the rotational splitting so that the vertical electron affinity of $CH_3I$ is actually negative. However, the anion stabilizes quickly with the increase of the C–I distance, and this gives rise to the VFR. Other examples of VFR observations are DEA to $CH_2Br_2$ (Schramm et al., 2002), $CH_3Br$ (Braun et al., 2007b), uracil (Hanel et al., 2003; Scheer et al., 2004), and thymine (Denifl et al., 2004) molecules. The width and the strength of this resonance in the DEA cross sections are determined by the coupling between the weakly bound anion state and the diabatic valence anion state leading, at large internuclear separations, to the final reaction products. If the vibrational energy lies well below the crossing between the neutral potential energy curve and the diabatic anion curve, this coupling is weak and the corresponding VFR is very narrow and weak as well. The strongest VFR usually appears near the VE threshold, which is closest to the crossing point; for higher thresholds the width increases strongly and the resonance disappears quickly. If the long-range interaction is not strong enough to support a bound state, a virtual-state cusp may appear at VE thresholds. Such cusps have been observed in DEA to $CFCl_3$ (Klar et al., 2001), $CF_3Br$ (Marienfeld et al., 2006a), $CF_3I$ (Marienfeld et al., 2006b), $CHCl_3$ (Kopyra et al., 2008), $SF_5CF_3$ (Graupner et al., 2008), and $CCl_2F_2$ (Graupner et al., 2010). Similar features are observed in the VE cross sections (Allan and Fabrikant, 2002; Gallup and Fabrikant, 2007). A detailed description of the VFR and threshold cusp phenomena may be found in the previous reviews (Fabrikant, 2016a; Hotop et al., 2003, 2004; Ruf et al., 2007). Here, we will present some examples of recently obtained results.

For illustration, we have chosen two molecules with relatively small dipole moments and moderate polarizabilities, $CF_3Br$ and $CCl_2F_2$. In the first case, we are looking for the process

$$e^- + CF_3Br \rightarrow CF_3 + Br^-$$

which is coupled with the vibrational excitation of the symmetric C–Br stretch $\nu_3$. Since only this stretch is active in these two processes, we treat the dynamics by incorporating only C–Br coordinate. In Fig. 7, we present potential energy curves of the neutral and anion along the reaction coordinate. The crossing between the neutral and anion curves occurs approximately at the turning point of the nuclear motion in the $\nu_3=2$ state; therefore, pronounced cusps are expected at the first three vibrational

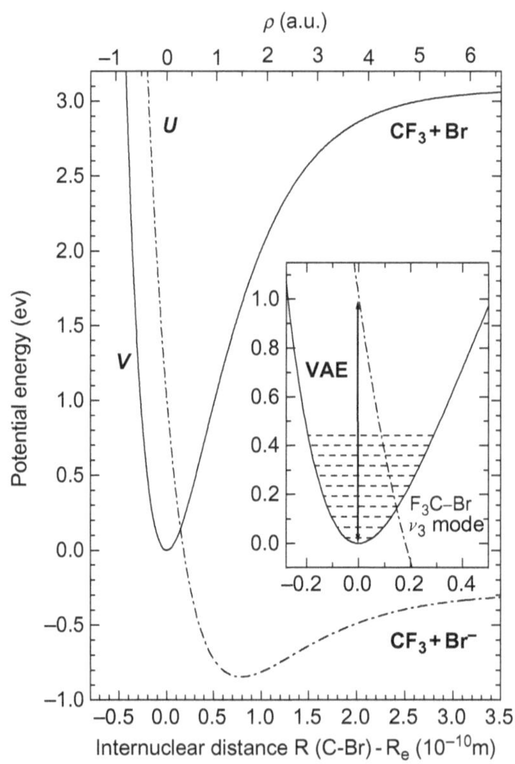


**Fig. 7** Potential curves for $CF_3Br$ and $CF_3Br^-$, adopting a vertical attachment energy (VAE) of 1.011 eV. *From Marienfeld, S., Sunagawa, T., Fabrikant, I.I., Braun, M., Ruf, M.-W., Hotop H., 2006a. The dependence of low-energy electron attachment to $CF_3Br$ on electron and vibrational energy. J. Chem. Phys. 124, 154316 (1–14).*

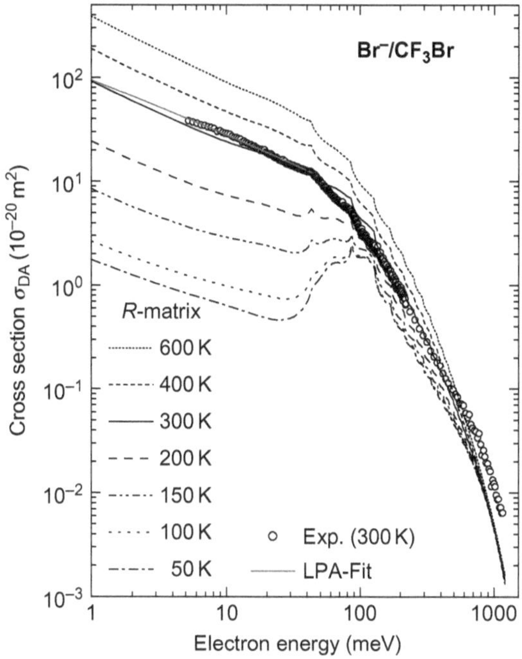


**Fig. 8** Vibrationally averaged absolute cross sections for $Br^-$ formation due to DEA to $CF_3Br$ for vibrational temperatures from 50K to 600K. For comparison, the experimental DEA cross section determined at 300K is shown as *open circles. From Marienfeld, S., Sunagawa, T., Fabrikant, I.I., Braun, M., Ruf, M.-W., Hotop H., 2006a. The dependence of low-energy electron attachment to $CF_3Br$ on electron and vibrational energy. J. Chem. Phys. 124, 154316 (1–14).*

excitation thresholds. This is illustrated in Fig. 8 by showing DEA cross sections averaged over vibrational energy distributions for several temperatures. High-resolution laser photoelectron attachment (LPA) measurements confirm these features, although they are less pronounced than in theory. This can be partly explained by the finite energy resolution and partly by interaction of the $\nu_3$ mode with other modes not accounted by the theory.

Although the dipole moment and polarizability of $CCl_2F_2$ are close to those of $CF_3Br$, the DEA mechanism of the low-energy reaction

$$e^- + CCl_2F_2 \rightarrow CClF_2 + Cl^-$$

is more complicated since there is no direct coupling between the DEA channel and VE of the symmetric $CCl_2$ vibrations. Therefore, for a rigorous calculation of the DEA process a multimode treatment, including the interaction between the vibrational modes, is necessary. To avoid the complexity of the multimode dynamics, the following simplification was used (Graupner et al., 2010). On the basis of the experimental finding of a substantial channel interaction between the primary attachment process and vibrational excitation of the symmetric $CCl_2$ mode $\nu_3$, it was assumed that the electron capture initially drives the symmetric stretch $\nu_3$ mode and that the primary temporary anion state is stabilized rapidly by motion along the $\nu_3$ coordinate. Then, mediated by intramolecular vibrational energy redistribution, the excess energy is channeled into the C–Cl stretch that eventually leads to the dissociation into the fragments $Cl^-$ and $CClF_2$. The attachment process is thus treated in a one-dimensional approximation, assuming coupling of the anion state with only the $\nu_3$ mode.

In Fig. 9, we present the result of such a calculation and its comparison with the LPA/EXLPA experiment (see Section 3.1). Experimental evidence

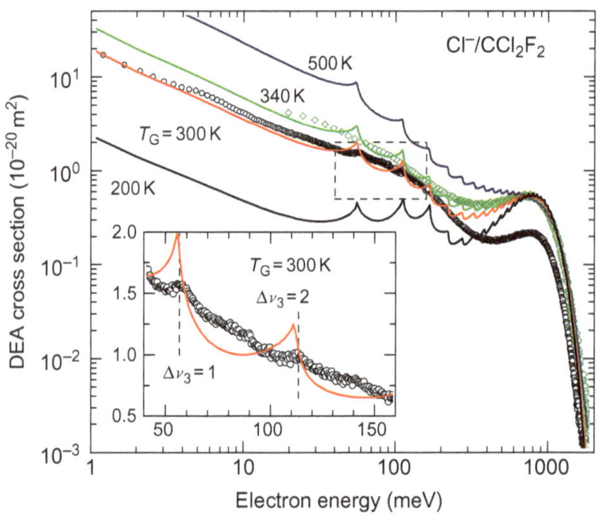


**Fig. 9** Calculated, thermally averaged DEA cross sections for $Cl^-$ formation in e–$CCl_2F_2$ collisions, obtained with initial vibrational state populations for temperatures of $T$=200K, 300K, 340K, and 500K, in comparison with the experimental data measured at $T$=340K (*open diamonds*, Aflatooni et al., 2001) and $T$=300K (*open circles*, LPA/EXLPA; Graupner et al., 2010). The *inset* presents a magnified view of the LPA data (*full circles*) in comparison with the calculated cross section (*full curve*, $T$=300K), both exhibiting cusp structure that occurs at the thresholds for VE of one and two quanta of the $\nu_3(a_1)$ mode ($\approx$56.7 and $\approx$113 meV, respectively) and which is due to interaction between the attachment and the VE channel. *From Graupner, K., Haughey, S.A., Field, T.A., Mayhew, C.A., Hoffmann, T.H., May, O., Fedor, J., Allan, M., Fabrikant, I.I., Illenberger, E., Braun, M., Ruf, M.-W., Hotop, H., 2010. Low-energy electron attachment to the dichlorodifluoromethane ($CCl_2F_2$) molecule. J. Phys. Chem. A 114, 1474–1484.*

for cusps at the $\nu_3$ thresholds confirms the assumption about the dominance of the $\nu_3$ mode at the initial stage of the DEA process.

## 2.5 Interaction Between Resonances

Theoretical treatment of resonance processes in atomic collisions is usually accomplished in the approximation of one resonance or in approximation of nonoverlapping resonances. However, interaction between two resonances can lead to rather interesting and unexpected effects. In particular a generalization (Devdariani et al., 1976) of the Landau–Zener model to two interacting decaying states leads to a coupling parameter whose imaginary part is proportional to $\sqrt{(\Gamma_1 \Gamma_2)}$. Physically, it corresponds to interaction through the continuum and leads to a strong (exponential) increase of the probability of nonadiabatic transitions between two states (Devdariani et al., 1976). This effect might be of great significance for DEA (Hazi, 1983) and the related process of dissociative recombination (Royal et al., 2004). If the resonance width is large, the survival probability of the intermediate negative-ion state is very small, which strongly reduces the DEA cross section. However, interaction of the dissociating state with another, long-lived, state can increase the survival probability. More detailed theory related to interaction of resonance coupling in electron–molecule collisions involving several nuclear degrees of freedom was developed by Estrada et al. (1986) and Feuerbacher et al. (2004) who generalized the theory of conical intersection to interaction between resonance states. This interaction can occur due to coupling through the continuum and due to vibronic coupling. Estrada et al. (1986) analyzed the latter mechanism in the local approximation using a generalization of the displaced oscillator model. Several more advanced models were analyzed by Feuerbacher et al. (2004). We outline here their results for conical intersection of resonances using what they call the minimal model. It represents the interaction in terms of a $2 \times 2$ matrix coupling totally symmetric vibrational modes with nontotally symmetric modes. The resonance width does not depend on the position of the nuclei, and the coupling between resonances is a real number. The first interesting result of this model is that, in contrast to the interaction of two bound states, there are two conical intersections connected by a straight line formed by the seam of intersections of the real part of the adiabatic energy surfaces. The seam of intersections of the imaginary part of the surfaces has two branches, which start from the two intersecting points and continue to $\pm$infinity in a straight line. The imaginary potential energy surfaces are discontinuous over the whole range of points, which form the real seam. Feuerbacher et al. (2004) also explored two other models, one with the linear dependence of the resonance width on the totally symmetric coordinate, and the other with the complex coupling parameter. They applied their results to investigate the resonance coupling in chloroethene. This is an example of unsaturated organic compounds where a halogen atom is connected to one of the carbon atoms involved in the double bond. These compounds represent a good example of the resonance interaction in DEA (Aflatooni et al., 2010; Burrow et al., 1981; Stricklett et al., 1986). They possess low-lying empty orbitals of $\pi^*$ symmetry associated with C=C double bonds, as well as those of C–Cl $\sigma^*$ character. In molecules such as allyl chloride and benzyl chloride, $\pi^*$ and $\sigma^*$ orbitals are coupled due to the nonplanar geometries of the molecules. In planar compounds such as vinyl chloride (chloroethene) and chlorobenzene, this interaction occurs due to vibronic mixing. The latter type of process was experimentally studied by Stricklett et al. (1986). Feuerbacher et al. (2004) found that their linear model describes well $\pi^*$–$\sigma^*$ coupling in chloroethene due to out-of-plane motion of the Cl atom. DEA calculations with inclusion of this coupling are very challenging and were not carried out for this system. However, DEA calculations involving this mechanism were more recently done by Chourou and Orel (2008) for acetylene and will be described in Section 7.3. A conical intersection related to DEA in water was explored by Haxton et al. (2005, 2007b), and will be described in Section 7.1.

Another interesting study of resonance coupling in DEA was carried out by Aflatooni et al. (2010) who reported experimental absolute total cross sections for a number of chlorinated compounds belonging to the nonplanar group and examined correlation between the cross sections and vertical attachment energies (VAEs). They compared this correlation with the dependence of the peak cross section as a function of VAE for approximately 40 saturated mono- and polychloroalkanes bearing only empty $\sigma^*$ orbital (Aflatooni and Burrow, 2000; Aflatooni et al., 1998). The latter dependence fits very well a universal curve over the range from 0.6 to 2.8 eV. Then they plotted the peak DEA for unsaturated chloroalkenes at both the VAE corresponding to the LUMO ($\pi^*$ resonance) and the LUMO + 1 ($\sigma^*$ resonance) as shown in Fig. 10.

As the bridge connecting the C=C and C–Cl groups increases in length, the peak DEA cross section was declining in an exponential manner as a function of LUMO VAE, falling below the chloroalkane line by steadily increasing amounts. The DEA cross section plotted at the LUMO + 1 VAEs exceeds the corresponding values for chloroalkanes. For the compounds

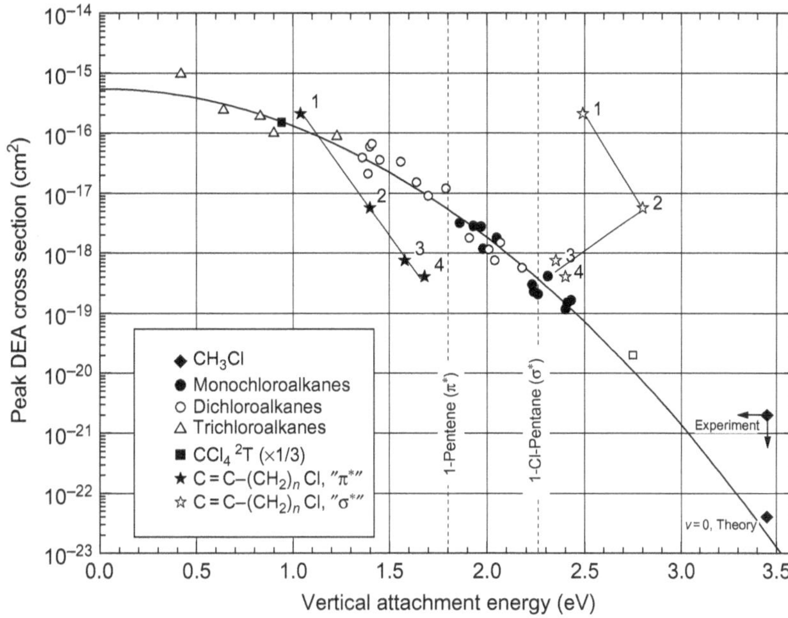


**Fig. 10** The peak DEA cross section for a series of mono- and polychloroalkanes as a function of VAE and a best fit line taken from Aflatooni and Burrow (2000). The *open* and *filled stars* show the DEA cross sections for the chloroalkenes measured by Aflatooni et al. (2010) and plotted as both the lower, primarily $\pi^*$ VAE, and upper, primarily $\sigma^*$ VAE. Open square represents DEA cross section to syn-7-chloro-2-norbornene where $\sigma^*$–$\pi^*$ coupling is forbidden by symmetry (Pearl et al. 1993). With increasing number of ($CH_2$) spacer groups, the DEA cross section is approaching the line characteristic of chloroalkanes. *Reproduced from Aflatooni, K., Gallup, G.A., Burrow, P.D., 2010. Dissociative electron attachment in nonplanar chlorocarbons with $\pi^*/\sigma^*$-coupled molecular orbitals. J. Chem. Phys. 132, 094306 (1–10).*

with the shortest bridge, allyl chloride and 4-chloro-1-butene, the cross sections are orders of magnitudes larger than those of saturated compounds with comparable values of VAE. However, for compounds with a longer bridge, 5-chloro-1-pentene and 6-chloro-1-hexene, cross sections lie only a factor of 3 above the line for chloroalkanes. A similar effect was observed for the series of 1-chloro-*n*-phenylalkanes.

## 3. EXPERIMENTAL METHODS

In this section, we discuss the experimental approaches to probe DEA based on the observation of product anions. *Classical* crossed-beam arrangements and methods for measuring absolute cross sections are described in Section 3.1. Several new experiments exploiting position-sensitive anion detection techniques are described in Section 3.2. We focus solely on experiments directly detecting DEA products (anions), but note that the important information about the relaxation dynamics of resonances can also be obtained by analyzing ejected electrons. This can be done via measuring cross sections for competitive electron scattering channels (Brunger and Buckman, 2002; Hotop et al., 2003) or by time- and angle-resolved photoelectron imaging (Anstoter et al., 2016). Furthermore, we note the experiments by Limão-Vieira and coworkers (Almeida et al., 2012; da Silva et al., 2013) comparing DEA with the dissociation of anions produced via electron transfer in atom– or anion–molecule collisions.

## 3.1 Crossed-Beam Experiments and DEA Cross-Section Measurements

A classical arrangement in DEA experiments involves crossing a controlled electron beam with either a molecular beam or a stagnant gas target. The anions resulting from the DEA process in the interaction region are either mass analyzed or collected for the purpose of a total DEA cross-section measurement. This section provides an overview of the most common experimental approaches. The experimental setups are arranged here based upon how the interacting electron beam is produced. We also refer the reader to an earlier review by Hotop et al. (2003) that includes descriptions of several important controlled electron beam experiments.

The source of electrons in DEA studies may be thermionic (typically a heated metal filament) or photoionization. Thermionic emission produces an energy distribution with a breadth of 0.5–1 eV, which is often too large to observe the fine structure in the resonance features typical of DEA; therefore, various types of electron monochromator have been used to select a portion of the beam and hence achieve a narrower energy distribution. Among the electrostatic devices, the most frequently used is a hemispherical electron monochromator (HEM), and DEA setups using HEM have been built, for example, in Innsbruck (Denifl et al., 1998), Orsay (Abouaf and Diaz-Tendero, 2009), and Fribourg (Allan, 1992). The major advantage of HEM is its good energy resolution when compared to other types of monochromator and DEA measurements with a resolution of 14 meV FWHM have been reported from last of the listed setups (Čížek et al., 2001). The maximum electron current achievable at such resolution is around 1 nA, while the typical values are in the order of hundreds of picoamperes.

A good balance between energy resolution (above 50 meV) and electron current (tens of nanoamperes) is provided by the use of a trochoidal electron monochromator (TEM). Its combination with a mass filter (usually a quadrupole) represents perhaps the most common experimental approach to DEA and examples include apparatus built in Fribourg (Stepanovic et al., 1999), Innsbruck (Cicman et al., 2007; Grill et al., 2001), Bratislava (Stano et al., 2003a,b), and Ufa (Khvostenko et al., 2008). The setup used for many years in Illenberger's group in Berlin (described, for example, in a review by Balog et al., 2004) has been recently transferred to Kopyra and coworkers' laboratory at Siedlce (Kopyra, 2012). More recent apparatus operating in this arrangement include the ones in Notre Dame (Dawley and Ptasinska, 2014) and Reyjkavik (Bjarnasón et al., 2014).

An alternative approach to the analysis of product ions is to couple the TEM with an anion time-of-flight (TOF) mass spectrometer. The difference to the setups listed earlier (where the ion yield has to be recorded at each mass separately) is that TEM + TOF combination records a complete mass spectrum at each electron energy. The spectra are thus usually recorded as two-dimensional maps from which the mass spectra and the energy dependences can be extracted simultaneously. Examples include the experiments in Belfast (Field et al., 2005) and Corvallis (Voinov et al., 2004). A similar setup built in Fribourg (May et al., 2008, 2009) is additionally quantitative, i.e., it provides absolute cross sections. It uses a stagnant gas target with the controlled local pressure, and the extraction, transmission, and detection efficiency of an anion are independent of its mass or initial kinetic energy. It is shown schematically in Fig. 11 and has been recently transferred to Prague.

The total ion collection setups represent a special class of instruments. They utilize the facts that TEM uses an axial magnetic field for its operation and that anions have much larger Larmor radii than electrons. The total anion current is measured on an electrode (or a system of electrodes) enclosing the electron beam trajectory, and examples of DEA cross sections measured this way are given by Aflatooni and Burrow (2000) and Fedor et al. (2008).

Several electron-collision setups are currently in use that do not employ any monochromator and use only a system of few electrodes (and, typically, a weak magnetic field) to collimate the electron beam. The broad energy distribution (0.5–1 eV from thermionic emission) is compensated by a high intensity of the beam. This enables probing of targets with low intensity of the molecular beam, for example, molecules embedded in

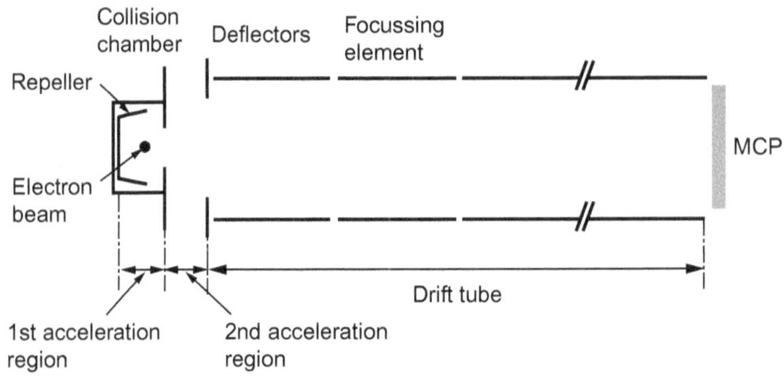


**Fig. 11** Scheme the TEM-TOF experiment to measure absolute partial cross sections for DEA (recently transferred from Fribourg to Prague). The orientation of electron beam is perpendicular to the paper plane. *Reproduced from May, O., Fedor, J., Allan, M., 2009. Isotope effect in dissociative electron attachment to acetylene. Phys. Rev. A 80, 012706 (1–5).*

helium nanodroplets (Denifl et al., 2006), certain biomolecules (Huber et al., 2006), various types of large clusters (Kočišek et al., 2016a), or microhydrated species (Kočišek et al., 2016b). Another advantage of high-intensity sources is that the produced ions can be further investigated for their metastable decay pathways and kinetic energy release in sector-field analyzers (Fedor et al., 2006; Gluch et al., 2003). Indeed, sources of this kind are also used in the DEA imaging experiments described in Section 3.2.

At the other end of the energy-resolution scale are the experiments where the source of the electron beam is not a thermionic emission but the photoionization of atoms. The Kaiserslautern group has carried out an extensive series of *LPA* measurements (Hotop et al., 2003), achieving sub-meV electron energy resolution extending to very low energies (200 to <1 meV). To study the electron attachment over an extended energy range (up to 2 eV), the group developed the *EXLPA* method in which photoelectrons are accelerated electrostatically in a guiding magnetic field. Although EXLPA has lower resolution than LPA, the group still achieved higher resolution (down to 14 meV) than filament-based systems (Braun et al., 2009). The Kaiserslautern facility has, however, been discontinued. An electron-collision setup working on the similar principle, but using a synchrotron light for photoionization, has been built at the ASTRID storage ring in Aarhus (Balog et al., 2009; Čurík et al., 2006). An alternative approach is to use photoionization of solid-state heterostructure crystals as the electron source (Schrotter et al., 2005).

An alternative way of studying the electron attachment properties of molecules is to utilize electron transport (swarm) experiments, described in depth, for example, by Christophorou (1984). They are conceptually quite different from all the above-mentioned approaches, since they use mixtures of the sample gas with buffer gas at considerably higher pressures. In this way, the effect of multiple collisions on the electron attachment, either dissociative or nondissociative, can be studied. In addition to several long-established setups (e.g., Shuman et al., 2013a,b), new experiments of this type have been recently built in Siedlce (Kopyra et al., 2007), Bratislava (Kučera et al., 2013), and ETH Zurich (Chachereau et al., 2016; Dahl et al., 2012).

Apart from producing and controlling the electron beam, another important characteristic of each DEA setup is the means of producing the molecular target. In principle, the experiments can be divided into three classes: those that use stagnant gas targets (collision cell), effusive molecular beams, or supersonic beams. The collision cell experiments are used mainly in setups where the absolute cross sections are measured, either by the total ion collection method (Aflatooni and Burrow, 2000) or by controlled extraction (May et al., 2009). The advantage of this arrangement is that the electron beam usually interacts with the target molecules along a long flight path, which leads to a large number of created anions (although achieving equivalent extraction/detection conditions for all anions is a challenge with this approach).

The use of effusive molecular beams is common in electron-collision experiments. They do not require high backing pressures; thus the beam can be (and in most cases is) composed of the pure target sample, without any buffer gas. The well-defined profile of an effusive beam enables absolute cross sections to be determined via calibration using the relative flow technique (e.g., Krishnakumar et al., 1997). The use of supersonic beams on the other hand brings the possibility of producing target molecules that are cold both rovibrationally and translationally (in the frame of reference of the beam). However, to achieve a suitably high number of collisions in the expansion, a mixture with the buffer gas (typically helium or argon) is typically required. Care has to be taken that the electron scattering on the buffer gas does not influence the impinging electron beam significantly.

Supersonic expansions are commonly used also for producing various types of molecular clusters. The traditional setups for probing electron attachment to clusters included those in Berlin (Balog et al., 2004), Kaiserslautern (Barsotti et al., 2002a), or Innsbruck (Matejcik et al., 1996).

Recently, DEA in more complex heterogeneous molecular clusters has been probed, and these were created by either coexpansion of two different species (Neustetter et al., 2015), pickup of molecules on larger clusters (Lengyel et al., 2016) or controlled humidification of the buffer gas (Kočišek et al., 2016a,b). Helium nanodroplets represent a very specific type of cluster that can be produced upon cooling the expansion nozzle to low temperatures and using high helium backing pressure (and, correspondingly, high pumping speed in the expansion chamber). In recent years, the Innsbruck group has paid a lot of attention to electron attachment to molecules embedded in such helium nanodroplets (Denifl et al., 2006).

All the molecular and cluster sources listed above require at some point the target molecule in the gas phase. This is not a problem with targets that are gaseous or have sufficiently high vapor pressure in standard laboratory conditions. However, in recent years a lot of interest has been dedicated to biomolecular targets that have low volatility and have to be heated in order to achieve utilizable vapor density. This brings a risk of thermally decomposing the target molecules prior to the interaction with electrons, especially for more complex biomolecules, such as nucleotides. Several techniques were used in order to verify that such decomposition does not occur, e.g., NMR spectroscopy (Kopyra, 2012). An alternative is to use a softer method for bringing nonvolatile samples to the gas phase, for example, using the laser desorption systems developed in Berlin (Bald et al., 2008b) and Belfast (Poully et al., 2015).

## 3.2 Velocity Map Imaging

Our understanding of the dynamics of DEA and the deconvolution of overlapping resonances in DEA cross sections has been greatly improved by the adoption of the method of velocity map imaging (VMI) in which the kinetic energy and the angular distributions of the product anions are measured using a combination of TOF spectroscopy and spatial detection using a position-sensitive detector. Such an ion imaging technique was first introduced by Chandler and Houston (1987) for multiphoton ionization products. This was followed by the development of the VMI technique by Eppink and Parker (1997) who exploited the technique to study photodissociation and reactive scattering using multiphoton ionization. Eppink and Parker used an inhomogeneous extraction field to focus the product ions onto the imaging plane; by projecting the expanding fragment distribution onto a two-dimensional position-sensitive detector (PSD) the original 3D distribution may be reconstructed. VMI was successfully adapted for the study of DEA by Krishnakumar and coworkers (Nandi et al., 2005) and has revolutionized our "view" of the DEA process revealing new resonances and quantum effects that challenge current theoretical methods.

The basic idea of VMI is that all the ions with the same velocity produced in a given interaction volume are mapped onto the same point on a two-dimensional PSD. The image position on the PSD and the time to arrival define all three velocity components of the anion fragments when they are formed. A Newton sphere is created when each dissociation event yields two partner fragments traveling with equal momentum in opposite directions in the center of mass frame. When the same dissociation is observed a number of times, the fragments build up a spherical distribution in velocity space (Fig. 12). The size of the image enables the kinetic energy of the ions to be derived, as the distance of the ion impact point on the detector from the center of the image is directly proportional to the square root of the kinetic energy of the ion (Whitaker, 2003).

A central slice through the Newton sphere contains the full angular distribution and translational kinetic energy information if an axis of symmetry exists for the process; thus the method is also commonly called *velocity slice imaging* (VSI). It is now standard for the electron beam momentum vector *k* to be the symmetry axis. The angular distribution in any plane containing the electron beam momentum vector *k* will be the same as there is azimuthal symmetry of the ions about *k*. The angular distribution of the anions produced by DEA to diatomic molecules can be fitted using the methodology

**Fig. 12** Schematic diagrams of Newton spheres. The figures show two events with equal and opposite momentum. Fragment A is heavier than Fragment B. By summing up a large number of events a spherical pattern emerges. *Adapted from Szymanska-Skolimowska, E., 2014. Dissociative electron attachment and dipolar dissociation processes in organic molecules found in the interstellar medium. PhD Thesis, The Open University, UK.*

derived by O'Malley and Taylor (1968) invoking the simple symmetry arguments described by Dunn (1962). The angular distribution is, in general, anisotropic and depends on the initial state of the neutral molecule and the TNI (resonance) state. Tronc et al. (1977) showed that measured angular distributions can be analyzed using the equation:

$$\sigma_{\mathrm{DA}}(\Omega, k) \propto \left| \sum_{l=\mu}^{\infty} i^{l} e^{i\delta_{l}} a(k)_{l\mu} Y_{l\mu}(\Omega) \right|^{2} \quad (1)$$

where $k$ is the incident electron wave vector, $\Omega$ the scattering direction of the anion, $a(k)_{l\mu}$ is the energy-dependent expansion coefficient, $Y_{l\mu}$ spherical parameters, $\delta_l$ a phase shift due to potential scattering, and $\mu = [\Lambda_r - \Lambda_I]$, where $\Lambda_I$ and $\Lambda_r$ are the target and resonance axial angular momenta, respectively.

Strictly the above equation only holds for diatomic molecules. However, it can be used to fit angular distributions for anions produced by DEA to polyatomic molecules if the molecular center of mass lies on the dissociative axis and the dissociation process is direct with the majority of the excess energy going into translational motion along the molecular axis. When these conditions are not met, we can use the theory of Azria et al. (1979) who extended that of O'Malley and Taylor to polyatomic molecules.

In principle, the same information as in VMI—kinetic energy release and angular distribution—can be obtained by an electrostatic setup with a mechanically rotating analyzer (e.g., Tronc et al., 1977). However, such angular measurements become challenging at low kinetic energies of the detected species: slow charged particles are sensitive to stray fields and these change with the mechanical configuration around the collision region. Since DEA fragments often have low kinetic energies (especially at the threshold region), such a method, commonly used in electron scattering, has been successfully applied only to several target molecules in the past. VMI enables the angular distribution of fragment anions to be derived over the entire $2\pi$ range. This greatly extends earlier measurements using the conventional table technique, which is often limited to angles between approximately 20 degrees and 160 degrees. VMI measurements have often contradicted measurements made by the turntable technique and by extending the angular range have provided new symmetry assignments to the measured resonances.

Fig. 13A shows an example of VMI (or more correctly VSI) images recorded for $O^-/OH^-$ from acetaldehyde, $CH_3CHO$ by Szymanska et al. (2013) (it should be noted that due to the relatively short TOF the mass

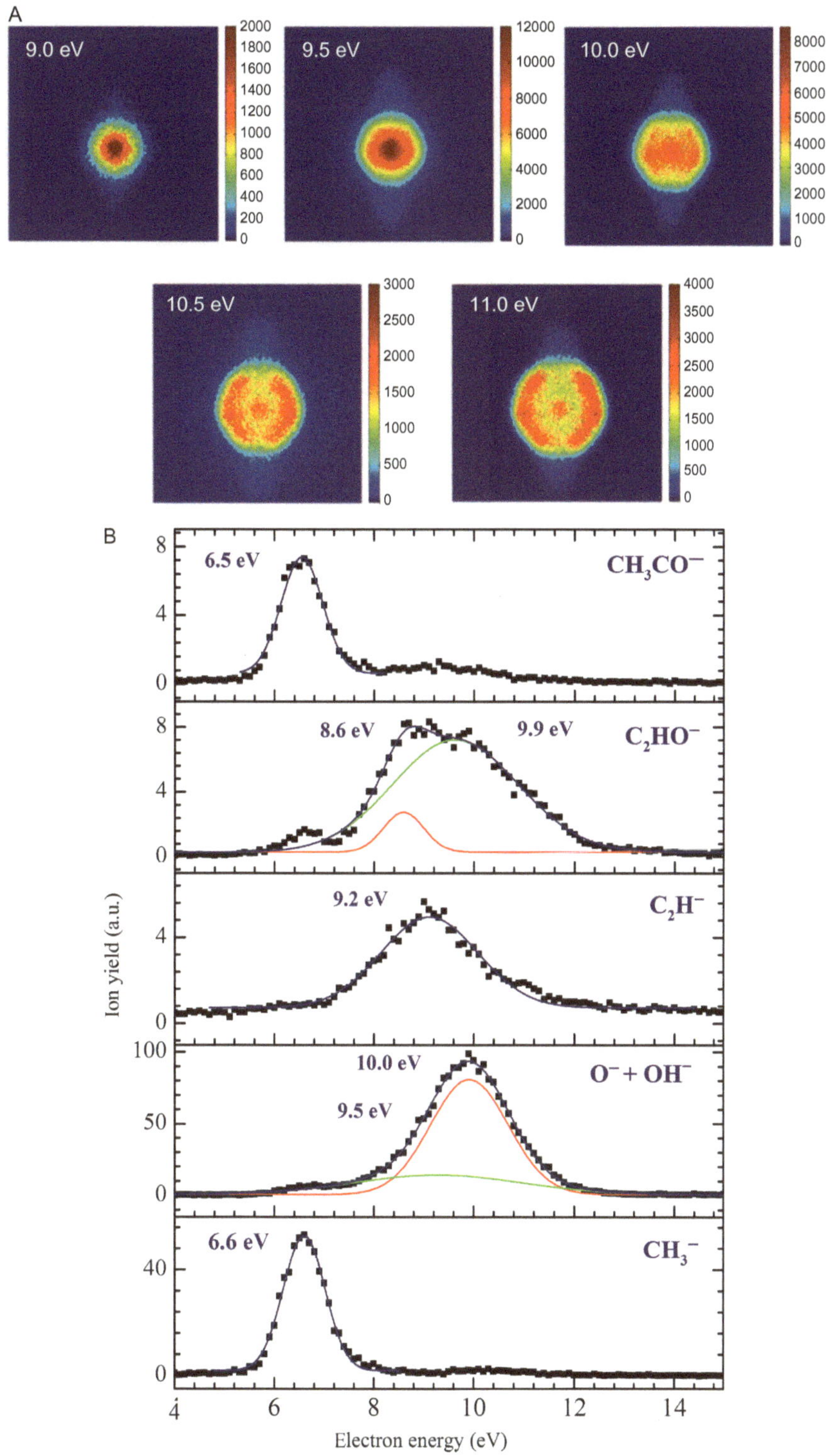
A
9.0 eV
9.5 eV
10.0 eV
10.5 eV
11.0 eV
B
6.5 eV
CH3CO−
8.6 eV
9.9 eV
C2HO−
9.2 eV
C2H−
10.0 eV
9.5 eV
O− + OH−
6.6 eV
CH3−
Ion yield (a.u.)
Electron energy (eV)

Fig. 13—Cont'd

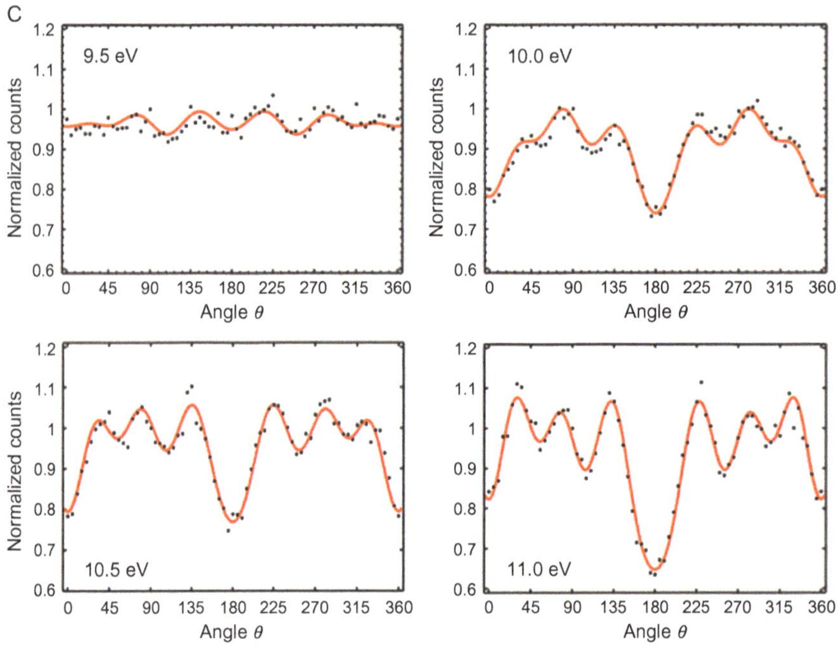


**Fig. 13** (A) Velocity map images recorded for DEA in acetaldehyde $CH_3CHO$ resulting in $O^-/OH^-$ anions. (B) Ion yields for different product anions recorded for DEA in acetaldehyde $CH_3CHO$ showing resonance formation. (C) Angular distributions derived from velocity map images recorded for DEA in acetaldehyde $CH_3CHO$ resulting in $O^-/OH^-$ anions. *Reproduced from Szymanska E., Prabhudesai V., Mason N. J., Krishnakumar E, 2013. Dissociative electron attachment to acetaldehyde, CH₃CHO. A laboratory study using the velocity map imaging technique. Phys. Chem. Chem. Phys. 15, 998–1005.*

resolution was insufficient to clearly delineate $O^-$ from $OH^-$). The advantages of VMI are clearly shown with the changing structure in the image as a function of energy, revealing that two resonances may contribute to the resonance structure observed in the $O^-/OH^-$ ion yield (Fig. 13B). $O^-$ anions are ascribed to the outer ring and found to have a detailed and complex angular distribution invoking many partial waves (Fig. 13C). The distinct angular distribution and kinetic energy release seen in the velocity image indicate that the dissociation is a two-body process, with $CH_3CH$ being the product neutral fragment and assuming no "hydrogen scrambling" occurs (see Szymanska et al., 2013 for further details).

VMI/VSI is now widely adopted by the DEA community having been pioneered by Krishnakumar and coworkers in India with groups now functioning in TIFR Mumbai and Kolkata (Nag and Nandi, 2016).

Krishnakumar built an upgraded version of this experiment at the Open University UK (Krishnakumar et al., 2011). Velocity imaging systems for DEA have since been developed in Heidelberg, Germany and Lawrence Berkley National Lab, USA (Rescigno et al., 2016; Slaughter et al., 2011, 2013) and by Tian and coworkers in China (Tian and Luo, 2015; Tian et al., 2013; Wang et al., 2015) and several other groups are currently developing their own systems. In the next decade, VMI/VSI may be adopted as the standard tool for DEA measurement and analysis. Further results using VMI/VSI are discussed later.

## 4. ARRHENIUS LAW FOR DEA

Many of the DEA reactions studied experimentally were found to obey the Arrhenius equation

$$K = A\exp(-E_a/T) \tag{2}$$

albeit over rather limited temperature ranges. Here, $K$ is the rate coefficient, $T$ is the temperature in energy units, $A$ is a constant, and $E_a$ is the so-called activation energy. In numerous cases, activation energies have thus been determined from plots of ln $K$ vs $1/T$. The thermal equilibrium is assumed, that is, the electron temperature equals the gas temperature. The variation of $K$ with temperature is determined by the dependence of the DEA cross section on the electron energy and vibrational state. Sometimes the rotational energy dependence plays a role too. A theoretical analysis (Fabrikant and Hotop, 2008) showed that the region of validity of the Arrhenius equation is limited. For endothermic DEA processes and favorable curve crossing, the Arrhenius equation has a rather wide range of validity limited by high temperatures. The upper temperature limit depends on the threshold energy $E_t$ and becomes quite low for low $E_t$ and low vibrational frequency. The activation energy in the Arrhenius equation is close to $E_t$. In the case of exothermic DEA processes with an intermediate barrier the range of validity of the Arrhenius equation is also limited toward low temperatures where the rate coefficient as a function of $T$ becomes flat. The activation energy $E_a$ is found to be lower than the reaction barrier $E_B$, and the relative difference $(E_B - E_a)/E_B$ decreases toward higher $E_B$. This behavior is partly due to the relatively slow growth of the DEA cross section with the vibrational quantum number $v$ for levels just below the barrier.

As an example, we present in Fig. 14 an Arrhenius plot of the DEA rate coefficient for the $CF_3Br$ molecule (Marienfeld et al., 2006a). One observes that the validity of the Arrhenius equation is certainly limited toward low temperatures and probably toward high temperatures as well. The slope calculated from the exponential part of the theoretical curve in Fig. 14 yields an activation energy of $E_a = 52$ meV, which is substantially lower than the reaction barrier $E_B = 120$ meV. A similar situation occurs for DEA to $CH_3Br$ (Braun et al., 2007b): the theoretical value of $E_a = 249$ meV, which agrees with several measurements, is substantially lower than the reaction barrier $E_B = 372$ meV.

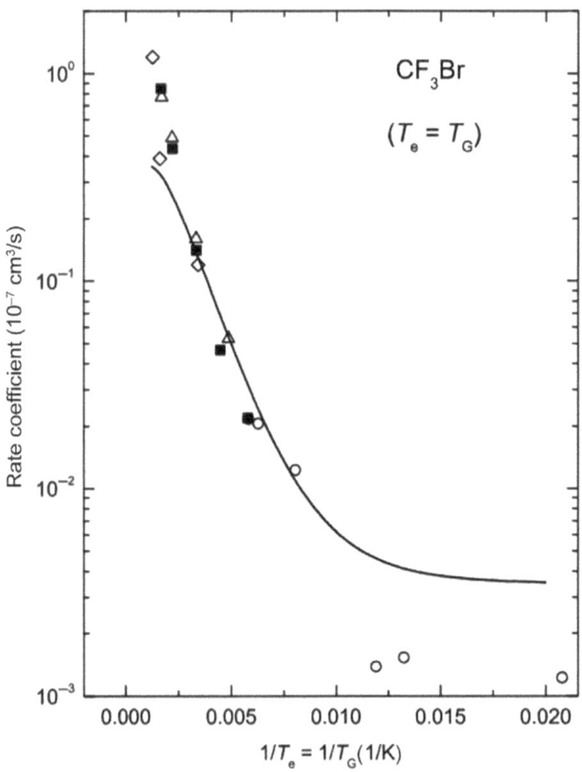


**Fig. 14** Dependence of the thermal rate coefficient on temperature for DEA to $CF_3Br$. *Full curve*: *R*-matrix theory (Marienfeld et al., 2006a). *Symbols* denote different swarm experiments: *open triangles*, Alge et al. (1984); *open diamonds*, Levy et al. (1994); *open circles*, Le Garrec et al. (1997); and *full squares*, Marienfeld et al. (2006a) and Fabrikant and Hotop (2008). *Reproduced from Fabrikant, I.I., Hotop, H., 2008. On the validity of the Arrhenius equation for electron attachment rate coefficient. J. Chem. Phys. 128, 124308 (1–8).*

Another, a more controversial, example is the thermal electron attachment to the $CH_2Cl_2$ molecule, methylene chloride. Experimental studies (Aflatooni and Burrow, 2000, 2001; Aflatooni et al., 1998) of DEA for many organic chlorides show a universal empirical relation between the peak value of the DEA cross section, $\sigma_{max}$, and the VAE. This relationship can be described by a curve giving the DEA cross section in terms of the VAE $E_{VA}$. The $CH_2Cl_2$ is the only compound that is substantially removed from the trend observed for all other compounds: its peak cross section is more than one order of magnitude lower than that predicted by the universal curve. Comparison of $CH_2Cl_2$ with $CCl_2F_2$ is particularly remarkable: in contrast to the $CH_2Cl_2$ case, the peak value of DEA for $CCl_2F_2$ fits perfectly into the universal dependence $\sigma_{max}(E_{VA})$ for chlorofluoromethanes. In an attempt to understand this anomaly, DEA cross sections for three dichloroalkanes, $CH_2Cl_2$, $CHCH_3Cl_2$ (1,1-dichloroethane), and $C(CH_3)_2Cl_2$ (2,2-dichloropropane), were calculated (Gallup and Fabrikant, 2011b). It was found that the cross section for the methylene chloride is anomalously small because of the relatively large resonance width for this compound. However, the Arrhenius plot for $CH_2Cl_2$ brought additional surprises.

In Fig. 15, we present Arrhenius plots for $CCl_2F_2$ and $CH_2Cl_2$ and compare them with the most recent swarm data of Miller et al. (2009) and some other measurements done in the past. The overall temperature dependence and the absolute value of the rate coefficient for $CCl_2F_2$ agree quite well with the experiment, although the theoretical coefficient grows more slowly with rising temperature than the experimental one. However, for the $CH_2Cl_2$ case, the situation is more uncertain: at room temperature, the calculated rate coefficient is substantially higher, by one order of magnitude, than the experimental value of the Oak Ridge National Lab (ORNL) group (Pinnaduwage et al., 1999). Miller et al. (2009) combined these data with their measurements at higher temperatures and obtained for the activation energy in the Arrhenius equation $E_a = 378$ meV, which is substantially higher than the theoretical value, 270 meV. On the other hand, the Boston College measurements (Burns et al., 1996) yield $E_a = 248$ meV in much better agreement with the calculations. In addition, earlier measurements (Fessenden and Bansal, 1970) produce the value of the rate coefficient at room temperature, which is only a factor of 2 higher than the calculation. The value of $K$ extracted by Gallup et al. (2003) from their beam measurements of the DEA cross sections is even higher. It is apparent that there is a certain inconsistency between the beam and swarm measurements and

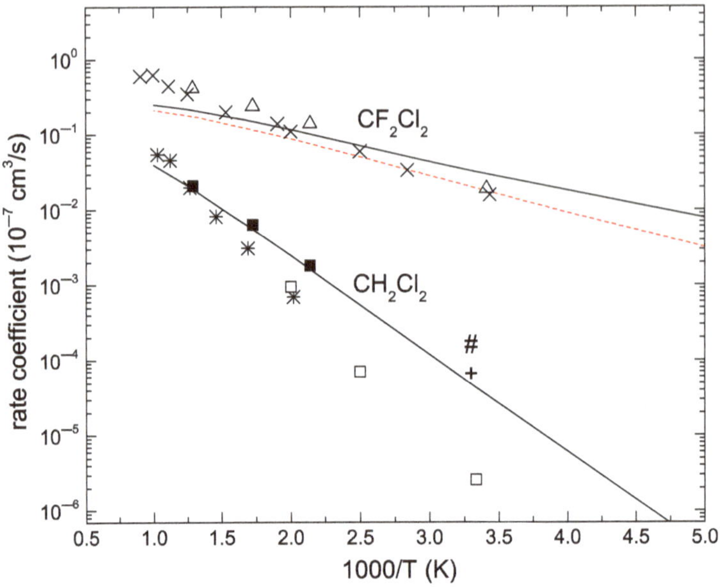


**Fig. 15** Arrhenius plots of the DEA rate coefficients. *Solid line*: theory for $CH_2Cl_2$ (Gallup and Fabrikant, 2011b). *Dashed line*: calculations for $CCl_2F_2$ based on the model described in Graupner et al. (2010). Experimental data for $CCl_2F_2$—*triangles*: Burns et al. (1996); *crosses*: Miller et al. (2009). Experimental data for $CH_2Cl_2$—*filled squares*: Burns et al. (1996); *asterisks*: Miller et al. (2009). *Open squares*: data of Pinnaduwage et al. (1999) extrapolated by Miller et al., 2009 to thermal energies; +: Fessenden and Bansal (1970); #: thermal average of the beam experiment (Gallup et al., 2003). *Reproduced from Gallup, G.A., Fabrikant, I.I., 2011b. Dissociative electron attachment to $CH_2Cl_2$, $CHCH_3Cl_2$, and $C(CH_3)_2Cl_2$. J. Chem. Phys. 135, 134316 (1–8).*

between different swarm measurements of DEA to $CH_2Cl_2$, and further experimental and theoretical studies are warranted.

Another interesting aspect of DEA is a relation between the thermal rate coefficient and the activation energy studied by Kopyra et al. (2007) and Hotop et al. (2011). The results of both model *R*-matrix calculations involving a simple anion potential shift model (activation energies between 0 and 0.61 eV) and *R*-matrix calculations for specific molecules (including $CF_3Br$, $CF_2Cl_2$, $CH_3Br$, $CH_2Cl_2$, $CF_3Cl$, $CH_3Cl$) closely follow the exponential relation $k(T=300\text{K}) \sim \exp(-E_a/E_0)$ with a slope parameter of $E_0 = 24.7$ meV, which is close to the thermal energy (25.85 meV at 300K) and in satisfactory agreement with the slope parameter of an exponential fit to critically assessed experimental data pairs $[k(T=300\text{K}), E_a]$ for 14 closed-shell molecules (halogenated alkanes, mainly methane and ethane derivatives).

# 5. DEA TO DIATOMICS

## 5.1 Hydrogen

DEA to molecular hydrogen is the classic DEA process and one that has important applications, for example, in the creation of neutral beams in next-generation nuclear fusion, and it is also perhaps one of the first electron molecule processes to occur in the early universe (Capitelli et al., 2007). Among the first measurements of the DEA cross section of $H_2$ are those of Schulz (1959) and Rapp et al. (1965), the former being widely quoted (and misquoted) in many databases and applications. These have recently been updated by Krishnakumar et al. (2011) and are shown in Fig. 16.

$H^-$ production from $H_2$ through DEA occurs as three peaks centered at 4 eV, 14 eV and a broad peak centered between 7 and 13 eV. The threshold for the formation of $H^-$ from $H_2$ is 3.724 eV since the dissociation energy of $H_2$ is 4.478 eV and the electron affinity of H is 0.7545 eV. The 4 eV resonance is known to be due to a "shape" resonance, which dissociates to give $H^- + H$. At this low energy, both H and $H^-$ are necessarily in their respective ground states. This is necessitated by the fact that the excited state of $H^-$ lies at 10.9 eV and the excited state of H atom at 10.2 eV above their respective ground states. The broad peak in the $H^-$ cross section between 7 and 13 eV is understood to be due to a purely repulsive negative-ion resonant state, which dissociates into the ground states of $H^-(1s^2)$ and H(1s). The excess electron energy in this case is converted into the kinetic energy of the two fragments. The broad 14 eV peak has been understood to be due

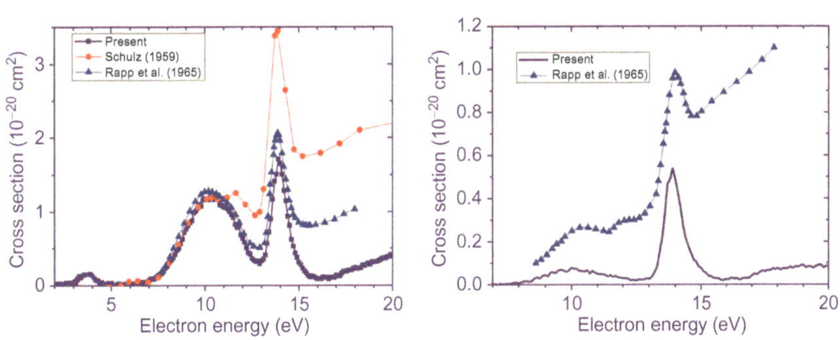


**Fig. 16** Cross sections for $H^-$ from $H_2$ (*left panel*) and $D^-$ from $D_2$ (*right panel*). *Red circles*, Schulz (1959); *blue triangles*, Rapp et al. (1965); and *purple squares and lines*, Krishnakumar et al. (2011). *Adapted from Krishnakumar, E., Denifl, S., Cadez, I., Markelj, S., Mason, N.J., 2011. Dissociative electron attachment cross sections for $H_2$ and $D_2$. Phys. Rev. Lett. 106, 243201.*

to a resonance that dissociates to give $H^-(1\ s^2)$ and the H atom in its first excited state ($n=2$) with a threshold of 13.92 eV (Yoon et al., 2010). Hence, most of the excess energy is transferred into electronic excitation of the neutral atom and the fragments are formed with very low kinetic energy similar to that of the 4 eV channel.

$H^-$ formation is also possible at higher electron energies, with the major contribution coming from dipolar dissociation. In this case, the electrons excite the molecule to high-lying doubly excited states and escape without getting captured. These super excited states can decay by autoionization to form a positive molecular ion and an electron or decay by dissociation to form a proton and an $H^-$ ion. This has a threshold of 17.75 eV. It has been shown that between 14 eV and the dipolar dissociation threshold, resonant attachment and subsequent dissociation to form $H^-$ and H ($n=3$, 4, etc.) in excited states do occur (Cadez et al., 2008).

A known feature of DEA in molecular hydrogen is that it displays a strong isotope effect (Celiberto et al. 2011, 2012; Xu and Fabrikant, 2001). This occurs due to the relatively larger time of dissociation in the case of $D_2^-$ transient state as compared to the $H_2^-$ transient state. The longer dissociation time affects the survival probability of transient negative ion against autodetachment resulting in lower cross section for the heavier isotope. This isotope effect is seen in all the three resonances, albeit in varying degrees. Recent measurements have also shown an isotope effect to be present in the dipolar dissociation process, as the dissociation process here is competing with the autoionization process (Krishnakumar et al., 2011). Similarly, DEA cross sections to $H_2$ and $D_2$ show a very strong rovibrational dependence (characterized as a temperature dependence) explained by the combined effect of the Franck–Condon factors and the survival factor appropriate to the calculated potential energy curves (Celiberto et al., 2001; Xu and Fabrikant, 2001; Xu et al., 2000). Angular distributions of $H^-$ from $H_2$ for the 4 and 10 eV resonances have been measured using the turntable technique, Tronc et al. (1977), but were only possible over a limited angular range since it is impossible to make measurements along the line of electron propagation. In contrast, using the VSI allows the anion yield to be determined over the entire angular range.

## 5.2 Oxygen

Molecular oxygen has been used as a standard for DEA for more than three decades, oxygen often being used to test and commission new DEA apparatus and calibrate the incident energy scale. DEA to $O_2$ also has many practical applications in atmospheric pressure discharges, industrial plasmas and occurs naturally in the D-layer of the Earth's ionosphere where it is the major source. There also exists a large body of experimental work on the $e^-$–$O_2$ scattering, revealing resonances in the elastic and vibrational excitation channels (see Brunger and Buckman, 2002).

Four resonances have been identified below 15 eV—$^2\Pi_g$, $^2\Pi_u$, $^4\Sigma^-_u$, and $^2\Sigma^-_g$ the first three playing a major role in the electronic and vibrational excitation of molecular oxygen. The DEA cross section from $O_2$ shows a single broad peak at 6.5 eV with early angular distribution measurements using the turntable method to measure $O^-$ angular yields between 23 degrees and 157 degrees (Van Brunt and Keiffer, 1970), indicating that the $^2\Pi_u$ resonance is responsible for this peak. However, the possibility of contributions from the $^4\Sigma^-_u$ resonance was proposed and, from symmetry arguments, is expected to manifest itself in contributions in the forward and backward directions. Prabhudesai et al. (2006) developed a new experiment employing VMI (Fig. 17) to obtain kinetic energy and angular distribution data simultaneously over the entire $2\pi$ angles, which was able to identify the presence of the $^4\Sigma^-_u$ in the DEA channel (Fig. 18). This suggests the need for further theoretical work and possibly a different interpretation of the new structures seen in the electron stimulated desorption (ESD) of the $O^-$ ions from $O_2$ clusters as well as in condensed phase.

There have been several theoretical calculations of DEA to molecular oxygen based upon knowledge of the structure and dynamics of negative-ion states of $O_2$. The most recent calculations, and ones that are extended to study DEA to vibrationally excited molecular oxygen, are those of Laporta et al. (2015) using ab initio resonance curves calculated with the *R*-matrix method. These calculations are in accord with the experiments of Prabhudesai et al. (2006), showing that when $O_2$ is in its vibrational ground state, the main contribution for both dissociative electron attachment and electron-impact dissociation comes from the $^2\Pi_u$ resonance state of $O^-_2$ but with a significant contribution from the $^4\Sigma^-_u$ resonant state. Furthermore, vibrational excitation leads to an increased contribution from the low-lying $^2\Pi_g$ resonance, greatly increasing the cross sections for both processes and moving the thresholds to lower energies. DEA experiments to vibrationally excited $O_2$ to compare with these calculations should be undertaken. However, our detailed understanding of DEA to molecular oxygen makes this the ideal standard for testing new theoretical calculations and experimental systems studying DEA.

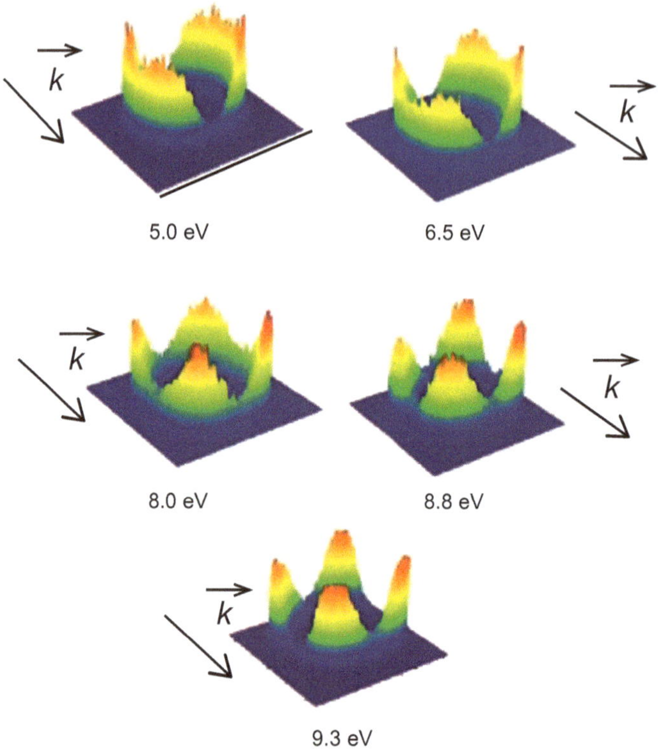


**Fig. 17** Velocity map images of $O^-$ from $O_2$ at different electron energies across the resonance. The *arrow* indicates the direction of electron momentum vector, $\vec{k}$ (Prabhudesai et al., 2006).

## 5.3 Carbon Monoxide

DEA to carbon monoxide (CO) has been extensively studied both experimentally and theoretically revealing the production of both $O^-$ and $C^-$ anions. Two different channels for $O^-$ formation were first identified by Chantry (1968) with the C atom being formed in the ground state $^3P$ (threshold = 9.62 eV) and the first excited state C ($^1D$) (threshold = 10.88 eV), respectively. $C^-$ production was found to be much weaker with a threshold of 10.2 eV and a peak cross section at 10.4 eV and a second threshold at 10.8 eV with a peak at 10.9 eV (Stamatovic and Schulz, 1970). The $C^-$ channel in DEA was investigated further by Abouaf et al. (1981) who reported three peaks of almost equal intensity at 10.27, 10.5, and 10.82 eV, as well as a smaller one at 11.5 eV. The first three peaks were attributed to the $C^-(^2D) + O(^3P)$ channel with a threshold of 9.824 eV and

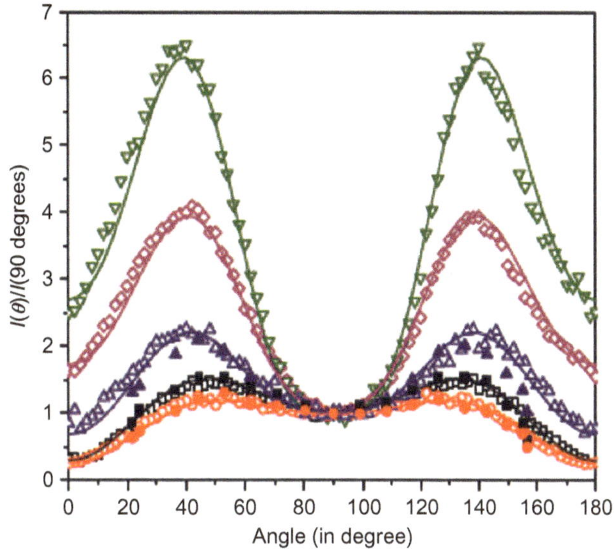


**Fig. 18** Angular distribution of $O^-$ at different electron energies. *Open symbols*: ○, −5; □, −6.5; △ −8 eV; ◊, −8.8; and ▽, −9.3 eV; *closed symbols*: ●, −5.75; ■, −6.7; and ▲, −7.8 eV. *Open and closed symbols: From Prabhudesai, V.S., Nandi, D., Krishnakumar, E., 2006. On the presence of the* $^4\Sigma^-{}_u$ resonance in dissociative electron attachment to $O_2$. J. Phys. B At. Mol. Opt. Phys. 39, L277–L283.

the weak peak at 11.5 eV assigned to the $C^-(^2D) + O(^3P)$ channel with a threshold of 11.057 eV. Abouaf et al. (1981) attributed the formation of the first two peaks as due to predissociation of the $^2\Sigma^+$ Feshbach resonance at 10.04 eV through another $^2\Sigma^+$ state and the third peak due to direct DEA by the latter state.

The symmetry of the underlying resonances was probed by Hall et al. (1977) who measured angular distributions of the $O^-$ yield. They showed that both the channels leading to $O^-$ formation have $^2\Pi$ symmetry. However, recent experiments using the VSI/VMI technique led to some controversy regarding the angular distribution data and the symmetry of the resonances involved in the DEA process. A set of papers by Tian and coworkers (Tian and Luo, 2015; Tian et al., 2013; Wang et al., 2015) claimed a distinctly different angular distribution for the ions than that reported by Hall et al. (1977), while Nag and Nandi (2015a,b) reported an angular distribution fairly close to that by Hall et al. The most striking issue in the discussion that ensued was whether or not the DEA process in CO involved coherent excitation of several resonances. The angular distribution reported by Tian and coworkers showed almost no intensity in the forward hemisphere, which was interpreted as due to coherent excitation of resonances of different symmetries, while Nag and Nandi (2015a,b) argued that the angular distribution was similar to that of Hall et al. (1977) and could be explained as due to an incoherent sum of $\Sigma$ and $\Pi$ states. These contradictory results were explored by Gope et al. (2016b) who observed that the angular distributions of the ions derived from VSI from the two dominant channels were very close to those measured by Hall et al. (1977). Their measured angular distributions are shown in Fig. 19 and reveal finite intensity for the $O^-$ ions in the forward hemisphere in stark contrast to those reported by Tian and coworkers (Tian and Luo, 2015; Tian et al., 2013; Wang et al., 2015). This could be explained as an incoherent sum of $\Sigma$ and $\Pi$ states. Though the $\Pi$ state contributions can be assigned to specific states observed in other scattering experiments, there are still unresolved questions on the states of $\Sigma$ symmetry involved in the DEA process and therefore new theoretical calculations are urgently needed to explain these results along with new measurements using better energy resolution. Gope et al. (2016b) also confirmed the presence of a third channel for the formation of $O^-$ with the C atom in the $^1S$ state.

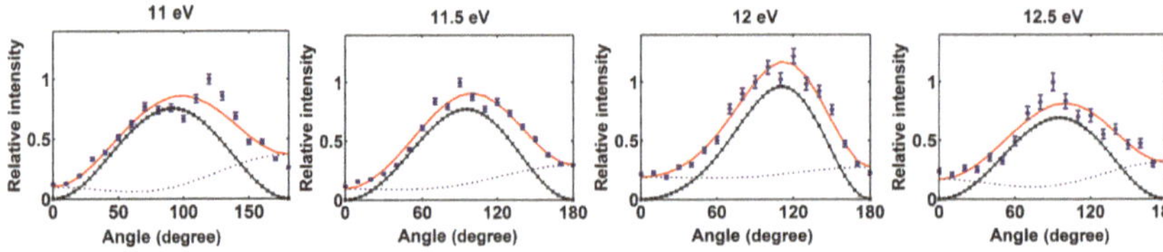


**Fig. 19** Angular distributions of $O^-$ produced by DEA from CO measured at different incident electron energies. *Top panel*: $O^-(^2P) + C(^3P)$ threshold = 9.62 eV; *bottom panel*: $O^-(^2P) + C(^1D)$ threshold = 10.88 eV. *Stars*—measured data by Gope et al. (2016b) solid line best fit using incoherent sum of $\Sigma$ (*dotted line*) state and $\Pi$ state (*line with crosses*) contributions. *From Gope, K., Tadsare, V., Prabhudesai, V.S., Mason, N.J., Krishnakumar, E., 2016b. Negative ion resonances in carbon monoxide. Eur. Phys. J. D 70, 134–141.*

## 5.4 Hydrogen Halides

Hydrogen halides (HCl, HF, HBr) represent an ideal system for testing the theoretical description of electron scattering because of a variety of phenomena encountered in various cross sections. These phenomena include threshold peaks in vibrational excitation cross sections, Wigner cusps, structures due to VFRs, and outer well resonances in the elastic and vibrationally inelastic scattering. These phenomena are much more complex than "simple" shape resonances, and their origin is closely linked to long-range electron–molecule interactions as described in Section 2.

The detailed account of successive experimental discoveries of these phenomena has been provided Hotop et al. (2003); here we will thus list only the most important references. The first surprising experimental observation in electron–halide collisions was the discovery of Rohr and Linder (1976) that the vibrational excitation cross section in HCl and HF shows sharp peaks at thresholds for the excitation of individual vibrational levels. These threshold peaks attained surprisingly large values and showed nearly isotropic angular distributions. Further experiments revealed additional boomerang structures (Schafer and Allan, 1991), cusps and outer well resonances (Allan et al., 2000) in elastic and vibrational excitation cross sections in HCl, similar structures in HBr (Čížek et al., 2001), and pronounced VFRs in vibrational excitation of HF (Sergenton et al., 2000). The DEA at low electron energies proceeds as:

$$e^- + HX \rightarrow X^- + H \quad (X = F, Cl, Br).$$

In all hydrogen halides, this DEA channel is endothermic and occurs at the energetic threshold. The shape of the $X^-$ band shows sharp step-like structures at openings of vibrational excitation channels (Allan and Wong, 1981; Čížek et al., 2001).

In order to describe the fine structures in the spectra, it is essential to utilize nonlocal treatment; the local complex potential approach completely fails to reproduce majority of the above-listed phenomena. The success of a number of nonlocal models to reproduce the features observed in various cross sections has been reviewed by Hotop et al. (2003). However, the comparison was to a large degree limited to the shape of the cross sections, since the absolute cross sections for many of the channels were not known. The tests of the theoretically predicted magnitudes of DEA cross sections have been enabled by the construction of the quantitative DEA spectrometer in Fribourg (Section 3.2). Fig. 20 shows

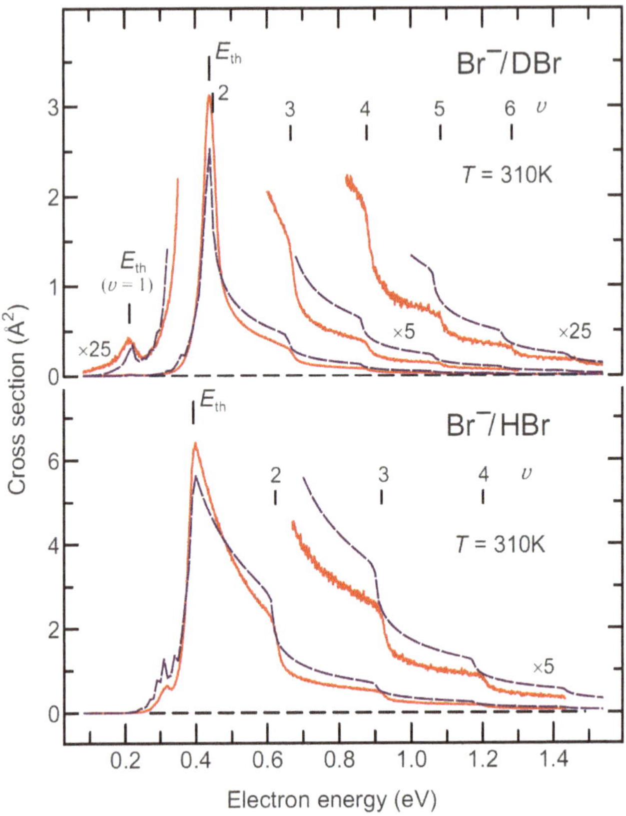


**Fig. 20** Absolute DEA cross sections for HBr and DBr. *Red lines*: experimental data of Fedor et al. (2008); *blue lines*: nonlocal resonance model of Čížek et al. (2001). The experimental and theoretical data sets are independently on absolute scales, without any mutual normalization or scaling. *From Fedor, J., May, O., Allan, M., 2008. Absolute cross sections for dissociative electron attachment to HCl, HBr and their deuterated analogs. Phys. Rev. A 78, 032701 (1–6).*

the DEA cross sections for HBr and DBr of Fedor et al. (2008). The experimental data (red lines) were obtained by scaling the high-resolution (10 meV) band shapes measured with an electrostatic spectrometer to the absolute values measured using a total ion collection setup. Scaling was done using the invariance of energy-integrated cross sections. The blue lines are predictions of the model of Čížek et al. (2001). The experiment and theory agree to an excellent degree, both in the shape and in the magnitude of the cross sections.

In view of this fact, it may seem somewhat surprising that a similarly constructed model for HCl by Čížek et al. (1999) overestimated the DEA cross-section magnitude in HCl and DCl by factors of 2.5 and 2.8, respectively. In both models for HBr and HCl, the parameters characterizing the negative-ion potential function were obtained from ab initio calculations (at the internuclear distances, where the anion is bound) and from fixed-nuclei scattering calculations (at the internuclear distances where the anion is embedded in continuum). For the second part, the HBr model of Čížek et al. (2001) used the *R*-matrix scattering calculations of Fandreyer et al. (1993), while the model for HCl used the scattering calculations of Padial and Norcross (1984), who used a free electron gas model for the exchange interaction. The generalized resonance width that results from the short-range part of the model and determines the detachment vs dissociation competition has been thus obtained using much older scattering data in the case of HCl than in the case of HBr.

Fedor et al. (2010) constructed an improved nonlocal resonance model for HCl by adjusting the short-range part of the model of Čížek et al. (1999) to new scattering calculations done by the Schwinger multichannel method. The improved model removed the discrepancy in the cross-section magnitude (the calculated cross sections agreed with the experimental ones within the experimental error) and reproduced the shapes and positions of the fine structures in all scattering cross sections to the same degree as the original model.

An independent test of the accuracy of various theories is represented by the DEA temperature dependence: the DEA cross sections for higher rovibrational states are much larger than for the ground state and even a relatively small thermal population of higher levels leads to appearance of hot bands at electron energies below the ground-state threshold. As can be seen in Fig. 20, for HBr these bands are clearly visible already at 310K. The DEA temperature dependence in HBr predicted by the model of Čížek et al. (2001) agreed very well with the measurements of Fedor et al. (2007) performed in Innsbruck. For HCl, the temperature dependence of DEA cross section has been measured by Allan and Wong (1981). The model for HCl of Čížek et al. (1999) reproduced the positions of the hot bands but could not reproduce their relative intensities. As demonstrated in Fig. 21, this discrepancy has also been resolved by the improved HCl model of Fedor et al. (2010); the new negative-ion potential and generalized width yielded cross sections that are in much better agreement with the experiment.

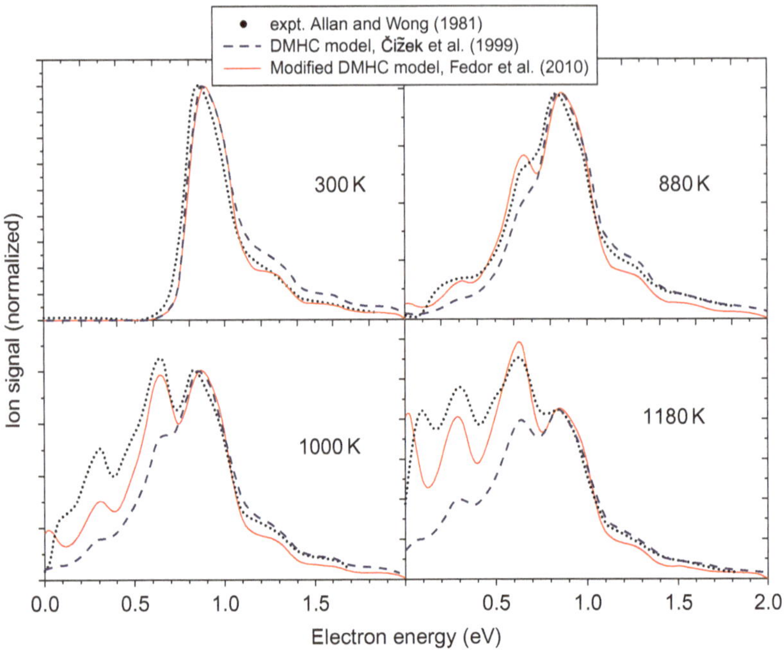


**Fig. 21** Temperature dependence of the $Cl^-$ ion yield form HCl. *Points*: experimental data of Allan and Wong (1981), *dashed blue lines*: nonlocal model of Čížek et al. (1999), and *full red lines*: modified nonlocal model of Fedor et al. (2010).

In the context of higher lying rovibrational levels, it is worth mentioning the associative detachment process, which occurs in the collision of a halogen anion with a hydrogen atom:

$$X^- + H \rightarrow HX + e^-$$

This is closely connected with dissociative attachment via microscopic reversibility; however, it probes a different range of rovibrational levels: since the detachment generally occurs early in the collision, at large internuclear distances, the population of the highest energetically available vibrational level of HX is preferred. The energy of the departed electron carries information about the final rovibrational state, and the product state distribution can be obtained from the electron spectra. Živanov et al. (2002, 2003) measured these distributions for X = Cl and Br. The spectra revealed not only the step-like structures associated with rovibrational onsets but also steep rises associated with interchannel coupling. From the theoretical point of view, the second type of the structure appears only in nonlocal treatments and is completely absent in the local complex potential theory.

## 5.5 Halogen Molecules

Due to the high electron affinity of halogen atoms, DEA reactions in halogen molecules are exothermic and occur with relatively high rates. Thermal rate coefficients for the diatomic halogens have been measured and are commonly used in plasma models (Ayala et al., 1981; Friedman et al., 2009; Shuman et al., 2013b; Sides et al., 1976), but details of dissociation dynamics remain poorly quantified and recent ion imaging experiments in molecular chlorine have highlighted discrepancies between experiment and theoretical calculations. Diatomic halogens are very reactive gases and can degrade an electron spectrometer by bleaching the surfaces and causing large changes in the so-called contact potential, which leads to a variation in the energy of the incident electron beam throughout the experiment. Hence, few DEA experiments have been conducted with halogens and it is particularly difficult to measure accurate DEA cross sections and resonance profiles at energies close to zero eV despite the large DEA cross sections.

### 5.5.1 $Cl_2$ and $F_2$

Low-energy DEA to homonuclear halogen molecules is dominated by the $\Sigma^+{}_u$ resonance with the p wave as a main component (Christophorou and Olthoff, 1999a; Tam and Wong, 1978). This leads, according to the Wigner threshold law (Wigner, 1948), to the $E^{1/2}$ behavior of the cross section as a function of electron energy $E$ at low $E$. Fabrikant et al. (2000) showed that the region of the validity of this law in the case of attachment to $Cl_2$ is very narrow due to the strong energy dependence of the Franck–Condon factor for the electron capture into the $\Sigma_u$ resonance state. Confirmation of this prediction became possible after the development of the LPA method (Klar et al., 1992a,b), generating electron beams with very high resolution in energy. The Wigner threshold law was observed for DEA to $Cl_2$ (Barsotti et al., 2002a; Ruf et al., 2004) and $F_2$ (Braun et al., 2007a). Ruf et al. (2004), using semiempirical $R$-matrix theory, calculated the contributions to DEA from the first three negative-ion resonances, ${}^2\Sigma^+{}_u(\sim 0\,\mathrm{eV})$, ${}^2\Pi_g(\sim 2.5\,\mathrm{eV})$, and ${}^2\Pi_u(\sim 5.6\,\mathrm{eV})$. According to calculations, these three contributions are well separated in energy. However, recent VSI results challenge this conclusion, as discussed later.

Nag and Nandi (2016) and Gope et al. (2016a) used VSI to measure the angular distributions of $Cl^-$, updating the earlier data of Kurepa (1978), Feketeova et al. (2003), and Azria et al. (1982). These experiments revealed that the ~2.5 eV peak, predominantly due to the transition $^1\Sigma^+$ to $^2\Pi_g$, has considerable intensity around 0 degrees and 180 degrees. This shows that there is contribution from the anion ground state $^2\Sigma^+{}_u$ to this peak that extends up to 3 eV. Theoretical studies predict that the $^2\Sigma^+{}_u$ resonance is expected to contribute to the ~0 eV DEA peak only, so its presence in the ~2.5 eV DEA peak requires a revision of the theoretical calculations. VSI also clearly reveals the presence of both $^2\Pi_g$ and $^2\Sigma^+{}_g$ resonances in the ~5.6 eV peak as can be seen by the changing shape of the image as a function of energy (Fig. 22). The $^2\Pi_g$ dominates in the rising part of the peak, whereas both $^2\Pi_g$ and $^2\Sigma^+{}_g$ play comparable roles in the falling part of the peak. Gope et al. (2016a) also explored dipolar dissociation of molecular chlorine. The momentum images indicate a strong cross section but no

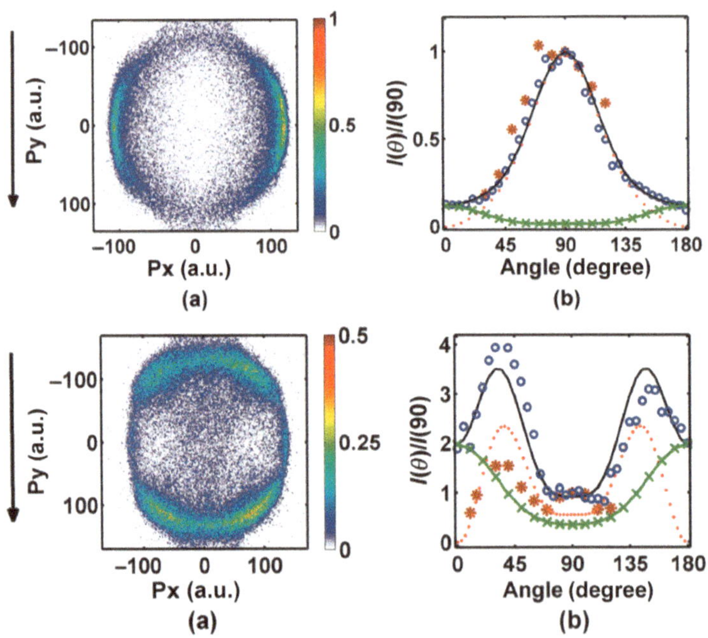


**Fig. 22** DEA to molecular chlorine ($Cl^-$ production) measured using velocity slice imaging (Gope et al., 2016a). *Top*, measured at incident energy 4.5 eV, *bottom*, at 6.5 eV. (A) VSI images. (B) Measured angular distributions (*blue open circles*), derived angular distributions (*solid line*) fitted as an incoherent sum of $^2\Pi_g$ (*red dotted line*) and $^2\Sigma^+{}_g$ (*crossed green line*). The earlier data of Azria et al. (1982) is shown with *maroon stars*.

signature of any resonant structures in the region up to 80 eV, in contrast to features reported by Kurepa (1978).

The situation with the threshold behavior of DEA to $F_2$ is not as clear as for $Cl_2$. An observation (Braun et al., 2007a) of a low-energy peak in DEA to $F_2$ at an energy of 31 meV suggested that perhaps earlier experiments (Chantry, 1982; Chutjian and Alajajian, 1987) suffered from insufficient resolution or other artifacts, and that the same theoretical description successful for $Cl_2$ would work for the attachment to $F_2$. However, all attempts to reproduce the low-energy peak in the DEA cross section observed by Braun et al. (2007a) have either failed to do so, or generated a very unrealistic anion potential energy curve (Fabrikant, 2010). A semiempirical semilocal calculation (Bardsley and Wadehra, 1983) (i.e., nonlocal for the imaginary portion of the complex potential and local for the real portion) gave a peak position at 70 meV. Previous fully nonlocal calculations (Brems et al., 2002; Hazi et al., 1981) gave the peak position at energies between 170 and 200 meV. These values seem reasonable in comparison to the $Cl_2$ peak at 50 meV as the $F_2$ molecule is lighter than $Cl_2$ and therefore the energy dependence of the Franck–Condon factor is weaker, but are well above the experimental result of Braun et al. (2007a).

Some clarification was achieved by comparison of calculated and experimental thermal rate coefficients for $Cl_2$ and $F_2$ (Friedman et al., 2009; Shuman et al., 2013b) obtained by a flowing afterglow-Langmuir probe (FALP) method. Theoretical calculations were carried out within the framework of the nonlocal complex potential theory (Domcke, 1991) and the equivalent resonance $R$-matrix theory (Fabrikant, 1991). This is essential for the threshold behavior since the local theory generates cross sections diverging as $1/E$ at $E \rightarrow 0$ (Hall, 1978). For both compounds an excellent agreement between theoretical and experimental rate coefficients is observed. In the case of $F_2$, the resonance width used in DEA calculations was taken from ab initio calculations of Hazi et al. (1981). The use of theoretical cross sections of Brems et al. (2002) results in rate coefficients, which are lower than the experiment by a factor of 5 and grow slower with the temperature. The question remains whether all theories developed so far are missing some physics allowing an explanation of the very-low-energy maximum and subsequent minimum in the attachment cross section observed by Braun et al. (2007a).

For both $Cl_2$ and $F_2$ the temperature dependence of the rate coefficients is strong, but it is not described by the Arrhenius law due to a conflict between the increase in attachment cross section with vibrational level (temperature) and the decline in the cross section with electron energy above the low-energy peak position.

### 5.5.2 $Br_2$ and $I_2$

Few experiments have been performed on electron attachment to molecular bromine with the most extensive study still that of Kurepa et al. (1981), which reported absolute cross sections for dissociative attachment between 0 and 10 eV, and ion-pair formation between 10 and 40 eV with a maximum in the anion production observed at 18.0 eV with a value of $5.26 \times 10^{-22}\,m^2$ were measured. Six dissociative attachment processes peaking at 0.00, 0.50, 1.45, 3.75, 5.30, and 8.50 eV with cross sections of 17.7, 1.50, 2.80, 4.94, 0.42, and 0.325 units of $10^{-22}\,m^2$, respectively, were detected. The symmetry and classification of these resonances remain speculative and new experiments exploiting VMI/VSI are needed.

To date, DEA to molecular iodine (and many iodine containing compounds) is largely unexplored. An early calculation was reported by Birtwistle and Modinos (1978) and a low-energy (0–8 eV) study was first carried out by Tam and Wong (1978) who reported three resonances: one at zero energy and two broader peaks at 2.5 and 3.5 eV assigned to $^2\Pi_g$, $^2\Pi_u$, and $^2\Sigma^+_g$ core-excited resonances, respectively. The most comprehensive experimental study was reported by Azria et al. (1988) whose kinetic energy and angular measurements of $I^-$ ions identified four $I_2^-$ resonant states with symmetry $(1/2)_u$, $(3/2)_g$, $(1/2)_g$, and $(3/2)_u$, leading to $I^- + I$ fragment formation at 0, 0.6, 1.3, and 2.3 eV respectively, whereas the $^2(1/2)_g$ and $^2(1/2)_u I_2^-$ resonant states are responsible for $I^- + I$ fragment formation at 2.4 and 3.5 eV, respectively. Once again new experiments utilizing ion imaging should be performed to explore these assigned symmetries and the dissociation dynamics.

### 5.5.3 Heteronuclear Halogen Diatomics

Experimental studies of DEA to heteronuclear halogen molecules (interhalogens) ClF, ICl, and IBr were performed recently by Wiens et al. (2016) using the FALP apparatus. Two product channels are energetically accessible in all cases, but $Cl^-$ is produced almost exclusively at thermal energies in the first two cases, whereas attachment to IBr was too slow to measure confidently. The exclusive production of $Cl^-$ was confirmed theoretically in the ClF case. In Fig. 23, we present potential energy curves for ClF and several electronic states of $ClF^-$. Each of the lowest curves of the

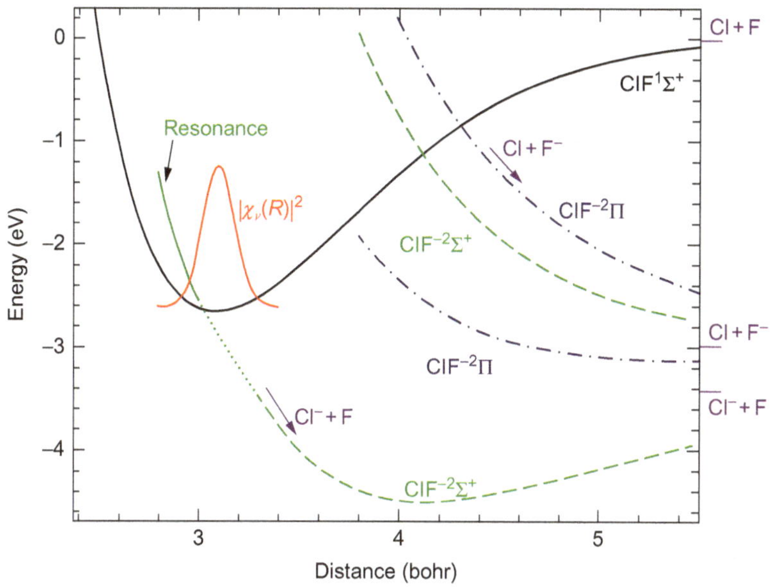


**Fig. 23** Potential energy curves of ClF and $ClF^-$. The *solid curve* is the ground electronic state of ClF. The two lowest electronic states of the $^2\Sigma^+$ and $^2\Pi$ symmetries of $ClF^-$ are shown by *dashed* (*green*) and *dot-dashed curves* (*blue*), respectively. The *curve* labeled $|\chi_\nu(R)|^2$ (*red*) shows the probability density of the ground vibrational level of ClF. The *thick line* (*green*) labeled "resonance" represents the energy of the lowest resonant state of $ClF^-$. The labels "Cl + F," "Cl + $F^-$," and "$Cl^-$ + F" with *short horizontal lines* refer to the corresponding calculated dissociation limits. *Reproduced from Wiens, J.P., Sawyer, J.C., Miller, T.M., Shuman, N.S., Viggiano, A.A., Khamesian, M., Kokoouline, V., Fabrikant, I.I., 2016. Electron attachment to the interhalogen compounds ClF, ICl, and IBr. Phys. Rev. A 93, 032706 (1–8).*

$^2\Sigma^+$ and $^2\Pi$ states dissociates only toward the F + $Cl^-$ limit. Out of all the $ClF^-$ curves, only the lowest $^2\Sigma^+$ state crosses the neutral state near the minimum of the ClF potential. In addition, the ground state does not approach any other curve to the right of the crossing. This suggests that if the ClF molecule is initially in one of the lowest vibrational levels, the DEA products could only be F + $Cl^-$.

Calculations of DEA cross sections for ClF (Wiens et al., 2016) were carried out in the local approximation. It works rather well in this case because of the contribution of the s wave at low energies. The $\nu = 1$ and 2 vibrational modes of ClF are predicted to attach about three to four times faster than $\nu = 0$ and 3 at energies lower than 0.2 eV. The resulting rate coefficients, calculated with inclusion of these modes, agree with measurements within the experimental error bars. In contrast to homonuclear diatomics, the temperature dependence of the rate coefficients is rather weak: it increases by a factor of 1.35 when the temperature grows from 300K to 1000K. An attempt to improve the local theory by using the Bardsley correction (Bardsley, 1979), which adjusts the threshold behavior, reduced the rate coefficient by the factor 1.5, but did not improve agreement with experiment. A more recent nonlocal calculation (Fabrikant, 2016b) has shown that the local theory for ClF is accurate within a factor of 2. Another concern is the accuracy of calculation of the resonance width. Analysis of uncertainties in the width calculation (Fabrikant, 2016b) shows that these uncertainties also can create a factor of 2 error.

In summary, our knowledge of DEA to heteronuclear halogens is sparse. In spite of their relative simplicity for the theory, not many DEA calculations have been performed so far. Furthermore, the theoretical predictions need to be tested by measuring DEA cross sections directly and using VMI/VSI to measure angular distributions that can explore both the symmetry of the resonances and the dominance of p-wave scattering at low (close to zero energy) or whether, as in the case of chlorine, there is evidence of contributions from the ground state in higher lying resonances.

## 6. HALOGEN-CONTAINING POLYATOMIC MOLECULES FOR PLASMA TECHNOLOGIES

As discussed in the introduction, Section 1.2, DEA plays an important role in the fluorocarbon chemistry that underpins the plasma etching and microchip fabrication industries. Typical fluorocarbons used by these industries include $CF_4$, $C_2F_6$, $C_3F_8$, $SiF_4$, $C_5F_{12}$, $CHF_3$, $CCl_2F_2$, $SF_6$, HF producing copious amounts of $F^-$, and associated anions. This review will not discuss DEA of all of these targets in detail but discuss two of the most extensively studied fluorocarbons that also demonstrate important properties and dissociation dynamics of such fluorocarbons, $CF_4$ and $SF_6$. For DEA of the other species, we refer the reader to the most recent papers and the references therein: for $C_2F_6$, Falcetta et al. (2000) and Hien et al. (2016); $C_3F_8$, Chantry and Chen (1989), Christophorou and Olthoff (1999b), and Jeon (2006); $C_5F_{12}$, Spyrou et al. (1985); $SiF_4$, Bjarnason et al. (2013); $C_5F_{12}$, Christophorou (1987); $CHF_3$, Christophorou and Olthoff (2012) (for HF see previous section; for $CCl_2F_2$ see also Section 2.4 and Fig. 9). Many of these studies are now relatively old and have not benefited yet from exploitation of modern techniques such as VMI/VSI (for a comprehensive review of perfluorocarbons pre-1995, see Weik and Illenberger, 1995).

## 6.1 Carbon Tetrafluoride

Carbon tetrafluoride is the benchmark fluorocarbon species for electron scattering studies. Its similarity to methane provides both experimentalists and theoreticians with comparative studies of elastic scattering and ionization processes, however, while $CF_4$ shows that a strong DEA cross-section methane has a very low anion production rate. Recent studies on $CF_4$ have exploited new ion imaging techniques (Fogle et al., 2015; Ómarsson et al., 2014a; Xia et al., 2013). These sophisticated experiments have revealed two major anionic fragments $F^-$, which is observed over a broad energy range from about 4 to 10 eV, and $CF_3^-$ (Fig. 24). The $F^-$ ion yield has been shown to consist of contributions from two overlapping resonances whose relative contributions can be clearly seen in the changing pattern observed in the VSI images (Fig. 25); indeed, this is an excellent example of how ion imaging can reveal the presence of one or more resonances in the broad anion yields that "traditional" ion counting and anion cross sections cannot. The angular distributions derived from VSI are shown in Fig. 25. A simple visual comparison of the $CF_3^-$ and $F^-$ velocity slice images (see Ómarsson et al., 2014a) and the angular distributions show that these are mirror images of one another with respect to a vertical axis through the center of the distribution and are therefore derived from the same resonance.

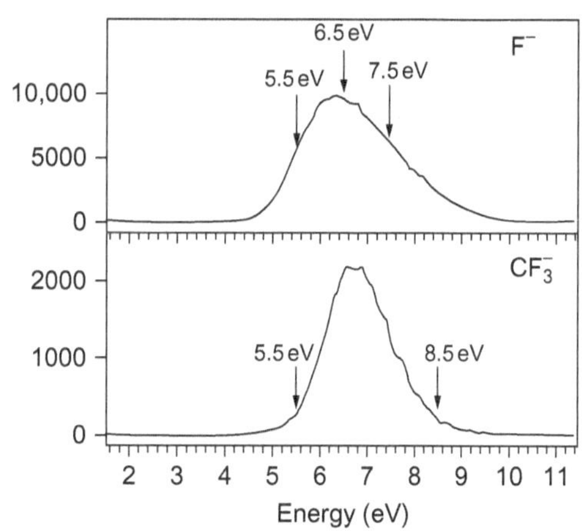


**Fig. 24** The ion yield of $F^-$ (*upper*) and $CF_3^-$ (*lower*) as a function of electron energy. The *vertical arrows* indicate where velocity slice images (VSI) were recorded by the Ómarsson et al. (2014a) and shown in Fig. 25.

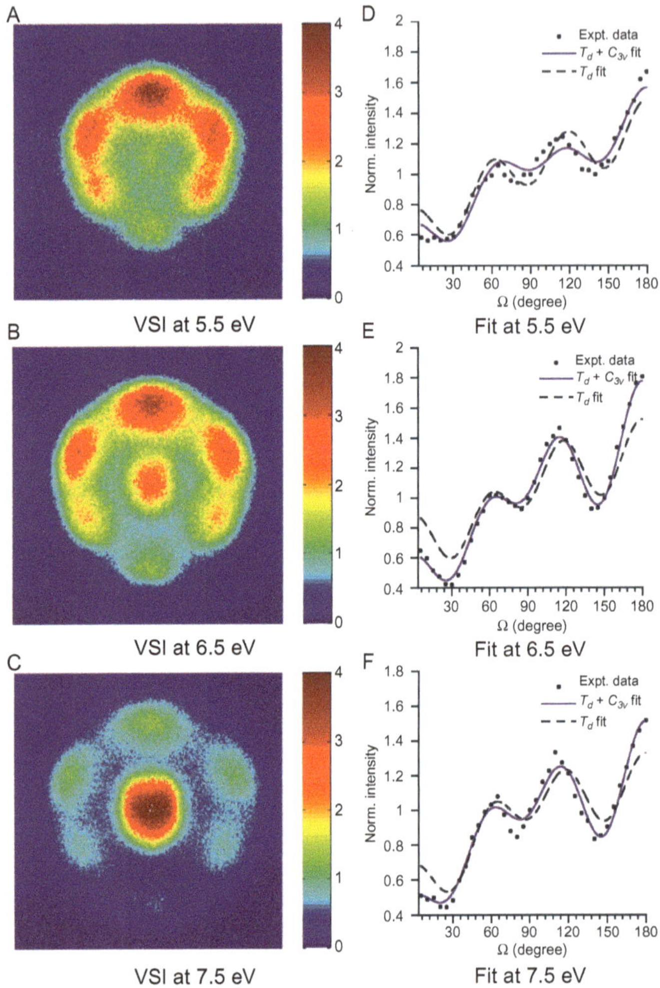


**Fig. 25** The velocity slice images of $F^-$ at (A) 5.5 eV, (B) 6.5 eV, and (C) 7.5 eV. The corresponding fitted angular distributions are shown in (D)–(F). *Copied from Ómarsson, F.H., Szymańska, E., Mason, N.J., Krishnakumar, E., 2014a. Velocity slice imaging study on dissociative electron attachment to CF4. Eur. Phys. J. D 68, 101.*

The lower energy resonance is found to result in high kinetic energy release (KER), while, in contrast, the higher energy resonance produces anions with close to thermal kinetic energies. The $CF_3^-$ anion yield is only observed to occur through the lower energy resonance, and the KER is relatively high as observed for corresponding $F^-$ produced through the same resonance. However, the $CF_3^-$ channel shows two distinct sections in the energy dependency of the KER. This indicates that the dissociation dynamics are not identical throughout the resonance. This was explained by the need of the $CF_3^-$ ion to relax toward its geometrical most stable configuration when the incident electron energy is below the C–F bond dissociation energy (BDE). Above the C–F BDE a contribution from the adiabatic electron affinity of $CF_3$ is no longer necessary and the channel for the formation of the unrelaxed ion becomes thermochemically accessible. The process yielding the unrelaxed $CF_3^-$ is, in turn, more direct and faster than the process yielding $CF_3^-$ in its geometrically relaxed configuration. Therefore, the slope of the line through the kinetic energy vs the electron energy becomes larger at higher electron energies.

These angular distributions and KER were best described due to the $CF^-$ anion being formed in two different geometrical configurations with a quantum superposition of the target state and product channels, respectively, of $T_d$ and $C_{3v}$ symmetry, providing the best fit to the measurements. These observations are attributed to a Jahn–Teller distortion of the molecule during the electron attachment process. A similar albeit less clear effect has been speculated for DEA to methane (Ram and Krishnakumar, 2011), and these experiments provide some of first evidence for the observation of Jahn–Teller effects in an electron scattering process (Omarsson et al., 2013) though this may be more common than is currently accepted. Since these fits are largely empirical, what is required is a formal detailed theoretical study, which can explore such dynamics and confirm the role of Jahn–Teller effects in low-energy electron-induced molecular fragmentation as it has in spectroscopic and photochemistry (Mckinlay et al., 2010).

## 6.2 Sulfur Hexafluoride

Sulfur hexafluoride is the classic molecule for electron attachment studies and has been labeled an "electron sponge" since it absorbs free electrons. This leads to it having excellent dielectric properties; indeed the dielectric strength of sulfur hexafluoride gas is about 2.5 times more than that of air. $SF_6$ has therefore been used as an insulating material in electrical systems, as a circuit breaker, in high voltage switch gear and as an arc-quenching medium. It has also been widely used in the semiconductor industry as an etchant $SF_6$ breaks down in the plasma into sulfur and fluorine, the fluorine plasma performing the etching.

Electron attachment to $SF_6$ creating long-lived $SF_6$ anions (>10 μs) occurs with low (thermal) energy electrons with an estimated cross section of electron attachment (EA) cross section of $>10^{-19}$ m$^2$; however, at energies above 0.3 eV DEA is observed to create $SF_5^-$ and $F^-$. The total electron attachment cross section is dominated by the formation of $SF_6^-$ below ~0.2 eV, by the formation of $SF_5^-$ between ~0.3 and 1.5 eV, and by the formation of $F^-$ beyond ~2.0 eV (see Fig. 26; Christophorou and Olthoff, 2001). Such a low-lying high cross section for anion formation has been widely used to calibrate energy scale of DEA experiments where the incident electron beam energy is such that the detection of $SF_6^-$ is assumed to be zero energy. However, very high resolution electron beams produced using photoionization sources (e.g., the LPA method of Hotop and coworkers) have more clearly defined the electron attachment cross section (Fig. 27) and allowed the complex attachment and dissociation dynamics to be explore both experimentally and theoretically.

$SF_6$ is also an excellent example of how the DEA cross section is strongly dependent upon temperature. As discussed earlier for $H_2$, DEA cross sections may be strongly influenced by vibrational excitation of the target and this is strongly observed in the temperature-dependent cross section of $SF_5^-$ (Fig. 27). As the temperature rises a larger proportion of the target molecules will be thermally excited into higher (ro)vibrational states (a fraction calculated by the Boltzmann distribution). However, the $\nu_3$ vibrational mode in $SF_6$ may also be pumped by a CO laser (117 meV photons) with a corresponding increase in $SF_5^-$ DEA cross section (Chen and Chantry, 1979) but a little difference in cross section for $SF_6^-$ production. A detailed discussion was presented first by Braun et al. (2006).

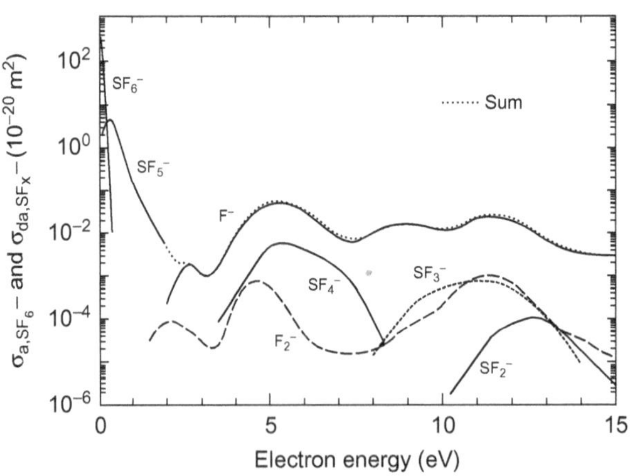


**Fig. 26** Recommended or suggested cross sections for the formation of anions to electron impact on $SF_6$ (Christophorou and Olthoff, 2001).

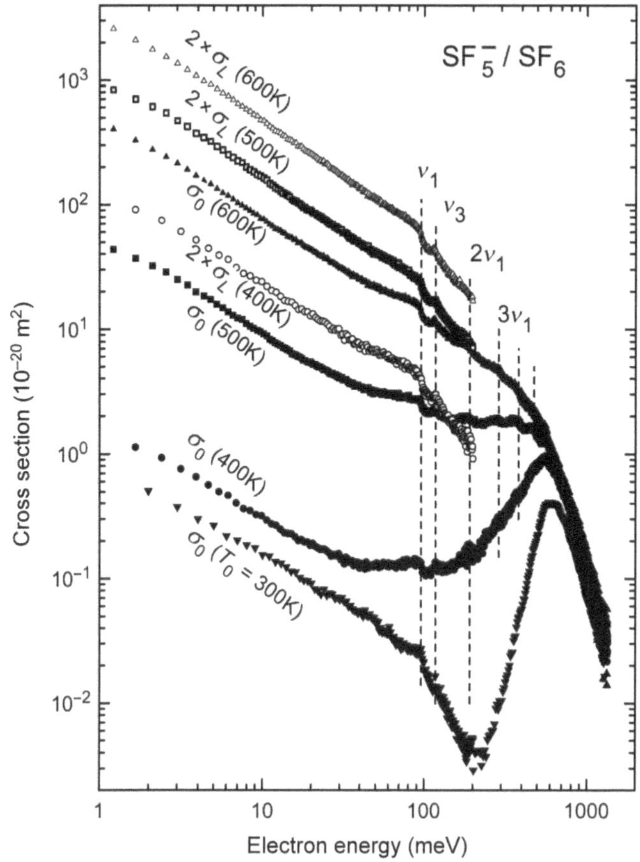


**Fig. 27** Temperature-dependent cross section of $SF_5^-$ produced by DEA from $SF_6$ (Braun et al., 2006).

It should be noted that due to the difficulty in producing high-intensity, very low energy (<1 eV) electron beams the modern ion imaging methods (VSI/VMI) have yet to be used to explore such complex EA and DEA dynamics in $SF_6$ or any vibrationally excited target. Theoretical work on electron attachment to $SF_6$ is very scarce due to complexity of inclusion of many vibrational degrees of freedom involved in the attachment process. Most of the theoretical work (Fabrikant et al., 2005; Gerchikov and Gribakin, 2008) concentrates on low-energy nondissociative attachment, which involves the mechanism of the intramolecular vibrational energy redistribution (IVR). According to this scenario (Gauyacq and

Herzenberg, 1984), at the first stage the incoming s-wave electron distorts the nuclear framework by coupling to the symmetric stretch motion with simultaneous capture. The symmetric $\nu_1$ mode strongly interacts with the spherically symmetric charge distribution due to the *s*-wave electron. The energy deposited by the electron is then distributed over all the nuclei in a chaotic long-lived anion state before the nuclear framework can oscillate back to its initial configuration. Completely ab initio description of this process is very challenging, but it can be described semiempirically by the multichannel effective range theory with complex boundary conditions (Fabrikant et al., 2005). The theory also produces cross sections for vibrational excitation of the Raman-active $\nu_1$ mode and the infrared-active $\nu_3$ mode, in a good agreement with experiments. Another, more phenomenological, approach to electron attachment to $SF_6$ is based on the Vogt and Wannier (1954) model for capture of a charged particle into a polarization well (Fabrikant and Hotop, 2001).

$SF_6$ was first produced on a commercial scale in 1948 and some 2000 $T_G$ of $SF_6$ were being produced annually in the early years of the 21st century. However, $SF_6$ is known to have the largest Global Warming Potential, GWP (23,900 times that of $CO_2$), of any known species, and thus it is required to reduce or phase out the use of sulfur hexafluoride by 2030. This has posed considerable challenges to the electronics and power industry since it is necessary to find a gas that has both electronegative properties while also being chemically inert and as nontoxic as $SF_6$. This has led to DEA properties of many new halogen-containing compounds being explored, for example, $CF_3I$ (Marienfeld et al., 2006b; Ómarsson et al., 2014b). The need to evaluate DEA cross sections, rate coefficients, and products for many halogenated compounds will therefore continue to be important for the development of next-generation technologies. Such experiments also provide new insights into fragmentation dynamics and the differences between photon- and electron-induced processes, for example, while DEA of $CF_3I$ proceeds from the same initial parent state as the corresponding photolysis process, in contrast to photolysis, which leads nearly exclusively to the (2)P(1/2) excited state of iodine, electron-induced dissociation leads predominantly to the (2)P(3/2) ground state. This has been explained by the changed spin state of the negative ion, allowing an adiabatic dissociation through a conical intersection, a path that is efficiently repressed by the required spin flip in the photolysis process (Ómarsson et al., 2014b).

## 7. MULTIDIMENSIONAL VIBRATIONAL DYNAMICS IN THE LOCAL APPROXIMATION

Multidimensional nonlocal theory of DEA has not been developed yet, and all calculations, including several nuclear degrees of freedom, performed so far were done in the local approximation. The processes described by the multidimensional dynamics can be separated into two types. In the first type, DEA occurs along a reaction coordinate corresponding, in a good approximation, to a certain vibrational mode, and other modes just modify dynamics. In the second class of DEA processes the nuclear dynamics is essentially multidimensional since there is no single vibrational mode corresponding to the dissociating coordinate. An example of a process belonging to the first class is low-energy DEA to methyl halides or perfluoromethyl halides

$$e + CY_3X \rightarrow CY_3 + X^-$$

where X stands for a halogen atom, and Y for H or F atom. The dynamics in this case is mostly driven by the C–X stretch vibrations, and other vibrational modes modify the process insignificantly. This is corroborated by numerous DEA calculations for molecules of this type with one active reaction coordinate, while all other coordinates are frozen.

An example of the second class of the DEA process is DEA to the water molecule. The vibrational dynamics is essentially six-dimensional involving three vibrational and three rotational degrees of freedom. The dynamics contains a rich physics including conical intersection, charge-transfer behavior at large internuclear separations, and Renner–Teller coupling between resonance surfaces (Haxton et al., 2005, 2007b). These calculations demonstrate that even in the case of a triatomic molecule as simple as water, the treatment of nuclear dynamics is extremely complicated. Certainly a development of new methods will be required for treating DEA to molecules with more than three atoms, even in case when the local complex potential approximation is valid.

For numerical treatment of nuclear dynamics on the local anion potential energy surface, two approaches have been developed. One is based on the numerical integration of the time-dependent Schroedinger equation. For the DEA problem, it was first explored by McCurdy and Turner (1983). The other reduces the stationary Schroedinger equation to a system of coupled ordinary differential equations describing different DEA channels (Tarana et al., 2009). The method is similar to that used in photodissociation problems (Shapiro and Bersohn, 1980).

In what follows we will present several case studies: the process of the first type, DEA to $CF_3Cl$, treated by a close coupling method, and the process of the second type, DEA to water, treated by a time-dependent wave packet propagation method of McCurdy and Turner (1983). We will briefly discuss then some other processes of the second type where the multidimensional treatment has been applied.

## 7.1 Perfluoromethyl Chloride

Tarana et al. (2009, 2011) studied the process

$$e + CF_3Cl(\nu_2, \nu_3) \rightarrow CF_3(\nu_2') + Cl^-$$

Here, $\nu_2$ and $\nu_3$ stand for the symmetric deformation vibrations (so-called umbrella mode) and the symmetric stretch vibrations, respectively, and $\nu_2'$ represents the umbrella mode of the free $CF_3$ radical. This choice of the modes can be justified by the same reasoning as that given by Shapiro and Bersohn (1980) in their studies of photodissociation of $CH_3I$. First, the intermediate anionic state responsible for the DEA at low energies has $A_1$ symmetry; therefore, degenerate vibrations $\nu_4$, $\nu_5$, and $\nu_6$ of the e type can be excited resonantly only in pairs. Second, the C–F bond length is very close to that for the anion and free $CF_3$ radical. Therefore, we can assume that the C–F symmetric stretching mode $\nu_1$ is unlikely to be excited as well. These assumptions are confirmed by measurements of Mann and Linder (1992) who observed strong electron-impact vibrational excitation of the $\nu_3$ and $\nu_2$ modes in the $A_1$ resonance region, whereas other modes were not significantly excited.

The use of the local theory for DEA to $CF_3Cl$ is justified by the fact that the $A_1$ shape resonance in $CF_3Cl$ appears at a relatively high energy (1.83 eV) and has rather narrow width. Also, the long-range interaction plays in this case an insignificant role. Indeed, comparison of local and non-local calculations in the one-dimensional case demonstrates good agreement between results of the two theories (Fabrikant, 2009; Tarana et al., 2009).

The potential energy surfaces for the $CF_3Cl$ anion were calculated by two methods. In the first publication (Tarana et al., 2009), quantum chemistry packages were used. This method, however, suffers from instabilities related to so-called variational collapse: an attempt to introduce a more diffuse basis set leads to larger contribution of the continuum states and the "collapse" of the anion energy to the neutral energy. The standard quantum chemistry packages also do not allow calculation of the resonance width, which was obtained by somewhat arbitrary extension of one-dimensional width to two-dimensional surface. In the subsequent calculations (Tarana et al., 2011) the real part of the potential energy surface and the width were obtained from scattering calculations employing *R*-matrix scattering code (Tennyson, 2010).

Turning to the nuclear motion, the Hamiltonian of the system can be written in terms of two nuclear coordinates: the distance $\rho$ between the center of mass of the $CF_3$ radical and the Cl atom, and the distance $r$ between the atom C and the plane formed by the F atoms. The latter coordinate describes the umbrella motion. After expanding the wave function for the nuclear motion in terms of the eigenstates of the vibrational Hamiltonian for the $CF_3$ radical, we obtain a system of coupled equations for the coefficient functions, which is solved then either by direct numerical integration or by a more efficient multichannel version of the exterior complex scaling method in the discrete variable representation basis set (Rescigno and McCurdy, 2000).

The total DEA cross sections calculated by different methods are plotted in Fig. 28. This graph compares two-dimensional results with one-dimensional nonlocal calculations of Wilde et al. (1999). Calculations with the model width (Tarana et al., 2009) give a significantly larger magnitude for the total DEA cross section than do the measurements (Aflatooni and Burrow, 2001; Underwood-Lemons et al., 1995) and the one-dimensional nonlocal semiempirical calculation (Wilde et al., 1999). This problem was attributed to an incorrect dependence of the width function (which was arbitrarily extended to two dimensions) on the coordinate $r$. Results of ab initio two-dimensional calculations are in very good agreement with the nonlocal calculations as well as with experimental results of Aflatooni and Burrow (2001), although they are smaller than the experimental results of Underwood-Lemons et al. (1995). The position of the peak agrees very well with the experimental value.

Fig. 29 shows the distribution of different final vibrational states of the $CF_3$ fragment. The low vibrationally excited states of the $CF_3$ fragment are more populated than the ground state and the positions of the peaks rise with increasing quantum number $\nu$. The *R*-matrix results of Tarana et al. (2011) give the highest peak for $\nu = 1$. The total DEA cross section is mainly determined by the complex potential energy surface in the region where

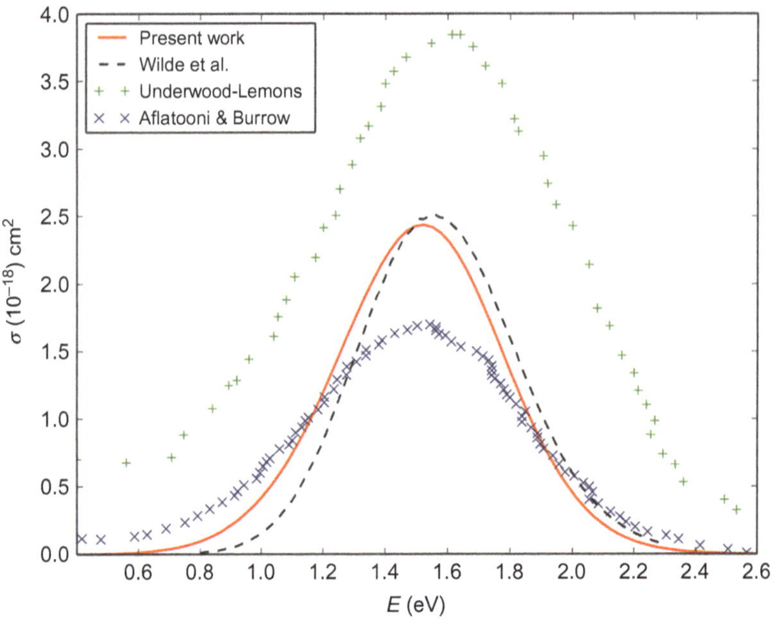


**Fig. 28** Total DEA cross section for $CF_3Cl$ calculated by Tarana et al. (2011) and compared with the experimental results of Underwood-Lemons et al. (1995) and Aflatooni and Burrow (2001), and with one-dimensional nonlocal calculations by Wilde et al. (1999). *Reproduced from Tarana, M., Houfek, K., Horacek, J., Fabrikant, I.I., 2011. Dissociative electron attachment and vibrational excitation of $CF_3Cl$: effect of two vibrational modes revisited. Phys. Rev. A 84, 052717 (1–9).*

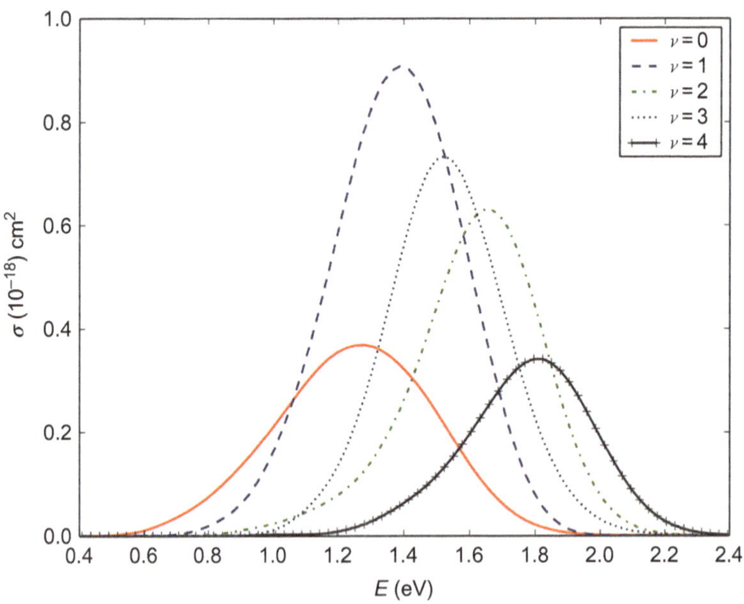


**Fig. 29** DEA cross sections for $CF_3Cl$ for $CF_3$ formation in different final vibrational states. *Reproduced from Tarana, M., Houfek, K., Horacek, J., Fabrikant, I.I., 2011. Dissociative electron attachment and vibrational excitation of $CF_3Cl$: effect of two vibrational modes revisited. Phys. Rev. A 84, 052717 (1–9).*

the anionic system is not bound. On the other hand, the distribution of vibrational states of the fragment is strongly influenced by the final-state interaction in the region where the anion is stable. This conclusion results from the additional model calculations (Tarana et al., 2009) whereby the vibrational coupling between the final states of $CF_3$ was switched off. This did not affect very much the total DEA cross section, but influenced substantially the final-state distribution.

## 7.2 Water

The main products of DEA in water are $H^-$ and $O^-$, with the production of $O^-$ being much smaller than that of $H^-$ at lower energies, but with $O^-$ dominating at higher electron-impact energies. A review of the earlier experimental work on DEA to water was given by Haxton et al. (2004a) and Itikawa and Mason (2005). In a series of papers, Haxton et al. (2004a,b, 2005, 2006, 2007a,b, 2008, 2011) presented detailed calculation of complex potential surfaces and DEA cross sections to water. The amount of material presented in these papers is enormous; therefore, we limit the present discussion to basic outlines and refer the reader for more details to the original papers.

Low-energy DEA to water goes through the three lowest Feshbach resonances, $^2B_1$ at 6.5 eV, $^2A_1$ at 8.4 eV, and $^2B_2$ at 11.8 eV. This classification assumes the $C_{2v}$ symmetry. Away from the equilibrium, this symmetry is broken and, according to the irreducible representation of the $C_s$ group, the resonances become $^2A''$, $1^2A'$, and $2^2A'$, respectively. Seng and Linder (1976) discovered a very broad low-energy resonance in electron-impact vibrational excitation of water, apparently associated with a temporary capture into the $4a_1$ virtual orbital. It is similar to the shape resonance in vibrational excitation of hydrogen halides (Rohr and Linder, 1976) and is responsible for large cross sections for the inverse process of associative detachment at low collision energies (Comer and Schulz, 1974). This resonance, however, does not appear in DEA to water. Another shape resonance with the configuration $[H_2O]2b_2$, which was also observed in vibrational excitation of water (El-Zein et al., 2000), is relevant to DEA. Haxton et al. (2005), using the complex Kohn method, located two resonance features in $^2B_2$ symmetry at certain distorted geometries and found a seam of degeneracy between the $^2B_2$ Feshbach and $^2B_2$ shape resonances. In addition, they found a conical intersection between the $^2A_1$ and $^2B_2$ Feshbach resonances. For an analytical description of the conical intersection, Haxton et al. (2005) used a model with the width exponentially dependent on one of the nuclear coordinates. This model produces a single intersection, in contrast to the linear model of Feuerbacher et al. (2004), which produces two conical intersections. Another important feature of the potential surfaces is the $^2A_1$–$^2B_1$ degeneracy at the linear geometry, leading to the Renner–Teller coupling between these states.

A full theoretical treatment of the problem of DEA to water, incorporating all of the features of the electronic states, contains rich physics. Haxton et al. (2007a,b) identified the reaction pathway for each resonance involved in the e–$H_2O$ collisions and calculated cross sections for the two-body breakup reactions

$$\begin{gathered} e + H_2O \rightarrow H + OH^- \quad (E_t = 3.27\,eV) \\ e + H_2O \rightarrow H_2 + O \quad (E_t = 3.56\,eV) \\ e + H_2O \rightarrow H^- + OH\left(X^2\Pi\right) \quad (E_t = 4.35\,eV) \\ e + H_2O \rightarrow H^- + OH\left(^2\Sigma\right) \quad (E_t = 8.38\,eV) \end{gathered}$$

and the three-body breakup reactions (Haxton et al., 2008)

$$\begin{gathered} e + H_2O \rightarrow H + H + O^- \quad (E_t = 8.04\,eV) \\ e + H_2O \rightarrow H^- + H + O \quad (E_t = 8.75\,eV) \end{gathered}$$

where values of $E_t$ indicate the reaction thresholds.

The experimental studies (Adaniya et al., 2009; Ram et al., 2009; Ram et al., 2011; Ram et al., 2012) employing the imaging technique confirmed theoretical results on the dynamics involving conical intersections. Haxton et al. (2004b) studied also the isotope effect whose physics in DEA to water is very different from that for the systems with a relatively large resonance width. In the latter case the isotope effect is controlled by the survival factor, which decreases very fast with the growth of the reduced mass. However, in systems with small resonance width like water, the magnitude of the cross section is controlled by the shape of the initial wave packet. Since the initial vibrational wave function for the heavier isotope is more sharply peaked and narrower, the cross sections are also more strongly peaked and narrower. In multidimensional systems, this argument is not that straightforward, but in any event, this mechanism produces a much weaker effect than the survival probability mechanism. In particular, Haxton et al. (2004b) found the peak values of the total DEA cross sections 5.99 and $7.04 \times 10^{-18}$ cm$^2$ for $H_2O$ and $D_2O$, respectively. Haxton et al. (2004b, 2007b) also investigated the dependencies of DEA cross sections on initial vibrational and rotational states and partial cross sections for producing molecular fragments in certain rovibrational states.

We conclude this section by emphasizing that the calculation of DEA to such a relatively simple triatomic molecule as water presents a big challenge to the theory even in the local approximation and involves very rich physics.

## 7.3 Acetylene and HCN

Acetylene is the smallest unsaturated hydrocarbon molecule. Although it has six vibrational degrees of freedom, only three of them are essential for the low-energy DEA process (Chourou and Orel, 2008)

$$C_2H_2\left(X^1\Sigma_g^{\ +},\nu\right)+e\rightarrow(C_2H_2)^-\left({}^2\Pi_g\right)\rightarrow C_2H^-\left({}^1\Sigma^+,\nu'\right)+H\left({}^2S\right)$$

going through the electron capture into the lowest ${}^2\Pi_g$ resonance. Chourou and Orel included $C_2H$–H stretching (the reaction coordinate), HC–C stretching, and CCH bending in their calculations. In the linear configuration the lowest anion potential energy curves ${}^2\Pi_g$ and ${}^2\Sigma$ exhibit a potential barrier toward the dissociation path. Upon bending the CCH configuration, the ${}^2\Sigma$ resonant state becomes the $2\,{}^2A'$ state and the doubly degenerate ${}^2\Pi_g$ resonant state splits into two states of ${}^2A'$ and ${}^2A''$ symmetry classified in terms of the $C_s$ point group. The bending of the CCH configuration provides a dissociation path without the barrier strongly increasing the DEA cross section. This phenomenon can also be interpreted in terms of the vibronic coupling between resonances discussed in Section 2.5.

After identifying this dissociation path, Chourou and Orel (2008) calculated DEA cross sections using the wave packet propagation technique. The results turned out to be in an excellent agreement with the measurements of May et al. (2008), both in shape and in the absolute magnitude. However, some disagreement was obtained initially in the magnitude of the isotope effect, that is, in the ratio of the $C_2H^-$ production to the $C_2D^-$ production in DEA to $C_2D_2$. Whereas the theoretical prediction for this ratio was 28.9, May et al. (2009) obtained the factor of 14.4. In their subsequent publication, Chourou and Orel (2009a) noted that at the room temperature, higher excited states corresponding to bending vibrations are substantially populated. When DEA from these states was incorporated in calculations, agreement with the experiment improved substantially. To summarize on acetylene, this molecule represents a nontrivial example of state-of-the-art ab initio calculations with the local nuclear dynamics resulting in a very impressive agreement with the experiment.

Similar calculations were performed for the HCN and HNC molecules (Chourou and Orel, 2009b, 2011). Both isomers are linear at the equilibrium and a low-energy $\pi^*$ resonance was observed in electron scattering calculations for both isomers. Accordingly, Chourou and Orel (2009b) studied the process

$$\mathrm{HCN}\left(X^1\Sigma_g{}^+, v\right) + \mathrm{e} \rightarrow (\mathrm{HCN})^-\left({}^2\Pi_g\right) \rightarrow \mathrm{CN}^-\left({}^1\Sigma^+, v'\right) + \mathrm{H}\left({}^2\mathrm{S}\right)$$

and a similar process for HNC. They found that the main resonance feature remains practically unchanged under stretching of the hydrogen bond. This implies a nondissociative process if the molecule is restrained to be linear. However, the dissociative channel opens if the bending coordinate is taken into account. Both HCN and HNC cross sections for DEA from the ground vibrational state peak at the energy about 3 eV with the HCN cross section higher by a factor of 2.7 than the HNC cross section. The isotope effect is similar to that in acetylene, but the effect of initial excitation in the bending mode is less significant. The peak cross section for HCN was predicted to be 370 times larger than in acetylene at 300K. This was basically confirmed by the experiment of May et al. (2010) whose measured cross section for HCN is 270 times larger than measured value in acetylene. However, May et al. observed differences in the shapes of DEA cross sections as functions of electron energy. Whereas the measured cross sections have a nearly vertical onset at the threshold (1.55 eV) and peak around 1.85 eV, the calculated cross sections have a very gradual onset and peak at much higher energy, around 3 eV.

New calculations by Chourou and Orel (2011) incorporating correlation and polarization effects substantially improved agreement with the experiment in terms of the position of the peak (new value 1.94 eV) and the absolute magnitude of the cross section. Much better agreement was also obtained for the isotope effect: the calculated ratio of the cross section for the H loss from HCN to the cross section for the D loss from DCN was found to be 3.2 at temperature $T=333$K (with the previous value about 13), whereas the measured ratio was 2.8. However, there was still some disagreement with regard to the threshold behavior of the DEA cross section. May et al. (2010) noted that the dependence of the resonance width on energy may be important. From the theoretical point of view, this means that the nonlocal effects in this process might be significant that is not surprising in view of a substantial dipole moment of the HCN. As was suggested by Allan (2016), HCN can be viewed as a case intermediate between the acetylene, where the local theory describes the process perfectly well, and hydrogen halides, where the nonlocal approach is crucial.

# 8. BIOLOGICALLY RELEVANT MOLECULES

## 8.1 Introduction

As described in Section 1.3, Sanche and coworkers from the University of Sherbrooke (Boudaïffa et al., 2000) demonstrated that subionization energy electrons are a significant source of DNA strand breaks. This result stimulated extensive research into low-energy free electron interactions with diverse biomolecules (for example, nucleobases, amino acids, and dipeptides (Puschnigg et al., 2013)), where the free electron can be considered to analogize an electron released in a local ionization event. These studies have provided indications of the most vulnerable DNA sites for DEA (Aflatooni et al., 2006; Denifl et al., 2004). Furthermore, a number of groups have studied low-energy electron interactions with larger subunits of DNA, notably nucleosides (a nucleobase coupled to a sugar unit (Ptasinska et al., 2006)) and oligonucleotides (several tens of sequenced base-pairs (Solomun et al., 2009)). Baccarelli et al. (2011) reviewed the research that explored low-energy electron-induced DNA damage over the 10 years that followed the Sherbrooke result (Boudaïffa et al., 2000) and we also recommend Bald et al.'s (2017) review chapter on DEA to biomolecules in *Nanoscale Insights Into Ion-Beam Cancer Therapy*. A further motivation to study DEA to biomolecular building blocks lies in assessing their radiostabilities and reactivities in diverse astronomical environments (for example, on dust particles in interstellar clouds and in early planetary atmospheres), with implications for understanding the molecular origins of life (Mason et al., 2014).

The following sections present case studies of DEA in biomolecular targets of increasing complexity. In Section 8.2, we discuss formic acid, which is a very good molecule for validating theoretical models. On one hand, it represents a target that is small enough to be described by various scattering approaches; on the other hand, its number of internal degrees of freedom gives rise to multidimensional effects. Section 8.3 discusses formic acid dimers and larger clusters with a particular focus on electron attachment-driven intracluster proton transfer. This mechanism is interesting from a radiation damage standpoint as Bowen and coworkers have since reported evidence for electron attachment-induced proton transfer in the cytosine–guanine pair based on the combination of photoelectron spectroscopy with ab initio calculations (Szyperska et al., 2010). Electron attachment to clusters is discussed further in Section 9. Section 8.4 addresses the role of VFRs and threshold cusps in DEA to fundamental biomolecules, focusing on the examples of uracil and thymine (the nucleobases that pair with adenine in RNA and DNA, respectively). Section 8.5 discusses progress in DEA experiments on relatively large biomolecules since Baccarelli et al.'s (2011) review.

## 8.2 Formic Acid

Formic acid, HCOOH, is the simplest organic acid. Due to its carboxylic functional group, it can serve as a model for amino acids that play critical roles in number of biochemical processes. DEA in gas-phase formic acid was initially probed experimentally by Pelc et al. (2002a,b) and the strongest dissociative channel was identified as simple dehydrogenation peaking at 1.3 eV:

$$\mathrm{HCOOH} + \mathrm{e} \rightarrow \mathrm{HCOO}^- + \mathrm{H}$$

The other DEA fragments ($OH^-$ and $O^-$) have cross sections that are orders of magnitude lower and appear at much higher electron energies (7.5 and 9.2 eV). The dehydrogenation process occurs at the energetic threshold for producing the formate anion $HCOO^-$ by the abstraction of H atom from the oxygen site. Weak structures on the high-energy flank of the $HCOO^-$ band were reported for the first time by Pelc et al. (2002b), which appear at the openings of vibrational channels. The absolute cross sections were measured by Prabudeshai et al. (2005).

As to the mechanism of dehydrogenation process, the structure of formic acid a priori allows for two quite different DEA mechanisms. On one hand, it has a double C=O bond and one can thus expect it to have a resonance associated with the low-lying unoccupied $\pi^*$ orbitals localized on this bond. Indeed the $\pi^*$ resonance has been observed in the electron transmission spectrum (Aflatooni et al., 2001) and in the spectra for vibrational excitation of HCOOH (Allan, 2006). As shown in Fig. 30, the vibrational excitation of C=O stretch shows this resonance at 1.9 eV: the excitation curve for this vibration has a form of a broad band centered around this energy with a clearly discernible boomerang structure (Allan, 2006). The shape of the

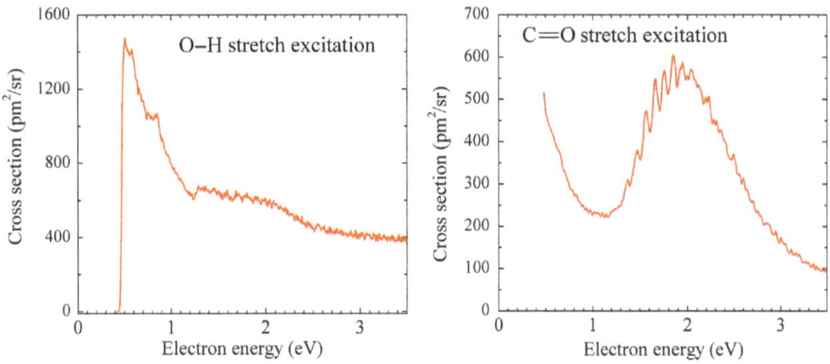


**Fig. 30** Vibrational excitation cross sections for C=O and O–H stretch vibrations in formic acid (Allan, 2006).

vibrational excitation cross section is typical of that of small molecules with double and triple bonds ($N_2$, CO, $C_2H_2$) and is characteristic of a $\pi^*$ shape resonance. This resonance thus can, in principle, serve as a doorway state for the DEA in spite of the fact that the resonance is initially localized on the C=O bond.

On the other hand the O–H bond that is actually cleaved is polar, which opens the possibility of long-range effects (described in Section 2.3) playing important roles. The vibrational excitation cross section for O–H stretch looks completely different: the $\pi^*$ resonance is barely discernible in its shape and the spectrum is dominated by a broad peak, which has clearly discernible structures at the thresholds for the excitation of higher O–H stretch quanta. This closely resembles the vibrational excitation cross sections observed in hydrogen halides, where the structures are due to Wigner cusps (like in HCl and HBr) or due to VFRs (like in HF, Section 5.5.3). This mechanism is clearly responsible for DEA in hydrogen halides, which opens the possibility that it also plays a role in formic acid.

These two possible mechanisms were reflected in the theoretical approaches describing the DEA in HCOOH. Fixed-nuclei scattering calculations in a local potential model (Gianturco and Luchesse, 2004) determined an isolated $\pi^*$ shape resonance at 3.5 eV. Rescigno et al. (2006) carried out a series of variational calculations employing the complex Kohn method, characterized this resonance at number of different geometries, and explored the potential pathways toward the dissociation of the resonant state. At the equilibrium geometry, they found the $\pi^*$ resonance to be located at 1.9 eV with a width of 0.2 eV. In the $C_s$ symmetry point group, this resonant state has an $A''$ symmetry. The fragments $HCOO^-(A')$ and

$H(^2S \equiv A')$ have $A'$ symmetry in planar geometry. Therefore, for the $\pi^*$ resonance to produce the $A'$ fragments, the planar geometry has to be broken along the reaction path allowing interaction with a second anion state that can dissociate into these fragments. Upon exploring the resonant potential energy surface in several directions, Rescigno et al. (2006) found that such geometry distortion should proceed via out-of-plane motion of the hydrogen atom that is bound to carbon and not of the H atom that is actually being abstracted. A similar mechanism was proposed earlier for description of DEA to 5-chlorouracil by Sommerfeld (2001). Unfortunately, calculating the actual DEA cross section for such a multidimensional problem is not feasible even in the local complex potential approach.

The second possible mechanism has been theoretically explored by Gallup et al. (2009a) using the $R$-matrix theory in a quasi-diatomic model allowing motion only along the OH-stretching coordinate. Even though this is significant simplification of an inherently multidimensional problem, it allows for inclusion of those nonlocal effects that were crucial for describing scattering and DEA in hydrogen halides. The calculated cross section is shown in Fig. 31 and reproduced very well both the shape and the magnitude of the absolute DEA cross section.

In order to resolve the theoretical dispute of Rescigno et al. (2009) and Gallup et al. (2009b), Janeckova et al. (2013) measured DEA cross sections for the H/D isotopomers of formic acid. The crucial part of the model involving the $\pi^*$ resonance is an out-of-plane motion of the H atom bound to carbon: in the calculation of Rescigno et al. (2006) the resonant state crosses the electronically bound anion state at the 35-degree out-of-plane rotation of the C–H bond. If this motion is indeed governing the dissociation, then deuteration at this site should lengthen the timescale of the whole dissociation process. Due to the competition of dissociation and electron autodetachment, the DEA cross section should drop significantly with mass. Fig. 31 shows the cross sections and energy-integrated cross sections (quantity independent on the electron-energy resolution) for four different formic acid isotopomers. Upon deuteration of the O-site, the cross section drops significantly, as expected in both mechanisms. However, the cross-section ratios HCOOH/DCOOH and HCOOD/DCOOD are very close to 1, within the limit of the stated experimental errors. The motion of the hydrogen atom bound to carbon thus seems to have very little influence on the detachment vs dissociation competition. Since its motion is crucial in the $\pi^*$ mechanism, it seems that this mechanism makes only a minor contribution to the DEA cross

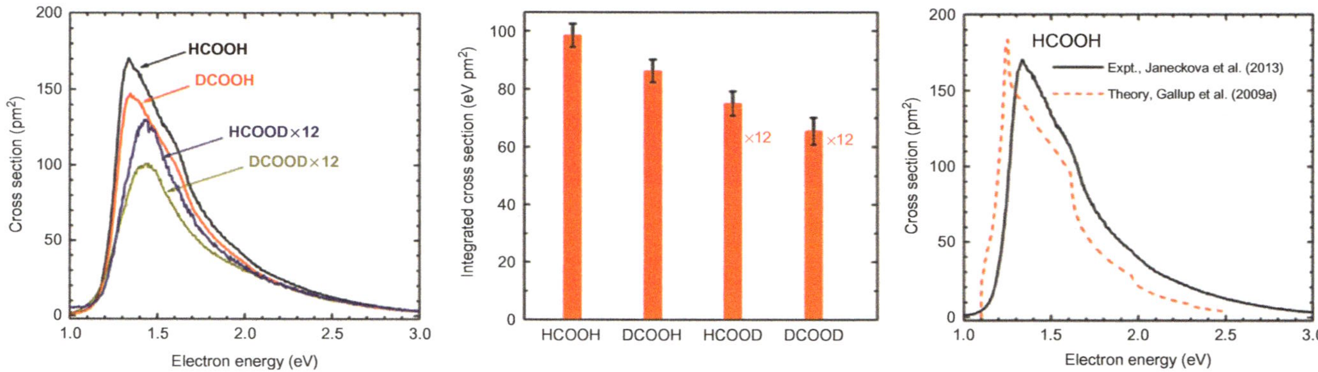


**Fig. 31** Cross sections (*left*) and energy-integrated cross sections (*middle*) for four different H/D isotopomers of formic acid (Janeckova et al., 2013). The *right panel* shows the comparison of experimental data with theoretical model of Gallup et al. (2009a).

section. Part of the reason why this contribution is weak may be that DEA occurs at an energy below the center of the $\pi^*$ resonance or at the edge of the Franck–Condon range for $\pi^*$ attachment.

The hydrogen halides-like mechanism of Gallup et al. (2009a) does not require any motion of the C–H bond and is thus consistent with the experimental findings of Janeckova et al. (2013). It also reproduces the magnitude and fine structures (associated with the opening of the O–H stretch vibrational excitation channels) of the experimental cross section. It should be noted at this point that a similar model also reproduced very well the structures in the DEA cross section to aminobutanoic acid (Vizcaino et al., 2012). However, the present state of the theory (see Section 2) is restricted to one-dimensional models and the treatment of nonlocal effects in more than one dimension remains an unresolved problem. The spectra of Janeckova et al. (2013) reveal weak structures in the DEA cross section that appear at the threshold for combination vibrations (O stretch with C=O stretch and O–C–O deformation). This demonstrates the need for a unified theoretical treatment that includes both long-range interaction with nonlocal effects and multidimensional phenomena.

## 8.3 Formic Acid Dimer and Larger Clusters

Molecules in living tissue are associated with neighboring molecules, often by means of hydrogen bridges. From an experimental point of view, formic acid appears to be a very good system for testing the influence of hydrogen bonding on the electron-induced processes, since its cyclic dimer is one of the most stable neutral complexes with the binding energy of 0.65 eV.

Perhaps the most intriguing effect of hydrogen bonding in formic acid dimers—its capacity to slow down and thermalize the electrons—has been shown by Allan (2007) by the means of electron energy loss spectroscopy. Fig. 32 shows the energy loss spectra for the formic acid monomer and dimer recorded at a constant incident energy of 2 eV. The $x$-axis shows how much energy the electrons lose upon the collision with the target. The peak at 0 eV is thus the elastic scattering peak corresponding to electrons that did not lose any energy, and the signal close to 2 eV corresponds to electrons that have lost nearly all the kinetic energy in the collision and are very slow. The sharp peaks between these two extremes correspond to selective excitation of specific vibrations. The slow electrons close to 2 eV have a rather broad distribution peaking at zero residual energy—they are quasithermal. This effect, where electrons leave nearly all of their energy in the internal modes of a polyatomic target without exciting any specific vibrational modes

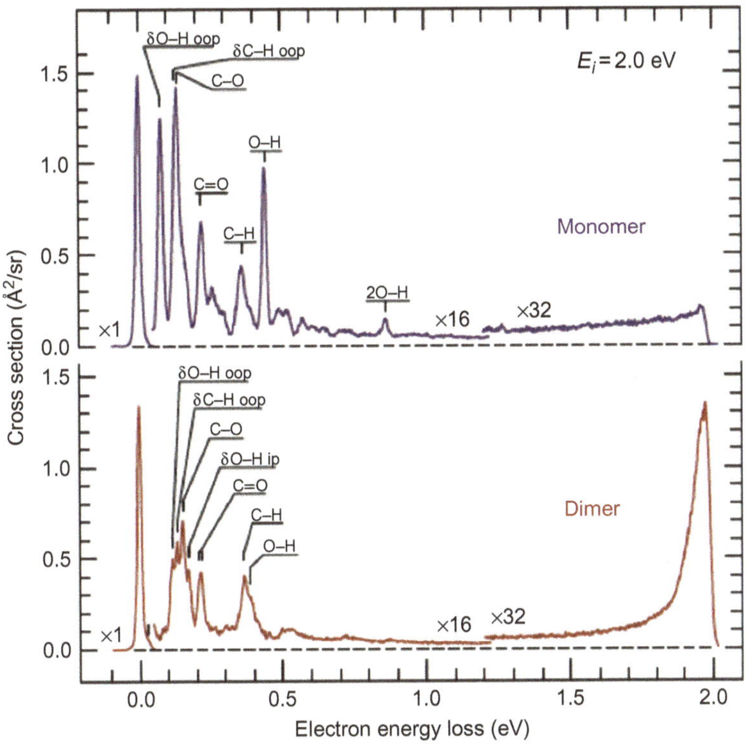


**Fig. 32** Kinetic energy distribution of electrons ejected after an impact of 2 eV electrons on the formic acid dimer and monomer. *Horizontal axis* shows the energy loss. The elastically scattered electrons have an energy loss of zero and the electrons with nearly zero kinetic energies are observed around 2 eV energy loss. The latter group is much more pronounced in the dimer than in the monomer. *From Allan, M., 2007. Electron collisions with formic acid monomer and dimer. Phys. Rev. Lett. 98, 123201 (1–4).*

(sometimes termed "unspecific excitation"), has been observed in a number of polyatomic molecules (Hotop et al., 2003). Upon the temporary attachment of an electron the excess energy is distributed among the nuclear degrees of freedom via IVR and the electrons are ejected via a process equivalent to thermionic emission. Fig. 32 clearly shows that the abundance of these quasithermal electrons is much higher in the formic acid dimer than in the monomer. At 1.3 eV, where this effect peaks, the difference between monomer and dimer in quasithermal electron production is a factor of 20. The explanation of this dramatic difference is demonstrated in the ab initio prediction of Bachorz et al. (2005) that the attachment of electron to the dimer initiates a spontaneous proton transfer: the charge initially localizes on one of the dimer constituents and this negatively charge unit attracts a proton from the second dimer constituent:

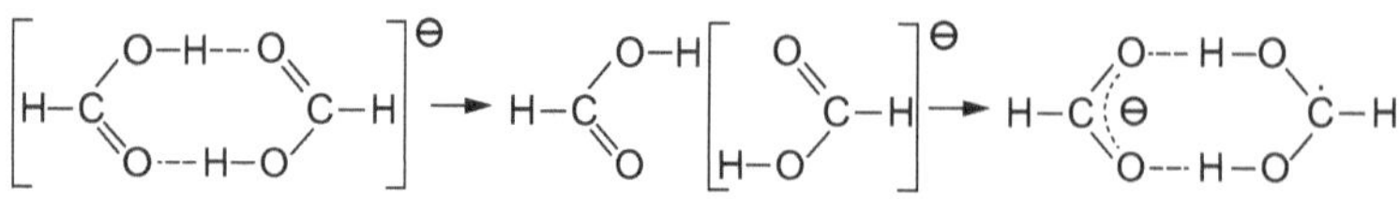

Such fast proton transfer quickly stabilizes the initial anion state, which would otherwise autodetach the electron, possibly upon selective vibrational excitation. Instead, the resulting complex of the formate anion and the methoxy radical is necessarily vibrationally hot (since it contains the stabilization energy of the complex) and can statistically emit the quasithermal electrons.

However, it is worth noting that an attempt to observe a similar thermalization effect in acetoacetic acid, a molecule with an intramolecular hydrogen bond, has failed and no increased abundance of slow electrons has been observed for this target (Keolopile et al., 2015). The reason is that the major gas-phase conformers of acetoacetic acid actually do not support the intramolecular hydrogen bond that would facilitate the fast proton transfer.

The production of negative ions in larger clusters of formic acid has been probed by Martin et al. (2005). A variety of negatively charged complexes have been detected, the most abundant being the homogeneous anions $(HCOOH)_n^-$ and dehydrogenated anions $(HCOOH)_n \cdot HCOO^-$. Apart from these, a rich variety of other fragments have been detected, indicating rich intracluster chemical reactions mediated by the hydrogen bonding. Perhaps the most interesting of these channels is a production of a water-containing $HCOO \cdot H_2O^-$ complex (identified using isotopically substituted formic acid) arising from the attachment to the cyclic dimer. It has been shown (Gianturco et al., 2005) that electron attachment to the dimer proceeds at lower electron energies than in monomer, with the experimental appearance energy close to 0 eV.

## 8.4 VFRs and Threshold Cusps in Electron Attachment to Biological Molecules

As in the case of formic acid the dominant negative-ion fragment formed by DEA to amino acids is $[M - H]^-$, the dehydrogenated anion of the original species. This process is strongly influenced by VFRs observed in DEA in hydrogen loss in uracil U (Aflatooni et al., 2005; Burrow et al., 2006; Feil et al., 2005; Hanel et al., 2003; Ptasinska et al., 2005a,c; Scheer et al., 2004, 2005), thymine T (Denifl et al., 2004) and adenine (Flosadottir et al., 2007). The lowest $^2\Sigma$ anion state in U and T compounds is strongly antibonding between $N_1$ and H. (The subscripts here and below refer to the atom location; see Fig. 33). Therefore, this state drives low-energy DEA in U (or T), leading to production of the $(U–H)^-$ anion and H atom. In addition, due to the large dipole moment of uracil, $\mu = 4.7$ D, the electron can be captured by the dipolar field with simultaneous vibrational excitation, leading to VFRs (Gallup and Fabrikant, 2011a).

Burrow et al. (2006) noticed that the $\sigma^*$ valence orbital is also strongly antibonding between $N_1$ and $C_6$ atoms, which leads to excitation of the $C_6$–H- and $N_1$–$C_6$-stretching vibrations. Indeed, additional structure corresponding to $\nu(N_1H)\nu(N_1C_6)$ overtones was observed in the DEA cross section. Uracil deuterated at the $N_1$ and $N_3$ positions was examined by Scheer et al. (2005) who reported the total anion yield, that is, without mass analysis. Two VFRs at 0.69 and 1.01 eV were observed in nondeuterated uracil. It was also noted that a broader peak at 1.7 eV was coincident with the location of the second $^2\Pi$ resonance. In N-deuterated uracil, on the other hand, no evidence for sharp VFRs was found and only a broad peak at 1.16 eV was observed. These data are evidence for a substantial isotope effect and imply that the reduced tunneling rate of a deuterium atom relative to that of hydrogen gives rise to VFRs that are too narrow to be observed with the resolution employed. Denifl et al. (2008a) studied N-deuterated thymine with high electron energy resolution and mass selectivity. They found a strong isotope effect in N-deuterated thymine, whereas the DEA cross section for the C-deuterated thymine was practically indistinguishable from that for thymine. Moreover, it was found that all narrow VFRs are due to the H loss from the $N_1$ site, whereas a broader asymmetric peak near 2 eV is due to the interaction between the second lowest $\pi^*$ resonance and the second $\sigma^*$ resonance. The isotope effect was the most pronounced in the H loss from the $N_1$ site. In this case the cross section for thymine was reduced by a factor of 40, whereas for the H loss from the $N_3$ site the cross section was reduced by a factor of 20.

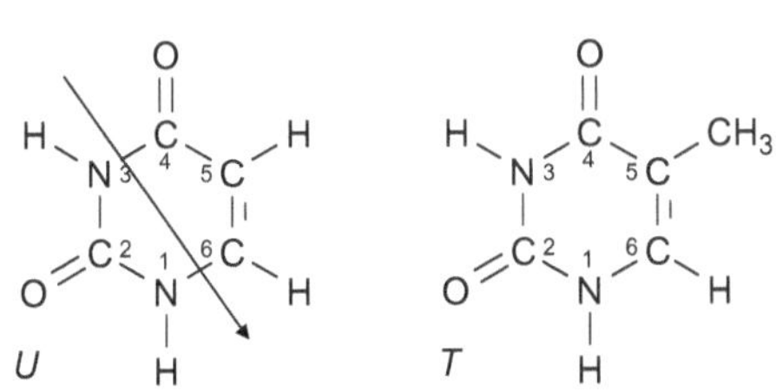


**Fig. 33** Structure of the uracil and thymine molecules. The *arrow* indicates the direction of the dipole moment of uracil (Gallup and Fabrikant, 2011a).

Theoretical description of these processes is a very challenging task because of the complex structure of these targets. A nonlocal resonance theory still has not been developed for polyatomic targets with several vibrational degrees of freedom; therefore, a "pseudo-diatomic" approach in these cases is used. This is similar to the treatment of the halogen-containing compounds and formic acid described (Sections 6 and 8.2), equations of the nonlocal complex potential theory incorporate only one reaction coordinate relevant to a specific bond cleavage, and all other coordinates are considered as "frozen." Surprisingly this simplification works quite well for many polyatomic halogen-containing molecules and has also been used in calculations of hydrogen loss process in the uracil molecule (Gallup and Fabrikant, 2011a)

$$\mathrm{e} + \mathrm{U} \rightarrow (\mathrm{U} - \mathrm{H})^{-} + \mathrm{H}$$

where the symbol U–H stands for the uracil molecule with the loss of a H atom. As was discussed earlier, there are three processes that can lead to the H loss: C–H bond breakage and N–H bond breaks at two sites (1 and 3, see Fig. 33) with the major contribution to the H loss coming from the N–H bond break. In order to separate these two processes, H coming from the site 1, and H coming from the site 3, two experiments were performed (Abdoul-Carime et al., 2004a; Ptasinska et al., 2005a): one with the normal uracil, and the other with the methylated uracil whereby the H atom is replaced by the methyl group at site 3. The measurements showed that the major signal is coming from the site 1 although there is a noticeable contribution from the site 3. Therefore, theoretical calculations (Gallup and Fabrikant, 2011a) were focused on the hydrogen loss from site 1.

The uracil molecule has a large dipole moment; therefore the major contribution to the binding force in the weakly bound state is due to the dipolar interaction. When this interaction is included in the calculation, a pronounced VFR appears below the first threshold for excitation of the N1–H stretch. In particular, the large dipole moment of the uracil allows theoretical detection of two resonances out of the theoretically predicted infinite progression (Fabrikant, 1983), but in practice, this infinite progression is limited by rotational splitting. We demonstrate this behavior by showing DEA cross section below the $\nu=3$ threshold in Fig. 34 where we plot it as a function of $-\log_{10}(E_t - E)$, where $E_t = 1.25778$ eV is the threshold energy. Apparently, this is the first case when two VFRs in DEA are produced near the same vibrational threshold. In Fig. 35, we present DEA to uracil and deuterated uracil over a broader energy range (Gallup and Fabrikant, 2011a).

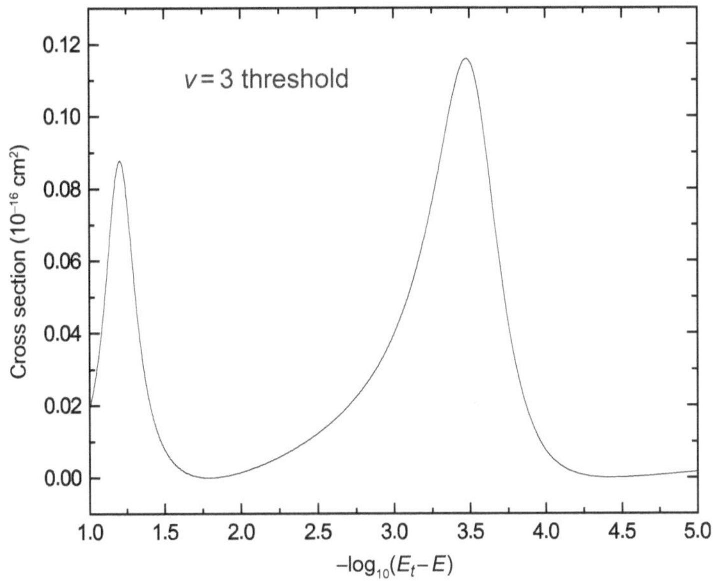


**Fig. 34** DEA cross sections for uracil below the $v=3$ threshold plotted as a function of $-\log_{10}(E_t-E)$, where $E_t=1.25778$ eV is the threshold energy and $E$ is the electron energy. *Reproduced from Gallup, G.A., Fabrikant, I.I., 2011a. Vibrational Feshbach resonances in dissociative electron attachment to uracil. Phys. Rev. A 83, 012706 (1–7).*

Due to the large value of the vibrational quantum corresponding to the N–H stretch vibrations, no significant temperature effect was found between $T=300$K and 450K in uracil. However, in a recent experimental paper on DEA to thymine (Kopyra and Abdoul-Carime, 2015) a strong temperature dependence of the production of dehydrogenated thymine was found within a relatively narrow temperature range. In particular, at 313K the DEA cross section decreased by two orders of magnitude in comparison to that at 423K. It is hard to explain such a strong dependence from the theoretical point of view, even if we recall that the theory in the approximation of one reaction coordinate is missing vibrational excitation of all modes other than N–H stretch. Note that the same authors (Kopyra and Abdoul-Carime, 2016) found that the cross section for hydrogen loss in 2-thiouracil is *decreasing* with increasing temperature, which is also hard to interpret theoretically.

Other theoretically investigated biologically relevant molecules are formic acid (see Section 8.2) and several isomers of aminobutanoic acid (Vizcaino et al., 2012). The dipole moments of these molecules are not high and threshold cusps, rather than VFRs, were predicted and experimentally

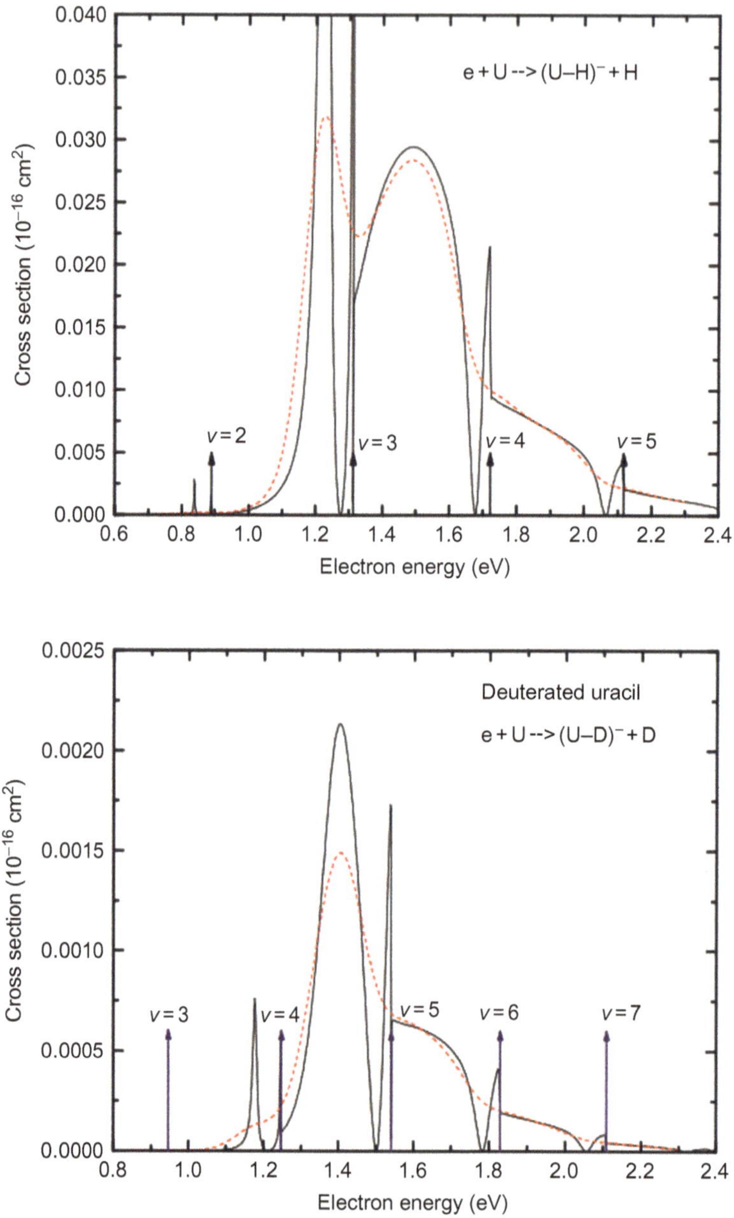


**Fig. 35** DEA cross sections for uracil and deuterated uracil (Gallup and Fabrikant, 2011a). *Solid curve*, nonaveraged cross sections; *dashed curve*, cross sections folded with a Gaussian distribution of width 0.07 eV. The vibrational excitation thresholds are indicated by *arrows*.

confirmed. There is also experimental evidence of cusp structures in DEA to glycine (Abouaf, 2008), which were confirmed by model calculations (Gallup et al., 2009a). Independent of the value of the dipole moment, the interaction between the first valence anion state and a weakly bound state due to the long-range electron–molecule interaction can produce a near-threshold peak in electron-impact dehydrogenation of many amino acids, particularly in proline (Abdoul-Carime and Illenberger, 2004; Sulzer et al., 2008), alanine (Ptasińska et al., 2005b), 5-nitrouracil (Ptasinska et al., 2008), and leucine (Abdoul-Carime et al., 2009).

Another demonstration of validity of calculations is the observation of deuterium loss in DEA to deuterated compounds. The isotope effect in uracil is similar to that in formic acid, as was discussed in Section 8.2. As an example, we present the isotope effect for uracil in Fig. 35. Observations of DEA to deuterated uracil (Scheer et al., 2005) and aminobutanoic acid (Vizcaino et al., 2012) confirm theoretical predictions of the isotope effects in these targets.

## 8.5 Extension to Larger Bio-/Macromolecules

The study of DEA in relatively large biomolecules (for example, larger than nucleobases) brings major challenges. From the theoretical perspective, the number of internal degrees of freedom rises very rapidly with the size of the system and the applicability of key approximations can break down. Experimentally, it becomes increasingly difficult to produce gas-phase targets of sufficient density without thermal decomposition or reactivity. Even some apparently *simple* biomolecules can present difficulties of this kind; for example, several groups have tried and failed to produce stable guanine targets via sublimation before and after Abdoul-Carime et al.'s (2005) successful DEA study. Naturally, the requirement of a neutral target for DEA experiments rules out the use of the electrospray methods that have revolutionized the study of many other important molecular processes in complex biomolecules (for example, VUV action spectroscopies of large molecules such as peptides (Rankovic et al., 2015)) since electrospray prepares charged targets.

Prior to Baccarelli et al.'s (2011) review, the most complex gas-phase DNA component to have been studied in DEA experiments was the nucleoside thymidine (Ptasinska et al., 2006). Since then, DEA experiments on nucleosides have been extended to the radiosensitizing agent gemcitabine (2′,2′-difluorocytidine) and the closely related molecules 2′-deoxycytidine and 2′-deoxy-5-fluorocytidine (Kopyra et al., 2014). We also note that electron transfer experiments have been carried out on uridine (Almeida et al., 2014). Only one experimental study has successfully probed the next step up in gas-phase biomolecular complexity—a nucleotide (a nucleoside bonded to phosphate group). Specifically, Kopyra (2012) studied low-energy electron interactions with deoxycytidine monophosphate (dCMP). By considering the structure of the nucleotide and its position within DNA, she concluded that DEA production of a phosphorus-containing fragment ion (notably $PO_4^-$) from dCMP within DNA would cause a single-strand break. Furthermore, by comparing the resonances observed in the isolated components of dCMP (the nucleobase, the sugar, and the phosphate unit), it was possible to link different features in the energy profile of $PO_4^-$ production to electron attachment to different sites on the nucleotide (see Fig. 36). The prominent part of the $PO_4^-$ profile close to zero eV was assigned to EA at the phosphate group and the shoulder at 1 eV (highlighted with an arrow in the figure) was assigned to EA at the cytosine component of the molecule. The latter case is particularly interesting as it indicates electron transfer from its initial attachment site on the nucleobase to the *vulnerable* (with respect to strand breaks) backbone of DNA. Indeed, this result appears to be consistent with earlier theoretical predictions that $\pi^*$ orbitals on the nucleobases act as "antennae" for low-energy electrons in DNA with subsequent transfer to the backbone and thence cleavage of $\sigma^*$ (C–O) bonds (Berdys et al., 2004). These experiments provide an elegant example of how studying DEA processes across sequential levels of complexity can lead to new insights into radiation damage mechanisms on the macromolecular scale.

The challenges in producing a viable target of a complex biomolecule can be overcome by studying such molecules on surfaces. In this review, we will not attempt to review the extensive subject DEA on surfaces, although we comment briefly on some of the associated theoretical challenges in Section 9.2. From the experimental standpoint, studying DEA on surfaces complicates the traditional approach of analyzing fragment anion production as a function of incident electron energy (for example, due to surface charging and the limited scope for analyzing the direction of anion production). However, this is offset by the great range of alternative analytical methods that are available in surface science. In particular, the Sherbrooke group's detection of low-energy electron-induced strand breaks in DNA depended on structural analysis postirradiation via agarose gel electrophoresis (Boudaïffa et al., 2000). Multiple replicates were required to

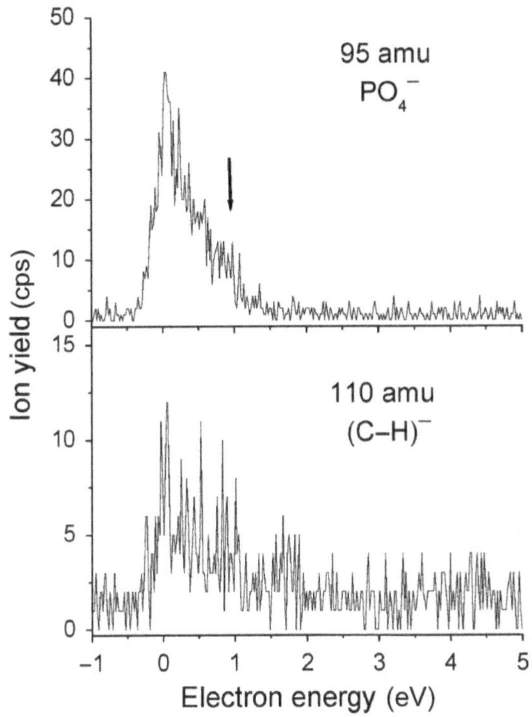


**Fig. 36** Electron-impact energy dependence of the production of $PO_4^-$ and (cytosine–H)$^-$ fragment ions from dCMP. The arrow indicates the contribution from initial electron localization at the cytosine base. *Reproduced from Kopyra, J., 2012. Low energy electron attachment to nucleotide deoxycitidine monophosphate: direct evidence for the molecular mechanism of electron-induced DNA strand breaks. Phys. Chem. Chem. Phys. 14, 8287–8287.*

attain adequate statistics and hence, despite its huge influence on the field and its applied importance, it was a number of years before another group repeated the key result that subionization energy electrons cause resonant DNA strand breaks. Orlando et al. (2008) showed that multiple scattering events play an important role in low energy-induced DNA damage to hydrated DNA. Kumar et al. (2012) irradiated DNA samples with relatively low electron currents and highlighted variations in the energies of the single-strand-break resonances observed by the three groups. This demonstrates the sensitivity of such experiments to specific target and irradiation conditions and hence the importance of further advancing our nanoscale understanding of DEA-induced DNA damage.

# 9. CLUSTERS

## 9.1 Electron Attachment to Molecular van der Waals Clusters

The work on highly energy-resolved electron attachment to van der Waals clusters and related VFR phenomenon was reviewed by Hotop et al. (2003). Here, after a brief introduction and a summary of earlier work, we present some recent experimental and theoretical updates.

Clusters serve as nanoscale objects for studying the effects of solvation due to the interaction between a solvated molecule or an ion and its surrounding solvent environment. Solvation effects also play a key role in the formation of negative ions by attachment of slow electrons to molecular clusters. The anion resonances, observed for single molecules, appear shifted toward lower energies in clusters due to the effects of solvation (Illenberger, 2000; Märk, 1991).

Using the LPA method at energy widths around 1 meV (Hotop et al., 2003), Hotop and coworkers in Kaiserslautern demonstrated that the cluster anion yield in low-energy (0–200 meV) electron collisions with $N_2O$ (Leber et al., 2000a; Weber et al., 1999) and $CO_2$ clusters (Barsotti et al., 2002b; Leber et al., 2000b) is, to a large extent, mediated by narrow VFRs, i.e., TNI states of the type $\left[(XY)_N(\nu_i \geq 1)\right]^-$, which occur at energies below those of the neutral cluster $\left[(XY)_{N-1}XY(\nu_i \geq 1)\right]$ (here the vibrational excitation $\nu_i$ in the VFR and the neutral cluster is assumed to be the same). It was thus shown for $N_2O$ and $CO_2$ clusters that the broad "zero energy resonance" observed in the early work at energy widths around 0.5 eV (Märk, 1991; Stamatovic, 1988) is due to the combined influence of the previously nonresolved, overlapping VFRs and a sharp s-wave type enhancement of the cross section toward zero energy.

The observation of VFRs requires the existence of weakly bound diffuse electron states residing at the surface of the molecule or cluster and corresponding to long-range binding of the electron, e.g., due to dipolar or polarization forces. VFRs are strongly influenced by the details of the electron–molecule interaction and by the overall strength of the electron–cluster attraction. In molecular van der Waals clusters the dipole moments of individual molecular units are typically randomly oriented so that the net dipole moment of the cluster is close to 0. However, isotropic polarizability is additive, meaning that the total polarizability of a cluster with $N$ molecular units is

$$\alpha_N = \alpha N \tag{3}$$

where $\alpha$ is the isotropic part of the polarizability of one molecular unit. This feature makes the long-range polarization interaction between an electron and a cluster very strong and makes it responsible for the VFR formation. While VFRs are still present in electron attachment yields for OCS clusters (Barsotti et al., 2004) (although less prominent than for $CO_2$ clusters), they are absent in those for $CS_2$ clusters (Barsotti et al., 2002b, 2004). The difference between $CO_2$ and $CS_2$ attachment was explained (Barsotti et al., 2002b) by the difference in their electron affinities: carbon dioxide monomer does not form a stable negative ion, while $CS_2$ does.

Processes of attachment to clusters typically involve intra- and intermolecular vibrational energy redistribution (IVR) (Uzer, 1991), the process whereby the initial vibrational excitation of one mode is followed by the redistribution of the corresponding energy among many modes in the cluster. A subsequent channeling of this energy into one reaction coordinate, leading to evaporation, might take a very long time, resulting to detection of the parent negative ion in experiments. The simplest system where IVR in electron collisions is observed is low-energy electron attachment to the $SF_6$ molecule. However, even in this case, an ab initio theory of attachment involving IVR still does not exist because of complexity of incorporating of many vibrational degrees of freedom (15 in the case of $SF_6$).

Some success in developing a theoretical description of electron attachment to large molecules and clusters has been achieved by application of the Vogt–Wannier model (Vogt and Wannier, 1954) for charged particle capture into a polarization well. This was developed further by Fabrikant and Hotop (2001) and was shown to work for molecules with strong EA cross sections such as $SF_6$ and $CCl_4$. However, this model accounts for only the polarization interaction and only provides a rough estimate of the capture rate coefficients; it does not provide a description of fine features in the energy dependence of the cross section like VFRs.

For a more detailed description, we need a model incorporating the essential physics of the electron attachment: long-range polarization interaction and vibrational energy spectrum of the attaching target. For the electron attachment to $SF_6$ a multichannel version of MERT was developed (Fabrikant et al., 2005), whereas for electron attachment to van der Waals clusters a model potential, consisting of a potential well (or a barrier) in the inner region and the polarization tail in the outer region, was used. In both cases the IVR process was incorporated phenomenologically by introducing a complex scattering length describing a loss of electron flux due to IVR.

This approach allowed a description of important experimentally observed features: cusps in elastic electron scattering and vibrational excitation of $SF_6$ and VFRs in electron attachment to $CO_2$ and $SO_2$ clusters. The subsequent decay channels include resonance elastic scattering, resonance vibrational excitation, and electron attachment. The position of the resonance is determined by the electron binding energy. The simple potential well plus polarization tail model is able to describe the dependence of the resonance position on the number of units $N$. With the growth of $N$ the resonances are red shifted because of the increase in the binding energy.

In the $CO_2$ clusters, there is an additional complication due to the existence of the Fermi resonance, leading to a strong mixture of the symmetric stretch $\nu_1$ and bending $\nu_2$ vibrations. The theory of electron attachment incorporating this phenomenon was developed by Fabrikant (2005). The theoretical results exhibit very sharp VFRs with the widths of 1.5–3.1 meV, which are associated with the (010) and (020)–(100) thresholds, where the numbers in parenthesis indicate vibrational quantum numbers $\nu_1$, $\nu_2$, and $\nu_3$ for symmetric stretch, bending, and asymmetric stretch vibrations, respectively. The symbol (020)–(100) represents the Fermi dyad corresponding to the mixture of one quantum of the symmetric stretch vibrations with two quanta of bending vibrations. The splitting between VFR dyad is somewhat higher than the splitting between dyad components for the neutral molecule, since the presence of an extra electron creates an additional interaction, leading to a repulsion between the vibrational levels. Generally the experimentally observed width of resonances is much larger than calculated (Fabrikant, 2005; Fabrikant and Hotop, 2005) and several mechanisms responsible for such resonance broadening have been proposed (Fabrikant, 2005).

The model developed for the process of electron attachment to $CO_2$ clusters was based on ab initio data for low-energy electron–$CO_2$ scattering at the equilibrium geometry, which is dominated by a virtual state. The same approach does not work for $SO_2$ clusters. Because of the large dipole moment of $SO_2$ (1.63 D), low-energy e–$SO_2$ scattering is dominated by the dipolar interaction that is not relevant in e–$SO_2$ cluster scattering since the dipole moment of a cluster is different from that of one molecule. On the other hand, the description of the vibrational spectrum of $SO_2$ is simpler because of the absence of the Fermi resonances that are present in the vibrational spectrum of $CO_2$. Also, the position of the first valence resonance states of the $SO_2^-$ anion at the equilibrium geometry of $SO_2$ (or the VAEs) is known from the electron transmission, scattering, and DEA (Krishnakumar et al., 1997) experiments. Based on these data for e–$SO_2$ collisions a model for electron attachment to $SO_2$ clusters was developed (Barsotti et al., 2014), which employs the displaced harmonic oscillator approximation for description of the vibrational motion in the cluster anion.

For illustration, we present here two figures showing experimental data collected for $SO_2$ clusters and compare these data with theoretical calculations. The measured VFR-related partial anion yields are summarized in Fig. 37. The vertical broken lines indicate the thresholds for vibrational excitation of the three fundamental modes of $SO_2$. The VFR peak associated with the bending vibration (010) is always the most prominent feature. These (010) peaks have widths (FWHM) from 14 meV for $q=2$ up to 29 meV for $q=10$, increasing, on average, by about 1.7 meV per added molecule. The (red) shift of the peak maximum of the (010) VFR is 1.7 (1) meV for $q=2$ and rises by about $-1.7(1)$ meV per added $SO_2$.

In Fig. 38, we present the results of model calculations of electron attachment to $SO_2$ clusters and compare them with the results of the Vogt–Wannier model for capture into the polarization well of clusters with size $N=4$. The choice of the model parameters corresponds to the electron–cluster interaction, which exhibits a barrier (rather than a well) at short distances making the $SO_2$ case similar to $N_2O$ (Leber et al., 2000a). The barrier's height varies from 0.619 eV at $N=4$ to 0.369 eV at $N=12$. The strength of VFR depends on the model coupling parameter, which can be adjusted to reproduce experimental data. For the weaker coupling, the VFRs are observed only at the first vibrational threshold (010) and are, in fact, substantially weaker than the VFRs observed experimentally. For the strong coupling, the resonances are much more pronounced and are observed at higher thresholds as well, i.e., at the overtones (020), (030) of the bending vibration. However, the predicted appearance of overtones is not compatible with the experimental data.

## 9.2 The Influence of Cluster Environments on DEA Processes

Theoretical descriptions of the influence of cluster, surface, and condensed-matter environments on DEA processes were recently reviewed by Fabrikant (2016a), and so here we give just a few outlines of the theory with a following detailed review of experimental work on DEA processes in cluster environments.

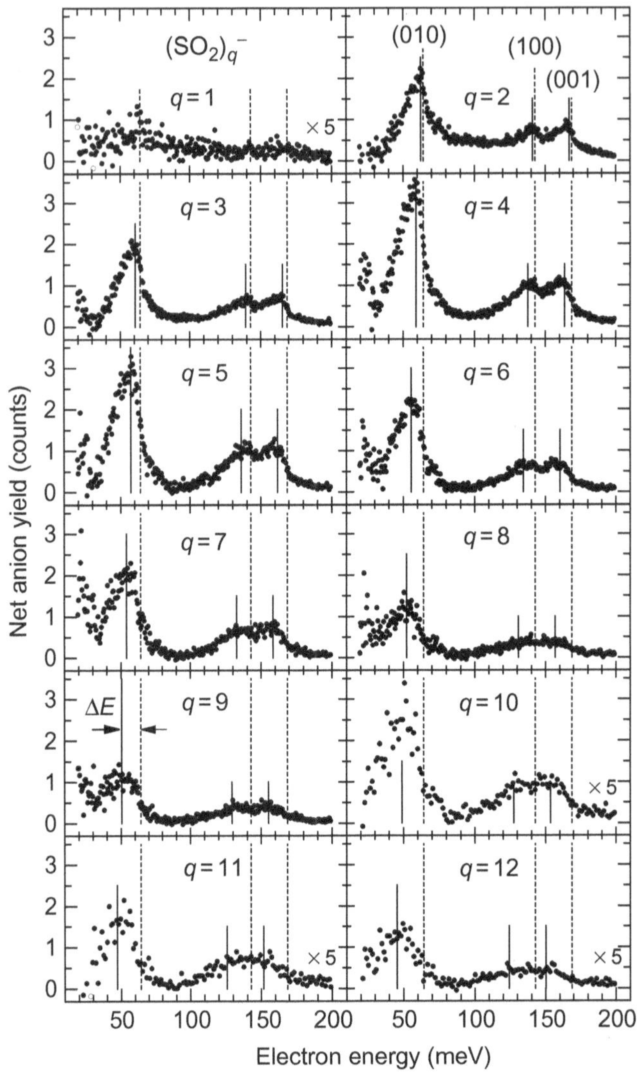


**Fig. 37** Linear plots of the net yield for the formation of $(SO_2)_q^-$ $(q = 1 - 12)$ anions via vibrational Feshbach resonances over the energy range of 20–200 meV. The *vertical broken lines* indicate the thresholds for vibrational excitation of the three fundamental modes of $SO_2$. The *vertical full lines* indicate the VFR peak positions, which are shifted by $\Delta E = -1.7 \times (q-1)$meV $(q \geq 2)$. *From Barsotti, S., Fabrikant, I.I., Ruf, M.-W., Hotop, H., 2014. High resolution low-energy electron attachment to molecular clusters of sulfur dioxide. Int. J. Mass Spectrom. 365–366, 301–310.*

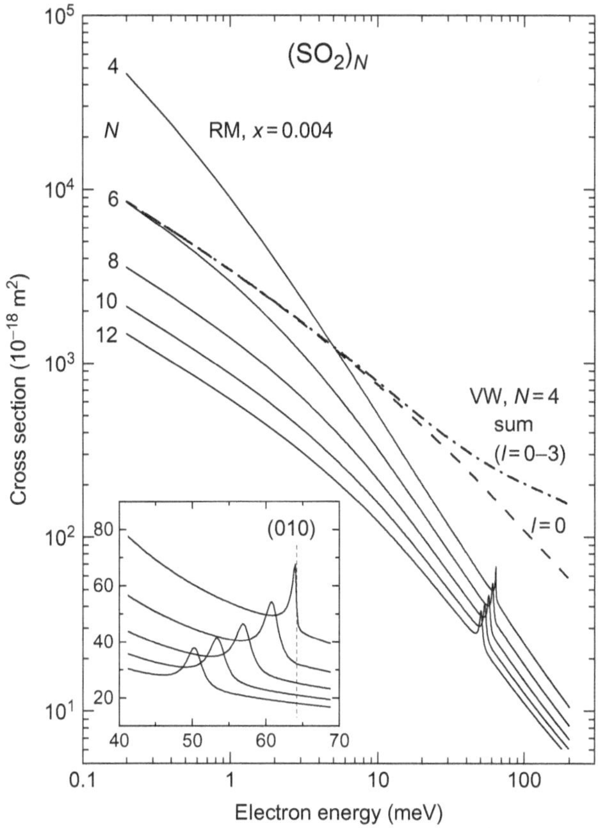


**Fig. 38** Model cross sections for electron attachment to $(SO_2)_N$ clusters for cluster sizes $N=4, 6, 8, 10, 12$. The Vogt–Wannier (VW) cross sections for the size $N=4$ ("sum" including orbital angular momenta $l=0$–$3$ in addition to $l=0$) are presented for comparison. *Inset*: double linear plot over the narrow range around the VFRs, showing in more detail their shifts from the threshold for vibrational excitation of the (010) mode (*vertical broken line*). *From Barsotti, S., Fabrikant, I.I., Ruf, M.-W., Hotop, H., 2014. High resolution low-energy electron attachment to molecular clusters of sulfur dioxide. Int. J. Mass Spectrom. 365–366, 301–310.*

The surface environment can lead both to enhancement and suppression of DEA cross sections. Enhanced cross sections are typically due to polarization interactions between the intermediate negative-ion state and the surface, which leads to increase in the capture cross section and in the survival probability (Nagesha et al., 2001; Sanche et al., 1995). Preferential surface orientation of $CF_3I$ molecules was used to explain the experimental observation (Le Coat et al., 1998) of an increase by more than two orders of magnitude in the cross section for ESD of $CF_3^-$ compared with the corresponding gas-phase DEA cross section. Decreases in the DEA cross section can arise if a substantial part of the signal is due to a VFR, then the surface environment suppresses the resonance and the DEA cross section is reduced. A typical example is the DEA to the methyl iodide on a surface of the rare-gas film or the water ice (Fabrikant, 2011; Miller et al., 2013).

As with surfaces, a cluster environment can also lead to DEA enhancement due to polarization interactions. In addition the resonance lifetime can be increased due to the multiscattering effect. This also leads to enhancement of the DEA cross section predicted for chlorofluorocarbons (Fabrikant et al., 2012) and uracil/thymine (Smyth et al., 2014). However, the developed theory does not follow the fate of the formed negative ion. It could be trapped by the cluster or surface environment due to caging effect (Nagesha et al., 2001), and in this case, the enhancement would not be observed experimentally. Indeed, as described later, a complete suppression of the dissociation has been experimentally observed in uracil and thymine surrounded by several water molecules (Kočišek et al., 2016b).

From an experimental standpoint, the question of how molecular DEA is modified by clustering has not been studied as extensively as might be expected in view of its relevance in applied fields such as atmospheric physics, plasma technology, and radiochemistry. We recommend the review by Illenberger and coworkers from the Freie Universität Berlin (Illenberger, 1993; Ingolfsson et al., 1996) for summaries of the relatively early work on this topic. Furthermore, Bald et al. (2008a) reported highlights of Illenberger's experimental contributions to the subject. Efficient quenching of low-energy DEA or autodetachment channels from low-energy resonances due to clustering is prevalent, with examples including $O_2$, $CF_3Cl$, $CF_3I$, 1,2-$C_2Cl_2F_4$, $C_2F_5I$, and $C_6F_5X$ (X=F, Cl, Br, I) (Illenberger, 1993; Ingolfsson et al., 1996; Langer et al., 2002, 2003). The converse effect—the enhancement of specific DEA channels due to clustering—has also been observed in a number of molecules (Bald et al., 2008a). For example, the conversion of a core-excited open channel resonance to a closed channel (Feshbach) resonance takes place in all forms of $C_6F_5X$ (X=F, Cl, Br, I) (Tegeder et al., 1996) under clustering conditions. This manifests itself in the appearance of a new channel for $X^-$ detection at about 6.5 eV.

Sanche's experiments on DNA (Boudaïffa et al., 2000) provided new impetus to explore the influence of clustering on molecular DEA processes, as clusters can be viewed as an intermediate stage between the gas phase and condensed biological environments. Accordingly, much of the recent work on this topic has focused on simple biological molecules. Denifl, Märk, Scheier, and coworkers (Denifl et al., 2006; Neustetter et al., 2015) at the University of Innsbruck and Kočišek, Farnik, and coworkers (Kočišek et al., 2016a,b) in Prague have played particularly important roles in advancing electron attachment experiments on biomolecules within clusters.

The first DEA experiments on DNA constituents within clusters were carried out in the special environment of helium nanodroplets (Denifl et al., 2006, 2008b). Helium droplets pick up atoms or molecules efficiently and provide an excellent environment for molecular cluster formation because long-range dipole interactions are not modified significantly by the superfluid helium matrix. Furthermore, evaporative cooling of the droplet provides an extremely efficient mechanism for quenching vibrational energy, maintaining a temperature of 0.37K and stabilizing even very weak intermolecular bonds. Fig. 39 compares DEA of a nucleobase (thymine in this case, while adenine behaves similarly) in the gas phase and within a helium droplet (Denifl, 2013; Denifl et al., 2008b). All fragment anions (e.g., $H^-$ and $NCO^-$) associated with the dissociation of an intermediate excited dehydrogenated nucleobase anion were strongly suppressed in the helium droplet. This was attributed to the superfluid helium environment's role as a sink for the internal energy of excited anion intermediates and can be viewed as an extreme case of the general effect within clusters of dissipating intramolecular vibrational excitation to intermolecular modes and evaporation. The counterpart of this result was a major enhancement in helium droplets of the single hydrogen loss DEA channels at electron energies above the prominent gas-phase resonance at 2.7 eV (Denifl et al., 2006, 2008a). This was observed in both single and clustered nucleobases, evidenced in the energy dependences of the signals for $(T_2\text{-H})^-$, $(A_2\text{-H})^-$, and $(AT\text{-H})^-$.

The Innsbruck group extended this work to probe electron attachment to clusters of amino acids (glycine, alanine, and serine) within helium nanodroplets (Ferreira da Silva et al., 2010), as well as nitromethane ($CH_3NO_2$ (Ferreira da Silva et al., 2011)) and the explosive trinitrotoluene (TNT) (Mauracher et al., 2009). In the latter case, the superfluid helium droplet completely suppressed the rich pattern of DEA products that were observed in the gas phase with only $TNT^-$ and cluster anions being observed. This extreme example of DEA suppression is traced to the complex

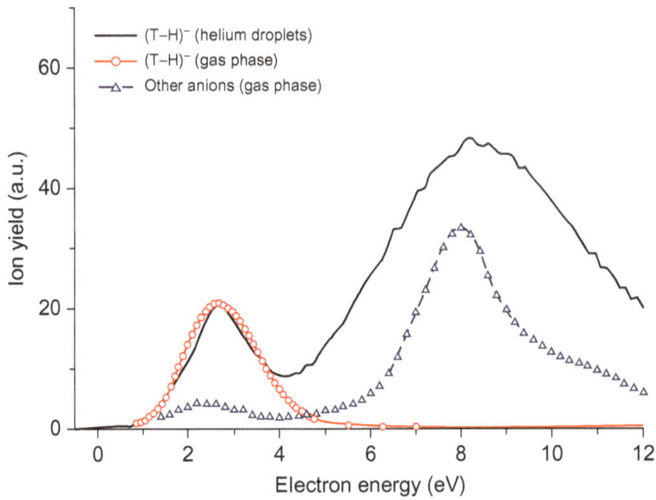


**Fig. 39** Ion yield of the dehydrogenated thymine anion $(T-H)^-$ formed upon electron attachment to single thymine in helium droplets (*full line*) and in the gas phase (*full line with open circles*). The comparison with the sum of all other gas-phase fragment anions (triangles) shows that at electron energies greater than ~4 eV, $(T-H)^-$ is a dissociation intermediate frozen in helium. *Reproduced from Denifl, S., 2013. Formation of cations and anions upon electron interaction with (doped) helium droplets. Eur. Phys. J. Spec. Top. 222, 2017–2033.*

rearrangement processes of the transient negative ion prior to decomposition with these metastable decay pathways being slow in comparison with the dissipation of internal energy in the helium droplet environment.

Further to their work with doped helium nanodroplets, the Innsbruck group has developed experiments probing DEA in clusters produced by more conventional supersonic expansion sources (Denifl et al., 2010; Vizcaino et al., 2010). Most recently, they studied pure and hydrated clusters of pyrimidine as a model system for nucleobases (Neustetter et al., 2015). Whereas EA to isolated pyrimidine does not produce the parent molecular anion, it was observed in clusters with resonance peaks at 6.1 eV (both in the pure and hydrated data) and 8.2 eV (in the pure case only). These resonances indicate that electronic excitation is required. The authors proposed that energy dissipation within the cluster and evaporation are required to access a metastable state of pyrimidine$^-$ in which spontaneous electron emission is restricted by Franck–Condon factors. By studying ion yields as a function of the expansion conditions the authors deduced that at least a neutral tetramer cluster is needed to achieve this stabilization. The main fragment anions associated with ring cleavage ($CN^-$ and $C_3H_2N^-$ from the isolated molecule) were strongly suppressed by clustering and only (pyrimidine—H)$^-$ remains. This may be partially due to weakening of the intramolecular C–H bond as a result of the hydrogen bonds in the pure and hydrated clusters. Fig. 40 shows the dependence of the (pyrimidine—H)$^-$ signal on the incident electron energy. The authors attributed the evident stabilization of the anion in clusters to the energy-dissipating environment, while the difference between the pure and hydrated case may be linked to the most stable calculated hydrated structures featuring water molecules surrounding the pyrimidine N sites.

A major limitation of experiments on neutral cluster targets is the lack of precise knowledge of the cluster conformation from which a detected anion originated. As an example indicating the potential significance of this limitation for DEA, we can consider Fabrikant et al.'s (2012) theoretical study of $Cl^-$ detachment from hydrated $CF_3Cl$ complexes. This showed strong dependence on cluster conformation, including an order of magnitude difference in the DEA cross section around 1.5 eV for two different geometries of $CF_3Cl(H_2O)_3$. One approach to reduce the range of neutral cluster conformations in an experiment is to probe targets comprising relatively small clusters. Kočišek et al. (2016a,b) have applied this rationale in their studies of DEA to 2-nitrophenol complexes and to *microhydrated* uracil and thymine

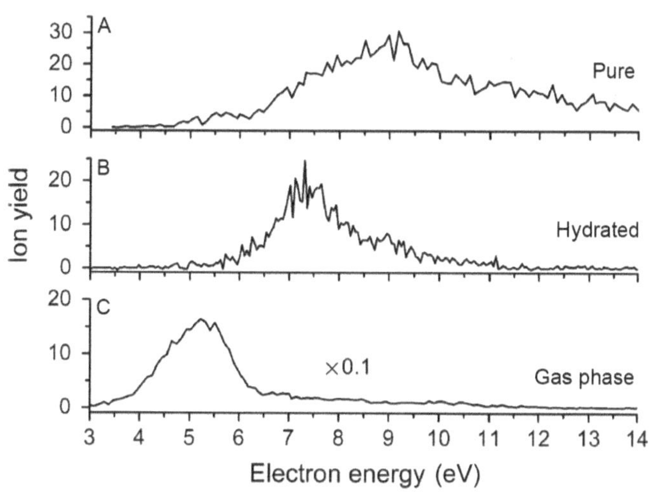


**Fig. 40** Ion yield of (pyrimidine–H)$^-$ resulting from DEA to pure pyrimidine clusters (A), hydrated pyrimidine clusters (B), and isolated pyrimidine (C). *Reproduced from Neustetter, M., Aysina, J., da Silva, FF., Denifl, S., 2015. The effect of solvation on electron attachment to pure and hydrated pyrimidine clusters. Angew. Chem. Int. Ed. 54, 9124–9126.*

(the first DEA experiments on DNA building blocks in an aqueous environment). Even in small hydrated clusters, the dissociative channels of the nucleobase TNIs were suppressed and only intact molecular and hydrated anions were detected, as shown in Fig. 41. The authors attributed this to caging-like effects and fast energy transfer to the solvent. However, this result appears to contradict the theoretical prediction of strongly enhanced DEA in uracil or thymine embedded within a cluster of five water molecules (Smyth et al., 2014). Possible reasons for the discrepancy include the fact that the calculations did not address whether or not DEA products escape the cluster and a single cluster conformation in the calculations was compared

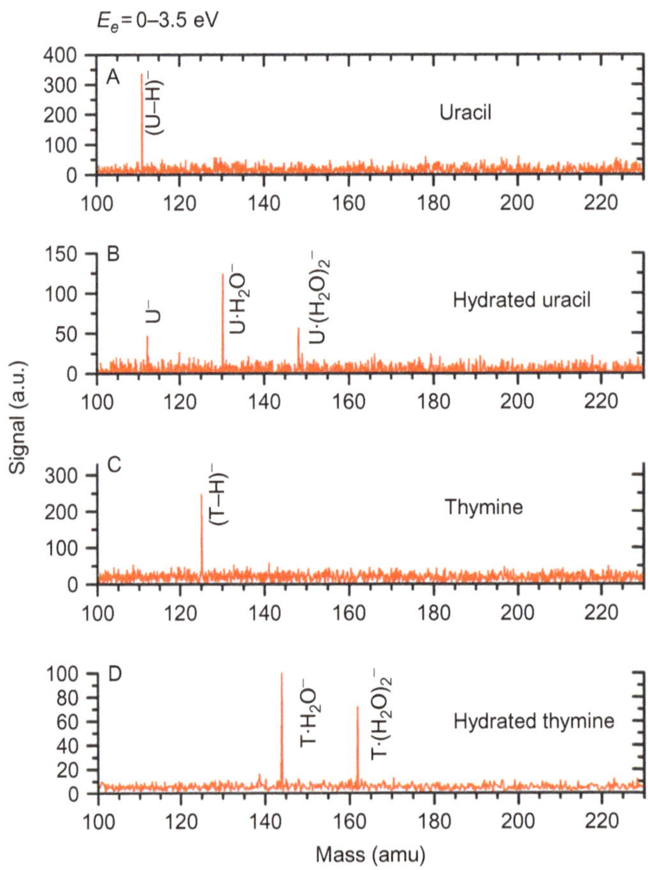


**Fig. 41** Negative ion mass spectra resulting from electron attachment to (A) gas-phase uracil, (B) small uracil–water clusters $U \cdot (H_2O)_n$, (C) gas-phase thymine, and (D) small thymine–water clusters $T \cdot (H_2O)_n$. Complete suppression of nucleobase fragmentation can be observed (Kočišek et al., 2016b).

with a range of small cluster targets in the experiment. This highlights the importance of further experimental and theoretical work in order to achieve the most direct comparisons possible.

## 10. CONCLUSIONS AND OUTLOOK

DEA is an active area of current research with significant progress having been made in the last decade since the review of Hotop et al. (2003) was published. On the theoretical front, the treatment of DEA has been further developed for diatomic molecules addressing such issues as the influence of long-range interactions on DEA processes, nonlocal dynamics, interactions between resonances, and the angular distribution of the products. Theoretical calculations of the angular distributions have been recently extended to polyatomic molecules (Haxton et al., 2006). This was particularly timely because of recent experimental developments where the VSI technique is revealing detailed angular measurements that pose considerable challenges to theoretical methods.

Theoretical studies of DEA in polyatomic molecules have included incorporation of multidimensional vibrational dynamics in the local approximation. Updated ab initio electron scattering codes, including complex Kohn and *R*-matrix, and improving computer speed and storage capacity, have allowed calculation of complex multidimensional potential energy surfaces, which serve as an input for subsequent calculation of dissociative dynamics. The latter is treated either by the wave packet propagation method or by solving time-independent coupled equations. Substantial progress has been achieved in theoretical studies of DEA to the water molecule through the lowest Feshbach resonances. The physics of these processes includes interesting phenomena such as interaction between resonances and related conical intersections, and Renner–Teller coupling.

Studies of DEA to polyatomic molecules, with the inclusion of nonlocal effects and the approximation of a single reaction coordinate, have demonstrated the ability of this approach to describe several interesting phenomena in DEA to polyatomics, notably VFRs and cusps at vibrational excitation thresholds. These studies have been extended to biological molecules, such as formic acid, uracil, and the conformers of aminobutanoic acid. Theoretical results, confirmed by experimental observations, have demonstrated that VFR is a very general phenomenon affecting DEA processes even in complex molecules such as uracil and thymine.

Theoretical descriptions of electron attachment to van der Waals clusters and to molecules in condensed-matter and cluster environments have also been significantly developed. Because of the complexity of these processes, these theories are mostly based on models capturing the major physics but containing empirical parameters. However, these model approaches are capable of describing important effects; in particular very pronounced VFRs are revealed in electron attachment to van der Waals clusters. The position of these resonances as a function of the cluster size is very well described by polarization interactions between the electron and the cluster. Several features observed in electron attachment to physisorbed molecules and molecules in condensed-matter and cluster environments were explained by theoretical models incorporating physical mechanisms involving band structure of solids, polarization interactions between the electron and the environment, and multiple scattering theory.

At the same time, many problems in the theoretical description of DEA remain unaddressed and need further work. The major challenge is the extension of existing ab initio methods to polyatomic molecules. As the number of vibrational degrees of freedom grows, completely ab initio calculations of complex multidimensional potential energy surfaces with the following treatment of dissociation dynamics in full dimensionality become infeasible. Semiempirical, semiclassical, and statistical methods should be incorporated if an understanding of the major physics involved in DEA within polyatomics is to be developed. A good example of such an approach is the theoretical treatment of enhanced positron annihilation in molecular gases (Gribakin et al., 2010) where both semiempirical and statistical methods allowed a description of VFRs in these processes. Statistical methods were proposed by Thoss and Domcke (1998) for a theoretical description of interactions between active vibration modes and a large number of inactive, so-called bath modes, and were also applied for the theory of electron attachment to van der Waals clusters (Fabrikant and Hotop, 2005). These should be further developed, as well as a theoretical description of intramolecular vibrational energy distribution in DEA processes. The theory of this phenomenon was well developed for the process of photoexcitation of molecules (Uzer, 1991), but very little is known about how to apply these methods for theoretical description of electron-impact vibrational excitation and DEA processes.

Another avenue for theoretical pursuits is further development of classical and semiclassical methods, which are much less challenging computationally for application to polyatomic molecules. Although the basic formulation of the classical approach was developed a long time ago (Goursaud et al., 1976), it has not been exploited yet using the full capacity of modern computational tools. Semiclassical and quasiclassical versions of the nonlocal theory (Fabrikant, 1991; Kalin and Kazansky, 1990) have been developed only for diatomic molecules. It is evident that for description of phenomenon such as VFRs in polyatomics, and particularly in biological molecules, further development of these methods is necessary. In the field of electron attachment to molecules in condensed-matter and cluster environments, further development of methods combining multiple scattering theory approach with condensed-matter physics approach (Fabrikant, 2007; Marinica et al., 2005; Michaud et al., 1993; Mills, 1992) is necessary.

Significant progress has been achieved in the experimental characterization of DEA over the past decade. This has been partially motivated by the progress in theoretical approaches outlined earlier; in particular the prediction of cross sections in small polyatomic molecules has created a need for accurate experimental data. However, the main impetus for modern DEA research has been the recognition that DEA is a fundamental process in many natural (atmospheric, astronomical, and biological radiation damage) and technological processes. This renaissance in DEA studies has fostered significant advances in experimental techniques as the range of targets to be studied has greatly expanded (e.g., large biomacromolecules including DNA) with the emergence of cluster and surface studies. The development of VMI/VSI has allowed new information on the molecular dynamics of DEA and has been adapted by several groups. The angular distributions of fragment ions from VMI/VSI experiments have provided new insights into the resonant states and relaxation dynamics of their TNI precursors.

The results acquired from the experimental programs described in this review help to identify a number of challenges for the DEA community in the near future. One is the quantification of absolute cross sections in targets where the neutral density is difficult to control. This is particularly challenging for targets with low volatility (e.g., complex biomolecules) or low stability (notably radicals). The same difficulty arises for targets that require a supersonic expansion for their production. To our knowledge, there are no experimental absolute DEA cross sections in clusters. Quantitative information on how aggregation influences bond-breaking processes in molecules is of great importance, especially for extrapolating knowledge obtained at the molecular level to bulk. A closely related challenge with respect to aggregation effects is preparing neutral targets (heterogeneous clusters) with well-defined size and composition. Experiments on clusters in supersonic coexpansions (or a combination of expansion and pickup) typically suffer from limited knowledge of the beam composition. A possible solution could be to use a beam diagnosis based on a soft ionization technique (e.g., sodium doping and photoionization introduced by Bobbert et al., 2002) or to preselect just a desired fraction of the neutral cluster beam using, for example, Stark deflection (Trippel et al., 2012).

Very little experimental data exist on electron collisions with excited molecules and DEA is no exception. These processes offer new fundamental insights and are also important for modeling and describing plasmas. Plasmas typically contain high concentrations of molecules in either vibrationally or electronically excited states, and electron interactions with molecules in long-lived metastable states are particularly important. The DEA cross sections for such targets can differ from the ground state by orders of magnitude. This has only been explored experimentally for thermally accessible rovibrational excitation and it is now well established that the DEA cross section rises dramatically for targets with increasing vibrational and rotational quantum numbers (e.g., Allan and Wong, 1978; Fedor et al., 2007). Electronic excitation of the target is expected to profoundly change the magnitude of the dissociative cross section, its electron energy dependence, and the resultant fragmentation pathways, but to date, there is little quantified data on such processes.

Another process that remains largely unquantified but whose importance is increasingly being recognized is ion-pair formation (IPF) or dipolar dissociation (DD) producing anions at energies in excess of the ionization energy

$$AB + e \rightarrow A^- + B^+ + e$$

DD/IPF has been shown to be the most important source of anions in hydrocarbon plasmas and a leading process in Titan's upper atmosphere. It is also believed to play an important role in the dissociation of precursor gases used in focused electron beam-induced processing (FEBIP). Developing experimental tools for lower energy DEA (e.g., VMI/VSI) and new theoretical methods to probe these DD/IPF processes is therefore an important challenge to the community.

Finally, the role of DEA in many technological processes provides the opportunity for the electron-collision community to provide valuable data and play a key role in next-generation industries. Since low-energy electrons are the most abundant secondary products of ionizing radiation and DEA is a bond-breaking process, it has attracted considerable attention as a mechanism for initiating structural and chemical damage in biological material and suggested new radiotherapy agents, e.g., suggestions of new radiosensitizers that have strong DEA cross sections to enhance DNA damage or use of nanoparticles, the irradiation of which releases an avalanche of lower energy secondary electrons to further DNA damage. Understanding DEA and being able to exploit its specificity of bond rupture provide a means to develop "electron-induced chemistry." FEBID and similar extreme ultraviolet processing can exploit such electron-driven chemistry to develop the next generation of (sub-10 nm) nanostructures, while the role of electron capture and DEA in combustion proves another area of transdisciplinary research.

## ACKNOWLEDGMENTS

We thank M. Allan, P.D. Burrow, G.A. Gallup (deceased), J. Gorfinkiel, H. Hotop, J. Kohanoff, T.M. Miller, M.-W. Ruf, and E. Krishnakumar for long-term collaborations and many stimulating discussions. I.I.F. was supported by the US National Science Foundation under Grant No. PHY-1401788. S.E. acknowledges the support the British EPSRC through the Grants Nos. EP/J002577/1 and EP/L002191/1. J.F. acknowledges the Czech Science Foundation Project No. 17-04844S. I.I.F.'s collaboration with the Open University was supported by a Marie Curie International Incoming Fellowship (FP7-PEOPLE-2009-IIF-252714).

## REFERENCES

Abdoul-Carime, H., Illenberger, E., 2004. Fragmentation of proline induced by slow electrons. Chem. Phys. Lett. 397, 309–313.

Abdoul-Carime, H., Huels, M.A., Illenberger, E., Sanche, L., 2001. Sensitizing DNA to secondary electron damage: resonant formation of oxidative radicals from 5-halouracils. J. Am. Chem. Soc. 123, 5354–5355.

Abdoul-Carime, H., Gohlke, S., Illenberger, E., 2004a. Site-specific dissociation of DNA bases by slow electrons at early stages of irradiation. Phys. Rev. Lett. 92. 168103 (1–4).

Abdoul-Carime, H., Limao-Vieira, P., Gohlke, S., Petrushko, I., Mason, N.J., Illenberger, E., 2004b. Sensitization of 5-bromouridine by slow electrons. Chem. Phys. Lett. 393, 442–447.

Abdoul-Carime, H., Langer, J., Huels, M.A., Illenberger, E., 2005. Decomposition of purine nucleobases by very low energy electrons. Eur. Phys. J. D 35, 399–404.

Abdoul-Carime, H., König-Lehmann, C., Kopyra, J., Farizon, B., Farizon, M., Illenberger, E., 2009. Dissociative electron attachment to amino-acids: the case of leucine. Chem. Phys. Lett. 477, 245–248.

Abouaf, R., 2008. Low energy electron impact in gas phase glycine, alanine and propanoic acid: electronic, vibrational excitations and negative ions. Chem. Phys. Lett. 451, 25–30.

Abouaf, R., Diaz-Tendero, S., 2009. Electron energy loss spectroscopy and anion formation in gas phase coronene. Phys. Chem. Chem. Phys. 11, 5686–5694.

Abouaf, R., Teillet-Billy, D., Goursaud, S., 1981. Predissociation process in C-/CO formation. J. Phys. B At. Mol. Phys. 14, 3517–3523.

Adaniya, H., Rudek, B., Osipov, T., Haxton, D.J., Weber, T., Rescigno, T.N., McCurdy, C.W., Belkacem, A., 2009. Imaging the molecular dynamics of dissociative electron attachment to water. Phys. Rev. Lett. 103. 233201 (1–4).

Aflatooni, K., Burrow, P.D., 2000. Total cross sections for dissociative electron attachment in dichloroalkanes and selected polychloroalkanes: the correlation with vertical attachment energies. J. Chem. Phys. 113, 1455–1464.

Aflatooni, K., Burrow, P.D., 2001. Dissociative electron attachment in chlorofluoromethanes and the correlation with vertical attachment energies. Int. J. Mass Spectrom. 205, 149–161.

Aflatooni, K., Gallup, G.A., Burrow, P.D., 1998. Dissociative electron attachment in chloroalkanes and the correlation with vertical attachment energies. Chem. Phys. Lett. 282, 398–402.

Aflatooni, A., Hitt, B., Gallup, G.A., Burrow, P.D., 2001. Temporary anion states of selected amino acids. J. Chem. Phys. 115, 6489–6494.

Aflatooni, K., Scheer, A.M., Burrow, P.D., 2005. Dissociative electron attachment in uracil: total anion yield. Chem. Phys. Lett. 408, 426–428.

Aflatooni, K., Scheer, A.M., Burrow, P.D., 2006. Total dissociative electron attachment cross sections for molecular constituents of DNA. J. Chem. Phys. 125, 054301.

Aflatooni, K., Gallup, G.A., Burrow, P.D., 2010. Dissociative electron attachment in nonplanar chlorocarbons with $\pi^*/\sigma^*$-coupled molecular orbitals. J. Chem. Phys. 132. 094306 (1–10).

Alge, E., Adams, N.G., Smith, D., 1984. Rate coefficients for the attachment reactions of electrons with c-C7F14, CH3Br, CF3Br, CH2Br2 and CH3I determined between 200 and 600 K using the FALP technique. J. Phys. B At. Mol. Phys. 17, 3827–3833.

Allan, M., 1992. Measurement of differential cross sections for excitation of helium by electron impact within the first 4 eV above threshold. J. Phys. B At. Mol. Phys. 25, 1559–1575.

Allan, M., 2006. Study of resonances in formic acid by means of vibrational excitation by slow electrons. J. Phys. B At. Mol. Phys. 39, 2939–2947.

Allan, M., 2007. Electron collisions with formic acid monomer and dimer. Phys. Rev. Lett. 98. 123201 (1–4).

Allan, M., 2016. Private communication.

Allan, M., Fabrikant, I.I., 2002. Threshold peaks and structures in vibrational excitation of $CH_3I$ by electron impact. J. Phys. B At. Mol. Opt. Phys. 35, 1025–1034.

Allan, M., Wong, S.F., 1978. Effect of vibrational and rotational excitation on dissociative attachment in hydrogen. Phys. Rev. Lett. 41, 1791–1794.

Allan, M., Wong, S.F., 1981. Dissociative electron attachment from vibrationally and rotationally excited HCl and HF. J. Chem. Phys. 74, 1687–1691.

Allan, M., Čížek, M., Horáček, J., Domcke, W., 2000. Electron scattering in cooled HCl: boomerang structures and outer-well resonances in elastic and vibrational excitation cross sections. J. Phys. B At. Mol. Opt. Phys. 33, L209–L213.

Almeida, D., Antunes, R., Garcia, G., McCullough, R.W., Eden, S., Limão-Vieira, P., 2012. Mass spectrometry of anions and cations produced in 1-4 keV H-, O-, and OH- collisions with nitromethane, water, ethanol, and methanol. Int. J. Mass Spectrom. 311, 7–16.

Almeida, D., Ferreira da Silva, F., Kopyra, J., García, G., Limão-Vieira, P., 2014. Anion formation in gas-phase potassium–uridine collisions. Int. J. Mass Spectrom. 365–366, 243–247.

Anstoter, C.S., Bull, J.N., Verlet, J.R.R., 2016. Ultrafast dynamics of temporary anions probed through the prism of photodetachment. Int. Rev. Phys. Chem. 35 (4), 509–538.

Ayala, J.A., Wentworth, W.E., Chen, E.C.M., 1981. Electron attachment to halogens. J. Phys. Chem. 85, 768–777.
Azria, R., Lecoat, Y., Lefevre, G., Simon, D., 1979. Dissociative attachment on $H_2S$—energy and angular distributions of $H^-$ ions. J. Phys. B At. Mol. Phys. 12, 679–687.
Azria, R., Abouaf, R., Teillet-Billy, D., 1982. Symmetry of Cl2- resonant states formed in dissociative electron attachment reaction on Cl2. J. Phys. B At. Mol. Phys. 15, L569–L574.
Azria, R., Abouaf, R., Teillet-Billy, D., 1988. Negative-ion states of molecular iodine. J. Phys. B At. Mol. Phys. 21, L213–L217.
Baccarelli, I., Bald, I., Gianturco, F.A., Illenberger, E., Kopyra, J., 2011. Electron-induced damage of DNA and its components: experiments and theoretical models. Phys. Rep. 508, 1–44.
Bachroz, R.A., Haranczyk, M., Dabkowska, I., Rak, J., Gutowski, M., 2005. Anion of the formic acid dimer as a model for intermolecular proton transfer induced by a $\pi^*$ excess electron. J. Chem. Phys. 122. 204304 (1–7).
Bald, I., Langer, J., Tegeder, P., Ingólfsson, O., 2008a. From isolated molecules through clusters and condensates to the building blocks of life. A short tribute to Prof. Eugen Illenberger's work in the field of negative ion chemistry. Int. J. Mass Spectrom. 277, 4–25.
Bald, I., Dabkowska, I., Illenberger, E., 2008b. Probing biomolecules by laser-induced acoustic desorption: electrons at near zero electron volts trigger sugar–phosphate cleavage. Angew. Chem. Int. Ed. 47, 8518–8520.
Bald, I., Curik, R., Kopyra, J., Tarana, M., 2017. Dissociative electron attachment to biomolecules. In: Solov'yov, A. (Ed.), Nanoscale Insights Into Ion-Beam Cancer Therapy. Springer, Cham, Switzerland, pp. 159–208.
Balog, R., Langer, J., Gohlke, S., Stano, M., Abdoul-Carime, H., Illenberger, E., 2004. Low energy electron driven reactions in free and bound molecules: from unimolecular processes in the gas phase to complex reactions in a condensed environment. Int. J. Mass Spectrom. 233, 267–291.
Balog, R., Cicman, P., Jones, N.C., Field, D., 2009. Spontaneous dipole alignment in films of $N_2O$. Phys. Rev. Lett. 102. 073033 (1–4).
Bardsley, J.N., 1968. Configuration interaction in the continuum states of molecules. J. Phys. B At. Mol. Phys. 1, 349–364.
Bardsley, J.N., 1979. Molecular resonance phenomena. In: Rescigno, T., McKoy, V., Schneider, B. (Eds.), Electron-Molecule and Photon-Molecule Collisions. Plenum, New York, pp. 267–280.
Bardsley, J.N., Wadehra, J.M., 1983. Dissociative attachment in HCl, $D_2$, and $F_2$. J. Chem. Phys. 78, 7227–7234.
Barsotti, S., Ruf, M.-W., Hotop, H., 2002a. Clear experimental evidence for p-wave attachment-threshold behavior in electron attachment to chlorine molecules. Phys. Rev. Lett. 89. 083201 (1–4).
Barsotti, S., Leber, E., Ruf, M.-W., Hotop, H., 2002b. High resolution study of cluster anion formation in low-energy electron collisions with molecular clusters of $CO_2$, $CS_2$, and $O_2$. Int. J. Mass Spectrom. 220, 313–330.
Barsotti, S., Sommerfeld, T., Ruf, M.-W., Hotop, H., 2004. High resolution study of cluster anion formation in low-energy electron collisions with OCS clusters. Int. J. Mass Spectrom. 233, 181–192.
Barsotti, S., Fabrikant, I.I., Ruf, M.-W., Hotop, H., 2014. High resolution low-energy electron attachment to molecular clusters of sulfur dioxide. Int. J. Mass Spectrom. 365–366, 301–310.
Berdys, J., Anusiewicz, I., Skurski, P., Simons, J., 2004. Damage to model DNA fragments from very low-energy (<1 eV) electrons. J. Am. Chem. Soc. 126, 6441–6447.

Birtwistle, D.T., Herzenberg, A., 1971. Vibrational excitation of $N_2$ by resonance scattering of electrons. J. Phys. B At. Mol. Phys. 4, 53–70.
Birtwistle, D.T., Modinos, A., 1978. Dissociative attachment in iodine. J. Phys. B At. Mol. Phys. 11, 2949–2955.
Bisettia, F., Morsli, M.E., 2014. Kinetic parameters, collision rates, energy exchanges and transport coefficients of non-thermal electrons in premixed flames at sub-breakdown electric field strengths. Combust. Theor. Model. 18, 148–184.
Bjarnason, E.H., Ómarson, F.H., Hoshino, M., Tanaka, H., Brunger, M.J., Limão-Vieira, P., Ingólfsson, O., 2013. Negative ion formation through dissociative electron attachment to the group IV tetrafluorides: carbon tetrafluoride, silicon tetrafluoride and germanium tetrafluoride. Int. J. Mass Spectrom. 339-340, 45–53.
Bjarnason, E.H., Omarsson, B., Engmann, S., Omarsson, F.H., Ingolfsson, O., 2014. Dissociative electron attachment to titanium tetrachloride and titanium tetraisopropoxide. Eur. Phys. J. D 68. 121 (1–8).
Bobbert, C., Schutte, S., Steinbach, C., Buck, U., 2002. Fragmentation and reliable size distributions of large ammonia and water clusters. Eur. Phys. J. D 19, 183–192.
Boudaïffa, B., Cloutier, P., Hunting, D., Huels, M.A., Sanche, L., 2000. Resonant formation of DNA strand breaks by low-energy (3 to 20 eV) electrons. Science 287, 1658–1660.
Braun, M., Gruber, F., Ruf, M.W., Kumar, S.V.K., Illenberger, E., Hotop, H., 2006. IR photon enhanced dissociative electron attachment to $SF_6$: dependence on photon, vibrational, and electron energy. Chem. Phys. 329, 148–162.
Braun, M., Ruf, M.-W., Fabrikant, I.I., Hotop, H., 2007a. Observation of p-wave threshold behavior in electron attachment to $F_2$ molecules. Phys. Rev. Lett. 99. 253202 (1–4).
Braun, M., Fabrikant, I.I., Ruf, M.-W., Hotop, H., 2007b. Low-energy electron collisions with $CH_3Br$: the dependence of elastic scattering, vibrational excitation, and dissociative attachment on the initial vibrational energy. J. Phys. B At. Mol. Phys. 40, 659–674.
Braun, M., Marienfeld, S., Ruf, M.-W., Hotop, H., 2009. High-resolution electron attachment to the molecules CCl4 and SF6 over extended energy ranges with the (EX)LPA method. J. Phys. B At. Mol. Opt. Phys. 42. 125202 (1–20).
Breit, G., Wigner, E.P., 1936. Capture of slow neutrons. Phys. Rev. 49, 519–531.
Brems, V., Beyer, T., Nestmann, B.M., Meyer, H.D., Cederbaum, L.S., 2002. Ab initio study of the resonant electron attachment to the $F_2$ molecule. J. Chem. Phys. 117, 10635–10647.
Brunger, M.J., Buckman, S.J., 2002. Electron-molecule scattering cross-sections. I. Experimental techniques and data for diatomic molecules. Phys. Rep. 357, 215–458.
Burns, S.J., Matthews, J.M., McFadden, D.L., 1996. Rate coefficients for dissociative electron attachment by halomethane compounds between 300 and 800 K. J. Phys. Chem. 100, 19436–19440.
Burrow, P.D., Modelli, A., Chiu, N.S., Jordan, K.D., 1981. Temporary $\Sigma$ and $\Pi$ anions of the chloroethylenes and chlorofluoroethylenes. Chem. Phys. Lett. 82, 270–276.
Burrow, P.D., Gallup, G.A., Scheer, A.M., Denifl, S., Ptasinska, S., Märk, T., Scheier, P., 2006. Vibrational Feshbach resonances in uracil and thymine. J. Chem. Phys. 124. 124310 (1–7).
Cadez, I., Markelj, S., Rupnik, Z., Pelicon, P., 2008. Processes with neutral hydrogen and deuterium molecules relevant to edge plasma in tokamaks. J. Phys. Conf. Ser. 133, 012029.
Capitelli, M., Coppola, C.M., Diomede, P., Longo, S., 2007. An evaluation of the effect of the dissociative attachment of vibrationally excited H2 on primordial universe chemistry. Astron. Astrophys. 470, 811–813.
Celiberto, R., Janev, R.K., Laricchiuta, A., Capitelli, M., Wadehra, J.M., Atems, D.E., 2001. Cross section data for electron-impact inelastic processes of vibrationally

excited molecules of hydrogen and its isotopes. At. Data Nucl. Data Tables 77, 161–213.
Celiberto, R., Janev, R.K., Wadehra, J.M., Laricchiuta, A., 2011. Cross sections for 14-eV e-$H_2$ resonant collisions: isotope effect in dissociative electron attachment. Phys. Rev. A 84. 012707 (1–7).
Celiberto, R., Janev, R.K., Wadehra, J.M., Tennyson, J., 2012. Dissociative electron attachment to vibrationally excited $H_2$ molecules involving the $^2\sum_g^+$ resonant Rydberg electronic state. Chem. Phys. 398, 206–213.
Celiberto, R., Armenise, I., Cacciatore, M., Capitelli, M., Esposito, F., Gamallo, P., Janev, R.K., Lagana, A., Laporta, V., Laricchiuta, A., Lombardi, A., Rutigliano, M., Sayos, R., Tennyson, J., Wadehra, J.M., 2016. Atomic and molecular data for spacecraft re-entry plasmas. Plasma Sour. Sci. Technol. 26, 033004.
Chachereau, A., Fedor, J., Janečková, R., Kočišek, J., Rabie, M., Franck, C.M., 2016. Electron attachment properties of c-$C_4F_8O$ in different environments. J. Phys. D Appl. Phys. 49. 375201 (1–12).
Chandler, D.W., Houston, P.L., 1987. Two-dimensional imaging of state-selected photodissociation products detected by multiphoton ionization. J. Chem. Phys. 87, 1445–1447.
Chantry, P.J., 1968. Dissociative attachment in CO and NO. Phys. Rev. 172, 125–136.
Chantry, P.J., 1982. Negative ion formation in gas lasers. In: Massey, H.S.W., McDaniel, E.W., Bederson, B. (Eds.), In: Applied Atomic Collision Physics: Gas Lasers, vol. 3. Academic Press, New York, pp. 35–70.
Chantry, P.J., Chen, C.L., 1989. Ionization and temperature dependent attachment cross section measurements in C3F8 and C2H3Cl. J. Chem. Phys. 90, 2585–2592.
Chen, C.L., Chantry, P.J., 1979. Photon-enhanced dissociative electron attachment in $SF_6$ and its isotopic selectivity. J. Chem. Phys. 71, 3897–3907.
Chen, H.-Y., Chen, H.-F., Kao, C.-L., Yang, P.-Y., Hsua, S.C.N., 2014. Interaction of electrons with cisplatin and the subsequent effect on DNA damage: a density functional theory study. Phys. Chem. Chem. Phys. 16, 19290–19297.
Chourou, S.T., Orel, A.E., 2008. Dissociative electron attachment to acetylene. Phys. Rev. A 77. 042709 (1–10).
Chourou, S.T., Orel, A.E., 2009a. Improved calculation on the isotope effect in dissociative electron attachment to acetylene. Phys. Rev. A 80. 034701 (1–2).
Chourou, S.T., Orel, A.E., 2009b. Dissociative electron attachment to HCN and HNC. Phys. Rev. A 80. 032709 (1–8).
Chourou, S.T., Orel, A.E., 2011. Isotope effect in dissociative electron attachment to HCN. Phys. Rev. A 83. 032709 (1–4).
Christophorou, L.G. (Ed.), 1984. In: Electron-Molecule Interactions and Their Applications, vols. 1 and 2. Academic Press, New York.
Christophorou, L.G., 1987. Electron attachment and detachment processes in electronegative gases. Contrib. Plasma Phys. 27, 237–281.
Christophorou, L.G., Olthoff, J.K., 1999a. Electron interactions with $Cl_2$. J. Phys. Chem. Ref. Data 28, 131–169.
Christophorou, L.G., Olthoff, J.K., 1999b. Electron interactions with plasma processing gases: an update for $CF_4$, $CHF_3$, $C_2F_6$, and $C_3F_8$. J. Phys. Chem. Ref. Data 28, 967–982.
Christophorou, L.G., Olthoff, J.K., 2000. Electron interaction with $SF_6$. J. Phys. Chem. Ref. Data 29, 267–330.
Christophorou, L.G., Olthoff, J.K., 2001. Electron attachment cross sections and negative ion states of $SF_6$. Int. J. Mass Spectrom. 205, 27–41.
Christophorou, L.G., Olthoff, J.K., 2012. Fundamental Electron Interactions with Plasma Processing Gases. Springer Science & Business Media.

Chutjian, A., Alajajian, S.H., 1987. Electron-attachment in $F_2$—conclusive demonstration of nonresonant, s-wave coupling in the limit of zero electron-energy. Phys. Rev. A 35, 4512–4514.
Cicman, P., Skalny, J.D., Fedor, J., Mason, N.J., Scheier, P., Illenberger, E., Märk, T.D., 2007. Dissociative electron attachment to ozone at very low energies revisited. Int. J. Mass Spectrom. 260, 85–87.
Čížek, M., Horáček, J., Domcke, W., 1999. Associative detachment, dissociative attachment, and vibrational excitation of HCl by low-energy electrons. Phys. Rev. A 60, 2873–2881.
Čížek, M., Horáček, J., Sergenton, A.-Ch., Popovic, D.B., Allan, M., Domcke, W., Leininger, T., Gadea, F.X., 2001. Inelastic low-energy electron collisions with the HBr and DBr molecules: experiment and theory. Phys. Rev. A 63. 062710 (1–14).
Comer, J., Schulz, G.J., 1974. Measurements of e1ectron-detachment cross sections from $O^-$ and $S^-$. Phys. Rev. A 10, 2100–2106.
Coulter, J.A., Hyland, W.B., Nicol, J., Currell, F.J., 2013. Radiosensitising nanoparticles as novel cancer therapeutics—pipe dream or realistic prospect? Clin. Oncol. 25, 593–603.
Čurík, R., Ziesel, J.P., Jones, N.C., Field, T.A., Field, D., 2006. Rotational excitation of $H_2O$ by cold electrons. Phys. Rev. Lett. 97. 123202 (1–4).
Dahl, D.A., Teich, T.H., Franck, C.M., 2012. Obtaining precise electron swarm parameters from a pulsed Townsend setup. J. Phys. D Appl. Phys. 45. 485201 (1–9).
da Silva, F.F., Matias, C., Almeida, D., Garcia, G., Ingolfsson, O., Flosadottir, H.D., Omarsson, B., Ptasinska, S., Puschnigg, B., Scheier, P., Limao-Vieira, P., Denifl, S., 2013. NCO-, a key fragment upon dissociative electron attachment and electron transfer to pyrimidine bases: site selectivity for a slow decay process. J. Am. Soc. Mass Spectrom. 24, 1787–1797.
Dawley, M.M., Ptasinska, S., 2014. Dissociative electron attachment to N-methylformamide. Int. J. Mass Spectrom. 365–366, 143–151.
Denifl, S., 2013. Formation of cations and anions upon electron interaction with (doped) helium droplets. Eur. Phys. J. Spec. Top. 222, 2017–2033.
Denifl, G., Muigg, D., Stamatovic, A., Märk, T.D., 1998. Dissociative electron attachment to CO and NO: the $O^-$ production threshold revisited. Chem. Phys. Lett. 288, 105–110.
Denifl, S., Ptasinska, S., Probst, M., Hrusak, J., Scheier, P., Märk, T.D., 2004. Electron attachment to the gas-phase DNA bases cytosine and thymine. J. Phys. Chem. A 108, 6562–6569.
Denifl, S., Zappa, F., Maehr, I., Lecointre, J., Probst, M., Märk, T.D., Scheier, P., 2006. Mass spectrometric investigation of anions formed upon free electron attachment to nucleobase molecules and clusters embedded in superfluid helium droplets. Phys. Rev. Lett. 97. 043201 (1–4).
Denifl, S., Sulzer, P., Zappa, F., Moser, S., Kräutler, B., Echt, O., Bohme, D.K., Märk, T.D., Scheier, P., 2008a. Isotope effects in dissociative electron attachment to the DNA base thymine. Int. J. Mass Spectrom. 277, 296–299.
Denifl, S., Zappa, F., Mauracher, A., Ferreira da Silva, F., Bacher, A., Echt, O., Märk, T.D., Bohme, D.K., Scheier, P., 2008b. Dissociative electron attachment to DNA bases near absolute zero temperature: freezing dissociation intermediates. ChemPhysChem 9, 1387–1389.
Denifl, S., Vizcaino, V., Mark, T.D., Illenberger, E., Scheier, P., 2010. High resolution electron attachment to $CO_2$ clusters. Phys. Chem. Chem. Phys. 12, 5219–5224.
Devdariani, A.Z., Ostrovskii, V.N., Sebyakin Yu, N., 1976. Crossing of quasistationary levels. Sov. Phys. JETP 44, 477–482.
Domcke, W., 1991. Theory of resonance and threshold effects in electron-molecule collisions: the projection-operator approach. Phys. Rep. 208, 97–188.

Dunn, G.H., 1962. Anistrotropies in angular distributions of molecular dissociation products. Phys. Rev. Lett. 8, 62–64.
El-Zein, A.A.A., Brunger, M.J., Newell, W.R., 2000. Excitation of vibrational quanta in water by electron impact. J. Phys. B At. Mol. Phys. 33, 5033–5044.
Engelking, P.C., 1982. Strong electron-dipole coupling in photodetachment of molecular negative ions: anomalous rotational thresholds. Phys. Rev. A 26, 740–745.
Eppink, A.T.J.B., Parker, D.H., 1997. Velocity map imaging of ions and electrons using electrostatic lenses: application in photoelectron and photofragment ion imaging of molecular oxygen. Rev. Sci. Instrum. 68, 3477–3484.
Estrada, H., Cederbaum, L.S., Domcke, W., 1986. Vibronic coupling of short-lived electronic states. J. Chem. Phys. 84, 152–169.
Fabrikant, I.I., 1977. Threshold behavior of the cross sections for scattering of electrons by polar molecules. Sov. Phys. JETP 46, 693–697.
Fabrikant, I.I., 1983. Frame transformation effective-range theory: application to interaction of polar molecules with electrons. J. Phys. B At. Mol. Phys. 16, 1253–1269.
Fabrikant, I.I., 1990. Resonance processes in e-HCl collisions: comparison of the R-matrix and the nonlocal-complex potential methods. Comments At. Mol. Phys. 24, 37–52.
Fabrikant, I.I., 1991. Quasiclassical R-matrix theory of inelastic processes in collisions of electrons with HCl molecules. Phys. Rev. A 43, 3478–3486.
Fabrikant, I.I., 1996. Electron scattering by neutral targets at milli- and submilli-electron-volt energies. Comments At. Mol. Phys. 32, 267–279.
Fabrikant, I.I., 2005. Theory of electron attachment to van der Waals clusters. J. Phys. B At. Mol. Phys. 38, 1745–1764.
Fabrikant, I.I., 2007. Dissociative electron attachment on surfaces and in bulk media. Phys. Rev. A 76. 012902 (1–12).
Fabrikant, I.I., 2009. Dissociative electron attachment: threshold phenomena and multimode effects. J. Phys. Conf. Ser. 192. 012002 (1–8).
Fabrikant, I.I., 2010. Recent progress in the theory of dissociative attachment: from diatomics to biomolecules. J. Phys. Conf. Ser. 204. 012004 (1–11).
Fabrikant, I.I., 2011. Dissociative electron attachment on surfaces and in bulk media: vibrational Feshbach resonance suppression. J. Phys. B At. Mol. Phys. 44. 225202 (1–7).
Fabrikant, I.I., 2016a. Long-range effects in electron scattering by polar molecules. J. Phys. B At. Mol. Phys. 49. 222005 (1–20).
Fabrikant, I.I., 2016b. Dissociative electron attachment to halogen molecules: angular distributions and nonlocal effects. Phys. Rev. A 94. 052707 (1–9).
Fabrikant, I.I., Hotop, H., 2001. Low-energy behavior of exothermic dissociative electron attachment. Phys. Rev. A 63. 022706 (1–10).
Fabrikant, I.I., Hotop, H., 2005. Electron attachment to $CO_2$ clusters. Phys. Rev. Lett. 94. 063201 (1–4).
Fabrikant, I.I., Hotop, H., 2008. On the validity of the Arrhenius equation for electron attachment rate coefficient. J. Chem. Phys. 128. 124308 (1–8).
Fabrikant, I.I., Leininger, T., Gadea, F.X., 2000. Low-energy dissociative electron attachment to $Cl_2$ molecules. J. Phys. B At. Mol. Phys. 33, 4575–4580.
Fabrikant, I.I., Hotop, H., Allan, M., 2005. Elastic scattering, vibrational excitation, and attachment in low-energy electron-SF6 scattering: experiment and effective range theory. Phys. Rev. A 71. 022712 (1–10).
Fabrikant, I.I., Caprasecca, S., Gallup, G.A., Gorfinkiel, J.D., 2012. Electron attachment to molecules in a cluster environment. J. Chem. Phys. 136. 184301 (1–8).
Falcetta, M.F., Choi, Y., Jordan, K.D., 2000. Ab initio investigation of the temporary anion states of perfluoroethane. J. Phys. Chem. A 104, 9605–9612.

Fandreyer, R., Burke, P.G., Morgan, L.A., Gillan, C.J., 1993. Low-energy electron scattering in HBr. J. Phys. B At. Mol. Opt. Phys. 26, 3625–3637.
Fano, U., 1961. Effects of configuration interaction on intensities and phase shifts. Phys. Rev. 124, 1866–1878.
Fedor, J., Cicman, P., Coupeir, B., Feil, S., Winkler, M., Gluch, K., Husarik, J., Jaksch, D., Farizon, B., Mason, N.J., Scheier, P., Märk, T.D., 2006. Fragmentation of transient water anions following low-energy electron capture by $H_2O/D_2O$. J. Phys. B At. Mol. Opt. Phys. 39, 3935–3944.
Fedor, J., Cingel, M., Skalny, J.D., Scheier, P., Märk, T.D., Cizek, M., Kolorenc, M., Horacek, J., 2007. Dissociative electron attachment to HBr: a temperature effect. Phys. Rev. A 75. 022703 (1–5).
Fedor, J., May, O., Allan, M., 2008. Absolute cross sections for dissociative electron attachment to HCl, HBr and their deuterated analogs. Phys. Rev. A 78. 032701 (1–6).
Fedor, J., Winstead, C., McKoy, W., Cizek, M., Houfek, K., Kolorenc, P., Horacek, J., 2010. Electron scattering in HCl: an improved nonlocal resonance model. Phys. Rev. A 81. 042702 (1–7).
Feil, S., Gluch, K., Matt-Leubner, S., Scheier, P., Limtrakul, J., Probst, M., Deutsch, H., Becker, K., Stamatovic, A., Märk, T.D., 2005. Partial cross sections for positive and negative ion formation following electron impact on uracil. J. Phys. B At. Mol. Phys. 37, 3013–3020.
Feketeova, L., Skalny, D.J., Hanel, G., Gstir, B., Francis, M., Märk, T.D., 2003. Production of Cl- via dissociative electron attachment to $Cl_2$. Int. J. Mass Spectrom. 223, 661–667.
Ferreira da Silva, F., Denifl, S., Märk, T.D., Ellis, A.M., Scheier, P., 2010. Electron attachment to amino acid clusters in helium nanodroplets: glycine, alanine, and serine. J. Chem. Phys. 132. 214306 (1–6).
Ferreira da Silva, F., Ptasinska, S., Denifl, S., Gschliesser, D., Postler, J., Matias, C., Märk, T.D., Limão-Vieira, P., Scheier, P., 2011. Electron interaction with nitromethane embedded in helium droplets: attachment and ionization measurements. J. Chem. Phys. 135. 174504 (1–7).
Feshbach, H., 1962. A unified theory of nuclear reactions II. Ann. Phys. 19, 287–313.
Fessenden, R.W., Bansal, K.M., 1970. Direct observation of electron disappearance in pulse irradiated fluorocarbon gases. J. Chem. Phys. 53, 3468–3473.
Feuerbacher, S., Sommerfeld, T., Cederbaum, L.S., 2004. Intersections of potential energy surfaces of short-lived states: the complex analogue of conical intersections. J. Chem. Phys. 120, 3201–3214.
Field, T.A., Slattery, A.E., Adams, D.J., Morrison, D.D., 2005. Experimental observation of dissociative electron attachment to $S_2O$ and $S_2O_2$ with a new spectrometer for unstable molecules. J. Phys. B At. Mol. Opt. Phys. 38, 255–264.
Flosadottir, H.D., Denifl, S., Zappa, F., Wendt, N., Mauracher, A., Bacher, A., Johnsson, H., Märk, T.D., Scheier, P., Ingolfsson, O., 2007. Combined experimental and theoretical study on the nature and the metastable decay pathways of the amino acid ion fragment $[M{-}H]^-$. Angew. Chem. Int. Ed. 46, 8057–8059.
Fogle, M., Landers, A.L., Moradmand, A., Sartor, J., Slaughter, D.S., Haxton, D.J., Rescignot, T.N., McCurdy, C.W., Weber, Th., Belkacem, A., Orel, A.E., Matsika, S., 2015. Ion-momentum imaging of dissociative-electron-attachment dynamics in $CO_2$, $N_2O$, HCCH and $CF_4$. J. Phys. Conf. Ser. 635, 072028.
Franck, J., Hertz, G., 1914. Uber Zusammenstöße zwischen Elektronen und Molekülen des Quecksilberdampfes und die Ionisierungsspannung desselben. Verhandlungen der Deutschen Physikalischen Gesellschaft (in German) 16, 457–467 and 512–517.
Frey, M.T., Hill, S.B., Ling, X., Smith, K.A., Dunning, F.B., Fabrikant, I.I., 1994. Inelastic electron-dipole-molecule scattering at sub-milli-electron-volt energies: possible role of dipole-supported states. Phys. Rev. A 50, 3124–3128.

Friedman, J.F., Miller, T.M., Schaffer, L.C., Viggiano, A.A., Fabrikant, I.I., 2009. Electron attachment to $Cl_2$ from 300-1100 K: experiment and theory. Phys. Rev. A 79. 032707 (1–6).
Gailitis, M., 1982. The influence of finite masses on the threshold behavior of scattering in a three charged particle system. J. Phys. B At. Mol. Phys. 15, 3423–3440.
Gailitis, M., Damburg, R., 1963. The influence of close coupling on the threshold behaviour of cross sections of electron-hydrogen scattering. Proc. Phys. Soc. 82, 192–200.
Gallup, G.A., Fabrikant, I.I., 2007. Resonances and threshold effects in low-energy electron collisions with methyl halides. Phys. Rev. A 75. 032719 (1–10).
Gallup, G.A., Fabrikant, I.I., 2011a. Vibrational Feshbach resonances in dissociative electron attachment to uracil. Phys. Rev. A 83. 012706 (1–7).
Gallup, G.A., Fabrikant, I.I., 2011b. Dissociative electron attachment to $CH_2Cl_2$, $CHCH_3Cl_2$, and $C(CH_3)_2Cl_2$. J. Chem. Phys. 135. 134316 (1–8).
Gallup, G.A., Aflatooni, K., Burrow, P.D., 2003. Dissociative electron attachment near threshold, thermal attachment rates, and vertical attachment energies of chloroalkanes. J. Chem. Phys. 118, 2562–2574.
Gallup, G.A., Burrow, P.D., Fabrikant, I.I., 2009a. Electron induced bond-breaking at low energies in HCOOH and glycine: the role of very short-lived $\sigma^*$ anion states. Phys. Rev. A 79. 042701 (1–7).
Gallup, G.A., Burrow, P.D., Fabrikant, I.I., 2009b. Reply to comment on "Electron-induced bond-breaking at low energies in HCOOH and glycine: the role of very short-lived $\sigma^*$ anion states". Phys. Rev. A 80. 046702 (1–2).
Gauyacq, J.P., Herzenberg, A., 1982. Nuclear-excited Feshbach resonances in e-HCl scattering. Phys. Rev. A 25, 2959–2967.
Gauyacq, J.P., Herzenberg, A., 1984. The attachment of very slow electrons to polyatomic molecules. J. Phys. B At. Mol. Phys. 17, 1155–1171.
Gerchikov, L.G., Gribakin, G.F., 2008. Electron attachment to $SF_6$ and lifetimes of $SF_6^-$ negative ions. Phys. Rev. A 77. 042724 (1–15).
Gianturco, F.A., Luchesse, R.R., 2004. Nanoscopic models for radiobiological damage: metastable precursors of dissociative electron attachment to formic acid. New J. Phys. 6. 66 (1–14).
Gianturco, F.A., Luchesse, R.R., Langer, J., Martin, I., Stano, M., Karwasz, G., Illenberger, E., 2005. Modelling electron-induced processes in "condensed" formic acid. Resonant states of $(HCOOH)_2^-$ at low energies. Eur. Phys. J. D 35, 417–428.
Gluch, K., Fedor, J., Matt-Leubner, S., Echt, O., Stamatovic, A., Probst, M., Scheier, P., Märk, T.D., 2003. Kinetic-energy release in Coulomb explosion of metastable $C_3H_5^+$. J. Chem. Phys. 118, 3090–3095.
Gope, K., Prabhudesai, V.S., Mason, N.J., Krishnakumar, E., 2016a. Probing the resonant states of $Cl_2$ using velocity slice imaging. J. Phys. B At. Mol. Phys. 49. 015201 (1–9).
Gope, K., Tadsare, V., Prabhudesai, V.S., Mason, N.J., Krishnakumar, E., 2016b. Negative ion resonances in carbon monoxide. Eur. Phys. J. D 70, 134–141.
Goursaud, S., Sizun, M., Fiquet-Fayard, F., 1976. Energy partitioning and isotope effects in the fragmentation of triatomic negative ions: Monte Carlo scheme for a classical trajectory study. J. Chem. Phys. 65, 5453–5461.
Graupner, K., Graham, L.M., Field, T.A., Mayhew, C.A., Fabrikant, I.I., Miller, T.M., Braun, M., Ruf, M.-W., Hotop, H., 2008. Highly-resolved absolute cross sections for dissociative electron attachment to $SF_5CF_3$. Int. J. Mass Spectrom. 277, 113–122.
Graupner, K., Haughey, S.A., Field, T.A., Mayhew, C.A., Hoffmann, T.H., May, O., Fedor, J., Allan, M., Fabrikant, I.I., Illenberger, E., Braun, M., Ruf, M.-W., Hotop, H., 2010. Low-energy electron attachment to the dichlorodifluoromethane ($CCl_2F_2$) molecule. J. Phys. Chem. A 114, 1474–1484.
Gribakin, G.F., Young, J.A., Surko, C.M., 2010. Positron-molecule interactions: resonant attachment, annihilation, and bound states. Rev. Mod. Phys. 82, 2557–2607.

Grill, V., Drexel, H., Sailer, W., Lezius, M., Märk, T.D., 2001. The working principle of trochoidal electron monochromator revisited. Int. J. Mass Spectrom. 205, 209–226.
Hall, R.J., 1978. Dissociative attachment and vibrational excitation of $F_2$ by slow electrons. J. Chem. Phys. 68, 1803–1807.
Hall, R.I., Cadez, I., Schermann, C., Tronc, M., 1977. Differential cross sections for dissociative attachment in CO. Phys. Rev. A 15, 599–610.
Hanel, G., Gstir, B., Denifl, S., Scheier, P., Probst, M., Farizon, B., Farizon, M., Illenberger, E., Märk, T.D., 2003. Electron attachment to uracil: effective destruction at subexcitation energies. Phys. Rev. Lett. 90. 188104 (1–4).
Haume, K., Rosa, S., Grellet, S., Śmiałek, M.A., Butterworth, K.T., Solov'yov, A.V., Prise, K.M., Golding, J., Mason, N.J., 2016. Gold nanoparticles for cancer radiotherapy: a review. Cancer Nanotechnol. 7, 1–20.
Haxton, D.J., Zhang, Z., McCurdy, C.W., Rescigno, T.N., 2004a. Complex potential surface for the $^2B_1$ metastable state of the water anion. Phys. Rev. A 69. 062713 (1–11).
Haxton, D.J., Zhang, Z., Meyer, H.-D., Rescigno, T.N., McCurdy, C.W., 2004b. Dynamics of dissociative attachment of electrons to water through the metastable state of the anion. Phys. Rev. A 69. 062714 (1–16).
Haxton, D.J., Rescigno, T.N., McCurdy, C.W., 2005. Topology of the adiabatic potential energy surfaces for the resonance states of the water anion. Phys. Rev. A 72. 022705 (1–12).
Haxton, D.J., McCurdy, C.W., Rescigno, T.N., 2006. Angular dependence of dissociative electron attachment to polyatomic molecules: application to the $^2B_1$ metastable state of the H2O and H2S anions. Phys. Rev. A 73. 062724 (1–15).
Haxton, D.J., McCurdy, C.W., Rescigno, T.N., 2007a. Dissociative electron attachment to the H2O molecule. I. Complex-valued potential-energy surfaces for the $^2B_1$, $^2A_1$, and $^2B_2$ metastable states of the water anion. Phys. Rev. A 75. 012710 (1–25); Erratum: 76, 049906 (E).
Haxton, D.J., Rescigno, T.N., McCurdy, C.W., 2007b. Dissociative electron attachment to the H2O molecule. II. Nuclear dynamics on coupled electronic surfaces within the local complex potential model. Phys. Rev. A 75. 012711(1–24); Erratum: 76, 049907 (E).
Haxton, D.J., Rescigno, T.N., McCurdy, C.W., 2008. Three-body breakup in dissociative electron attachment to the water molecule. Phys. Rev. A 78. 040702 (R) (1–4).
Haxton, D.J., Adaniya, H., Slaughter, D.S., Rudek, B., Osipov, T., Weber, T., Rescigno, T.N., McCurdy, C.W., Belkacem, A., 2011. Observation of the dynamics leading to a conical intersection in dissociative electron attachment to water. Phys. Rev. A 84. 030701 (R) (1–4).
Hazi, A.U., 1983. Effects of resonance-resonance coupling on dissociative electron attachment. J. Phys. B At. Mol. Phys. 16, L29–L34.
Hazi, A.U., Orel, A.E., Rescigno, T.N., 1981. Ab initio study of dissociative attachment of low-energy electrons to $F_2$. Phys. Rev. Lett. 46, 918–922.
Hien, P.X., Tuan, D.A., Jeon, B.-H., 2016. Determination of the electron collision cross section set for the C2F6 molecule. J. Korean Phys. Soc. 69, 1755–1761.
Hill, S.B., Frey, M.T., Dunning, F.B., Fabrikant, I.I., 1996. Electron-hydrogen fluoride scattering at ultralow electron energies: possible role of dipole-supported states. Phys. Rev. A 53, 3348–3357.
Hotop, H., Ruf, M.-W., Allan, M., Fabrikant, I.I., 2003. Resonance and threshold phenomena in low-energy electron collisions with molecules and clusters. Adv. At. Mol. Opt. Phys. 49, 86–216.
Hotop, H., Ruf, M.-W., Fabrikant, I.I., 2004. Resonance and threshold phenomena in low-energy electron collisions with molecules and clusters. Phys. Scr. T110, 22–31.
Hotop, H., Ruf, M.-W., Kopyra, J., Miller, T.M., Fabrikant, I.I., 2011. On the relation between the activation energy for electron attachment reactions and the size of their thermal rate coefficients. J. Chem. Phys. 134. 064303 (1–7).

Huber, D., Beikircher, M., Denifl, S., Zappa, F., Matejcik, S., Bacher, A., Grill, V., Märk, T.D., Scheier, P., 2006. High resolution dissociative electron attachment to gas phase adenine. J. Chem. Phys. 125. 084304 (1–7).

Illenberger, E., 1993. Anion formation in low energy electron impact to gaseous and condensed molecules. In: Christophorou, L.G., Illenberger, E., Schmidt, W.F. (Eds.), Linking the Gaseous and Condensed Phases of Matter. Plenum, New York, pp. 49–74.

Illenberger, E., 2000. Electron capture processes by free and bound molecules. In: Ng, C.-Y. (Ed.), Photoionization and Photodetachment (Part II). In: Advanced Series in Physical Chemistry, vol. 10B. World Scientific, Singapore, pp. 1063–1160.

Illenberger, E., Scheunemann, H.U., Baumgärtel, H., 1979. Negative ion formation in CF2Cl2, CF3Cl and CFCl3 following low energy (0–10 eV) impact with near monoenergetic electrons. Chem. Phys. 37, 21–31.

Ingolfsson, O., Weik, F., Illenberger, E., 1996. The reactivity of slow electrons with molecules at different degrees of aggregation: gas phase, clusters and condensed phase. Int. J. Mass Spectrom. 155, 1–68.

Itikawa, Y., Mason, N., 2005. Cross sections for electron collisions with water molecules. J. Phys. Chem. Ref. Data 34 (1), 1–22.

Janeckova, R., Kubala, D., May, O., Fedor, J., Allan, M., 2013. Experimental evidence on the mechanism of the dissociative electron attachment in formic acid. Phys. Rev. Lett. 111. 213201 (1–5).

Jeon, B.H., 2006. Determination of electron collision cross-sections for the C3F8 molecule by using an electron swarm study. J. Korean Phys. Soc. 49, 2321–2326.

Kalin, S.A., Kazansky, A.K., 1990. The semiclassical version of the non-local resonance theory of electron-molecule collisions. J. Phys. B At. Mol. Phys. 23, 4377–4400.

Keolopile, Z.G., Gutowski, M., Buonaugurio, A., Collins, E., Zhang, X., Erb, J., Lectka, T., Bowen, K.H., Allan, M., 2015. Importance of time scale and local environment in electron-driven proton transfer. The anion of acetoacetic acid. J. Am. Chem. Soc. 137, 14329–14340.

Khvostenko, O.G., Shchukin, P.V., Tuimedov, G.V., Muftakhov, M.V., Tseplin, E.E., Tseplina, S.N., Mazunov, V.A., 2008. Negative ion mass spectrum of the resonance electron capture by molecules of p-benzoquinone. Int. J. Mass Spectrom. 273, 69–77.

Klar, D., Ruf, M.-W., Hotop, H., 1992a. Attachment of electrons to molecules at submillielectronvolt resolution. Chem. Phys. Lett. 189, 448–454.

Klar, D., Ruf, M.-W., Hotop, H., 1992b. Attachment of electrons to molecules at meV resolution. Aust. J. Phys. 45, 263–291.

Klar, D., Ruf, M.-W., Fabrikant, I.I., Hotop, H., 2001. Dissociative electron attachment to dipolar molecules at low energies with meV resolution: $CFCl_3$, 1,1,1-$C_2Cl_3F_3$, and HI. J. Phys. B At. Mol. Phys. 34, 3855–3878.

Knoth, G., Gote, M., Rädle, M., Jung, K., Ehrhardt, H., 1989. Nuclear-excited Feshbach resonances in the electron scattering from hydrogen halides. Phys. Rev. Lett. 62, 1735–1737.

Kočišek, J., Grygoryeva, K., Lengyel, J., Fárník, M., Fedor, J., 2016a. Effect of cluster environment on the electron attachment to 2-nitrophenol. Eur. Phys. J. D 70. 98 (1–8).

Kočišek, J., Pysanenko, A., Fárník, M., Fedor, J., 2016b. Microhydration prevents fragmentation of uracil and thymine by low-energy electrons. J. Phys. Chem. Lett. 7, 3401–3405.

Kopyra, J., 2012. Low energy electron attachment to nucleotide deoxycitidine monophosphate: direct evidence for the molecular mechanism of electron-induced DNA strand breaks. Phys. Chem. Chem. Phys. 14, 8287.

Kopyra, J., Abdoul-Carime, H., 2015. Temperature dependence of the cross section for the fragmentation of thymine via dissociative electron attachment. J. Chem. Phys. 142. 174303 (1–4).

Kopyra, J., Abdoul-Carime, H., 2016. Unusual temperature dependence of the dissociative electron attachment cross section of 2-thiouracil. J. Chem. Phys. 144. 034306 (1–7).

Kopyra, J., Wnorowska, J., Forys, M., Szamrej, I., 2007. A new apparatus for measuring rate constants and activation energies of thermal electron capture processes in the gas phase. Int. J. Mass Spectrom. 268, 60–65.

Kopyra, J., Szamrej, I., Graupner, K., Graham, L.M., Field, T.A., Sulzer, P., Denifl, S., Maerk, T.D., Scheier, P., Fabrikant, I.I., Braun, M., Ruf, M.-W., Hotop, H., 2008. Low-energy electron attachment to chloroform ($CHCl_3$) molecules: a joint experimental and theoretical study. Int. J. Mass Spectrom. 277, 130–141.

Kopyra, J., Keller, A., Bald, I., 2014. On the role of fluoro-substituted nucleosides in DNA radiosensitization for tumor radiation therapy. RSC Adv. 4, 6825–6829.

Krishnakumar, E., Kumar, S.V.K., Rangwala, S.A., Mitra, S.K., 1997. Dissociative-attachment cross sections for excited and ground electronic states of SO2. Phys. Rev. A 56, 1945–1953.

Krishnakumar, E., Denifl, S., Cadez, I., Markelj, S., Mason, N.J., 2011. Dissociative electron attachment cross sections for $H_2$ and $D_2$. Phys. Rev. Lett. 106, 243201.

Kučera, M., Stano, M., Wnorowska, J., Barszczewska, W., Loffhagen, D., Matejčík, Š., 2013. Electron attachment to oxygen in nitrogen buffer gas at atmospheric pressure. Eur. Phys. J. D 67. 234 (1–8).

Kumar, S.V.K., Pota, T., Peri, D., Dongre, A.D., Raob, B.J., 2012. Low energy electron induced damage to plasmid DNA pQE30. J. Chem. Phys. 137, 045101.

Kurepa, M.V., 1978. Electron-chlorine molecule total ionisation and electron attachment cross sections. J. Phys. B At. Mol. Phys. 11, 3719–3730.

Kurepa, M.V., Babic, D.S., Belic, D.S., 1981. Electron-bromine-molecule total ionization and electron attachment cross sections. J. Phys. B At. Mol. Phys. 14, 375–384.

Lane, A.M., Thomas, R.G., 1958. R-matrix theory of nuclear reactions. Rev. Mod. Phys. 30, 257–353.

Langer, J., Matejcik, S., Illenberger, E., 2002. Electron attachment to C2F5I molecules and clusters. Int. J. Mass Spectrom. 220, 211–220.

Langer, J., Stano, M., Gohke, S., Rosa, A., Barszczewska, W., Matejcík, S., EIllenberger, E., 2003. Low energy (0-15 eV) electron stimulated reactions in single 1,2-C2F4Cl2 molecules and clusters. Int. J. Mass Spectrom. 223, 193–204.

Laporta, V., Celiberto, R., Tennyson, J., 2015. Dissociative electron attachment and electron-impact resonant dissociation of vibrationally excited $O_2$ molecules. Phys. Rev. A 91, 012701.

Leber, E., Barsotti, S., Bömmels, J., Weber, J.M., Fabrikant, I.I., Ruf, M.-W., Hotop, H., 2000a. Vibrational Feshbach resonances in electron attachment to nitrous oxide clusters: decay into heterogeneous and homogeneous cluster anions. Chem. Phys. Lett. 325, 345–353.

Leber, E., Barsotti, S., Fabrikant, I.I., Weber, J.M., Ruf, M.-W., Hotop, H., 2000b. Vibrational Feshbach resonances in electron attachment to carbon dioxide clusters. Eur. Phys. J. D 12, 125–131.

Le Coat, Y., Azria, R., Tronc, M., Ingólfsson, O., Illenberger, E., 1998. Medium enhanced, electron stimulated desorption of CF3- from condensed CF3I. Chem. Phys. Lett. 296, 208–214.

Le Garrec, J.-L., Sidko, O., Queffelec, J.L., Hamon, S., Mitchell, J.B.A., Rowe, B.R., 1997. Experimental studies of cold electron attachment to $SF_6$, $CF_3Br$, and $CCl_2F_2$. J. Chem. Phys. 107, 54–63.

Lengyel, J., Kočišek, J., Fárník, M., Fedor, J., 2016. Self-scavenging of electrons in $Fe(CO)_5$ aggregates deposited on argon nanoparticles. J. Phys. Chem. C 120, 7397–7402.

Levy, R.G., Burns, S.J., McFadden, D.L., 1994. Temperature dependence of rate coefficients for thermal electron attachment reactions of $CH_3Br$, $CF_3Br$, and $CF_3I$. Chem. Phys. Lett. 231, 132–138.

Makabe, T., Petrovic, Z.Lj., 2014. Plasma Electronics: Applications in Microelectronic Device Fabrication, second ed. CRC Press, Boca Raton, USA.
Mann, A., Linder, F., 1992. Low-energy electron scattering from halomethanes. III. e-$CF_3Cl$. J. Phys. B At. Mol. Phys. 25, 1621–1632.
Marienfeld, S., Sunagawa, T., Fabrikant, I.I., Braun, M., Ruf, M.-W., Hotop, H., 2006a. The dependence of low-energy electron attachment to $CF_3Br$ on electron and vibrational energy. J. Chem. Phys. 124. 154316 (1–14).
Marienfeld, S., Fabrikant, I.I., Braun, M., Ruf, M.-W., Hotop, H., 2006b. High-resolution low-energy electron attachment to $CF_3I$. J. Phys. B At. Mol. Phys. 39, 105–126.
Marinica, D.C., Teillet-Billy, D., Gauyacq, J.-P., 2005. Inelastic electron scattering by a molecule embedded in a host medium: role of the electronic structure of the medium. Phys. Rev. B 71. 115438 (1–10).
Märk, T.D., 1991. Free electron attachment to van der Waals clusters. Int. J. Mass Spectrom. Ion Process. 107, 143–163.
Martin, I., Skalicky, T., Langer, J., Abdoul-Carime, H., Karwasz, G., Illenberger, E., Stano, M., Matejcik, S., 2005. Low energy electron driven reactions in single formic acid molecules (HCOOH) and their homogeneous clusters. Phys. Chem. Chem. Phys. 7, 2212–2216.
Mason, N.J., Nairn, B., Jheeta, S., Szymanska, E., 2014. Electron induced chemistry: a new Frontier in astrochemistry. Faraday Discuss. 168, 235–247.
Matejcik, S., Cicman, P., Kiendler, A., Skalny, J.D., Illenberger, E., Stamatovic, A., Märk, T.D., 1996. Low-energy electron attachment to mixed ozone/oxygen clusters. Chem. Phys. Lett. 261, 437–442.
Mauracher, A., Schobel, H., Ferreira da Silva, F., Edtbauer, A., Mitterdorfer, C., Denifl, S., Mark, T.D., Illenberger, E., Scheier, P., 2009. Electron attachment to trinitrotoluene (TNT) embedded in He droplets: complete freezing of dissociation intermediates in an extended range of electron energies. Phys. Chem. Chem. Phys. 11, 8240.
May, O., Fedor, J., Ibănescu, B.C., Allan, M., 2008. Absolute cross sections for dissociative electron attachment to acetylene and diacetylene. Phys. Rev. A 77. 040701 (R) (1–4).
May, O., Fedor, J., Allan, M., 2009. Isotope effect in dissociative electron attachment to acetylene. Phys. Rev. A 80. 012706 (1–5).
May, O., Kubala, D., Allan, M., 2010. Absolute cross sections for dissociative electron attachment to HCN and DCN. Phys. Rev. A 82. 010701 (R) (1–4).
McCarthy, M.C., Gottlieb, C.A., Gupta, H., Thaddeus, P., 2006. Laboratory and astronomical identification of the negative molecular ion $C6H^-$. Astrophys. J. Lett. 652, L141–L144.
McCurdy, C.W., Turner, J.L., 1983. Wave packet formulation of the boomerang model for resonant electron-molecule scattering. J. Chem. Phys. 78, 6773–6779.
McCurdy, C.W., Isaacs, W.A., Meyer, H.-D., Rescigno, T.N., 2003. Resonant vibrational excitation of CO2 by electron impact: nuclear dynamics on the coupled components of the resonance. Phys. Rev. A 67, 042708–042719.
Mckinlay, R.G., Z¨ure, J.M., Paterson, M.J., 2010. Vibronic coupling in inorganic systems. Photochemistry, conical intersections, and the Jahn-Teller and pseudo-Jahn-Teller effects. Adv. Inorg. Chem. 62, 351–390.
Michaud, M., Cloutier, P., Sanche, L., 1993. Direct observation of a Feshbach-type electron resonance in solid argon. Phys. Rev. B 47, 4131–4133.
Miller, T.M., Friedman, J.F., Schaffer, L.C., Viggiano, A.A., 2009. Electron attachment to halomethanes at high temperature: $CH_2Cl_2$, $CF_2Cl_2$, $CH_3Cl$, and $CF_3Cl$ attachment rate constants up to 1100 K. J. Chem. Phys. 131. 084302 (1–7).
Miller, E.R., Muirhead, G.D., Jensen, E.T., 2013. Mechanisms for the near-UV photodissociation of $CH_3I$ on $D_2O/Cu(110)$. J. Chem. Phys. 138. 084702 (1–11).

Mills, R.M., 1968. Flame inhibition with electron attachment as the first step. Combust. Flame 12, 513–520.
Mills, D.L., 1992. Resonant scattering of slow electrons in molecular solids: suppression of the elastic beam. Phys. Rev. B 45, 36–42.
Nag, P., Nandi, D., 2015a. Comment on "Coherent interference in the resonant dissociative electron attachment to carbon monoxide". Phys. Rev. A 91. 056701 (1–2).
Nag, P., Nandi, D., 2015b. Fragmentation dynamics in dissociative electron attachment to CO probed by velocity slice imaging. Phys. Chem. Chem. Phys. 17, 7130–7137.
Nag, P., Nandi, D., 2016. Identification of overlapping resonances in dissociative electron attachment to chlorine molecules. Phys. Rev. A 93. 012701 (1–7).
Nagesha, K., Fabrikant, I.I., Sanche, L., 2001. Electron attachment to $CF_3Cl$ and $CH_3Cl$ on the surface and in the bulk of solid Kr. J. Chem. Phys. 114, 4934–4944.
Nandi, D., Prabhudesai, V.S., Krishnakumar, E., Chatterjee, A., 2005. Velocity slice imaging for dissociative electron attachment. Rev. Sci. Instrum. 76. 053107 (1–8).
Neustetter, M., Aysina, J., Ferreira da Silva, F., Denifl, S., 2015. The effect of solvation on electron attachment to pure and hydrated pyrimidine clusters. Angew. Chem. Int. Ed. 54, 9124–9126.
O'Malley, T.F., 1966. Theory of dissociative attachment. Phys. Rev. 150, 14–29.
O'Malley, T.F., Taylor, H.S., 1968. Angular dependence of scattering products in electron-molecule resonant excitation and in dissociative attachment. Phys. Rev. 176, 207–221.
Omarsson, F.H., Szymanska, E., Mason, N.J., Krishnakumar, E., Ingolfsson, O., 2013. Quantum superposition of target and product states in reactive electron scattering from $CF_4$ revealed. Phys. Rev. Lett. 111, 063201.
Ómarsson, F.H., Szymańska, E., Mason, N.J., Krishnakumar, E., 2014a. Velocity slice imaging study on dissociative electron attachment to CF4. Eur. Phys. J. D 68, 101.
Ómarsson, F.H., Mason, N.J., Krishnakumar, E., Ingólfsson, O., 2014b. State selectivity and dynamics in dissociative electron attachment to $CF_3I$ revealed through velocity slice imaging. Angew. Chem. Int. Ed. Engl. 53, 12051–12054.
Orlando, T.M., Oh, D., Chen, Y., Aleksandrov, A.B., 2008. Low-energy electron diffraction and induced damage in hydrated DNA. J. Chem. Phys. 128, 195102.
Padial, N.T., Norcross, D.W., 1984. Ro-vibrational excitation of HCl by electron impact. Phys. Rev. A 29, 1590–1593.
Pearl, D.M., Burrow, P.D., Nash, J.J., Morrison, H., Jordan, K.D., 1993. Dissociative attachment as a probe of intramolecular electron transfer. JACS 115, 9876–9877.
Pelc, A., Sailer, W., Scheier, P., Mason, N.J., Märk, T.D., 2002a. Low energy electron attachment to formic acid. Eur. Phys. J. D 20, 441–444.
Pelc, A., Sailer, W., Scheier, P., Probst, M., Mason, N.J., Illenberger, E., Märk, T.D., 2002b. Dissociative electron attachment to formic acid (HCOOH). Chem. Phys. Lett. 361, 277–284.
Pinnaduwage, L.A., Tav, C., McCorkle, D.L., Ding, W.X., 1999. Temperature dependence of electron attachment to methylene chloride. J. Chem. Phys. 110, 9011–9016.
Porcel, E., Liehn, S., Remita, H., Usami, N., Kobayashi, K., Furusawa, Y., Sech, C., Lacombe, S., 2010. Platinum nanoparticles: a promising material for future cancer therapy? Nanotechnology 21. 085103 (1–7).
Posseme, N., 2015. Plasma Etching Processes for Interconnect Realization, in VLSI Elsevier Science & Technology. ISTE Press Ltd, London.
Potter, E.D., Herek, J.L., Pedersen, S., Liu, Q., Zewail, A.H., 1992. Femtosecond laser control of a chemical-reaction. Nature 355, 66–68.
Poully, J.-C., Miles, J., De Camillis, S., Cassimi, A., Greenwood, J.B., 2015. Proton irradiation of DNA nucleosides in the gas phase. Phys. Chem. Chem. Phys. 17, 7172–7180.
Prabhudesai, V.S., Nandi, D., Krishnakumar, E., 2006. On the presence of the $^4\Sigma^-{}_u$ resonance in dissociative electron attachment to $O_2$. J. Phys. B At. Mol. Opt. Phys. 39, L277–L283.

Prabudeshai, V.S., Nandi, D., Kelkar, A.H., Parajuli, R., Krishakumar, E., 2005. Dissociative electron attachment to formic acid. Chem. Phys. Lett. 405, 172–175.
Ptasinska, S., Denifl, S., Scheier, P., Illenberger, E., Märk, T.D., 2005a. Bond- and site-selective loss of H atoms by very-low-energy electrons (<3 eV). Angew. Chem. Int. Ed. 44, 6941–6943.
Ptasińska, S., Denifl, S., Candori, P., Matejcik, S., Scheier, P., Märk, T.D., 2005b. Dissociative electron attachment to gas phase alanine. Chem. Phys. Lett. 403, 107–112.
Ptasińska, S., Denifl, S., Mróz, B., Probst, M., Grill, V., Illenberger, E., Scheier, P., Märk, T.D., 2005c. Bond selective dissociative electron attachment to thymine. J. Chem. Phys. 123. 124302 (1–8).
Ptasinska, S., Denifl, S., Gohlke, S., Scheier, P., Illenberger, E., Märk, T.D., 2006. Decomposition of thymidine by low-energy electrons: implications for the molecular mechanisms of single-strand breaks in DNA. Angew. Chem. Int. Ed. 45, 1893–1896.
Ptasinska, S., Alizadehb, E., Sulzer, P., Abouaf, R., Mason, N.J., Märk, T.D., Scheier, P., 2008. Negative ion formation by low energy electron attachment to gas-phase 5-nitrouracil. Int. J. Mass Spectrom. 277, 291–295.
Puschnigg, B., Huber, S.E., Probst, M., Tanzer, K., Vizcaino, V., Ferreira da Silva, F., Scheier, P., Limao-Vieira, P., Denifl, S., 2013. Electron attachment to the dipeptide dialanine: influence of methylation on site selective dissociation reactions. Phys. Chem. Chem. Phys. 15, 3834–3840.
Ram, N.B., Prabhudesai, V.S., Krishnakumar, E., 2009. Resonances in Dissociative Electron Attachment to Water. J. Phys. B: At. Mol. Opt. Phys. 42. 225203 (1–13).
Ram, N.B., Prabhudesai, V.S., Krishnakumar, E., 2011. Comment on "Imaging the Molecular Dynamics of Dissociative Electron Attachment to Water" Phys. Rev. Lett. 106. 049301 (1).
Ram, N.B., Prabhudesai, V.S., Krishnakumar, E., 2012. Dynamics of the dissociative attachment in H2O and D2O: the A1 resonance and axial recoil approximation. J. Chem. Sci. 124, 271–279.
Ram, N.B., Krishnakumar, E., 2011. Dissociative electron attachment to methane probed using velocity slice imaging. Chem. Phys. Lett. 511, 22–27.
Rankovic, M.L., Canon, F., Nahon, L., Giuliani, A., Milosavljevic, A.R., 2015. VUV action spectroscopy of protonated leucine-enkephalin peptide in the 6-14 eV range. J. Chem. Phys. 143, 244311.
Rapp, D., Sharp, T.E., Briglia, D.D., 1965. Large isotope effect in formation of H- or D- by electron impact on $H_2$, HD and $D_2$. Phys. Rev. Lett. 14, 533–535.
Rescigno, T.N., McCurdy, C.W., 2000. Numerical grid methods for quantum-mechanical scattering problems. Phys. Rev. A 62. 032706 (1–8).
Rescigno, T.N., Trevisan, C.S., Orel, A.E., 2006. Dynamics of low-energy electron attachment to formic acid. Phys. Rev. Lett. 96. 21320 (1–4).
Rescigno, T.N., Trevisan, C.S., Orel, A.E., 2009. Comment on "Electron-induced bond breaking at low energies in HCOOH and glycine: the role of very short-lived sigma* anion states". Phys. Rev. A 80. 046701 (1–2).
Rescigno, T.N., Trevisan, C.S., Orel, A.E., Slaughter, D.S., Adaniya, H., Belkacem, A., Weyland, M., Dorn, A., McCurdy, C.W., 2016. Dynamics of dissociative electron attachment to ammonia. Phys. Rev. A 93. 052704 (1–10).
Rohr, K., Linder, F., 1976. Vibrational excitation of polar molecules by electron impact. I. Threshold resonances in HF and HCl. J. Phys. B 9, 2521–2537.
Royal, J., Larson, A., Orel, A.E., 2004. Effects of couplings in the resonance continuum. J. Phys. B At. Mol. Phys. 37, 3075–3083.
Ruf, M.W., Barsotti, S., Braun, M., Hotop, H., Fabrikant, I.I., 2004. Dissociative attachment and vibrational excitation in low-energy electron collisions with chlorine molecules. J. Phys. B At. Mol. Phys. 37, 41–62.

Ruf, M.-W., Braun, M., Marienfeld, S., Fabrikant, I.I., Hotop, H., 2007. High resolution studies of dissociative electron attachment to molecules: dependence on electron and vibrational energy. J. Phys. Conf. Ser. 88. 012013 (1–10).
Samukawa, S., 2006. Ultimate top-down etching processes for future nanoscale devices: advanced neutral-beam etching. Jpn. J. Appl. Phys. 45, 2395–2407.
Samukawa, S., Hori, M., Rauf, S., Tachibana, K., Bruggeman, P., Kroesen, G., Whitehead, J.Ch., Murphy, A.B., Gutsol, A.F., Starikovskaia, S., 2012. The 2012 plasma roadmap. J. Phys. D Appl. Phys. 45. 253001 (1–37).
Sanche, L., Bass, A.D., Ayotte, P., Fabrikant, I.I., 1995. The effect of the condensed phase on dissociative electron attachment: CH3Cl condensed on a Kr surface. Phys. Rev. Lett. 75, 3568.
Schafer, O., Allan, M., 1991. Measurement of near-threshold vibrational excitation of HCl by electron impact. J. Phys. B At. Mol. Opt. Phys. 24, 3069–3076.
Scheer, A.M., Aflatooni, K., Gallup, G.A., Burrow, P.D., 2004. Bond breaking and temporary anion states in uracil and halouracils: implications for the DNA bases. Phys. Rev. Lett. 92. 068102 (1–4).
Scheer, A.M., Silvernail, C., Belot, J.A., Aflatooni, K., Gallup, G.A., Burrow, P.D., 2005. Dissociative electron attachment to uracil deuterated at the N1 and N3 positions. Chem. Phys. Lett. 411, 46–50.
Schneider, B.I., LeDourneuf, M., Burke, P.G., 1979. Theory of vibrational excitation and dissociative attachment: an R-matrix approach. J. Phys. B At. Mol. Phys. 12, L365–L369.
Schramm, A., Fabrikant, I.I., Weber, J.M., Leber, E., Ruf, M.-W., Hotop, H., 1999. Vibrational resonance and threshold effects in inelastic electron collisions with methyl iodide molecules. J. Phys. B At. Mol. Phys. 32, 2153–2171.
Schramm, A., Ruf, M.-W., Stano, M., Matejcik, S., Fabrikant, I.I., Hotop, H., 2002. High resolution study of dissociative electron attachment to dipolar molecules at low energies: $CH_2Br_2$ and $CCl_3Br$. J. Phys. B At. Mol. Phys. 35, 4179–4202.
Schrotter, C.D., Rudenko, A., Dorn, A., Moshammer, R., Ullrich, J., 2005. Status of the pulsed photoelectron source for atomic and molecular collision experiments. Nucl. Inst. Methods Phys. Res. A 536, 312–318.
Schulz, G.J., 1959. Formation of H- ions by electron impact on $H_2$. Phys. Rev. 113, 816–819.
Schulz, G.J., 1963. Resonance in the elastic scattering of electrons in helium. Phys. Rev. Lett. 10, 104–106.
Schulz, G.J., 1973. Resonances in electron impact on atoms. Rev. Mod. Phys. 45, 378–486.
Seng, G., Linder, F., 1976. Vibrational excitation of polar molecules by electron impact. II. Direct and resonant excitation in $H_2O$. J. Phys. B At. Mol. Phys. 9, 2539–2551.
Sergenton, A.-Ch., Jungo, L., Allan, M., 2000. Excitation of vibrational levels of HF up to v=4 by electron impact. Phys. Rev. A 61. 062702 (1–6).
Shapiro, M., Bersohn, R., 1980. Vibrational energy distribution of the $CH_3$ radical photodissociated from $CH_3I$. J. Chem. Phys. 73, 3810–3818.
Shuman, N.S., Miller, T.M., Friedman, J.F., Viggiano, A.A., 2013a. Electron attachment to $Fe(CO)_n$ (n=0-5). J. Phys. Chem. A 117, 1102–1109.
Shuman, N.S., Miller, T.M., Viggiano, A.A., Fabrikant, I.I., 2013b. Thermal electron attachment to $F_2$. Phys. Rev. A 88. 062708 (1–6).
Sides, G.D., Tiernan, T.O., Hanrahan, R.J., 1976. Measurement of thermal electron dissociative attachment rate constants for halogen gases using a flowing afterglow technique. J. Chem. Phys. 65, 1966–1975.
Slaughter, D.S., Adaniya, H., Rescigno, T.N., Haxton, D.J., Orel, A.E., McCurdy, C.W., Belkacem, A., 2011. Dissociative electron attachment to carbon dioxide via the 8.2 eV Feshbach resonance. J. Phys. B At. Mol. Phys. 44. 205203 (1–5).

Slaughter, D.S., Haxton, D.J., Adaniya, H., Weber, T., Rescigno, T.N., McCurdy, C.W., Belkacem, A., 2013. Ion-momentum imaging of resonant dissociative-electron-attachment dynamics in methanol. Phys. Rev. A 87. 052711 (1–7).
Smyth, M., Kohanoff, J., Fabrikant, I.I., 2014. Electron-induced hydrogen loss in uracil in a water cluster environment. J. Chem. Phys. 140. 184313 (1–7).
Solomun, T., Seitz, H., Sturm, H., 2009. DNA damage by low-energy electron impact: dependence on guanine content. J. Phys. Chem. B 113, 11557–11559.
Sommerfeld, T., 2001. Electron-induced chemistry of 5-chlorouracil. Chem. Phys. Chem. 2, 677–679.
Spyrou, S.M., Hunter, S.R., Christophorou, L.G., 1985. A study of the isomeric dependence of low-energy (<10 eV) electron attachment: perfluoroalkanes. J. Chem. Phys. 83, 641–654.
Stamatovic, A., 1988. Electron attachment to van der Waals clusters: the zero energy resonance. In: Gilbody, H.B., Newell, W.R., Read, F.H., Smith, A.C.H. (Eds.), Electronic and Atomic Collisions. North Holland, Amsterdam, pp. 729–735.
Stamatovic, A., Schulz, G.J., 1970. Dissociative attachment in CO and formation of $C^-$. J. Chem. Phys. 53, 2663–2667.
Stano, M., Foltin, V., Matejcik, S., Langer, J., Gohlke, S., Illenberger, E., 2003a. A study of dissociative electron attachment to $CH_3Br$ and $CH_3I$. J. Phys. B At. Mol. Opt. Phys. 36, 443–452.
Stano, M., Matejcik, S., Skalny, J.D., Märk, T.D., 2003b. Electron impact ionization of $CH_4$: ionization energies and temperature effects. J. Phys. B At. Mol. Opt. Phys. 36, 261–271.
Stepanovic, M., Pariat, Y., Allan, M., 1999. Dissociative electron attachment in cyclopentanone, gama-butyrolactone, ethylene carbonate and ethylene carbonate-d4: role of dipole-bound resonances. J. Chem. Phys. 110, 11376–11382.
Stricklett, K.L., Chu, S.C., Burrow, P.D., 1986. Dissociative attachment in vinyl and allyl chloride, chlorobenzene and benzyl chloride. Chem. Phys. Lett. 131, 279–284.
Sulzer, P., Alizadeh, E., Mauracher, A., Märk, T.D., Scheier, P., 2008. Detailed dissociative electron attachment studies on the amino acid proline. Int. J. Mass Spectrom. 277, 274–278.
Szymanska, E., Prabhudesai, V., Mason, N.J., Krishnakumar, E., 2013. Dissociative electron attachment to acetaldehyde, $CH_3CHO$. A laboratory study using the velocity map imaging technique. Phys. Chem. Chem. Phys. 15, 998–1005.
Szyperska, A., Rak, J., Leszczynski, J., Li, X., Ko, Y.J., Wang, H., Bowen, K.H., 2010. Low-energy-barrier proton transfer induced by electron attachment to the guanine—cytosine base pair. ChemPhysChem 11, 880–888.
Tam, W.-C., Wong, S.F., 1978. Dissociative attachment of halogen molecules by 0–8 eV electrons. J. Chem. Phys. 68, 5626–5630.
Tarana, M., Wielgus, P., Roszak, S., Fabrikant, I.I., 2009. Effects of two vibrational modes in the dissociative electron attachment to $CF_3Cl$. Phys. Rev. A 79. 052712 (1–11).
Tarana, M., Houfek, K., Horacek, J., Fabrikant, I.I., 2011. Dissociative electron attachment and vibrational excitation of $CF_3Cl$: effect of two vibrational modes revisited. Phys. Rev. A 84. 052717 (1–9).
Tegeder, P., Lehmann, L., Ingólfsson, O., Illenberger, E., 1996. Strong solvation effects in the reactivity of C(6)F(5)X(-) anions (X=Cl, Br, I). Competition between dissociative and associative processes following electron capture. Z. Phys. Chem. 195, 217–236.
Tennyson, J., 2010. Electron–molecule collision calculations using the R-matrix method. Phys. Rep. 491, 29–76.
Thoss, M., Domcke, W., 1998. Theory of vibrational relaxation processes in resonant collisions of low-energy electrons with large molecules. J. Chem. Phys. 109, 6577–6595.

Tian, S.X., Luo, Y., 2015. Reply to "Comment on 'Coherent interference in the resonant dissociative electron attachment to carbon monoxide'" Phys. Rev. A 91. 056702 (1–2).
Tian, S.X., Wu, B., Xia, L., Wang, Y.F., Li, H.K., Zeng, X.J., Luo, Y., 2013. Coherent interference in the resonant dissociative electron attachment to carbon monoxide. Phys. Rev. A 88. 012708 (1–6).
Trippel, S., Chang, Y.-P., Stern, S., Mullins, T., Holmegaard, L., Küpper, J., 2012. Spatial separation of state- and size-selected neutral clusters. Phys. Rev. A 86, 033202.
Tronc, M., Fiquet-Fayard, F., Schermann, C., Hall, R.I., 1977. Differential cross sections and angular distributions of H from dissociative electron attachment to $H_2$ between 3.75 and 13 eV. J. Phys. B At. Mol. Phys. 10, 305–332.
Underwood-Lemons, T., Gergel, T.J., Moore, J.H., 1995. Dissociative electron attachment cross sections for halofluoromethanes. J. Chem. Phys. 102, 119–123.
Uzer, T., 1991. Theories of intramolecular vibrational energy transfer. Phys. Rep. 199, 73–146.
Van Brunt, R.J., Keiffer, L.J., 1970. Angular distribution of $O^-$ from dissociative electron attachment to $O_2$. Phys. Rev. A 2, 1899–1905.
Vizcaino, V., Denifl, S., Märk, T.D., Illenberger, E., Scheier, P., 2010. Low energy (0–4 eV) electron impact to $N_2O$ clusters: dissociative electron attachment, ion-molecule reactions, and vibrational Feshbach resonances. J. Chem. Phys. 133. 154512 (1–6).
Vizcaino, V., Puschnigg, B., Huber, S.E., Probst, M., Fabrikant, I.I., Gallup, G.A., Illenberger, E., Scheier, P., Denifl, S., 2012. Hydrogen loss in aminobutanoic acid isomers by the $\sigma^*$ resonance formed in electron capture. New J. Phys. 14. 043017 (1–12).
Vogt, E., Wannier, G.H., 1954. Scattering of ions by polarization forces. Phys. Rev. 95, 1190–1198.
Voinov, V.G., Vasilev, Y.V., Ji, H., Figard, B., Morre, J., Egan, T.F., Barovsky, D.F., Deinzer, M.L., 2004. A gas chromatograph/resonant electron capture-TOF mass spectrometer for four dimensions of negative ion analytical information. Anal. Chem. 76, 2951–2957.
Vuitton, V., Lavvas, P., Yelle, R.V., Galand, M., Wellbrock, A., Lewis, G.R., Coates, A.J., Wahlund, J.E., 2009. Negative ion chemistry in Titan's upper atmosphere. Planet. Space Sci. 57, 1558–1572.
Wang, X.D., Xuan, C.J., Luo, Y., Tian, S.X., 2015. Note: coherent resonances observed in the dissociative electron attachment to carbon monoxide. J. Chem. Phys. 143, 066101.
Weber, J.M., Leber, E., Ruf, M.-W., Hotop, H., 1999. Nuclear-excited Feshbach resonances in electron attachment to molecular clusters. Phys. Rev. Lett. 82, 516–519.
Weik, F., Illenberger, E., 1995. Interaction of low energy electrons with gaseous and condensed perfluoro compounds. J. Chem. Phys. 103, 1406–1412.
Whitaker, B.J., 2003. Imaging in Molecular Dynamics. Technology and Applications. Cambridge University Press, Cambridge, UK.
Wiens, J.P., Sawyer, J.C., Miller, T.M., Shuman, N.S., Viggiano, A.A., Khamesian, M., Kokoouline, V., Fabrikant, I.I., 2016. Electron attachment to the interhalogen compounds ClF, ICl, and IBr. Phys. Rev. A 93. 032706 (1–8).
Wigner, E.P., 1948. On the behavior of cross sections near thresholds. Phys. Rev. 73, 1002–1009.
Wilde, R.S., Gallup, G.A., Fabrikant, I.I., 1999. Semiempirical R-matrix theory of low energy electron–$CF_3Cl$ inelastic scattering. J. Phys. B At. Mol. Phys. 32, 663–673.
Wilde, R.S., Gallup, G.A., Fabrikant, I.I., 2000. Comparative studies of dissociative electron attachment to methyl halides. J. Phys. B At. Mol. Phys. 33, 5479–5492.
Xia, L., Zeng, X.J., Li, H.K., Wu, B., Tian, S.X., 2013. Orientation effect in the low-energy electron attachment to the apolar carbon tetrafluoride molecule. Angew. Chem. Int. Ed. 52, 1013–1016.

Xu, Y., Fabrikant, I.I., 2001. Dissociative electron attachment rates for $H_2$ and its isotopes. Appl. Phys. Lett. 78, 2598–2600.

Xu, Y., Gallup, G.A., Fabrikant, I.I., 2000. Dissociative electron attachment to vibrationally and rotationally excited $H_2$ and HF molecules. Phys. Rev. A 61. 052705(1–7).

Yoon, J.-S., Kim, Y.-W., Kwon, D.-C., Song, M.-Y., Chang, W.-S., Kim, C.-G., Kumar, V., Lee, B., 2010. Electron-impact cross sections for deuterated hydrogen and deuterium molecules. Rep. Prog. Phys. 73, 116401.

Živanov, S., Allan, M., Čížek, M., Horáček, J., Thiel, F.A.U., Hotop, H., 2002. Effects of interchannel coupling in associative detachment: electron spectra for H + Cl- and H + Br- collisions. Phys. Rev. Lett. 89. 073201 (1–4).

Živanov, S., Čížek, M., Horáček, J., Allan, M., 2003. Electron spectra for associative detachment in low-energy collisions of Cl- and Br- with H and D. J. Phys. B At. Mol. Opt. Phys. 36, 3513–3531.

# INDEX

Note: Page numbers followed by "*f*" indicate figures, and "*t*" indicate tables.

## C

## D

## E

## N

## Q

## V



## W



## Z